Handbook of Graphene

Scrivener Publishing
100 Cummings Center, Suite 541J
Beverly, MA 01915-6106

Publishers at Scrivener
Martin Scrivener (martin@scrivenerpublishing.com)
Phillip Carmical (pcarmical@scrivenerpublishing.com)

Handbook of Graphene comprises 8 volumes:

Volume 1: Growth, Synthesis, and Functionalization
Edited by Edvige Celasco and Alexander Chaika
ISBN 978-1-119-46855-4

Volume 2: Physics, Chemistry, and Biology
Edited by Tobias Stauber
ISBN 978-1-119-46959-9

Volume 3: Graphene-Like 2D Materials
Edited by Mei Zhang
ISBN 978-1-119-46965-0

Volume 4: Composites
Edited by Cengiz Ozkan
ISBN 978-1-119-46968-1

Volume 5: Energy, Healthcare, and Environmental Applications
Edited by Cengiz Ozkan and Umit Ozkan
ISBN 978-1-119-46971-1

Volume 6: Biosensors and Advanced Sensors
Edited by Barbara Palys
ISBN 978-1-119-46974-2

Volume 7: Biomaterials
Edited by Sulaiman Wadi Harun
ISBN 978-1-119-46977-3

Volume 8: Technology and Innovation
Edited by Sulaiman Wadi Harun
ISBN 978-1-119-46980-3

Handbook of Graphene

Volume 4: Composites

Edited by
Cengiz Ozkan
*Department of Materials Science & Engineering,
University of California,
Riverside, USA*

This edition first published 2019 by John Wiley & Sons, Inc., 111 River Street, Hoboken, NJ 07030, USA and Scrivener Publishing LLC, 100 Cummings Center, Suite 541J, Beverly, MA 01915, USA
© 2019 Scrivener Publishing LLC
For more information about Scrivener publications please visit www.scrivenerpublishing.com.

All rights reserved. No part of this publication may be reproduced, stored in a retrieval system, or transmitted, in any form or by any means, electronic, mechanical, photocopying, recording, or otherwise, except as permitted by law. Advice on how to obtain permission to reuse material from this title is available at http://www.wiley.com/go/permissions.

Wiley Global Headquarters
111 River Street, Hoboken, NJ 07030, USA

For details of our global editorial offices, customer services, and more information about Wiley products visit us at www.wiley.com.

Limit of Liability/Disclaimer of Warranty
While the publisher and authors have used their best efforts in preparing this work, they make no representations or warranties with respect to the accuracy or completeness of the contents of this work and specifically disclaim all warranties, including without limitation any implied warranties of merchantability or fitness for a particular purpose. No warranty may be created or extended by sales representatives, written sales materials, or promotional statements for this work. The fact that an organization, website, or product is referred to in this work as a citation and/or potential source of further information does not mean that the publisher and authors endorse the information or services the organization, website, or product may provide or recommendations it may make. This work is sold with the understanding that the publisher is not engaged in rendering professional services. The advice and strategies contained herein may not be suitable for your situation. You should consult with a specialist where appropriate. Neither the publisher nor authors shall be liable for any loss of profit or any other commercial damages, including but not limited to special, incidental, consequential, or other damages. Further, readers should be aware that websites listed in this work may have changed or disappeared between when this work was written and when it is read.

Library of Congress Cataloging-in-Publication Data

ISBN 978-1-119-46968-1

Cover image: Pixabay.Com
Cover design by Russell Richardson

Set in size of 11pt and Minion Pro by Manila Typesetting Company, Makati, Philippines

10 9 8 7 6 5 4 3 2 1

Contents

Preface — xvii

1 Graphene Composites — 1
Xiao-Jun Shen, Xiao-Ling Zeng and Chen-Yang Dang
1.1 Introduction — 1
1.2 History of Graphene — 2
1.3 Synthesis of Graphene — 2
    1.3.1 Top-Down Approach — 3
        1.3.1.1 Exfoliation and Cleavage — 3
        1.3.1.2 Chemically Derived Graphene — 4
    1.3.2 Bottom-Up Approach — 4
        1.3.2.1 Chemical Vapor Deposition — 4
        1.3.2.2 Epitaxial Growth — 5
    1.3.3 Other Methods — 6
1.4 Characterization and Properties — 6
    1.4.1 Characterization — 6
        1.4.1.1 Optical Imaging of Graphene Layers — 6
        1.4.1.2 Atomic Force Microscopy (AFM) — 7
        1.4.1.3 Transmission Electron Microscopy (TEM) — 7
        1.4.1.4 Raman Spectroscopy — 8
    1.4.2 Properties — 8
        1.4.2.1 Electrical Transport Property — 8
        1.4.2.2 Optical Properties — 9
        1.4.2.3 Mechanical Properties — 9
        1.4.2.4 Thermal Properties — 10
    1.4.3 Application — 10
1.5 Graphene-Based Composites — 11
    1.5.1 Graphene–Polymer Composites — 11
        1.5.1.1 Synthesis of Graphene-Reinforced Polymer Composite — 12
        1.5.1.2 Mechanical Properties — 12
        1.5.1.3 Electrical Properties — 13
        1.5.1.4 Thermal Conductivity — 14
        1.5.1.5 Other Properties — 14
        1.5.1.6 Application — 15
    1.5.2 Graphene–Nanoparticle Composites — 15
        1.5.2.1 Synthesis of Graphene–Nanoparticle Composites — 15
        1.5.2.2 Properties — 16

1.6	Future Prospects	17
	Acknowledgment	18
	References	18

2 Graphene-Reinforced Advanced Composite Materials — 27
Xiaochao Ji, Shaojun Qi, Rajib Ahmed and Ahmmed A. Rifat

- 2.1 Introduction — 27
- 2.2 Graphene–Metal Matrix Composites (MMCs) — 29
  - 2.2.1 Processing of MMCs — 30
    - 2.2.1.1 Powder Metallurgy — 30
    - 2.2.1.2 Melting and Solidification — 41
    - 2.2.1.3 Electrochemical Deposition — 42
    - 2.2.1.4 Thermal Spray — 44
    - 2.2.1.5 Other Techniques — 44
  - 2.2.2 Properties of the Graphene-Reinforced MMCs — 45
    - 2.2.2.1 Mechanical Properties — 45
    - 2.2.2.2 Corrosion Properties — 48
    - 2.2.2.3 Tribological Properties — 49
    - 2.2.2.4 Other Properties — 50
- 2.3 Graphene–Reinforced Polymer Matrix Composites (PMCs) — 50
  - 2.3.1 Preparation of Graphene Polymer Composites — 51
    - 2.3.1.1 Melt Blending — 51
    - 2.3.1.2 Solution Compounding — 54
    - 2.3.1.3 *In Situ* Polymerization — 54
    - 2.3.1.4 Other Methods — 55
  - 2.3.2 Properties of Graphene-Reinforced PMCs — 56
    - 2.3.2.1 Electrical Properties — 56
    - 2.3.2.2 Mechanical Properties — 59
    - 2.3.2.3 Thermal Properties — 59
    - 2.3.2.4 Corrosion Properties — 61
- 2.4 Graphene-Reinforced Ceramic Matrix Composites (CMCs) — 61
  - 2.4.1 Processing Methods — 61
    - 2.4.1.1 Types of Graphene Fillers — 61
    - 2.4.1.2 Powder Processing — 65
    - 2.4.1.3 Densification — 66
    - 2.4.1.4 Thermal/Cold/Plasma Spraying — 71
    - 2.4.1.5 Electrophoretic Deposition (EPD) — 71
  - 2.4.2 Performance — 72
    - 2.4.2.1 Mechanical Properties — 72
    - 2.4.2.2 Electrical Properties — 74
- 2.5 Applications of Graphene-Reinforced Composites — 74
  - 2.5.1 Low Friction and Wear Components — 74
  - 2.5.2 Intelligent Interfaces and Anti-Corrosion Coatings — 75
  - 2.5.3 Antibacterial and Biocompatible Implants — 77
  - 2.5.4 Flame-Retardant Materials — 77

	2.6	Conclusion	79
		References	80
3	**Graphene-Based Composite Materials**	91
	Munirah Abdullah Almessiere, Kashif Chaudhary, Jalil Ali and Muhammad Sufi Roslan	
	3.1	Introduction	92
	3.2	Graphene Composites	92
		3.2.1 Graphene Filled Polymer Composites	92
		3.2.1.1 Graphene Filled Polymers	94
		3.2.1.2 Layered Graphene Polymers	94
		3.2.1.3 Polymer-Functionalized Graphene Nanosheets	94
		3.2.2 Graphene Nanostructure Composites	95
		3.2.3 Hybrid Graphene/Microfiber Composites	95
		3.2.4 Graphene Colloids and Coatings	97
		3.2.5 Graphene Bioactive Composites	99
	3.3	Processing Routes for Graphene Composites	99
		3.3.1 Melt Bending/Mixing	100
		3.3.2 Solution Blending/Mixing	101
		3.3.3 *In Situ* Polymerization/Crystallization	102
		3.3.4 Layer-by-Layer Assembly	103
		3.3.5 Other Processing Techniques	105
		3.3.5.1 Chemical Reduction	105
		3.3.5.2 Sol-Gel Methods	105
		3.3.5.3 Colloidal Processing	105
		3.3.5.4 Powder Processing	106
	3.4	Summary	106
		References	106
4	**Interfacial Mechanical Properties of Graphene/Substrate System: Measurement Methods and Experimental Analysis**	115
	Chaochen Xu, Hongzhi Du, Yilan Kang and Wei Qiu	
	4.1	Methodology of Raman Mechanical Measurements of Graphene	115
		4.1.1 Theory of Graphene Strain Measurement	117
		4.1.2 Characterization of Graphene Strain Using *In Situ* Raman Spectroscopy	120
	4.2	Experimental Investigations of Interfacial Mechanical Behaviors of Graphene	120
		4.2.1 Raman-Spectroscopy-Based Investigations of Interfacial Properties of Graphene	121
		4.2.2 Influencing Factors of Experimental Measurements on Interfacial Properties	122
	4.3	Experimental Investigation of Mechanical Behavior of Graphene/Substrate Interface	123
		4.3.1 Graphene/Substrate Specimen and Raman Experiments	123
		4.3.2 Interfacial Strain Transfer of the Graphene/Substrate Interface	124
		4.3.3 Interfacial Shear Stress of Graphene/Substrate Interface	127

4.4	Size Effect on Mechanical Behavior of Graphene/Substrate Interface	128
	4.4.1 Serial Experiments on Graphene/Substrate Interface	129
	4.4.2 Size Effect of Graphene/Substrate Interface	131
4.5	Effect of Cyclic Loading on Mechanical Behavior of Graphene/Substrate Interface	134
	4.5.1 Initial Strain of Graphene	135
	4.5.2 Release of Initial Strain by Cyclic Loading Treatment	136
	4.5.3 Improvement of Interfacial Mechanical Properties	138
	4.5.4 Discussion	140
4.6	Conclusion	141
	Acknowledgments	143
	References	143

5 Graphene-Based Ceramic Composites: Processing and Applications 147
Kalaimani Markandan and Jit Kai Chin

5.1	Introduction	147
	5.1.1 Technical Ceramics	147
	5.1.2 Graphene	148
5.2	Processing of GCMC	148
	5.2.1 Powder Processing	148
	5.2.2 Colloidal Processing	151
	5.2.3 Sol-Gel Processing	151
	5.2.4 Polymer-Derived Ceramics	152
	5.2.5 Molecular-Level Mixing	153
	5.2.6 Compaction and Consolidation	154
5.3	Properties of GCMC	155
	5.3.1 Mechanical Properties and Toughening Mechanisms	155
	5.3.2 Electrical Properties	158
	5.3.3 Tribological Behavior	162
5.4	Application of GCMC	163
	5.4.1 Anode Materials for Li-Ion Batteries	163
	5.4.2 Supercapacitors	164
	5.4.3 Engine Components/Bearings/Cutting Tools	164
5.5	Conclusion	165
	References	166

6 *Ab Initio* Design of 2D and 3D Graphene-Based Nanostructure 171
Andrei Timoshevskii, Sergiy Kotrechko, Yuriy Matviychuk and Eugene Kolyvoshko

6.1	Introduction	171
6.2	The Subject and the Methods of Simulation	172
	6.2.1 1D Modeling	173
	6.2.2 2D Modeling	173
	6.2.3 3D Modeling	173
6.3	*Ab Initio* Modeling of the Atomic Structure and Mechanical Properties	174
	6.3.1 Atomic Structure and Strength of Carbynes	174
	6.3.2 Atomics of Instability and Break of a Contact Bond in 2D Structures	179

	6.4	Modeling of 3D Crystal Structures	184
	6.5	Thermomechanical Stability	192
		6.5.1 Fluctuation Model	192
		6.5.2 Lifetime Prediction	195
	6.6	Conclusions	198
		Funding	200
		References	200
7	**Graphene-Based Composite Nanostructures: Synthesis, Properties, and Applications**		**203**
	Mashkoor Ahmad and Saira Naz		
	7.1	Introduction	203
	7.2	Carbon Nanomaterials	204
	7.3	Graphene	205
		7.3.1 Graphene Structure	205
		7.3.2 Graphene Synthesis	207
		7.3.2.1 Exfoliation of Graphite	207
		7.3.2.2 CVD Synthesis	208
		7.3.2.3 Epitaxial Growth	210
		7.3.2.4 Chemical Method	210
		7.3.3 Graphene Properties	211
		7.3.3.1 Physicochemical Properties	211
		7.3.3.2 Thermal and Electrical Properties	212
		7.3.3.3 Optical Properties	212
		7.3.3.4 Mechanical Properties	213
		7.3.3.5 Biological Properties	213
	7.4	Carbon-Based Nanocomposites	214
		7.4.1 Graphene-Based Composites	214
		7.4.2 Graphene-Based Composite Synthesis	215
		7.4.2.1 Solution Mixing Method	216
		7.4.2.2 Sol–Gel Method	216
		7.4.2.3 Hydrothermal/Solvothermal Method	217
		7.4.2.4 Self-Assembly	217
		7.4.2.5 Other Methods	217
		7.4.3 Graphene-Based Composite Properties	218
	7.5	Applications	219
		7.5.1 Gas Sorption and Storage	219
		7.5.2 Hydrogen Storage	220
		7.5.3 Energy Storage Devices	221
		7.5.4 Antibacterial Activity	222
		7.5.5 Bioimaging	222
		7.5.6 Biosensing	224
		7.5.7 Photocatalysis	224
		Acknowledgments	225
		References	225

8 Graphene-Based Composites with Shape Memory Effect—Properties, Applications, and Future Perspectives 233
André Espinha, Ana Domínguez-Bajo, Ankor González-Mayorga and María Concepción Serrano

    List of Abbreviations 234
8.1 Introduction 234
    8.1.1 Graphene 235
    8.1.2 Shape Memory Polymers 236
    8.1.3 Shape Memory Polymer Composites 238
8.2 Graphene-Doped SMP Composites 238
    8.2.1 Morphological Properties 242
    8.2.2 Optical Properties 244
    8.2.3 Mechanical Properties 244
    8.2.4 Electrical Properties 246
    8.2.5 Shape Memory Characterization 248
8.3 Applications 252
8.4 Future Perspectives 254
    Acknowledgments 255
    References 255

9 Graphene-Based Scroll Structures: Optical Characterization and Its Application in Resistive Switching Memory Devices 261
Janardhanan R. Rani and Jae-Hyung Jang

9.1 Graphene-Based Scroll Structures 262
    9.1.1 Introduction 262
    9.1.2 Reduced Graphene-Oxide-Based Scroll Fabrication: Iron Oxide Intercalation with rGO Powder 262
    9.1.3 Reduced Graphene-Oxide-Based Scroll Fabrication: Scrolls Formed due to Phosphor Intercalation 263
    9.1.4 Optical Properties of rGO–Phosphor Hybrid Scrolls 265
    9.1.5 Raman Spectra of the Scrolls 266
9.2 Reduced Graphene-Oxide-Based Resistive Switching Devices 270
    9.2.1 Resistive Switching in GO–Phosphor Hybrid Scrolls 272
    9.2.2 Resistive Switching in Graphene-Oxide–Iron Oxide Hybrid Thin Films 275
    References 280

10 Fabrication and Properties of Copper–Graphene Composites 285
Vladimir G. Konakov, Ivan Yu. Archakov and Olga Yu. Kurapova

10.1 Introduction 285
10.2 Powder Metallurgy Technique 286
    10.2.1 Hot Pressing Technique 288
    10.2.2 Microwave Heating 290
    10.2.3 Spark Plasma Synthesis 291
10.3 Electrochemical Deposition 293
    10.3.1 Deposition in the Direct Current Regime 294
    10.3.2 Deposition of Cu–Gr Composites in a Pulse Regime 297

		10.3.3 Electrochemical Deposition of Nanotwinned Copper–Graphene Composites	304
	10.4	Electroless Deposition	308
	10.5	Molecular-Level Mixing (MLM) Technique	309
	10.6	Chemical Vapor Deposition (CVD) Technique	313
	10.7	Functionalization of Copper Powder Surface	316
	10.8	Conclusions	316
		References	318

11 Graphene–Metal Oxide Composite as Anode Material in Li-Ion Batteries — 323
Sanjaya Brahma, Shao-Chieh Weng and Jow-Lay Huang

11.1	Introduction		324
11.2	Type of Anode Materials		325
11.3	Metal Oxides as Anode Materials in Lithium Ion Battery		325
11.4	Graphene/Graphene–Metal Oxide as Anode in Li-Ion Battery		328
	11.4.1	Graphene as Anode Materials in Lithium Ion Battery	328
	11.4.2	Graphene–MnO_2 as Anode in Li-Ion Battery	329
	11.4.3	Graphene–SnO_2 as Anode in Li-Ion Battery	337
	11.4.4	Graphene–Co_3O_4 as Anode in Li-Ion Battery	343
	11.4.5	Graphene–Fe_2O_3 as Anode in Li-Ion Battery	345
11.5	Conclusion		346
	Acknowledgment		347
	References		347

12 Graphene/TiO_2 Nanocomposites: Synthesis Routes, Characterization, and Solar Cell Applications — 353
Chin Wei Lai, Foo Wah Low, Siti Zubaidah Binti Mohamed Siddick and Joon Ching Juan

12.1	Introduction		354
12.2	History of Solar Cells		356
12.3	DSSC Structure and Working Operation		358
	12.3.1	Transparent Conductive Films	361
	12.3.2	Semiconductor Film Electrodes	361
	12.3.3	TiO_2	361
	12.3.4	rGO	363
	12.3.5	rGO-TiO_2 NC	364
	12.3.6	Dye Sensitizer	367
	12.3.7	Liquid Electrolyte	369
	12.3.8	Cathode Electrodes	372
12.4	rGO-TiO_2 NC Properties		373
	12.4.1	Mechanism of rGO-TiO_2 NC	374
	12.4.2	Mechanism of rGO-TiO_2 NC in DSSCs	374
12.5	rGO-TiO_2 NC Synthesis		375
	12.5.1	Sol–Gel Synthesis	376
	12.5.2	Solution Mixing Synthesis	377
	12.5.3	*In Situ* Growth Synthesis	377

	12.6	Fabrication Technique of rGO-TiO$_2$ NC-Based Photoanode in DSSC Application	378
		12.6.1 PVD Methods—rGO-TiO$_2$ NC (Liquid-Phase Processes)	380
		12.6.1.1 Spin-Coating Technique	380
		12.6.1.2 Doctor Blade Printing Technique	381
		12.6.1.3 Electrohydrodynamic Deposition Technique	381
		12.6.2 PVD Methods—rGO-TiO$_2$ NC (Gas-Phase Processes)	382
		12.6.2.1 Physical Vapor Deposition (PVD) Technique	382
		12.6.2.2 Thermal Evaporation Technique	382
		12.6.2.3 Electron Beam Evaporation (EBE) Technique	383
		12.6.2.4 Sputtering Technique	384
		12.6.2.5 Pulsed DC Sputtering Power Technique	384
		12.6.2.6 DC and RF Magnetron Sputtering Technique	386
	Acknowledgments		387
	References		387
13	**Role of Reduced Graphene Oxide Nanosheet Composition with ZnO Nanostructures in Gas Sensing Properties**		**395**
	A.S.M. Iftekhar Uddin and Hyeon Cheol Kim		
	13.1	Introduction	395
	13.2	Experimental	397
		13.2.1 Synthesis of ZnO Nanostructures	397
		13.2.1.1 Nanoparticles (NPs)	397
		13.2.1.2 Nanocapsules (NCs)	398
		13.2.1.3 Nanograins (NGs)	398
		13.2.1.4 Nanorods (NRs)	398
		13.2.1.5 Nanoflakes (NFls)	398
		13.2.1.6 Microflowers (MFs)	398
		13.2.1.7 Microurchins (MUs)	399
		13.2.1.8 Microspheres (MSs)	399
		13.2.2 Synthesis of Bare Reduced Graphene Oxide (rGO) Nanosheets	399
		13.2.3 Synthesis of ZnO NSs–rGO Hybrids	400
		13.2.4 Device Fabrication	400
		13.2.5 Characterization and Sensor Test	400
	13.3	Results and Discussions	401
		13.3.1 Morphological Studies of the ZnO NSs	402
		13.3.2 Morphological and Elemental Studies of GO and rGO	403
		13.3.3 Chemical Composition Studies of GO and rGO	404
		13.3.4 Morphological and Elemental Analysis of the ZnO NSs–rGO Hybrids	405
		13.3.5 Structural Studies of GO, rGO, ZnO NSs, and ZnO NSs–rGO Hybrids	407
		13.3.6 Gas Sensing Mechanism	409
		13.3.7 Gas Sensor Studies	410
	13.4	Conclusions	414
	References		414

14 Functional Graphene Oxide/Epoxy Nanocomposite Coatings with Enhanced Protection Properties   419
H. Alhumade, R.P. Nogueira, A. Yu, L. Simon and A. Elkamel

- 14.1 Introduction   419
- 14.2 Experimental   421
  - 14.2.1 Materials   421
  - 14.2.2 Composite Synthesis   422
  - 14.2.3 Composites Characterization   422
  - 14.2.4 Adhesion   424
  - 14.2.5 Electrochemical Measurements   424
  - 14.2.6 Gravimetric Analysis   424
  - 14.2.7 Thermal Analysis and UV Degradation   425
  - 14.2.8 Impact Resistance   425
- 14.3 Results and Discussion   426
  - 14.3.1 Composite Characterization   426
  - 14.3.2 Adhesion   429
  - 14.3.3 Gravimetric Analysis   430
  - 14.3.4 Impedance   431
  - 14.3.5 Potentiodynamic Polarization   435
  - 14.3.6 Thermal Stability and UV Degradation   437
  - 14.3.7 Impact Resistance   440
- 14.4 Conclusion   441
- References   441

15 Supramolecular Graphene-Based Systems for Drug Delivery   443
Sandra M.A. Cruz, Paula A.A.P. Marques and Artur J.M. Valente

- 15.1 Introduction   444
- 15.2 Graphene Oxide and Cyclodextrin: Entities Applied in Drug Delivery   444
  - 15.2.1 Graphene Oxide   444
  - 15.2.2 Cyclodextrin   451
- 15.3 GO-CD Nanocomposites as Drug Delivery Systems   456
  - 15.3.1 Strategies of Preparation of GO-CD   456
  - 15.3.2 Biocompatibility   460
  - 15.3.3 Drug Release Profiles   462
- 15.4 Concluding Remarks   468
- References   468

16 Polymeric Nanocomposites Including Graphene Nanoplatelets   481
Ismaeil Ghasemi and Sepideh Gomari

- 16.1 Introduction   481
- 16.2 Functionalization of Graphene Nanosheets   482
  - 16.2.1 Covalent Modification   483
  - 16.2.2 Noncovalent Modification   484
- 16.3 Preparation Methods of Polymeric Nanocomposites   484
- 16.4 Crystallization Behavior of Polymer/Graphene Nanocomposites   485
  - 16.4.1 Isothermal Crystallization Kinetics   486

		16.4.2 Non-Isothermal Crystallization Kinetics	489

	16.4.2 Non-Isothermal Crystallization Kinetics	489
16.5	Electrical Conductivity	492
16.6	Mechanical Properties	496
16.7	Gas Barrier Properties	498
16.8	Thermal Conductivity	502
16.9	Rheology	503
16.10	Hybrid Nanocomposites Including Graphene and Other Nanofillers	507
16.11	Applications of Polymer/Graphene Nanocomposites	509
	References	510

17 Graphene Oxide–Polyacrylamide Composites: Optical and Mechanical Characterizations 517

Gülşen Akın Evingür and Önder Pekcan

17.1	Introduction	517
17.2	Theoretical Considerations	521
	17.2.1 Universality	521
	17.2.2 Fractal Analysis	522
	17.2.3 Optical Energy Band Gap	522
	17.2.3.1 Tauc's Model	522
	17.2.3.2 Tail of Absorption Edge	523
	17.2.4 Elasticity	523
17.3	Experiment	524
	17.3.1 Preparation of PAAm–GO Composites	524
	17.3.2 Fluorescence Measurement	524
	17.3.3 UV Measurement	524
	17.3.4 Mechanical Measurement	524
17.4	Results and Discussion	525
17.5	Conclusion	535
	References	538

18 Synthesis, Characterization, and Applications of Polymer/Graphene Oxide Composite Materials 541

Carmina Menchaca-Campos, César García-Pérez, Miriam Flores-Domínguez, Miguel A. García-Sánchez, M.A. Hernández-Gallegos, Alba Covelo and Jorge Uruchurtu-Chavarín

18.1	Introduction	541
18.2	Graphene Oxide Synthesis	544
	18.2.1 Synthesis of GO by the Modified Hummers Method	544
	18.2.2 Synthesis of Graphene Oxide by Exfoliation Method	545
	18.2.3 Electro-Reduction of GO	545
18.3	Graphene Oxide Characterization	546
18.4	Applications of Polymer/Graphene Oxide Composite Materials	548
	18.4.1 Corrosion Protection Coating Application	548
	18.4.1.1 Corrosion Protection Properties of Sol-Gel Coatings Reinforced with Graphene Nanoparticles on Aluminum	549
	18.4.1.2 Graphene Oxide–Nylon Coating System for Steel	553

18.4.2	Storage Energy Applications	555
	18.4.2.1 Graphene Oxide–Nylon–Porphyrin System	556
	18.4.2.2 Nylon/$H_2T(p-NH_2)PP$ System Preparation	557
	18.4.2.3 Nylon/$H_2T(p-NH_2)PP$/GO Compound Preparation	557
	18.4.2.4 Characterization	558
	18.4.2.5 Electrochemical Evaluation	561
18.4.3	Water Solar Heater Application	564
	18.4.3.1 Graphene Oxide–Polypropylene (GO/PP) Composite	565
References		568

Index 575

Preface

Despite being just a one-atom-thick sheet of carbon, graphene is one of the most valuable nanomaterials. Initially discovered through scotch-tape-based mechanical exfoliation, graphene can now be synthesized in bulk using various chemical techniques. Counted among the contrasting properties of this remarkable material are its lightweight, thinness, flexibility, transparency, strength, and resistance, along with superior electrical, thermal, mechanical and optical properties. Due to these novel traits, graphene has attracted attention for use in cutting-edge applications in almost every area of technology, which are projected to change the world.

The *Handbook of Graphene* is presented in a unique eight-volume format covering all aspects relating to graphene—its development, synthesis, application techniques and integration methods; its modification and functionalization, its characterization tools and related 2D materials; physical, chemical and biological studies of graphene and related 2D materials; graphene composites; use of graphene in energy, healthcare and environmental applications (electronics, photonics, spintronics, bioelectronics and optoelectronics, photovoltaics, energy storage, fuel cells and hydrogen storage, graphene-based devices); and its large-scale production and characterization as well as graphene-related 2D material innovations and their commercialization.

This fourth volume of the handbook is solely focused on *graphene composite materials*. Some of the important topics include but are not limited to graphene composites; graphene-reinforced advanced composite materials; interfacial mechanical properties of graphene/substrate system: measurement methods and experimental analysis; graphene-based ceramic composites; *ab initio* design of 2D and 3D graphene-based nanostructure; graphene-based composite nanostructures; graphene-based composites with shape memory effect; graphene-based scroll structures: optical characterization and its application in resistive switching memory devices; fabrication and properties of copper–graphene composites; graphene–metal oxide composite as anode material in li-ion batteries; graphene/TiO_2 nanocomposites: synthesis routes, characterization, and solar cell applications; role of reduced graphene oxide nanosheet composition with ZnO nanostructures in gas sensing properties; functional graphene oxide/epoxy nanocomposite coatings with enhanced protection properties; supramolecular graphene-based systems for drug delivery; polymeric nanocomposites including graphene nanoplatelets; graphene oxide–polyacrylamide composites: optical and mechanical characterizations; and synthesis, characterization, and applications of polymer/graphene oxide composite materials.

In conclusion, thank you to all the authors whose expertise in their respective fields have contributed to this book as well as a sincere appreciation to the International Association of Advanced Materials.

February 15, 2019

1

Graphene Composites

Xiao-Jun Shen*, Xiao-Ling Zeng and Chen-Yang Dang

Materials and Textile Engineering College, Jiaxing University, Jiaxing, Zhejiang Province, China

Abstract

At present, the preparation and application of graphene and its composites have become the focus of the material industry. As a hexagonal honeycomb structure planar material composed of all the carbon atoms by sp^2 hybridization, graphene has many excellent physical and chemical properties; hence, its application prospect is broad. In this chapter, based on the brief introduction of four kinds of preparation methods, namely, graphene exfoliation and cleavage method, chemical vapor deposition method, epitaxial growth method, and chemically derived graphene method, the structure and properties of graphene and its composites are reviewed. This paper reviews the research and application of graphene and related polymer-based composites and elaborates the unique advantages of graphene nanocomposites in electronic devices, microwave absorption, bioengineering, etc.

Keywords: Graphene, graphene composites preparation, application

1.1 Introduction

Graphene is a two-dimensional carbon nanomaterial that is predicted to be nonexistent and shows a hexagonal honeycomb structure by sp^2 hybrid orbitals. Once reported, it triggered a great deal of global attention. Since 2004, Geim and Novoselov [1] have found an extremely simple micromechanical exfoliation method for the preparation of graphene, which disproves the theory that the perfect two-dimensional structure cannot exist on non-absolute zero degree stability. At the same time, it has shown unique advantages in the fields of electricity [2], light [3–5], machine [6, 7], thermodynamics [2], and biomedicine [8]. It has been favored by researchers and has become a hot topic for research.

First, the local superconductivity and high carrier mobility of graphene can be used in plasma [9, 10]. Many composite materials based on its properties, such as optical modulators [11–13], plasma-excited components [14–16], and broadband photodetectors [17], have also been applied in the past few years. Recently, an array of graphene monolayers covering the underlying metal layer has also been proposed, enabling efficient excitation of multielectron resonant modes under terahertz (THz) waves. This mode can be applied to displays [18], multichannel sensors [19], etc. Second, the advantages of graphene, such as high coefficient of thermal conductivity (about 5000 $Wm^{-1} K^{-1}$), excellent carrier mobility

Corresponding author: sxj908@163.com

Cengiz Ozkan (ed.) Handbook of Graphene: Volume 4, (1–26) © 2019 Scrivener Publishing LLC (about 200,000 cm$^2 \cdot$V$^{-1} \cdot$s^{-1}), and high specific surface area (about 2600 m$^2 \cdot$g^{-1}) [20], make it an ideal carrier, namely, as photocatalyst support [21], photonic crystal [22], or microwave absorption support [23]. Moreover, its chemical derivatives contain rich functional groups like hydroxyl, carboxyl, and epoxy groups. It is beneficial to the interfacial bonding of modified materials, and at the same time, its strength is high and it is also beneficial to the modified materials when combined with other materials [24]. Furthermore, its excellent thermal conductivity can be used to increase the thermal conductivity of solar cells, which can enhance their latent heat storage capacity and thermal conductivity [25]. In addition, its outstanding biocompatibility and solubility are good for biomaterials. Graphene and its derivatives have shown great potential as biosensor and bioimaging materials [26], and it shows great prospects for the application of membrane separation owing to its splendid selective performance [27].

1.2 History of Graphene

Actually, graphene exists naturally in nature, but it is difficult to peel off a single-layer structure. In fact, graphite is layered by graphene and graphite with a thickness of 1 mm contains approximately 3 million layers of graphene [28]. While the pencil strokes gently on the paper, the traces left may be several layers or even only one layer of graphene.

In 2004, two scientists from the University of Manchester, Andre Geim and Konstantin Novoselov, peeled off the graphite sheet from the highly oriented pyrolytic graphite. Then, they stuck both sides of the sheet to a special type of tape. The graphite sheet can be divided into two by tearing the tape. The sheets become much thinner with the repeated operations. Finally, they got a sheet made of only one layer of carbon atoms, which is named graphene.

After this, different new methods of preparing graphene have emerged. In 2009, Andre Geim and Konstantin Novoselov received the 2010 Nobel Prize in physics for the quantum Hall (QH) effect. Integer quantum Hall effect was found in monolayer and bilayer graphene systems. Furthermore, they discovered quantum Hall effect under normal temperature conditions [29–31]. Most physicists believed that thermodynamic fluctuations did not allow any two-dimensional crystals to exist at a finite temperature before the discovery of graphene. Although both theoretical and experimental communities believe that a perfect two-dimensional structure cannot be stable at non-absolute zero degrees, monolayer graphene can be prepared in experiments [32–34]. Therefore, the discovery of graphene immediately shocked the academic community of agglomeration physics.

On March 31, 2018, China's first fully automated mass production graphene organic solar optoelectronic device production line was launched in Heze, Shandong. The project mainly produces graphene organic solar cells (OSCs) that can generate electricity under low light [35]. It solves three major solar power problems: application limitations, angle sensitivity, and difficulty in modeling.

1.3 Synthesis of Graphene

Graphene has been on the research spotlight for its excellent properties and its rapidly developed production technology. Furthermore, the outstanding properties of graphene and its great potential applications have promoted the rapid development of graphene preparation technology. There are two main categories of graphene synthesis: the top-down approach and the bottom-up approach. The top-down approach is devoted to the decomposition of graphene precursor (graphite) from the stack into atomic layers, while the bottom-up approach enforces carbon molecules as building blocks, which are gained from alternative sources [36].

While the top-down approach includes mechanical exfoliation [37], ball milling [38], sonication [39], and electrochemical exfoliation [40], the bottom-up approach contains chemical vapor deposition [41], epitaxial growth on silicon carbide (SiC) [42], growth from metal-carbon melts [43], deposition [44], etc. Examples of top-down and bottom-up approaches on graphene synthesis are provided in the following text.

1.3.1 Top-Down Approach

1.3.1.1 Exfoliation and Cleavage

1.3.1.1.1 Mechanical Exfoliation

Graphene was first separated by the mechanical stripping method. The mechanical exfoliation method has been a turning point in the history of graphene. The mechanical stripping method uses transparent tape to press the highly oriented pyrolytic graphite (HOPG) sheet on the other surface and peel it off repeatedly to obtain a single layer or several layers of graphene. In 2004, Andre Geim and Konstantin Novoselov obtained monolayer graphene for the first time through this method [45]. They used a very simple method named "the scotch tape", or what we often call peel off method, to repeatedly split graphite crystals into increasingly thinner pieces. They proved that a two-dimensional crystal structure can exist at room temperature.

The mechanical or micro-mechanical exfoliation method is still the primary means to receive high-quality and defect-free graphene. The mechanical stripping method is easy to use and can obtain high-quality samples, and it is currently the main method for preparing single-layer high-quality graphene. However, its controllability is poor, and the produced graphene using this method has a small size and great uncertainty. At the same time, it has low efficiency and high cost, so it is not suitable for large-scale production.

1.3.1.1.2 Graphite Intercalation

Graphene can be synthesized by graphite intercalation. Graphite intercalation can be conducted in two ways: one is by adding small molecules into the layers of graphite; the other is by attaching molecules or polymers onto the sheets by noncovalent bonds, and hence graphite intercalation compounds (GICs) are formed. Sulfuric acid and hydrogen peroxide were used respectively as the intercalant and oxidant for the GIC formation in An's [46] team. The spheroidized natural graphite (SNG) was successfully converted into both expanded graphite and graphene nanoplatelets (GNPs). Bae *et al.* [47] have studied the effect of polymer intercalation on sound absorption and reported that the sound transmission loss was improved.

In GICs, the graphite layers remain unaltered with guest molecules located in the interlayer galleries [48]. Different intercalants could lead to GICs having different properties, which greatly benefit applications focusing on electrical [49, 50], thermal [51], chemical

[52, 53], and magnetic [49] performance. Other intercalants have some special functions: Horie *et al.* [54] have studied the conversion and selectivity of cinnamaldehyde (CAL) between graphite layers by platinum nanosheet intercalants.

1.3.1.2 Chemically Derived Graphene

At present, chemical conversion of graphite to graphene oxide (GO) has been a workable route [55]. Large quantities of graphene-based single sheets can be obtained by this way; namely, graphene nanoflakes/powder can be obtained by chemical reduction. Graphene that is obtained by this method is more viable than the bottom-up approach for the large-scale production of graphene. The reason is that the exfoliation degree of graphite and expandable graphite is lower using this method [36]. GO can be obtained by exfoliating graphite oxide easily via sonication. The reduction of GO method is also considered to be one of the best methods for preparing graphene currently. This method is easy to use and inexpensive to prepare. The graphene can be prepared in a large scale. Another advantage of this method is that it can produce functionalized graphene and GO, which also have broad application prospects.

There are many ways to synthesize GO [56–59]. The most popular synthesis approach to graphene is Hummers' method, and it has been modified by many researchers [60, 61]. The specific operation process of a classical preparation is as follows [62]: (a) First, oxidize graphite into graphite oxide by strong oxidizers such as concentrated sulfuric acid, concentrated nitric acid, and potassium permanganate. During this oxidation process, the oxygen functional groups are interpenetrated between the graphite layers, thus the space between graphite layers. With the increase in graphite layer spacing, the formation of graphene sheets in the next stage will be easy. (b) After being subjected to ultrasonic treatment for a period of time, single or multiple layers of GO can be formed in this stage. (c) Finally, GO is reduced to graphene by strong reductants like hydrazine hydrate aqueous sodium borohydride ($NaBH_4$).

However, some problems arise with this method. The conductivity and specific surface area of graphene will decline due to the graphene aggregations because of its low thickness, which has an effect on its application for optoelectronic equipment. In addition, crystal structure defects like the loss of carbon atoms on the carbon ring will occur during this process.

1.3.2 Bottom-Up Approach

1.3.2.1 Chemical Vapor Deposition

Chemical vapor deposition (CVD) is a vapor-depositing method of preparing a graphene film while it reacts with a carbon-containing organic gas used as a raw material [63]. This is the most effective method for producing graphene films. Thus, CVD is considered as the method with the highest potential for large-scale production. The CVD method is the most promising method for producing high-quality, large-area graphene, which is most ideal for industrially producing graphene films [64]. CVD can be classified into two main types [65]: thermal CVD and plasma CVD. The difference between them is the means of reducing the growth temperature. Plasma-enhanced chemical vapor deposition (PECVD) offers another route of graphene synthesis of a lower temperature [36], unlike thermal CVD.

The specific process of a typical CVD approach is as follows [66]: (a) passing gases such as hydrocarbon methane and ethanol into the surface of Cu and Ni heated on high-temperature metal substrates; (b) cooling the materials when the continuous reaction is done. Several layers or monolayers of graphene are formed on the surface of the substrate during the cooling process. This process involves two parts: the dissolution and diffusion of carbon atoms on the substrate.

Researchers continue to study this method. Rybin *et al.* [67] have obtained more details on the CVD method used in a real-time regimen for graphene film fabrication on nickel foils. Films of various thicknesses (from 3 to 53 layers and more) have been obtained in their report. Dong's group [68] focused on grain boundaries (GBs) in a graphene film for application and some parameters are defined for formation of overlapping grain boundaries (OLGBs), which leads to a deep insight into graphene CVD growth. It can be applied to the fabrication of capacitors with top electrodes of high-quality graphene by the introduction of an ultrathin Ti catalytic layer [69].

Graphene preparation using this method has the advantages of having a large area and being high quality, but the cost is high and the process conditions need to be further improved at this stage. The large-area graphene films cannot be used separately because of its low thickness. On account of this, it must be attached to macro-devices for use, such as touch screens, heating devices, etc.

1.3.2.2 Epitaxial Growth

Epitaxial growth includes two main approaches: epitaxial growth on silicon carbide (SiC) and metal-catalyzed epitaxial growth.

1.3.2.2.1 Epitaxial Growth on Silicon Carbide (SiC)

The method of epitaxial growth on SiC is that silicon atoms evaporated from the SiC surface escaping from the surface during the heating of SiC single crystals at high temperatures. The remaining carbon atoms are reconstructed in a self-assembled form, resulting in graphene-based SiC substrates. Graphene can be epitaxially grown on SiC substrates ideal to be used in transistors and circuits because thin graphene films with a size larger than 50 μm can be obtained through this method. This method can lead to epitaxial graphene, but the size of graphene flakes still depend on the size of SiC wafers. It is very important to the preparation of a certain graphene. High-quality graphene can be obtained through it, but this method has serious equipment requirements.

1.3.2.2.2 Growth from Metal–Carbon Melts

This method of epitaxial growth reacts with the surface of the catalytically active transition metal. This method is similar to the CVD approach. The metal-catalyzed epitaxial growth method is under ultra-high vacuum conditions, passing a hydrocarbon into a catalytically active transition metal substrate such as Pt, Ir, Ru, and Cu. The graphene is prepared by catalytically dehydrogenating the adsorbed gas by heating. Comparing these two methods, the advantage of the CVD method is that it can be performed at a lower temperature, so that the energy consumption of the preparation process can be reduced, and the graphene and the substrate can be easily separated by a chemical corrosion metal method, which is favorable for subsequent processing of the graphene.

The metal-catalyzed epitaxial growth needs a condition with ultra-high vacuum to compare the epitaxial growth of SiC. The metal-catalyzed epitaxial growth must be stable in high-temperature surroundings. The gas can fill in the entire metal substrate during the adsorption process and the growth process is a self-limiting process. It means that the substrate will not repeat absorption after adsorbing gas. Therefore, the graphene prepared by this method is mostly monolayer. Moreover, the uniform graphene can be prepared for a large scale.

1.3.3 Other Methods

Other methods of preparing graphene include carbon nanotube cutting method, graphite intercalation method [70], ion implantation method, high-pressure and high-temperature (HPHT) growth method, explosion method, and organic synthesis method.

In general, the existing methods cannot satisfy the requirements for the industrialization of graphene. In particular, industrialization requires the graphene production technology to produce graphene with large area and high purity under stable and low-cost conditions. This technical problem has not been solved yet.

The preparation method restricts the industrialization of graphene. The various top properties of graphene can be demonstrated when the graphene is of high quality. With the increase in layers and the accumulation of internal defects, many superior properties of graphene will be reduced. Only when a suitable industrialization of graphene processes emerges can the industrialization on the application of graphene be realized.

1.4 Characterization and Properties

1.4.1 Characterization

With the boom in graphene research, scientists have developed many new and important analysis testing technologies to study the surface morphology, chemical structure, and properties of graphene. Furthermore, it lays an experimental basis of inferring the physicochemical properties of graphene, as well as the interface formation and interfacial function in the composite material. At present, analytical techniques for the surface morphology and chemical structure of graphene have developed relatively well.

Graphene characterization is mainly divided into image and graph categories. The image category was mainly studied by optical microscopy, transmission electron microscopy (TEM), scanning electron microscopy (SEM), and atomic force microscopy (AFM). The spectra are represented by Raman spectrum, infrared spectroscopy (IR), X-ray photoelectron spectroscopy (XPS), and ultraviolet and visible spectrum (UV). Among them, TEM, SEM, Raman spectrum, AFM, and optical microscope are generally used to judge the number of graphene layers, while IR, XPS, and UV can be used to characterize the structure of graphene and monitor the synthesis of graphene.

1.4.1.1 Optical Imaging of Graphene Layers

At present, analytical techniques for the surface morphology and chemical structure of graphene have developed relatively well: AFM, TEM, and SEM. The single-layer graphene has a thickness of 0.335 nm and has about 1 nm undulation in the vertical direction [71]. The graphene prepared by different processes has large differences in morphology and in the number and structure of the graphene. However, the final products obtained by any method are more or less mixed with multilayer graphene sheets. The identification of monolayer graphene will be hindered in this situation. How to effectively identify the number and structure of graphene is one of the key steps in obtaining high-quality graphene.

After the discovery of graphene, optical microscopy is mainly used for imaging because it is the cheapest means, is nondestructive, and is readily available in the laboratory [72]. However, a combination of two or more techniques for complete imaging is often used to find different layers of graphene.

1.4.1.2 Atomic Force Microscopy (AFM)

There is not much to say regarding AFM. At present, there are main three operating modes of AFM [73]: contact mode [74], tapping mode [75], and noncontact mode [76]. Each mode of operation has its own characteristics, which are suitable for different experimental needs.

Atomic force characterization of graphene generally adopts the tapping mode [77]: the tapping mode is between the contact mode and the noncontact mode, which is a hybrid concept. AFM can be used to understand the fine morphology and precise thickness information of graphene. It uses the interaction force between the tip and the sample to sense the micro-cantilever, and then the laser reflection system detects the cantilever bending deformation. This indirectly measures the force between the needle tip samples to reflect the sample surface topography. Therefore, the characterization method mainly characterizes the thickness of the sheet, surface undulations, and topography, as well as the measurement of the height difference between layers.

AFM is the best way to determine whether it is graphene, because it can be used directly to observe the surface morphology of graphene. At the same time, the thickness of the graphene can be measured and then compared with the thickness of the single layer of graphene to determine whether there is a single layer of graphene. However, AFM has the disadvantage of low efficiency. This is because there are often some adsorbates on the surface of graphene, which makes the measured graphene thickness slightly larger than its actual thickness.

1.4.1.3 Transmission Electron Microscopy (TEM)

The TEM [78] electron beam passes through the ultrathin sample and reaches the imaging lens and detector. TEM has higher resolution imaging capabilities. It can observe the microscopic morphology of the graphene surface and can measure the clear structure and atomic-scale details of suspended graphene. At the same time, single-layer and multilayer graphene can be identified by using an electron diffraction pattern.

Using TEM, the number and size of the graphene sheets can be estimated by means of high-resolution electron micrographs at the edges or folds of the graphene. This method is relatively simple and fast. TEM seems to be the only tool that can resolve the atomic characteristics of graphene because graphene and reduced graphene oxide (rGO) are atomically thick layers.

1.4.1.4 Raman Spectroscopy

For the study of graphene, it is crucial to determine the number of layers and quantify the disorder. Laser Raman spectroscopy [79, 80] is precisely a standard ideal analytical tool for characterizing these two properties. Measuring the graphene by Raman spectrum, we can determine the structure and properties of the graphene layers, stacking modes, defects, edge structures, tension, and doping states. In addition, Raman spectroscopy also plays an important role in understanding the electronic phonon behavior of graphene [81].

The difference in electronic dispersion between multi-layer and single-layer graphene leads to significant differences in Raman spectra. A large number of studies have shown that graphene contains some second-order sum and frequency doubled Raman peaks. These Raman signals are often overlooked due to their weaker intensity. If these weak signals of the Raman spectra are analyzed, the electron-electron and electron-phonon interactions and the Raman scattering processes of graphene can be systematically studied.

As is well known, graphene is a two-dimensional honeycomb carbon lattice and is a zero gap material [82]. In order to adapt to its rapid application, people have developed a series of methods to open the band gap between graphene, for example, drilling, doping with boron or nitrogen, and chemical modification. This will introduce defects into the graphene, which will have a great impact on its electrical properties and device performance. Raman spectroscopy has unique advantages in characterizing graphene material defects. In summary, Raman spectroscopy is a very effective tool for judging the defect type and defect density of graphene.

When some molecules are adsorbed on the surface of a specific material (such as gold and silver), the signal intensity of the Raman spectrum of the molecule will increase significantly. We call this Raman scattering enhancement phenomenon the surface-enhanced Raman scattering (SERS) effect [83]. The SERS technology overcomes the weakness of the traditional Raman signal and can increase the Raman intensity by several orders of magnitude. Of course, you need to get a good base first if you want to get a strong boost signal. As a new type of two-dimensional ultra-thin carbon material, graphene can easily adsorb molecules and can easily meet the needs of a natural substrate. When certain molecules are adsorbed on the surface of graphene, the Raman signal of the molecule is manifestly enhanced.

1.4.2 Properties

Graphene possesses similar mechanical properties compared to carbon nanotubes (CNTs). However, graphene has excellent electrical and thermal properties and a large specific surface area thanks to its two-dimensional crystal structure. It possesses many excellent properties.

1.4.2.1 Electrical Transport Property

Each carbon atom of graphene is sp^2 hybridized and contributes one of the remaining p orbital electrons to form a large bond, and the electrons can move freely, giving graphene excellent electrical conductivity. When electrons are transported in graphene, they are less likely to scatter, and the mobility can reach 200,000 $cm^2/(V*s)$ [48], which is approximately

140 times that of electrons in silicon. Its conductivity is up to 104 S/m, which is the best conductivity at room temperature.

There are many studies about the electrical transport property of graphene. In the study of Bang et al. [84], the effect of ribbon width on electrical transport properties of graphene nanoribbons (GNRs) has been found, which broadens the application of graphene. There are also many researchers that studied the modification of graphene to enhance the properties of new materials [85, 86].

Graphene's stable lattice structure gives carbon atoms excellent electrical conductivity. When electrons in graphene move in orbit, they do not scatter due to lattice defects or the introduction of foreign atoms. Because the interaction between atoms is very strong, at room temperature, even if the surrounding carbon atoms collide, the interference in the electrons in the graphene is very small.

1.4.2.2 *Optical Properties*

The k-point energy and kinetic energy of the Brillouin zone have a linear relationship, and the effective mass of the carrier is zero [87]. Different from the traditional material electronic structure, it has quantum Hall effect and carrier near-ballistic transmission at room temperature. The monolayer graphene has a high light absorption, and the linear distribution of Dirac electrons causes each layer of graphene to absorb 2.3% of light from visible to terahertz wide band [88]. The ultra-fast kinetics of Dirac electrons and the presence of Pauli groups in the cone-shaped band structure endow graphene excellent nonlinear optical properties.

Graphene has excellent optical and electrical properties. It has compatibility with silicon-based semiconductor processes, unique two-dimensional atomic crystal materials, ultra-high thermal conductivity and carrier mobility, and ultra-wide bandwidth optical response spectrum of strong nonlinear optical properties. New graphene-based optoelectronic devices have been developed in the field of new optical and optoelectronic devices. Often, a combination of its optical properties and electrical properties was used in many fields [89–92].

1.4.2.3 *Mechanical Properties*

Graphene is the highest known substance to humans, harder than diamonds, and 100 times more powerful than the best steel in the world. Graphene's Young's modulus, Poisson's ratio, tensile strength and other basic mechanical properties are the main parameters of graphene's mechanical properties in recent years. It should be pointed out that the Young's modulus and other mechanical property parameters belong to the mechanical concept under the continuous medium framework. Thus, its thickness must be calculated by using the continuum hypothesis, and its mechanical properties are only meaningful because of the monolayer carbon atoms of graphene.

In terms of experimental detection, it is difficult to obtain the effective mechanical properties of graphene through traditional macroscopic material testing methods and techniques due to the two-dimensional structure of graphene. Therefore, an atomic force nano-indentation experimental system has been applied further [93]. Graphene is usually employed to improve the mechanical properties of polymer-based composites.

1.4.2.4 Thermal Properties

Graphene is a layered structural material and its thermal properties are mainly caused by lattice vibrations. It has been reported that there are six types of polar phonons in graphene by calculating the dispersion curve of the optical phonon and the acoustic phonon in graphene, which are as follows [94]:

1. Out-of-plane acoustic phonons (ZA mode phonons) and optical phonons (ZO mode phonons);
2. In-plane transverse acoustic phonons (TA mode phonons) and transverse optical phonons (TO mode phonons);
3. In-plane longitudinal acoustic phonons (LA mode phonons) and longitudinal optical phonons (LO mode phonons).

At present, among the commonly used thermal conductive materials, the thermal conductivity of aluminum foil is 160 W/mK, the thermal conductivity of copper is 380 W/mK, the thermal conductivity of single-walled CNTs is 3500 W/mK, and that of multiwalled carbon nanotubes (MWCNTs) is 3000 W/mK. The thermal conductivity of diamond is between 1000 and 2200 W/mK. The results show that monolayer graphene has a thermal conductivity of up to 5000 W/mK [48].

1.4.3 Application

Graphene has a wide range of applications, from electronic products to body armor and paper, and even future space elevators can use graphene as raw material.

The unique two-dimensional structure of graphene makes it a bright application prospect in the field of sensors. The huge surface area makes it very sensitive to the surrounding environment. Even the adsorption or release of a gas molecule can be detected. This test can be divided into direct detection and indirect detection. The single atom adsorption and release process can be directly observed by a TEM. The measurement of the Hall effect method can indirectly detect the adsorption and release of single atoms. When a gas molecule is adsorbed on the graphene surface, a local change in resistance occurs at the adsorption site. Graphene's good electrical and optical properties make it a very good candidate for the usage of transparent conductivity electrodes. Touch screens, liquid crystal displays, organic photovoltaic cells, organic light-emitting diodes, and the like all require good transparent conductive electrode materials. In particular, graphene is superior in mechanical strength and flexibility to indium tin oxide, which is a common material. Due to the high brittleness of indium tin oxide, it is relatively easy to damage. The graphene film in solution can be deposited in a large area. Using the CVD method, a large-area, continuous, transparent, high-conductivity, few-layered graphene film can be fabricated. It is mainly used for anodes of photovoltaic devices and its energy conversion efficiency is as high as 1.71%. It is about 55.2% of its energy conversion efficiency compared with those made of indium tin oxide material.

As we all known, graphene nano-walls perpendicular to the substrate surface were successfully prepared in 2002, and it is considered to be a very good field emission electron source material. In 2011, scholars of the Georgia Institute of Technology first reported the application of a vertical three-dimensional structure of functionalized multilayer graphene in thermal interface materials and its ultra-high equivalent thermal conductivity and ultra-low interface thermal resistance.

Graphene can be used as a conductive electrode for supercapacitors because of the particularly high surface area to mass ratio of it [95]. Scientists believe that such supercapacitors have a higher storage energy density than existing capacitors. Due to its modifiability in chemical functions, large contact area, atomic thickness, molecular gate structure, and other characteristics, graphene is an option for bacteria detection and diagnostic devices. Scientists believe that graphene is a material with this potential.

According to US researchers, one of the biggest obstacles to "space elevators" is making a 23,000-mile-long, strong cable that is connected to the space satellites. American scientists have confirmed that being the most intense substance on the earth, graphene is entirely suitable for use in manufacturing space elevator cables.

Some researchers have shown that stacking monolayers precisely will create a large number of new materials and equipment. Graphene and related monoatomic thickness crystals provide a broad choice for this purpose. The single-atomic layer crystals of graphene and boron nitride are stacked (one on top of another) to build a "multilayer cake" that can be used as a nano-scale transformer.

1.5 Graphene-Based Composites

Various polymers and nanoparticle composites have been exploited based on the peculiar properties of graphene. Graphene possesses similar mechanical properties to CNTs. However, graphene has excellent electrical and thermal properties and a large specific surface area thanks to its two-dimensional crystal structure. The polymer-based composites material is an important research direction for the application of graphene. It has broad application prospects because it exhibits excellent performance in the fields of energy storage [96, 97], liquid crystal devices [98, 99], electronic devices [100, 101], biomaterials [102, 103], sensing materials [104, 105], and catalyst supports [106, 107]. At present, the study of graphene-based composites mainly focuses on the graphene–polymer composites and graphene–nanoparticle composites.

1.5.1 Graphene–Polymer Composites

Graphene and its derivatives as fillers for polymer matrix composites have shown a great potential in many fields [108–110]. In the past few years, researchers have made successful attempts on synthesis graphene-based composites. But several challenges should be overcome for the wide use of graphene or GO based polymer composites:

1. Functionalization of graphene sheets [48]
2. Efficient mixing and homogeneous dispersion of materials [108]
3. Establish good interaction/interfacial bonding between graphene sheets and polymer matrices
4. Understanding the interfacial structure and properties [48]

This part is devoted to graphene and graphene-based polymer composites and discusses their properties and applications.

1.5.1.1 Synthesis of Graphene-Reinforced Polymer Composite

The synthesis of the graphene-reinforced polymer composite method is similar to CNTs. The most common synthesis methods of polymer matrix composites include *in situ* polymerization method, solution mixing method, and melt mixing method. These methods will be reviewed as follows.

1.5.1.1.1 *In Situ* Polymerization

The *in situ* polymerization method is a preparation method that forms a composite material by polymerizing monomers with graphene sheets and polymer monomers. The method can also open bonds of graphene sheets under the action of an initiator or make the functional groups on the surface of graphene sheets participate in the polymerization to obtain a polymer/graphene-based composite material [111].

This approach has produced a variety of composites, such as carboxyl-functionalized graphene oxide-polyaniline (CFGO-PANI) composite [112], poly(ethylene succinate) (PES)/graphene nanocomposites [113], and so on.

1.5.1.1.2 Solution Mixing Method

In the solution mixing method, the graphene sheets and polymers are respectively added to a solvent or the graphene sheets are directly added to the liquid polymer to be uniformly mixed. At the same time, the graphene sheets are dispersed during this phase. Finally, the composite material is prepared by removing the solvent by evaporation or precipitation.

1.5.1.1.3 Melt Mixing Method

A polymer/graphene composite material can be obtained by melt mixing method, which is melting the polymer matrix and then mixing, dispersing, and curing with graphene sheets.

An important feature of the melt mixing method is the use of high temperature and high shear forces to disperse graphene, and it is compatible with existing industrial application equipment. The equipment required is generally an extruder, an injection molding machine, and the like. Its advantages are as follows: (a) the preparation method is simple and (b) no surfactant or solvent was added to it during the preparation. Thus, the prepared composite material will not be polluted by the addition of a solvent or a surfactant.

1.5.1.2 Mechanical Properties

Polymer matrix composites are widely used in many industries for their unique properties and performance, especially their mechanical property. The excellent mechanical property of graphene has attracted the attention of researchers. They have studied the introduction of graphene into a polymer matrix to markedly improve its mechanical properties and electrical conductivity [89]. The mechanical properties depend on the concentration and distribution of the reinforcing phase in the host matrix, the interfacial adhesion, the reinforcement aspect ratio, etc.

Exploiting graphene's exceptional physical properties well in polymer composites remains a challenge because it is hard to control the dispersion of graphene. Controlling it well not only is beneficial for the mechanical properties of the composite but is also good for other properties [114]. One study showed the reinforcing effect of graphene in enhancing impact properties of epoxy composites. It provided details on how the cryogenic tensile and impact strength of the composites can be improved by graphene addition at a certain content [115].

1.5.1.3 Electrical Properties

The most attractive property of graphene is its electrical conductivity. Many researchers have studied its high conductivity. Its dielectric properties, electrochemical performances, electromagnetic wave absorption property, and other properties have been studied for many years [116–119].

With the development of supercapacitors, many researchers work at synthesizing a new electrode material with graphene that possesses high carrier mobility, thermal conductivity, elasticity, and stiffness. The main reason is that graphene has a theoretical specific capacitance of 2630 $m^2 \cdot g^{-1}$ during their experiment [120]. In the Boothroyd *et al.* study [114], they not only provided critical insights into understanding and controlling GNP orientation and dispersion within composites but also enhanced the electrical conductivity of the composite. A novel and facile synthesis approach of graphene composite is proposed, and this approach helps in the growth of high-performance supercapacitors [121]. As an effective nano-filler, GNP influenced the electromechanical responses of the MG/PLA(polylactic acid)/DBP(dibutyl phthalate) composites. During the temporal response experiment, it demonstrated that such composites have good recoverability under the electric field [122].

The reduced GO is suitable as a filler for composites even though GO is electrically insulating. The thermal reduction can eliminate the oxygen functional groups so that the electrical conductivity can be restored partially. It is recommendable and expedient to make rGO-based composite materials. Pham *et al.* [123] have reported a simple, environmentally friendly approach for preparing poly(methyl methacrylate)-reduced graphene oxide (PMMA-rGO) composites. The obtained PMMA-rGO composites possess excellent electrical properties. This highly conductive composite material was first prepared via self-assembly of the electrostatic interactions of positively charged PMMA latex particles and negatively charged graphene oxide sheets. Then, it was reduced by hydrazine.

It is certain that the NiO nanoparticles can disperse uniformly between graphene layers and the reduction of GO happened simultaneously [124]. They put forward an approach to improve stiffness for epoxy (EP) composites and showed great promise for the application of carbon/polymer composites [125].

In a word, graphene may greatly enhance the electrical conductivity of the composites while it is used as a filler to the insulating polymer matrix.

1.5.1.4 Thermal Conductivity

The coefficient of thermal conductivity (λ/κ) is controlled by the lattice vibrations (phonon). The graphene has shown a high coefficient of thermal conductivity (about 5000 W·m^{-1}·K^{-1}) [48], making it an outstanding filler to enhance the heat transport materials. Researchers want to obtain materials with high thermal conductivity and good mechanical properties to solve the heat dissipation problem of current electronic products. The excellent thermal stability of graphene makes it attractive filler for the fabrication of thermally stable composites. One study showed the improvement in thermal stability of C-P covalent bonds with GO phosphonic and phosphinic acids (GOPAs) [126].

Based on the unique properties of graphene, high-performance graphene-based composites are prepared by researchers from many fields. It shows great potential as a thermal conductive material in the future [127–130]. The thermal conductivity of the new form-stable composite phase change materials (PCMs) was highly improved from 0.305 to 0.985 (W/mK) [131]. Poly(vinylidene fluoride) (PVDF) is a thermoplastic polymer with excellent corrosion resistance, electrochemical stability, and thermal property. They doped the graphene in PVDF matrix in preparing graphene/PVDF composite membranes. It demonstrated that the thermal conductivity of composite membranes was significantly improved with the addition of graphene [132]. Zhang and coworkers [133] have studied the graphene-aligned composites in areas that need high thermal conductive materials. Furthermore, they discussed the properties of aligned composites and have achieved some results. The effects of interfacial interactions between graphene and polymer are also discussed.

With the rapid development of science and technology, more and more polymer materials and nano-materials are emerging. The exploration of the thermal conductivity of various new materials will be a new and captivating field, which will eventually lead to advancement in multifunctional composite materials.

1.5.1.5 Other Properties

The electromagnetic absorption properties and high thermal stability of the graphene-reinforced polymer have also been demonstrated. Thanks to the improved impedance matching and multi-interfacial polarization, the electromagnetic absorption properties are enhanced. Huang and coworkers [134] have successfully prepared the quaternary composites of CoNi@SiO$_2$ @graphene @PANI with enhanced electromagnetic absorption properties. Furthermore, the structures, the morphologies, and the electromagnetic parameters of obtained composites are analyzed in detail. In a word, the stronger electromagnetic absorption properties can be observed with the addition of graphene.

Similar to carbon nanotube, graphene sheets show excellent selectivity in terms of tribological performance [135]. Hassan's group has performed a detailed study on low-density polyethylene (LDPE)/GNPs prepared by paraffin oil (PO). The results show that the composites have lower coefficient of friction (COF) and wear rates when compared to the pure LDPE [136].

1.5.1.6 Application

Graphene-based polymer composite materials have solved many problems due to their widespread applications such as in aerospace, automobiles, coatings, and packaging materials. Graphene/polymer composites have proven to have limited use in energy storage, conductive polymers, antistatic coatings, and electromagnetic interference shielding.

Other potential applications of graphene polymer composites, such as being an effective photocatalyst for the photocatalytic degradation of Rose Bengal (RB) dye, have also been explored [137]. Inspired by natural nacre, Chen *et al.* have designed the synthesis of graphene oxide-polydopamine (GO-PDA) nanocomposites. In their study, the polymer composites have increased electrical conductivity, and the highlight of this work is that they proposed the most stable chemical connection between PDA and GO [138].

1.5.2 Graphene–Nanoparticle Composites

Roy [139] first proposed the concept of nanocomposites in 1984. This nanocomposite is a composite material having at least one dispersed phase and having a one-dimensional size of less than 100 nm. The synthesis and application of nanoparticles are relatively mature. New frontiers of nanoparticle (NP) composites have been opened with the addition of graphene. Recently, various metals [140], metal oxides [141], and semiconducting NPs [142] have been incorporated to graphene 2-D structures to realize the splendid properties of the composites. The outstanding properties of these composites indicated that the graphene has an effect on blocking dislocation propagation. In other words, graphene may prevent additional trap states along the sheets.

The NPs are directly decorated on the graphene sheets, and there are no molecular linkers to bridge the NPs and the graphene. Therefore, many types of NPs are deposited on graphene sheets to impart new functionality for their application in different fields.

1.5.2.1 Synthesis of Graphene–Nanoparticle Composites

Graphene-based nanocomposites exhibit many excellent properties in energy storage, liquid crystal devices, electronic devices, biomaterials, sensing materials, and catalyst support, and have broad application prospects. The main preparation methods include direct dispersion method, simultaneous formation, and *in situ* polymerization. These methods will be reviewed in this section.

1.5.2.1.1 *In Situ* Polymerization

In situ polymerization means that a suitable polymer is prepared first and then the nanoparticles are generated *in situ* through a chemical reaction under a controlled environment provided by the polymer (a nano-template or a nano-reactor) [143]. The polymers can provide nano-templates with strong polar groups into the molecular structure, such as sulfonic acid groups, carboxylic acid groups, hydroxyl groups, amine groups, and nitrile groups [144]. The strong interactions such as ionic bonds and complex coordination bonds can be formed between these strong polar groups and metal ions in the strongly polar inorganic nanoparticles. Thus, the probability of collisions between the particles can be reduced. At the same time, the polymer chains can prevent the excessive aggregation of the particles and facilitate the formation of nanoparticles. These polar polymers may be ionomers, ion exchange resins, homopolymers containing polar groups, copolymeric (random copolymers, block copolymers) polymer compounds and blends thereof, dendrimers, and so on. The polyester/rGO composites were prepared via *in situ* polymerization with terephthalic acid (PTA) and ethylene glycol containing well-dispersed GO [89].

1.5.2.1.2 Direct Dispersion

The direct dispersion method means that the nanoparticles are first prepared by a certain method and then the polymer-based inorganic nanocomposite materials are prepared by a suitable method for the NP and the polymer component (monomer or polymer). This method is one of the most widely used methods for preparing polymer-based inorganic nanocomposites, and most of the nanoparticles can be prepared into corresponding polymer-based nanocomposites by this method.

NPs have a strong tendency to agglomerate, and the nano-scale distribution cannot be restored by using conventional processing methods once the agglomeration appears. Therefore, the primary problem in the direct dispersion method is maintaining the nano-scale of the particles. At the same time, the polymer-based inorganic nanocomposite material can be prepared by uniformly dispersing the polymer components.

1.5.2.1.3 Simultaneous Formation

Simultaneous formation means that the nanoparticles as the dispersed phase and the polymer as the matrix are produced in the same preparation process, but the nanoparticles are preferentially formed when the monomers are polymerized. This differs from the direct dispersion method in which inorganic nanoparticles are previously prepared and then dispersed and polymerized in the monomer. There are only a few examples of this method, but it has its own distinguishing feature.

1.5.2.2 Properties

Composite materials with unique thermal, electrical, and antibacterial properties or other special properties will be valuable in science and in the industry field. Graphene-based nanocomposites have shown superior performance in thermal, electrical, photoelectric, and other potential fields.

1.5.2.2.1 Thermal Properties

The composite material consisting of graphene (GN as shorthand in the reference) and silver nanoparticles (AgNPs) may have weak thermal properties due to the nanofluids. With this in mind, Myekhlai *et al.* [145] used a facile and environment-friendly method for synthesizing GN-AgNPs composite material and studied its thermal conductivity. They found that the GN-AgNPs composite material considerably enhanced thermal conductivity.

1.5.2.2.2 Electrical

There are many researchers that study graphene sheets and NPs. Kholmanov and coworkers [146] found that the rGO/Cu NW hybrid films have improved electrical conductivity, oxidation resistance, substrate adhesion, and stability in harsh environments, which made it more useful. New lightweight flexible dielectric composites have been produced by Dimiev et al. [147]. They studied the permittivity and loss tangent values of the composites, and these parameters have been changed by the conductive filler type and content. $BaTiO_3$ nanoparticles have been added into the graphene-polyimide (PI) system as a filler [148]. Thus, a flexible three-phase (PI–graphene–$BaTiO_3$) composite material with enhanced dielectric permittivity was prepared. The composites possess good permittivity and lower dielectric loss thanks to the unique dimensional structure of graphene.

1.5.2.2.3 Other Properties

The graphene-based nanocomposites also perform well in friction [149]. It is beneficial to the development of the composite material if researchers can adjust the wear coefficient of the composite material by changing the relevant parameters.

In another aspect, the increasing energy demands have motivated researchers to study more materials. Researchers continue research on energy storage materials. There are two main aspects for energy storage in this section: ion batteries and supercapacitors.

With the in-depth study of graphene, the application of graphene reinforcements in composites has also received increasing attention. Multifunctional NP composites with graphene and high-strength porous ceramic materials [150] enhance the special properties of the composites.

1.6 Future Prospects

Many problems still exist even though graphene has excellent electrical, optical, thermal, and mechanical performance. In the development of graphene application process, the dispersion caused by the surface properties greatly limits the display of the excellent performance of graphene. Achieving a good dispersion of graphene is an effective way of solving the current research dilemma in the application of graphene.

Although graphene has a rich and interesting history, people are paying more attention to its future development. Integrating the research results of the previous researchers on the properties of graphene, we put forward the following suggestions:

1. In order to obtain 3D network relationships of surface structure, surface properties, and properties, we can start from the surface structure of graphene; a variety of analytical techniques are used to comprehensively and quantitatively study the relationship between the surface structure and surface properties of graphene, and it should be combined with practical application performance.
2. In the preparation process of graphene, realizing the precise control of surface structure, size, and number of graphene will play a key role in the in-depth

study of the surface properties of graphene and will provide an accurate template for the functionalization of graphene.
3. In terms of interface interactions, the interface between graphene and its derivatives and composite materials or the substrates can be further studied.

In short, an in-depth study of the properties of graphene will play a positive role in promoting the application of graphene in composite materials, nano-coatings, and electronic devices, and at the same time, it is important in expanding the potential properties of graphene.

Acknowledgment

This work is financially supported by the National Natural Science Foundation of China (No. 11502096).

References

1. Geim, A.K. and Novoselov, K.S., The rise of graphene. *Nat. Mater.*, 6, 183–191, 2007.
2. Verma, D., Gope, P.C., Shandilya, A. *et al.*, Mechanical-thermal-electrical and morphological properties of graphene reinforced polymer composites: A review. *T. Indian I. Metals*, 67, 6, 803–816, 2014.
3. Diez-Pascual, A.M., Sanchez, J.A.L., Capilla, R.P. *et al.*, Recent developments in graphene/polymer nanocomposites for application in polymer solar cells. *Polymers*, 10, 2, 217, 2018.
4. Liu, J., Zhang, L., Ding, Z. *et al.*, Tuning work functions of graphene quantum dot-modified electrode for polymer solar cells application. *Nano*, 9, 10, 3524, 2017.
5. Kim, H. and Yong, K., A highly efficient light capturing 2D (nanosheet)-1D (nanorod) combined hierarchical ZnO nanostructure for efficient quantum dot sensitized solar cells. *Phys. Chem. Chem. Phys.*, 15, 6, 2109–2116, 2013.
6. Li, X., Sun, M., Shan, C. *et al.*, Mechanical properties of 2D materials studied by *in situ* microscopy techniques. *Adv. Mater Interfaces*, 1701246, 2018.
7. Argentero, G., Mittelberger, A., Reza, M., Monazam, A. *et al.*, Unraveling the 3D atomic structure of a suspended graphene/hBN van der Waals Heterostructurep. *Nano Lett.*, 17, 1409–1416, 2017.
8. Mohan, V.B., Lau, K.T., Hui, D. *et al.*, Graphene-based materials and their composites: A review on production, applications and product limitations. *Compos. Part B-Eng.*, 142, 200–220, 2018.
9. Novoselov, K.S., Fal'Ko, V.I., Colombo, L. *et al.*, A roadmap for graphene. *Nature*, 490, 7419, 192–200, 2012.
10. Koppens, F.H.L., Chang, D.E., Abajo, F.J.G.D., Graphene plasmonics: A platform for strong lightmatter interactions. *Nano Lett.*, 11, 8, 3370, 2011.
11. Liu, M., Yin, X., Ulin-Avila, E., Geng, B., Zentgraf, T., Ju, L. *et al.*, A graphene-based broadband optical modulator. *Nature*, 474, 7349, 64–67, 2011.
12. He, X., Zhao, Z.-Y., Shi, W., Graphene-supported tunable near-IR metamaterials. *Opt. Lett.*, 40, 2, 178–181, 2015.
13. Xiao, S., Wang, T., Liu, T., Yan, X., Li, Z., Xu, C., Active modulation of electromagnetically induced transparency analogue in terahertz hybrid metal-graphene metamaterials. *Carbon*, 126, 271–278, 2018.

14. Arezoomandan, S., Quispe, H.O.C., Ramey, N., Nieves, C.A., Sensale-Rodriguez, B., Graphene based reconfigurable terahertz plasmonics and metamaterials. *Carbon*, 112, 177–184, 2017.
15. He, X., Gao, P., Shi, W., A further comparison of graphene and thin metal layers for plasmonics. *Nano*, 8, 19, 10388–10397, 2016.
16. He, X., Lin, F., Liu, F. *et al.*, Terahertz tunable graphene Fano resonance. *Nanotechnology*, 27, 48, 485202, 2016.
17. Pospischil, A., Humer, M., Furchi, M.M. *et al.*, CMOS-compatible graphene photodetector covering all optical communication bands. *Nat. Photonics*, 7, 11, 892–896, 2013.
18. Tuteja, S.K., Ormsby, C., Neethirajan, S., Noninvasive label-free detection of cortisol and lactate using graphene embedded screen-printed electrode. *Nano-Micro Lett.*, 10, 3, 41, 2018.
19. Chen, X., Fan, W., Song, C., Multiple plasmonic resonance excitations on graphene metamaterials for ultrasensitive terahertz sensing. *Carbon*, 133, 416–422, 2018.
20. Qi, L., Yu, J., Jaroniec, M., Preparation and enhanced visible-light photocatalytic H_2-production activity of CdS-sensitized Pt/TiO_2 nanosheets with exposed (001) facets. *Phys. Chem. Chem. Phys.*, 13, 19, 8915, 2011.
21. Qian, X.F., Ren, M., Fang, M.Y. *et al.*, Hydrophilic mesoporous carbon as iron(III)/(II) electron shuttle for visible light enhanced Fenton-like degradation of organic pollutants. *Appl. Catal. B-Environ.*, 231, 108–114, 2018.
22. Likodimos, V., Photonic crystal-assisted visible light activated TiO_2 photocatalysis. *Appl. Catal. B-Environ.*, 230, 269–303, 2018.
23. Li, C., Huang, Y., Chen, J., Dopamine-assisted one-pot synthesis of grapheme @Ni@C composites and their enhanced microwave absorption performance. *Mater. Lett.*, 154, 136–139, 2015.
24. Imtiaz, S., Siddiq, M., Kausar, A. *et al.*, A review featuring fabrication, properties and applications of carbon nanotubes (cnts) reinforced polymer and epoxy nanocomposites. *Chinese J. Polym. Sci.*, 36, 4, 445–461, 2018.
25. Atinafu, D.G., Dong, W.J., Huang, X.B. *et al.*, One-pot synthesis of light-driven polymeric composite phase change materials based on N-doped porous carbon for enhanced latent heat storage capacity and thermal conductivity. *Sol. Energ. Mat. Sol. C*, 179, 392–400, 2018.
26. Banerjee, A.N., Graphene and its derivatives as biomedical materials: Future prospects and challenges. *J. R. Soc. Interface*, 6, 8, 3, 2018.
27. Song, N., Gao, X.L., Ma, Z. *et al.*, A review of graphene-based separation membrane: Materials, characteristics, preparation and applications. *Desalination*, 437, 59–72, 2018.
28. Fan, Y., Jiao, W., Huang, C., Effect of the noncovalent functionalization of graphite nanoflakes on the performance of MnO_2/C composites. *J. Appl. Electrochem.*, 48, 2, 187–199, 2018.
29. Gerstner, E., Nobel Prize 2010: Andre Geim & Konstantin Novoselov. *Nat. Phys.*, 6, 11, 836, 2010.
30. Ledwith, P., Kort-Kamp, W.J.M., Dalvit, D.A.R., Topological phase transitions and quantum hall effect in the graphene family. *Front. Optics*, 97, 16, 165426, 2017.
31. Zhu, J., Li, J., Wen, H., Gate-controlled tunneling of quantum Hall edge states in bilayer graphene. *Phys. Rev. Lett.*, 120, 5, 057701, 2018.
32. Sun, X., Mu, Y., Zhang, J. *et al.*, Tuning the self-assembly of oligothiophenes on chemical vapor deposition graphene: Effect of functional group, solvent, and substrate. *Chem. Asian J.*, 9, 7, 1888–1894, 2014.
33. Zhang, H., Yin, H.F., Zhang, K.B. *et al.*, Progress of surface plasmon research based on time-dependent density functional theory. *Acta Phys. Sin-Ch. Ed.*, 64, 7, 2015.
34. Ariga, K., Li, M., Richards, G.J. *et al.*, Nanoarchitectonics: A conceptual paradigm for design and synthesis of dimension-controlled functional nanomaterials. *J. Nanosci. Nanotechnol.*, 11, 1, 1–13, 2011.

35. Dong, H.S., Sang, W.S., Kim, J.M. et al., Graphene transparent conductive electrodes doped with graphene quantum dots-mixed silver nanowires for highly-flexible organic solar cells. *J. Alloy Compd.*, 744, 1–6, 2018.
36. Saqib Shams, S., Zhang, R., Zhu, J. et al., Graphene synthesis: A review. *Mater. Sci. Poland*, 33, 3, 566–578, 2015.
37. Martinez, A., Fuse, K., Yamashita, S., Mechanical exfoliation of graphene for the passive mode-locking of fiber lasers. *Appl. Phys. Lett.*, 99, 12, 3077, 2011.
38. Jeon, I.Y., Shin, Y.R., Sohn, G.J. et al., Edge-carboxylated graphene nanosheets via ball milling. *Proc. Natl. Acad. Sci. USA*, 109, 15, 5588–5593, 2012.
39. Lin, Z., Karthik, P., Hada, M. et al., Simple technique of exfoliation and dispersion of multilayer graphene from natural graphite by ozone-assisted sonication. *Nanomaterials*, 7, 6, 125, 2017.
40. Dai, W., Chung, C.Y., Hung, T.T. et al., Superior field emission performance of graphene/carbon nanofilament hybrids synthesized by electrochemical self-exfoliation. *Mater. Lett.*, 205, 223–225, 2017.
41. Habib, M.R., Liang, T., Yu, X. et al., A review of theoretical study of graphene chemical vapor deposition synthesis on metals: Nucleation, growth, and the role of hydrogen and oxygen. *Rep. Prog. Phys.*, 81, 3, 036501, 2018.
42. Forti, S., Rossi, A., Büch, H. et al., Electronic properties of single-layer tungsten disulfide on epitaxial graphene on silicon carbide. *Nano*, 9, 42, 16412–16419, 2017.
43. Carreño, N.L.V., Barbosa, A.M., Duarte, V.C. et al., Metal-carbon interactions on reduced graphene oxide under facile thermal treatment: Microbiological and cell assay. *J. Nanomater.*, 2017, 4, 6059540, 2017.
44. Trudeau, C., Dionbertrand, L.I., Mukherjee, S. et al., Electrostatic deposition of large-surface graphene. *Materials*, 11, 1, 2018.
45. Novoselov, K.S., Geim, A.K., Morozov, S.V., Jiang, D., Zhang, Y., Dubonos, S.V., Grigorieva, I.V., Firsov, A.A., Electric field effect in atomically thin carbon films. *Science*, 306, 666–669, 2004.
46. An, J.C., Lee, E.J., Hong, I., Preparation of the spheroidized graphite-derived multi-layered graphene via GIC (graphite intercalation compound) method. *J. Ind. Eng. Chem.*, 47, 56–61, 2016.
47. Bae, Y.H., Kwon, T.S., Yu, M.J. et al., Acoustic characteristics and thermal properties of polycarbonate/(graphite intercalation compound) composites. *Polym-Korea*, 41, 2, 189, 2017.
48. Singh, V., Joung, D., Lei, Z. et al., Graphene based materials: Past, present and future. *Prog. Mater. Sci.*, 56, 8, 1178–1271, 2011.
49. Ovsiienko, I., Matzui, L., Berkutov, I. et al., Magnetoresistance of graphite intercalated with cobalt. *J. Mater. Sci.*, 53, 1, 716–726, 2018.
50. Maruyama, S., Fukutsuka, T., Miyazaki, K. et al., Observation of the intercalation of dimethyl sulfoxide-solvated lithium ion into graphite and decomposition of the ternary graphite intercalation compound using *in situ*, Raman spectroscopy. *Electrochim. Acta*, 265, 41–46, 2018.
51. Poláková, L., Sedláková, Z., Ecorchard, P. et al., Poly(meth)acrylate nanocomposite membranes containing *in situ*, exfoliated graphene platelets: Synthesis, characterization and gas barrier properties. *Eur. Polym. J.*, 94, 431–445, 2017.
52. Rozmanowski, T. and Krawczyk, P., Influence of chemical exfoliation process on the activity of $NiCl_2$-$FeCl_3$-$PdCl_2$-graphite intercalation compound towards methanol electrooxidation. *Appl. Catal. B-Environ.*, 224, 53–59, 2017.
53. Jeon, I., Yoon, B., He, M. et al., Hyperstage graphite: Electrochemical synthesis and spontaneous reactive exfoliation. *Adv. Mater.*, 30, 3, 1704538, 2018.
54. Horie, M., Takahashi, K., Nanao, H. et al., Selective hydrogenation of cinnamaldehyde over platinum nanosheets intercalated between graphite layers. *J. Nanosci. Nanotechnol.*, 18, 1, 80–85, 2018.

55. Marcano, D.C., Kosynkin, D.V., Berlin, J.M. et al., Improved synthesis of graphene oxide. *ACS Nano*, 12, 2, 2078, 2018.
56. Justh, N., Berke, B., László, K. et al., Thermal analysis of the improved Hummers' synthesis of graphene oxide. *J. Therm. Anal. Calorim.*, 1–6, 2017.
57. Wu, X., Ma, L., Sun, S. et al., A versatile platform for the highly efficient preparation of graphene quantum dots: Photoluminescence emission and hydrophilicity-hydrophobicity regulation and organelle imaging. *Nano*, 10, 3, 1532–1539, 2017.
58. Mandal, P., Naik, M.J.P., Saha, M., Room temperature synthesis of graphene nanosheets. *Cryst. Res. Technol.*, 53, 2, 1700250, 2018.
59. Mazanek, V., Matejkova, S., Sedmidubsky, D. et al., One step synthesis of B/N co-doped graphene as highly efficient electrocatalyst for oxygen reduction reaction - synergistic effect of impurities. *Chemistry*, 24, 4, 928–936, 2017.
60. Hack, R., Correia, C.H.G., Zanon, R.A.D.S. et al., Characterization of graphene nanosheets obtained by a modified Hummer's method. *Matéria*, 23, 1, 2018.
61. Sinitsyna, O.V., Meshkov, G.B., Grigorieva, A.V. et al., Blister formation during graphite surface oxidation by Hummers' method. *Beilstein J. Nanotech.*, 9, 407–414, 2018.
62. Yuan, R., Yuan, J., Wu, Y. et al., Graphene oxide-monohydrated manganese phosphate composites: Preparation *via*, modified Hummers method. *Colloid Surface A*, 547, 56–63, 2018.
63. Li, X., Cai, W., An, J. et al., Large-area synthesis of high-quality and uniform graphene films on copper foils. *Science*, 324, 5932, 1312, 2009.
64. Ani, M.H., Kamarudin, M.A., Ramlan, A.H. et al., A critical review on the contributions of chemical and physical factors toward the nucleation and growth of large-area graphene. *J. Mater. Sci.*, 53, 10, 7095–7111, 2018.
65. Naghdi, S., Rhee, K.Y., Park, S.J., A catalytic, catalyst-free, and roll-to-roll production of graphene *via* chemical vapor deposition: Low temperature growth. *Carbon*, 127, 1–12, 2018.
66. Tu, R., Liang, Y., Zhang, C. et al., Fast synthesis of high-quality large-area graphene by laser CVD. *Appl. Surf. Sci.*, 445, 204–210, 2018.
67. Rybin, M.G., Kondrashov, I.I., Pozharov, A.S. et al., *In situ* control of CVD synthesis of graphene film on nickel foil. *Phys. Status Solidi.*, 255, 1700414, 2017.
68. Dong, J., Wang, H., Peng, H. et al., Formation mechanism of overlapping grain boundaries in graphene chemical vapor deposition growth. *Chem. Sci.*, 8, 3, 2209, 2017.
69. Park, B.J., Choi, J.S., Eom, J.H. et al., Defect-free graphene synthesized directly at 150°C *via* chemical vapor deposition with no transfer. *ACS Nano*, 12, 2, 2008–2016, 2018.
70. Lobiak, E.V., Shlyakhova, E.V., Gusel'Nikov, A.V. et al., Carbon nanotube synthesis using Fe-Mo/MgO catalyst with different ratios of CH_4 and H_2 gases. *Phys. Status Solidi*, 255, 1, 1700274, 2018.
71. Wang, C.Y., Jing, W.X., Jiang, Z.D. et al., The measurement of single-layer thickness of graphene materials by high resolution transmission electron microscopy. *Acta Metrologica Sinica*, 38, 2, 145–148, 2017.
72. Duong, D.L., Gang, H.H., Lee, S.M. et al., Probing graphene grain boundaries with optical microscopy. *Nature*, 490, 7419, 235, 2012.
73. Almeida, C.M., Carozo, V., Prioli, R. et al., Identification of graphene crystallographic orientation by atomic force microscopy. *J. Appl. Phys.*, 110, 8, 666–301, 2011.
74. Lindvall, N., Kalabukhov, A., Yurgens, A., Cleaning graphene using atomic force microscope. *J. Appl. Phys.*, 111, 6, 666, 2012.
75. Nemes-Incze, P., Osváth, Z., Kamarás, K. et al., Anomalies in thickness measurements of graphene and few layer graphite crystals by tapping mode atomic force microscopy. *Carbon*, 46, 11, 1435–1442, 2008.

76. Sun, Z., Hamalainen, S.K., Sainio, J. et al., Topographic and electronic contrast of the graphene moiré on Ir(111) probed by scanning tunneling microscopy and noncontact atomic force microscopy. *Phys. Rev. B Condens. Matter.*, 83, 8, 210–216, 2010.
77. Chu, L., Korobko, A.V., Bus, M. et al., Fast and controlled fabrication of porous graphene oxide: Application of AFM tapping for mechano-chemistry. *Nanotechnology*, 29, 18, 185301, 2018.
78. Ziatdinov, M., Dyck, O., Maksov, A. et al., Deep learning of atomically resolved scanning transmission electron microscopy images: Chemical identification and tracking local transformations. *ACS Nano*, 12742–12752, 2018.
79. Xia, M., A review on applications of two-dimensional materials in surface enhanced Raman spectroscopy. *Internet. J. Spectro.*, 2018-1-1, 2018, 2018.
80. Wu, J.B., Lin, M.L., Cong, X. et al., Raman spectroscopy of graphene-based materials and its applications in related devices. *Chem. Soc. Rev.*, 47, 5, 1822–1873, 2018.
81. Ferrari, A., Ferrante, C., Virga, A. et al., Raman spectroscopy of graphene under ultrafast laser excitation. *Nat. Commun.*, 9, 1, 2018.
82. Dass, D., Structural analysis, electronic properties, and band gaps of a graphene nanoribbon: A new 2D materials. *Superlattice Microst.*, 115, 88–107, 2018.
83. Cai, Q., Mateti, S., Yang, W. et al., Boron nitride nanosheets improve sensitivity and reusability of surface enhanced Raman spectroscopy. *Angew. Chem. Int. Ed.*, 128, 29, 8597–8597, 2016.
84. Bang, K., Chee, S.S., Kim, K. et al., Effect of ribbon width on electrical transport properties of graphene nanoribbons. *Nano Converg.*, 5, 1, 7, 2018.
85. Shcherbakov, D., Stepanov, P., Watanabe, K. et al., *Electrical Transport in hybrid graphene/CrI$_3$ junctions*, APS March Meeting. American Physical Society, 2018.
86. Nazir, G., Khan, M.F., Aftab, S. et al., Gate tunable transport in graphene/MoS$_2$/(Cr/Au) vertical field-effect transistors. *Nanomaterials*, 8, 1, 14, 2018.
87. Warmbier, R. and Quandt, A., Brillouin zone grid refinement for highly resolved *ab initio* THz optical properties of graphene. *Comput. Phys. Commun.*, 228, 96–99, 2018.
88. Lin, I.T., Liu, J.M., Shi, K.Y. et al., Terahertz optical properties of multilayer graphene: Experimental observation of strong dependence on stacking arrangements and misorientation angles. *Phys. Rev. B Condens. Matter.*, 86, 23, 278–281, 2012.
89. Liu, K., Chen, L., Chen, Y. et al., Preparation of polyester/reduced graphene oxide composites via *in situ* melt polycondensation and simultaneous thermo-reduction of graphene oxide. *J. Mater. Chem.*, 21, 24, 8612–8617, 2011.
90. Anishmadhavan, A., Kalluri, S., Kchacko, D. et al., Electrical and optical properties of electrospun TiO2-graphene composite nanofibers and its application as DSSC photo-anodes. *RSC Adv.*, 2, 33, 13032–13037, 2012.
91. Yuan, J., Ma, L.P., Pei, S. et al., Tuning the electrical and optical properties of graphene by ozone treatment for patterning monolithic transparent electrodes. *ACS Nano*, 7, 5, 4233–4241, 2013.
92. Ho, X., Lu, H., Liu, W. et al., Electrical and optical properties of hybrid transparent electrodes that use metal grids and graphene films. *J. Mater. Res.*, 28, 4, 620–626, 2013.
93. Roos, W.H., How to perform a nanoindentation experiment on a virus. *Methods Mol. Biol.*, 783, 251–264, 2011.
94. Zou, J.H., Ye, Z.Q., Cao, B.Y., Phonon thermal properties of graphene from molecular dynamics using different potentials. *J. Chem. Phys.*, 145, 13, 134705, 2016.
95. Qiu, B., Li, Q., Shen, B. et al., Stöber-like method to synthesize ultradispersed Fe$_3$O$_4$ nanoparticles on graphene with excellent Photo-Fenton reaction and high-performance lithium storage. *Appl. Catal. B-Environ.*, 183, 216–223, 2016.

96. Huo, P., Zhao, P., Wang, Y. *et al.*, A roadmap for achieving sustainable energy conversion and storage: Graphene-based composites used both as an electrocatalyst for oxygen reduction reactions and an electrode material for a supercapacitor. *Energies*, 11, 1, 167, 2018.
97. Kim, M., Hwang, H.M., Park, G.H. *et al.*, Graphene-based composite electrodes for electrochemical energy storage devices: Recent progress and challenges. *Flatchem*, 6, 48–76, 2017.
98. Wang, H., Liu, B., Wang, L. *et al.*, Graphene glass inducing multidomain orientations in cholesteric liquid crystal devices toward wide viewing angles. *ACS Nano*, 12(7), 2018.
99. Zhang, J., Seyedin, S., Gu, Z. *et al.*, Liquid crystals of graphene oxide: A route towards solution-based processing and applications. *Part. Part. Syst. Char.*, 34, 9, 1600396, 2017.
100. Shtein, M., Nadiv, R., Buzaglo, M. *et al.*, Graphene-based hybrid composites for efficient thermal management of electronic devices. *ACS Appl. Mater. Interfaces*, 7, 42, 23725–23730, 2015.
101. Wan, S.J., Wei, H.U., Jiang, L. *et al.*, Bioinspired graphene-based nanocomposites and their application in electronic devices. *Chinese Sci. Bull.*, 62, 27, 3173–3200, 2017.
102. Ryan, A.J., Kearney, C.J., Shen, N. *et al.*, Electroconductive biohybrid collagen/pristine graphene composite biomaterials with enhanced biological activity. *Adv. Mater.*, 1706442, 2018.
103. Weng, W., Nie, W., Zhou, Q. *et al.*, Controlled release of vancomycin from 3D porous graphene-based composites for dual-purpose treatment of infected bone defects. *RSC Adv.*, 7, 5, 2753–2765, 2017.
104. D'Elia, E., Barg, S., Ni, N. *et al.*, Self-healing graphene-based composites with sensing capabilities. *Adv. Mater.*, 27, 32, 4788, 2015.
105. Samaddar, P., Son, Y.S., Tsang, D.C.W. *et al.*, Progress in graphene-based materials as superior media for sensing, sorption, and separation of gaseous pollutants. *Coordin. Chem. Rev.*, 368, 93–114, 2018.
106. Cao, P., Huang, C., Zhang, L. *et al.*, One-step fabrication of RGO/HNBR composites *via* selective hydrogenation of NBR with graphene-based catalyst. *RSC Adv.*, 5, 51, 41098–41102, 2015.
107. Gürsel, S.A., Layer-by-layer polypyrrole coated graphite oxide and graphene nanosheets as catalyst support materials for fuel cells. *Fullerene Sci. Technol.*, 21, 3, 233–247, 2013.
108. Silva, M., Alves, N.M., Paiva, M.C., Graphene-polymer nanocomposites for biomedical applications. *Polym. Adv. Technol.*, 29, 2, 687–700, 2017.
109. He, D., Xue, W., Zhao, R. *et al.*, Reduced graphene oxide/Fe-phthalocyanine nanosphere cathodes for lithium-ion batteries. *J. Mater. Sci.*, 53, 12, 9170–9179, 2018.
110. Embrey, L., Nautiyal, P., Loganathan, A. *et al.*, Three-dimensional graphene foam induces multifunctionality in epoxy nanocomposites by simultaneous improvement in mechanical, thermal, and electrical properties. *ACS Appl. Mater. Interfaces*, 9, 45, 39717–39727, 2017.
111. Yoo, Y., Choi, H.S., Kim Y.S. *et al.*, Polyamide based polymer compositions comprising cyclic compound and polymer based composite material using the same. US 20180030220, 2018.
112. Liu, Y., Deng, R., Wang, Z. *et al.*, Carboxyl-functionalized graphene oxide-polyaniline composite as a promising super capacitor material. *J. Mater. Chem.*, 22, 27, 13619–13624, 2012.
113. Zhao, J., Wang, X., Zhou, W. *et al.*, Graphene-reinforced biodegradable poly(ethylene succinate) nanocomposites prepared by *in situ* polymerization. *J. Appl. Polym. Sci.*, 130, 5, 3212–3220, 2014.
114. Boothroyd, S.C., Johnson, D.W., Weir, M.P. *et al.*, Controlled structure evolution of graphene networks in polymer composites. *Chem. Mater.*, 30, 5, 1524–1531, 2018.
115. Shen, X.J., Liu, Y., Xiao, H.M. *et al.*, The reinforcing effect of graphene nanosheets on the cryogenic mechanical properties of epoxy resins. *Compos. Sci. Technol.*, 72, 13, 1581–1587, 2012.

116. Rubrice, K., Castel, X., Himdi, M. et al., Dielectric characteristics and microwave absorption of graphene composite materials. *Materials*, 9, 10, 825, 2016.
117. Almadhoun, M.N., Alshareef, H.N., Bhansali, U.S. et al., Graphene-based composite materials, method of manufacture and applications thereof: US, WO 2014058860 A1, 2014.
118. Bica, I., Anitas, E.M., Averis, L.M.E. et al., Magnetodielectric effects in composite materials based on paraffin, carbonyl iron and graphene. *J. Ind. Eng. Chem.*, 21, 1, 1323–1327, 2014.
119. Li, H., Xu, C., Chen, Z. et al., Graphene/poly(vinylidene fluoride) dielectric composites with polydopamine as interface layers. *Sci. Eng. Compos. Mater.*, 24, 3, 327–333, 2017.
120. Kandasamy, S.K. and Kandasamy, K., Recent advances in electrochemical performances of graphene composite (graphene-polyaniline/polypyrrole/activated carbon/carbon nanotube) electrode materials for supercapacitor: A review. *J. Inorg. Organomet P.*, 3, 559–584, 2018.
121. Liu, T., Zhang, X., Liu, K. et al., A novel and facile synthesis approach of porous carbon/graphene composite for the supercapacitor with high performance. *Nano*, 29, 9, 2018.
122. Thummarungsan, N., Paradee, N., Pattavarakorn, D. et al., Influence of graphene on electromechanical responses of plasticized poly(lactic acid). *Polymer*, 138, 169–179, 2018.
123. Pham, V.H., Dang, T.T., Hur, S.H. et al., Highly conductive poly(methyl methacrylate) (PMMA)-reduced graphene oxide composite prepared by self-assembly of PMMA latex and graphene oxide through electrostatic interaction. *ACS Appl. Mater. Interfaces*, 4, 5, 2630, 2012.
124. Sun, X., Lu, H., Liu, P. et al., A reduced graphene oxide-NiO composite electrode with a high and stable capacitance. *Sustain. Energ. Fuels*, 2, 3, 673–678, 2017.
125. Tschoppe, K., Beckert, F., Beckert, M. et al., Thermally reduced graphite oxide and mechanochemically functionalized graphene as functional fillers for epoxy nanocomposites. *Macromol. Mater. Eng.*, 300, 2, 140–152, 2015.
126. Li, J., Song, Y., Ma, Z. et al., Preparation of polyvinyl alcohol graphene oxide phosphonate film and research of thermal stability and mechanical properties. *Ultrason. Sonochem.*, 43, 1, 2018.
127. Xin, F., Fan, L.W., Ding, Q. et al., Increased thermal conductivity of eicosane-based composite phase change materials in the presence of graphene nanoplatelets. *Energ. Fuel.*, 27, 7, 4041–4047, 2013.
128. Han, P., Fan, J., Jing, M. et al., Effects of reduced graphene on crystallization behavior, thermal conductivity and tribological properties of poly(vinylidene fluoride). *J. Compos. Mater.*, 48, 6, 659–666, 2013.
129. Choi, J.Y., Lee, J.H., Mi, R.K. et al., Graphene attached on microsphere surface for thermally conductive composite material. *Clean Technol. Environ.*, 19, 3, 243–248, 2013.
130. Kumar, P., Yu, S., Shahzad, F. et al., Ultrahigh electrically and thermally conductive self-aligned graphene/polymer composites using large-area reduced graphene oxides. *Carbon*, 101, 120–128, 2016.
131. Mehrali, M., Latibari, S.T., Mehrali, M. et al., Shape-stabilized phase change materials with high thermal conductivity based on paraffin/graphene oxide composite. *Energ. Convers. Manage.*, 67, 3, 275–282, 2013.
132. Guo, H., Li, X., Li, B. et al., Thermal conductivity of graphene/poly(vinylidene fluoride) nanocomposite membrane. *Mater. Des.*, 114, 355–363, 2016.
133. Zhang, Z., Qu, J., Feng, Y. et al., Assembly of graphene-aligned polymer composites for thermal conductive applications. *Com. Commun.*, 9, 33–41, 2018.
134. Huang, Y., Yan, J., Zhou, S. et al., Preparation and electromagnetic wave absorption properties of CoNi@SiO$_2$ microspheres decorated graphene-polyaniline nanosheets. *J. Mater. Sci. Mater. El.*, 29, 1, 1–10, 2017.
135. Shen, X.J., Pei, X.Q., Fu, S.Y. et al., Significantly modified tribological performance of epoxy nanocomposites at very low graphene oxide content. *Polymer*, 54, 3, 1234–1242, 2013.

136. Hassan, E.S.M., Eid, A.I., El-Sheikh, M. et al., Effect of graphene nanoplatelets and paraffin oil addition on the mechanical and tribological properties of low-density polyethylene nanocomposites. *Ara. J. Sci. Eng.*, 11, 1–9, 2017.
137. Ameen, S., Seo, H.K., Akhtar, M.S. et al., Novel graphene/polyaniline nanocomposites and its photocatalytic activity toward the degradation of rose Bengal dye. *Chem. Eng. J.*, 210, 6, 220–228, 2012.
138. Chen, C.T., Martinmartinez, F.J., Ling, S. et al., Nacre-inspired design of graphene oxide-polydopamine nanocomposites for enhanced mechanical properties and multi-functionalities. *Nano Futures*, 1, 1, 011003, 2017.
139. Roy, C.S., Introduction to nursing: An adaptation model. *Am. J. Nurs.*, 84, 10, 1331, 1984.
140. Kim, Y., Lee, J., Yeom, M.S. et al., Strengthening effect of single-atomic-layer graphene in metal-graphene nanolayered composites. *Nat. Commun.*, 4, 2114, 2013.
141. Zhang, J., Xiong, Z., Zhao, X.S., Graphene–metal–oxide composites for the degradation of dyes under visible light irradiation. *J. Mater. Chem.*, 21, 11, 3634–3640, 2011.
142. Xiang, H., Zhang, K., Ji, G. et al., Graphene/nanosized silicon composites for lithium battery anodes with improved cycling stability. *Carbon*, 49, 5, 1787–1796, 2011.
143. Iqbal, N., Khan, I., Yamani, Z.H.A. et al., Corrigendum to "A facile one-step strategy for *in-situ* fabrication of WO3-BiVO4, nanoarrays for solar-driven photoelectrochemical water splitting applications" [Solar Energy 144 (2017) 604-611]. *Sol. Energy*, 160, 604–611, 2018.
144. Wang, G., Chen, R., Zhao, S. et al., Efficient synthesis of 1,2,4-Oxadiazine-5-ones via [3+3] cycloaddition of *in situ* generated aza-oxyallylic cations with nitrile oxides. *Tetrahedron Lett.*, 59, 21, 2018–2020, 2018.
145. Myekhlai, M., Lee, T., Baatar, B. et al., Thermal conductivity on the nanofluid of graphene and silver nanoparticles composite material. *J. Nanosci. Nanotechnol.*, 16, 2, 1633, 2016.
146. Kholmanov, I.N., Domingues, S.H., Chou, H. et al., Reduced graphene oxide/copper nanowire hybrid films as high-performance transparent electrodes. *ACS Nano*, 7, 2, 1811–1816, 2013.
147. Dimiev, A., Zakhidov, D., Genorio, B. et al., Permittivity of dielectric composite materials comprising graphene nanoribbons. The effect of nanostructure. *ACS Appl. Mater. Interfaces*, 5, 15, 7567, 2013.
148. Liu, J., Tian, G., Qi, S. et al., Enhanced dielectric permittivity of a flexible three-phase polyimide-graphene-$BaTiO_3$ composite material. *Mater. Lett.*, 124, 124, 117–119, 2014.
149. Zhai, W., Shi, X., Wang, M. et al., Effect of graphene nanoplate addition on the tribological performance of Ni_3Al matrix composites. *J. Compos. Mater.*, 48, 30, 3727–3733, 2014.
150. Zhou, M., Lin, T., Huang, F. et al., Highly conductive porous graphene/ceramic composites for heat transfer and thermal energy storage. *Adv. Funct. Mater.*, 23, 18, 2263–2269, 2013.

2

Graphene-Reinforced Advanced Composite Materials

Xiaochao Ji[1]*, Shaojun Qi[1], Rajib Ahmed[2] and Ahmmed A. Rifat[3]

[1]*School of Metallurgy and Materials, University of Birmingham, Birmingham, UK*
[2]*Nanotechnology Laboratory, School of Engineering, University of Birmingham, Birmingham, UK*
[3]*Nonlinear Physics Centre, Research School of Physics and Engineering, The Australian National University, Canberra, ACT, Australia*

Abstract
Graphene is one of the most promising two-dimensional (2D) materials, which has attracted much attention during the past decade due to its exciting mechanical, electronic, and thermal properties. Graphene-reinforced composites are being projected for structural materials as well as functional materials for the unique characteristics of graphene. There are several critical issues of graphene-reinforced composites, including manufacturing techniques, dispersion, bondings, strengthen mechanisms, and mechanical properties. In this chapter, processes applied for synthesis of graphene-reinforced composites (polymer, metal and ceramic) have been critically reviewed with an aim to generate homogeneous distribution of graphene in the matrix. Mechanical properties of the composites are summarized, and the important factors that determine the strength and interface have been reviewed. Applications of the graphene-reinforced composite materials with different matrices are also summarized. Future work that needs to promote the graphene-reinforced composites is addressed.

Keywords: Graphene, composite material, polymer matrix, metal matrix, ceramic matrix

2.1 Introduction

Composite materials provide designers flexibility in designing shapes, strength, and properties, which are widely needed due to their versatile properties, and enable them to be used in different fields. Generally, the composite material is defined as a material formed with more than one distinct phases physically or chemically, arranged or distributed. The continuous phase is named as a matrix, and the distributed phase is referred to as the reinforcement. Typically, composite materials can be classified on the basis of the type of matrix or reinforcement, such as metal matrix composites (MMCs), ceramic matrix composites (CMCs), polymer matrix composites (PMCs), particle-reinforced composites, short-fiber-reinforced composites, continuous-fiber-reinforced composites, and laminate

---

Corresponding author: jixiaochao@gmail.com

composites. High-strength and lightweight composite materials are designed to meet the requirement in industry, which can lead to advantages like the improvement of payload and fuel efficiency in the case of aerospace and automobiles.

Besides the conventional composite materials, the emergence of nanotechnology makes it possible to deposit nanocomposites, which have wide ranges of applications due to their special structural features. Thus far, tons of research have been carried out to study the nanofillers and their nanocomposites, such as carbon nanotubes (CNTs), carbon nanofibers (CNFs), carbon black, etc. Unlike the traditional composite materials that contain a lot of fillers, low volume fraction of nanofillers can dramatically change the properties of the composites, because of the large surface area per unit the nanofillers have.

Graphene is the most promising two-dimensional material first synthesized by A. K. Geim and K. S. Novoselov through the mechanical exfoliation method in 2004 [1]. This pioneering work started the tendency of studies on graphene. Both theoretical and experimental results indicate that this single layer of carbon has a lot of attractive properties [2–5]. Apart from the excellent mechanical properties, high electrical and thermal conductivity, large surface area, and inertness surface properties all make graphene a promising material in the field of electronic devices, thermal sensors, energy storages, and biomedical applications. Graphene is formed by carbon atoms arranged in a hexagonal lattice, which is responsible for its high elastic modulus (~1 TPa) and tensile strength (~130 GPa) [2–5]. Graphene has some inherent advantages over other carbon allotropes like graphite, carbon black, and carbon nanotubes. The large surface area of graphene makes it an ideal nanofiller for composites because of greater interaction area with the matrix. This can be helpful for the transmission of phonons, electrons, thermal energy, or mechanical stress. Graphene can be synthesized through bottom-up approaches or top-down approaches. Pure graphene can be achieved by chemical vapor deposition process or mechanical exfoliation of graphite. Chemical exfoliation of graphite is a relatively simple and cheap way to synthesize graphene oxide, but subsequent reduction is required to obtain graphene. Schematic diagrams of graphene and graphene oxide are shown in Figure 2.1. Studies on the graphene-reinforced composites developed very fast after the deposition of graphene in 2004. More explorations have been carried out on the polymer matrix composites than on the metal and ceramic matrix composites due to the simple deposition process, because high pressure and high temperature are needed for the formation of metal and ceramic matrix composites.

Due to the low mechanical strength, polymers cannot be applied in structural applications, while metal and ceramic matrix composites are widely used as structural materials for lightweight, high-strength, and long-life components. There is increasing interest in the graphene-reinforced metal and ceramic matrix composite due to the potential of these promising composites.

In this chapter, several critical issues of the graphene-reinforced composite materials are summarized, including manufacturing techniques, dispersion, bondings, strengthen mechanisms, and mechanical properties. Processes applied for synthesis of graphene-reinforced composites (polymer, metal and ceramic) have been critically reviewed with an aim to generate homogeneous distribution of graphene in the matrix. Mechanical properties of the composites studied to date have been summarized, and important factors that determine the strength and interface have been reviewed. The interfacial reaction and stability of graphene in different material systems are reviewed. Along with the chemical stability for the homogeneous distribution of graphene, the dispersion behavior of graphene in the matrix is also discussed. Properties of the composites that were affected by graphene have

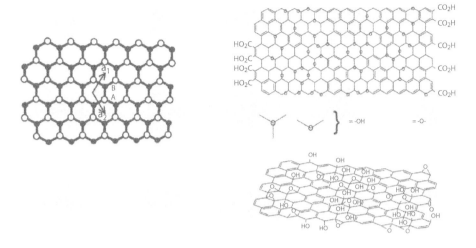

Figure 2.1 Schematic diagrams of graphene and graphene oxide [6].

been summarized, such as electrical, thermal, corrosion, and catalytic properties. Potential applications of the graphene-reinforced composite materials with different matrices are summarized. Directions of future researches on the graphene-reinforced composites are addressed based on the previous discussion.

2.2 Graphene–Reinforced Metal Matrix Composites (MMCs)

Graphene is recognized as the most promising reinforcement for metal matrix composites. Recently, studies on graphene-reinforced metal matrix composites have been reported. The number of papers published in the past years has been summarized in Figure 2.2, which indicates that this field has not received enough attention compared to the polymer matrix composites. The reason can be attributed to the difficulties in the dispersion process and other processing problems, such as interfacial chemical reactions or bonding issues. The defects and porosity level caused by graphene during the reinforce process directly affect the mechanical properties of the composites.

To date, several techniques have been applied for the fabrication of the graphene-reinforced metal matrix composites, such as powder metallurgy, melting and solidification, electrochemical deposition, and some novel deposition techniques. Different metal-based composites have been processed, and their microstructures and interfacial reactions have been studied to get a clear understanding of the graphene-reinforced metal matrix composites. Here, the recent published papers on graphene-reinforced metal-based composites are categorized according to the matrices, which are summarized in Table 2.1. Their mixing methods, processing techniques, reinforcements, and properties are also listed. Though graphene has excellent electrical, thermal, and mechanical properties, it should be noticed that all these properties were the ideal properties of single-layer graphene. In most cases shown in Table 2.1, the nanofillers were multilayer graphene or graphene oxides, which may not be as good as the single-layer graphene. Thus, the specific graphene used in the

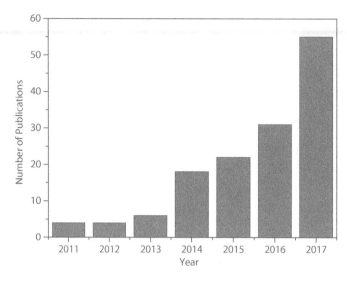

Figure 2.2 Trends of publications on graphene-reinforced metal matrix composites. (From Scopus.)

experiments have been listed in Table 2.1 for an overview of the state of graphene applied for metal matrix composites.

2.2.1 Processing of MMCs

The great challenge in fabrication of graphene-reinforced MMCs is to achieve a homogeneous dispersion of the nanofillers in the metal matrix because of the high difference of surface energies between graphene and metal matrix. Various fabrication routes have been developed to process the graphene-reinforced MMCs, such as powder metallurgy, melting and solidification, and electrodeposition.

2.2.1.1 Powder Metallurgy

Powder metallurgy is widely used for the deposition of graphene-reinforced metal matrix composites. Basically, there are mainly two steps for the powder metallurgy processing of graphene-reinforced metal matrix composites. First, graphene and metal powders are mixed by grinding or mechanical alloying, and then the mixed powders are consolidated by different types of techniques, such as sintering, spark plasma sintering, cold isostatic pressing, and hot isostatic pressing. The mixing process plays an important role in ensuring the homogeneous mixture of the graphenes and metal powders. High mixing energy is needed to overcome the high surface energy of graphene, which can cause agglomeration of graphene. The compact processes are subjected to the secondary deformation processes, such as rolling and extrusion. All the methods are focused on achieving homogeneous distribution of graphene within the metal matrix and well-bonded interfaces. Agglomeration of the nanofillers may significantly influence the mechanical and electrical properties of the composites. It is shown in Table 2.2 that most of the metal matrix composites can be fabricated by the powder metallurgy routes, such as Al-, Cu-, Mg-, Ni-, Ag-, and Ti-based composite materials.

Table 2.1 Summary of processing methods, matrices, and reinforcements of graphene-reinforced metal matrix composites.

Author	Mixing method	Method	Matrix	Reinforcement	Properties
Bastwros et al. [7]	Ball milling	Semi-solid sintering	Al6061	1.0 wt.% graphene	47% increase in flexural strength
Bartolucci et al. [8]	Ball milling	Hot isostatic pressing and extrusion	Al	0.1 wt.% graphene	lower hardness and tensile strength
Yang et al. [9]	Ball milling	Pressure infiltration method	Al	0.06 wt.%, 0.14 wt.%, 0.21 wt.%, 0.54 wt.% graphene	228% increase in yield strength; 93% increase in tensile strength
Gao et al. [10]	Electrostatic self-assembly	Hot pressing	Al	0.1, 0.3, and 0.5 wt.% GO	Tensile strength 110 MPa with 0.3 wt.% GO
Li et al. [11]	High-energy ball milling	Sintering	Al	0.25, 0.5, and 1 wt.% graphene	Hardness 81 Hv; yield strength and tensile strength increased by 38.27% and 56.19%
Kumar et al. [12]	Magnetic stirring	Hot extrusion	Al	Graphene	Hardness ~75 HV; 46% improvement in tensile strength; elongation decrease
Wang et al. [13]	Mechanical stirring	Hot extrusion	Al	GO	249 MPa tensile strength;
Li et al. [14]	Mechanical stirring	Hot pressing	Al	0.3 wt.% GO	18% and 17% increase in elastic modulus and hardness

(Continued)

Table 2.1 Summary of processing methods, matrices, and reinforcements of graphene-reinforced metal matrix composites. (*Continued*)

Author	Mixing method	Method	Matrix	Reinforcement	Properties
Jeon et al. [15]	Solvent mixing	Friction stir processing	Al	GO	15% improvement in thermal conductivity
Pérez et al. [16]	High-energy ball milling	Sintering	Al	0.25, 0.50, and 1.0 wt.% graphene	~138% increase in hardness
Zhang et al. [17]	Ball milling	Hot extrusion	Al 5083	0.5 and 1.0 wt.% graphene	Yield strength and tensile strength are 332 MPa and 470 MPa with 1.0 wt.% graphene
Khodabakhshi et al. [18]	Wet mixing	Friction stir processing	Al-Mg	Graphene	84 Hardness; 300% yield strength
Kavimani et al. [19]	Wet mixing	Sintering	AZ31	GO	64 Hv hardness; lower corrosion rate
Liu et al. [20]	Solvent mixing	Pulse electrodeposition	Co	GO	430 Hv hardness; lower friction coefficients ~0.55; lower corrosion rate
Rekha et al. [21]	Solvent mixing	Electrodeposition	Cr	graphene	Enhance corrosion resistance
Jiang et al. [22]	Solvent mixing	Spark plasma sintering	Cu	GO	90% and 81% increase in yield strength and compression strength; high electrical conductivity

(*Continued*)

Table 2.1 Summary of processing methods, matrices, and reinforcements of graphene-reinforced metal matrix composites. (*Continued*)

Author	Mixing method	Method	Matrix	Reinforcement	Properties
Kim et al. [23]	Ball milling	High-ratio differential speed rolling	Cu	0.5 and 1 vol.% MLG	Optimizing grain size; 360 MPa and 425 MPa of yield strength and tensile strength with 1 vol.% MLG
Hwang et al. [24]	Molecular-level mixing	Spark plasma sintering	Cu	GO	Elastic modulus and yield strength are 131 GPa and 284 MPa
Tang et al. [25]	Solvent mixing	Spark plasma sintering	Cu	Graphene/Ni	Elastic modulus and yield strength are 132 GPa and 268 MPa with 1.0 vol.%
Raghupathy et al. [26]	Solvent mixing	Electrodeposition	Cu	GO	Lower corrosion rate
Akbulut et al. [27]	Solvent mixing	Electrophoretic deposition	Cu	WC/graphene	Low friction coefficient 0.2 and low wear rate
Hu et al. [28]	Solvent mixing	Laser additive manufacturing	Cu	GO	Elastic modulus and yield strength are 118.9 GPa and 3 GPa
Luo et al. [29]	Ball milling	Hot-press sintering	Cu	GO/silver	89.1 HV; high thermal and electrical performance
Jagannadham et al. [30]	Solvent mixing	Electrochemical deposition	Cu	Graphene	Thermal conductivity

(*Continued*)

Table 2.1 Summary of processing methods, matrices, and reinforcements of graphene-reinforced metal matrix composites. (*Continued*)

Author	Mixing method	Method	Matrix	Reinforcement	Properties
Jagannadham et al. [31]	Solvent mixing	Electrochemical deposition	Cu	Graphene	Thermal conductivity
Chu et al. [32]	Ball milling	Hot pressing	Cu	3, 5, 8, and 12 vol.% graphene	114% and 37% increases of yield strength and Young's modulus with 8 vol.%
Pavithra et al. [33]	Solvent mixing	Electrochemical deposition	Cu	GO	~2.5 GPa hardness; ~137 GPa elastic modulus; comparable electrical conductivity
Xie et al. [34]	Solvent mixing	Electrochemical deposition	Cu	GO	Electroactivity
Li et al. [35]	Solvent mixing	Spark plasma sintering	Cu	0.8 vol.% GO/Ni	42 % improvement in tensile strength
Peng et al. [36]	Solvent mixing	Electroless plating	Cu	GO/Sn/Pd	Sandwich-like structure
Zhao et al. [37]	Solvent mixing	Electroless plating	Cu	GO	107% and 21% increase in tensile strength and Young's modulus with 1.3 wt.%
Dutkiewicz et al. [38]	Ball milling	Hot pressing	Cu	1 wt.% and 2 wt.% of GO or graphene	Hardness higher by 20 Hv with 2 wt.% GO
Xiong et al. [39]	Solvent mixing	Hot pressing	Cu	0.3 vol.% and 1.2 vol.% GO	Yield strength 233 MPa, and tensile stress 308 MPa with 1.2 vol.%

(*Continued*)

Table 2.1 Summary of processing methods, matrices, and reinforcements of graphene-reinforced metal matrix composites. (*Continued*)

Author	Mixing method	Method	Matrix	Reinforcement	Properties
Zhao et al. [40]	Solvent mixing	Clad rolling	Cu-Al	3 wt.% GO	Tensile strength and hardness improved by 77.5% and 29.1%
Liu et al. [41]	Solvent mixing	Vacuum cold spraying	HA	0.1 wt.% and 1 wt.% graphene	Fracture property enhanced
Turan et al. [42]	Solvent mixing	Hot press sintering	Mg	0.1, 0.25, and 0.5 wt.% of graphene	Hardness and wear resistance improved, lower corrosion rate
Rashad et al. [43]	Solvent mixing	Hot extrusion	Mg-Al	0.5 wt.% graphene/0.1 wt.% CNT	63 Hv hardness; Improvement in tensile and compressive strength
Qi et al. [44]	Solvent mixing	Electro-brush plating	Ni	GO	8.65 GPa hardness; lower corrosion rate
Algul et al. [45]	Solvent mixing	Pulse electroplating	Ni	Graphene	High hardness and low friction coefficient 0.2
Jabbar et al. [46]	Solvent mixing	Electrochemical deposition	Ni	Graphene	Enhanced corrosion resistance
Zhou et al. [47]	Solvent mixing	Reverse pulse electrodeposition	Ni	Ce/GO	Excellent anticorrosion property
Kumar et al. [48]	Solvent mixing	Electrodeposition	Ni	GO	Low corrosion rate
Kuang et al. [49]	Solvent mixing	Electrodeposition	Ni	GO	Lower hardness, higher thermal conductivity

(*Continued*)

Table 2.1 Summary of processing methods, matrices, and reinforcements of graphene-reinforced metal matrix composites. (*Continued*)

Author	Mixing method	Method	Matrix	Reinforcement	Properties
Szeptycka *et al.* [50]	Solvent mixing	Electrodeposition	Ni	Graphene	Better corrosion resistance
Chen *et al.* [51]	Solvent mixing	Pulse electrodeposition	Ni	Graphene	High hardness ~223 Hv; low friction coefficient
Ren *et al.* [52]	Solvent mixing	Electrochemical deposition	Ni	Multilayer graphene	Hardness 4.6 GPa; elastic modulus 240 GPa
Khalil *et al.* [53]	Solvent mixing	Electrodeposition	Ni	GO/TiO_2	Low corrosion rate
Jiang *et al.* [54]	Solvent mixing	Electrochemical deposition	Ni	Graphene	Smaller grain sizes; low corrosion rate
Zhai *et al.* [55]	Ball milling	Spark plasma sintering	Ni_3Al	1 wt.% graphene	Low friction coefficients at low temperature
Qiu *et al.* [56]	Solvent mixing	Pulse current deposition	Nickel hydroxide	GO	Better corrosion resistance
Zhang *et al.* [57]	Solvent mixing	Electrodeposition	Ni-Fe	Graphene	Threefold hardness and 14.9% improvement in elastic modulus
Gao *et al.* [58]	Solvent mixing	Electrochemical deposition	PtNi	GO	Sensitive to glucose
Berlia *et al.* [59]	Solvent mixing	Electrodeposition	Sn	Graphene	Better corrosion resistance
Song *et al.* [60]	Ball milling	Spark plasma sintering	Ti	0.5 wt.% and 1.5 wt.% multilayer graphene	~15 GPa hardness; ~264 GPa elastic modulus; ~918 MPa yield strength; ~24 GPa scratch resistance

(*Continued*)

Table 2.1 Summary of processing methods, matrices, and reinforcements of graphene-reinforced metal matrix composites. (*Continued*)

Author	Mixing method	Method	Matrix	Reinforcement	Properties
Hu et al. [61]	Solvent mixing	Laser sintering	Ti	1 wt.%, 2.5 wt.% and 5 wt.% GO	~11 GPa Hardness
Zhang et al. [62]	Mechanical stirring	Spark plasma sintering	Ti	3.0 vol. and 7 vol.% GO	2.64 GPa and 1.93 GPa of compressive strength and yield strength with 7 vol.% GO
Cao et al. [63]	Mechanical blending	Hot isostatic pressing	Ti	0.5 wt.% graphene	125 GPa elastic modulus; 1.06 GPa tensile strength; 1.02 GPa yield strength
Mu et al. [64]	Solvent mixing	Spark plasma sintering	Ti	Graphene	54.2% increase in tensile strength with 0.1 wt.%
Xu et al. [65]	Ball milling	Spark plasma sintering	TiAl	3.5 wt.% multilayer graphene	Low friction coefficient and wear rate
Kumar et al. [66]	Solvent mixing	Electrodeposition	Zn	Graphene	Better corrosion resistance
Lin et al. [67]	Solvent mixing	Laser sintering	Fe	2 wt.% GO	93.5% increase of hardness; 167% increase of fatigue life
Rashad et al. [68]	Mechanical stirring	Stir-casting	Mg alloy	1.5 and 3.0 wt.% graphene	195 MPa yield strength and 299 MPa tensile strength with 3 wt.%
Liu et al. [69]	Solvent mixing	Electrochemical deposition	Au	GO	Conductivity improvement
Shin et al. [70]	Ball milling	Hot pressing	Ti, Al	0.3, 0.5, and 0.7 vol.% graphene	~148 GPa elastic modulus; ~1.5 GPa yield stress

Ball milling is widely used for powder mixing, which is a grinding process with high energy to impair agglomeration and increase adhesion between graphene nanofillers and metal powders. The milling processes can be carried out in either dry or liquid environments, with or without grinding balls. Liquid such as acetone or water can improve deagglomeration and prohibit the growth of metal grains. The milling balls can provide shearing forces to break the agglomeration of graphenes. Besides, the wet mixing process is also preferred because it is relatively simple. The graphene slurry and metal powders are mixed using ultrasonic vibration. The ultrasonic waves can generate cavitation in liquid to agitate the mixtures. The vibration force is not as high as the ball milling process and longer time is needed to decrease agglomeration. The solvent should not react with graphene and should be easy to evaporate during the baking process. Wang *et al.* presented a four-step powder metallurgy route for the fabrication of graphene-oxide-reinforced aluminum composites. Few layered graphene oxides were dispersed in deionized water before mixing with the aluminum flakes. Ball milling was applied to modify the aluminum flakes with a hydrophilic PVA membrane. The aluminum powder slurry was added into the graphene oxide aqueous dispersion and was stirred until the brown slurry became transparent. The dry and consolidated composite powders were sintered in an ambient of argon and then deformed by hot extrusion.

It is reported that better dispersion can be achieved by employing high-energy ball milling or mechanical alloying mixing processes. Mechanical alloying provides a solid-state way for the mixing of powders with homogeneous dispersion and fine structure of the composite powder. The tensile properties of the composite and the pure Al are shown in Figure 2.3a. Figure 2.3b shows the fracture surface of the graphene/Al composite, while graphene was pulled out during the tensile process [13].

As shown in Figure 2.4, there are some hydroxyl and epoxy groups on the surface of graphene oxide that are beneficial for its dispersion in solutions. This makes graphene oxide a better reinforcement than the pure graphene nanosheets [13, 71, 72]. Mina *et al.* reported a semi-solid processing technique for the fabrication of few-layered graphene-oxide-reinforced aluminum alloy composite. The Raman spectra shown in Figure 2.5 indicate that the structure of the stress within graphene can be changed related to the variation in the wavenumber

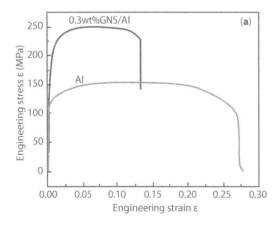
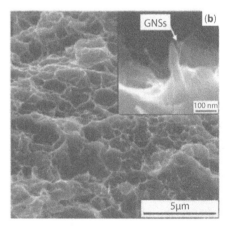

Figure 2.3 Tensile properties of the graphene-reinforced Al composite and the pure Al sample (a); (b) fracture surface of the graphene/Al composite [13].

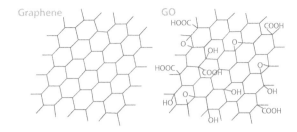

Figure 2.4 Schematic structure of graphene and graphene oxide with functional groups [13].

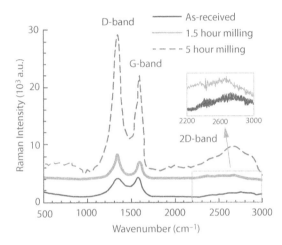

Figure 2.5 Raman spectra of graphene and the graphene/Al6061 milled under different conditions [7].

shifts. The graphene/aluminum composites were synthesized by pressure-assisted sintering. It was found that the sizes of the composite particles increased with the ball milling times, and the graphene sheets were repeatedly enclosed, folded, and embedded into the aluminum particles. It is hard to distinguish graphene from the composite particles, and their shape changed from flakes to particulate shape [7].

Instead of sintering, hot pressing is also widely used for consolidation of the mixed powders. Bartolucci *et al.* blended the thermal reduction graphene with aluminum powders with ball milling, and stearic acid was used as control agent to impair agglomeration of the graphene. Mixed powders were processed with hot isostatic pressing and hot extrusion. The defect nature of the graphene led to the formation of aluminum carbide, which lowered the hardness and tensile strength of the composites [8]. Chu *et al.* obtained good dispersion of graphene sheets in Cu matrix powder by ball milling in an argon atmosphere; the as-milled powders were compacted to a density of 75% and consolidated using a hot isostatic pressing technique. The composite powders were compacted at a pressure of 40 MPa and sintered at 800°C for 15 min [32]. Luo *et al.* obtained the GO/silver-reinforced Cu matrix composites by hot-press sintering at various pressure. It was verified that microhardness, electronic conductivity, and thermal conductivity were correlated with press pressures. The properties and morphologies of the Ag-Cu/GO composites are shown in Figure 2.6. The composite pressed with 50 MPa showed 18.6% and 21.8% improvements in thermal and

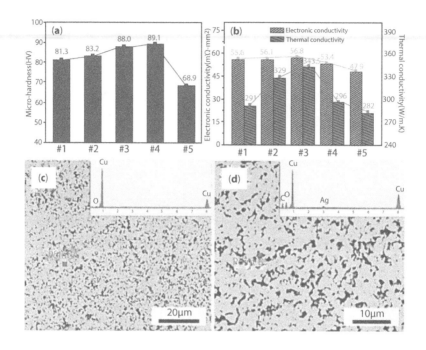

Figure 2.6 Properties and morphologies of the Cu/GO/Ag composites. (a) Hardness; (b) electronic and thermal conductivity; (c) low-powered composite; (d) high-powered composite [29].

electrical properties compared to pure Cu [29]. Hot pressing is a technique suitable for production of large-scale composite parts with limited damage to the graphene nanofillers, but the process is relatively long, which may lead to high grain growth.

Spark plasma sintering technique is a relatively new sintering method that uses a pulsed direct current to generate rapid heating. The fast sintering rate makes it suitable for consolidate nano powders. The rapid consolidation process can limit the growth of grains, but the high energy can cause graphitization of the graphene nanofillers. The processing quality is based on the conductivity of the materials, and size is limited. Jiang et al. prepared the graphene/Cu and graphene oxide/Cu composite with a similar process that is shown in Figure 2.7. Both yield strength and compression strength were improved due to the strength of the two-dimensional structures that also constrained the movement of dislocations [22].

Xu et al. mixed the multilayer graphenes with the TiAl composite powders by ball milling in vacuum, and the mixed powders were hot pressed in vacuum and formed by spark plasma sintering at a temperature of 1100°C with a loading pressure of 50 MPa. The ball milling process can crack and decrease the size of the multilayer graphene, and introduction of graphene sheets can refine the grain, because the nanofillers can decrease the diffusion during the hot pressing process [65]. Song et al. reported the fabrication of the graphene-reinforced titanium-based composites using the spark plasma sintering method. The composite powders were mixed in ethanol and ball milled using Si_3N_4 milling balls for 12 h. The dried powders were then loaded into the graphite die. SEM observation and Raman spectra indicated that multilayer graphene survived during harsh spark plasma sintering [60].

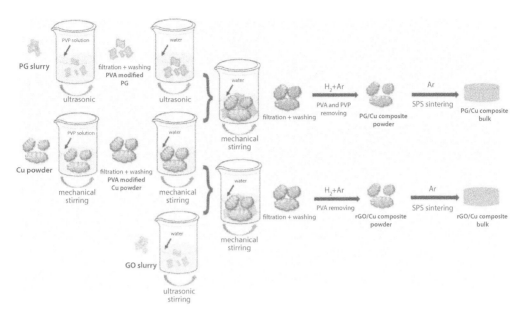

Figure 2.7 Schematic diagram of the processing procedures of the graphene/Cu and graphene oxide/Cu composites by spark plasma sintering [22].

Zhai *et al.* found that the graphene nanosheets could refine the grains of Ni$_3$Al. The raw composite powders were mixed by ball milling for 6 h and then sintered by spark plasma sintering technique [55].

2.2.1.2 *Melting and Solidification*

Melting and solidification are traditional ways to fabricate metal matrix composites, which are also utilized for the processing of graphene-reinforced metal composite materials. Few studies are available for this process because the high temperature for melting that can damage graphene and chemical reactions will be generated at the graphene/metal interfaces. Besides, carbon clusters may formed due to the surface tension forces. Infiltration method was applied to fabricate the graphene-reinforced Al matrix composite materials. Figure 2.8 shows the schematic diagram of the fabrication processes for the graphene-reinforced Al composites.

Yang *et al.* indicated that no Al$_4$C$_3$ phase was formed during the infiltrate process, and both yield strength and tensile strength were significantly improved before and after hot extrusion. The composite with 0.54 wt.% graphene had the best mechanical properties as shown in Figure 2.9 [9]. Hu *et al.* reported on the single-layer graphene-oxide-reinforced Ti-based nanocomposite using the laser sintering method. The survival of graphene oxide in the composite was confirmed by XRD, EDS, and Raman spectrum [61].

A laser-based additive manufacturing method was utilized for the sintering of iron-based composite material, and it was observed that the fast laser heating process could prevent the aggregation of graphene oxide powders. Schematic diagram and TEM images of the cross-section view of the composite coating are shown in Figure 2.10. Tensile strength,

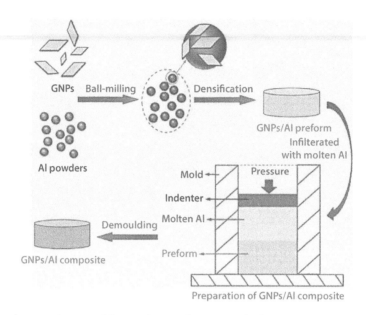

Figure 2.8 The schematic diagram of the powder metallurgy route for fabrication of graphene-reinforced Al composites [9].

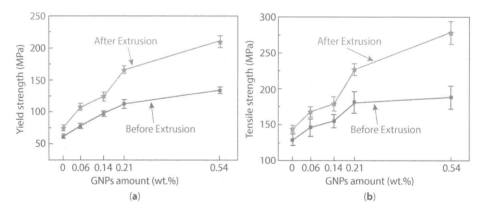

Figure 2.9 Yield strength (a) and tensile strength (b) of the graphene-reinforced Al matrix composites with different loading fractions of graphene [9].

Young's modulus, and surface hardness of the graphene/iron composites were all improved [67]. Stir-casting method was applied for the fabrication of graphene-reinforced magnesium alloy. The AZ31 alloy was melted first and then the graphene powders were added at 740°C. The mixture was poured into a mold and solidified. The ingots were extruded as rods with a ratio of 5.2:1 at 350°C. Rashad *et al.* found that the presence of graphene affected the formation of the intermetallic phases during the solidification process and increased the fracture strain of the AZ31 matrix [68].

2.2.1.3 Electrochemical Deposition

Electrochemical deposition is a popular route for the deposition of graphene-reinforced composite coatings projected for applications such as electrodes and sensors. This route

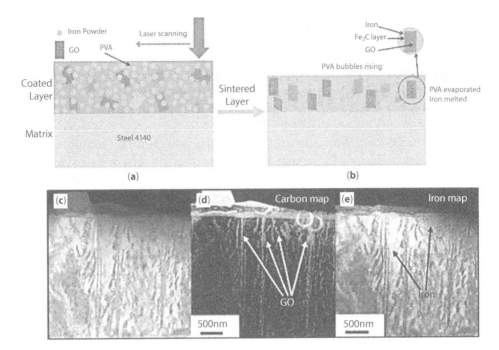

Figure 2.10 Schematic diagram and TEM images of the iron-based composite coating before and after laser sintering [67]. (a) and (b) show the laser sintering process and the schematic of the composite coating; (c) is the TEM image of the composite structure, and (d) and (e) are the EDS mappings of carbon and iron.

is flexible to mix the graphene slurry and the metal ion solutions for the deposition of the composite coating. A pulse electroplating method was applied to produce the graphene/Ni composite at a fixed current density of 5 A/dm². The incorporation of graphene can reduce the grain size of Ni and change the microstructure of the composite coating. The schematic diagram of the electrodeposition process is shown in Figure 2.11 [45].

Kuang *et al.* used the mixing solution of graphene oxide sheets and nickel sulfamate for the deposition of graphene/Ni nanocomposite coating. Nickel ions and graphene oxide sheets were reduced by electrochemical processes [49]. Liu *et al.* prepared a graphene/Au

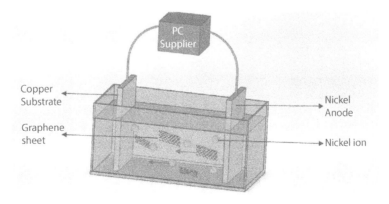

Figure 2.11 Schematic diagram of the electrodeposition process [45].

nanocomposite film via a direct electrodeposition method using the mixture of graphene oxide and $HAuCl_4$ solution. The nano Au particles were embedded on the surface of the graphene, which can improve the conductivity of the composite structure [69]. Gao et al. synthesized the graphene-reinforced PtNi nanocomposite for the detection of glucose, and the PtNi nanoparticles were electrochemically deposited in a solution mixed with Na_2SO_4, H_2PtCl_6, and $NiSO_4$. The composite coating is more sensible to glucose compared to the monolithic graphene [58].

Electro-brush plating has been applied for fabrication of graphene-oxide/nickel nanocomposite coating; the graphene oxide suspension was added into the nickel plating solution and mixed by ultrasonic vibration. The nanocomposite coatings were then deposited by wiping of the plating pen at a voltage of 14 V [44]. Jabbar et al. fabricated a Ni/graphene composite coating via an electrochemical deposition method at different temperatures; coarser surfaces were induced by the addition of graphene, and the grain sizes of the Ni were refined [46]. Graphene and Ce reinforced Ni-based composite coating by a reverse pulse electrodeposition, and the microstructure of the Ni deposits was refined by the incorporation of the reinforcements [47].

2.2.1.4 Thermal Spray

Thermal spray technology is widely used for deposition of thick coatings for engines, turbine blades, and rollers. Molten or semi-molten powders were sprayed through a hot jet, and the coatings were formed with the splat droplets on the substrates. A fast cooling rate (10^8 K s^{-1}) during the solidification process can lead to the formation of nanocrystalline. Thermal spray can be categorized into plasma spray, flame spray, electric arc spray, high-velocity oxyfuel spraying (HVOF), or cold spraying based on the heat source. Plasma spray and HVOF are widely used for the deposition of nanocomposite coatings. Heat source of the plasma spray is generated by the ionization of the feeding gas by arc between the cathode and anode. Powders absorb heats and accelerated to impact on the substrate with a speed of ~1000 m s^{-1}, which can form a dense coating with good adhesion. The source of heat in HVOF is the high-pressure combustion of a mixture of fuel (methane, hydrogen, or kerosene) and oxygen, and the powders can be projected at a very high velocity (~1500 m s^{-1}), which can improve the density of the coating. Thermal spray provides a near net shape way for the fabrication of graphene-reinforced metal-based composite coatings. Several studies have been carried out to fabricate the CNT-reinforced metal matrix composites [73–75], while limited work has been done on the graphene-reinforced MMCs by thermal spray techniques. Liu et al. reported the fabrication of graphene-reinforced hydroxyapatite composite coating by vacuum cold spraying technique for biomedical applications. The composite coating presented good adhesive strength and fracture toughness because the graphene sheets were uniformly embedded in the hydroxyapatite matrix [41, 76].

2.2.1.5 Other Techniques

Some studies have been carried out to explore unique processing routes for the graphene-reinforced metal-based composites; while some of them are an updated version of the conventional processes, others are novel methods to fabricate the composites. The high-ratio differential speed rolling method fabricated the multilayer graphene-reinforced Cu composites, and obvious strength improvement was gained due to the homogeneous dispersion of the graphene sheets [23]. Zhao et al. first utilized the high-pressure torsion technique for the fabrication of graphene-reinforced Al matrix composites. The Al powders and the graphene fillers were uniformly mixed by the planetary mill for 5 h. The milled powders were then compacted into discs pressed by the tablet press machine under 60 MPa, and then the discs were consolidated by high-pressure torsion at a pressure of 3 GPa. High-pressure torsion is typically used for severe plastic deformation for decreasing the grain sizes. The process can be carried out at room temperature, which can impair the structure damage of the reinforcements [9]. Zhang et al. processed the graphene-reinforced Cu/Al composite through a novel clad-forming technique. The $CuCl_2$ solution was mixed with the oxidized graphene and then $N_2H_4H_2O$ was added to obtain the graphene-Cu. Al powders were blended with the graphene-Cu slurry and then rolled to form the final composite [40]. Multi-pass friction stir processing was reported for the formation of graphene-reinforced AA5052 aluminummagnesium alloy nanocomposites, and negligible deterioration of the graphene was observed [18]. Graphene-reinforced Cu nanocomposite was fabricated by a molecular-level mixing process and spark plasma sintering process. This process was reported to reduce the issues of dispersion and thermal damages of the nanofillers. During the molecular-level mixing process, functional groups were attached on the surface of graphene. Then, the fast heating and cooling process was applied to limit the growth of grains [24].

2.2.2 Properties of the Graphene-Reinforced MMCs

2.2.2.1 Mechanical Properties

Yang et al. added 0.54 wt.% graphene in the Al matrix, and 116% and 45% improvement of the yield and tensile strength were achieved before extrusion, respectively. After extrusion, the increases of the yield and tensile strength were 228% and 93%, respectively [9]. Gao et al. indicated that the ultimate tensile strength of the graphene/Al composite increases first and then declines with increasing of the fraction of the graphene fillers, and the maximum ultimate tensile strength was achieved when 0.3 wt.% graphene was added. However, the elongation to fracture decreases with the increase of the fraction of graphene. TEM images of the Al matrix are shown in Figure 2.12, and the grain boundaries can be clearly viewed. The increase of graphene contents will change the fracture mode of the Al composite from ductile mode to brittle mode [10]. The Al_4C_3 phase was observed in graphene-reinforced Al5083 composite fabricated by ball milling, hot pressing, and hot extrusion.

Few graphene was reacted with the Al matrix during the consolidation process, and Al_4C_3 was formed. Both yield and tensile strength of the composite with 1 wt.% graphene increased by 50% compared with the Al matrix [17]. Good interfacial bonding was formed in the graphene/Al composite fabricated by high-energy ball milling and vacuum hot pressing, while rod-like aluminum carbide Al_4C_3 was found at the interface, as shown in Figure 2.13. The amount of aluminum carbides increases with the increase of the content of the reinforcements. The yield strength and ultimate tensile strength of the composite with 0.25 wt.% presented an incremental of 38.27% and 56.19%, respectively [11]. Clad rolled graphene-reinforced Cu/Al composites presented an enhancement of tensile strength and hardness with increases by 77.5% and 29.1%. The synergistic action of stress transfer and dispersion strengthening were applied to explain the mechanical property improvement [40]. The severe friction stir process can cause deterioration of the planar structure of the graphene, but the reinforced graphene/Al/Mg composite shows an increase of hardness of 53% and more than three times improvement in yield strength. A mixed ductile-brittle fracture behavior was observed, and its preserving ductility enhanced by 20% [18]. Latief et al.

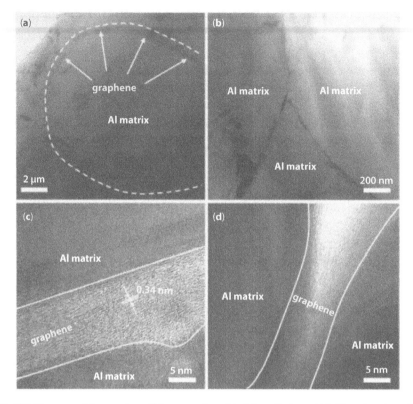

Figure 2.12 TEM images of the graphene/Al composite with 0.5 wt.%. (a and b) Al grains; (c and d) grain boundaries [10].

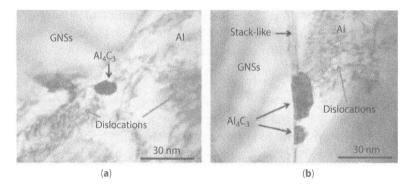

Figure 2.13 TEM images of graphene/Al composite with aluminum carbides at interfaces [11]. (a) Shows Al4C3 is formed within the Al matrix; (b) shows the carbides formed at the interface.

employed the powder metallurgy process to fabricate the graphene-reinforced Al matrix composite with a different percentage of graphene. It was found that the hardness and compression strength were enhanced with the increase of the loading fraction of graphene, while the density decreased.

Tang et al. reported a Cu matrix composite reinforced with graphene/Ni nanoparticles by spark plasma sintering. A strong interface interaction was achieved, which led to a 61%

improvement in Young's modulus and a 94% increase in yield strength, and a load transfer mechanism was used to explain the enhancement [25]. Hwang *et al.* fabricated the graphene-oxide-reinforced Cu matrix composite by a novel molecular-level mixing process that prohibited the agglomeration of the nanofillers and enhanced the adhesion between the graphene and the Cu matrix. Yield strength of the composite material was increased by 80% when 2.5 vol.% graphene oxide was added in the Cu matrix compared with that of the pure Cu [24]. Kim *et al.* pointed that the high-ratio differential speed rolling technique can achieve a graphene-reinforced Cu matrix composite with a higher density of nanosized graphene particles compared with that of the conventionally rolled Cu composite materials. The Orowan strengthening mechanism can be used to explain the improvements in strength of the composite structure, and TEM images of the microstructure graphene/Cu composite are shown in Figure 2.14, and the multilayer graphene is homogeneously dispersed within the Cu matrix [23].

Low volume fraction of 0.7 vol.% graphene-reinforced titanium composite presents a strength of ~1.5 GPa, which is superior to that of the pure titanium. The strength was attributed to several factors, such as the large surface area of the reinforcement and the interfacial features between the matrix and the reinforcements [70]. The graphene sheets survived after the laser sintering for the deposition of the graphene/Ti composites due to the fast

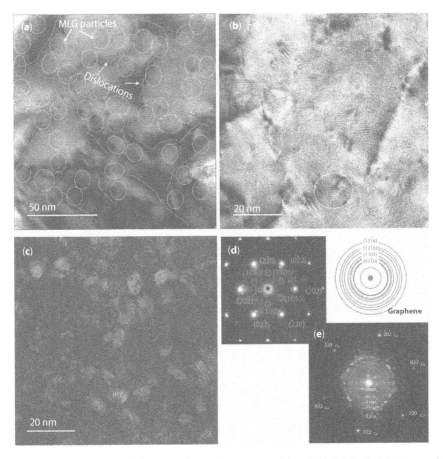

Figure 2.14 TEM images of the multilayer graphene/Cu composite (a) and (b). (c) Dark-field image of the area shows in (b); (d) diffraction pattern of the yellow circle area; (e) FFT image of the lattice image in (b) [23].

heating and cooling process. Hardness of the graphene/Ti composite is doubled to that of the barely sintered Ti. The dispersion of graphene, interfaces, and defects during the sintering are three main factors that influence the performance of the composite [62]. TiC particles were *in situ* formed at the interfaces of the graphene-reinforced Ti composite during the isothermal forging process at 970°C. With 0.5 wt.% incorporation of graphene sheets, the yield strength increased from 850 MPa to 1021 MPa and the ultimate tensile strength increased from 942 MPa to 1058 MPa, and no obvious loss of ductility was observed [63]. Mu *et al.* achieved a 54.2% increase in ultimate tensile strength of the Ti-based composite with only 0.1 wt.% nanofillers. The graphene can block slipping so that compressive twining was generated in matrix during the rolling process with the increase of nanofiller content. They attributed the strengthen reason to three main factors, namely, load transfer, grain refinement, and texturing strengthen [64]. Graphene oxides survived during the laser sintering process, which was verified by XRD and Raman spectroscopy, and the nano-hardness results of the graphene/Ti composite were significantly improved by around three times that of the bare Ti substrate. An optimal content was achieved for Vickers hardness improvement because large amounts of graphene may cause pores in the composite during the laser sintering process [61].

2.2.2.2 Corrosion Properties

Metallic coatings are commonly utilized to protect steels from corrosion when they are exposed to some corrosive environments such as high temperature, humidity, and pH. It is reported that the inclusion of nano reinforcements can improve the corrosion resistance. Few layered graphene can act as a corrosion protection barrier to decrease the corrosion rate [77]. Kumar *et al.* reported the deposition of graphene/Ni composite by electrodeposition process, and the surface morphologies of the Ni and graphene/Ni coatings are shown in Figure 2.15. A more noble performance of the graphene/Ni composite coating was achieved, and the decrease in corrosion current indicated higher corrosion resistance compared to that of the pure Ni coating [48]. The phase structure and morphologies of the graphene/Co composite coating can be

Figure 2.15 Morphologies of the (a) Ni coating and (b) graphene/Ni composite coating [48].

affected by the incorporation of the graphene oxide sheet during the electrodeposition process. A lower corrosion current was achieved by the composite coating compared to that of the bare Co coating [20]. The graphene-oxide-reinforced Cu composite coating was electrodeposited on the mild steel to improve its corrosion resistance. The composite had a fine granular morphology and the Cu matrix exhibited a <220> preferred orientation. A decrease in the susceptibility of the coated samples to the attacks from Cl^{-1} indicated the corrosion protective performance of the graphene oxide/Cu composite coating. The optimal sample showed 88% reduction in corrosion rate compared to the bare mild steel. Passive films were formed on Cu and promoted by the nanocrystalline microstructure affected by the graphene oxides [26].

Qiu *et al.* reported a $GO/Ni(OH)_2$ composite coating by a pulse current deposition technique on the 316 stainless steel. The graphene oxides can be easily dispersed in polar solvent because they are hydrophilic, and a compact $GO/Ni(OH)_2$ composite coating was formed with $Ni(OH)_2$ and graphene oxide particles. The corrosion inhibition efficiency of the GO/$Ni(OH)_2$ reached 97.1%, indicating that the composite coating can act as a barrier to prevent the permeation of the corrosive medium [56]. Kavimani *et al.* reported a magnesium-based metal matrix composite reinforced with graphene oxide nanosheets through powder metallurgy route. The composite coating with 0.3 wt.% of graphene oxide posed an excellent corrosion inhibition efficiency of 96% and a low corrosion rate of 3.57×10^{-7} mpy [19]. Corrosion properties of the graphene/Ni composite coating were studied by polarization tests and electrochemical impedance spectroscopy. The results indicated that the composite coating deposited at a specific temperature showed a better corrosion resistance [46].

2.2.2.3 Tribological Properties

Liu *et al.* found that the graphene-oxide-reinforced Co coating presented a friction coefficient around 0.33, which is lower than that of the bare Co coating around 0.8. This significant decrease of friction coefficient and wear rate was due to the self-lubricating property of the graphene oxide [20]. Zhai *et al.* indicated that the graphene-nanosheet-reinforced composite can act as a solid lubricant due to the existence of graphene. A low loading fraction of graphene is able to significantly reduce the friction coefficient and wear rate at an elevated temperature. It was found that the grain size of Ni_3Al can be refined by the graphene oxide and stress can be dissipated through the slippage of the graphene sheets, leading to the improvement of tribological properties. The schematic diagram of the wear mechanism is shown in Figure 2.16 [55].

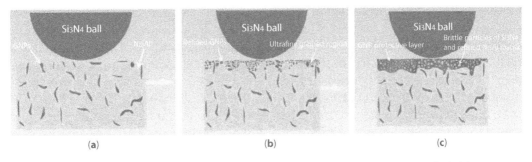

Figure 2.16 Schematic diagram of the wear mechanism of the graphene/Ni_3Al composites [55]. (a) Pristine composite; (b) refined GNPs formed; (c) protective layer formed.

There were obvious reductions in friction coefficients by 0.4 and wear rate by about 4-9 times the magnitude of the graphene/TiAl composite. The reductions were attributed to the formation of an anti-wear protective layer on the contact interfaces [65]. Micro-tribological tests on the graphene-related composite coatings were carried out by AFM. The GO/Ni(OH)$_2$ composite coating presented a lower friction coefficient compared to that of the bare Ni(OH)$_2$ coating [56]. Kavimani *et al.* studied the influence of the loading fraction of graphene oxide on the wear performance of the GO/Mg composite. Hardness of the composite coating was improved by the addition of strong graphene sheets, and it was found that the wear performances of the composite coating were affected by the loading fraction of the graphene oxide nanofillers. Wear loss reduction can be attributed to the self-lubricating layer formed at the interfaces [19].

2.2.2.4 Other Properties

Graphene-reinforced nickel composite has been prepared by a electrodeposition process, which has a reduction effect on the graphene oxide. The nickel growth orientation changed from (200) to (111) during the electrodeposition process. The graphene/nickel composite presented a 15% increase in thermal conductivity compared with that of the monolithic nickel [49]. Hybrid graphene/silver particle fillers were synthesized as a thermal interface material. The thermal conductivity of the composite was improved significantly with a low loading fraction of graphene fillers. About fivefold increase in thermal conductivity was achieved by the composite reinforced with 5 vol.% of graphene, and the enhancement of thermal conductivity was attributed to the excellent intrinsic thermal conductivity of the graphene/silver composite fillers [78].

2.3 Graphene–Reinforced Polymer Matrix Composites (PMCs)

Polymers are widely used in today's society due to their broad range of properties. Both natural polymers and synthetic polymers are relatively cheap and simple to fabricate. Some of the synthetic polymers play important and ubiquitous roles in our everyday life, such as polyurethane, polyimide, polyethylene, polypropylene, polycarbonate, poly(methyl methacrylate), and poly(vinyl alcohol). However, due to the limitation of their physical properties, some polymer-based composites are developed for their applications in some specific areas. Graphene and graphene oxide are broadly studied in the past decades due to their excellent mechanical, electrical, and thermal properties. These superior properties arise from the single carbon layer 2D structure. The incorporation of these carbon nanofillers into the polymeric chains may significantly enhance the mechanical strength and the electrical or thermal conductivity of the composites. The processing of graphene-reinforced polymer matrix composites (PMC) is relatively simple and a large number of studies have been carried out and a rising trend on this topic can be observed from the number of publications in the past several years (Figure 2.17); summaries of the preparation methods and properties of the graphene-reinforced PMCs are provided in Table 2.2.

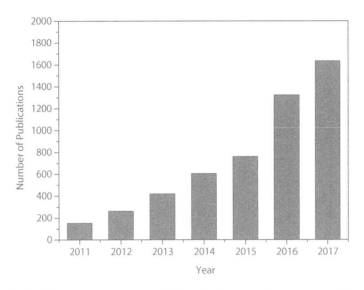

Figure 2.17 Trends of publications on graphene-reinforced polymer matrix composites. (From Scopus.)

2.3.1 Preparation of Graphene Polymer Composites

The use of graphene-related materials for processing of the polymer-based composites is widespread in academia. Different methods have been carried out for the preparation since the first isolation of graphene in 2004. The main problem during the manufacturing of the graphene-reinforced polymer-based composite is to ensure the homogeneous dispersion of the nanofillers. The mechanical properties of the composite structure are directly influenced by the dispersion of the graphenes. Moreover, the interfacial features between the fillers and the matrix are important for the processing of the graphene-reinforced composites. A number of fabrication techniques have been developed to prepare the composites, such as melt blending, *in-situ* polymerization, solution compounding, and some novel processes.

2.3.1.1 *Melt Blending*

The melt blending process is widely used in the industry for the production of thermoplastic composites. The melt blending method is relatively simple, fast, and cost-effective. First, the polymer is melted at elevated temperature and the graphene flakes are added and mixed. Then, the composite mixture is extruded through a single-, twin-, or quad-screw extruder. The working temperature should be carefully controlled because high temperature may cause degradation of polymer. Besides, high shearing force needed for the mixing can cause damage on the graphene sheets. Even though there are some limitation of the melt blending process, it is suitable for large-scale fabrication of the graphene-based nanocomposite with adequate properties. Shen *et al.* studied the influence of melt blending on the interaction between graphene and PS through different melt blending time. The π-π stacking

Table 2.2 Summary of preparation methods, matrix, and reinforcements of the graphene/polymer matrix composites.

Author	Method	Matrix	Reinforcement	Properties
Shen et al. [79]	Melt blending	Polystyrene	5 wt.% graphene	Formation of π-π stacking
Bao et al. [80]	Melt blending	Poly(lactic acid)	0.08 to 2 wt.% graphene	Storage modulus 4.04 GPa; tensile strength 60 MPa
Zhang et al. [81]	Melt blending	Polyethylene terephthalate	1 to 7 vol.% graphene	2.11 S/m electrical conductivity with 3.0 vol.% graphene
Pang et al. [82]	Melt blending	Polystyrene	0.25, 0.5, 0.75, and 1 wt.% graphene or CNT	Activation energy for graphene-reinforced conductive network is 80 kJ/mol
Zeng et al. [83]	Solution compounding	poly(methyl methacrylate)	0.1, 0.5, 1, and 2 wt.% GO	0.037 S/m electrical conductivity with 2.0 wt.% GO
He et al. [84]	Solution compounding	Poly(vinylidene fluoride)	0.4 to 3 vol.% GO	108 dielectric permittivity with 2.5 vol.%
Stankovich et al. [85]	Solution compounding	Polystyrene	0.1 to 2.4 vol.% GO	Electrical conductivity ~ 1 S m^{-1} at 2.5 vol.% of GO
Kim et al. [86]	Solution compounding	Low-density polyethylene	5, 7, and 12 wt.% multilayer graphene	Higher thermal stability
Wang et al. [87]	In situ polymerization	Polyurethane	0.5, 1, and 2 wt.% GO	Tensile strength and storage modulus increased by 239% and 202% with 2 wt.%
Aidan et al. [88]	In situ polymerization	Polyamide 6	0.1, 0.25, 0.5, 0.75, and 1 wt.% GO	62 MPa tensile strength; 51.2 MPa yield strength with 0.75 wt.% GO

(Continued)

Table 2.2 Summary of preparation methods, matrix, and reinforcements of the graphene/polymer matrix composites. (*Continued*)

Author	Method	Matrix	Reinforcement	Properties
Yu et al. [89]	*In situ* polymerization	Epoxy	0.1, 10, and 25 vol.% multilayer graphene	3000% improvement in thermal conductivity; thermal conductivity 6.44 W/mK
Min et al. [90]	*In situ* polymerization	Epoxy	0.270 and 2.703 vol.% multilayer graphene	Thermal conductivity 0.72 W/mK; 240% improvement with loading of 2.703 vol.%
Ren et al. [91]	*In situ* polymerization	Cyanate ester–epoxy	0.1, 0.3, 0.5, 0.7, 0.9, and 1 wt.% GO	Flexural strength 128.1 MPa; impact strength 11.5 kJ/m^2
Zhao et al. [92]	Layer by layer assembly	Poly(vinyl alcohol)	GO	98.7% improvement of elastic and 240.4% increase of hardness

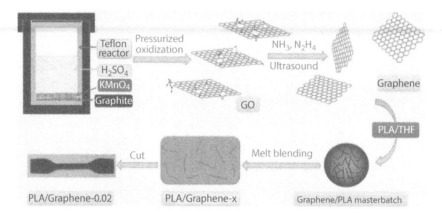

Figure 2.18 Schematic diagram of the preparation of graphene and graphene-reinforced PLA composite by the melt blending technique [80].

between PS and graphene oxide was formed during melt blending, which can improve the interaction between PS and graphene [79]. As shown in Figure 2.18, graphene-reinforced poly(lactic acid) composite was prepared by melt blending. Graphene was deposited from graphite by pressurized oxidation and reduction process, and then dispersed into PLA. The characterization results indicated that the percolation threshold was 0.08 wt.% and the interactions among the polymer matrix were cut down by the reinforcement of graphene leading to reduction of mechanical properties [80].

2.3.1.2 Solution Compounding

The solution compounding method is based on the mixing of the graphene suspension and the polymer in a solution. The mixtures are commonly stirred or ultrasonic vibrated to achieve a homogeneous dispersion of the nanofillers. Then, the mixed solution can be cast into a mold to remove the solvent. This process is relatively versatile, and a variety of solvents can be used for the mixing. One disadvantage of the process is that the removal of the solvent may cause restacking or aggregation of the nanofiller. He *et al.* used a solution mixing process to fabricate graphene-reinforced PVDF composites. The percolation threshold of graphene loading fraction was 1 vol.% and the dielectric constant was 200 around the percolation threshold at 1 kHz [84]. Graphene/poly(methyl methacrylate) (PMMA) nanocomposites were deposited by a simple solution mixing method: a certain volume of GO suspension was mixed with PMMA and continuously stirred and sonicated for 2 h, and then the mixtures were poured into a beaker with methanol. Precipitates were filtered and dried for further characterization [83] (Figure 2.19).

2.3.1.3 In Situ Polymerization

The process makes it possible to graft the nanofillers on the polymer for the improvement of compatibility between the elements in the composites. However, the viscosity of the mixture is increased during the polymerization process, which limits the loading fraction of the nanofillers. Functionalized graphene-oxide-reinforced polyimide composite was

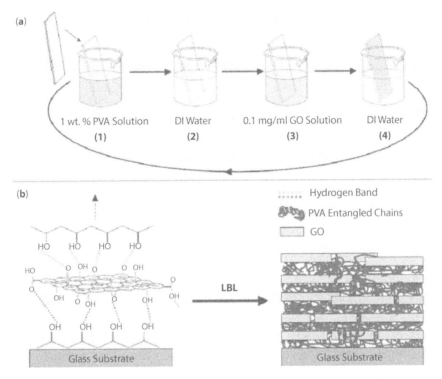

Figure 2.19 Schematic diagram of the deposition processes for the GO/PVA composite film using layer-by-layer assembly. (a) Basic sequences; (b) schematic cross-section view of the composite film [92].

fabricated by the *in situ* polymerization method. Amine groups were added to the surface of the graphene oxide to improve the dispersion and enable the formation of strong bonds between the graphene and the polymers [93]. Bielawski *et al.* indicated that the graphene oxide reinforcement can play roles in catalyzing the dehydrative polymerization process and graphene-like flakes were formed after the reaction [94]. Yu *et al.* embedded the graphene into epoxy using the *in situ* cross-linking method. Graphene slurry was shearing mixed with epoxy in acetone to ensure homogeneous dispersion and prevent aggregation. The mixture was then mixed with cure agent [89]. Aidan *et al.* prepared the GO/PA6 nanocomposite via *in situ* polymerization of ε-caprolactam and single-layer graphene oxide. The nanocomposite was relatively thermal stable, but the induction of graphene oxide can directly affect the molecular weight and the crystallinity of the nanocomposite [88].

2.3.1.4 Other Methods

The layer-by-layer assembly process has been carried out for the graphene-reinforced polymer-based composites. Various nanomaterials can be prepared for the formation of multilayer films with specific thickness and nanostructures by alternating the phases. Various hydroxyl and epoxy groups on the surface of the graphene oxide make it interactive during the assembly process. The functional composite film can be used for a wide range of applications, such as supercapacitor, Li-ion battery, and electrodes. Graphene-oxide-reinforced

PVA composite thin film was deposited using the layer-by-layer assembly method with a bilayer thickness of 3 nm. The modulus of the composite film was doubled compared to that of the pure film [92]. Singh *et al.* reported the processing of a graphene-reinforced hydroxy functional acrylic adhesive composite coating by aqueous cathodic electrophoretic deposition (EPD) to protect the Cu substrate from electrochemical degradation [95].

2.3.2 Properties of Graphene-Reinforced PMCs

2.3.2.1 Electrical Properties

Graphene exhibits excellent electrical conductivity, which makes it an ideal filler in polymers for the fabrication of flexible sensors, conductive films, and microwave absorbers. A continuous conductive network can form within the composite due to the interactions between the graphene fillers [1]. The large aspect ratio of graphene makes it possible to enable the insulator polymer to conducting composites at significantly low graphene loading compared to the electrical percolation thresholds for other carbon materials [4, 96].

Several graphene-reinforced polymer composites have been reported, and the matrices included epoxy, polyolefin, polyamide, polyester, vinyl, PS, PU, and synthetic rubbers [97–100]. The percolation thresholds depend on the loading fraction of the electrical fillers and typically their electrical conductivity increases nonlinear with the increasing of the electrical fillers. Stankovich *et al.* reported that the percolation threshold for the PS/GO nanocomposite was 0.1 vol.%, which is comparable with other carbon fillers, such as SWCNT and MWCNTs. The relationship between the electrical conductivity and the loading fraction of graphene oxide is shown in Figure 2.20 [85]. Xie *et al.* modeled graphene fillers that have better conductivity than the cylinder-like CNTs [101]. Kalaitzidou *et al.* indicated that 0.1 to 0.3 vol.% loading of graphene fillers can reach the electrical percolation threshold of PP [102]. Steurer *et al.* reported that the percolation thresholds of the graphene-reinforced thermoplastic nanocomposites varied from 1.3 to 3.8 vol.% [98].

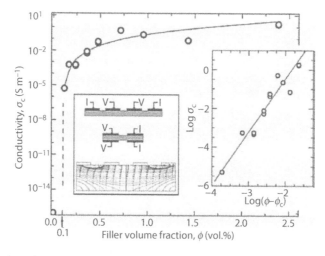

Figure 2.20 Electrical conductivity of the graphene/polystyrene composites as a function of the loading fraction of graphene [85].

It was summarized that there are several factors that affect the electrical conductivity and electrical percolation threshold of the graphene-reinforced polymer composite materials, such as fabrication methods, concentration of the graphene fillers, functionalization of the graphene fillers, and internal distribution of graphene in the matrices.

The processing methods directly influence the distribution of graphene fillers within the polymer matrices and thus affect their electrical properties. Better electrical conductivity can be achieved through solvent blending process and the *in situ* polymerized process than the melt mixing process [103], because solution-based processes can provide better dispersion of the nanofillers. On the other hand, the melt mixing can generate an annealing effect, which is useful to adjust the junctions between the fillers. Lower electrical percolation threshold was achieved for the solvothermal reduced graphene-reinforced PVDF than that composite processed by direct blending of PVDF and graphene [104, 105]. The graphene sheets were homogeneously dispersed and relatively stable during the solvent-based process, which led to a low percolation threshold. The large aspect ratio of graphene can be maintained because graphene cannot easily fold in solution. However, poor distribution of the graphene was generated during the mixing process, which leads to a notably low electrical conductivity even with high fraction of the graphene filler. The nanocomposite was deposited using an *in situ* polymerization process that is sufficient to obtain nanosized distribution without other pretreatments.

In order to achieve current flows within the polymer matrix composites, there should be a conductive network within the polymer based on the fillers. However, direct contacts of the graphene fillers are unnecessary because the conduction can generate through the tunneling between the fillers and the thin polymer layers around them. Thus, there is no need to add high concentration of nanofillers to create the percolation. Besides, it was reported that percolation thresholds for polymers are different and the concentration of the nanofillers plays an important role for their electrical conductivity. Pang *et al.* reported the electrical conductivity of the graphene-reinforced UHMWPE. Figure 2.21 shows that the percolation threshold of the electrical conductivity was 0.07 vol.% [106].

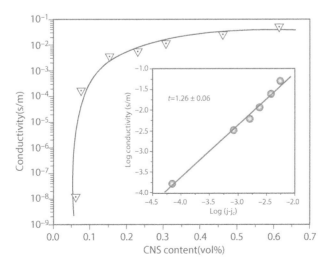

Figure 2.21 Relationship between electrical conductivity and loading fraction of graphene in UHMWPE [106].

Zhang *et al.* fabricated the graphene/PET nanocomposite, and its percolation threshold of electrical conductivity is 0.47 vol.% [81]. Liang *et al.* indicated that solution with the graphene/epoxy nanocomposite has a lower percolation of 0.1 vol.% compared to that of the neat nanocomposite [107]. Pang *et al.* pointed out that the conductive nanosheets can overcome the matrix barrier films and overlap with each other. Figure 2.22 shows the formation of the conductive network of the graphene-reinforced polystyrene. Conducting channels are formed in the "conducting square," which can generate responses to the electric field [82].

Modification of the graphene is a feasible way to improve the electrical conductivity of the graphene-based polymer composites. Graphene can be functionalized to be covalent or noncovalent in order to promote its dispersion and make the graphene more stable. To prevent the agglomeration of the nanofillers during the mixing process [108, 109]. Functional groups can be attached to graphene, which can improve the dispersion and enhance the interactions within the polymers. Some chemical processes were also applied to functionalize the graphene, such as amination and esterification [110, 111]. Liu *et al.* utilized an ionic-liquid electrochemical process to modify graphene [112]. Park *et al.* obtained a stable suspension of graphene by chemically reducing graphene oxide in a stabilized medium [113]. Li *et al.* provide a facile way for large-scale fabrication of aqueous graphene suspension by electrostatic stabilization [114].

Stankovich *et al.* reported that the wettability of the GO sheets could be modified by different chemical agents, such as organic amines and isocyanates [115]. It was also reported by Stankovich *et al.* that low percolation threshold at 0.1 vol.% could be achieved by using GO modified by isocyanation as nanofillers [85]. The modified GO presented high electrical conductivity and can be homogenously dispersed in the polymer. Chen *et al.* studied the electrical conductivity of the chemical reduction of GO-reinforced PDMS composites and the directly deposited graphene/PDMS using the CVD method.

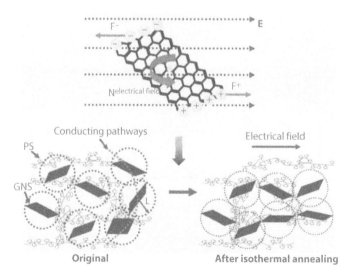

Figure 2.22 Schematic diagram of the conductive network of graphene before and after the inducement of the electric field. The electrons are circulated in the blue circle [82].

The modified samples exhibited higher electrical conductivity due to the interconnected graphene sheets [116]. Stankovich *et al.* modified GO by attaching long-chain aliphatic amines that made it easier for the dispersion of the modified GO in the organic solvents [117]. Zhang *et al.* reported that graphene cannot be well dispersed in water-soluble polymer using the ultrasonic method, and polymeric anions were applied to modify the graphene for a stable dispersion [81]. Kim *et al.* indicated that graphene treated with pyrene could significantly improve the electrical conductivity of graphene/epoxy nanocomposite, because pyrene can be tightly attached on the surface of graphene through the π–π bonds [118].

2.3.2.2 Mechanical Properties

Wang *et al.* reported polyurethane composites reinforced with graphene nanofillers by an *in situ* polymerization process. It was found that the graphene sheets were well dispersed in the polyurethane matrix due to the chemical bonds. The tensile strength of the graphene oxide/polyurethane composite increased by 239% with the addition of 2.0 wt.% of the nanofillers [87]. Song *et al.* achieved a well-dispersed graphene-reinforced polypropylene composite material by a two-step mixing process. The graphene was coated with the polypropylene latex first and then melt-blending with the polypropylene matrix. With the incorporation of 0.42 vol.% of graphene, there is a ~75% and ~74% increase in yield strength and Young's modulus of the polypropylene composite. The cross-section images of the composite after tensile test are shown in Figure 2.23 [119].

The graphene-oxide-reinforced PVA film prepared using the LBL process presented a 98.7% improvement of elastic modulus and a 240.4% increase of hardness. The improvements were attributed to uniform dispersion and the orientation of the graphene oxide sheets that can maximize the interaction between GO and matrix and constrain the motion of the polymer chain [92]. Aidan *et al.* found a linear increase of the yield strength of the GO/PA6 composite with the increase of the loading content of GO. However, tensile strength increases are not linear with the increase of GO contents. The use of water during the *in situ* polymerization process reduced the mechanical properties of the composite [88].

2.3.2.3 Thermal Properties

Alexander *et al.* reported the excellent thermal conductivity of the single-layer graphene sheet, which ranges from ~$(4.84 \pm 0.44) \times 10^3$ to ~$(5.30 \pm 0.48) \times 10^3$ W/mK at room temperature. This extremely high thermal conductivity makes graphene an ideal filler for thermal conductive composite materials [4]. Kim *et al.* reported that the multilayer graphene-reinforced low-density polyethylene nanocomposite showed better thermal stability using TGA and TMA tests. The weight percentage of the nanofiller directly affects the thermal expansion coefficient [86]. Yu *et al.* measured the thermal conductivity of the graphene/epoxy composites with different loading fractions of graphene. It was confirmed that graphene is efficient to enhance the thermal conductivity of the epoxy matrix (Figure 2.24). The highest enhancement of thermal conductivity was more than 3000% with ~25% vol.% of nanofillers and correspond to about 100% increase per vol.% [89].

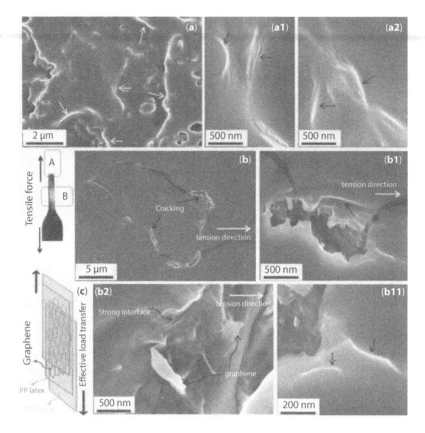

Figure 2.23 SEM images of the cross-section view of the graphene/PP composites with 1 wt.% of graphene after tensile tests; black arrows point to the graphene sheets [119]. (a) Graphene/PP composite; (b) fracture surface of the composite; (c) schematic of tensile test.

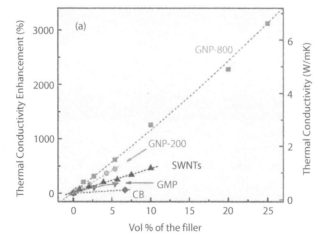

Figure 2.24 Thermal conductivity of the epoxy matrix composites with different carbon fillers [89].

2.3.2.4 Corrosion Properties

Chang *et al.* prepared a hydrophobic graphene/epoxy composite coating as a corrosion inhibitor to protect the cold rolled steel. The graphene sheets embedded in the epoxy matrix can prevent the corrosion due to its large aspect ratio [120]. Functionalized graphene-like sheets presented a better dispersion in polyaniline. The composite coating acted as a barrier against O_2 and H_2O and exhibited better corrosion resistance with a higher corrosion voltage and lower corrosion current compared with that of the bare steel [121]. A well-dispersed graphene oxide/polystyrene nanocomposite was prepared using an *in situ* miniemulsion polymerization process that exhibited excellent anti-corrosion performances. With 2 wt.% incorporation of the modified graphene oxide, the corrosion protection efficiency was improved from 37.90% to 99.53% [122].

Singh *et al.* fabricated a graphene-oxide-reinforced hydroxy functional acrylic adhesive composite coating on Cu to improve its corrosion resistance. The composite coating was deposited by aqueous cathodic electrophoretic deposition with thickness around 40 nm. Tafel analysis results indicated that the corrosion rate of the composite coating was an order of magnitude lower than that of the untreated Cu [95]. Qiu *et al.* functionalized 316 stainless steel surface with a graphene-oxide/polyaniline composite coating via a pulse current deposition method. The composite coating displayed high corrosion inhibition efficiency and protection efficiency, which were 98.4% and 99.3%, respectively. The wettability and porosity of the composite coating can be affected by the deposition parameters [123].

2.4 Graphene-Reinforced Ceramic Matrix Composites (CMCs)

2.4.1 Processing Methods

2.4.1.1 Types of Graphene Fillers

Since its first successful isolation in 2004 via mechanical exfoliation (i.e., so-called "Scotch-tape" method) [1], many other top-down graphene synthesis methods, including liquid exfoliation [124] and reduced graphene oxide [125], and bottom-up methods, such as chemical vapor deposition (CVD) [126] and epitaxial growth on SiC [127], have been developed. Although the original mechanical exfoliation route gives high-quality graphene in terms of pristine electronic properties and mechanical strength, its low yield withdraws itself from the selection list as the reinforcement phase of composite materials. The latest development on CVD-grown graphene shows that continuous growth of high-quality graphenes up to 100 m long has been possible [128] and there is no debate that CVD is the most promising route for the mass production of high-quality graphene. However, a complicated separation and transfer process from the growth substrate (usually Cu or Ni foils) is still inevitable for this type of graphene products, which makes them unsuitable for composites [128].

In the literature so far on graphene-reinforced ceramic composites, few-layer graphene (FLG, 1–5 layers), graphene/graphite nanoplatelets (GNPs, normally in the range of 5–50 layers), and graphene oxide/reduced graphene oxide (GO/rGO) are the mainstream choices due to not only their wide availability and competitive prices on the market but also the compatibility with many ceramic processing methodologies. Table 2.3 lists the types

Table 2.3 Different graphene-related nanofillers used in previous studies.

G type	Synthesis method	Layer no.	Lateral sizes (μm)	Composite matrix	Mixing method	Properties desired	Ref.
FLG	Electrochemical expansion	<5	10–20	Al_2O_3	Ball milling, 24 h	Wear resistance, toughness	[129]
	Liquid exfoliation	<3	1.5	Al_2O_3	Sonication, ball milling	Toughness, hardness, elastic modulus	[131]
MLG	High-energy milling	10–20	–	Si_3N_4	Ball milling, 600 rpm 30 min	Elasticity, bending strength	[134]
GNP	Commercially sourced	~60	2	Si_3N_4	Sonication, blade mixing	Toughness	[141]
	Commercially sourced	–	–	Si_3N_4	Sonication	Tribology	[142]
	Commercially sourced	–	–	Si_3N_4	Sonication for 1 h	Electrical	[143]
	Commercially sourced	6–8 nm	5	ZrB_2	Colloidal, ball milling	Toughness, flexural strength	[144]
	Mechanical milling; commercially sourced	5–50 nm	1	Si_3N_4	Ball milling	Toughness	[145]
	Commercially sourced	<32	4–12	ZrB_2, Si_3N_4	Colloidal, stirring	Hardness, toughness	[146]
	Chemical exfoliation and thermal reduction	3–4	–	Si_3N_4	Sonication	Toughness	[147]
	Thermal expansion	6–8 nm	15–25	Al_2O_3	Sonication, ball milling	Toughness	[132]
	Thermal exfoliation, ball milling	2.5–20 nm	–	Al_2O_3	Ball milling	Electrical conductivity	[130]

(*Continued*)

Table 2.3 Different graphene-related nanofillers used in previous studies. (*Continued*)

G type	Synthesis method	Layer no.	Lateral sizes (μm)	Composite matrix	Mixing method	Properties desired	Ref.
GO	Hummers	–	–	Al_2O_3	Colloidal	Wear resistance	[148]
	Hummers	~10	3	Si_3N_4	Sonication, blade mixing	Toughness	[141]
	Hummers	1	–	Al_2O_3	Colloidal	Toughness, electrical conductivity	[149]
rGO	Hummers, *in situ* reduction	–	–	Hydroxyapatite	Hydrothermal synthesis	Biomedical, toughness, hardness, elasticity	[150]
	Hummers, chemical reduction	–	–	YSZ	Colloidal	Electrical conductivity, toughness	[151]
	Hummers, *in situ* reduction	–	–	Al_2O_3	Colloidal	Electrical conductivity, toughness	[152]
	Hummers, *in situ* reduction	–	10	ZrB_2-SiC	Ball milling	Toughness	[153]

of graphene fillers in related studies. It is worth noting that FLG and GNP here refer to graphene structures that have been subject to no or very small chemical modification, thus retaining more pristine properties of graphene compared with those synthesized via chemical routes, i.e., GO or rGO.

Few-layer graphene was employed in a few studies. Literally, FLGs are normally much thinner than their multilayer counterparts, thus enabling higher flexibility without sacrificing the great mechanical strength of graphene. Considering the normally high strength yet very low ductility of common ceramics, such graphene structures are desirable for toughness improvements. Another important advantage of FLG over GO or rGO is that the former usually did not experience any severe chemical oxidation and reduction process as with GO and rGO, thus containing much fewer structural defects. To obtain FLGs, different approaches of energy input can be adapted to exfoliate thicker graphite apart into thin sheets. For instance, Kim et al. electrochemically expanded graphite in a Li-containing electrolyte and further exfoliated with the aid of sonication [129]. In their work on graphene-alumina composites, Fan et al. thermally expanded graphite by heating to 1000°C in nitrogen atmosphere. The following mechanical exfoliation was integrated with the powder mixing step in a planetary mill together with α-Al_2O_3 powder [130]. To improve the quality and homogeneity of the resulting graphene, Porwal and coworkers employed a liquid phase exfoliation in *n*-methyl pyrrolidone (NMP) under sonication [131].

From Table 2.3, it is clear that graphene nanoplatelet (GNP) is particularly popular among the graphene-CMC researchers. This is mainly due to the very low price (GBP ~0.5, still decreasing) and stable mechanical, thermal, and electrical properties. In the literature, it may be named graphene nanoplate, graphene nanosheets (GNS), or graphene platelet (GPL), but all refer to a graphene structure of up to 100 nm, usually unoxidized [132]. GNP synthesis methods include but are not limited to thermal expansion, liquid exfoliation, mechanical milling, etc. The starting materials are normally natural graphite flakes or expandable graphite powders (pre-expanded via thermal shocks or chemical intercalation). In a typical powder process, Tapasztó et al. milled thermally expanded graphite in a high-energy attritor mill for 3 h and ended up with few-layer to multilayer (1–30 layers) graphene flakes in dimensions of a few square microns [133]. Kun et al. [134] prepared MLGs using a similar mechanical milling technique prior to the fabrication of silicon-nitride-based composites. A few commercially available GNPs were also involved as comparison groups. They found that the MLGs even outperformed the commercial GNPs in terms of the elastic modulus and bending strength of the composites. Of particular interest is the process used by Fan et al., in which expandable graphite and Al_2O_3 powder were milled together in a planetary mill for 30 h in the presence of NMP as the dispersing solvent. By this method, the exfoliation of graphite and the powder mixing were integrated into one single stage. A planetary mill rather than a vibratory mill was used as the former favors the cleavage of the particles [130]. The thickness of the resulting graphene sheets was in the 2.5–20 nm range.

It is well known that pristine graphene shows an elastic modulus of astonishingly 1 TPa and an intrinsic strength of 130 GPa, while GO and rGO show effective Young's modulus of ~200 and 250 GPa, respectively, suggesting the inverse effect of structural disorders on the mechanical properties of the graphene product [135]. This is mainly due to the functional groups and structural defects on the graphene plane originated from the oxidation process. Nevertheless, some advantages remain for GO and rGO, such as easy dispersion, likely enhanced interactions with the ceramic matrix owing to the functional groups on GO sheets, and the potential for improved mechanical and electrical properties via the reduction of GO, etc. GO can be easily synthesized in a large quantity (1–5 g per batch in the laboratory and much higher in the industry) via Hummers' method [136] and its derivatives [137–140] using natural graphite flakes or expandable graphite as the starting material. After oxidation, mild sonication is sufficient to split the expanded graphitic structure further into few-layer or even monolayer graphene sheets.

2.4.1.2 Powder Processing

For graphene-reinforced ceramic and metal composites, it is crucial to ensure the homogeneous distribution of the graphene reinforcements in the targeted matrix in order to take advantage of the extraordinary mechanical, thermal, and electrical properties of graphene. The powder mixing step thus plays a significant role in the whole composite fabrication process. However, due to the difference in surface energies of carbonaceous materials and most ceramic powders, graphene materials naturally tend to aggregate during processing, as with CNT-reinforced composites [154, 155]. Therefore, in the published studies, authors tailored the powder processing recipes carefully to improve this situation.

Ball milling has been commonly used to minimize the agglomeration of graphene-related materials and improve the dispersion of these materials within ceramic powder precursors. Important parameters include the milling media (liquid or dry), selection of milling balls, and the ball-to-powder ratio, as well as the milling time. To overcome the high surface energy of graphene sheets that causes agglomeration, a high-energy input is usually required. Another advantage of using ball milling is that the shear forces introduced during the ball milling can also break the van der Waals forces that hold the graphene layers together, and exfoliate thicker graphene platelets into few-layer sheets, as mentioned in the last section. But what is inevitable is that the average particle size of the final powder will be reduced, and contaminations may be introduced during the milling process (the use of ultra-hard ceramic milling jar and balls like WC). The solvents involved in the ball milling process vary from case to case. Surfactants help with the agglomeration issue. It has been reported that CTAB [147] and polyethylene glycol (PEG) [134] are both effective in improving the graphene dispersion in the final composite matrix. On the other hand, based on their quantitative analysis of the liquid exfoliated graphene sheets, Coleman *et al.* predict that any solvent with a surface tension of 40–50 mJ/m^2 is desirable, as the surface energies of such solvents match that of graphene [156]. Thus, it is beneficial to choose such a proper solvent for mixing and ball milling of the graphene nanofillers and the ceramic powder [130]. Typical solvents involved in the ball milling process are deionized water plus surfactants [134, 145], isopropanol [142, 143], DMF [129, 131, 132], ethanol [146], and NMP [130], to name a few.

Colloidal processing is also popular for CNT-based and graphene-based CMCs [131, 144, 146]. Alternative energy inputs other than ball milling, such as ultrasonication and mechanical stirring, are adapted to help with the agglomerations of GNPs. Unlike ball milling, the particle size will not be reduced during the agitation due to absence of strong shear forces. The contamination issue associated with ball milling is also avoided for colloidal processing routes. The problems are the attachment of the liquid media to the final particles, which may result in poor densification, and the density difference between the ceramic powder and the graphene, which increases the risk of inhomogeneous composite products. The liquid media can be baked out at the end of the mixing process [143, 145].

2.4.1.3 Densification

Following powder mixing and compacting, the green composite needs to be sintered to go through a densification or consolidation stage. Consolidation methods include hot pressing (HP), hot isostatic pressing (HIP), spark plasma sintering (SPS) and pressure-less sintering. These techniques, together with typical processing conditions and references, are summarized in Table 2.4.

As a conventional sintering technique, pressure-less sintering is cost-effective and environment friendly. But in order to reach full densification, it requires high temperatures and prolonged sintering times compared with other techniques. Kim *et al.* prepared unoxidized graphene/Al_2O_3 composites via pressure-less sintering and investigated toughness, strength, and wear resistance. The ball-milled dried powder mixture was initially shaped in a uniaxial press and subsequent cold isostatic press at 200 MPa. The formed bars were then subject to sintering in an electrical furnace in an Ar atmosphere for 3 h. Without applied pressure, the sintering temperatures varied from 1450 to 1700°C according to the sample composition.

Hot pressing (HP) and hot isostatic pressing (HIP) techniques introduce either uniaxial (for HP) or isostatic (for HIP) pressures, thus enabling large ceramics to be fully densified. While maintaining the high sintering temperature, normally a uniaxial pressure, like 20 MPa, is applied between the mold and die. A typical holding time of a few hours is necessary. It should be noted that, during long-time sintering, the ceramic grains may grow continuously to a large extent, resulting in a "softening" effect in terms of the mechanical properties.

Unlike conventional sintering routes, spark plasma sintering (SPS) is a rapid sintering technique. With the assistance of pressure and an electric field, the sintering time can be reduced dramatically from hours to a matter of minutes. Gutierrez-Gonzalez *et al.* [148] synthesized a GNP/alumina composite by SPS. The green powder compact was sintered at 1500°C with a heating rate of 100°C/min under a pressure of 80 MPa. The dwelling time was merely 1 min. Compared with the other studies on graphene-based CMCs using HP, HIP, or pressure-less sintering, it is clear that not only is the sintering efficiency of SPS significantly higher, but the sintering temperature can be remarkably reduced.

It has been found that graphene promotes the densification by particle re-arrangement in the early stages of sintering [144]. For some specific ceramics, carbon species can help remove the oxide impurities on the surfaces of the ceramic powder (such as ZrO_2 and B_2O_3 on ZrB_2), thus promoting the densification [146]. Yadhukulakrishnan *et al.* [144] investigated the densification behavior of monolithic ZrB_2 and ZrB_2/GNP composites by monitoring the punch displacement during the sintering process. It was found that the GNPs not only enhanced the early-stage compaction of the powder mixture but also clearly promoted the final-stage densification by shifting the thermal expansion-densification transition point in the sintering process to an earlier time (Figure 2.25). In this sense, the GNPs have acted as a sintering aid like conventional ones (Al_2O_3 and Y_2O_3) [134, 141]. The enhanced densification process can in turn contribute to an improved strength, as the denser the final composite is, the higher hardness it could possess.

Table 2.4 Deposition techniques, properties, and conditions of the graphene-reinforced composites.

Densification method	Conditions	Matrix	Filler type and content	Improvements	Ref.
HP	Heating 15°C/min, 15 min dwelling @ 1000°C, 60 min @ 1850°C, uniaxial pressure 20 MPa, in vacuum	ZrB_2-SiC	GNP, 5 wt.%	Densification improved with a relative density of >99%, Vickers hardness +30%, indentation fracture toughness +250%	[146]
HIP	Dry pressing @ 220 MPa for green samples, pre-heating @ 400°C, HIP @1700°C in N_2, 20 MPa, 3 h	Al_2O_3	MLG, GNP, 1–3 wt.%	Young's modulus +14%, bending strength +20%	[134]
	1700°C in N_2, heating <25°C/min, 20 MPa, 3 h	α-Si_3N_4	MLG, GNP	The best toughness improvement (9.9 MPa vs. 6.9 MPa for the monolithic) achieved on the composite with the smallest graphene fillers	[145]
	1700°C, 20 MPa	α-Si_3N_4	CNT, FLG, 3 wt.%	FLG composite outperformed CNT composite by 10–50% in mechanical properties, CNTs aggregate in the matrix while FLGs disperse well according to the neutron scattering analysis	[133]

(*Continued*)

Table 2.4 Deposition techniques, properties, and conditions of the graphene-reinforced composites. (*Continued*)

Densification method	Conditions	Matrix	Filler type and content	Improvements	Ref.
SPS	1350°C (100°C/min), 50 MPa, 5 min	α-Al_2O_3, 200 nm	FLG, 0.2–5 vol.%	Fracture toughness +40%	[131]
	Pre-sintering in vacuum (<6 Pa), heating 140°C/min, 1300°C, 60 MPa	α-Al_2O_3, 100 nm	FLG, 0.8–15 vol.%	Electrical conductivity percolation threshold 3 vol.%, with 15 vol.% addition 170% more conductive than the optimized CNT/Al_2O_3	[130]
	Heating 100°C/min, 1500–1550°C, 50 MPa, 3 min, in a vacuum of 5 Pa	α-Al_2O_3, 150 nm	GNP, 0–1.33 vol.%	Flexural strength +30.75%, fracture toughness +27.20%	[132]
	Heating 100°C/min, 1500°C, 80 MPa, 1 min	Al_2O_3	GO, 0.22 wt.%	Friction coefficient −10%, wear rate halved	[148]
	1625°C, 50 MPa, 5 min, in a vacuum of 5 Pa	Si_3N_4	GNP, 4–24 vol.%	Bulk electrical conductivity shows strong anisotropic effect; the preferential *ab*-plane of the GNPs perpendicular to the pressing axis is the most conductive	[143]

(*Continued*)

Table 2.4 Deposition techniques, properties, and conditions of the graphene-reinforced composites. (*Continued*)

Densification method	Conditions	Matrix	Filler type and content	Improvements	Ref.
SPS	1625°C, 50 MPa, 5 min, in a vacuum of 4–6 Pa	α-Si_3N_4	GNP, 1–5 wt.% GO, 1–5 wt.%	Harmony between the experimental and modeling results on the toughening effect, graphene bridging mechanism dominating	[141]
	1625°C, 50 MPa, 5 min, under 6 Pa	Si_3N_4	GNP, 4.4 vol.%	Friction −11% under high loads, wear resistance +56%	[142]
	Heating 100°C/min, 1900°C, 70 MPa, 15 min, in Ar	ZrB_2, 1–2 µm	GNP, 2–6 vol.%	Biaxial flexural strength +100%, fracture toughness +83%	[144]
Pressureless	Mixture shaped by uniaxial and cold isostatic presses (200 MPa), sintered in Ar for 3 h, heating 10°C/min, 1450–1700°C	Al_2O_3	FLG, 0.25–1.5 vol.%	Fracture toughness +75%, flexural strength +25%, wear resistance increased 1 order of magnitude	[129]

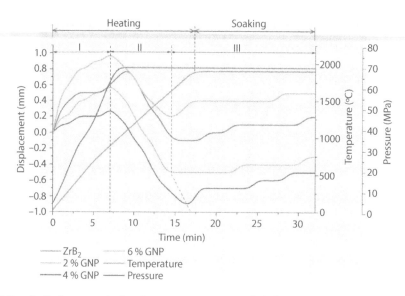

Figure 2.25 Punch displacement during the sintering process with different nanofillers [144].

Porwal et al. [131] investigated the effect of the reinforcement content on the microstructure of the graphene/alumina nanocomposite. No appreciable agglomeration was found when graphene concentration was smaller than 2%, but further increase in the graphene concentration led to the overlapping of graphene sheets within the composite matrix, as evidenced by both a decrease in I_D/I_G in the Raman spectrum and SEM observation [131]. Interestingly, however, it has been reported by multiple authors that graphene materials are superior to carbon nanotubes in terms of the homogeneous distribution in the matrices [130, 133]. In the work by Tapasztó et al., small-angle neutron scattering (SANS) was employed to mathematically reveal the distribution of the nanoscale fillers within the whole volume of the silicon nitride matrices. By detailed analysis of both the neutron scattering spectra and the SEM observations of the CNT-reinforced and the FLG-reinforced Si_3N_4 composites, it was concluded that CNTs tend to form dense aggregations within the ceramic composite, while graphene nanofillers appear as individual 2D platelets throughout the entire volume of the matrix [133].

During the pressure-assisted sintering process, a preferred orientation of the graphene nanofillers may take place [143]. Upon their measurements of the electrical conductivity of the GNP/Si_3N_4 composites fabricated by spark plasma sintering, Ramirez et al. found that the electrical conductivity along the direction perpendicular to the compression axis was one order of magnitude higher than that of the parallel direction, arguing that the *ab*-plane of graphene nanosheets may take the perpendicular direction as the preferential orientation is an effect of the applied pressure during SPS.

It is worth noting and has been reported by a number of groups that graphene introduces pores in the composite matrix [129, 132, 145, 148]. The origin of the porosity could be the insufficient bonding between graphene sheets and ceramic matrix, which leads to inhomogeneous residual stress during a cooling process due to the different thermal expansion coefficients of graphene fillers and the matrix [132]. In cases of GO, the porosity can be attributed to the gas evolution during the GO reduction at lower temperatures [129]. The porosity leads to a degraded mechanical strength, which could be even lower than the monolithic ceramic [145, 157]. This graphene-induced porosity problem can be mediated by using smaller starting graphene fillers or further improving the dispersion of the graphene sheets [134].

As mentioned above, in the cases of using GO nanosheets as reinforcements, GO can be reduced to rGO during the sintering process [141, 149, 152, 153]. It is well known that a relatively low temperature of 200–250°C is sufficient for the oxygen-containing groups on GO to thermally decompose, thus resulting in the reduction of the GO [158]. At high temperatures of up to 1900°C as with most of the sintering techniques, the thermal reduction of GO is very likely, and the effect can be immediate [152]. This reduction effect, which is usually associated with the evolution of gas, can cause porosity in the sintered body. One possible solution to this problem is to preheat the green ceramic composite compact at lower temperatures before final sintering at high temperatures with or without pressure.

2.4.1.4 Thermal/Cold/Plasma Spraying

Liu et al. [159] were able to fabricate graphene nanosheet (GN)-reinforced zirconia ceramic coatings on Ti-6Al-4V substrates using an atmospheric plasma spraying technique. To enhance the adhesion, a Ni-Cr bonding layer was applied. With a relatively high current of 630 A, a voltage of 67 V, and a powder feed rate of 20 g/min from a spraying distance of 120 mm, the ZrO_2/GNs mixing powder was fabricated with a high homogeneity. The graphene nanosheets survived the high temperature during the plasma spraying process. Compared with the ZrO_2/graphite composite prepared under the otherwise same conditions that showed a large number of voids, pores, and graphite aggregates, the graphene-reinforced composite showed a denser structure. It was found that an addition of 1 wt.% GNs reduced the wear rate by ~50%, and the friction coefficient was reduced from 0.27 to 0.19 when the normal load increased from 10 to 100 N. The improvement by GNs, especially under high load, can be attributed to the formation of a continuous GN-reinforced transfer layer, which effectively prevents the substrate from further damage. In contrast to this, graphite was less effective since the transfer layer was discontinuous.

A hydroxyapatite (HA)/graphene oxide composite coating has been successfully fabricated via vacuum cold spraying by Liu et al. [159, 160]. Unlike conventional thermal spraying, which is usually conducted at high temperatures in order to melt the particulate coating precursors and achieve a fine coating with sufficient adhesion and cohesion, vacuum cold spraying is a method based on shock-loading solidification and can be fulfilled at room temperature, without sacrificing the deposition efficiency. Liu and coworkers found that the coatings prepared this way not only retained the fine nanostructure of both HA and the graphene sheets but also demonstrated good biocompatibility with the human osteoblast cells, indicating the promising biomedical application.

2.4.1.5 Electrophoretic Deposition (EPD)

Li et al. reported their work on the fabrication and characterization of GO/HA nanocomposite coatings [198, 199]. In their study, the coatings were fabricated on titanium substrates from a co-suspension of GO and HA nanoparticles by electrophoretic deposition (EPD). TEM observation confirmed a uniform distribution of HA particles on GO sheets, and SEM images showed dense morphology and much less cracks for the GO-reinforced

HA coating. Apart from the strengthened mechanical properties, the GO containing coatings showed improved corrosion resistance in simulated body fluid (SBF), and superior *in vitro* biocompatibility (~95% cell viability for 2 wt.% GO). In a very recent report, Janković *et al.* prepared similar GO/HA composite coatings and evaluated their bio-activity as well as the corrosion behavior in simulated body fluid [245]. The hardness, elastic modulus, and thermal stability were all found increased, and an appetite layer was freshly formed in SBF, indicating the good biocompatibility of the composite coating. The GO/HA composite coatings demonstrated an improved corrosion resistance as confirmed by EIS measurements, although no antibacterial activity was observed.

2.4.2 Performance

2.4.2.1 Mechanical Properties

Conceptually, mechanical strength and fracture toughness are mutually exclusive. The strong bonding and low plasticity limit in a high-strength material usually lead to brittle fracture under high and continuous stress, known as low fracture toughness. The unusual structure of graphene determines that it can offer not only extraordinary mechanical strength but also extraordinary flexibility due to its strong C–C bonds and high aspect ratio. This is beneficial for ceramic processing as most ceramics tend to fatigue and crack under repetitive and persistent stress as a consequence of their low fracture toughness.

It has been shown that only a small fraction of graphene additives is sufficient to reinforce the composite. 0.25–0.5 vol.% addition in the Al_2O_3 matrix resulted in improvements in the fracture toughness by ~75% and in the flexural strength by ~25% compared with those of the pure Al_2O_3 sintered ceramic [129]. It can be ascribed to the high aspect ratio of graphene, compared with conventional reinforcement species such as carbon nanotubes and fibers, which usually require a relatively high content of 1–10 vol.% [129]. Higher graphene contents, however, may limit the fracture toughness improvements [131, 132]. It is echoed by Kim *et al.*, who reported the decreasing toughening effect with increasing graphene concentration in the Al_2O_3 matrix, regardless of the nature of graphene (unoxidized graphene, GO, rGO) [129].

Porwal *et al.* investigated the microstructures of the alumina composites with increasing graphene concentrations. No significant difference in the grain sizes of the material groups was found, and the hardness values were similar. Although the fracture toughness for the composite with 0.8 vol.% FLGs was almost 40% higher than that of the pure alumina, the Young's modulus did not change until a reinforcement content of 2 vol.%. The toughness and elastic modulus then both decreased significantly when the FLG content increased to 5 vol.%. The deterioration of the mechanical properties is likely due to the increase in density of the interconnected graphene network [131]. Dusza *et al.* investigated the effects of different graphene geometries (thickness and lateral sizes) on the toughening mechanisms of the GNP/Si_3N_4 composites. They found that the highest toughness enhancement was obtained from the composites with the lowest average lateral size and the narrowest size distribution of the GNPs. In contrast, the lowest fracture toughness improvement was associated with the largest average size and the widest size distribution of the GNPs. The strengthening mechanism could be the inhibition of dislocation motions due to the presence of graphenes along the ceramic grain boundaries [129]. However, graphene sheets in larger sizes may appear as structural defects that lower the toughness of the composite.

Despite the reinforcement concentration and the densification techniques, the reinforcing efficiency is mainly governed by several factors [133]: (1) the intrinsic mechanical properties of the nanofillers, (2) the efficiency for the load transfer in between the fillers and the composite matrix, and (3) the distribution homogeneity of the nanofillers throughout the whole volume of the matrix. Although apparently mechanically exfoliated graphene sheets are superior to those by chemical routes like GO and rGO in many cases, for the same type of graphene nanofillers, their intrinsic mechanical strength should be identical. On the other hand, considering the small reinforcement concentration, the mechanical properties of the final product should not be mostly dominated by the intrinsic mechanical properties of the nanofillers (Factor 1), but more likely by the interactions between the fillers and the matrix (Factors 2 and 3). It is not surprising that any factors that may lead to insufficient bonding between the two phases or introduce voids, localized defects, or graphene agglomerations will result in unexpected poor mechanical performance of the composite.

The interactions between graphene fillers and the ceramic matrix are thus crucial for any strengthening and toughening. It has been observed that the I_D/I_G ratio in the Raman spectra of the GNP/ZrB_2 composites is larger than that of the monolithic ceramic, likely due to interfacial interactions between the GNPs and the ZrB_2 matrix [144].

Toughness improvement mechanisms proposed in the literature include intergranular to transgranular, interconnecting graphene network, graphene pull-out, crack bridging, crack deflection, and crack branching [131, 144]. According to the SEM observations by various authors, crack bridging could be the dominant mechanism, as with common reinforcement fillers such as whiskers and fibers. The advantage of graphene fillers is that they promote the fracture toughness isotropically due to the homogeneous distribution of graphene nanosheets along the grain boundaries [129]. It has also been evidenced that graphene nanoplatelets are able to wrap the composite grains, which may contribute to the toughening of the composite [146, 147]. Graphene embedded within the grains can potentially improve the strength and toughness due to its intrinsic mechanical strength and flexibility [132].

There has been also increasing interests in exploiting the tribology of graphene for CMCs. Belmonte and coworkers fabricated GNP/Si_3N_4 composites and found that GNPs lubricated the tribosystem under high loads, reaching a minimum friction coefficient of 0.16, which is 11% lower than that of the monolithic ceramic. The GNP-reinforced composite demonstrated higher wear resistance regardless of the normal load. Under the highest load, the composite was 56% more wear resistant compared to the monolithic, which was attributed to the continuous exfoliation of the GNPs and thus the form of an adhered tribofilm between the tribopairs [142]. In the work on GO/alumina composites by Gutierrez-Gonzalez *et al.*, it was found that under dry sliding conditions, the GO containing composite showed a 10% less friction coefficient and only half the wear rate in comparison to the monolithic ceramic [148]. The friction reduction was ascribed to the lubricant nature of the GO sheets that smoothed the sliding contact. The suppression in wear was argued to stem from the relief of the intergrain tensile tensions enabled by the incorporated graphene platelets. Because of this strain relief, the pullout of alumina grains was reduced, which in turn reduced the formation of wear debris and the associated severe third body abrasion.

2.4.2.2 Electrical Properties

The percolation threshold for the graphene concentration in the graphene/alumina composites was found to be higher than those of CNT-reinforced counterparts [130]. However, the conductivity value increased rapidly even over the percolation threshold, which differs from the behavior of CNT-reinforced CMCs, as for these composites the electrical conductivity usually levels off after the percolation threshold. The desired conductivity behavior of the graphene/alumina composites was attributed to the agglomeration-free distribution of graphene nanofillers within the alumina matrix, thanks to their improved powder processing method, as well as the high aspect ratio 2D geometry of graphene, which can form a homogeneous network within the matrix, rather than "bundle"- or "rope"-like agglomerations usually associated with CNTs. Moreover, the area-to-area contact in a graphene network is believed to be more electrically effective than the point-to-point contact between CNTs [130].

Shin *et al.* [151] fabricated YSZ/rGO ceramic composites by spark plasma sintering, using hydrazine reduced GO nanosheets as the reinforcements. With the increasing reinforcement concentration, the indentation hardness of the composite gradually decreased while the fracture toughness was improved from 4.4 MPa to 5.9 MPa, a trade-off similar to the findings by others. More importantly, the electrical conductivity of the YSZ/rGO composite can be at least one order of magnitude higher than that of the monolithic, showing a percolation threshold of ~2.5 vol.%. These improvements were again attributed to the formation of a conductive, interlinked 3D graphene network.

In the work by Centeno *et al.* [152], Al_2O_3/GO composites were prepared via SPS. As a consequence of the thermal reduction of GO during the sintering state and due to the good dispersion of the GO sheets throughout the matrix, the electrical conductivity of the composite was drastically improved by 8 orders of magnitude with a GO concentration of merely 0.22 vol.%. A preferential orientation of the graphene sheets, which is the plane perpendicular to the compression axis, was suggested, consistent with the findings by Ramirez *et al.* [143].

2.5 Applications of Graphene-Reinforced Composites

2.5.1 Low Friction and Wear Components

Too often, graphene is cited as the strongest material ever, outperforming structural steel by some 200 times [161]. Recently, the tribology of graphene has also attracted much research interest. Atomic-level studies have proved that it can demonstrate an extremely low friction coefficient of ~0.03, far more superior to graphite (~0.1) [162–164]. Given its thickness of 0.34 nm, graphene could be the thinnest solid lubricant ever discovered [163]. Moreover, it has been reported that unlike graphite that lubricates well in the presence of moistures but is less effective in dry atmosphere, graphene's excellent lubricity is valid regardless of the environment humidity [165]. These unusual properties indicate graphene's great potential for tribological applications. However, the lubricity of single-layer graphene is affected heavily by not only out-of-plane deformations but also the weak bonding with the underlying support [166–168]. Because of that, and the complexity of real application conditions, it would be more practical to design graphene-based composites in order to exploit the tribology of graphene.

Tai *et al.* [169] reported the tribological performance of the GO/ultrahigh molecular weight polyethylene (UHMWPE) composites prepared by hot pressing (HP). Both the hardness and wear resistance of the GO-reinforced composites increased with the GO content up to 1.0 wt.%. The wear rate was reduced remarkably by 40% with a GO content of 3%. A transfer layer mechanism was proposed to explain the wear reduction, which is in agreement with other reports [170–173]. The friction increased slightly with increasing GO loading, consistent with reported phenomenon of CNT/UHMWPE composites [169]. In contrast, Lahiri *et al.* [174] demonstrated that the friction coefficient of UHMWPE-matrix composites was decreased with the increasing reinforcement content of graphene nanoplatelet (GNP). The discrepancy can be attributed to the difference in the graphene material types, testing conditions, and more likely the graphene-matrix interactions. Kandanur *et al.* [175] investigated the tribological behavior of the GO-reinforced polytetrafluoroethylene (PTFE) composites under a high normal load of 50 N. The wear was significantly reduced by 10-fold with merely 0.32 wt.% GO and dramatically by 4000-fold with 10 wt.% GO reinforcements. As a comparison, graphite-reinforced PTFE resulted in 10–30 times more wearable.

The tribological improvements for graphene-reinforced metallic and ceramic composites have also been demonstrated [170, 172, 176, 177]. Xu *et al.* fabricated multilayer graphene (MLG)-reinforced TiAl composites by spark plasma sintering. Upon the confirmation of homogeneous distribution of MLGs within the matrix and the resulting improvements in the mechanical properties, it was also found that the graphene nanofillers decreased the friction coefficient by a factor of 4 and the wear rate by 4–9 orders of magnitude.

More recently, Berman *et al.* reported their significant finding on the superlubricity (friction coefficient ~0.004) for graphene wrapped nanodiamond spheres [178]. Likely inspired by this, another group studied the friction reduction mechanism of amorphous carbon films in detail and revealed that upon sliding contact, a large number of graphene-like nano scrolls with inner amorphous carbon hard cores were developed in the tribofilms, thus reducing the friction coefficients impressively (see Figure 2.26).

2.5.2 Intelligent Interfaces and Anti-Corrosion Coatings

Graphene's unique 2D structure gives rise to some unique properties, including its impermeability. Graphene, although only one atom thick, can block all molecules, including helium (the smallest), from passing through, given that it is free of defect [180]. In addition, the overlapping π-electron cloud in the vicinity of graphene's basal plane renders a repelling field to foreign atoms and molecules, while remaining transparent to electrons [181]. These characteristics imply that graphene can be the world's thinnest separation membrane [180] and corrosion barrier [182]. Graphene-based materials, such as GO in particular, have been incorporated into polymers [95, 183–196], inorganic/ceramic materials [197–199], and metallic matrices [200, 201] to form composites for corrosion protection.

Meanwhile, graphene, as well as graphite, is commonly seen as a hydrophobic material, although recently some authors have pointed out that the wettability of graphene should be affected by the ambient environment (i.e., contaminations and hydrocarbon species that could be adsorbed onto graphene surface) [202, 203] or the liquid-graphene and liquid-substrate interactions [204]. The hydrophobicity of graphene has been employed for

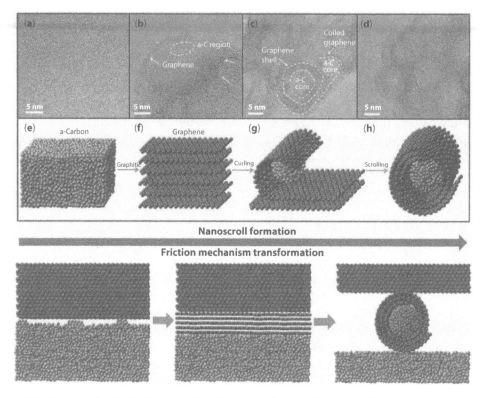

Figure 2.26 Schematic for the development of low-friction graphene scrolls on top of the amorphous carbon material [179]. (a) to (d) are TEM images of the formation of nanoscrolls; (e) to (h) schematic of the formation process.

self-cleaning composite coatings [205]. In this study, a mixture of diatomaceous earth (DE), reduced graphene oxide (rGO), and a portion of TiO_2 nanoparticles were made into a composite. The composite showed a water contact angle of 170 ± 2°, and hence a long-lasting self-cleaning behavior (see Figure 2.27). The self-cleaning feature would not be destroyed even after sand blasting or crosscut scratching and could be applied conveniently on literally any substrates by spraying, brush painting, or dip coating.

Graphene's different wetting behavior with water and oil media can lead to the applications of graphene-based intelligent interfaces. In a very interesting work, Nguyen *et al.* processed a commercially available sponge, which is normally highly hydrophilic, by dipping in a dispersion of graphene nanosheets at 100°C for 2 h. The final composite product ended

Figure 2.27 The fabrication of DE/rGO/TiO_2 composite coating by spraying and the superhydrophobic behavior of the composite [205].

Figure 2.28 The graphene-coated sponge repels water perfectly while exhibiting strong adsorption to oil [206]. (a) Graphene based sponge in water; (b) water and oil droplets on the graphene coated sponge.

up to show a water contact angle of 162°, while keeping a high wettability to oil. That is, superhydrophobic yet superoleophilic (see Figure 2.28).

2.5.3 Antibacterial and Biocompatible Implants

The high aspect ratio of graphene facilitates various biological reactions. Unlike CNTs, which can cause lesions in a live body if without specific functionalization [207], graphene-based materials usually demonstrate satisfactory biocompatibility. Human cell culturing studies showed that GO did not cause cell toxicity [208, 209] or minor concentration-dependent toxicity [210], although there may be some doubts for this as for a few specific cells it showed toxicity [211]. In the meantime, it has been reported that GO promoted the differentiation, growth, and proliferation of stem cells (see Figure 2.29), likely due to the electrostatic interaction and hydrogen bonds between the GO sheets and the cells [212].

Interestingly, while showing good biocompatibility to human cells, some graphene-based materials are reported to be antibacterial, and this concept has been applied to a few composites containing graphene derivatives. Kulshrestha *et al.* [213] prepared graphene/zinc oxide nanocomposites (GZNC) via a simple colloidal process and investigated the activities of *Streptococcus mutans* (*S. mutans*), a cariogenic bacterium commonly seen in dental practices on this nanocomposite. The results clearly showed that the formation of *S. mutans* biofilm was essentially inhibited on the GZNC-coated acrylic tooth surfaces (see Figure 2.30).

2.5.4 Flame-Retardant Materials

Conventionally, a large family of flame-retardant fillers are inorganic materials, including hydroxide, metal oxide, phosphate, and silicate. These materials have good thermal stability, low toxicity and pollution, and low cost, but usually require a high load in the composite in order to enhance the efficiency [214]. The organic group of flame-retardant fillers shows higher efficiency and better compatibility with polymeric matrices, but inevitably contain halogen or phosphorus nitrogen species, which may cause very high health risks due to the toxic gases formed during combustion.

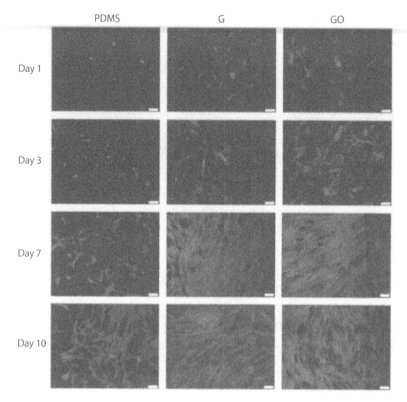

Figure 2.29 Fluorescent images of proliferation of human mesenchymal stem cells (MSCs) on PDMS (reference), CVD-grown graphene, and GO. Scale bar, 100 μm [212].

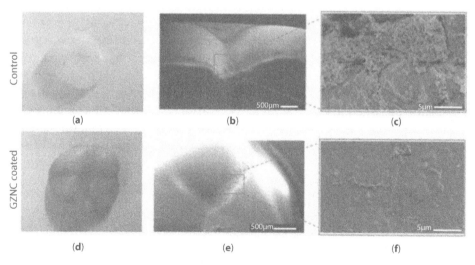

Figure 2.30 Photographs and SEM images of (a–c) noncoated and (d–f) GZNC-coated acrylic teeth. (c) Shows well-defined biofilm generation on noncoated teeth, while (f) suggests almost negligible biofilm on GZNC-coated teeth [213].

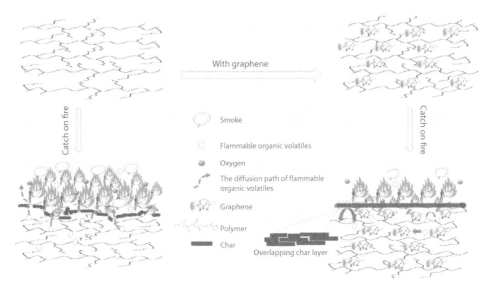

Figure 2.31 A retarding mechanism of graphene-based flame retardant proposed by Sang *et al.* [214].

Graphene-based materials could be alternatives to these conventional retardant fillers owing to (1) their excellent physical barrier effect, which insulates the heat and fuel upon fire emergencies; (2) large specific area, which can effectively absorb flammable vapors and provide catalytic and carbonization platforms for other materials like metal oxides [215]; and (3) high thermal stability, which prevents the self-propagating decomposition of the flame-retardant composite. An illustration of the graphene-reinforced polymeric composite and the fire-retarding mechanism is shown in Figure 2.31 [214].

2.6 Conclusion

Graphene is one of the most exciting scientific discovery in the 21st century. It has presented a large number of gifts and possibilities and the magic is still not fully unveiled. It is believed that graphene and related derivatives will boost a new class of composite materials. Nevertheless, it should be noted that there are several technical issues that must be overcome in order to fully realize the potential of graphene: (1) The quality of graphene, in terms of layer number, lateral dimensions, and structural defects, needs to be stable and tunable in order to fulfill the varying requirements for different applications. (2) The dispersion of graphene fillers in composite matrices need to be further improved and fully formulated to ensure a homogeneous composite microstructure and thus reliable properties. (3) The interactions of graphene fillers with the matrices, either mechanically or electrically, must be enhanced accordingly to avoid structural voids and weakness and guarantee desired mechanical stability and electrical and thermal conductivity. Regarding the applications of graphene-reinforced composites, it is beneficial to develop new possibilities such as energy harvesting devices and MEMS devices.

References

1. Novoselov, K.S. et al., Electric field effect in atomically thin carbon films. *Science*, 306, 5696, 666–669, 2004.
2. Lee, C. et al., Measurement of the elastic properties and intrinsic strength of monolayer graphene. *Science*, 321, 5887, 385–388, 2008.
3. Balog, R. et al., Bandgap opening in graphene induced by patterned hydrogen adsorption. *Nat. Mater.*, 9, 4, 315–319, 2010.
4. Balandin, A.A. et al., Superior thermal conductivity of single-layer graphene. *Nano Lett.*, 8, 3, 902–907, 2008.
5. Park, S. and Ruoff, R.S., Chemical methods for the production of graphenes. *Nat. Nanotechnol.*, 4, 4, 217–224, 2009.
6. Zhu, Y.W. et al., Graphene and graphene oxide: Synthesis, properties, and applications. *Adv. Mater.*, 22, 35, 3906–3924, 2010.
7. Bastwros, M. et al., Effect of ball milling on graphene reinforced Al6061 composite fabricated by semi-solid sintering. *Compos. Part B Eng.*, 60, 111–118, 2014.
8. Bartolucci, S.F. et al., Graphene-aluminum nanocomposites. *Mater. Sci. Eng. A Struct. Mater.*, 528, 27, 7933–7937, 2011.
9. Yang, W. et al., Microstructure and mechanical properties of graphene nanoplates reinforced pure Al matrix composites prepared by pressure infiltration method. *J. Alloys Compd.*, 732, 748–758, 2018.
10. Gao, X. et al., Preparation and tensile properties of homogeneously dispersed graphene reinforced aluminum matrix composites. *Mater. Des.*, 94, 54–60, 2016.
11. Li, G. and Xiong, B., Effects of graphene content on microstructures and tensile property of graphene-nanosheets/aluminum composites. *J. Alloys Compd.*, 697, 31–36, 2017.
12. Kumar, S.J.N. et al., Mechanical properties of aluminium-graphene composite synthesized by powder metallurgy and hot extrusion. *T. Indian I. Metals*, 70, 3, 605–613, 2017.
13. Wang, J.Y. et al., Reinforcement with graphene nanosheets in aluminum matrix composites. *Scr. Mater.*, 66, 8, 594–597, 2012.
14. Li, Z. et al., Uniform dispersion of graphene oxide in aluminum powder by direct electrostatic adsorption for fabrication of graphene/aluminum composites. *Nanotechnology*, 25, 32, 2014.
15. Jeon, C.H. et al., Material properties of graphene/aluminum metal matrix composites fabricated by friction stir processing. *Int. J. Precis. Eng. Man.*, 15, 6, 1235–1239, 2014.
16. Perez-Bustamante, R. et al., Microstructural and hardness behavior of graphene-nanoplatelets/aluminum composites synthesized by mechanical alloying. *J. Alloys Compd.*, 615, S578–S582, 2014.
17. Zhang, H. et al., Enhanced mechanical properties of Al5083 alloy with graphene nanoplates prepared by ball milling and hot extrusion. *Mater. Sci. Eng. A*, 658, 8–15, 2016.
18. Khodabakhshi, F. et al., Fabrication of a new Al-Mg/graphene nanocomposite by multi-pass friction-stir processing: Dispersion, microstructure, stability, and strengthening. *Mater. Charact.*, 132, 92–107, 2017.
19. Kavimani, V., Prakash, K.S., Pandian, M.A., Influence of r-GO addition on enhancement of corrosion and wear behavior of AZ31 MMC. *Appl. Phys. A Mater. Sci. Process.*, 123, 8, 2017.
20. Liu, C.S., Su, F.H., Liang, J.Z., Producing cobalt-graphene composite coating by pulse electrodeposition with excellent wear and corrosion resistance. *Appl. Surf. Sci.*, 351, 889–896, 2015.
21. Rekha, M.Y., Kumar, M.K.P., Srivastava, C., Electrochemical behaviour of chromium-graphene composite coating. *RSC Adv.*, 6, 67, 62083–62090, 2016.
22. Jiang, R.R. et al., Copper-graphene bulk composites with homogeneous graphene dispersion and enhanced mechanical properties. *Mater. Sci. Eng. A Struct. Mater.*, 654, 124–130, 2016.

23. Kim, W.J., Lee, T.J., Han, S.H., Multi-layer graphene/copper composites: Preparation using high-ratio differential speed rolling, microstructure and mechanical properties. *Carbon*, 69, 55–65, 2014.
24. Hwang, J. et al., Enhanced mechanical properties of graphene/copper nanocomposites using a molecular-level mixing process. *Adv. Mater.*, 25, 46, 6724–6729, 2013.
25. Tang, Y.X. et al., Enhancement of the mechanical properties of graphene-copper composites with graphene-nickel hybrids. *Mater. Sci. Eng. A Struct. Mater.*, 599, 247–254, 2014.
26. Raghupathy, Y. et al., Copper-graphene oxide composite coatings for corrosion protection of mild steel in 3.5% NaCl. *Thin Solid Films*, 636, 107–115, 2017.
27. Akbulut, H. et al., Co-deposition of Cu/WC/graphene hybrid nanocomposites produced by electrophoretic deposition. *Surf. Coat. Technol.*, 284, 344–352, 2015.
28. Hu, Z.R. et al., Laser additive manufacturing bulk graphene-copper nanocomposites. *Nanotechnology*, 28, 44, 2017.
29. Luo, H.B. et al., Copper matrix composites enhanced by silver/reduced graphene oxide hybrids. *Mater. Lett.*, 196, 354–357, 2017.
30. Jagannadham, K., Orientation dependence of thermal conductivity in copper-graphene composites. *J. Appl. Phys.*, 110, 7, 2011.
31. Jagannadham, K., Thermal conductivity of copper-graphene composite films synthesized by electrochemical deposition with exfoliated graphene platelets. *Metall. Mater. Trans. B*, 43, 2, 316–324, 2012.
32. Chu, K. and Jia, C.C., Enhanced strength in bulk graphene-copper composites. *Phys. Status Solidi A*, 211, 1, 184–190, 2014.
33. Pavithra, C.L.P. et al., A new electrochemical approach for the synthesis of copper-graphene nanocomposite foils with high hardness. *Sci. Rep.*, 4, 2014.
34. Xie, G.X., Forslund, M., Pan, J.S., Direct electrochemical synthesis of reduced graphene oxide (rgo)/copper composite films and their electrical/electroactive properties. *ACS Appl. Mater. Interfaces*, 6, 10, 7444–7455, 2014.
35. Li, M.X. et al., Highly enhanced mechanical properties in Cu matrix composites reinforced with graphene decorated metallic nanoparticles. *J. Mater. Sci.*, 49, 10, 3725–3731, 2014.
36. Peng, Y.T. et al., Ultrasound-assisted fabrication of dispersed two-dimensional copper/reduced graphene oxide nanosheets nanocomposites. *Compos. Part B Eng.*, 58, 473–477, 2014.
37. Zhao, C. and Wang, J., Fabrication and tensile properties of graphene/copper composites prepared by electroless plating for structrual applications. *Phys. Status Solidi A*, 211, 12, 2878–2885, 2014.
38. Dutkiewicz, J. et al., Microstructure and properties of bulk copper matrix composites strengthened with various kinds of graphene nanoplatelets. *Mater. Sci. Eng. A Struct. Mater.*, 628, 124–134, 2015.
39. Xiong, D.B. et al., Graphene-and-copper artificial nacre fabricated by a preform impregnation process: Bioinspired strategy for strengthening-toughening of metal matrix composite. *ACS Nano*, 9, 7, 6934–6943, 2015.
40. Zhao, Z.Y. et al., Microstructures and properties of graphene-Cu/Al composite prepared by a novel process through clad forming and improving wettability with copper. *Adv. Eng. Mater.*, 17, 5, 663–668, 2015.
41. Liu, Y. et al., Hydroxyapatite/graphene-nanosheet composite coatings deposited by vacuum cold spraying for biomedical applications: Inherited nanostructures and enhanced properties. *Carbon*, 67, 250–259, 2014.
42. Turan, M.E. et al., The effect of GNPs on wear and corrosion behaviors of pure magnesium. *J. Alloys Compd.*, 724, 14–23, 2017.

43. Rashad, M. et al., Synergetic effect of graphene nanoplatelets (GNPs) and multi-walled carbon nanotube (MW-CNTs) on mechanical properties of pure magnesium. *J. Alloys Compd.*, 603, 111–118, 2014.
44. Qi, S.J. et al., Fabrication and characterisation of electro-brush plated nickel-graphene oxide nano-composite coatings. *Thin Solid Films*, 644, 106–114, 2017.
45. Algul, H. et al., The effect of graphene content and sliding speed on the wear mechanism of nickel-graphene nanocomposites. *Appl. Surf. Sci.*, 359, 340–348, 2015.
46. Jabbar, A. et al., Electrochemical deposition of nickel graphene composite coatings: Effect of deposition temperature on its surface morphology and corrosion resistance. *RSC Adv.*, 7, 49, 31100–31109, 2017.
47. Zhou, P.W. et al., Fabrication and corrosion performances of pure ni and ni-based coatings containing rare earth element ce and graphene by reverse pulse electrodeposition. *J. Electrochem. Soc.*, 164, 2, D75–D81, 2017.
48. Kumar, C.M.P., Venkatesha, T.V., Shabadi, R., Preparation and corrosion behavior of Ni and Ni-graphene composite coatings. *Mater. Res. Bull.*, 48, 4, 1477–1483, 2013.
49. Kuang, D. et al., Graphene-nickel composites. *Appl. Surf. Sci.*, 273, 484–490, 2013.
50. Szeptycka, B., Gajewska-Midzialek, A., Babul, T., Electrodeposition and corrosion resistance of Ni-graphene composite coatings. *J. Mater. Eng. Perform.*, 25, 8, 3134–3138, 2016.
51. Chen, J.J. et al., Preparation and tribological behavior of Ni-graphene composite coating under room temperature. *Appl. Surf. Sci.*, 361, 49–56, 2016.
52. Ren, Z. et al., Mechanical properties of nickel-graphene composites synthesized by electrochemical deposition. *Nanotechnology*, 26, 6, 2015.
53. Khalil, M.W. et al., Electrodeposition of Ni-GNS-TiO2 nanocomposite coatings as anticorrosion film for mild steel in neutral environment. *Surf. Coat. Technol.*, 275, 98–111, 2015.
54. Jiang, K., Li, J.R., Liu, J., Electrochemical codeposition of graphene platelets and nickel for improved corrosion resistant properties. *RSC Adv.*, 4, 68, 36245–36252, 2014.
55. Zhai, W.Z. et al., Grain refinement: A mechanism for graphene nanoplatelets to reduce friction and wear of Ni3Al matrix self-lubricating composites. *Wear*, 310, 1-2, 33–40, 2014.
56. Qiu, C.C. et al., Corrosion resistance and micro-tribological properties of nickel hydroxide-graphene oxide composite coating. *Diam. Relat. Mater.*, 76, 150–156, 2017.
57. Zhang, L. et al., Preparation and mechanical properties of (Ni-Fe)-Graphene composite coating. *Adv. Eng. Mater.*, 18, 10, 1716–1719, 2016.
58. Gao, H.C. et al., One-step electrochemical synthesis of ptni nanoparticle-graphene nanocomposites for nonenzynnatic amperometric glucose detection. *ACS Appl. Mater. Interfaces*, 3, 8, 3049–3057, 2011.
59. Berlia, R., Kumar, M.K.P., Srivastava, C., Electrochemical behavior of Sn-graphene composite coating. *RSC Adv.*, 5, 87, 71413–71418, 2015.
60. Song, Y. et al., Microscopic mechanical properties of titanium composites containing multi-layer graphene nanofillers. *Mater. Des.*, 109, 256–263, 2016.
61. Hu, Z.R. et al., Laser sintered single layer graphene oxide reinforced titanium matrix nanocomposites. *Compos. Part B Eng.*, 93, 352–359, 2016.
62. Zhang, X.J. et al., Microstructural and mechanical characterization of *in-situ* TiC/Ti titanium matrix composites fabricated by graphene/Ti sintering reaction. *Mater. Sci. Eng. A Struct. Mater.*, 705, 153–159, 2017.
63. Cao, Z. et al., Reinforcement with graphene nanoflakes in titanium matrix composites. *J. Alloys Compd.*, 696, 498–502, 2017.
64. Mu, X.N. et al., Microstructure evolution and superior tensile properties of low content graphene nanoplatelets reinforced pure Ti matrix composites. *Mater. Sci. Eng. A Struct. Mater.*, 687, 164–174, 2017.

65. Xu, Z.S. *et al.*, Preparation and tribological properties of TiAl matrix composites reinforced by multilayer graphene. *Carbon*, 67, 168–177, 2014.
66. Kumar, M.K.P., Singh, M.P., Srivastava, C., Electrochemical behavior of Zn-graphene composite coatings. *RSC Adv.*, 5, 32, 25603–25608, 2015.
67. Lin, D., Liu, C.R., Cheng, G.J., Single-layer graphene oxide reinforced metal matrix composites by laser sintering: Microstructure and mechanical property enhancement. *Acta Mater.*, 80, 183–193, 2014.
68. Rashad, M. *et al.*, High temperature formability of graphene nanoplatelets-AZ31 composites fabricated by stir-casting method. *J. Magnesium Alloys*, 4, 4, 270–277, 2016.
69. Liu, C.B. *et al.*, Direct electrodeposition of graphene enabling the one-step synthesis of graphene-metal nanocomposite films. *Small*, 7, 9, 1203–1206, 2011.
70. Shin, S.E. *et al.*, Strengthening behavior of carbon/metal nanocomposites. *Sci. Rep.*, 5, 2015.
71. Shukla, A.K. *et al.*, Processing copper-carbon nanotube composite powders by high energy milling. *Mater. Charact.*, 84, 58–66, 2013.
72. Zhou, T.N. *et al.*, A simple and efficient method to prepare graphene by reduction of graphite oxide with sodium hydrosulfite. *Nanotechnology*, 22, 4, 2011.
73. Laha, T., Liu, Y., Agarwal, A., Carbon nanotube reinforced aluminum nanocomposite via plasma and high velocity oxy-fuel spray forming. *J. Nanosci. Nanotechnol.*, 7, 2, 515–524, 2007.
74. Bakshi, S.R. *et al.*, Aluminum composite reinforced with multiwalled carbon nanotubes from plasma spraying of spray dried powders. *Surf. Coat. Technol.*, 203, 10-11, 1544–1554, 2009.
75. Balani, K. *et al.*, Plasma-sprayed carbon nanotube reinforced hydroxyapatite coatings and their interaction with human osteoblasts *in vitro*. *Biomaterials*, 28, 4, 618–624, 2007.
76. Liu, Y., Huang, J., Li, H., Nanostructural characteristics of vacuum cold-sprayed hydroxyapatite/graphene-nanosheet coatings for biomedical applications. *J. Therm. Spray Technol.*, 23, 7, 1149–1156, 2014.
77. Kirkland, N.T. *et al.*, Exploring graphene as a corrosion protection barrier. *Corros. Sci.*, 56, 1–4, 2012.
78. Goyal, V. and Balandin, A.A., Thermal properties of the hybrid graphene-metal nano-micro-composites: Applications in thermal interface materials. *Appl. Phys. Lett.*, 100, 7, 2012.
79. Shen, B. *et al.*, Melt blending *in situ* enhances the interaction between polystyrene and graphene through pi-pi stacking. *ACS Appl. Mater. Interfaces*, 3, 8, 3103–3109, 2011.
80. Bao, C.L. *et al.*, Preparation of graphene by pressurized oxidation and multiplex reduction and its polymer nanocomposites by masterbatch-based melt blending. *J. Mater. Chem.*, 22, 13, 6088–6096, 2012.
81. Zhang, H.B. *et al.*, Electrically conductive polyethylene terephthalate/graphene nanocomposites prepared by melt compounding. *Polymer*, 51, 5, 1191–1196, 2010.
82. Pang, H.A. *et al.*, The effect of electric field, annealing temperature and filler loading on the percolation threshold of polystyrene containing carbon nanotubes and graphene nanosheets. *Carbon*, 49, 6, 1980–1988, 2011.
83. Zeng, X.P., Yang, J.J., Yuan, W.X., Preparation of a poly(methyl methacrylate)-reduced graphene oxide composite with enhanced properties by a solution blending method. *Eur. Polym. J.*, 48, 10, 1674–1682, 2012.
84. He, F. *et al.*, High Dielectric Permittivity and Low Percolation Threshold in Nanocomposites Based on Poly(vinylidene fluoride) and Exfoliated Graphite Nanoplates. *Adv. Mater.*, 21, 6, 710, 2009.
85. Stankovich, S. *et al.*, Graphene-based composite materials. *Nature*, 442, 7100, 282–286, 2006.
86. Kim, S., Do, I., Drzal, L.T., Thermal Stability and Dynamic Mechanical Behavior of Exfoliated Graphite Nanoplatelets-LLDPE Nanocomposites. *Polym. Compos.*, 31, 5, 755–761, 2010.

87. Wang, X. et al., In situ polymerization of graphene nanosheets and polyurethane with enhanced mechanical and thermal properties. *J. Mater. Chem.*, 21, 12, 4222–4227, 2011.
88. O'Neill, A. et al., Polymer nanocomposites: In situ polymerization of polyamide 6 in the presence of graphene oxide. *Polym. Compos.*, 38, 3, 528–537, 2017.
89. Yu, A.P. et al., Graphite nanoplatelet-epoxy composite thermal interface materials. *J. Phys. Chem. C*, 111, 21, 7565–7569, 2007.
90. Min, C. et al., A graphite nanoplatelet/epoxy composite with high dielectric constant and high thermal conductivity. *Carbon*, 55, 116–125, 2013.
91. Ren, F. et al., In situ polymerization of graphene oxide and cyanate ester-epoxy with enhanced mechanical and thermal properties. *Appl. Surf. Sci.*, 316, 549–557, 2014.
92. Zhao, X. et al., Alternate multilayer films of poly(vinyl alcohol) and exfoliated graphene oxide fabricated via a facial layer-by-layer assembly. *Macromolecules*, 43, 22, 9411–9416, 2010.
93. Wang, J.Y. et al., Preparation and properties of graphene oxide/polyimide composite films with low dielectric constant and ultrahigh strength via in situ polymerization. *J. Mater. Chem.*, 21, 35, 13569–13575, 2011.
94. Dreyer, D.R., Jia, H.P., Bielawski, C.W., Graphene oxide: A convenient carbocatalyst for facilitating oxidation and hydration reactions. *Angew. Chem. Int. Ed.*, 49, 38, 6813–6816, 2010.
95. Singh, B.P. et al., The production of a corrosion resistant graphene reinforced composite coating on copper by electrophoretic deposition. *Carbon*, 61, 47–56, 2013.
96. Gudarzi, M.M. and Sharif, F., Enhancement of dispersion and bonding of graphene-polymer through wet transfer of functionalized graphene oxide. *Express Polym. Lett.*, 6, 12, 1017–1031, 2012.
97. Vermant, J. et al., Quantifying dispersion of layered nanocomposites via melt rheology. *J. Rheol.*, 51, 3, 429–450, 2007.
98. Steurer, P. et al., Functionalized graphenes and thermoplastic nanocomposites based upon expanded graphite oxide. *Macromol. Rapid Commun.*, 30, 4–5, 316–327, 2009.
99. Jang, J.Y. et al., Graphite oxide/poly(methyl methacrylate) nanocomposites prepared by a novel method utilizing macroazoinitiator. *Compos. Sci. Technol.*, 69, 2, 186–191, 2009.
100. Kim, H. and Macosko, C.W., Processing-property relationships of polycarbonate/graphene composites. *Polymer*, 50, 15, 3797–3809, 2009.
101. Xie, S.H., Liu, Y.Y., Li, J.Y., Comparison of the effective conductivity between composites reinforced by graphene nanosheets and carbon nanotubes. *Appl. Phys. Lett.*, 92, 24, 2008.
102. Kalaitzidou, K. et al., The nucleating effect of exfoliated graphite nanoplatelets and their influence on the crystal structure and electrical conductivity of polypropylene nanocomposites. *J. Mater. Sci.*, 43, 8, 2895–2907, 2008.
103. Lotya, M. et al., High-concentration, surfactant-stabilized graphene dispersions. *ACS Nano*, 4, 6, 3155–3162, 2010.
104. Hussain, F. et al., Review article: Polymer-matrix nanocomposites, processing, manufacturing, and application: An overview. *J. Compos. Mater.*, 40, 17, 1511–1575, 2006.
105. Kalaitzidou, K., Fukushima, H., Drzal, L.T., A new compounding method for exfoliated graphite-polypropylene nanocomposites with enhanced flexural properties and lower percolation threshold. *Compos. Sci. Technol.*, 67, 10, 2045–2051, 2007.
106. Pang, H. et al., An electrically conducting polymer/graphene composite with a very low percolation threshold. *Mater. Lett.*, 64, 20, 2226–2229, 2010.
107. Liang, J.J. et al., Electromagnetic interference shielding of graphene/epoxy composites. *Carbon*, 47, 3, 922–925, 2009.
108. Geng, Y., Wang, S.J., Kim, J.K., Preparation of graphite nanoplatelets and graphene sheets. *J. Colloid Interface Sci.*, 336, 2, 592–598, 2009.

109. An, J.C., Kim, H.J., Hong, I., Preparation of Kish graphite-based graphene nanoplatelets by GIC (graphite intercalation compound) via process. *J. Ind. Eng. Chem.*, 26, 55–60, 2015.
110. Niyogi, S. et al., Solution properties of graphite and graphene. *J. Am. Chem. Soc.*, 128, 24, 7720–7721, 2006.
111. Ling, C. et al., Electrical transport properties of graphene nanoribbons produced from sonicating graphite in solution. *Nanotechnology*, 22, 32, 2011.
112. Liu, N. et al., One-step ionic-liquid-assisted electrochemical synthesis of ionic-liquid-functionalized graphene sheets directly from graphite. *Adv. Funct. Mater.*, 18, 10, 1518–1525, 2008.
113. Park, S. et al., Aqueous suspension and characterization of chemically modified graphene sheets. *Chem. Mater.*, 20, 21, 6592–6594, 2008.
114. Li, D. et al., Processable aqueous dispersions of graphene nanosheets. *Nat. Nanotechnol.*, 3, 2, 101–105, 2008.
115. Stankovich, S. et al., Synthesis and exfoliation of isocyanate-treated graphene oxide nanoplatelets. *Carbon*, 44, 15, 3342–3347, 2006.
116. Chen, Z.P. et al., Three-dimensional flexible and conductive interconnected graphene networks grown by chemical vapour deposition. *Nat. Mater.*, 10, 6, 424–428, 2011.
117. Stankovich, S. et al., Stable aqueous dispersions of graphitic nanoplatelets via the reduction of exfoliated graphite oxide in the presence of poly(sodium 4-styrenesulfonate). *J. Mater. Chem.*, 16, 2, 155–158, 2006.
118. Kim, S.C. et al., Effect of pyrene treatment on the properties of graphene/epoxy nanocomposites. *Macromol. Res.*, 18, 11, 1125–1128, 2010.
119. Song, P.G. et al., Fabrication of exfoliated graphene-based polypropylene nanocomposites with enhanced mechanical and thermal properties. *Polymer*, 52, 18, 4001–4010, 2011.
120. Chang, K.C. et al., Room-temperature cured hydrophobic epoxy/graphene composites as corrosion inhibitor for cold-rolled steel (vol 66, pg 144, 2014). *Carbon*, 82, 611–611, 2015.
121. Chang, C.H. et al., Novel anticorrosion coatings prepared from polyaniline/graphene composites. *Carbon*, 50, 14, 5044–5051, 2012.
122. Yu, Y.H. et al., High-performance polystyrene/graphene-based nanocomposites with excellent anti-corrosion properties. *Polym. Chem.*, 5, 2, 535–550, 2014.
123. Qiu, C.C. et al., Electrochemical functionalization of 316 stainless steel with polyaniline-graphene oxide: Corrosion resistance study. *Mater. Chem. Phys.*, 198, 90–98, 2017.
124. Coleman, J.N., Liquid exfoliation of defect-free graphene. *Acc. Chem. Res.*, 46, 1, 14–22, 2013.
125. Eda, G. and Chhowalla, M., Chemically derived graphene oxide: Towards large-area thin-film electronics and optoelectronics. *Adv. Mater.*, 22, 22, 2392–415, 2010.
126. Li, X. et al., Large-area synthesis of high-quality and uniform graphene films on copper foils. *Science*, 324, 5932, 1312–1314, 2009.
127. Norimatsu, W. and Kusunoki, M., Epitaxial graphene on SiC {0001}: Advances and perspectives. *Phys. Chem. Chem. Phys.*, 16, 8, 3501–3511, 2014.
128. Kobayashi, T. et al., Production of a 100-m-long high-quality graphene transparent conductive film by roll-to-roll chemical vapor deposition and transfer process. *Appl. Phys. Lett.*, 102, 2, 023112, 2013.
129. Kim, H.J. et al., Unoxidized graphene/alumina nanocomposite: Fracture- and wear-resistance effects of graphene on alumina matrix. *Sci. Rep.*, 4, 5176, 2014.
130. Fan, Y. et al., Preparation and electrical properties of graphene nanosheet/Al2O3 composites. *Carbon*, 48, 6, 1743–1749, 2010.
131. Porwal, H. et al., Graphene reinforced alumina nano-composites. *Carbon*, 64, 359–369, 2013.
132. Liu, J., Yan, H., Jiang, K., Mechanical properties of graphene platelet-reinforced alumina ceramic composites. *Ceram. Int.*, 39, 6, 6215–6221, 2013.

133. Tapasztó, O. et al., Dispersion patterns of graphene and carbon nanotubes in ceramic matrix composites. *Chem. Phys. Lett.*, 511, 4–6, 340–343, 2011.
134. Kun, P. et al., Determination of structural and mechanical properties of multilayer graphene added silicon nitride-based composites. *Ceram. Int.*, 38, 1, 211–216, 2012.
135. Zheng, Q. et al., Molecular dynamics study of the effect of chemical functionalization on the elastic properties of graphene sheets. *J. Nanosci. Nanotechnol.*, 10, 11, 7070–7074, 2010.
136. Hummers, W.S. and Offeman, R.E., Preparation of graphitic oxide. *JACS*, 1958.
137. Marcano, D.C., Kosynkin, D.V., Berlin, J.M., Improved synthesis of graphene oxide. *ACS Nano*, 4, 4806–4814, 2010.
138. Park, S. et al., Aqueous suspension and characterization of chemically modified graphene sheets. *Chem. Mater.*, 20, 6592–6594, 2008.
139. Zhao, J. et al., Efficient preparation of large-area graphene oxide sheets for transparent conductive films. *ACS Nano*, 4, 5245–5252, 2010.
140. Kovtyukhnova, N.I. et al., Layer-by-layer assembly of ultrathin composite films from micron-sized graphite oxide sheets and polycations. *Chem. Mater.*, 11, 771–778, 1999.
141. Ramirez, C. and Osendi, M.I., Toughening in ceramics containing graphene fillers. *Ceram. Int.*, 40, 7, 11187–11192, 2014.
142. Belmonte, M. et al., The beneficial effect of graphene nanofillers on the tribological performance of ceramics. *Carbon*, 61, 431–435, 2013.
143. Ramirez, C. et al., Graphene nanoplatelet/silicon nitride composites with high electrical conductivity. *Carbon*, 50, 10, 3607–3615, 2012.
144. Yadhukulakrishnan, G.B. et al., Spark plasma sintering of graphene reinforced zirconium diboride ultra-high temperature ceramic composites. *Ceram. Int.*, 39, 6, 6637–6646, 2013.
145. Dusza, J. et al., Microstructure and fracture toughness of Si3N4+graphene platelet composites. *J. Eur. Ceram. Soc.*, 32, 12, 3389–3397, 2012.
146. Shahedi Asl, M. and Ghassemi Kakroudi, M., Characterization of hot-pressed graphene reinforced ZrB2–SiC composite. *Mater. Sci. Eng. A*, 625, 385–392, 2015.
147. Walker, L.S. et al., Toughening in graphene ceramic composites. *ACS Nano*, 5, 4, 3182–3190, 2011.
148. Gutierrez-Gonzalez, C.F. et al., Wear behavior of graphene/alumina composite. *Ceram. Int.*, 41, 6, 7434–7438, 2015.
149. Wang, K. et al., Preparation of graphene nanosheet/alumina composites by spark plasma sintering. *Mater. Res. Bull.*, 46, 2, 315–318, 2011.
150. Baradaran, S. et al., Mechanical properties and biomedical applications of a nanotube hydroxyapatite-reduced graphene oxide composite. *Carbon*, 69, 32–45, 2014.
151. Shin, J.-H. and Hong, S.-H., Fabrication and properties of reduced graphene oxide reinforced yttria-stabilized zirconia composite ceramics. *J. Eur. Ceram. Soc.*, 34, 5, 1297–1302, 2014.
152. Centeno, A. et al., Graphene for tough and electroconductive alumina ceramics. *J. Eur. Ceram. Soc.*, 33, 15–16, 3201–3210, 2013.
153. Zhang, X. et al., Graphene nanosheet reinforced ZrB2–SiC ceramic composite by thermal reduction of graphene oxide. *RSC Adv.*, 5, 58, 47060–47065, 2015.
154. Puertolas, J.A. and Kurtz, S.M., Evaluation of carbon nanotubes and graphene as reinforcements for UHMWPE-based composites in arthroplastic applications: A review. *J. Mech. Behav. Biomed. Mater.*, 39, 129–45, 2014.
155. Lahiri, D., Ghosh, S., Agarwal, A., Carbon nanotube reinforced hydroxyapatite composite for orthopedic application: A review. *Mater. Sci. Eng. C*, 32, 7, 1727–1758, 2012.
156. Hernandez, Y. et al., High-yield production of graphene by liquid-phase exfoliation of graphite. *Nat. Nanotechnol.*, 3, 9, 563–568, 2008.

157. Kvetková, L. *et al.*, Fracture toughness and toughening mechanisms in graphene platelet reinforced Si3N4 composites. *Scr. Mater.*, 66, 10, 793–796, 2012.
158. Stankovich, S. *et al.*, Synthesis of graphene-based nanosheets via chemical reduction of exfoliated graphite oxide. *Carbon*, 45, 7, 1558–1565, 2007.
159. Liu, Y., Huang, J., Li, H., Nanostructural characteristics of vacuum cold-sprayed hydroxyapatite/graphene-nanosheet coatings for biomedical applications. *J. Therm. Spray Technol.*, 23, 7, 1149–1156, 2014.
160. Liu, Y. *et al.*, Hydroxyapatite/graphene-nanosheet composite coatings deposited by vacuum cold spraying for biomedical applications: Inherited nanostructures and enhanced properties. *Carbon*, 67, 250–259, 2014.
161. *Columbia Engineers Prove Graphene is the Strongest Material.* [Internet] 2008 [cited 2016 August 03]; Available from: http://www.columbia.edu/cu/news/08/07/graphene.html.
162. Marchetto, D. *et al.*, Friction and wear on single-layer epitaxial graphene in multi-asperity contacts. *Tribol. Lett.*, 48, 1, 77–82, 2012.
163. Kim, K.S. *et al.*, Chemical vapor deposition-grown graphene: The thinnest solid lubricant. *ACS Nano*, 5, 6, 5107–5114, 2011.
164. Penkov, O. *et al.*, Tribology of graphene: A review. *Int. J. Precis. Eng. Man.*, 15, 3, 577–585, 2014.
165. Berman, D., Erdemir, A., Sumant, A.V., Reduced wear and friction enabled by graphene layers on sliding steel surfaces in dry nitrogen. *Carbon*, 59, 167–175, 2013.
166. Lee, C. *et al.*, Elastic and frictional properties of graphene. *Phys. Status Solidi B*, 246, 11–12, 2562–2567, 2009.
167. Lee, C. *et al.*, Frictional characteristics of atomically thin sheets. *Science*, 328, 5974, 76–80, 2010.
168. Cho, D.H. *et al.*, Effect of surface morphology on friction of graphene on various substrates. *Nanoscale*, 5, 7, 3063–3069, 2013.
169. Tai, Z. *et al.*, Tribological behavior of UHMWPE reinforced with graphene oxide nanosheets. *Tribol. Lett.*, 46, 1, 55–63, 2012.
170. Li, H. *et al.*, Microstructure and wear behavior of graphene nanosheets-reinforced zirconia coating. *Ceram. Int.*, 40, 8, 12821–12829, 2014.
171. Li, Y. *et al.*, Preparation and tribological properties of graphene oxide/nitrile rubber nanocomposites. *J. Mater. Sci.*, 47, 2, 730–738, 2011.
172. Xu, Z. *et al.*, Preparation and tribological properties of TiAl matrix composites reinforced by multilayer graphene. *Carbon*, 67, 168–177, 2014.
173. Xu, Z. *et al.*, Formation of friction layers in graphene-reinforced TiAl matrix self-lubricating composites. *Tribol. Trans.*, 58, 4, 668–678, 2015.
174. Lahiri, D. *et al.*, Nanotribological behavior of graphene nanoplatelet reinforced ultra high molecular weight polyethylene composites. *Tribol. Int.*, 70, 165–169, 2014.
175. Kandanur, S.S. *et al.*, Suppression of wear in graphene polymer composites. *Carbon*, 50, 9, 3178–3183, 2012.
176. Xu, S. *et al.*, Mechanical properties, tribological behavior, and biocompatibility of high-density polyethylene/carbon nanofibers nanocomposites. *J. Compos. Mater.*, 2014.
177. Dorri Moghadam, A. *et al.*, Mechanical and tribological properties of self-lubricating metal matrix nanocomposites reinforced by carbon nanotubes (CNTs) and graphene – A review. *Compos. Part B Eng.*, 77, 402–420, 2015.
178. Berman, D. *et al.*, Macroscale superlubricity enabled by graphene nanoscroll formation. *Science*, 348, 6239, 1118–1122, 2015.
179. Gong, Z. *et al.*, Graphene nano scrolls responding to superlow friction of amorphous carbon. *Carbon*, 116, 310–317, 2017.

180. Bunch, J.S. et al., Impermeable atomic membranes from graphene sheets. *Nano Lett.*, 8, 8, 2458–2462, 2008.
181. Berry, V., Impermeability of graphene and its applications. *Carbon*, 62, 1–10, 2013.
182. Raman, R.S. and Tiwari, A., Graphene: The thinnest known coating for corrosion protection. *JOM*, 66, 4, 637–642, 2014.
183. Merisalu, M. et al., Graphene–polypyrrole thin hybrid corrosion resistant coatings for copper. *Synth. Met.*, 200, 16–23, 2015.
184. Chang, C.-H. et al., Novel anticorrosion coatings prepared from polyaniline/graphene composites. *Carbon*, 50, 14, 5044–5051, 2012.
185. Mayavan, S., Siva, T., Sathiyanarayanan, S., Graphene ink as a corrosion inhibiting blanket for iron in an aggressive chloride environment. *RSC Adv.*, 3, 47, 24868–24871, 2013.
186. Sahu, S.C. et al., A facile electrochemical approach for development of highly corrosion protective coatings using graphene nanosheets. *Electrochem. Commun.*, 32, 22–26, 2013.
187. Singh, B.P. et al., Development of oxidation and corrosion resistance hydrophobic graphene oxide-polymer composite coating on copper. *Surf. Coat. Technol.*, 232, 475–481, 2013.
188. Park, J.H. and Park, J.M., Electrophoretic deposition of graphene oxide on mild carbon steel for anti-corrosion application. *Surf. Coat. Technol.*, 254, 167–174, 2014.
189. Krishnamoorthy, K. et al., Graphene oxide nanopaint. *Carbon*, 72, 328–337, 2014.
190. Yu, Y.-H. et al., High-performance polystyrene/graphene-based nanocomposites with excellent anti-corrosion properties. *Polym. Chem.*, 5, 2, 535–550, 2014.
191. Sun, W. et al., Synthesis of low-electrical-conductivity graphene/pernigraniline composites and their application in corrosion protection. *Carbon*, 79, 605–614, 2014.
192. Sun, W. et al., Inhibiting the corrosion-promotion activity of graphene. *Chem. Mater.*, 27, 7, 2367–2373, 2015.
193. Ramezanzadeh, B. et al., Covalently-grafted graphene oxide nanosheets to improve barrier and corrosion protection properties of polyurethane coatings. *Carbon*, 93, 555–573, 2015.
194. Yu, Z. et al., Fabrication of graphene oxide–alumina hybrids to reinforce the anti-corrosion performance of composite epoxy coatings. *Appl. Surf. Sci.*, 351, 986–996, 2015.
195. Ramezanzadeh, B. et al., Enhancement of barrier and corrosion protection performance of an epoxy coating through wet transfer of amino functionalized graphene oxide. *Corros. Sci.*, 103, 283–304, 2016.
196. Chang, K.-C. et al., Room-temperature cured hydrophobic epoxy/graphene composites as corrosion inhibitor for cold-rolled steel. *Carbon*, 66, 144–153, 2014.
197. Janković, A. et al., Bioactive hydroxyapatite/graphene composite coating and its corrosion stability in simulated body fluid. *J. Alloy. Compd.*, 624, 148–157, 2015.
198. Li, M. et al., Electrophoretic deposition and electrochemical behavior of novel graphene oxide-hyaluronic acid-hydroxyapatite nanocomposite coatings. *Appl. Surf. Sci.*, 284, 804–810, 2013.
199. Li, M. et al., Graphene oxide/hydroxyapatite composite coatings fabricated by electrophoretic nanotechnology for biological applications. *Carbon*, 67, 185–197, 2014.
200. Jiang, K., Li, J., Liu, J., Electrochemical codeposition of graphene platelets and nickel for improved corrosion resistant properties. *RSC Adv.*, 4, 68, 36245–36252, 2014.
201. Kumar, C.M.P., Venkatesha, T.V., Shabadi, R., Preparation and corrosion behavior of Ni and Ni–graphene composite coatings. *Mater. Res. Bull.*, 48, 4, 1477–1483, 2013.
202. Editorial, Not so transparent. *Nat. Mater.*, 12, 10, 865–865, 2013.
203. Li, Z. et al., Effect of airborne contaminants on the wettability of supported graphene and graphite. *Nat. Mater.*, 12, 10, 925–931, 2013.
204. Shih, C.-J., Strano, M.S., Blankschtein, D., Wetting translucency of graphene. *Nat. Mater.*, 12, 10, 866–869, 2013.

205. Nine, M.J. et al., Robust superhydrophobic graphene-based composite coatings with self-cleaning and corrosion barrier properties. *ACS Appl. Mater. Interfaces*, 7, 51, 28482–28493, 2015.
206. Nguyen, D.D. et al., Superhydrophobic and superoleophilic properties of graphene-based sponges fabricated using a facile dip coating method. *Energ. Environ. Sci.*, 5, 7, 7908, 2012.
207. Castranova, V., Schulte, P.A., Zumwalde, R.D., Occupational nanosafety considerations for carbon nanotubes and carbon nanofibers. *Acc. Chem. Res.*, 46, 3, 642–649, 2013.
208. Chang, Y. et al., In vitro toxicity evaluation of graphene oxide on A549 cells. *Toxicol. Lett.*, 200, 3, 201–210, 2011.
209. Wang, K. et al., Biocompatibility of graphene oxide. *Nanoscale Res. Lett.*, 6, 1–8, 2011.
210. Hu, W. et al., Protein Corona-mediated mitigation of cytotoxicity of graphene oxide. *ACS Nano*, 5, 5, 3693–3700, 2011.
211. Liao, K.H. et al., Cytotoxicity of graphene oxide and graphene in human erythrocytes and skin fibroblasts. *ACS Appl. Mater. Interfaces*, 3, 7, 2607–2615, 2011.
212. Lee, W.C. et al., Origin of enhanced stem cell growth and differentiation on graphene and graphene oxide. *ACS Nano*, 5, 9, 7334–7341, 2011.
213. Kulshrestha, S. et al., A graphene/zinc oxide nanocomposite film protects dental implant surfaces against cariogenic Streptococcus mutans. *Biofouling*, 30, 10, 1281–1294, 2014.
214. Sang, B. et al., Graphene-based flame retardants: A review. *J. Mater. Sci.*, 51, 18, 8271–8295, 2016.
215. Shi, Y. and Li, L.-J., Chemically modified graphene: Flame retardant or fuel for combustion? *J. Mater. Chem.*, 21, 10, 3277–3279, 2011.

3
Graphene-Based Composite Materials

Munirah Abdullah Almessiere[1]*, Kashif Chaudhary[2], Jalil Ali[2] and Muhammad Sufi Roslan[3]

[1]*Department of Physics, College of Science, Imam Abdulrahman Bin Faisal University, Dammam, Saudi Arabia*
[2]*Laser Center, Ibnu Sina Institute for Scientific & Industrial Research (ISI-SIR), Universiti Teknologi Malaysia (UTM), Johor Bahru, Malaysia*
[3]*Center for Diploma Studies (CeDS), Universiti Tun Hussein Onn Malaysia (UTHM), Johor, Malaysia*

Abstract

Graphene, a single-layer carbon sheet with hexagonal lattice structure, has shown several unique characteristics such as the quantum Hall effect, high carrier mobility, large theoretical specific surface area, good optical transparency, high Young's modulus, and excellent thermal conductivity. A single square-meter sheet of graphene of weight 0.0077 g can support up to 4 kg. The two-dimensional structure, large surface area, and extraordinary mechanical characteristics make graphene a potential candidate as a nanofiller in a variety of composite materials. The significant advantages are as follows: the possibility to enhance the mechanical properties even with a relatively low nanofiller (graphene), effective dispersion, interface chemistry, and nanoscale morphology. The exceptional properties and the variety of additional incredible traits make graphene one of the important materials in the future. It is predicted that graphene will revolutionize every industry known to man. To exploit these characteristics of graphene, different reliable synthetic techniques have been developed to fabricate graphene and its derivatives, ranging from the bottom-up epitaxial growth to the top-down exfoliation of graphite through oxidation, intercalation, and/or sonication. The increase in production of graphene and its derivatives, such as graphene oxide (GO) and reduced graphene oxide (rGO), offers numerous possibilities to fabricate graphene-based functional materials for a variety of technological applications. The incredible improvement has been achieved with graphene-based composites, where the performance depends on the inherent characteristics of the nanofiller (as graphene and its derivatives). The addition of graphene and its derivatives to the host matrix allows the improvements in the composite characteristics to be used in a variety of applications such as electric, optics, electrochemical energy conversion, storage, etc. The graphene as nanofiller has been successfully added to inorganic nanostructures, organic crystals, polymers, biomaterials, metal-organic frameworks, etc. The modifications allow interactions or chemical bond formation between the nanocrystals (host material) and/or graphene sheets. Significant enhancements in the electrical conductivity, thermal stability, and mechanical characteristics are achieved by the addition of graphene and its derivatives to nanocomposites, which are explored for applications such as batteries, supercapacitors, fuel cells, photovoltaic devices, photo-catalysis, sensing platforms, and so on.

Keywords: Graphene composites, graphene filled polymers, graphene nanostructure, hybrid graphene, microfiber composites, graphene colloids and coatings, graphene bioactive, processing graphene composites

Corresponding author: malmessiere@iau.edu.sa

3.1 Introduction

Graphene can introduce or mix any to metals, polymers, and ceramics to create composites for specific application with controlled conductive and resistant characteristics. The applications seem endless, as one graphene–polymer proves to be light, flexible, and an excellent electrical conductor, while another dioxide–graphene composite is found to have interesting photocatalytic efficiencies, with possibilities of numerous types of coupling with other materials to make all kinds of composites. In this chapter, different types of graphene composites and processing routes are discussed as illustrated in Figure 3.1.

3.2 Graphene Composites

Remarkable developments have been accomplished with graphene-based composites, where the properties and performance depend on the inherent characteristics of the nanofillers such as graphene and its derivatives. The presence of graphene or its derivatives to the host matrix allows enhancements in the composite characteristics that are employed in a variety of technological applications such as electric, optics, electrochemical energy conversion, storage, etc. Graphene and its derivatives as nanofiller has been successfully added to different types of host materials such as inorganic nanostructures, organic crystals, polymers, biomaterials and metal-organic frameworks, metals and their oxides, etc. [1].

3.2.1 Graphene Filled Polymer Composites

In recent years, the development of dispersion of graphene particles polymer matrix at nanoscale has opened a new area in materials science. These polymer nanohybrid materials have shown incredible improvement in properties. The extent of the improvement depends on the degree of nanofiller dispersion in the polymer matrix [2, 3]. Overall, graphene-based polymer nanocomposites have shown impressive functional characteristics and record mechanical properties, electrical conductivity, unique optical transportation, anisotropic transport, and low permeability for a variety of demanding applications [4–6]. It has been reported that even the addition of a small fraction of a graphene component to the host material can dramatically enhance the performance of the variety of the polymeric

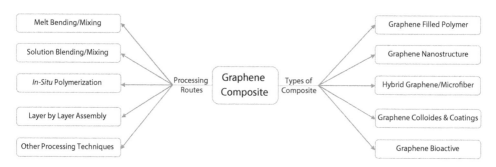

Figure 3.1 Organization of the chapter.

matrices and provide extraordinary reinforcing and functional characteristics [7]. Different types of nano graphite forms have also been used to grow conducting nanocomposites with enhanced physicochemical characteristics such as epoxy, [8, 9] PMMA, polypropylene [10], LLDPE [11], polystyrene [12], Nylon [13], polyaniline [14], phenylethynyl-terminated polyimide [15], and silicone rubber [16]. The conductivity and percolation thresholds of different graphite forms or graphene/its derivative composites depend on the processing route, polymer matrix, and filler type [17].

Interfacial interactions between graphene-based materials and polymers play a major role in the performance and integrity of the corresponding nanocomposite. Due to the homogeneous composition of carbon atoms in graphene, the molecular interaction of polymers are limited to weak van der Waals forces, stacking, and hydrophobic–hydrophobic interactions [18]. Van der Waals forces are weak attractive interactions developed between molecules by the transient or permanent dipoles. These forces contribute a major part in developing interfacial strength between graphene-based materials and host polymers due to intimate contact and large specific surface area [19]. In case of hydrophobic polymer matrices, the hydrophobic–hydrophobic interactions are a dominant means for binding graphene with the host matrix. Such interactions are more dominant for polymers with electron-rich aromatic rings such as phenyl rings. The stacking can adapt to different structural arrangements and significantly improve the bonding in graphene nanocomposites [20].

As compared to pure graphene, graphene oxide develops oxygen-containing polar functionalities such as epoxide, carbonyl, hydroxyl, and carboxyl groups [21]. The interaction of graphene oxide with different polymers is more versatile due to functionalization. Furthermore, the covalent grafting of polymer chains on graphene oxide surfaces can provide better blending of polymer matrix and graphene oxide components. The covalent bonds possess high mechanical strength among the intermolecular interactions and compatibility to grafted graphene oxide is higher due to the possibility of replacement of exposed functionalities. Polymers with hydroxyl functional groups directly cross-link the graphene oxide sheets with their carboxyl groups through esterification. This interfacial cross-linking significantly enhances the modulus of the nanocomposite; however, due to the covalent cross-linking, the compliance can be compromised [22]. The presence of polar functionalities in certain polymers leads to the development of the electrostatic interactions, which are stronger and restorable alternatives to covalent bonds for graphene oxides. As a result of these electrostatic interactions, the nanocomposites are much stronger and tougher as compared to their counterparts without graphene oxide fillers [23]. For graphene oxides, hydrogen bonding between highly polarized donor and acceptor groups also play a key role. The epoxide, hydroxyl, carbonyl, and carboxyl functional groups on graphene oxide are all highly polarized, where oxygen atom acts as a negative center [24], which enables graphene oxide to establish hydrogen bonds with different polar polymers, particularly polyelectrolytes and proteins [25]. The high density of the highly polar functionalities develops strong interfacial interactions through hydrogen bonding network between polymer–graphene nanocomposites. However, the interaction through the hydrogen bond network is less stronger than covalently cross-linked nanocomposites [7]. Based on the spatial arrangement and the kind of interaction between graphene material and polymers, the graphene-based polymer composites can be classified into three types, i.e., graphene-filled polymer composites, layered graphene–polymer films, and polymer-functionalized graphene nanosheets [1].

3.2.1.1 Graphene Filled Polymers

Conventionally, carbon-based materials, such as carbon nanotubes (CNTs), amorphous carbon, and graphitic particles, are used as fillers to enhance the electronic, mechanical, and thermal characteristics of polymer matrices. CNTs are one of the effective filler materials, but with a relatively high cost. Graphene-based materials are expected to be a promising replacement or supplement to CNT filler. In order to achieve the optimal characteristics of composite and lower the content of the graphene filler, the dispersity and its bonding with the host polymer matrix are the key factors. In common practice, the graphene-based material polymer composites are prepared via solution mixing, melt blending, or *in situ* polymerization [1].

3.2.1.2 Layered Graphene Polymers

In layered graphene polymer composites, graphene derivatives are composited with polymer matrices in layered structures, whereas in graphene filler composites, the fillers are randomly distributed in the polymer matrices. The layered graphene composites are prepared for specific applications, such as the directional load-bearing membranes and thin films for photovoltaic applications. Layer-by-layer (LbL) assembling of graphene derivatives and polymers is deposited via the Langmuir–Blodgett (LB) technique onto films of polyelectrolyte poly(allylamine hydrochloride) (PAH) and poly(sodium 4-styrene sulfonate) (PSS) multilayers [1, 23]. A similar approach is used to fabricate multilayer PVA–graphene oxide films with better elastic modulus and hardness [26]. Composite films are grown using sequential spin-coating of functional components in the device configurations. Li *et al.* have grown graphene oxide and poly(3-hexylthiophene) (P3HT)/phenylC61-butyric acid (PCBM) film by depositing layer after layer on ITO substrate for photovoltaic applications [27].

3.2.1.3 Polymer-Functionalized Graphene Nanosheets

In functionalized graphene nanosheet polymers, graphene derivatives are used as templates for polymer decoration through covalent and noncovalent functionalizations, instead of being used as filler. On the other hand, the polymer coating is used to enhance the solubility of the graphene derivatives, which offers additional functionality to the grown hybrid nanosheets. For example, in case of graphene oxide–PVA composite sheets, the carboxylic groups on graphene oxide is attached with the hydroxyl groups in PVA [28]. Similarly, the carboxylic groups on graphene oxide involves the carbodiimide-catalyzed amide formation process to bind with the six-armed polyethylene glycol (PEG)-amine stars [29]. However, the carboxylic groups usually confine at the periphery of the graphene oxide sheets and also the grafting of certain polymers needs the presence of nonoxygenated functional groups such as amine or chloride on graphene sheet. Therefore, alternative strategies with additional chemical reactions are developed to alter graphene oxide sheet surface with appropriate functional groups prior to grafting of polymers. Due to the rich surface chemistry, there are versatile possibilities of covalent functionalization of polymers on graphene-based sheets. The noncovalent functionalization only relies on the van der Waals force, electrostatic interaction, or p–p stacking [30], which are easier to develop without altering the chemical structure, and serves as an effective way to fabricate the electronic/optical property and solubility of the nanosheets. For example, *in situ* reduction of graphene oxide sheet with hydrazine in the presence of poly(sodium 4-styrene sulfonate) (PSS), where the hydrophobic backbone of PSS stabilizes the graphene oxide sheet and hydrophilic sulfonate side groups maintain a good dispersion in water [1, 31].

3.2.2 Graphene Nanostructure Composites

Graphene-based nanocomposites have received tremendous attention and lead to the development of a broad new class [32]. The fabrication of nanoparticles is well established and implemented in a variety of applications. The composite of nanoparticles with graphene-based materials has opened new avenues to realize synergistic effects of individual components [33]. To enhance the electronic, optical, and electrochemical energy conversion characteristics, the inorganic nanostructures and metals such as Au, Ag, Pd, Pt, Ni, Cu, Ru and Rh [25, 34–36]; oxides like TiO_2, ZnO, SnO_2, MnO_2, Co_3O_4, Fe_3O_4, NiO, Cu_2O, RuO_2, and SiO_2; and chalcogenides like CdS and CdSe have been composited with graphene and its derivatives [1, 37–41]. Nanoparticles can directly be decorated on the graphene sheets without establishing molecular linkers to bridge the nanoparticles and graphene. Therefore, numerous types of second phase can incorporate on graphene in the form of nanoparticles to introduce new functionalities for different applications such as catalytic, energy storage, photocatalytic, sensor, and optoelectronics applications [33]. Chao Xu *et al.* synthesized graphene–metal (Au, Pt, and Pd) nanoparticles using graphene oxide sheets as a precursor in the solution approach. Graphene incorporates specific properties to the functional particles, useful for technological applications [42]. In graphene-based metal nanocomposites, the noble metals are used in majority as second components. The incorporation of graphene not only reduces the consumption of noble metal but also significantly enhances electronic interaction [43]. Other metal nanomaterials/nanoparticles such as bimetals [44] and alloys [45] are also composited with graphene to fabricate inorganic graphene composites. The combination of graphene and inorganic material produces carbon materials with large specific surfaces, high electrical conductivities, and unique mechanical properties. The charge-transfer electronic as well as magnetic interactions are developed between graphene (and its derivatives) and the attached semiconducting oxide or magnetic nanomaterials [46] Graphene as an electron transport channel enhances the performance of metal compound nanomaterials in various applications. Nonmetallic materials such as S, Si, SiO_2, Si_3N_4, SiOC, CN, and C_3N_4 [47, 48] have also been prepared to develop metal-free catalysts to substitute metal catalysts. Graphene-based C_3N_4 nanocomposites have been prepared to enhance the performance catalyst to activate molecular oxygen for selective oxidation of secondary C–H bonds of saturated alkanes with good conversion and high selectivity to the corresponding ketones [32].

3.2.3 Hybrid Graphene/Microfiber Composites

A promising approach to prepared graphene-based composites is the use of hybrid fillers, which includes graphene-based materials and an inorganic material. The hybrid combination gives an advantage such as the ultimate properties resulting from additive or synergistic effects between the fillers. The hybridization procedure will encounter some of the filler disadvantages and improves its interactions with the host matrix, depending on the functionalization process. Multifunctionality is a key parameter to fabricate hybrid composites as the collective characteristics of the different fillers result in the material with different sets of properties as compared to properties of individual materials. Additionally, the cost of the final product can be lowered by using well-established microscale reinforcements such as carbon and glass fibers with small amounts of graphene where the stress can shift from micro to nanoscale reinforcement, enhancing the ultimate mechanical and other characteristics [49].

Hybrid graphene materials and other inorganic family members have shown exceptional features at nano or atomic scale or they may also revealed synergy in addition to complementarity. Hybridizing graphene and other 2D materials improvise the electronic properties of graphene, and more essentially, it could be used to adjust the bandgap of graphene [50–52]. Zero bandgap of original graphene is not perfect for semiconductor devices [53, 54], but this characteristic is addressed by its composites, which is a major breakthrough for the construction of graphene-based electronics [55–57]. The bandgap tuning of graphene (zero bandgap) with h-boron nitride (BN) (B5.8 eV), e.g., broadens the scope for semiconducting applications. An in-plane modification of h-BN on graphene (at 64% C) widens the bandgap up to 4 eV. On the contrary, carbon addition to h-BN domains increases electron conductivity, which is an insulator in its pristine form. Induced bandgap in graphene can efficiently be used to fabricate field effect transistors of graphene/h-BN films with carrier motilities ranging from 190 to 2000 cm^2/V/s [58]. Composites of chemically cross-linked graphene and boron nitride materials can improve energy storage efficiency. The graphene/h-BN supercapacitors show a maximum specific capacitance of B240 F/g [59, 60].

The bonding interaction between graphene and the individual fillers plays a vital role on the final physicochemical characteristics of resultant reinforced composite. Different routes have been explored to enhance the affinity among the fillers such as chemical functionalization, solution blending, growth on top of each other, mechanical mixing, etc. Due to van der Waals interaction and high specific surface area, the graphene agglomerates very easily. The decoration of graphene is the most pursued approach to avoid the agglomeration with other nanomaterials. The functionalization of graphene (graphene derivatives) increased the reactive sites on graphene, which enhances the bonding interaction with other nanomaterials and leads to the development of significantly fewer covalent bonds between the hybrid filler and host matrix [33]. Yang et al. prepared unfunctionalized and amine-functionalized multiwall carbon nanotube (MWCNT) and multi-graphene platelet (MGP) epoxy hybrid reinforced composites to improve mechanical and thermal properties. The functionalization of MWCNTs developed strong bonding between the fillers and MGPs. The 2D filler (MGPs) and 1D filler (MWCTNs) combination results in the formation of a 3D structure with a high surface area. The thermal conductivity of functionalized MWCNTs has increased more than 50% as compared to the pristine hybrid filler [61]. Lin et al. reported the fabrication of silica/reduced graphene oxide (SiO_2/rGO) hybrids via electrostatic route followed by mechanical blending of the hybrid filler with a styrene butadiene rubber matrix. Better dispersion has been observed in the case of hybrid materials as compared to the individual fillers, with strong interaction with the matrix [62].

Fiber-reinforced composites have replaced the conventional metals and are used in a variety of applications such as aerospace, automotive, marine, and construction industries. The ultimate properties of fiber composites attribute to the strong interface between the fiber and the host matrix. The addition of graphene-based materials to the fiber matrix can make a real difference. The long, micron-sized fibers are capable of confining nanoparticles in the interfacial region. The localized presence of nanoparticles in the stress transfer or electron/phonon conduction area can significantly reduce the quantity of nanoparticles required for reinforcement. Yavari *et al.* directly sprayed graphene on glass fibers and prepared graphene/glass-fiber/epoxy composites. Enhancement in the flexural bending fatigue magnitude by order of three has been observed for the hybrid filler with very low fractions of graphene [63]. Pathak *et al.* fabricated hybrid carbon-fiber/graphene-oxide/epoxy composite by reinforced epoxy with graphene oxide followed by impregnating carbon fiber fabric to modified epoxy resin. The 66% increase in flexural strength has been obtained, whereas the modulus has increased to 72%. Due to the development of hydrogen bonding, interlaminar shear strength has increased by 25% [64]. To enhance the toughness of the hybrid composites, Mannov *et al.* laminated cross-ply with glass fibers and carbon fibers and introduced thermally reduced graphene oxide. The results revealed an increase in residual compressive strength and toughness due to the presence of graphene oxide in the matrix and led to less impact damage and an increased residual compressive strength [65]. Knoll *et al.* investigated the fatigue performance of carbon-fiber-reinforced epoxy mixed with layer graphene and multiwalled carbon nanotubes. The 15-fold improvement in fatigue load and the significant decrease in fatigue degradation lifetime have been observed [66]. Wang *et al.* combined graphene nanoplatelets with glass fibers to reinforce polypropylene matrix. Three times higher tensile modulus has been observed as compared to the original modulus in the presence of graphene nanoplatelets, whereas the tensile strength also improved significantly [33, 67].

3.2.4 Graphene Colloids and Coatings

Surface coating is a vital approach to modify the surface quality and provide protection for a substrate in numerous applications. Graphene is a promising new-generation material to deposit advanced coating on different types of substrates due to its exceptional characteristics attributed to its unique structure. Somani *et al.* deposited graphene on Ni-foil substrate via the chemical vapor deposition technique where camphor has been used as precursor [68]. Chemical vapor deposition is a promising route to deposit graphene on substrate in large scale. Different approaches are used to grow graphene on substrates, which depends on the nature of substrates such as the substrates with medium-high carbon solubility like Ni, Cu, etc. [33]. The concentration of dissolved carbon atoms depends on the type and substrate thickness and the concentration of the hydrocarbon gas (precursor). In turn, the thickness of the graphene layers is dependent on the dissolved carbon atoms and substrate cooling rate [69]. In case of low carbon solubility substrates, the development of graphene layer on substrate surface does not follow the diffusion process [70].

The electrical characteristics of graphene are the most promising in industrial applications. The graphene-based transparent conductive coating for the practical touch panel application is fabricated on copper substrate. Ishikawa *et al.* deposited graphene film on glass substrate. Apart from the electrical conductivity, hierarchical graphene oxide/MnO_2 nanostructure sponge has been grown and used as a supercapacitor for batteries due to high specific capacitance, wide operation range, good energy and power density, and excellent cycling stability [71]. Jeon *et al.* have demonstrated that the moderately reduced graphene oxide (RGO) can be utilized as the hole transporting layer (HTL) in polymer solar cells [72].

RGO coating on aluminum film is used to generate high-pressure and high-frequency ultrasounds for optoacoustic transmitters [73]. The aluminum transmitter coated with RGO enhances optoacoustic pressure 64 times as compared to the aluminum-based transmitter. The TiO2 thin film coated with graphene-based material possess better photocatalytic activity due to the presence of a giant π-conjunction system and high specific surface area. However, large quantity of graphene-based material reduces the photocatalytic activity of the TiO2 material due to absorbance and scattering of photons [74]. Graphene offers an effective barrier toward oxidation and corrosion due to its inertness to oxidizing gas and liquid solutions; e.g., GO/poly(ethylene imide) (PEI) coating has been deposited through the layer-by-layer method to form a GO/PEI bilayer for oxygen barrier. However, the oxidation resistance of graphene is limited to pressure [75, 76]. Singh Raman et al. investigated the anti-corrosion behavior of graphene coating on various metal substrates [77].

Graphene-based materials are also a promising candidate in sensing and absorbent applications [78]. The graphene-coated super-hydrophobic and super-hydrophilic sponge exhibited high absorption capacities up to 165 times of its weight, high selectivity, good recyclability, lightweight, robustness, and inertness to corrosive environments [79]. Apart from the use of graphene and its derivative to grow new sensing structures, graphene-based material coatings are also used on commercially available equipment to enhance their performance and efficiency. Zhang et al. demonstrated the growth of graphene coating on a plunger-in-needle micro-syringe by the sol-gel method for a solid-phase micro-extraction (SPME) device as a sorbent material to use as UV filter [80].

Signal enhancement is also an area that can benefit from graphene-based materials. Graphene coating is also employed on conventional metallic surface enhanced Raman scattering (SERS) substrates to enhance the sensitivity of SERS detection [81]. Wang et al. reported the coating of sulfur particles by graphene oxide used to modify the capacity and cycling stability of rechargeable lithium-sulfur battery cathode materials [82]. SnO_2-SiC/G nanocomposites have been grown using the ball mill approach. The graphene-based material coating on polymer not only significantly enhances the properties of a polymer but also improves several characteristics simultaneously such as electrical and mechanical properties. Liao et al. coated polyurethane acrylate (PUA) with graphene to reinforce the polymer using *in situ* polymerization on a TEFLON plate [83]. An increase in the electric conductivity of the composite is observed with an increase in graphene loading [84].

The large delocalized π-electron system of graphene makes it a good candidate as an adsorbent to adsorb benzenoid base compounds. The graphene-based solid-phase micro-extraction fibers are used for the extraction of organic pollutants [85]. Colloids are the formation stage of crystallization, where atoms form clusters in different types of solutions or melts; during the nucleation process, a colloid is formed and promoted to crystal growth. Their strong size- and shape-dependent physical characteristics, ease in fabrication, and processing make colloidal nanocrystals potential building blocks for future materials with desired design functions. The colloidal form of graphene is capable of providing better performances as compared to the other forms of graphene. Colloidal graphene can provide a versatile platform to fabricate devices through solution route. However, the preparation of colloidal graphene solution from graphite materials is a critical point for commercial applications [86]. Wu et al. coated SPME fiber with graphene and applied to extract four triazine herbicides (atrazine, prometon, ametryn, and prometryn) in water samples [85]. Chen et al. prepared a graphene-coated SPME fiber to extract pyrethroid pesticides from water [87]. Lee *et al.* grew graphene-based SPME films to extract polybrominated diphenyl ethers from water samples [88].

3.2.5 Graphene Bioactive Composites

Different biosensors based on graphene material for diverse sensing applications include different mechanisms of optical and electrochemical signaling [89]. Graphene, due to its high sensitivity, low cost, quick response, excellent electrochemical properties, and easy operation, is a potential candidate for biomolecule detection [90–92]. Zhou *et al.* investigated the electrochemical behavior of graphene-modified electrode toward hydrogen peroxide. Significant increase in the electron transfer rate has been observed in case of graphene-based electrodes as compared to bare electrodes [93]. Shao *et al.* reported that N-doped graphene (N-graphene) possesses better electrocatalytic activity toward hydrogen peroxide reduction as compared to graphene, which is attributed to the presence of nitrogen functional groups, oxygen-containing groups, and structural defects [94, 95]. Graphene oxide exhibits fluorescence over a wide range of wavelength (200 nm to 1200 nm) [96] and effectively quenches the fluorescence of other fluorescent dyes [97]. These optical characteristics make graphene oxide a potential material to fabricate fluorescence resonance energy transfer (FRET) sensors. Numerous graphene-based FRET sensors have been fabricated for target ssDNA monitoring. Researchers also introduced fluorescent molecules to graphene oxide and resultant functionalized graphene is employed as an *in vitro* and *in vivo* imaging probe. Liu *et al.* have conjugated nano-graphene sheets (NGS) with Cy7 using polyethylene glycol [98]. Aminodextran coated ferrite nanoparticles immobilized onto graphene oxide are employed to improve magnetic resonance imaging (MIR). Significant improvement is observed in cellular MRI signal in case of graphene-coated ferrite nanoparticles as compared to pure ferrites [99]. The advantage of graphene-based material over other nanoparticles is ultrahigh surface area and *sp*2 hybridized carbon atoms, which make it a potential drug carrier to load drug molecules in large amounts on both sides of the single atom layer sheet [29].

Due to its remarkable optical properties, graphene has been used in photothermal therapy (PTT). Zhang *et al.* have grown DOX-loaded PEGylated nano-graphene oxide [NGO-PEG (polyethylene glycol)-DOX], which is capable of delivering heat and drug to the tumorigenic region to facilitate the combined chemotherapy and photothermal treatment in a signal system [100]. Graphene-based field-effect transistors are applied for the sensitive detection of both biomolecules and electrophysiological signals. Graphene derivatives, such as graphene oxide (GO), are widely applied in biological imaging, drug delivery, and cancer therapy, due to their large surface area and ease of preparation and functionalization [92].

3.3 Processing Routes for Graphene Composites

A single, purely sp^2-hybridized graphene layer free of heteroatomic defects (pristine graphene) is fabricated via different techniques such as chemical vapor deposition, micro-mechanical exfoliation of graphite, and growth on crystalline silicon carbide. However, the synthesis of powdered graphene samples in large enough quantities for use as composite is still a challenging task [101]. The ultimate characteristic of graphene-based composites critically depends on the processing routes and conditions [102]. The functionality of graphene plays a critical role in lowering filler loading rate, dispersion, and organization of sheets within the host matrix to improve the overall performance of composites. For example, the mechanical properties of graphene composites depend on the aspect ratio, organization of sheets, specific surface area, and loading content of graphene materials, whereas the interfacial strength, dispersion, spatial organization, and affinity of components determine the final strength, stiffness, toughness, and elongation of composites [103]. The pretreatment process and synthesis route define the morphology and physiochemical characteristics of resulting composites. In several graphene-based polymer composites, the dispersion and exfoliation of graphene layers are controlled through shear force, temperature, and solvent polarity. The development of high-performance composite requires effective control of restacking, wrinkling, and aggregation of graphene sheets. As, the flexibility and high-aspect ratio of graphene components prone to random wrinkling, buckling, or folding during processing, which dramatically disturbs the ultimate performance of composite. Therefore, the selection of synthesis route is determined by the surface functionalization of integrated graphitic sheets. In general, the traditional synthesis techniques include melt-based processing and solution-based processing [104]. The interaction mechanism in graphene-based composites depends on the polarity, molecular weight, hydrophobicity, reactive groups, etc. [105]. The most popular approaches used for chemical modification and assembly are *in situ* polymerization, chemical grafting, blending, layer-by-layer (LbL) assembly, and directed assembly [7].

3.3.1 Melt Bending/Mixing

The melt-based mixing technique is a solvent-free route for preparation of graphene composites. The melt mixing process involves high temperature and mechanical shear forces to distribute the reinforcement phase (filler) in the matrix through a screw extruder or blending mixer. Graphene or modified graphene derivatives are mixed with the host matrix in the molten state. The high temperature liquefies the polymer and permits easy dispersion and intercalation of graphene sheets and its derivatives. The process is applicable to both types of polar and nonpolar matrices. This technique allows graphene or reduced graphene oxide sheets to be exfoliated into a viscous matrix by pressing unfavorable interactions and inducing component dispersion. Melt mixing is considered a practical approach to prepare graphene-based polymer nanocomposites. However, the high localized mechanical stress due to thermal heating can disturb the stability of components, shape of flakes, and reduction state of the graphene oxide sheets [7]. Melt mixing is a fast, relatively inexpensive process and is widely used in industry for thermoplastic nanocomposite production. The polymer is melted at elevated temperatures and the graphene-based material in powder form is mixed using a single-, twin-, triple-, or even quad-screw extruder. The process does not involves the use of any toxic solvents. A number of graphene-based nanocomposites are prepared using this technique [106–109]. In general, the composites with an adequate degree of dispersion can be prepared via a melt mixing procedure. However, mixing temperature should be selected carefully as elevated temperature can cause a degradation of the host material. In certain cases, high shear forces are required to mix the polymer with the graphene flakes, which can cause buckling or breakage of the graphene sheets. In spite of poor dispersion, this process can be adopted to synthesize graphene-based nanocomposites with adequate properties. Adjustments of the melt mixing process is required to improve the dispersion and subsequent enhancement of the composite properties. Li *et al.* introduced graphene nanoplatelets into poly(methyl methacrylate)/polystyrene (PMMA/PS) and PMMA/PMMA multilayer films using force assembly. The filler planar orientation led to a higher degree of reinforcement in prepared composites [49, 110]. A variety of polymer nanocomposites, such as PP/EG [111], HDPE/EG, [112] PPS/EG, [113] PA6/EG, [114] etc., have been synthesized using the melt mixing technique [17].

The melt mixing route is often considered more cost effective and compatible in the context of industrial practices [115]. However, this process does not provide dispersion of the filler as a solvent mixing technique or *in situ* polymerization process [116]. The low bulk density of dry powders is difficult and poses processing challenges such as processing equipment (melt extruder) [117]. In a different approach, the host matrix and filler are subject to premixing as sonicated in a nonsolvent prior to melt mixing, which lowers the electrical percolation threshold of GNP/polypropylene composites [111]. In the case of GO platelets as a filler, melt processing and molding can lead to substantial reduction of the platelets due to their thermal instability [101, 118]. The thermal reduction can lead to the loss of the functional group, which is another hurdle to obtain homogeneous dispersion in a polymeric matrix particularly in nonpolar polymers. Kim and Macosko did not observe any significant enhancement in mechanical properties of graphene composite with polycarbonate due to the removal of the oxygen functional groups, which affects the interfacial bonding [33].

3.3.2 Solution Blending/Mixing

Solution blending is the most widely used method to synthesize polymer-based graphene composites provided the polymer is soluble in aqueous or organic solvents such as water, acetone, DMF, chloroform, DCM, and toluene. The solution-based process includes the blending of graphene platelet colloidal suspensions or other graphene-based materials with the desired polymer, in solution or by dissolving the polymer in the graphene platelet suspension by stirring or shear mixing or ultrasonication. The polymers are dissolved in a suitable solvent and blended with dispersed graphene suspension. The polar polymers such as PMMA, PAA, PAN, and polyesters are effectively blended with graphene-based material. The surface of the graphene material is usually functionalized by isocyanates, alkylamine, alkylchlorosilanes, etc., prior to blending in order to enhance the dispersion in organic solvent [33]. The extent of dispersion of graphene platelets in the composite is mainly governed by the level of exfoliation of the platelets achieved prior to, or during, mixing. Therefore, solution mixing provides a simple potential approach to disperse platelets into a polymer matrix. Lyophilization methods [119], phase transfer techniques [120, 121], and surfactants [122] are used to facilitate dispersion in solution mixing of graphene-based composites. However, the use of surfactants can lead to the attenuation of the thermal conductivity [101, 123]. The final product can be precipitated from blended suspension using a nonsolvent for the polymer, causing the polymer chains to encapsulate the filler upon precipitation. The precipitated composite is then filtered, dried, and processed for application. In an alternative process, the suspension is directly casted into mold and dried. However, this process can lead to aggregation of the filler in composite [101]. The solution processing route maximizes the dispersion of filler (graphene sheets) in a polymer matrix.

This technique is widely exploited to fabricate the polymer composites due to high dispersion efficiency, facile and fast fabrication, and control on component behavior. The finding of common solvents, toxic solvent utilization, thin-film limitation, difficulties in solvent removal, and common aggregation issues are the key challenges [7]. In general, the solution blending technique offers an adequate dispersion of the flakes or sheets and it is quite versatile due to the availability of a number of different solvents to dissolve the matrix and disperse the filler [49]. For the synthesis of the PVA–GO composite, the esterificated GO is blended with the solution of PVA dissolved in DMSO [28]. To obtain homogenized dispersion of graphene sheets, ultrasonication is used. However, the long exposure to high-power ultrasonication may lead to defects in graphene sheets, which in turn can be detrimental to the composite properties. The higher loading of graphene sheets with better dispersion in water and other organic solvents can be attained via functionalization of graphene. In a blending process, the polymer is coated on an individual sheet that interconnects each sheet as solvents are removed. During solution mixing, GO and RGO sheets tend to agglomerate in case of slow solvent evaporation, which results in inhomogeneous disbursement of sheets in matrix. The distribution of filler in the matrix can be controlled by adjusting the evaporation time using drop casting or spin coating [33]. The gain in entropy due to desorption of solvent molecules is the driving force for polymer intercalation. The gained entropy is compensated by a decrease in conformational entropy of the intercalated polymer chains. Therefore, to accommodate the incoming polymer chains, a relatively large number of solvent molecules are required for desorption. The melt blending allows the fabrication of intercalated nanocomposites for polymers with low or even no polarity [124]. Liao *et al.* prepared aqueous-reduced graphene thermoplastic polyurethane composites via a solution mixing approach using a co-solvent process. An organic solvent (dimethylformamide, DMF) is added before the removal of water to avoid restacking and aggregation of filler [49].

3.3.3 *In Situ* Polymerization/Crystallization

In situ polymerization synthesis of graphene nanocomposite involves mixing of filler in neat monomer or multiple monomers, or a solution of monomer, followed by polymerization. The graphene or modified graphene sheets are swollen into the liquid monomer and then a suitable initiator is added to the solution. The polymerization is initiated either by heat or by radiation [101]. *In situ* polymerization is one of the low-cost thermal processes with high moldability into any shape and better optical properties [125]. Similar to the solution mixing method, functionalization of graphene sheets is performed to improve the initial dispersion in the monomer solution and subsequently in the composites. *In situ* polymerization exfoliates the graphene layered structure into nano plates by the intercalation of monomers that produce polymers with well-dispersed graphene in a polymer matrix [33]. During the *in situ* polymerization process, graphene flakes are first dispersed into monomers or pre-polymers and then polymerization is performed, leading to the crystallization/precipitation of composites with good dispersion and strong interactions between the matrix and the filler. The *in situ* polymerization process allows grafting of the filler on the polymer with or without functionalization to enhance the compatibility between the components of the system. However, the increase in viscosity of the system during the polymerization process limits the loading fraction and the processing of the composites

[49]. High-level graphene-based filler dispersion can be achieved via the *in situ* polymerization synthesis route without a prior exfoliation step as compared to solution mixing and melting techniques. In this technique, monomer is intercalated between the graphene layers followed by polymerization to isolate the layers. This method, also referred as intercalation polymerization, is mostly used to investigate and test graphene-based derived polymer composites [49].

An alkali metal and a monomer (e.g., isoprene or styrene) can be used to intercalate graphene and are polymerized by the negatively charged graphene sheets. The covalent linkages are developed between the matrix and filler fabricated during the *in situ* polymerization process. However, a variety of polymers such as poly(ethylene), PMMA, and poly(pyrrole) can also be prepared through development of noncovalent linkages via the *in situ* polymerization process. The larger spacing in GO layers facilitates intercalation by both monomers and polymers as compared to graphite, whereas polar functional groups support direct intercalation of hydrophilic molecules as the interlayer spacing increased with uptake of monomer or polymer [101]. The functionalization routes enable the development of strong interfaces between the filler and the matrix. The oxygen-containing functional groups of graphene provide sufficient active sites to develop bonds with the matrix or secondary filler, which significantly enhances the ultimate properties of the composites [49]. Extensive research have been conducted to produce epoxy-based nanocomposites by the *in situ* polymerization process where fillers are first dispersed into resin followed by curing and adding hardener [126]. A variety of composites, such as PANI–GO/PANI–graphene [127], graphene nanosheet/carbon nanotube/polyaniline [128], and PANI–GO [129], are prepared through this technique [33]. In a recent study, PE chains between the graphitic layers are grown through metallocene-mediated polymerization of poly(ethylene) in the presence of dispersed GNPs [101]. Wang *et al.* synthesized graphene oxide/polyimide (PI/GO) composites using *in situ* polymerization. GO is functionalized with amine (ANH2) (ODA-GO) groups to enhance flake dispersion [130]. Bielawski *et al.* have proposed that during the polymerization procedure, GO performs two different functions [131, 132]. Initially, it catalyzes the dehydrative polymerization, whereas the residual carbon from the GO catalyst undergoes a dehydrogenation during the reaction and serves as an additive in the composite [49].

3.3.4 Layer-by-Layer Assembly

Layer-by-layer (LbL) assembly is a versatile technique used for the fabrication of graphene-based composites. LbL assembly is one of the efficient deposition routes to fabricate ultrastrong and robust coating and thin films, membranes with controlled adhesion, flexibility, and environmental stability [133, 134]. In LbL assembly, a variety of nano-architectures can be deposited in the form of multilayer thin films with specific thickness or hierarchy by alternating cationic and anionic phases on a substrate. In the LbL approach, the stacking assembly provides precise control on distribution and content of graphene to engineer a graphene–polymer interface on the molecular level by alternating deposition of complementary components (polymer solution and graphene filler suspension) [135]. The LbL fabrication route also enables fine-tuning of morphology of the nanocomposite films by manipulating deposition mode (dipping or spin and spray), solvent removal procedure, or applied shear force. In, vacuum-assisted deposition techniques micro-flow controller at the filter/solution interface is employed to deposit layers. The LbL approach is capable of producing uniform and large-area thin films with precise controlled thickness on a variety of substrates [5]. However, the vacuum-assisted method might not control the precise deposition (layer assembly) of different complementary components and is challenging in vacuum-assisted techniques. The high structural uniformity and chemical stability can be achieved through chemical and electrochemical post-reduction for conductive nanocomposite films [7]. Novel functional composites can be fabricated by manipulation of the deposition sequence, for a wide variety of applications, such as membranes, Li-ion batteries, field-effect transistors, anodes, and supercapacitors. During the LbL deposition process, the experimental parameters such as temperature, ionic strength, pH, and the actual polyelectrolyte play a significant role and can affect the interactions such as hydrogen bonding, covalent bonding, electrostatic, charge transfer, and coordination chemistry interactions [49].

Zhao *et al.* deposited multilayer thin films of exfoliated graphene oxide and PVA using the hydrogen bonding LbL technique and measured their mechanical properties [26]. Zhu *et al.* deposited PVA and graphene oxide nanocomposites via the vacuum-assisted technique and the dip-assisted LbL technique and compared the electrical and mechanical properties. It has been reported that morphology (layered structure) determines the mechanical behavior, whereas the electrical conductivities depend on the dispersion of nanostructures as electron transportation is dependent on the tunneling barrier among the finely distributed conductive components [136]. Li *et al.* used the dip-assisted electrostatic LbL technique to deposit hybrid multilayered films using negatively charged graphene oxide nanosheets and polyoxometalate clusters with cationic polyelectrolytes [25]. Kulkarni *et al.* fabricated ultrathin graphene oxide/polyelectrolyte multilayers using spin-assisted LbL assembly in a combination with Langmuir–Blodgett. The combination of LbL with Langmuir–Blodgett facilitates the growth of a highly integrated nanocomposite membrane with large dimensions by suppressing folding and wrinkling of graphene sheets. Dramatic enhancement in mechanical properties and elastic modulus has been observed [23]. In another study, Hu *et al.* incorporated graphene oxide sheets into silk fibroin matrix using spin-assisted LbL assembly through heterogeneous surface interactions. Incredible mechanical properties of prepared LbL membranes have been observed, which are attributed to the effective coupling of the silk fibroin matrix graphene filler [5]. Choi *et al.* fabricated nanocomposite films with a graphene conductive network by pressing graphene-wrapped PS microspheres into thin films. The PS polymer latex has been used to facilitate the uniform filler dispersion in the polymer matrix. The electrically conductive graphene/PS nanocomposites are fabricated using a combination of latex technology and LbL assembly, which offers a facile, efficient, and environmentally friendly route. Significant enhancement in the interfacial adhesion has observed due to diverse chemical functionality [137]. Kesong *et al.* exploited an interface-mediated assembly technique to fabricate micelle-decorated graphene oxide sheets. Amphiphilic heteroarm star copolymers [PSnP2VPn and PSn(P2VP-b-PtBA)n (n = 28 arms)] have been adsorbed on the pre-suspended graphene oxide sheets at the air–water interface. The high-order, discrete assemblies of micelles of amphiphilic star uniformly covered with flat graphene oxide sheets in pancake conformation have been obtained. The resulting morphology is attributed to the strong affinity between positively charged pyridine groups of star polymers and negatively charged basal plane of graphene oxide [7, 137].

3.3.5 Other Processing Techniques

3.3.5.1 Chemical Reduction

Graphene derivatives such as graphene oxide (GO) platelets have a surface rich in reactive functional groups, and numerous techniques have been developed to develop covalent linkages between GO platelets and polymers. For example, both grafting-to and grafting-from methods are introduced for the attachment of wide range of polymers. Lee *et al.* attached covalent atom transfer radical polymerization (ATRP) initiators with the alcohols present in GO platelet through esterification [138]. Polymer brushes have been fabricated in a controlled manner by adding an ATRP-compatible monomer (such as styrene, butylacrylate, or methyl methacrylate) and a source of copper iodide. Similar studies using such ATRP-based methods have reported an increased scope of monomer reactivity [139]. Significant improvements in thermal and mechanical properties have been reported for polymer-grafted CMG platelets into a polymer matrix as compared to the neat matrix polymer [140]. The grafting-based methods have also been used for heterogeneous blending of polymer-functionalized GO in matrices composed of conducting polymers, such as poly(3-hexylthiophene) (P3HT) and a triphenylamine-based poly(azomethine) [141]. The grafting-to approach includes grafting of poly(styrene) (PS) chains to the alkyne-functionalized GO platelets through CuI-catalyzed 1,3-dipolar cycloaddition [142], and PVA to GO platelets through carbodiimide-activated esterification. The selection between grafting-to and grafting-from approach depends on the resultant polymer. However, a grafting-to method may lower the grafting density of chains to the platelet surface [143], which in turn can reduce the dispersion of these polymer-grafted platelets [144]. In some polymers, covalent linkage between GO platelets and the matrix is developed during polymerization without performing any prior functionalization. In case of epoxy matrix composite, curing with an amine hardener may lead to the incorporation of GO platelets directly into the cross-linked network [145]. Xu *et al.* grafted polyamide brushes to GO platelets through condensation reactions between carboxylic acid groups of GO platelets and amine containing monomer [101, 146].

3.3.5.2 Sol-Gel Methods

The sol-gel processing route is used to synthesize graphene glass/ceramic composites. The process involves the preparation of a precursor that undergoes condensation to grow a substance with well-dispersed graphene. In this method, a stable suspension of well-dispersed graphene is prepared using ultrasonic bath in the first step. In the second step, a catalyst such as acidic water is added to promote hydrolysis and formation of gel after condensation at room temperature. Silica nanocomposites are mainly prepared using the sol-gel technique, as well as CNT-silica composites [147, 148].

3.3.5.3 Colloidal Processing

Colloidal processing is used to prepare ceramic suspensions on the basis of colloidal chemistry. This technique is also utilized to fabricate graphene–ceramic mixtures by mixing graphene colloidal suspensions and ceramic powders. In common practice, the same solvent is preferred to prepare suspension in order to obtain uniform dispersion, prepared by slowing mixing using magnetic stirring/ultrasonication. Colloidal processing also requires surface modification of both host matrix and graphene, which is achieved either by direct functionalization (i.e., oxidation) or heterocoagulation (using surfactants by generating same/opposite electric charges between graphene and ceramic particles) [149, 150].

3.3.5.4 Powder Processing

The powder processing technique is commonly used to prepare CNT–ceramic composites with different matrices such as alumina, zirconia [151, 152], silicon nitride, silica [153], and borosilicate glass [154]. In the powder processing method, the filler material (graphene or CNTs) de-agglomerates through either ultrasonication followed by mixing with the ceramic powder in a solvent or by using conventional ball milling to produce slurries of well-dispersed ceramic composites. Kun *et al.* have synthesized well-dispersed graphene–ceramic composites using NMP/ethanol as the dispersing media obtained by attritor ball milling. Graphene is easier to process via powder processing as compared to CNTs to fabricate well-dispersed composites [155].

3.4 Summary

Graphene can introduce or mix in a variety of materials such as metals, polymers, and ceramics to fabricate composites for specific application with controlled conductive and resistant characteristics. The incredible improvement has been achieved with graphene-based composites, where the performance of the composite depends on the inherent characteristics of the nanofiller (such as graphene and its derivatives). The addition of graphene and its derivatives in the host matrix allows enhancement in the composite characteristics, which can be implemented in a variety of technological applications, such as electric, optics, electrochemical energy conversion, and storage. The different types of graphene-based nanocomposite and their fabrication routes have been presented in this chapter. Graphene-based nanocomposites have endless possibilities and capabilities for advanced engineering applications attributed to multifunctional graphene-based nanofillers. The details of different synthesis techniques have been described, whereas the literature has been analyzed for the evaluation of different types of graphene composites in different ranges of matrices. The efficiency of reinforcement of nanocomposites is dependent on the synthesis route. For the successful synthesis of advanced graphene-based composites, homogeneous dispersion of the filler in the matrix and strong bonding between the matrix and the filler are key factors for the ultimate properties of the resultant composite. In spite of intensive research, there are still several challenges that need to be addressed and tackled before industrial-scale mass production of graphene nanocomposites.

References

1. Huang, X., Qi, X., Boey, F., Zhang, H., Graphene-based composites. *Chem. Soc. Rev.*, 41, 666–686, 2012.
2. Stankovich, S., Dikin, D.A., Dommett, G.H., Kohlhaas, K.M., Zimney, E.J., Stach, E.A., Piner, R.D., Nguyen, S.T., Ruoff, R.S., Graphene-based composite materials. *Nature*, 442, 282–286, 2006.

3. Dikin, D.A., Stankovich, S., Zimney, E.J., Piner, R.D., Dommett, G.H., Evmenenko, G., Nguyen, S.T., Ruoff, R.S., Preparation and characterization of graphene oxide paper. *Nature*, 448, 457–460, 2007.
4. El-Kady, M.F. and Kaner, R.B., Scalable fabrication of high-power graphene micro-supercapacitors for flexible and on-chip energy storage. *Nat. Commun.*, 4, 1475, 2013.
5. Hu, K., Gupta, M.K., Kulkarni, D.D., Tsukruk, V.V., Ultra-robust graphene oxide-silk fibroin nanocomposite membranes. *Adv. Mater.*, 25, 2301–2307, 2013.
6. Shahil, K.M. and Balandin, A.A., Thermal properties of graphene and multilayer graphene: Applications in thermal interface materials. *Solid State Commun.*, 152, 1331–1340, 2012.
7. Hu, K., Kulkarni, D.D., Choi, I., Tsukruk, V.V., Graphene-polymer nanocomposites for structural and functional applications. *Prog. Polym. Sci.*, 39, 1934–1972, 2014.
8. Park, J.K., Do, I.-H., Askeland, P., Drzal, L.T., Electrodeposition of exfoliated graphite nanoplatelets onto carbon fibers and properties of their epoxy composites. *Compos. Sci. Technol.*, 68, 1734–1741, 2008.
9. Ye, L., Meng, X.-Y., Ji, X., Li, Z.-M., Tang, J.-H., Synthesis and characterization of expandable graphite–poly (methyl methacrylate) composite particles and their application to flame retardation of rigid polyurethane foams. *Polym. Degrad. Stab.*, 94, 971–979, 2009.
10. Wakabayashi, K., Pierre, C., Dikin, D.A., Ruoff, R.S., Ramanathan, T., Brinson, L.C., Torkelson, J.M., Polymer–graphite nanocomposites: Effective dispersion and major property enhancement via solid-state shear pulverization. *Macromolecules*, 41, 1905–1908, 2008.
11. Kim, S., Seo, J., Drzal, L.T., Improvement of electric conductivity of LLDPE based nanocomposite by paraffin coating on exfoliated graphite nanoplatelets. *Compos. Part A Appl. Sci. Manuf.*, 41, 581–587, 2010.
12. Kim, H., Hahn, H.T., Viculis, L.M., Gilje, S., Kaner, R.B., Electrical conductivity of graphite/polystyrene composites made from potassium intercalated graphite. *Carbon*, 45, 1578–1582, 2007.
13. Scully, K. and Bissessur, R., Decomposition kinetics of nylon-6/graphite and nylon-6/graphite oxide composites. *Thermochim Acta*, 490, 32–36, 2009.
14. Du, X., Xiao, M., Meng, Y., Facile synthesis of highly conductive polyaniline/graphite nanocomposites. *Eur. Polym. J.*, 40, 1489–1493, 2004.
15. Du, X., Xiao, M., Meng, Y., Synthesis and characterization of polyaniline/graphite conducting nanocomposites. *J. Polym. Sci. Part B: Polym. Phys.*, 42, 1972–1978, 2004.
16. Cho, D., Lee, S., Yang, G., Fukushima, H., Drzal, L.T., Dynamic mechanical and thermal properties of phenylethynyl-terminated polyimide composites reinforced with expanded graphite nanoplatelets. *Macromol. Mater. Eng.*, 290, 179–187, 2005.
17. Kuilla, T., Bhadra, S., Yao, D., Kim, N.H., Bose, S., Lee, J.H., Recent advances in graphene based polymer composites. *Prog. Polym. Sci.*, 35, 1350–1375, 2010.
18. Israelachvili, J.N., *Intermolecular and Surface Forces*, Second Edition, Academic Press, University of California, Santa Barbara, USA, 2011.
19. Jiang, L.Y., Huang, Y., Jiang, H., Ravichandran, G., Gao, H., Hwang, K., Liu, B., A cohesive law for carbon nanotube/polymer interfaces based on the van der Waals force. *J. Mech. Phys. Solids*, 54, 2436–2452, 2006.
20. Shen, B., Zhai, W., Chen, C., Lu, D., Wang, J., Zheng, W., Melt blending *in situ* enhances the interaction between polystyrene and graphene through π–π stacking. *ACS Appl. Mater. Interfaces*, 3, 3103–3109, 2011.
21. Pei, S. and Cheng, H.-M., The reduction of graphene oxide. *Carbon*, 50, 3210–3228, 2012.
22. Cheng, Q., Wu, M., Li, M., Jiang, L., Tang, Z., Ultratough artificial nacre based on conjugated cross-linked graphene oxide. *Angew. Chem. Int. Ed.*, 52, 3750–3755, 2013.
23. Kulkarni, D.D., Choi, I., Singamaneni, S.S., Tsukruk, V.V., Graphene oxide–polyelectrolyte nanomembranes. *ACS Nano*, 4, 4667–4676, 2010.

24. Compton, O.C. and Nguyen, S.T., Graphene oxide, highly reduced graphene oxide, and graphene: Versatile building blocks for carbon-based materials. *Small*, 6, 711–723, 2010.
25. Liu, J., Fu, S., Yuan, B., Li, Y., Deng, Z., Toward a universal "adhesive nanosheet" for the assembly of multiple nanoparticles based on a protein-induced reduction/decoration of graphene oxide. *J. Am. Chem. Soc.*, 132, 7279–7281, 2010.
26. Zhao, X., Zhang, Q., Chen, D., Lu, P., Enhanced mechanical properties of graphene-based poly (vinyl alcohol) composites. *Macromolecules*, 43, 2357–2363, 2010.
27. Li, S.-S., Tu, K.-H., Lin, C.-C., Chen, C.-W., Chhowalla, M., Solution-processable graphene oxide as an efficient hole transport layer in polymer solar cells. *ACS Nano*, 4, 3169–3174, 2010.
28. Salavagione, H.J., Gomez, M.A., Martínez, G., Polymeric modification of graphene through esterification of graphite oxide and poly (vinyl alcohol). *Macromolecules*, 42, 6331–6334, 2009.
29. Liu, Z., Robinson, J.T., Sun, X., Dai, H., PEGylated nanographene oxide for delivery of water-insoluble cancer drugs. *J. Am. Chem. Soc.*, 130, 10876–10877, 2008.
30. Björk, J., Hanke, F., Palma, C.-A., Samori, P., Cecchini, M., Persson, M., Adsorption of aromatic and anti-aromatic systems on graphene through π-π stacking. *J. Phys. Chem. Lett.*, 1, 3407–3412, 2010.
31. Stankovich, S., Piner, R.D., Chen, X., Wu, N., Nguyen, S.T., Ruoff, R.S., Stable aqueous dispersions of graphitic nanoplatelets via the reduction of exfoliated graphite oxide in the presence of poly (sodium 4-styrenesulfonate). *J. Mater. Chem.*, 16, 155–158, 2006.
32. Bai, S. and Shen, X., Graphene–inorganic nanocomposites. *RSC Adv.*, 2, 64–98, 2012.
33. Singh, V., Joung, D., Zhai, L., Das, S., Khondaker, S.I., Seal, S., Graphene based materials: Past, present and future. *Prog. Mater. Sci.*, 56, 1178–1271, 2011.
34. Zhou, X., Huang, X., Qi, X., Wu, S., Xue, C., Boey, F.Y., Yan, Q., Chen, P., Zhang, H., In situ synthesis of metal nanoparticles on single-layer graphene oxide and reduced graphene oxide surfaces. *J. Phys. Chem. C*, 113, 10842–10846, 2009.
35. Hassan, H.M., Abdelsayed, V., Abd El Rahman, S.K., AbouZeid, K.M., Terner, J., El-Shall, M.S., Al-Resayes, S.I., El-Azhary, A.A., Microwave synthesis of graphene sheets supporting metal nanocrystals in aqueous and organic media. *J. Mater. Chem.*, 19, 3832–3837, 2009.
36. Marquardt, D., Vollmer, C., Thomann, R., Steurer, P., Mülhaupt, R., Redel, E., Janiak, C., The use of microwave irradiation for the easy synthesis of graphene-supported transition metal nanoparticles in ionic liquids. *Carbon*, 49, 1326–1332, 2011.
37. Nethravathi, C., Nisha, T., Ravishankar, N., Shivakumara, C., Rajamathi, M., Graphene–nanocrystalline metal sulphide composites produced by a one-pot reaction starting from graphite oxide. *Carbon*, 47, 2054–2059, 2009.
38. Lin, Y., Zhang, K., Chen, W., Liu, Y., Geng, Z., Zeng, J., Pan, N., Yan, L., Wang, X., Hou, J., Dramatically enhanced photoresponse of reduced graphene oxide with linker-free anchored CdSe nanoparticles. *ACS Nano*, 4, 3033–3038, 2010.
39. Yang, X., Zhang, X., Ma, Y., Huang, Y., Wang, Y., Chen, Y., Superparamagnetic graphene oxide–Fe$_3$O$_4$ nanoparticles hybrid for controlled targeted drug carriers. *J. Mater. Chem.*, 19, 2710–2714, 2009.
40. Williams, G. and Kamat, P.V., Graphene–semiconductor nanocomposites: Excited-state interactions between ZnO nanoparticles and graphene oxide. *Langmuir*, 25, 13869–13873, 2009.
41. Zhang, Y., Tang, Z., Fu, X., Xu, Y., TiO$_2$-graphene nanocomposites for gas-phase photocatalytic degradation of volatile aromatic pollutant: Is TiO$_2$-graphene truly different from other TiO$_2$-carbon composite materials? *ACS Nano*, 4, 7303–7314, 2010.
42. Xu, C., Wang, X., Zhu, J., Graphene–metal particle nanocomposites. *J. Phys. Chem. C*, 112, 19841–19845, 2008.
43. Subrahmanyam, K., Manna, A.K., Pati, S.K., Rao, C., A study of graphene decorated with metal nanoparticles. *Chem. Phys. Lett.*, 497, 70–75, 2010.

44. Bian, J., Wei, X.W., Wang, L., Guan, Z.P., Graphene nanosheet as support of catalytically active metal particles in DMC synthesis. *Chin. Chem. Lett.*, 22, 57–60, 2011.
45. Bai, S., Shen, X., Zhu, G., Xu, Z., Liu, Y., Reversible phase transfer of graphene oxide and its use in the synthesis of graphene-based hybrid materials. *Carbon*, 49, 4563–4570, 2011.
46. Das, B., Choudhury, B., Gomathi, A., Manna, A.K., Pati, S., Rao, C., Interaction of inorganic nanoparticles with graphene. *Chem. Phys. Chem.*, 12, 937–943, 2011.
47. Cao, Y., Li, X., Aksay, I.A., Lemmon, J., Nie, Z., Yang, Z., Liu, J., Sandwich-type functionalized graphene sheet-sulfur nanocomposite for rechargeable lithium batteries. *Phys. Chem. Chem. Phys.*, 13, 7660–7665, 2011.
48. Xiang, Q., Yu, J., Jaroniec, M., Preparation and enhanced visible-light photocatalytic H2-production activity of graphene/C3N4 composites. *J. Phys. Chem. C*, 115, 7355–7363, 2011.
49. Papageorgiou, D.G., Kinloch, I.A., Young, R.J., Mechanical properties of graphene and graphene-based nanocomposites. *Prog. Mater. Sci.*, 90, 75–127, 2017.
50. Geim, A.K. and Grigorieva, I.V., Van der Waals heterostructures. *Nature*, 499, 419–425, 2013.
51. Chen, X., Wu, B., Liu, Y., Direct preparation of high quality graphene on dielectric substrates. *Chem. Soc. Rev.*, 45, 2057–2074, 2016.
52. Wang, H., Liu, F., Fu, W., Fang, Z., Zhou, W., Liu, Z., Two-dimensional heterostructures: Fabrication, characterization, and application. *Nanoscale*, 6, 12250–12272, 2014.
53. Niu, T. and Li, A., From two-dimensional materials to heterostructures. *Prog. Surf. Sci.*, 90, 21–45, 2015.
54. Zeng, Q., Wang, H., Fu, W., Gong, Y., Zhou, W., Ajayan, P.M., Lou, J., Liu, Z., Band engineering for novel two-dimensional atomic layers. *Small*, 11, 1868–1884, 2015.
55. Tan, C. and Zhang, H., Two-dimensional transition metal dichalcogenide nanosheet-based composites. *Chem. Soc. Rev.*, 44, 2713–2731, 2015.
56. Koppens, F., Mueller, T., Avouris, P., Ferrari, A., Vitiello, M., Polini, M., Photodetectors based on graphene, other two-dimensional materials and hybrid systems. *Nat. Nanotechnol.*, 9, 780–793, 2014.
57. Xie, C., Mak, C., Tao, X., Yan, F., Photodetectors based on two-dimensional layered materials beyond graphene. *Adv. Funct. Mater.*, 27, 1–41, 2017.
58. Ci, L., Song, L., Jin, C., Jariwala, D., Wu, D., Li, Y., Srivastava, A., Wang, Z., Storr, K., Balicas, L., Atomic layers of hybridized boron nitride and graphene domains. *Nat. Mater.*, 9, 430–435, 2010.
59. Chang, C.-K., Kataria, S., Kuo, C.-C., Ganguly, A., Wang, B.-Y., Hwang, J.-Y., Huang, K.-J., Yang, W.-H., Wang, S.-B., Chuang, C.-H., Band gap engineering of chemical vapor deposited graphene by *in situ* BN doping. *ACS Nano*, 7, 1333–1341, 2013.
60. Rao, C. and Gopalakrishnan, K., Borocarbonitrides, B x C y N z: Synthesis, characterization, and properties with potential applications. *ACS Appl. Mater. Interfaces*, 9, 19478–19494, 2016.
61. Jiang, X. and Drzal, L.T., Multifunctional high density polyethylene nanocomposites produced by incorporation of exfoliated graphite nanoplatelets 1: Morphology and mechanical properties. *Polym. Compos.*, 31, 1091–1098, 2010.
62. Compton, O.C., Kim, S., Pierre, C., Torkelson, J.M., Nguyen, S.T., Crumpled graphene nanosheets as highly effective barrier property enhancers. *Adv. Mater.*, 22, 4759–4763, 2010.
63. Liu, N., Luo, F., Wu, H., Liu, Y., Zhang, C., Chen, J., One-step ionic-liquid-assisted electrochemical synthesis of ionic-liquid-functionalized graphene sheets directly from graphite. *Adv. Funct. Mater.*, 18, 1518–1525, 2008.
64. Wang, S., Tambraparni, M., Qiu, J., Tipton, J., Dean, D., Thermal expansion of graphene composites. *Macromolecules*, 42, 5251–5255, 2009.

65. Lape, N.K., Nuxoll, E.E., Cussler, E., Polydisperse flakes in barrier films. *J. Membr. Sci.*, 236, 29–37, 2004.
66. Wang, H., Hao, Q., Yang, X., Lu, L., Wang, X., A nanostructured graphene/polyaniline hybrid material for supercapacitors. *Nanoscale*, 2, 2164–2170, 2010.
67. Wang, D.-W., Li, F., Zhao, J., Ren, W., Chen, Z.-G., Tan, J., Wu, Z.-S., Gentle, I., Lu, G.Q., Cheng, H.-M., Fabrication of graphene/polyaniline composite paper via in situ anodic electropolymerization for high-performance flexible electrode. *ACS Nano*, 3, 1745–1752, 2009.
68. Somani, P.R., Somani, S.P., Umeno, M., Planer nano-graphenes from camphor by CVD. *Chem. Phys. Lett.*, 430, 56–59, 2006.
69. Kim, K.S., Zhao, Y., Jang, H., Lee, S.Y., Kim, J.M., Kim, K.S., Ahn, J.-H., Kim, P., Choi, J.-Y., Hong, B.H., Large-scale pattern growth of graphene films for stretchable transparent electrodes. *Nature*, 457, 706–710, 2009.
70. Li, X., Cai, W., Colombo, L., Ruoff, R.S., Evolution of graphene growth on Ni and Cu by carbon isotope labeling. *Nano Lett.*, 9, 4268–4272, 2009.
71. Yamada, T., Ishihara, M., Hasegawa, M., Large area coating of graphene at low temperature using a roll-to-roll microwave plasma chemical vapor deposition. *Thin Solid Films*, 532, 89–93, 2013.
72. Jeon, Y.-J., Yun, J.-M., Kim, D.-Y., Na, S.-I., Kim, S.-S., High-performance polymer solar cells with moderately reduced graphene oxide as an efficient hole transporting layer. *Sol. Energy Mater. Sol. Cells*, 105, 96–102, 2012.
73. Hwan Lee, S., M.-a. Park, J.J., Song, H., Yun Jang, E., Hyup Kim, Y., Kang, S., Seop Yoon, Y., Reduced graphene oxide coated thin aluminum film as an optoacoustic transmitter for high pressure and high frequency ultrasound generation. *Appl. Phys. Lett.*, 101, 241909, 2012.
74. Yoo, D.-H., Cuong, T.V., Pham, V.H., Chung, J.S., Khoa, N.T., Kim, E.J., Hahn, S.H., Enhanced photocatalytic activity of graphene oxide decorated on TiO_2 films under UV and visible irradiation. *Curr. Appl. Phys.*, 11, 805–808, 2011.
75. Yu, L., Lim, Y.-S., Han, J.H., Kim, K., Kim, J.Y., Choi, S.-Y., Shin, K., A graphene oxide oxygen barrier film deposited via a self-assembly coating method. *Synth. Met.*, 162, 710–714, 2012.
76. Nayak, P.K., Hsu, C.-J., Wang, S.-C., Sung, J.C., Huang, J.-L., Graphene coated Ni films: A protective coating. *Thin Solid Films*, 529, 312–316, 2013.
77. Raman, R.S., Banerjee, P.C., Lobo, D.E., Gullapalli, H., Sumandasa, M., Kumar, A., Choudhary, L., Tkacz, R., Ajayan, P.M., Majumder, M., Protecting copper from electrochemical degradation by graphene coating. *Carbon*, 50, 4040–4045, 2012.
78. Novoselov, K.S. and Geim, A., The rise of graphene. *Nat. Mater.*, 6, 183–191, 2007.
79. Nguyen, D.D., Tai, N.-H., Lee, S.-B., Kuo, W.-S., Superhydrophobic and superoleophilic properties of graphene-based sponges fabricated using a facile dip coating method. *Energ. Environ. Sci.*, 5, 7908–7912, 2012.
80. Zhang, H. and Lee, H.K., Simultaneous determination of ultraviolet filters in aqueous samples by plunger-in-needle solid-phase microextraction with graphene-based sol–gel coating as sorbent coupled with gas chromatography–mass spectrometry. *Anal. Chim. Acta*, 742, 67–73, 2012.
81. Hao, Q., Wang, B., Bossard, J.A., Kiraly, B., Zeng, Y., Chiang, I.-K., Jensen, L., Werner, D.H., Huang, T.J., Surface-enhanced Raman scattering study on graphene-coated metallic nanostructure substrates. *J. Phys. Chem. C*, 116, 7249–7254, 2012.
82. Wang, H., Yang, Y., Liang, Y., Robinson, J.T., Li, Y., Jackson, A., Cui, Y., Dai, H., Graphene-wrapped sulfur particles as a rechargeable lithium–sulfur battery cathode material with high capacity and cycling stability. *Nano Lett.*, 11, 2644–2647, 2011.
83. Liao, K.-H., Qian, Y., Macosko, C.W., Ultralow percolation graphene/polyurethane acrylate nanocomposites. *Polymer*, 53, 3756–3761, 2012.

84. Tong, Y., Bohm, S., Song, M., Graphene based materials and their composites as coatings. *Austin J. Nanomed. Nanotechnol.*, 1, 1003, 2013.
85. Wu, Q., Feng, C., Zhao, G., Wang, C., Wang, Z., Graphene-coated fiber for solid-phase microextraction of triazine herbicides in water samples. *J. Sep. Sci.*, 35, 193–199, 2012.
86. Yin, Y. and Alivisatos, A.P., Colloidal nanocrystal synthesis and the organic–inorganic interface. *Nature*, 437, 664, 2004.
87. Chen, J., Zou, J., Zeng, J., Song, X., Ji, J., Wang, Y., Ha, J., Chen, X., Preparation and evaluation of graphene-coated solid-phase microextraction fiber. *Anal. Chim. Acta*, 678, 44–49, 2010.
88. Zhang, H. and Lee, H.K., Plunger-in-needle solid-phase microextraction with graphene-based sol–gel coating as sorbent for determination of polybrominated diphenyl ethers. *J. Chromatogr. A*, 1218, 4509–4516, 2011.
89. Liu, Y., Dong, X., Chen, P., Biological and chemical sensors based on graphene materials. *Chem. Soc. Rev.*, 41, 2283–2307, 2012.
90. Liu, G., Riechers, S.L., Mellen, M.C., Lin, Y., Sensitive electrochemical detection of enzymatically generated thiocholine at carbon nanotube modified glassy carbon electrode. *Electrochem. Commun.*, 7, 1163–1169, 2005.
91. Liu, G. and Lin, Y., Electrochemical stripping analysis of organophosphate pesticides and nerve agents. *Electrochem. Commun.*, 7, 339–343, 2005.
92. Yang, Y., Asiri, A.M., Tang, Z., Du, D., Lin, Y., Graphene based materials for biomedical applications. *Mater. Today*, 16, 365–373, 2013.
93. Zhou, M., Zhai, Y., Dong, S., Electrochemical sensing and biosensing platform based on chemically reduced graphene oxide. *Anal. Chem.*, 81, 5603–5613, 2009.
94. Shao, Y., Zhang, S., Engelhard, M.H., Li, G., Shao, G., Wang, Y., Liu, J., Aksay, I.A., Lin, Y., Nitrogen-doped graphene and its electrochemical applications. *J. Mater. Chem.*, 20, 7491–7496, 2010.
95. Wu, P., Qian, Y., Du, P., Zhang, H., Cai, C., Facile synthesis of nitrogen-doped graphene for measuring the releasing process of hydrogen peroxide from living cells. *J. Mater. Chem.*, 22, 6402–6412, 2012.
96. Loh, K.P., Bao, Q., Eda, G., Chhowalla, M., Graphene oxide as a chemically tunable platform for optical applications. *Nat. Chem.*, 2, 1015–1024, 2010.
97. Liu, Z., Liu, Q., Huang, Y., Ma, Y., Yin, S., Zhang, X., Sun, W., Chen, Y., Organic photovoltaic devices based on a novel acceptor material: Graphene. *Adv. Mater.*, 20, 3924–3930, 2008.
98. Yang, K., Zhang, S., Zhang, G., Sun, X., Lee, S.-T., Liu, Z., Graphene in mice: Ultrahigh *in vivo* tumor uptake and efficient photothermal therapy. *Nano Lett.*, 10, 3318–3323, 2010.
99. Chen, W., Yi, P., Zhang, Y., Zhang, L., Deng, Z., Zhang, Z., Composites of aminodextran-coated Fe3O4 nanoparticles and graphene oxide for cellular magnetic resonance imaging. *ACS Appl. Mater. Interfaces*, 3, 4085–4091, 2011.
100. Zhang, W., Guo, Z., Huang, D., Liu, Z., Guo, X., Zhong, H., Synergistic effect of chemo-photothermal therapy using PEGylated graphene oxide. *Biomaterials*, 32, 8555–8561, 2011.
101. Potts, J.R., Dreyer, D.R., Bielawski, C.W., Ruoff, R.S., Graphene-based polymer nanocomposites. *Polymer*, 52, 5–25, 2011.
102. Ramanathan, T., Abdala, A., Stankovich, S., Dikin, D., Herrera-Alonso, M., Piner, R., Adamson, D., Schniepp, H., Chen, X., Ruoff, R., Functionalized graphene sheets for polymer nanocomposites. *Nat. Nanotechnol.*, 3, 327–331, 2008.
103. Xu, Y., Wang, Y., Liang, J., Huang, Y., Ma, Y., Wan, X., Chen, Y., A hybrid material of graphene and poly (3, 4-ethyldioxythiophene) with high conductivity, flexibility, and transparency. *Nano Res.*, 2, 343–348, 2009.
104. Liang, J., Huang, Y., Zhang, L., Wang, Y., Ma, Y., Guo, T., Chen, Y., Molecular-level dispersion of graphene into poly (vinyl alcohol) and effective reinforcement of their nanocomposites. *Adv. Funct. Mater.*, 19, 2297–2302, 2009.

105. Zhang, H.-B., Zheng, W.-G., Yan, Q., Yang, Y., Wang, J.-W., Lu, Z.-H., Ji, G.-Y., Yu, Z.-Z., Electrically conductive polyethylene terephthalate/graphene nanocomposites prepared by melt compounding. *Polymer*, 51, 1191–1196, 2010.
106. Istrate, O.M., Paton, K.R., Khan, U., O'Neill, A., Bell, A.P., Coleman, J.N., Reinforcement in melt-processed polymer–graphene composites at extremely low graphene loading level. *Carbon*, 78, 243–249, 2014.
107. Vasileiou, A.A., Kontopoulou, M., Docoslis, A., A noncovalent compatibilization approach to improve the filler dispersion and properties of polyethylene/graphene composites. *ACS Appl. Mater. Interfaces*, 6, 1916–1925, 2014.
108. Maio, A., Fucarino, R., Khatibi, R., Rosselli, S., Bruno, M., Scaffaro, R., A novel approach to prevent graphene oxide re-aggregation during the melt compounding with polymers. *Compos. Sci. Technol.*, 119, 131–137, 2015.
109. Vallés, C., Abdelkader, A.M., Young, R.J., Kinloch, I.A., Few layer graphene–polypropylene nanocomposites: The role of flake diameter. *Faraday Discuss.*, 173, 379–390, 2014.
110. Li, X., McKenna, G.B., Miquelard-Garnier, G., Guinault, A., Sollogoub, C., Regnier, G., Rozanski, A., Forced assembly by multilayer coextrusion to create oriented graphene reinforced polymer nanocomposites. *Polymer*, 55, 248–257, 2014.
111. Kalaitzidou, K., Fukushima, H., Drzal, L.T., A new compounding method for exfoliated graphite–polypropylene nanocomposites with enhanced flexural properties and lower percolation threshold. *Compos. Sci. Technol.*, 67, 2045–2051, 2007.
112. Kim, S., Do, I., Drzal, L.T., Thermal stability and dynamic mechanical behavior of exfoliated graphite nanoplatelets-LLDPE nanocomposites. *Polym. Compos.*, 31, 755–761, 2010.
113. Chen, G., Wu, C., Weng, W., Wu, D., Yan, W., Preparation of polystyrene/graphite nanosheet composite. *Polymer*, 44, 1781–1784, 2003.
114. Weng, W., Chen, G., Wu, D., Transport properties of electrically conducting nylon 6/foliated graphite nanocomposites. *Polymer*, 46, 6250–6257, 2005.
115. Anwar, Z., Kausar, A., Rafique, I., Muhammad, B., Advances in epoxy/graphene nanoplatelet composite with enhanced physical properties: A review, *Polymer-Plastics Technology and Engineering*, 55, 643–662, 2015.
116. Kim, H., Miura, Y., Macosko, C.W., Graphene/polyurethane nanocomposites for improved gas barrier and electrical conductivity. *Chem. Mater.*, 22, 3441–3450, 2010.
117. Steurer, P., Wissert, R., Thomann, R., Mülhaupt, R., Functionalized graphenes and thermoplastic nanocomposites based upon expanded graphite oxide. *Macromol. Rapid Commun.*, 30, 316–327, 2009.
118. Jeong, H.-K., Lee, Y.P., Jin, M.H., Kim, E.S., Bae, J.J., Lee, Y.H., Thermal stability of graphite oxide. *Chem. Phys. Lett.*, 470, 255–258, 2009.
119. Cao, Y., Feng, J., Wu, P., Preparation of organically dispersible graphene nanosheet powders through a lyophilization method and their poly (lactic acid) composites. *Carbon*, 48, 3834–3839, 2010.
120. Choi, E.-Y., Han, T.H., Hong, J., Kim, J.E., Lee, S.H., Kim, H.W., Kim, S.O., Noncovalent functionalization of graphene with end-functional polymers. *J. Mater. Chem.*, 20, 1907–1912, 2010.
121. Wei, T., Luo, G., Fan, Z., Zheng, C., Yan, J., Yao, C., Li, W., Zhang, C., Preparation of graphene nanosheet/polymer composites using *in situ* reduction–extractive dispersion. *Carbon*, 47, 2296–2299, 2009.
122. Lee, H.B., Raghu, A.V., Yoon, K.S., Jeong, H.M., Preparation and characterization of poly (ethylene oxide)/graphene nanocomposites from an aqueous medium. *J. Macromol. Sci. Part B*, 49, 802–809, 2010.

123. Bryning, M., Milkie, D., Islam, M., Kikkawa, J., Yodh, A., Thermal conductivity and interfacial resistance in single-wall carbon nanotube epoxy composites. *Appl. Phys. Lett.*, 87, 161909, 2005.
124. Kim, J.K., Yang, S.Y., Lee, Y., Kim, Y., Functional nanomaterials based on block copolymer self-assembly. *Prog. Polym. Sci.*, 35, 1325–1349, 2010.
125. Tripathi, S.N., Saini, P., Gupta, D., Choudhary, V., Electrical and mechanical properties of PMMA/reduced graphene oxide nanocomposites prepared via *in situ* polymerization. *J. Mater. Sci.*, 48, 6223–6232, 2013.
126. Rafiee, M.A., Rafiee, J., Srivastava, I., Wang, Z., Song, H., Yu, Z.Z., Koratkar, N., Fracture and fatigue in graphene nanocomposites. *Small*, 6, 179–183, 2010.
127. Yan, X., Chen, J., Yang, J., Xue, Q., Miele, P., Fabrication of free-standing, electrochemically active, and biocompatible graphene oxide–polyaniline and graphene–polyaniline hybrid papers. *ACS Appl. Mater. Interfaces*, 2, 2521–2529, 2010.
128. Yan, J., Wei, T., Fan, Z., Qian, W., Zhang, M., Shen, X., Wei, F., Preparation of graphene nanosheet/carbon nanotube/polyaniline composite as electrode material for supercapacitors. *J. Power Sources*, 195, 3041–3045, 2010.
129. Wang, H., Hao, Q., Yang, X., Lu, L., Wang, X., Graphene oxide doped polyaniline for supercapacitors. *Electrochem. Commun.*, 11, 1158–1161, 2009.
130. Wang, J.-Y., Yang, S.-Y., Huang, Y.-L., Tien, H.-W., Chin, W.-K., Ma, C.-C.M., Preparation and properties of graphene oxide/polyimide composite films with low dielectric constant and ultrahigh strength via *in situ* polymerization. *J. Mater. Chem.*, 21, 13569–13575, 2011.
131. Dreyer, D.R., Jarvis, K.A., Ferreira, P.J., Bielawski, C.W., Graphite oxide as a carbocatalyst for the preparation of fullerene-reinforced polyester and polyamide nanocomposites. *Polym. Chem.*, 3, 757–766, 2012.
132. Dreyer, D.R., Jarvis, K.A., Ferreira, P.J., Bielawski, C.W., Graphite oxide as a dehydrative polymerization catalyst: A one-step synthesis of carbon-reinforced poly (phenylene methylene) composites. *Macromolecules*, 44, 7659–7667, 2011.
133. Decher, G. and Schlenoff, J.B., *Multilayer Thin Films: Sequential Assembly of Nanocomposite Materials*, 2nd Edition, Wiley-VCH Verlag, Berlin, John Wiley & Sons, 2006.
134. Ariga, K., *Organized Organic Ultrathin Films: Fundamentals and Applications*, Wiley-VCH Verlag, Berlin, John Wiley & Sons, 2012.
135. Yang, M., Hou, Y., Kotov, N.A., Graphene-based multilayers: Critical evaluation of materials assembly techniques. *Nano Today*, 7, 430–447, 2012.
136. Zhu, J., Zhang, H., Kotov, N.A., Thermodynamic and structural insights into nanocomposites engineering by comparing two materials assembly techniques for graphene. *ACS Nano*, 7, 4818–4829, 2013.
137. Choi, I., Kulkarni, D.D., Xu, W., Tsitsilianis, C., Tsukruk, V.V., Star polymer unimicelles on graphene oxide flakes. *Langmuir*, 29, 9761–9769, 2013.
138. Massoumi, B., Ghandomi, F., Abbasian, M., Eskandani, M., Jaymand, M., Surface functionalization of graphene oxide with poly (2-hydroxyethyl methacrylate)-graft-poly (ε-caprolactone) and its electrospun nanofibers with gelatin. *Appl. Phys. A*, 122, 1000, 2016.
139. Fang, M., Wang, K., Lu, H., Yang, Y., Nutt, S., Single-layer graphene nanosheets with controlled grafting of polymer chains. *J. Mater. Chem.*, 20, 1982–1992, 2010.
140. Layek, R.K., Samanta, S., Chatterjee, D.P., Nandi, A.K., Physical and mechanical properties of poly (methyl methacrylate)-functionalized graphene/poly (vinylidine fluoride) nanocomposites: Piezoelectric β polymorph formation. *Polymer*, 51, 5846–5856, 2010.
141. Zhuang, X.D., Chen, Y., Liu, G., Li, P.P., Zhu, C.X., Kang, E.T., Noeh, K.G., Zhang, B., Zhu, J.H., Li, Y.X., Conjugated-polymer-functionalized graphene oxide: Synthesis and nonvolatile rewritable memory effect. *Adv. Mater.*, 22, 1731–1735, 2010.

142. Park, S., Dikin, D.A., Nguyen, S.T., Ruoff, R.S., Graphene oxide sheets chemically cross-linked by polyallylamine. *J. Phys. Chem. C*, 113, 15801–15804, 2009.
143. Coleman, J.N., Khan, U., Gun'ko, Y.K., Mechanical reinforcement of polymers using carbon nanotubes. *Adv. Mater.*, 18, 689–706, 2006.
144. Akcora, P., Kumar, S.K., Moll, J., Lewis, S., Schadler, L.S., Li, Y., Benicewicz, B.C., Sandy, A., Narayanan, S., Ilavsky, J., "Gel-like" mechanical reinforcement in polymer nanocomposite melts. *Macromolecules*, 43, 1003–1010, 2009.
145. Yang, H., Li, F., Shan, C., Han, D., Zhang, Q., Niu, L., Ivaska, A., Covalent functionalization of chemically converted graphene sheets via silane and its reinforcement. *J. Mater. Chem.*, 19, 4632–4638, 2009.
146. Xu, Z. and Gao, C., *In situ* polymerization approach to graphene-reinforced nylon-6 composites. *Macromolecules*, 43, 6716–6723, 2010.
147. Zheng, C., Feng, M., Zhen, X., Huang, J., Zhan, H., Materials investigation of multi-walled carbon nanotubes doped silica gel glass composites. *J. Non-Cryst. Solids*, 354, 1327–1330, 2008.
148. Hongbing, Z., Wenzhe, C., Minquan, W., Chunlin, Z., Optical limiting effects of multi-walled carbon nanotubes suspension and silica xerogel composite. *Chem. Phys. Lett.*, 382, 313–317, 2003.
149. Cho, J., Inam, F., Reece, M.J., Chlup, Z., Dlouhy, I., Shaffer, M.S., Boccaccini, A.R., Carbon nanotubes: Do they toughen brittle matrices? *J. Mater. Sci.*, 46, 4770–4779, 2011.
150. Lewis, J.A., Colloidal processing of ceramics. *J. Am. Ceram. Soc.*, 83, 2341–2359, 2000.
151. Inam, F., Yan, H., Reece, M.J., Peijs, T., Dimethylformamide: An effective dispersant for making ceramic–carbon nanotube composites. *Nanotechnology*, 19, 195710, 2008.
152. Yang, Y., Wang, Y., Tian, W., Z.-q. Wang, Y., Wang, L., Bian, H.-M., Reinforcing and toughening alumina/titania ceramic composites with nano-dopants from nanostructured composite powders. *Mater. Sci. Eng. A*, 508, 161–166, 2009.
153. Dusza, J., Blugan, G., Morgiel, J., Kuebler, J., Inam, F., Peijs, T., Reece, M.J., Puchy, V., Hot pressed and spark plasma sintered zirconia/carbon nanofiber composites. *J. Eur. Ceram. Soc.*, 29, 3177–3184, 2009.
154. Guo, S., Sivakumar, R., Kitazawa, H., Kagawa, Y., Electrical Properties of Silica-Based Nanocomposites with Multiwall Carbon Nanotubes. *J. Am. Ceram. Soc.*, 90, 1667–1670, 2007.
155. Kun, P., Tapasztó, O., Wéber, F., Balázsi, C., Determination of structural and mechanical properties of multilayer graphene added silicon nitride-based composites. *Ceram. Int.*, 38, 211–216, 2012.

4

Interfacial Mechanical Properties of Graphene/Substrate System: Measurement Methods and Experimental Analysis

Chaochen Xu, Hongzhi Du, Yilan Kang* and Wei Qiu[†]

Department of Mechanics, School of Mechanical Engineering, Tianjin University, Tianjin, People's Republic of China

Abstract

Graphene and its composites have wide application prospects in the microelectronics field. However, in general, graphene must be attached to a substrate to realize its function because of its atomic thickness. Therefore, it is critical to fully understand the mechanical properties of the interface between graphene and a flexible substrate. This chapter introduces the progress made by our research group in experimental investigations of the graphene/substrate interface using Raman spectroscopy as well as other relevant literature, including the theory of Raman-spectroscopy-based mechanical measurements on graphene, experimental measurement of the interfacial mechanical parameters under tensile loading, and quantitative characterization of the mechanical behavior of the graphene/substrate interface. The effect of cyclic loading regulation and the size effect on the interfacial mechanical properties of graphene are also discussed. Finally, several phenomena and problems reported in previous papers, such as the scattering of experimental data and the initial strain in graphene, are summarized and analyzed.

Keywords: Graphene, Raman spectroscopy, interfacial behavior, interfacial shear stress, critical length, size effect, initial strain, cyclic loading regulation

4.1 Methodology of Raman Mechanical Measurements of Graphene

Graphene, an ideal two-dimensional atomic crystal [1–3], is a flexible and transparent material with the lowest resistance rate [4, 5] and highest strength [6, 7] and heat conduction [8, 9] ever reported. Thus, graphene has important application prospects for flexible electronics [10, 11], fiber-reinforced composites [12–14], and photoelectric energy storage components [15, 16].

Micro-Raman spectroscopy is one of the most effective methods to study the properties of graphene because it is nondestructive, noncontact, and rapid with high spatial resolution

Corresponding author: tju_ylkang@tju.edu.cn
[†]Corresponding author: qiuwei@tju.edu.cn

(~1 μm) and enables quantitative measurement of mechanical parameters to be performed [17–19]. Micro-Raman spectroscopy is based on the principle of stimulated Raman scattering, which is an inelastic scattering effect of photons on the molecules of tested materials [20–22]. When a molecule is exposed to exciting light of a certain frequency, the frequency of some part of the scattered light equals that of the incident light, which leads to an elastic collision between the molecule and photon with no energy exchange; this scattering is called Rayleigh scattering. The frequency of the other part of the scattered light unequal to the incident light leads to an inelastic collision called Raman scattering, as shown in Figure 4.1. Graphene has a truly two-dimensional structure, consisting of sp² carbon hexagonal networks with strong covalent bonds [23]. The lattice vibration of graphene is closely related not only to Raman scattering but also to the optical, electrical, and thermal properties and structural phase transition; therefore, Raman spectroscopy can provide fingerprint information of the structure and properties of graphene.

In the Raman spectrum of graphene, two main peaks are related to the mechanical information: the G peak at ~1580 cm^{-1} and the 2D peak at ~2650 cm^{-1}, as shown in Figure 4.2a. The G peak corresponds to the stretching vibration among the sp² carbon atoms, corresponding to the doubly degenerate in-plane vibrating mode, E_{2g}, in the center of the Brillouin zone. The 2D peak corresponds to the double resonance transition of two phonons

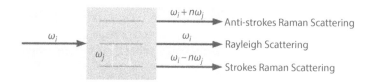

Figure 4.1 Schematic diagram of the Raman scattering.

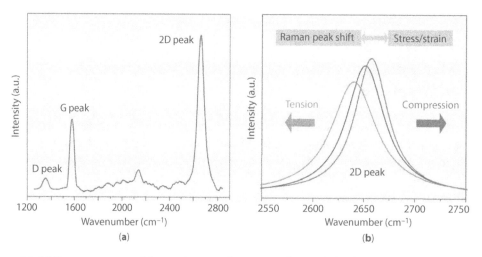

Figure 4.2 (a) Raman spectrum of the graphene on the copper substrate before being transferred onto the PET substrate. (b) Schematic diagram of the Raman 2D peak position shift of graphene under strain-free, tension and compression conditions.

with reverse momentum in carbon atoms [24]. The D peak at 1350 cm^{-1} will appear in the Raman spectrum if the graphene structure has defects or is near the fringe of graphene [25].

4.1.1 Theory of Graphene Strain Measurement

Raman spectroscopy strain measurement is used to determine the material strain or stress by detecting and analyzing spectral characteristic peaks. When a material with Raman activity like graphene is subjected to strain, i.e., the atomic bond length in the crystal lattice changes, the original atomic vibration frequency changes, thereby causing the frequency of the scattered photon to change. As a result, the position of the characteristic peak shifts. Therefore, the graphene strain can be determined from the change in the position of the Raman characteristic peak of the material [26–29]. In the presence of strain, the dynamical equations that describe phonon modes in the solid have the following form [30]:

$$m\ddot{u}_i = -\sum_k K_{ik} u_k = -\left(m\omega_0^2 u_i + \sum_{klm} K^{(1)}_{iklm} \varepsilon_{lm} u_k \right), \tag{4.1}$$

where u_i is the ith component of the relative displacement of two atoms in one unit cell, m is the mass of two atoms, ω_0 is the vibration frequency under free strain, and the second term describes the change in the phonon frequency under the applied strain. Because of the hexagonal lattice symmetry of the graphene sheet, the symmetric tensor $K^{(1)}$ only has three non-zero components:

$$\begin{aligned} K_{1111} &= K_{2222} = m\tilde{K}_{11} \\ K_{1122} &= m\tilde{K}_{12} \\ K_{1212} &= \frac{m}{2}\left(\tilde{K}_{11} - \tilde{K}_{12}\right). \end{aligned} \tag{4.2}$$

According to the conditions given by Equation 4.2, the secular equation of Equation 4.1 is

$$\begin{vmatrix} \tilde{K}_{11}\varepsilon_{xx} + \tilde{K}_{12}\varepsilon_{yy} - \lambda & \left(\tilde{K}_{11} - \tilde{K}_{12}\right)\varepsilon_{xy} \\ \left(\tilde{K}_{11} - \tilde{K}_{12}\right)\varepsilon_{xy} & \tilde{K}_{11}\varepsilon_{yy} + \tilde{K}_{12}\varepsilon_{xx} - \lambda \end{vmatrix} = 0, \tag{4.3}$$

where ε_{xx} and ε_{yy} are the strain in the x and y directions, respectively. In general, the change of the peak position shift due to strain, $\Delta\omega$, is small compared with ω_0; therefore, the following approximate relationship holds:

$$\lambda = \omega^2 - \omega_0^2 \approx 2\omega_0 \Delta\omega, \tag{4.4}$$

where ω represents the phonon vibration frequency under strain. Solving the characteristic Equation 4.3 results in

$$\Delta\omega = -\omega_0 \gamma (\varepsilon_{xx} + \varepsilon_{yy}) \pm \frac{1}{2}\omega_0 \beta (\varepsilon_{xx} - \varepsilon_{yy}). \quad (4.5)$$

The Grüneisen parameter γ and shear deformation potential β are defined as follows:

$$\gamma = -\frac{\tilde{K}_{11} + \tilde{K}_{12}}{4\omega_0^2}, \quad \beta = \frac{\tilde{K}_{11} - \tilde{K}_{12}}{2\omega_0^2}. \quad (4.6)$$

The G peak of graphene is related to the tensile vibration between sp² carbon atoms, which corresponds to the vibration of the E_{2g} optical phonon in the center of the Brillouin zone. Therefore, for the G peak of graphene, the solution of the secular equation of the E_{2g} vibration mode can be written as follows [31]:

$$\begin{aligned}\Delta\omega_G^\pm &= \Delta\omega_G^h \pm \frac{1}{2}\Delta\omega_G^s \\ &= -\omega_G^0 \gamma_G (\varepsilon_{xx} + \varepsilon_{yy}) \pm \frac{1}{2}\omega_G^0 \beta_G (\varepsilon_{xx} - \varepsilon_{yy}),\end{aligned} \quad (4.7)$$

where $\Delta\omega_G^+$ and $\Delta\omega_G^-$ are the peak position shifts of the two sub-peaks generated when the G peak of graphene is bi-directionally strained, $\Delta\omega_G^h$ is the peak position shift caused by the hydrostatic pressure component of the strain, $\Delta\omega_G^s$ is the mode splitting caused by the shearing component of the strain, ω_G^0 is the initial peak position for the G peak, γ_G is the Grüneisen parameter, and β_G is the shear deformation potential. Using Equation 4.7, we can obtain

$$\varepsilon_{xx} = \frac{\Delta\omega_G^+}{2\omega_G^0}\left(\frac{1}{\beta_G} - \frac{1}{2\gamma_G}\right) - \frac{\Delta\omega_G^-}{2\omega_G^0}\left(\frac{1}{\beta_G} + \frac{1}{2\gamma_G}\right) \quad (4.8)$$

$$\varepsilon_{yy} = -\frac{\Delta\omega_G^+}{2\omega_G^0}\left(\frac{1}{\beta_G} + \frac{1}{2\gamma_G}\right) + \frac{\Delta\omega_G^-}{2\omega_G^0}\left(\frac{1}{\beta_G} - \frac{1}{2\gamma_G}\right). \quad (4.9)$$

Considering the generalized Hooke's law of planar biaxial stress, the stress–strain relationship satisfies

$$\begin{bmatrix} \sigma_{xx} \\ \sigma_{yy} \end{bmatrix} = \begin{bmatrix} Q_{11} & Q_{12} \\ Q_{12} & Q_{22} \end{bmatrix} \begin{bmatrix} \varepsilon_{xx} \\ \varepsilon_{yy} \end{bmatrix}$$

$$Q_{11} = Q_{22} = \frac{E}{1-v^2}, \quad Q_{12} = Q_{21} = \frac{vE}{1-v^2}. \quad (4.10)$$

The simultaneous biaxial stress of graphene can be determined using Equations 4.10–4.12, where $E = 1$ TPa is the Young's modulus of graphene and v is the Poisson's ratio of graphene.

When the strain of graphene is unknown, the initial peak position ω_G^0 for the G peak and peak position shifts of the two sub-peaks G⁺ and G⁻ can be experimentally measured using the Grüneisen parameter γ_G and shear deformation potential β_G given in the literature, and the strain of graphene can be calculated using Equations 4.8 and 4.9 [31]. When the strain is known, for example, when a uniaxial tensile test of graphene is performed, and the deformation is transmitted from the substrate to the graphene, the lateral strain of the graphene is caused by the Poisson ratio of the substrate; therefore, $\varepsilon_{yy} = -\nu \varepsilon_{xx}$. The initial peak position ω_G^0 for the G peak and peak position shifts of the two sub-peaks G⁺ and G⁻ with applied known tensile strain ε_{xx} in uniaxial tension can be experimentally measured. Then, γ_G and β_G of graphene can be calculated using Equation 4.7:

$$\gamma_G = \frac{\Delta\omega_G^+ + \Delta\omega_G^-}{2\omega_G^0(1-\nu)\varepsilon_{xx}} \quad (4.11)$$

$$\beta_G = \frac{\Delta\omega_G^+ - \Delta\omega_G^-}{2\omega_G^0(1+\nu)\varepsilon_{xx}}. \quad (4.12)$$

The 2D peak of graphene is associated with two phonon double resonance transitions with opposite momentum in the carbon atom, corresponding to the single degenerate mode. For pure A_{1g} symmetry and small strain, the uniaxial peak position shift is given by the hydrostatic pressure component of the stress:

$$\Delta\omega_{2D} = -\omega_{2D}^0 \gamma_{2D}(\varepsilon_{xx} + \varepsilon_{yy}), \quad (4.13)$$

where ω_{2D}^0 and γ_{2D} are the initial peak position of the 2D peak and Grüneisen parameter, respectively.

For the graphene film under uniaxial strain, $\varepsilon_{yy} = -\nu\varepsilon_{xx}$,

$$\Delta\omega_{2D} = -\omega_{2D}^0 \gamma_{2D}(1-\nu)\varepsilon_{xx}. \quad (4.14)$$

Then, the relationship between the graphene stress and 2D peak position shift of the Raman spectrum can be determined using

$$\sigma_{xx} = \frac{E\Delta\omega_{2D}}{-\omega_{2D}^0 \gamma_{2D}(1-\nu)}. \quad (4.15)$$

In our experiments, the width of the specimens was large; therefore, the effect of Poisson's ratio could be neglected, i.e., $\varepsilon_{xx} = \varepsilon$, $\varepsilon_{yy} = 0$, and Equation 4.15 can be simplified as

$$\Delta\omega_{2D} = -\omega_{2D}^0 \gamma_{2D}\varepsilon, \quad (4.16)$$

where $-\omega_{2D}^0 \gamma_{2D}$ is typically called the Raman 2D shift to strain coefficient (RSS_{2D}). In a uniaxial tensile test of graphene, the peak position shift $\Delta\omega_{2D}$ of the graphene 2D peak position with applied tensile strain can be measured *in situ*, and RSS_{2D} of graphene can be calibrated using Equation 4.16.

4.1.2 Characterization of Graphene Strain Using *In Situ* Raman Spectroscopy

Based on the theory above, the strain of graphene can be effectively measured by monitoring the shift of the G and 2D peak positions in the Raman spectra. The peak positions associated with the strain were determined using the Lorentzian function to fit the Raman spectrum, as shown in Figure 4.2b. When graphene is subjected to tensile deformation, the C–C bonds are stretched in the tensile direction, causing the peak position to shift to the left and the wavenumber to decrease linearly with increasing applied strain, termed a "red shift", as shown for the red curve in Figure 4.2b. Based on the linear relationship between the Raman peak position shift and applied strain, the shift rate of the peak position can be determined from the corresponding slope, i.e., the Raman shift to strain coefficient (RSS). In particular, the 2D peak position shift is highly sensitive to the strain of graphene because its RSS_{2D} can reach up to -64 cm^{-1}/% [31]. Under the condition of small strain, the effect of the strain on the Raman shift of graphene is reversible; when the external strain is released, the G and 2D peaks will return to their unstressed original positions. In contrast, when the graphene is subjected to compressive deformation, the C–C bonds are constricted in the compressive direction, causing the peak positions to shift to the right and the wavenumber to increase linearly with increasing applied strain, termed a "blue shift", as shown for the blue curve in Figure 4.1b.

It is noteworthy that the Raman shift to strain coefficient, RSS, can vary greatly in different graphene/substrate systems. This parameter is affected by several internal and external factors, including the type of graphene and substrate, the doping effect, and the surrounding temperature. For example, the RSS_{2D} of graphene prepared using mechanical exfoliation ranges between -17 [30] and -64 cm^{-1}/% [31], whereas that of graphene prepared using chemical vapor deposition ranges from -19.4 [32] to -36 cm^{-1}/% [33]. Therefore, before each Raman experiment, a calibration test must be performed to determine the RSS of the specific graphene/substrate system that will be used in the subsequent experiments and to determine the relationship between the graphene peak position and its strain.

The well-defined Raman spectra in all the experiments were obtained using a Renishaw InVia system with a 633-nm He–Ne laser as the excitation source. The spot size of the laser was approximately 1 μm in diameter after being focused through a 50× objective lens (numerical aperture = 0.75). A low laser power of 0.85 mW was used to avoid a local heating effect or damage to graphene. The spectrum of graphene inside the sampled spot area could then be obtained. Using the Raman mapping scanning method, spectral collection from point to surface for large areas of graphene could be realized, and real-time contour maps of the peak position (graphene strain field) were constructed.

4.2 Experimental Investigations of Interfacial Mechanical Behaviors of Graphene

Because graphene has only atomic thickness, in general, it must be attached to a substrate to realize its function. Graphene/substrate microstructures have wide application prospects for nanocomposite materials, wearable sensing devices [34, 35], and micro-electromechanical systems, and in all these microstructures, interfacial interactions occur between all the materials. At the macroscopic scale, the interaction of the interface is relatively weak [36, 37]; however, as the structure is reduced to the nanometer scale, the interfacial interaction dominated by van der Waals forces should not be neglected and even directly determines the mechanical properties of the entire structure [38]. Graphene is very sensitive to the interfacial force because of its extremely large specific surface area [39]. During loading deformation, the micro-nanoscale mechanical behaviors of the graphene/substrate interface, such as adhesion, sliding, and debonding, can control the performance and service life of microelectronic devices [40–42]. Therefore, the deformation of graphene and performance of the interface are the key scientific problems hindering the application of graphene in microelectronic devices; research on the interfacial mechanical behavior of graphene is urgently needed to provide guidance for these applications.

4.2.1 Raman-Spectroscopy-Based Investigations of Interfacial Properties of Graphene

During the last decade, significant progress has been made in experimental investigations characterizing the interfacial mechanical properties of graphene mainly using the double cantilever beam fracture method, blister test method, direct loading method, and nanoindentation. The double cantilever beam fracture [43–46] and blister test methods [47–49] are often employed to measure the adhesion energy of the normal interface, whereas the mechanical properties of the tangential interface are mainly studied using the direct loading method [32, 33, 50–56]. The nanoindentation method has advantages for examining the friction behaviors between graphene and the substrate [57–60].

Yoon *et al.* performed double cantilever beam fracture testing to obtain a direct measurement of the adhesion energy of the normal interface between graphene and a substrate. Large-area monolayer graphene synthesized on copper was peeled off the seed copper, and the adhesion energy of graphene and copper was determined from the force–displacement curve generated from the peeling process [43]. Motivated by this work, Na *et al.* used fracture mechanics analyses to determine the adhesion energy between graphene and its seed copper foil and between graphene and epoxy. They also developed a very fast and dry selective mechanical transfer method to transfer graphene from the seed copper foil to a specific target substrate using rate effects [44]. Bunch *et al.* performed a pressurized blister test by creating a pressure difference across the graphene membrane on microcavities to directly measure the adhesion energy of graphene of different layers with a silicon oxide substrate [47, 48]. Motivated by this work, Zhang *et al.* used the blister test method to measure the adhesion energy of bilayer graphene with a silicon oxide substrate and then calculated the interfacial shear stress between two graphene layers [49].

For experimental investigation of the tangential interface between graphene and a specific substrate, the direct tensile loading of graphene on a flexible substrate was first realized by Ni [50] and Mohiuddin [31] *et al.* They performed Raman experiments to investigate the interfacial stress transfer of the graphene/substrate structure. Motivated by their work, for the first time, Young *et al.* introduced the traditional shear lag model typically used in the study of fiber-reinforced composites to analyze the graphene/flexible substrate interface. They measured several mechanical parameters that can describe interfacial behaviors, such as the interfacial shear strength, by analyzing the in-plane strain distribution of graphene [51, 52]. Jiang *et al.* developed this shear lag theory into a nonlinear shear lag theory by considering the sliding of graphene relative to the substrate during the deformation process. They calculated the interfacial shear strength from the linear slope of the sliding region of graphene during the loading process and the adhesion energy from the wrinkle morphology formed during the unloading process [53]. In addition, our research group used the direct loading method combined with *in situ* Raman spectroscopy to evaluate the mechanical properties of the tangential interface of graphene with a flexible plastic substrate. We performed a series of experiments, and our findings are presented in Sections 4.3 to 4.5 of this chapter [33, 54–56].

4.2.2 Influencing Factors of Experimental Measurements on Interfacial Properties

Currently, a prominent problem in research on the interfacial properties of nanomaterials such as graphene is the large discrepancy between the predicted data given by theory and simulation and the experimentally measured results as well as that between experimentally measured interfacial mechanical parameters from similar experimental studies, with the difference reaching one to three orders of magnitude. Theory and numerical simulation are usually based on ideal materials and an ideal interface between graphene and the substrate. In contrast, experimental results are affected by many factors, including the quality of the nanomaterial and its geometry (such as the appearance of any intrinsic ripples), the properties and surface roughness of the substrate, and any wrinkles or residual strain produced during the transfer process. Recognition and analysis of these influencing factors has posed new challenges to the quantitative characterization of nanomaterials such as graphene.

Some studies have been performed to determine these influencing factors. The geometry stability of graphene, roughness of substrates, wrinkles, and residual strain were investigated by means of theoretical modeling, numerical simulation, and experimental measurement. To study the surface undulation of graphene itself, Kusminskiy *et al.* studied the pinning of a two-dimensional membrane to a patterned substrate using elastic theory and found that both the in-plane strains and bending rigidity can lead to depinning [61]. In addition, numerical simulation results obtained by Zhang *et al.* indicated that topological defects such as declinations and dislocations can induce graphene wrinkling [62]. Xu *et al.* studied a new class of corrugations ubiquitous in exfoliated graphene using scanning tunneling microscopy [63]. Besides the corrugations of graphene itself, new wrinkles and residual strain can be introduced in the processes of transfer and the recombination with target substrate. Lanza *et al.* observed that graphene grown on a copper substrate using chemical vapor deposition has wrinkles and that the subsequent transfer process also produces new wrinkles [64]. Robinson *et al.* reported that the strain distribution of epitaxial graphene grown on SiC is inhomogeneous, which is correlated to the physical topography of the substrate [65]. The studies by Raju *et al.* and Du *et al.* showed that cyclic loading can improve the inhomogeneous strain distribution of graphene [66].

We speculate that these main influencing factors are often intertwined. In addition, the comprehensive effect of these factors causes the formation of initial strain in graphene.

Currently, there remains a lack of comprehensive and systematic research on these factors. In addition, problems with experimental measurements and effective characterization of graphene interfacial properties have also been encountered. In summary, a generally applicable testing method that can accurately characterize the mechanical properties of the graphene interface needs to be developed, and multiscale measurement and multiparameter modeling description need to be realized. In addition, the accuracy and reliability of the obtained mechanical parameters require further improvement.

4.3 Experimental Investigation of Mechanical Behavior of Graphene/Substrate Interface

In this section, we focused on a large-sized monolayer of graphene produced by the chemical vapor deposition method and investigated the mechanical properties of the interface between the large-sized graphene and the PET substrate in the tangential direction to explore the interfacial mechanical behavior of the graphene. With *in situ* Raman spectroscopy measuring the whole-field deformation of graphene subjected to a uniaxial tensile load, the process of interfacial stress/strain transfer from the PET surface to the graphene was analyzed and the evolution of the bonding state existing at the interface during loading was discussed. The mechanical parameters of the interface, such as the graphene limit strain and interfacial shear strength, were also provided.

4.3.1 Graphene/Substrate Specimen and Raman Experiments

The graphene used in the specimens was monolayer polycrystalline graphene synthesized on copper foil using chemical vapor deposition method with dimensions of 10 mm long by 3 mm wide. The monolayer graphene film was transferred to the top surface of the PET substrate using the poly(methyl methacrylate) (PMMA)-assisted wet transfer method. The composite samples haven't received any chemical modification, physical, and glue-bonding treatments. Therefore, the graphene monolayer was physically adsorbed on the substrate by van der Waals forces at the interface. The substrate was polyethylene terephthalate (PET), which is a flexible large-deformation material with good light transmission, creep resistance, and fatigue resistance. The size of the substrate was 20 mm long, 3 mm wide, and 0.1 mm thick. In the experiment, the substrate was stretched using a micro-loading device, and the stress–strain curve that was measured by a tensile testing machine is shown in Figure 4.3a. To ensure linear loading and uniform deformation throughout the substrate, the entire loading process was conducted in the elastic region (the gray area in Figure 4.3b in the strain range from 0% to 3%), and the loading step was set to a strain of 0.25%, marked as red dots, to obtain the Raman spectrum.

Figure 4.3c shows the Raman spectrum of the monolayer graphene on PET before loading. The initial position of the G and 2D Raman peak was 1580 and 2651 cm^{-1}, respectively. Since the G peak of graphene was easily overlapped by the Raman peak of PET around 1615 cm^{-1}, the 2D peak position was selected as the measuring target because its shift was highly sensitive to the strain of graphene (hereafter, the 2D peak position will be referred to as "peak position").

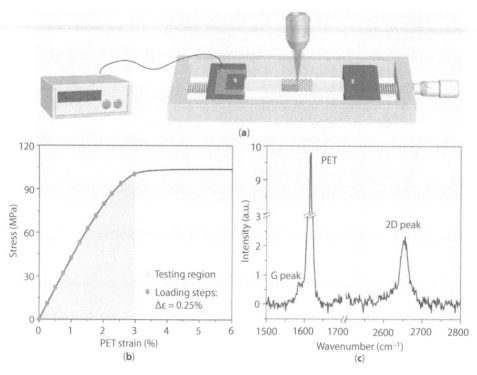

Figure 4.3 (a) Schematic diagram of the experimental setup (micro-Raman system, loading device, and graphene/PET specimen, not to scale). (b) Stress–strain curve of PET substrate by uniaxial tension test. (c) Raman spectrum of graphene on PET substrate before loading (the characteristic peak at 1615 cm^{-1} is from the PET background).

4.3.2 Interfacial Strain Transfer of the Graphene/Substrate Interface

Based on Section 4.1, before each Raman experiment, a calibration test must be performed to determine the RSS of the specific graphene/substrate system and to build the relationship between the graphene peak position and its strain. Therefore, the central region of the graphene was chosen as an observation spot first to analyze the strain of this region versus the PET strain and calibrate the RSS$_{2D}$. Figure 4.4a depicts the evolution of the 2D Raman peak of graphene at the central region with increasing PET strain. It can be clearly determined that the position of the Raman peak started to red shift linearly from the initial 2651 cm^{-1} position at a rate of −36 cm^{-1} per % until approximately 2633 cm^{-1}, which corresponded to the applied PET strain of 0.5%, and then continued to red shift to 2614 cm^{-1}. The rate of −36 cm^{-1} per % was the RSS$_{2D}$, which can be used to convert the Raman peak position shift to graphene strain hereafter. Figure 4.4b shows the peak position/strain in graphene as a function of PET strain during the loading process, and the peak position data were obtained from the statistical average of 100 measurement points at the center of graphene. When the PET was stretched up to 3%, the process of graphene strain can be divided into four stages called the initial stage, the linear stage, the nonlinear stage, and the stable stage. During the initial stage of loading, the peak position of graphene consistently fluctuated by approximately 2651 cm^{-1} and the average strain of graphene at the beam measuring points hardly changed. This phenomenon has also been reported in previous studies.

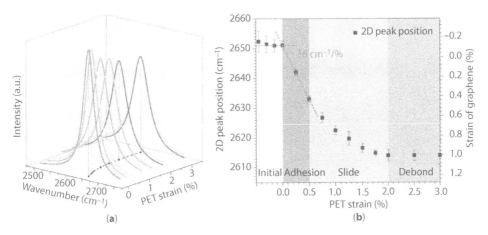

Figure 4.4 (a) Schematic diagram of the 2D Raman peaks of graphene for PET tensile strains of 0%–3% during the loading process along the tensile direction at a PET strain of 2.5%. (All of the Raman peaks are normalized in intensity.) (b) The 2D Raman band position/strain in graphene as a function of PET strain during the loading process. The shaded regions in (b) indicate the initial (white), adhesion (red), sliding (gray), and debonding (blue) stages.

This phenomenon and its mechanism will be investigated and analyzed in the Section 4.5 by a series of experiments. Since the parameters obtained from mechanical measurements were relative quantities, here we considered the corresponding PET strain where the graphene started the linear peak position shifts as the PET zero strain ($\Delta\varepsilon_s = 0\%$). The demarcation points between the latter three stages were the PET strain values of 0.5 and 2%. In the linear stage ($\Delta\varepsilon_s \leq 0.5\%$), the graphene strain was equal to the PET strain, which means that the deformation in the substrate was completely transferred to the graphene on its surface since the graphene tightly adhered to the PET by the van der Waals force. In the nonlinear stage, the graphene strain was less than the PET strain, which means that only part of the deformation in the substrate was transferred and interfacial sliding occurred between the graphene and PET. In the stable stage ($\Delta\varepsilon_s \geq 2\%$), the graphene strain did not change even as the PET strain kept increasing, which means that the deformation in the substrate was not transferred and that the graphene and PET totally debonded in the tangential direction because the van der Waals force was not strong enough to keep them together. Therefore, by comparing the relative strains of the graphene ($\Delta\varepsilon_g$) and the PET substrate ($\Delta\varepsilon_s$), the bonding states of the interface between them can be classified into the three stages of interfacial adhesion ($\Delta\varepsilon_g = \Delta\varepsilon_s$), interfacial slide in the tangential direction ($\Delta\varepsilon_g < \Delta\varepsilon_s$), and interfacial debond in the tangential direction ($\Delta\varepsilon_g = 0$) or, in short, adhesion, slide, and debond. In the adhesion stage, the Raman peak linearly shifted versus applied PET strain, and the slope can be used to calibrate the RSS_{2D} as the red dashed line shows in Figure 4.4b.

Because of the van der Waals forces at the interface, the graphene deformed as the PET substrate was stretched by the micro tensile device, and the Raman mapping method was used to scan the whole-field distribution of the graphene strain during the loading process. Considering the symmetry of the specimen, the mapping area (5000 × 1500 μm²) was a quarter of the entire graphene area, which can be seen as the shaded region in Figure 4.5a. The parameters set for Raman mapping were a horizontal step length of 50 μm, a vertical step length of 75 μm, and a scanning time of 5 s. Figure 4.5a shows the contour maps

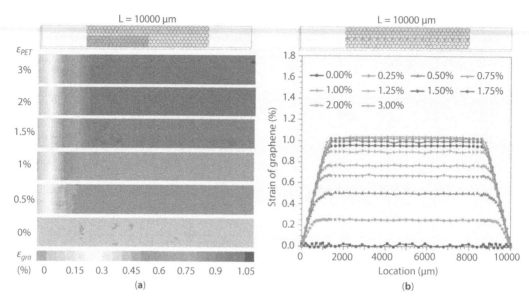

Figure 4.5 (a) Contour maps of strain over a quarter of the 10,000-μm-long graphene areas at six different levels of PET tensile strain. The list of numbers (left) shows the six different levels of PET strain from 0% to 3%, and the bar legend (bottom) plots the relationship between contour colors and graphene strain. (b) Strain distributions of graphene along the tensile direction at 10 levels of PET strain up to 3% in the loading process. Schematic diagram above shows the locations of the sampling points along the centerline of graphene.

of strain over a quarter of the graphene areas at six different levels of PET tensile strain. Numbers on the left are the six strain states of PET and the legend bar across the bottom reveals the relationship between the contour colors and the strain of graphene. Before loading, the main color of the contour map was orange, representing zero strain. Some individual regions were red, indicating that a minor local initial strain was induced in several regions of the graphene during the chemical transfer process. The strain field of graphene in the vertical direction was uniform during the loading process, which means the interfacial edge effect upon the deformation caused by the top and bottom edges of graphene was relatively small and the Poisson effect of the substrate can be ignored. However, the strain field in the horizontal direction was not uniform at each level of PET strain. After loading, the strain distributions of graphene in each PET strain consist of the fringe region and the central region, and the relative size of the two regions changed in the loading process.

To quantitatively characterize the mechanical parameters of the graphene/PET interface, Figure 4.5b provides the strain distribution along the centerline of the 10,000-μm-long graphene for PET tensile strains of 0%–3% during the loading process. The strain distribution of graphene in each PET strain can be divided into two parts: the two side fringe regions and the center strain stable region. There was a strain gradient from the relative zero strain at the edges to a steady value, and the strain value reached a maximum in the central region. First, we focus on the change of strain in the stable region at the center of graphene. During the increase of the PET substrate strain from 0% to 2%, the strain of graphene in the central region increased from 0% to around 1%. When the substrate strain was greater than 2%, the strain of graphene reached a stable maximum value and no longer continued to change with that of the substrate; at this point, the interface between graphene and substrate was considered to be critically debonding tangentially. Therefore, the bonding state of the central region underwent different stages from adhesion to debond as previously stated. Herein, during the loading process, when an interface was critically debonding, we defined the maximum strain that graphene could reach by the van der Waals force transferred from the graphene/substrate interface as the graphene limit strain (ε_{gmax}), and the corresponding strain of the substrate at this point was defined as the substrate debonding strain (ε_s). For a 10,000-μm-long graphene in Figure 4.5b, the ε_s was 2% and the ε_{gmax} was 1%. Next, the edges of the strain curve were analyzed. The fringe regions had the strain gradient from the relative zero strain at the edges to a steady value, and the length of the gradient region in the two edges of graphene gradually increased during the loading process. Until the PET strain was stretched to the substrate debonding strain, the whole graphene/substrate interface debonded, the distribution of the graphene strain did not change any more, and the corresponding length of the edges stabilized. Herein, we defined the length of edges when the interface critically debonded and the graphene strain was maximized as the critical length (L_c). The ratio of the critical length to the total length of graphene (L_c/L) is defined as the relative critical length (δ). This dimensionless parameter can be used to characterize the transfer efficiency of interfacial loads and the quality of the interface between the graphene and the substrate. In other words, a smaller relative critical length corresponded to a higher transfer efficiency of interfacial loads and a stronger interface. Figure 4.5b shows that the critical length for the 10,000-μm-long graphene was 2000 μm and the corresponding relative critical length was 20%.

4.3.3 Interfacial Shear Stress of Graphene/Substrate Interface

Here, the interfacial stress transfer between the graphene and the substrate was explored based on the force analysis of an element of graphene. Using a force balance of the shear forces at the interface and the tensile forces in a flake element, as shown in Figure 4.6a, the relationship between the interfacial shear stress (ISS) and the normal stress can be determined as:

$$\begin{cases} \sigma = E\varepsilon \\ \dfrac{d\sigma}{dx} = -\dfrac{\tau}{t} \end{cases} \tag{4.17}$$

$$\tau = -Et\dfrac{d\varepsilon}{dx}, \tag{4.18}$$

where τ is the ISS, and σ, ε, E, and t are the normal stress, normal strain, Young's modulus, and thickness of graphene, respectively. In Equation 4.18, we take $E = 1$ TPa and $t = 0.34$ nm. The force owing to the shear stress at the interface was balanced by the force owing to the variation of the normal stress in the graphene that was deduced from the force balance equation. The bonding states of the graphene/substrate interface in different regions can also be judged by the ISS, as illustrated in Figure 4.6a. In the adhesion stage, the ISS was equal to zero, indicating the common deformation of graphene and substrate and no normal strain difference. The interfacial slide stage began when ISS increased over zero and

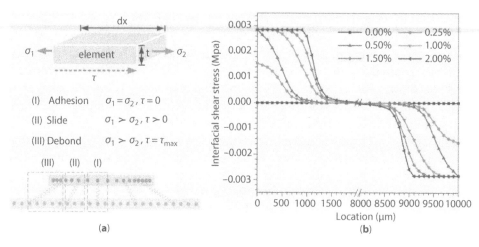

Figure 4.6 (a) Schematic of the force balance of an element of graphene illustrating the bonding state of the interface as a function of the interfacial stress. (b) Distributions of interfacial shear stress (ISS) along the tensile direction at six levels of PET strain in the loading process.

the normal strain difference occurred, indicating the interfacial behavior like static friction between the graphene and the substrate. The ISS kept increasing as the normal strain difference increased, and when the ISS reached a critical maximum value that the interface could stand, debond occurs in tangential direction at the interface, and interfacial behavior between the graphene and the substrate resembled the kinetic friction. Using Equation 4.18, the distribution of ISS along the tensile direction at six levels of PET strain was determined, as shown in Figure 4.6b. The curves indicate that the ISS increased with increasing substrate strain. Above a PET strain of 0.5%, the ISS of the edge on both sides reached the critical value of interfacial debond, and the interfacial edge began to debond. Upon further increasing the substrate strain, the debonding region at the edge moved toward the center, and the ISS in more areas reached a constant maximum value. The maximum ISS of the interface was defined as the interfacial shear strength (τ_{max}). Figure 4.6b shows that the maximum ISS of graphene was 0.0028 MPa; that is, the interfacial shear strength τ_{max} = 0.0028 MPa.

In this section, five important interfacial mechanical parameters were quantitatively determined, including the graphene limit strain (ε_{gmax}), substrate debonding strain (ε_s), critical length (L_c), relative critical length (δ), and interfacial shear strength (τ_{max}). It is noteworthy that the maximum strain that can be transferred to graphene was around 1%, which was similar to the value found in previous works. However, the interfacial shear strength was found to be around 0.003 MPa, which was two orders of magnitude smaller than that reported in the literature. Therefore, a question should be put forward, which was why the results obtained from similar experiments of graphene interface were so different.

4.4 Size Effect on Mechanical Behavior of Graphene/Substrate Interface

Based on the analysis in Section 4.3, the outstanding problem was that the experimental data obtained by different reports were scattered; the difference between the interfacial mechanical parameters of similar graphene/polymer substrate interface in different papers can reach two orders of magnitude. By comprehensive analysis on the difference between the specimens, the size of graphene was hypothesized to be the main factor affecting the experiment data. Currently, there remained a lack of comprehensive and systematic research on the size effect of the graphene/substrate interface. Therefore, our group designed eight composite specimens, containing PET substrate and eight different sizes of graphene. The specimens were studied by a series of experiments to explore how the mechanical properties of the tangential interface between graphene and the substrates can be influenced by the size of graphene. Micro-Raman spectroscopy was employed to measure the full-field strain of graphene subjected to a uniaxial tensile loading process, based on which, the evolution of the bonding states of the interface was obtained. The existence of a size effect in the interfacial strain transfer process at the graphene/PET interface was observed, and this phenomenon was characterized by a size threshold and the relative critical length. Combined with previous experimental results on the tangential interface of graphene, we discussed the size effect of the interfacial shear stress of graphene and the main cause for the inconformity of experimental data published in previous reports.

4.4.1 Serial Experiments on Graphene/Substrate Interface

Eight graphene/PET substrate specimens were designed. The lengths of graphene were 10,000, 5000, 2000, 800, 200, 100, 50, and 20 μm as shown in Figure 4.7. The width of graphene was identical, but the difference between the shortest and longest graphene was almost three orders of magnitude. A series of eight experiments was conducted on these specimens to explore the size effect of the interface. The experimental conditions and materials for these specimens were identical.

One of the eight specimens, where the length of graphene was 50 μm, was selected here to analyze the mechanical behavior of the graphene/PET interface. Owing to the van der Waals forces at the interface, graphene deformed simultaneously as the PET substrate was stretched by a micro-loading device in the loading process, and the strain of graphene in the mapping area was monitored by the *in situ* Raman system. Figure 4.8 shows the contour maps of strain over a quarter of the graphene areas and the strain distribution of graphene along the centerline of the 50-μm-long graphene at different levels of PET tensile strain. The color variance in the contour maps suggested the nonuniform distribution of strain along the tensile direction and the existence of a strain gradient in the fringe region. First, we focused on the central region, during the increase of the PET substrate strain from 0% to 2%, the strain of graphene in the central region increased from 0% to around 1%. When the substrate strain was greater than 2%, the strain of graphene reached a stable maximum value

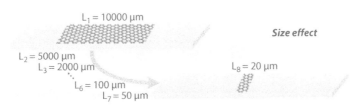

Figure 4.7 Sketch of the eight graphene/PET specimens to investigate the size effect of the graphene/substrate interface.

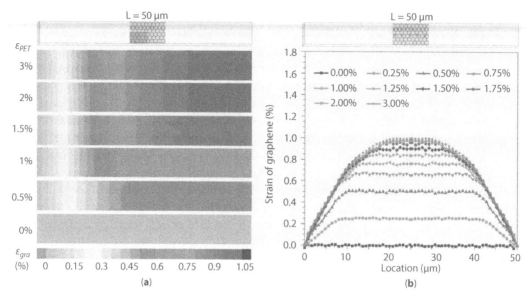

Figure 4.8 (a) Contour maps of strain over a quarter of the 50-μm-long graphene areas at six different levels of PET tensile strain. The list of numbers (left) shows the six different levels of PET strain from 0% to 3%, and the bar legend (bottom) plots the relationship between contour colors and graphene strain. (b) Strain distributions of graphene along the tensile direction at 10 levels of PET strain up to 3% in the loading process. Schematic diagram above shows the locations of the sampling points along the centerline of the 50-μm-long graphene.

and no longer continued to change with that of the substrate. This evolution of the bonding states of the 50-μm-long interface was similar to that of the 10,000-μm-long interface in Section 4.3. For a 50-μm-long graphene in Figure 4.8b, the ε_s was 2% and the ε_{gmax} was 1%. Then, the fringe region was analyzed, showing that the length of the fringe region and the slope was apparently different between the 50-μm-long and 10,000-μm-long graphene.

To quantitatively compare the difference among the strain distributions, Figure 4.9 shows the strain distributions of the graphene specimens with three different lengths (L = 10,000, 50, and 20 μm), which were the most representative for the whole range. For comparison, the

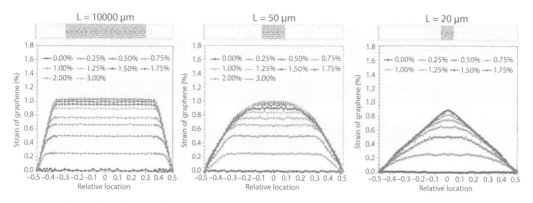

Figure 4.9 Strain distributions of the 10,000-μm-long graphene, 50-μm-long graphene, and 20-μm-long graphene along the tensile direction in the loading process. The lengths of graphene are normalized, and the fractional coordinates X are used.

lengths of graphene are normalized so that the locations along the graphene were expressed as a fractional coordinate, $X = x/L$, where L is the total length of specific graphene and $X = \pm 0.5$ represents the left and right edges of graphene. The distributions of the strain field were approximately the same for all three sizes of graphene; they all exhibited a strain gradient region at both edges of the interface, where the strain gradually rose from zero to a constant maximum value, and the strain of graphene increased under the external loading. However, the length and the ratio of the strain gradient region varied, i.e., the critical length and the relative critical length differed among graphene of different lengths. The relative critical length can be used to characterize the transfer efficiency of interfacial loads and the quality of the interface. Therefore, the transfer efficiency and the interface strength were varied. In Figure 4.9, the critical length for the 10,000-μm-long graphene was 2000 μm and the corresponding relative critical length was 20%, while L_c for the 50-μm-long graphene was 40 μm and δ was 80%; therefore, the 50-μm-long graphene/PET interface was weaker and the transfer efficiency of it was relatively low. It is worthwhile to focus on the 20-μm-long graphene. The maximum strain was only 0.91%, which failed to peak at the limit value of 1%, and the central zone disappeared. As determined by the slope of the fringe region, the critical length should be 21 μm to enable the midpoint of graphene to reach the limit strain of 1%. Therefore, if the total length of graphene was shorter than the critical length, δ > 1, the deformation transferred by the tangential interface could not enable the graphene to reach its limit strain.

4.4.2 Size Effect of Graphene/Substrate Interface

Figure 4.10a shows the strain distributions of graphene of eight different lengths after interfacial debond. The limit strain of the 20-μm-long graphene was 0.91%, obviously lower than that of the other samples, since the length was shorter than its critical length. The limit strains of the other seven graphene samples with a length longer than 20 μm were around 1%. This figure also reveals that graphene of varied lengths had different relative

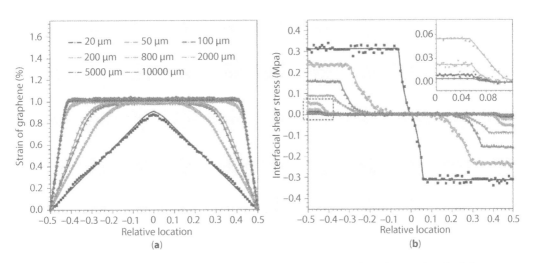

Figure 4.10 (a) Distributions of strain (a) and interfacial shear stress (b) of graphene with eight different lengths after debond of the interface. Inset shows the partial enlarged details for the dashed boxes in (b). The lengths of graphene are normalized, and the fractional coordinates X are used.

critical lengths. The shorter the graphene was, the bigger the relative critical length was, and then, the deeper the degree that the interface was affected by the edge. However, when the length was longer than 1000 μm, that is, reached the macroscopic millimeter level, the strain curves were almost overlapped, suggesting that the influence of the edges on the graphene was stable. Figure 4.10b shows the corresponding ISS distributions. The maximum ISS for the longest 10,000-μm-long graphene was 0.003 MPa, while for the shortest 20-μm-long graphene, it was 0.314 MPa. The results reveal that the maximum ISS dramatically increased with a decrease of the graphene length, and the difference between the ISS of micro-sized and macro-sized graphene could reach up to two orders of magnitude.

Table 4.1 presents the experimental results of the eight specimens, including five interfacial mechanical parameters, which are the substrate debonding strain (ε_s), graphene limit strain (ε_{gmax}), critical length (L_c), relative critical length (δ), and interfacial shear strength (τ_{max}). The parameters in the first two columns were size-independent, while the parameters in the last three columns were size-dependent. Figure 4.11 shows the variations of relative critical length and interfacial shear strength as a function of the graphene length, where the horizontal axis is the logarithmic coordinate of the graphene length. Data reveal that the two parameters increased rapidly with a slight decrease of the graphene length, especially in the micro-size region, as the variation gradient increased. Figure 4.11 also shows that these two parameters had a similar change tendency.

Based on the experimental data in Figure 4.11 and the analysis above, the fitting equation of the relative critical length and the graphene length thresholds are proposed as:

$$\delta \triangleq L_c/L = \begin{cases} \succ 1, & L \leq 20 \ \mu m \\ 1.47 - 0.38 \lg L, & 20 \ \mu m \prec L \prec 1000 \ \mu m. \\ 0.2, & L \geq 1000 \ \mu m \end{cases} \quad (4.19)$$

Table 4.1 The interfacial mechanical parameters of graphene of eight different lengths.

Graphene length L (μm)	Substrate debonding strain ε_s (%)	Graphene limit strain ε_{gmax} (%)	Critical length L_c (μm)	Relative critical length δ (%)	Interfacial shear strength τ_{max} (MPa)
20	1.5	0.9	21	105	0.314
50	2	1	40	80	0.237
100	2	1	70	70	0.158
200	2	1	116	58	0.089
800	2	1	280	35	0.055
2000	2	1.01	400	20	0.022
5000	2	1.01	1000	20	0.009
10,000	2	1.01	2000	20	0.003

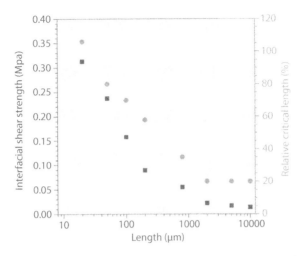

Figure 4.11 Variations of relative critical length and interfacial shear strength as a function of the graphene length.

For the graphene/PET interface, the two threshold values of 20 μm and 1000 μm were given by the equation above, and the lengths of graphene were divided into three ranges. The first range was micro-sized, where the relative critical length was larger than one, and the strain of graphene cannot be maximized to ε_{gmax} by the transfer load from the interface. The mechanical parameters, such as graphene strain and ISS, were extremely sensitive to the graphene length, and the size effect played the prominent role in this range. A size between 20 and 1000 μm was in the second range. The fitting equation shows how the relative critical length was affected by graphene length. The shorter the graphene length was, the larger the relative critical length was, and thus, the more the mechanical properties of the graphene/PET interface were influenced by the size effect. The upper threshold was 1000 μm, beyond which was the third range, where the relative critical length remained constant and the interfacial edges were no longer size-dependent.

The influence of the size effect on the interfacial mechanical properties of graphene was further discussed below. We analyzed the experimental results in this chapter combined with previous reports of similar graphene/substrate interfaces, and the integrated detail data are shown in Figure 4.12, where the horizontal axis is the logarithmic coordinate of the graphene length and the vertical axis is the maximum ISS, that is, interfacial shear strength. The red data in Figure 4.12 were extracted from papers of our group and others were extracted from papers of Young [51, 52], Jiang [53], Wang [67], and Galiotis [68, 69]. Figure 4.12 reveals that the data of maximum ISS exhibited their unanimity with the size effect. Based on the two graphene length thresholds and three ranges in Equation 4.19, 0–20 μm was concluded to be in the first range of micro-sized graphene, where the size effect was most apparent, which was observed by a surge of the interfacial strength with the decrease of graphene length. Ten lengths in the previous reports and the 20-μm-long graphene in this chapter belonged to this range. The data were not uniform and showed some scattering, which arose because of the different graphene/substrate materials and the treatment of interfaces; for example, the mechanically exfoliated (GE) graphene and a PMMA substrate were used in Young's experiment, and the interface was formed by the SU8 epoxy resin

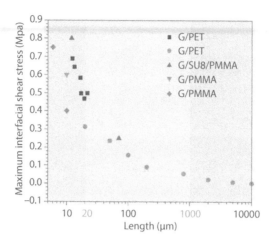

Figure 4.12 Variations of interfacial shear strength as a function of the graphene length. Red data are extracted from papers of our group and others are extracted from papers of Young [51, 52], Jiang [53], Wang [67], and Galiotis [68, 69]. The two threshold values of 20 and 1000 μm are marked by red.

rather than pure van der Waals forces. As a result, the data were relatively higher than others. The 20- to 1000-μm-long graphene was concluded to be in the second range shown in Figure 4.12, where the relative transfer length was less than 1. Data reveal that the size effect was still apparent, although its varying gradient was relatively milder than that observed in the first range. Macro-sized graphene, with a length larger than 1000 μm, was in the third range shown in Figure 4.12, where the relative critical length was a constant value. The maximum ISS of graphene with three different lengths shows that the interfacial shear strength was close to constant and the size effect in this range was too slight that it can be neglected. Therefore, the interfacial shear strength in a series of experiments on graphene with varied lengths were analyzed by Equation 4.19 and Figure 4.12, and the ranges based on two graphene length thresholds verified the existence of the size effect, which was the main cause of the scattering of experimental data on the graphene/substrate interface reported in previous reports, and the question raised in Section 4.3 was answered.

4.5 Effect of Cyclic Loading on Mechanical Behavior of Graphene/Substrate Interface

In the previous experiments on the graphene/substrate interface, we found that the initial strain always formed in graphene, which caused the appearance of the nonlinear initial stage in the subsequent loading process, as the Figure 4.4b in Section 4.3 shows. The initial strain directly produced inaccuracy in the experimental measurement. Therefore, in this section, the initial strain of graphene on a PET substrate and its effect on the interfacial properties were studied. First, the graphene/PET specimens were exposed to different types of cyclic loading treatment. Raman spectroscopy was used to experimentally measure the whole-field deformation of graphene and analyze the effect of different strain amplitudes and loading modes on the strain distribution and initial strain of graphene. Then, uniaxial tensile tests were performed to characterize the interfacial properties of the graphene/PET specimens. The strain distributions of the graphene upon uniaxial stretching were measured. In addition, the mechanical parameters of the graphene interface that received the different treatments were quantitatively characterized and compared. Finally, the effect and mechanism of cyclic loading on the graphene/PET interface was discussed. This section provides a reference for engineering application of graphene interface that improved via strain regulation.

4.5.1 Initial Strain of Graphene

The graphene and the substrates used in the specimens here were identical to those in Sections 4.3 and 4.4. Because the surface roughness of substrate affects the interfacial properties, a series of 5 μm × 5 μm surface areas on the PET substrate were measured using an FM-Nanoview 100 atomic force microscope, and then the root mean square (RMS) and roughness average (Ra) values (~10 and 6.5 nm, respectively) and maximum height difference (~49.7 nm) were calculated by mathematical statistics. In the experiment, the substrate was loaded using a micro-loading device. To ensure linear loading and uniform deformation throughout the substrate, the entire loading process was conducted in the strain range from 0% to 2.5% to obtain the Raman spectrum. The graphene length of all the specimens was 100 μm, and the experiments were conducted under the same conditions.

In this section, the initial Raman peak position of graphene on top of PET without any cyclic treatments was analyzed. Figure 4.13a shows the statistical distribution of the peak positions of 2500 points in the entire 50 μm × 50 μm graphene area. The inset presents a contour map of all the measured peak positions, which shows that the strain of the original graphene was not uniform. The statistical result of the peak positions indicates a normal distribution, where the peak positions range from 2633 to 2649 cm^{-1}. The central peak of the normal distribution curve was approximately 2641 cm^{-1}; however, the maximum difference between the peak positions was 14 cm^{-1}, which was an evidence of the existence of initial strain in the original graphene specimen. Assuming 2641 cm^{-1} is the approximate

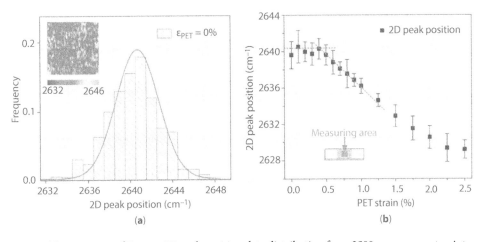

Figure 4.13 (a) Histogram of Raman 2D peak position data distribution from 2500 measurement points of graphene/PET specimen before treatment. (b) Statistical average of 2D peak position in central area as a function of PET strain during the loading process. The error bars are the standard deviation.

zero strain in the initial state of graphene, this finding indicates that the graphene without cyclic treatment has initial strain consisting of tensile strain and compressive strain. Initial strain mainly originates from the preparation and transfer processes, material factors such as the structural stability of the material itself, polycrystallinity, defects, and the substrate surface, which is inevitable. The existence of initial strain has a substantial effect on the strain transfer of the graphene/substrate microstructure. Figure 4.13b shows the peak position of graphene of the original specimen without any cyclic loading as a function of substrate strain under uniaxial tensile loading; the peak position data were obtained from the statistical average of 100 measurement points at the center of graphene. During the initial stage of loading (substrate strain of less than 0.5%), the peak position of graphene consistently fluctuated by approximately 2641 cm^{-1}. Therefore, the average strain of graphene at the measuring area hardly changed, indicating that the substrate strain could not be effectively transferred to the graphene, thus providing further evidence of the existence of initial strain. However, the initial strain directly produced experimental measurement error and the dispersion of the experimental results during the initial stage. This problem has also been reported in Section 4.3 and previous studies [70, 71].

4.5.2 Release of Initial Strain by Cyclic Loading Treatment

To study the effect of initial strain on the initial strain state of graphene, four types of cyclic loadings with either no loading or amplitudes of 0.3%, 0.7%, and 1.0% conducted in ascending-loading and direct-loading mode were designed. Then, the entire field of the strain distribution of graphene before and after cyclic loading was measured using Raman spectroscopy. Figure 4.14 shows the cyclic loading modes and statistical experimental results of the peak position contour of 1000 measurement points in a 100 μm × 40 μm area before and after cyclic loading, including the mapping contour and histogram with normal distribution. Figure 4.14a presents the results obtained before and after loading and unloading the specimen with the strain amplitude of 0.3%. Comparing the contour map and histogram of the peak position before and after cyclic loading, it is apparent that cyclic loading at this magnitude had little effect on the data. Figure 4.14b presents the results obtained before and after loading and unloading the specimen in 0%–0.3%–0.7% ascending-loading mode. Comparing the contour map and histogram of the peak position before and after cyclic loading, the strain homogeneity of graphene in the region was improved under this cyclic loading condition. Figure 4.14c presents the results obtained before and after loading and unloading the specimen in 0%–0.3%–0.7%–1% ascending-loading mode. The strain homogeneity of graphene in the region was also significantly improved with the appearance of a small blue shift. The statistical results indicate that the standard deviations of the normal distribution curve decreased and the average peak was located at approximately 2643 cm^{-1}.

The experimental data presented in Figure 4.14 and Table 4.2 demonstrate that the effect of cyclic loading is related to the strain amplitude. The effect of cyclic loading with a small amplitude is not detectable; however, the effect of cyclic loading with an amplitude greater than 0.7% is visible. These types of cyclic loading treatments can significantly improve the strain homogeneity of graphene and partly release the initial stain in the original graphene specimen.

INTERFACIAL MECHANICAL PROPERTIES OF GRAPHENE 137

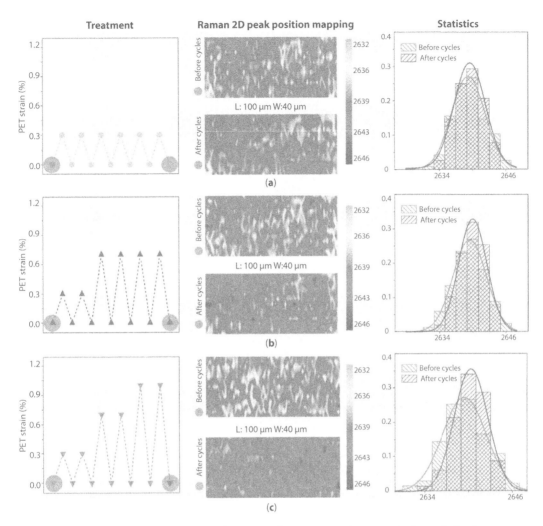

Figure 4.14 Contour maps and experimental statistics of Raman 2D peak position before and after three different cycles in ascending mode. The left column shows the three cyclic loading methods. The middle column shows the contour maps at 100 μm × 40 μm area before and after cyclic loading. The red dot represents the specimen before the cycles, and the blue point represents the specimen after the cycles. The right column shows the histograms of the Raman 2D peak position data before and after cyclic loadings with normal distribution.

Table 4.2 Statistical deviation of specimens after different cyclic loading treatments.

Amplitude of cyclic strain (%)	Before cyclic loading	Two cyclic loadings	Four cyclic loadings	Six cyclic loadings
0	1.985	—	—	—
0.3	1.932	1.872	1.865	1.868
0.7	1.896	1.851	1.789	1.761
1	1.910	1.885	1.755	1.748

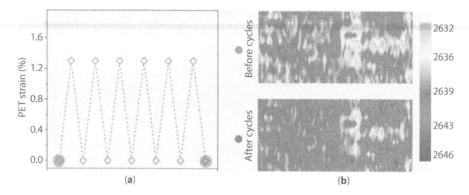

Figure 4.15 (a) Cyclic deformation with 1.3% amplitude. (b) Contour maps of 2D peak position before and after cyclic loading in direct-loading mode

Figure 4.15 shows the direct-loading mode and a contour map of the peak position before and after cyclic loading with a large strain amplitude of 1.3%. The peak positions in part of the regions clearly become concentrated, whereas those in the other part of the regions are more scattered. The load on the specimen in the direct-loading mode is considered to be significantly affected by the roughness of the substrate. Because of the inhomogeneity of the initial graphene strain distribution, direct loading would cause a local stress concentration and destroy the graphene where it was floating on the surface of the substrate. Therefore, the ascending-loading mode should be used to effectively improve the initial strain distribution of graphene and release the initial strain energy between the graphene and substrate.

4.5.3 Improvement of Interfacial Mechanical Properties

The effect of cyclic loading on the interfacial properties of graphene/PET was further studied. Three specimens with different strain amplitudes of 0.3%, 0.7%, and 1.0% in ascending-loading mode and one specimen without any cyclic loading treatment were subjected to tensile testing to measure the properties of the interface between graphene and PET. Figure 4.16a presents strain distribution of the graphene in the direction of the stretching axis when the four samples were critically debonded at the interface. A strain gradient area from the zero strain at both sides of the edge to a maximum constant strain value in the central area is observed. Two differences can be observed upon comparing the peak position distributions of the 2D peaks along graphene in the direction of the stretching axis of the three specimens with three different amplitudes and the specimen without any cyclic loading. First, the graphene limit strain (ε_{gmax}) differs. The ε_{gmax} was 1.31% and 1.01% after cyclic loading with an amplitude of 1% and 0.3%, respectively, indicating that the effect of cyclic loading is related to the cyclic loading amplitude. When the specimen was subjected to cyclic loading with a small strain amplitude of 0.3%, the strain distribution of graphene was almost the same as that of the graphene without any cyclic loading, and the two curves almost coincide. Second, the relative critical length differs, suggesting that the transfer efficiencies between the four graphene/substrate systems differed. Upon increasing the strain amplitude of the cyclic loading from 0.3% to 1%, the relative critical length of the specimen decreased from 70% to 40%. Thus, the load-transfer efficiency of the latter was significantly

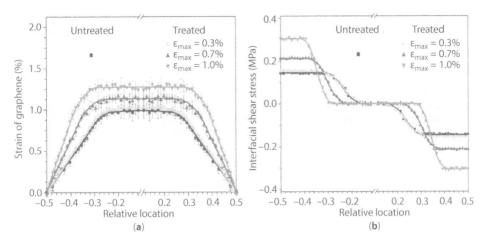

Figure 4.16 The strain distribution (a) and interfacial shear stress (b) of graphene in the direction of the stretching axis when the interface was critically debonded. Graphene subjected to the three cyclic loading methods are compared to that without any treatment.

improved. Therefore, there was an obvious difference in the slope, i.e., the ISS, due to the comprehensive effect of the two differences mentioned above, which will be analyzed below. Therefore, cyclic loading with different strain amplitudes can affect the load transfer and interfacial mechanical properties.

The distribution of the ISS along the tensile direction of the different specimens could be obtained, and the results are presented in Figure 4.16b. The ISS was largest at the edge of both sides and gradually decreased toward the center, where the ISS was zero. The distributions are consistent with the results of the model of shear lag theory. However, there is a significant difference between the four specimens. The maximum ISS values of the specimen without any cyclic loading and that with cyclic loading with a strain amplitude of 0.3% were approximately 0.15 MPa, whereas that of the specimen after cyclic loading with a strain amplitude of 1.0% reached 0.305 MPa, indicating that the interfacial shear strength more than doubled. Therefore, the effect of cyclic loading on interface improvement is apparent.

Based on the results in Figure 4.16, the mechanical parameters of the four specimens were obtained, and the results are presented in Table 4.3. In the ascending-loading mode,

Table 4.3 Interfacial mechanical parameters of four graphene samples prepared with and without cyclic loading.

Amplitude of cyclic strain ε_{max}(%)	Graphene length L (μm)	Graphene limit strain ε_{gmax}(%)	Critical length L_c(μm)	Relative critical length δ (%)	Interfacial shear strength τ_{max}(MPa)
0	100	1.01	70	70	0.152
0.3	100	1.03	70	70	0.155
0.7	100	1.13	50	50	0.212
1	100	1.31	40	40	0.305

a larger strain amplitude led to increased release of the initial strain existing in the initial state, a higher load-transfer efficiency, a greater interfacial shear strength between graphene and the PET substrate, and improved effect of improvement on the graphene interface. For cyclic loading with a small strain amplitude (ε_{max} <0.3%), the interfacial properties between graphene and the PET substrate hardly improved. However, the interfacial properties between graphene and the PET substrate were maximized by cyclic loading with a strain amplitude of 1% in ascending mode.

4.5.4 Discussion

Figure 4.17 shows the deformation evolution of the midpoints of graphene with different treatments as a function of PET strain during the loading process. When the substrate strain was less than 0.5%, because of the presence of initial strain, the increase of substrate strain resulted in an improvement of the bonding of the graphene/substrate interface and the release of initial strain. Therefore, the graphene strain only slightly increased during this stage. When the substrate strain was greater than 0.5%, the effect of the initial strain in the initial stage was gradually eliminated, and the graphene became primarily bonded to the substrate. In addition, the strain generated by the continuous stretching of the substrate could be completely transferred to the graphene. The strain of graphene increased linearly during this stage. By extending the linear section backward using its slope, the value of the initial strain can be quantified. The untreated graphene had an initial strain of 0.36%. After cyclic loading training, the initial strain was gradually released. The inset of Figure 4.17 shows that the initial strains of the specimen treated with 0.7% and 1% amplitude cyclic loading were 0.25% and 0.16%, respectively. It is observed that 0.20% of the initial strain was released in these cases. Therefore, the essence of cyclic loading treatment is improvement of the bonding state and release of the initial strain of graphene. This result also indicates that the amplitude of cyclic loading is an important factor. For the transferred graphene, the effect of the cyclic loading treatment with small amplitude on the interfacial properties was minimal. In addition, use of the ascending mode with large strain amplitude

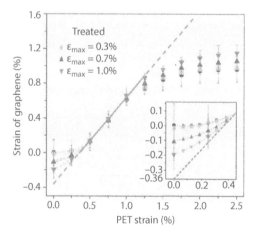

Figure 4.17 Deformation evolution of the midpoints of graphene of samples prepared using different treatments as a function of PET strain during the loading process. The inset shows an enlarged view of the part of the plot in the black box.

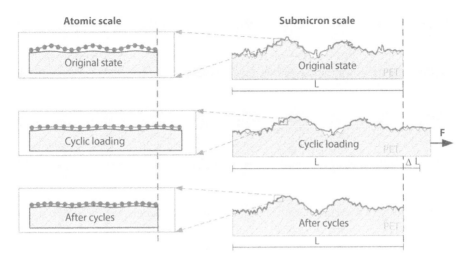

Figure 4.18 Schematic diagram of the improvement effect of cyclic loading on the bonding state.

will improve the initial strain distribution of graphene, leading to gradual homogenization and the effective release of the initial strain. The experimental results clearly demonstrate that the ultimate maximum strain of graphene treated with 1% cyclic loading can be increased by approximately 25% compared with that of the untreated graphene, indicating improvement of the interfacial properties.

Why do the cyclic loading treatments enable improvement of the interfacial properties of graphene? Figure 4.18 presents a schematic diagram of the deformation evolution of the graphene/substrate system during the cyclic loadings to explain the mechanism driving this behavior. Because of the roughness of the substrate, intrinsic fluctuation of graphene, folds induced by the production/transfer processes, and other factors, graphene cannot ideally bond with the substrate surface. There must be some suspending or wrinkled areas. The cyclic loading treatment for which repeated interfacial tangential deformation and load transfer occur causes the suspending graphene to partially conform with the substrate; in addition, some of the folds are also stretched to conform with the substrate. Therefore, the treatment changed the bonding degree and increased the graphene/substrate contact area at the sub-microscale. The insets present schematic views of the nanostructures. The bonding between graphene and substrate is due to van der Waals forces, which is inversely proportional to the equilibrium distance between pairs of atoms; therefore, an increase of the equilibrium distance will result in an abrupt reduction of the van der Waals energy between graphene and the substrate. Therefore, the bonding degree after cyclic loading treatment is improved, and as a result, the interfacial adhesion energy is increased, and the interfacial properties are improved.

4.6 Conclusion

In this chapter, the mechanical properties of the tangential interface between a polycrystalline graphene monolayer attached to a flexible PET substrate through van der Waals forces were systematically investigated from an experimental viewpoint. The micro-loading devices and *in situ* Raman spectroscopy technique were employed to measure the whole-field deformation of graphene subjected to a uniaxial tensile load. The process of interfacial strain transfer from PET to graphene was analyzed, and some experimental phenomena were observed. In particular, three stages of adhesion, sliding, and debonding of the graphene/substrate interface were studied. The bonding state of the interface was observed to evolve from adhesion to sliding and finally to debonding during the loading process. In addition, five important interfacial mechanical parameters were quantitatively determined: the graphene limit strain (ε_{gmax}), substrate debonding strain (ε_s), critical length (L_c), relative critical length (δ), and interfacial shear strength (τ_{max}). The interfacial mechanical properties of graphene/substrate can be effectively characterized using these five parameters, especially the relative critical length, which reflects the transfer efficiency of interfacial loads and the performance of the interface. Therefore, a stronger interface exhibits larger ε_{gmax} and ε_s and smaller L_c and δ and thus larger τ_{max}. The maximum strain that can be transferred to graphene was observed to be approximately 1%, which is similar to the values reported by Young et al. and Jiang et al. [51–53]. However, the maximum interfacial shear strength was observed to be 0.0028 MPa, which is two order of magnitude smaller than the values reported in the literature.

In addition, the new phenomenon of a size effect was observed in the interfacial behavior of graphene based on a series of experiments on graphene specimens with eight different lengths. The eight lengths transitioned from the micro- to macro-levels. The experimental results revealed that of the five interfacial mechanical parameters, ε_{gmax} and ε_s were size-independent, and L_c, δ, and τ_{max} were size-dependent. δ and τ_{max} dramatically increased with decreasing graphene length, especially for the micro-sized graphene, and the difference between the interfacial shear strength of the micro- and macro-sized graphene could reach two orders of magnitude. Based on the results of a series of experiments, two graphene length thresholds were proposed to characterize the effect of the size effect on the graphene/substrate interface. Graphene with a length of less than 20 μm belongs to the micro-sized range, where the interfacial mechanical parameters are extremely sensitive to the graphene length and the size effect plays a prominent role. The upper length threshold before reaching the macro-sized range is 1000 μm; in the macro-sized range, the interfacial mechanical parameters are size-independent and the relative transfer length remains constant, suggesting that the transfer efficiency of interfacial loads is stable. The experimental data in this chapter combined with previous reports on similar graphene interfaces were integrated, revealing that the data for the interfacial shear strength show good agreement with the size effect. The existence of the two graphene length thresholds (that is, the existence of a size effect) is the main reason for the scattering of experimental data for graphene/substrate interfaces reported in different studies.

Finally, the initial strain of graphene reported in this chapter and in previous similar experiments was systematically studied from an experimental viewpoint. The initial strain of graphene on a PET substrate was quantitatively characterized. To explore the effect of the initial strain on the interfacial properties of graphene/PET, three specimens prepared using cyclic loading treatments with different strain amplitudes and one without any treatment were designed. The 2D peak positions before and after treatment indicated that cyclic loadings with appropriate amplitude, where repeated interfacial tangential deformation and load transfer occur, can effectively improve the interfacial bonding state of graphene/PET, resulting in release of the initial strain of graphene. Using Raman spectroscopy and a micro-tensile device, the interfacial properties of the specimens with and without treatment were experimentally investigated. The total strain distributions were measured, and the interfacial properties including the interfacial strength and stiffness were compared.

The results indicated that the conformability of graphene/PET and the interfacial properties can be improved using a cyclic loading amplitude greater than 0.7%. Finally, based on the experimental measurements, the micro-mechanism of the cyclic loading for the improvement of the mechanical properties of the tangential interface was analyzed.

In summary, Raman spectroscopy is an effective method to study the mechanical and interfacial properties of graphene. The findings reported in this chapter demonstrate that the mechanical properties of the interface between graphene and a substrate are affected by many factors such as the size of graphene, texture of the substrate material, and roughness of the substrate. Therefore, the interfacial properties and interfacial strain transfer process of graphene should be considered in practical applications.

Acknowledgments

This work was supported by the National Natural Science Foundation of China (11372217 and 11672203) and the Science and Technology Supporting Major Project of Tianjin City (No. 16YFZCSY00850).

References

1. Geim, A.K. and Novoselov, K.S., The rise of graphene. *Nat. Mater.*, 6, 183, 2007.
2. Geim, A.K., Graphene: Status and prospects. *Science*, 324, 1530–1534, 2009.
3. Novoselov, K.S., Fal, V., Colombo, L., Gellert, P., Schwab, M., Kim, K., A roadmap for graphene. *Nature*, 490, 192, 2012.
4. Bolotin, K.I., Sikes, K., Jiang, Z., Klima, M., Fudenberg, G., Hone, J., Kim, P., Stormer, H., Ultrahigh electron mobility in suspended graphene. *Solid State Commun.*, 146, 351–355, 2008.
5. Kim, K.S., Zhao, Y., Jang, H., Lee, S.Y., Kim, J.M., Kim, K.S., Ahn, J.-H., Kim, P., Choi, J.-Y., Hong, B.H., Large-scale pattern growth of graphene films for stretchable transparent electrodes. *Nature*, 457, 706, 2009.
6. Daniels, C., Horning, A., Phillips, A., Massote, D.V., Liang, L., Bullard, Z., Sumpter, B.G., Meunier, V., Elastic, plastic, and fracture mechanisms in graphene materials. *J. Phys. Condens. Matter*, 27, 373002, 2015.
7. Lee, C., Wei, X., Kysar, J.W., Hone, J., Measurement of the elastic properties and intrinsic strength of monolayer graphene. *Science*, 321, 385–388, 2008.
8. Balandin, A.A., Thermal properties of graphene and nanostructured carbon materials. *Nat. Mater.*, 10, 569, 2011.
9. Balandin, A.A., Ghosh, S., Bao, W., Calizo, I., Teweldebrhan, D., Miao, F., Lau, C.N., Superior thermal conductivity of single-layer graphene. *Nano Lett.*, 8, 902–907, 2008.
10. Kim, H. and Ahn, J.-H., Graphene for flexible and wearable device applications. *Carbon*, 120, 244–257, 2017.
11. Park, J.J., Hyun, W.J., Mun, S.C., Park, Y.T., Park, O.O., Highly stretchable and wearable graphene strain sensors with controllable sensitivity for human motion monitoring. *ACS Appl. Mater. Interfaces*, 7, 6317–6324, 2015.
12. Porwal, H., Grasso, S., Reece, M., Review of graphene–ceramic matrix composites. *Adv. Appl. Ceram.*, 112, 443–454, 2013.
13. Das, T.K. and Prusty, S., Graphene-based polymer composites and their applications. *Polym. Plast. Technol. Eng.*, 52, 319–331, 2013.

14. Tjong, S.C., Recent progress in the development and properties of novel metal matrix nanocomposites reinforced with carbon nanotubes and graphene nanosheets. *Mater. Sci. Eng. R Rep.*, 74, 281–350, 2013.
15. La Notte, L., Villari, E., Palma, A.L., Sacchetti, A., Giangregorio, M.M., Bruno, G., Di Carlo, A., Bianco, G.V., Reale, A., Laser-patterned functionalized CVD-graphene as highly transparent conductive electrodes for polymer solar cells. *Nanoscale*, 9, 62–69, 2017.
16. Lee, S., Lee, S.H., Kim, T.H., Cho, M., Yoo, J.B., Kim, T.-I., Lee, Y., Geometry-controllable graphene layers and their application for supercapacitors. *ACS Appl. Mater. Interfaces*, 7, 8070–8075, 2015.
17. Deng, W., Qiu, W., Li, Q., Kang, Y., Guo, J., Li, Y., Han, S., Multi-scale experiments and interfacial mechanical modeling of carbon nanotube fiber. *Exp. Mech.*, 54, 3–10, 2014.
18. Qiu, W., Li, Q., Lei, Z.-K., Qin, Q.-H., Deng, W.-L., Kang, Y.-L., The use of a carbon nanotube sensor for measuring strain by micro-Raman spectroscopy. *Carbon*, 53, 161–168, 2013.
19. Qiu, W. and Kang, Y.-L., Mechanical behavior study of microdevice and nanomaterials by Raman spectroscopy: A review. *Chin. Sci. Bull.*, 59, 2811–2824, 2014.
20. Ferrari, A.C. and Basko, D.M., Raman spectroscopy as a versatile tool for studying the properties of graphene. *Nat. Nanotechnol.*, 8, 235, 2013.
21. Ling, X. and Zhang, J., Interference phenomenon in graphene-enhanced Raman scattering. *J. Phys. Chem. C*, 115, 2835–2840, 2011.
22. Gupta, A., Chen, G., Joshi, P., Tadigadapa, S., Eklund, P., Raman scattering from high-frequency phonons in supported n-graphene layer films. *Nano Lett.*, 6, 2667–2673, 2006.
23. Abergel, D., Apalkov, V., Berashevich, J., Ziegler, K., Chakraborty, T., Properties of graphene: A theoretical perspective. *Adv. Phys.*, 59, 261–482, 2010.
24. Malard, L., Pimenta, M., Dresselhaus, G., Dresselhaus, M., Raman spectroscopy in graphene. *Phys. Rep.*, 473, 51–87, 2009.
25. Das, A., Chakraborty, B., Sood, A., Raman spectroscopy of graphene on different substrates and influence of defects. *Bull. Mater. Sci.*, 31, 579–584, 2008.
26. Havener, R.W., Zhuang, H., Brown, L., Hennig, R.G., Park, J., Angle-resolved Raman imaging of interlayer rotations and interactions in twisted bilayer graphene. *Nano Lett.*, 12, 3162–3167, 2012.
27. Del Corro, E., Taravillo, M., Baonza, V.G., Nonlinear strain effects in double-resonance Raman bands of graphite, graphene, and related materials. *Phys. Rev. B*, 85, 033407, 2012.
28. Yoon, D., Son, Y.-W., Cheong, H., Strain-dependent splitting of the double-resonance Raman scattering band in graphene. *Phys. Rev. Lett.*, 106, 155502, 2011.
29. Zabel, J., Nair, R.R., Ott, A., Georgiou, T., Geim, A.K., Novoselov, K.S., Casiraghi, C., Raman spectroscopy of graphene and bilayer under biaxial strain: Bubbles and balloons. *Nano Lett.*, 12, 617–621, 2012.
30. Huang, M., Yan, H., Chen, C., Song, D., Heinz, T.F., Hone, J., Phonon softening and crystallographic orientation of strained graphene studied by Raman spectroscopy. *Proc. Natl. Acad. Sci.*, 106, 7304–7308, 2009.
31. Mohiuddin, T., Lombardo, A., Nair, R., Bonetti, A., Savini, G., Jalil, R., Bonini, N., Basko, D., Galiotis, C., Marzari, N., Uniaxial strain in graphene by Raman spectroscopy: G peak splitting, Grüneisen parameters, and sample orientation. *Phys. Rev. B*, 79, 205433, 2009.
32. Bousa, M., Anagnostopoulos, G., del Corro, E., Drogowska, K., Pekarek, J., Kavan, L., Kalbac, M., Parthenios, J., Papagelis, K., Galiotis, C., Stress and charge transfer in uniaxially strained CVD graphene. *Phys. Status Solidi B*, 253, 2355–2361, 2016.
33. Xu, C., Xue, T., Guo, J., Qin, Q., Wu, S., Song, H., Xie, H., An experimental investigation on the mechanical properties of the interface between large-sized graphene and a flexible substrate. *J. Appl. Phys.*, 117, 164301, 2015.

34. Rahimi, R., Ochoa, M., Yu, W., Ziaie, B., Highly stretchable and sensitive unidirectional strain sensor via laser carbonization. *ACS Appl. Mater. Interfaces*, 7, 4463–4470, 2015.
35. Boland, C.S., Khan, U., Binions, M., Barwich, S., Boland, J.B., Weaire, D., Coleman, J.N., Graphene-coated polymer foams as tuneable impact sensors. *Nanoscale*, 10, 5366–5375, 2018.
36. DelRio, F.W., de Boer, M.P., Knapp, J.A., Reedy, E.D., Jr., Clews, P.J., Dunn, M.L., The role of van der Waals forces in adhesion of micromachined surfaces. *Nat. Mater.*, 4, 629, 2005.
37. Maboudian, R. and Howe, R.T., Critical review: Adhesion in surface micromechanical structures. *J. Vac. Sci. Technol. B Microelectron. Nanometer Struct. Process. Meas. Phenom.*, 15, 1–20, 1997.
38. Israelachvili, J.N., *Intermolecular and surface forces*, Academic Press, 2011.
39. Sarabadani, J., Naji, A., Asgari, R., Podgornik, R., Many-body effects in the van der Waals–Casimir interaction between graphene layers. *Phys. Rev. B*, 84, 155407, 2011.
40. Cranford, S., Sen, D., Buehler, M.J., Meso-origami: Folding multilayer graphene sheets. *Appl. Phys. Lett.*, 95, 123121, 2009.
41. Li, S., Li, Q., Carpick, R.W., Gumbsch, P., Liu, X.Z., Ding, X., Sun, J., Li, J., The evolving quality of frictional contact with graphene. *Nature*, 539, 541, 2016.
42. Annett, J. and Cross, G.L., Self-assembly of graphene ribbons by spontaneous self-tearing and peeling from a substrate. *Nature*, 535, 271, 2016.
43. Yoon, T., Shin, W.C., Kim, T.Y., Mun, J.H., Kim, T.-S., Cho, B.J., Direct measurement of adhesion energy of monolayer graphene as-grown on copper and its application to renewable transfer process. *Nano Lett.*, 12, 1448–1452, 2012.
44. Na, S.R., Suk, J.W., Tao, L., Akinwande, D., Ruoff, R.S., Huang, R., Liechti, K.M., Selective mechanical transfer of graphene from seed copper foil using rate effects. *ACS Nano*, 9, 1325–1335, 2015.
45. Na, S., Rahimi, S., Tao, L., Chou, H., Ameri, S., Akinwande, D., Liechti, K., Clean graphene interfaces by selective dry transfer for large area silicon integration. *Nanoscale*, 8, 7523–7533, 2016.
46. Na, S.R., Suk, J.W., Ruoff, R.S., Huang, R., Liechti, K.M., Ultra long-range interactions between large area graphene and silicon. *ACS Nano*, 8, 11234–11242, 2014.
47. Boddeti, N.G., Koenig, S.P., Long, R., Xiao, J., Bunch, J.S., Dunn, M.L., Mechanics of adhered, pressurized graphene blisters. *J. Appl. Mech.*, 80, 040909, 2013.
48. Koenig, S.P., Boddeti, N.G., Dunn, M.L., Bunch, J.S., Ultrastrong adhesion of graphene membranes. *Nat. Nanotechnol.*, 6, 543, 2011.
49. Wang, G., Dai, Z., Wang, Y., Tan, P., Liu, L., Xu, Z., Wei, Y., Huang, R., Zhang, Z., Measuring interlayer shear stress in bilayer graphene. *Phys. Rev. Lett.*, 119, 036101, 2017.
50. Ni, Z.H., Yu, T., Lu, Y.H., Wang, Y.Y., Feng, Y.P., Shen, Z.X., Uniaxial strain on graphene: Raman spectroscopy study and band-gap opening. *ACS Nano*, 2, 2301–2305, 2008.
51. Gong, L., Kinloch, I.A., Young, R.J., Riaz, I., Jalil, R., Novoselov, K.S., Interfacial stress transfer in a graphene monolayer nanocomposite. *Adv. Mater.*, 22, 2694–2697, 2010.
52. Young, R.J., Gong, L., Kinloch, I.A., Riaz, I., Jalil, R., Novoselov, K.S., Strain mapping in a graphene monolayer nanocomposite. *ACS Nano*, 5, 3079–3084, 2011.
53. Jiang, T., Huang, R., Zhu, Y., Interfacial sliding and buckling of monolayer graphene on a stretchable substrate. *Adv. Funct. Mater.*, 24, 396–402, 2014.
54. Xu, C., Xue, T., Guo, J., Kang, Y., Qiu, W., Song, H., Xie, H., An experimental investigation on the tangential interfacial properties of graphene: Size effect. *Mater. Lett.*, 161, 755–758, 2015.
55. Xu, C., Xue, T., Qiu, W., Kang, Y., Size effect of the interfacial mechanical behavior of graphene on a stretchable substrate. *ACS Appl. Mater. Interfaces*, 8, 27099–27106, 2016.
56. Du, H., Xue, T., Xu, C., Kang, Y., Dou, W., Improvement of mechanical properties of graphene/substrate interface *via* regulation of initial strain through cyclic loading. *Opt. Lasers Eng.* Accepted.

57. Li, S., Li, Q., Carpick, R.W., Gumbsch, P., Liu, X.Z., Ding, X., Sun, J., Li, J., The evolving quality of frictional contact with graphene. *Nature*, 539, 541, 2016.
58. Huang, Y., Yao, Q., Qi, Y., Cheng, Y., Wang, H., Li, Q., Meng, Y., Wear evolution of monolayer graphene at the macroscale. *Carbon*, 115, 600–607, 2017.
59. Lee, C., Li, Q., Kalb, W., Liu, X.-Z., Berger, H., Carpick, R.W., Hone, J., Frictional characteristics of atomically thin sheets. *Science*, 328, 76–80, 2010.
60. Gong, P., Li, Q., Liu, X.-Z., Carpick, R.W., Egberts, P., Adhesion mechanics between nanoscale silicon oxide tips and few-layer graphene. *Tribol. Lett.*, 65, 61, 2017.
61. Kusminskiy, S.V., Campbell, D., Neto, A.C., Guinea, F., Pinning of a two-dimensional membrane on top of a patterned substrate: The case of graphene. *Phys. Rev. B*, 83, 165405, 2011.
62. Zhang, Z. and Li, T., Determining graphene adhesion *via* substrate-regulated morphology of graphene. *J. Appl. Phys.*, 110, 083526, 2011.
63. Xu, K., Cao, P., Heath, J.R., Scanning tunneling microscopy characterization of the electrical properties of wrinkles in exfoliated graphene monolayers. *Nano Lett.*, 9, 4446–4451, 2009.
64. Lanza, M., Wang, Y., Bayerl, A., Gao, T., Porti, M., Nafria, M., Liang, H., Jing, G., Liu, Z., Zhang, Y., Tuning graphene morphology by substrate towards wrinkle-free devices: Experiment and simulation. *J. Appl. Phys.*, 113, 104301, 2013.
65. Robinson, J.A., Puls, C.P., Staley, N.E., Stitt, J.P., Fanton, M.A., Emtsev, K.V., Seyller, T., Liu, Y., Raman topography and strain uniformity of large-area epitaxial graphene. *Nano Lett.*, 9, 964–968, 2009.
66. Raju, A.P.A., Lewis, A., Derby, B., Young, R.J., Kinloch, I.A., Zan, R., Novoselov, K.S., Wide-area strain sensors based upon graphene-polymer composite coatings probed by raman spectroscopy. *Adv. Funct. Mater.*, 24, 2865–2874, 2014.
67. Wang, G., Dai, Z., Liu, L., Hu, H., Dai, Q., Zhang, Z., Tuning the interfacial mechanical behaviors of monolayer graphene/PMMA nanocomposites. *ACS Appl. Mater. Interfaces*, 8, 22554–22562, 2016.
68. Anagnostopoulos, G., Androulidakis, C., Koukaras, E.N., Tsoukleri, G., Polyzos, I., Parthenios, J., Papagelis, K., Galiotis, C., Stress transfer mechanisms at the submicron level for graphene/polymer systems. *ACS Appl. Mater. Interfaces*, 7, 4216–4223, 2015.
69. Polyzos, I., Bianchi, M., Rizzi, L., Koukaras, E.N., Parthenios, J., Papagelis, K., Sordan, R., Galiotis, C., Suspended monolayer graphene under true uniaxial deformation. *Nanoscale*, 7, 13033–13042, 2015.
70. Tsoukleri, G., Parthenios, J., Papagelis, K., Jalil, R., Ferrari, A.C., Geim, A.K., Novoselov, K.S., Galiotis, C., Subjecting a graphene monolayer to tension and compression. *Small*, 5, 2397–2402, 2009.
71. Srivastava, I., Mehta, R.J., Yu, Z.-Z., Schadler, L., Koratkar, N., Raman study of interfacial load transfer in graphene nanocomposites. *Appl. Phys. Lett.*, 98, 063102, 2011.

5

Graphene-Based Ceramic Composites: Processing and Applications

Kalaimani Markandan[1*] and Jit Kai Chin[2]

[1]*Department of Chemical and Environmental Engineering, Faculty of Engineering, University of Nottingham Malaysia Campus, Semenyih, Malaysia*
[2]*Department of Chemical Sciences, University of Huddersfield, Huddersfield, UK*

Abstract

Research on carbonaceous nanofillers such as graphene has been developing at a relentless pace as it holds the promise of creating novel materials with meliorated properties. In particular, properties of graphene have been envisaged as an ideal filler material in monolithic ceramics. This is because despite the fact that monolithic ceramics have high stiffness, strength, and stability at high temperatures, they are still susceptible to brittleness, mechanical unreliability, and poor electrical conductivity. Thanks to graphene's exceptional properties (Young's modulus of 1 TPa, breaking strength of 42 N/m, and in-plane electrical conductivity of 10^7 S/m), incorporating graphene into ceramics has great potential to produce tough and electrically conductive composites. However, due to the high surface area and exhibiting strong van der Waals forces of graphene, they have the tendency to stick together and form agglomerates. This leads to inefficient load transfer from matrix to fillers that affects properties of the resulting composite. To avert these problems, processing routes need to be modified carefully before producing graphene-based ceramic composites (GCMC). This chapter aims to report the current understanding of GCMC with two particular topics: (i) various processing routes of GCMC and (ii) application prospective of GCMC.

Keywords: Graphene, ceramic, composites, dispersion, processing, application

5.1 Introduction

5.1.1 Technical Ceramics

Ceramic materials produced from raw material modification and refinement or synthesized to form new ceramic materials are called technical, advanced, engineering, or fine ceramics. Monolithic ceramics have attractive properties such as high mechanical strength and stability at high temperatures that make them useful for electronic, biomedical, automotive, and space applications. Despite the fact that monolithic ceramics are promising structural materials; they are susceptible to brittleness, mechanical unreliability, and poor electrical conductivity [1, 2]. Technical ceramics can be classified into oxide ceramics, non-oxide

---

Corresponding author: kebx3kaa@nottingham.edu.my

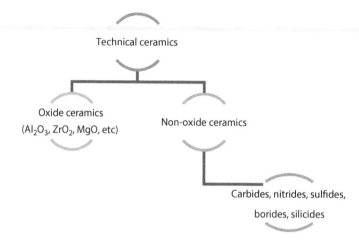

Figure 5.1 Classification of technical ceramics.

ceramics, and ceramic matrix composites (Figure 5.1). Generally, oxide ceramics are oxidation resistant, chemically inert, and very hard. Non-oxide ceramics can be classified based on their chemical components. In view of these limitations, ceramic matrix composites (CMCs) have been developed. CMCs have excellent mechanical stability and relatively low density in comparison to metals that make them potential material for aerospace industry.

5.1.2 Graphene

Graphene is a single layer form of graphite. It is a one-atom-thick planar sheet of sp^2 bonded carbon atoms that are densely packed in honeycomb crystal lattice. In view of its chemical structure, the 2s orbital interacts with the $2p_x$ and $2p_y$ orbitals to form 3σ bonds that are the strongest covalent bond that can exist [3]. This is the source of the excellent mechanical properties of graphene. A single layer of graphene has a thickness of less than 1 nm, but despite this thinness, the theoretical Young's modulus of a single layer of graphene has been reported as 1.06 TPa [4]. On the other hand, the $2p_z$ orbital is uniformly distributed between the two carbon atoms creating a π bond. The p_z electrons have very weak interaction with the nuclei leading to enhanced electrical properties of graphene [5]. The excellent properties of graphene alongside with its low density, high surface area, and high aspect ratio enables graphene to be a desirable material for reinforcement in composite materials. A summary on the properties of various carbon materials is provided in Table 5.1. It can be concluded that graphene precludes the problems associated with other carbon nanomaterials while outperforming their properties in many ways.

5.2 Processing of GCMC

5.2.1 Powder Processing

The traditional method of processing ceramics involves grinding, pressing, and heating of powder particles in the micrometer range. Powder processing can be considered as one of

Table 5.1 Properties of carbon nanomaterials as reported in literatures.

Carbon allotropes	Density (g/cm^3)	Specific surface area (m^2/g)	Thermal conductivity (W/mK)	Electrical conductivity (S/cm)	Young's modulus (TPa)	Hardness (GPa)
Graphene	2.2	2630 [6]	4840–5300 [7]	2000 [7, 8]	1 [9]	1.13 [10]
Graphite	2.2	10–20 [11]	1500–2000 [12]	20,000–30,000 [5]	0.795 [4]	0.2–0.9 [13]
Carbon nanotubes	>1	1300 [14]	3500 [15]	Structure dependent	2.8–3.6 [16]	0.5 [17]
Graphene oxide	0.5–1	736.6 [18]	18 [19]	Nonconducting	0.2076 [20]	NA
Reduced graphene oxide (rGO)	1.91	422–500 [21]	NA	6667 [20, 21]	NA	NA

the most common and pioneer processing route, in which ultrasonication and ball milling are involved. Ultrasonication technique employs energy to agitate graphene in solvent or dispersant, in which ultrasound propagates through series of compression and hence induces attenuated waves in the molecule of the medium it passes. The shear force due to this shockwave will "peel off" the individual nanoparticle (graphene) located at the outer part of nanoparticle bundle or agglomerates, thus resulting in separation of individualized nanoparticles of the bundle [22]. The most common dispersant for graphene are ethanol or n-methyl-2-pyrrolidone (NMP) whereas milling time ranges from 3 to 30 h to produce well-dispersed composites. On the other hand, the ball milling process maximizes load sharing and pull-out effects of graphene, which strengthens interfaces between ceramic and graphene. It is a proven technique to disperse and reduce the number of stacked graphene layers within the matrix. Figure 5.2 shows the flow chart of powder processing route to obtain GCMC composite.

For example, Kun et al. synthesized Si_3N_4-graphene composites by powder processing route [23]. The milling was performed at a high rotation speed of 3000 rpm for 4.5 h in which the graphene particles conferred a cumulative effect in improving the mechanical attributes of the composites and decreased the agglomeration quotient of graphene. A similar technique was demonstrated by Miranzo et al. using SiC powder and graphene where composite was milled in ethanol for 2 h. In 2014, Michalkova et al. compared homogenization of graphene in Si_3N_4 matrix using various methods such as attritor milling, ball milling, and planetary ball milling and reported that best results were obtained for ceramic composites prepared using planetary ball milling [24]. This is because planetary ball milling is able to produce the highest degree of fineness required in comparison to other milling techniques. Besides, extremely high centrifugal forces of planetary ball mill results in high pulverization energy and hence shortens the grinding time. The powder processing route offers excellent opportunities to reduce complexity, cost, and time to synthesize ceramic composites. Besides, this technique has successfully created homogenous dispersion of second phase (graphene) in ceramic composites. However, it should be noted that distribution

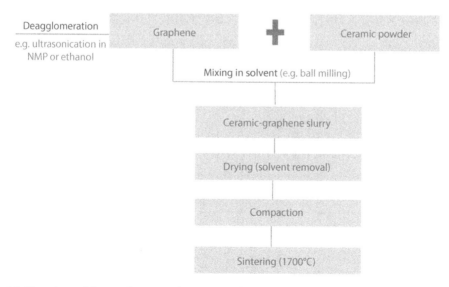

Figure 5.2 Flow chart of the powder processing route to obtain GCMC composite.

of high surface area and high aspect ratio filler in the absence of driving force impedes graphene to deagglomerate and distribute from ceramic powder particle surface into bulk of the mixture.

5.2.2 Colloidal Processing

Colloidal processing is a wet chemistry precipitation method in which solutions of different ions are mixed under controlled temperature and pressure to form insoluble precipitates. This technique produces composites with homogenous microstructure and controllable properties on the basis of colloidal chemistry. In GCMC, colloidal suspensions are used to coat graphene with ceramic particles by modifying the surface chemistry, stabilizing suspensions, and reducing repulsion between graphene, which facilitates homogenous dispersion of graphene throughout ceramic matrix grains.

Dispersion of graphene via this route is established by manipulating the surface chemistry of both phases. Hence, dispersion of graphene in the ceramic matrix will be retained even after sintering. A single solvent for both graphene and ceramic is always preferred in the colloidal processing route to ensure uniformity of dispersing medium. Besides, slow mixing by magnetic stirring or ultrasonication is essential to favor homogenous dispersion and prevent defects of graphene in the ceramic matrix.

Another requirement for colloidal processing is surface modification of both graphene and ceramic matrix. This is achievable by direct functionalization (i.e., oxidation) or using surfactants that generate electric charges. The common surface modification method involves generation of charges between ceramic powders and graphene, which is known as heterocoagulation. Many studies have reported heterocoagulation as an effective route to produce well-dispersed GCMC [25–30]. For example, Wang et al. used the heterocoagulation process to produce Al_2O_3–graphene composites [30]. In their work, graphene oxide (GO) and alumina suspension was prepared by ultrasonication in water separately. GO was then added dropwise into the alumina suspension under stirring conditions. A similar technique was repeated by Centeno et al. where graphene–alumina-based ceramic composites were fabricated by adding GO dropwise into alumina suspension under mechanical stirring while pH at 10 was maintained [25]. Fan et al. prepared Al_2O_3–GNS (graphene nanosheets) composites via colloidal processing route, where GO colloid and alumina colloid were added dropwise into each other [29]. In another study by Walker et al. Si_3N_4–graphene composites were successfully produced via the colloidal processing route [27], in which cetyl trimethyl ammonium bromide (CTAB) was used as the cationic surfactant to produce positive charges on both ceramic and graphene surfaces. 1 wt.% CTAB was used to develop electrostatic repulsive forces on the surfaces of graphene and Si_3N_4 and thus achieve dispersion of graphene within the ceramic matrix. Successful application of the technique on various ceramic composites shows that the method is versatile, nondestructive, and shall be recommended for preliminary work into ceramic composite.

5.2.3 Sol-Gel Processing

Alternative routes based on solution processing were developed to gain better control over the structure of nanosized particles and allow synthesis of ceramics at low processing temperatures. Sol-gel process is an example of such low-temperature processing routes for producing ceramic materials. It is a wet chemical process that uses colloidal chemistry technology, involving creation of a network by polycondensation reaction of molecular precursors in a liquid medium.

Some researchers have defined sol-gel as a chemical transformation process of liquid (sol) into the gel state and with subsequent posttreatment into solid material [31]. This process allows rapid solidification, powder-free synthesis, and soft chemical approach to synthesize stable or metastable oxide materials. By definition, "sol" refers to the stable suspension of colloidal solid particles or polymers in a liquid where the particles can be amorphous or crystalline. On the other hand, "gel" refers to the porous, three-dimensional (3-D) continuous solid network supporting the liquid phase. Sol particles can be connected by covalent bonds, van der Waals forces, or hydrogen bonds, while gels can be formed due to the entanglement of polymer chains. In most cases, gelation occurs due to the formation of covalent bonds, and this process is irreversible. For GCMC prepared by the sol-gel method, the filler is dispersed in a molecular precursor solution [e.g., tetra methyl ortho silicate (TMOS)] that undergoes condensation reaction to generate a green body for subsequent consolidation. Later, suspension of TMOS and graphene will be sonicated to obtain a uniformly dispersed sol. Gelation is initiated by adding catalyst, which promotes hydrolysis and leads to formation of composite gels upon condensation at room temperature.

In 2015, Markandan et al. used methacrylamide (MAM) and N,N'-methylenebisacrylamide (MBAM) as the monomer and cross-linker at a ratio of 6:1 to prepare YSZ–graphene composites via the polycondensation reaction [32]. They added polyvinylpyrrolidone (PVP) to assist the polymerization of acrylamide-based system to occur in ambient air. Before casting the suspension on PDMS soft molds, the authors added ammonium persulfate (APS) as initiator and N,N,N',N'-tetramethyl ethylenediamine as catalyst to initiate polymerization. Heating of the casted suspension further assisted in polymerization reaction.

Initial emphasis of this processing route was to improve the synthesis procedure and product quality of ceramics, which were only attainable via high-temperature processing. Today, new classes of sol-gel materials such as a variety of hybrid materials (inorganic matrices incorporating organic compounds, polymers, and biological moieties) are developed with focus on high-purity precursors and low processing temperatures.

Although the sol-gel process is ought to provide a route to good dispersions, agglomeration in the precursor suspensions has been problematic. Nevertheless, this technique only requires liquid precursors that eases the preparation of doped materials or well-dispersed composites by dissolving or suspending materials in liquid phase [33].

5.2.4 Polymer-Derived Ceramics

The polymer-derived ceramic (PDC) route is useful to produce GCMCs that are difficult to be processed via traditional powder technology. In this PDC route, common preceramic polymers such as poly(silazanes), poly(siloxanes), and poly(carbosilanes) are processed and shaped using well-established polymer-forming techniques such as polymer infiltration pyrolysis (PIP), injection molding, coating from solvent, extrusion, and resin transfer molding (RTM) [34]. After processing, materials made from preceramic polymers can be converted into ceramic components by heating to temperatures that are required to produce tough and densified ceramic composites. One of the advantages of the PDC route is the versatility to produce materials that can be shaped into the form of fibers or bulk composites. Besides, PDCs exhibit excellent thermo-mechanical properties that the materials can remain stable in temperatures up to 1500°C. In fact, recent studies shows that temperature stability up to 2000°C can be achieved if the preceramic polymers contain boron [35].

PDCs can be considered as additive free ceramic materials that have excellent oxidation and creep resistance. In particular, the PDC technique is suitable for GCMC since desired dispersion of nanofiller can be produced in liquid-phase precursors prior to pyrolysis [2, 36]. One of the earliest study utilizing the PDC route was reported by Ji *et al.* where GO was dispersed in polysiloxane (PSO) precursor liquid and SiOC followed by cross-linking and pyrolysis at 1000°C in argon gas to produce SiOC-GNS [37]. Another technique in the PDC route involves the infiltration of PDC resin from solution into pre-existing fabricated ceramic green body, which was demonstrated by Cheah *et al.* in 2013 [38]. They first fabricated green bodies of zirconia by gel casting the ceramic suspension on PDMS soft molds. These green bodies were then infiltrated with PDC resin (RD-212a) before sintering at 1200°C. In their study, infiltration of the pre-ceramic resin into pre-sintered ceramic has successfully sealed the pores.

Despite the advantages of the PDC route, there are several concerns in using this technique, such as considerable shrinkage ratio and volume decrease due to material change and gas loss during thermal treatment [39]. Shrinkage causes cracking while gas loss may leave poorly distributed pores behind. In some cases, retaining the as-formed shape throughout the thermal treatment stage becomes difficult since polymers become less viscous.

5.2.5 Molecular-Level Mixing

Molecular-level mixing is able to produce GCMC by utilizing a molecular-level mixing approach. In this route, functionalized graphene in a solvent will be mixed with ceramic salt, which will then be converted into ceramic particles by heat treatment or other processing methods [9, 40]. This route enables molecular-level coating between ceramic particles and graphene.

For example, Lee *et al.* in 2014 reported molecular-level mixing approach to produce Al_2O_3-GO composite with different wt.% of GO [40]. In their study, GO was dispersed in distilled water by sonication to form a GO suspension. Alumina nitrate precursor salt [$Al(NO_3)_3 \cdot 9H_2O$] was added and stirred for 12 h. The solution was then vaporized at 100°C and dried powders were oxidized at 350°C hot air to produce alumina particles. The powder was further processed by ball milling for 12 h to obtain well-dispersed Al_2O_3-GO powders. In the first stage, aluminum nitrate was thermally decomposed to Al^{3+} ions while hydroxyl and carboxylic groups present on the surface of GO react with Al^{3+} ions at the molecular level. This results in heterogeneous nucleation of Al ions on GO surface. Coating of Al^{3+} ions on the surface of GO avoids agglomeration of GO flakes. Interestingly enough, the authors examined Al–O–C bonding via FT-IR analysis and interface area of reduced Al_2O_3-GO matrix using TEM analysis, which are strong evidences of the molecular-level mixing process. Due to the characteristic microstructure, the GO–alumina composite shows enhanced strength, hardness, and fracture toughness superior to monolithic. The schematic representation for fabricating reduced GO–Al_2O_3 composite by the molecular-level mixing process is depicted in Figure 5.3.

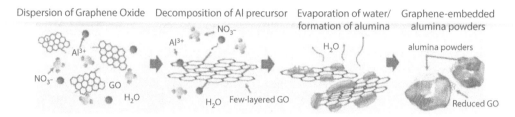

Figure 5.3 Schematic representation for fabricating reduced GO–alumina composite by the molecular-level mixing process [40].

The significant advantage of the molecular-level mixing process is the excellent dispersion of graphene in the ceramic matrix and strong interfacial bonding between ceramic particles and graphene at the molecular level. Molecular-level combination of ceramic particles and graphene required to enhance the property of GCMC may be relatively easier to achieve owing to this strong interfacial bonding.

Previous works on ceramic composites were mostly based on conventional powder metallurgy route resulting in lower than expected mechanical properties since graphene is prone to agglomeration due to van der Waals forces [30]. Although the sol-gel process is able to disperse graphene within ceramic matrix, the interface between graphene and ceramic matrix was not strong [41]. Hence, it is nonetheless plausible to claim that molecular-level processing is one of the most promising approaches to obtain homogenous dispersion of graphene with strong interfacial strength.

5.2.6 Compaction and Consolidation

In early stages, researches on GCMC were limited due to the low thermal stability of graphene (>600°C) [42]. While ceramics start to densify above 1500°C, challenges arise to incorporate graphene, which has low thermal stability. Insights into conventional sintering technique revealed that it requires long processing time and high temperature to prepare fully dense ceramics. This leads to grain growth and simultaneous degradation of graphene in the ceramic matrix [43]. Hence, to overcome these limitations in GCMC, novel sintering techniques such as spark plasma sintering (SPS), high-frequency induction heat sintering (HFIHS), and flash sintering are employed. These techniques are aimed to deliver extremely high temperature into the ceramic matrix in a short interval, reducing the damage on the filler, especially graphene, which is susceptible to elevated temperature.

The SPS technique can be considered as a new, high temperature with low dwell time powder consolidation technique, which is used to create fully dense ceramics [44–46]. This technique involves simultaneous application of pressure and electric current through a graphite die containing the ceramic powders to be sintered. The pulsed current assists in densification of ceramics via creep mechanism unlike conventional sintering techniques that relies on diffusion and mass transport phenomena across grain boundaries during long periods of dwelling time. In particular, SPS has been useful for investigating the sintering behavior of carbon-based fillers (graphene)-reinforced ceramic composites; isothermal conditions can be achieved rapidly, enabling densification to be studied over wide range of densities [47]. Other key advantages of the SPS technique include the *in situ* reduction of GO to graphene in a single step without additional steps and alignment of graphene in host matrix [25].

Meanwhile, HFIHS focuses on sintering ceramics over very short sintering times (<2 min) through the simultaneous application of induced current and high pressure (Figure 5.3). The role of the current in HFIHS is twofold: (i) intrinsic contribution of current to mass transport and (ii) fast heating attributed by Joule heating at contact points. The composites produced via this technique were fairly dense, with relative density as high as 96%. In 2015, Kwon *et al.* demonstrated that HFIHS sinter and densify ZrO_2–graphene ceramic composites [48]. Recently, Ahmad *et al.* reported a similar sintering technique (HFIHS) under processing conditions of 1500°C sintering at 60 MPa and 3 min holding time, which produced Al_2O_3–graphene composites with near-theoretical densities (>99%) [49]. While this sintering technique is not earth shattering, it is still plausible as it represents a new sintering approach to that of SPS, HP, or HIP. By careful modification or optimization of the process parameters, it may be possible to push the relative density values to near 100% by increasing the heating rates. The challenge is to avoid property deterioration of filler (graphene) within the ceramic composite and in an economical approach.

A more recent technique of sintering ceramics is via flash sintering. It occurs when an electrical field is applied to a heated ceramic compact. At a critical combination of field and temperature, a power surge ("flash event") results in sintering completion in a few seconds [50]. The growing interest in flash sintering arose after a publication from Cologna *et al.* in 2010 [51]. In their study, an initial voltage was applied to zirconia powder compact while it is slowly heated in a conventional furnace. At 850°C, "flash event" occurs within few seconds when specimens are sintered to near full density. This event is due to local Joule heating of grain boundaries, in which on one hand promotes grain-boundary diffusion (kinetic effect) and simultaneously restricts grain growth (thermodynamic effect). The smaller grain size and higher temperature at grain boundaries work synergistically to enhance rate of sintering. In a recent study, Grasso *et al.* used flash sintering to sinter ZrB_2 ceramics [52]. The ceramic was densified up to 95% in 35 s under an applied pressure of 16 MPa. In comparison to the conventional SPS technique, the newly developed flash sintering resulted in unprecedented energy and time savings of approximately 95% and 98%, respectively.

5.3 Properties of GCMC

5.3.1 Mechanical Properties and Toughening Mechanisms

Higher mechanical strength can be achieved when a carbonaceous nanofiller such as graphene is added into the ceramic materials. Besides, addition of graphene can transform the nonconducting ceramic materials to electrically conductive ceramic composites. Therefore, this section attempts to summarize the main mechanical properties, electrical properties, and tribological behavior of GCMC.

The intensity of reinforcement effects significantly depends on interfacial bonding of ceramic–graphene and graphene dispersion throughout the ceramic matrix. Toughening mechanisms proposed in GCMC are crack deflection, crack bridging, and graphene pullouts. All these mechanisms are crucial to be microstructurally characterized for fruitful discussions.

In 2013, Centeno *et al.* reported that 0.22 wt.% of graphene loading in alumina prepared by colloidal method has led to 50% increment in fracture toughness, due to crack bridging phenomena [25]. In another work, Dusza *et al.* prepared composites of 1 wt.% Si_3N_4–graphene by hot isostatic pressing and studied the effects of different types of graphene (multilayer graphene, exfoliated GNP, and GNP) on the mechanical properties [53]. According to the authors, the graphene platelets induced porosity leading to lower hardness and fracture toughness values in comparison to composites reinforced with multilayer graphene.

In a similar vein, Kun *et al.* prepared Si_3N_4 composites reinforced with multilayer graphene, graphene nanoplatelets, and nano-graphene platelets [23]. Their findings were in agreement with the studies performed by Dusza *et al.* where graphene platelets induced porosity in samples, leading to lower bending strength and modulus of elasticity values in comparison to composites reinforced with multilayer graphene. Most of the studies on graphene-based ceramic composites have shown toughening mechanisms that originated from pull-outs, crack deflection, crack branching, and crack bridging (Figure 5.4). Ramirez *et al.* discussed the toughening mechanism in Si_3N_4–graphene using a well-established model for reinforcement in ceramic composites [54]. The following assumptions were made by the authors:

(i) GNP/GO were aligned in a direction perpendicular to the pressing direction o f SPS.
(ii) Graphene in ceramic matrix are in residual tension due to the mismatch in thermal expansion coefficient of Si_3N_4 and graphene.
(iii) Fracture toughness due to graphene pull-out was not considered since graphene was in residual tension with Si_3N_4 matrix.

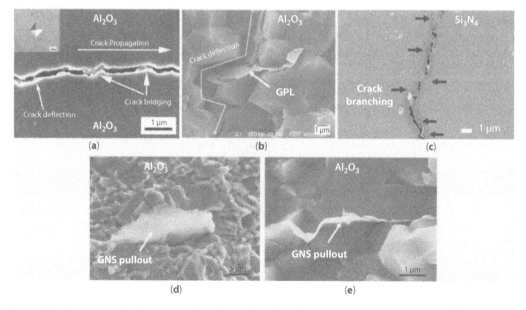

Figure 5.4 Various toughening mechanisms in graphene-based ceramic composites: (a) crack deflection and bridging [49], (b) crack deflection [55], (c) crack branching [56], and (d and e) GNS pull-outs [57].

However, the above assumption contradicted many experimental findings as many authors have reported improvement in fracture due to graphene pull-out. The improvement in toughness due to failure of graphene in wake zone can be calculated using Equation 5.1:

$$\Delta G_c = 2f \int_0^{t=S} t\,du + \frac{4 f \Gamma_i\, d}{(1-f)R}$$

$$= \frac{fS^2 R[(\lambda_1 + \lambda_2 (d/R))^2 - (E_F\, e_T/S)^2 ((\lambda_3 + \lambda_4 (d/R)))^2]}{Ef(\lambda_1 + \lambda_2 (d/R))} + \frac{4 f \Gamma_i\, d}{(1-f)R} \quad (5.1)$$

where f is filler volume, S is the strength of filler, R is the filler radius, E_f is the elastic modulus of fiber, e_T is the misfit strain and Γ_i is the interface fracture energy. λ_i coefficients depend on filler volume fraction and ratio between fiber and elastic modulus of fiber.

The first term in the equation is closely related to toughening due to crack bridging while the second term is related to debond surface energy. The toughness was converted into the critical strain energy release rate (G_c), using the equation $G_I = \frac{K_I^2}{E}$ and compared to data plotted using Equation 5.1. Experimental and theoretical data for GNP composites have shown to be in good correlation. Crack bridging is the dominant toughening mechanism in Si_3N_4–graphene composites, although it becomes invalid for high loading of graphene. This is due to the formation of three-dimensional interconnected network of graphene platelets leading to failure of the composites.

Walker et al. reported significant improvement of 235% in fracture toughness with 1.5 vol.% loading of graphene nanosheets (GNS) in Si_3N_4 matrix prepared by the aqueous colloidal processing method [27]. Unexpected toughening mechanisms such as cracks are not able to propagate through graphene walls and arrested. In such cases, cracks deviate around the graphene sheets. This toughening mechanism is new in comparison to previous studies.

A newer approach to incorporate graphene into Al_2O_3 was proposed by Porwal et al. Graphene was prepared using liquid-phase exfoliation and dispersed dropwise into Al_2O_3 via ultrasonication and powder processing route, which resulted in 40% increment in fracture toughness with the addition of 0.8 vol.% graphene [58]. This route appears advantageous in comparison to Hummer's method since it solves the problem of producing good-quality graphene without affecting its properties.

A breakthrough was reported in a work by Yazdani et al. where Al_2O_3 composites reinforced with graphene nanoplatelets (GNP) and carbon nanotubes (CNT) were prepared by wet dispersion and probe sonication technique [59]. Fracture toughness increased from 3.5 MPam$^{0.5}$ to 5.7 MPam$^{0.5}$ at hybrid addition of 0.5 wt.% GNP and 1 wt.% CNTs, respectively. The 63% improvement was related to CNTs that were attached to GNP surfaces and edges that assisted in deagglomeration and homogenous dispersion. Toughening mechanisms such as change in fracture mode from intergranular in monolithic Al_2O_3 to blurry and glaze-like transgranular mode in GCMC were observed.

Kim et al. investigated the mechanical properties of Al_2O_3 reinforced with unoxidized graphene, GO, and reduced GO [60]. The authors reported that the optimum values were obtained for Al_2O_3 reinforced with unoxidized graphene due to the lower defect concentration within the unoxidized graphene. Crack bridging was noticed to be the major toughening mechanism. In addition to that, there was an investigation into the effect of graphene size ($\approx$100, 20, and 10 μm) on fracture toughness of the Al_2O_3–graphene composite, and the best results were achieved for graphene flakes with a lateral size of 20 μm. Graphene flakes of 100 μm produced structural defects while toughening mechanism such as crack bridging is insignificant with the use of graphene flakes of size 10 μm.

In best-case scenarios, graphene should take a major share of the load when external force is applied to GCMC. Efficiency of load transfer to graphene depends on interfacial bonding between ceramic and graphene. In GCMC, the high strength of graphene is important because once crack starts, the load will be transferred from the ceramic to graphene. When ceramic–graphene adhesion is weak, the initiated crack will be deflected along the matrix–filler interface, leaving the fillers intact, thus toughening the composite material. However, when the matrix is too strong, crack penetrates through the ceramic particles, resulting in brittle composites as exemplified by monolithic ceramics.

Furthermore, it is noteworthy that in almost all studies, the mechanical properties of GCMC do not show proportional enhancement with the increase in graphene loading. The reasons for this behavior in GCMC are twofold:

(i) Increase in porosity with the increase of graphene loading, and these pores act as fracture initiation sites upon indentation load.
(ii) Overlapping/agglomeration of graphene at higher graphene loading.

When filler loading is above the optimum amount, agglomeration of graphene will be dominant in GCMC. Hence, formation of pores between agglomerated graphene platelets and the ceramic matrix interface is not uncommon. The presence of these pores reduces the contact area of ceramic matrix with graphene platelets. When crack is initiated, stresses will be released in an inefficient manner due to the presence of pores. For example, if crack propagates and meets the graphene platelets, it is arrested and deflected in plane. This mechanism develops a complex pathway to release stress, which assists in increasing the toughness and strength of GCMC. In the presence of the pores, interfacial friction during graphene pull-out from the ceramic matrix is weakened. Hence, agglomeration of graphene platelets leads to degradation of mechanical properties and inefficient toughening mechanisms in their host ceramic matrix. Table 5.2 summarizes the mechanical properties of selected GCMC reported in literatures.

5.3.2 Electrical Properties

Electrically conductive composites with volume conductivity higher than 10^{-10} S/cm are important materials that are useful for various engineering applications. From the moment of its discovery, graphene was expected to showcase superlative electrical and thermal properties by analogy to graphite. This is because it had been long known that graphite had an in-plane electrical conductivity of 10^7 S/m and a thermal conductivity of 5300 W/mK [7, 8].

Table 5.2 Effect of graphene addition on mechanical properties of ceramic composites.

Matrix	Optimum filler composition (wt.%)	Mechanical properties				Ref.
		Flexural strength (MPa)	Young's modulus (GPa)	Hardness (GPa)	Fracture toughness ($MPa^{\frac{1}{2}}$)	
Si_3N_4	7	740	–	–	–	[24]
Si_3N_4	1	–	–	16.38 ± 0.48	9.92 ± 0.38	[56]
Si_3N_4	1	–	–	16.4 ± 0.4	9.9 ± 0.3	[61]
Si_3N_4	1	876 ± 53	–	12.2 ± 0.1	8.6 ± 0.4	[62]
Si_3N_4	3	–	–	15.6 ± 0.2	4.2 ± 0.1	[63]
Si_3N_4	0.03*	–	290 ± 4	–	6.6 ± 0.1	[54]
Si_3N_4–ZrO_2	1	–	–	16.4 ± 0.4	9.9 ± 0.4	[64]
Al_2O_3	1	440	–	17	5.7	[59]
Al_2O_3	0.2	542	–	–	6.6	[57]
Al_2O_3	0.5	–	–	≈18.5	5.7	[49]
Al_2O_3	0.45*	–	373	21.6 ± 0.55	3.9 ± 0.13	[58]
Al_2O_3–3YTZP	1.1*	–	373.9 ± 3.1	23.5 ± 0.3	–	[28]
Al_2O_3–ZrO_2	0.43*	–	–	16.13 ± 0.53	9.05 ± 55	[65]
ZrB_2	4	219 ± 23	–	15.9 ± 0.84	2.15 ± 0.24	[26]
YSZ	1.63*	–	–	10.8	5.9	[66]

*When loading was reported in volume percent, the density of bulk graphite (2.2 g/cm³) was used to convert to a weight percent loading.

As such, graphene is also expected to have high electrical and thermal conductivities. In monolithic ceramics, addition of graphene transforms the nonconducting ceramic into electrically conductive composites.

Percolation theory can be used to explain the electrically conductive behavior of GCMC as shown in Figure 5.5. The percolation theory refers to critical filler loading where electrical conductivity increases notably to several orders of magnitude because of the formation of continuous electron and conducting paths. Electron paths will not exist below the percolation transition range. Hence, concentration of graphene must be above the percolation threshold so that a conducting network can be achieved in the ceramics. Electrical conductivity experiences saturation plateau when multiple electron paths exist above the percolation transition range. This phenomenon can be explained by the change in nanofiller concentration based on the scaling law (Equation 5.2):

$$\sigma_c = \sigma_o (\varnothing - \varnothing_c)^t \qquad (5.2)$$

where $\varnothing_c$ and σ_o are the conductivities of composite material and nanofiller (graphene), respectively. $\varnothing$ is the volume fraction of graphene, $\varnothing_c$ is the percolation threshold, and t is the universal critical exponent revealing the dimensionality of the conducting system. In comparison to spherical (graphene agglomerates) conducting fillers, the onset of percolation in fiber or "stick-like" (flakes and platelets) systems takes place at a lower $\varnothing$ of conducting filler. Even at low filler concentrations, graphene fillers may be in direct contact

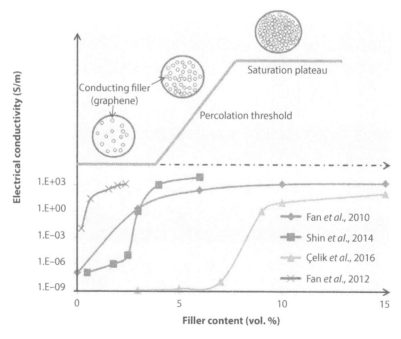

Figure 5.5 Electrical conductivity and percolation phenomenon as a function of filler volume fraction in graphene-based ceramic composites [29, 66, 68, 69].

with each other owing to their high aspect ratios [67]. This results in macroscale conductive pathways through the entire ceramic composite. For uniformly dispersed particles, ∅ for onset of percolation threshold decreases with increasing aspect ratio (L/D) of graphene.

In 2010, Fan et al. for the first time reported on the percolation threshold phenomena in GCMC [68]. Al_2O_3-graphene composites were prepared by varying graphene content from 0 to 15 vol.% and sintered using spark plasma sintering technique. They reported that the percolation threshold for the GCMC was at 3 vol.%. Electrical conductivity increased with graphene loading and was 5709 S/m at 15 vol.% of graphene. This is due to the increased number of charge carriers throughout the ceramic matrix.

A similar group of authors prepared Al_2O_3 reinforced with GO composites via the colloidal processing route and used SPS technique to reduce GO to graphene [29]. They reported an improvement from their previous study where percolation threshold was only at 0.38 vol.% in comparison to 3 vol.% reported previously. Furthermore, authors reported electrical conductivity of 1000 S/m at graphene loading of 2.35 vol.%. An important discovery was reported by the authors where the Hall coefficient reverses from positive to negative with the increment in graphene loading, which indicates a change in major charge carrier. Along with the Hall coefficient, Seebeck coefficient values also changed from positive to negative. The enhancement in electrical conductivity was due to the excellent dispersion of high-quality graphene, whereas a positive Hall coefficient was related to the doping of graphene by the Al_2O_3 matrix.

In another study by Centeno et al., the electrical conductivity of Al_2O_3-graphene composites had increased up to eight orders of magnitude with low percolation thresholds of 0.22 wt.% [25]. The significant improvement in electrical conductivity was due to the increase in graphene loading, which increased the intersheet connections along the graphene planes. Electrical conductivity of an insulator–conductor binary mixture (σ_m) can be expressed as a function of filler volume fraction (V_h) by the general effective media (GEM):

$$\frac{(1-V_h)\left(\sigma_l^{1/t} - \sigma_m^{1/t}\right)}{\sigma_l^{1/t} + A\sigma_m^{1/t}} + \frac{V_h\left(\sigma_h^{1/t} - \sigma_m^{1/t}\right)}{\sigma_h^{1/t} + A\sigma_m^{1/t}} = 0 \text{ where } A = (1-V_{h,c})V_{h,c}^{-1} \quad (5.3)$$

where σ_l and σ_h are the conductivities of low and high conductivity phases, respectively, and $V_{h,c}$ is the percolation threshold. The exponent t is a parameter dependent on shape and orientation of filler. Hence, t can be expressed as a phenomenological parameter typical of conductivity of a given composite.

Ramirez et al. used conductive scanning force microscopy and the GEM equation to study the effect of graphene loading above their percolation threshold (12 wt.% and 15 wt.%) on electrical conductivity of GCMC [70]. The graphene concentration is directly proportional to the conductivity of ceramic composite. Besides, the electronic response and final microstructure of the composite were due to the stiffness and aspect ratio that lead to self-orientation of graphene and lying on the a-b plane during SPS. The highly anisotropic nature of GCMC leads to differences in transport activation energy, which determined the different current values measured under the same conditions for parallel and perpendicular orientations. In 2012, a similar group of researchers extended their study using Si_3N_4 matrix reinforced with 20 vol.% graphene platelets [71]. An electrical conductivity of 40 S/m with preferential orientation of graphene platelets in the ceramic matrix was reported. In the direction perpendicular to SPS pressing axis, electrical conductivity was an order of magnitude higher than in the parallel direction. Different mechanisms of charge transport were reported for these directions such as the variable range hopping mechanism in perpendicular direction and metallic-type transition in parallel direction. The authors also reported percolation threshold in the range of 7–9 vol.%, which depends on the direction of conductivity measurement.

Shin *et al.* in 2014 fabricated fully densified YSZ ceramics reinforced with reduced GO by spark plasma sintering where GO was reduced to graphene by thermal treatment during SPS processing and reported a percolation threshold value of 2.5 vol.% [66]. The authors reported a percolation threshold value of 2.5 vol.%, which was comparable to studies performed by Fan *et al.* in the Al_2O_3-graphene composite. Furthermore, the electrical conductivity of the GCMC increased significantly and reached an optimum value of 12,000 S/m at an addition of 4.1 vol.% GO. Similar to previous studies, the improvement in electrical conductivity of GCMC was because of the effective distribution of GO, which created an interconnected electron pathway.

5.3.3 Tribological Behavior

A recent emphasis of GCMC has been on the tribological behavior such as wear and friction properties. Wear can be defined as the progressive loss of materials due to relative motion between a surface and the contacting substance. Wear damages can be in the form of microcracks or localized plastic deformation. The examples of wear are adhesive wear, abrasion wear, surface fatigue wear, erosion, and cavitation wear.

The excellent lubricating properties of graphene have been long known and atomic-scale studies have pointed out the low friction attained by graphene. It should be noted that friction monotonically decreases as the number of graphene layers increases, where four layers of graphene show similar behavior to bulk graphite [72]. Tribological properties of GCMC are expected to be improved in comparison to monolithic ceramics since graphene is an excellent lubricant due to its hexagonal structure. Ball-on-disc apparatus is the most common technique to study the tribological behavior of GCMC. Wear rate is calculated using the equation $W = \dfrac{V}{LF}$, where W is the specific wear rate, V is the worn volume, L is the sliding distance, and F represents the loading force during experiment. Coefficient of friction will be calculated by measuring the tangential forces during the test.

In a study by Hvizdos *et al.*, the tribological properties of Si_3N_4-graphene composites were measured using the ball-on-disc method [64]. They reported a 60% improvement in wear resistance of the Si_3N_4 matrix reinforced with 3 wt.% graphene. The three main findings reported by the authors are as listed below:

1. Coefficient of friction is independent on the type of graphene used (exfoliated graphene platelets, nano-graphene platelets, and multilayer graphene).

2. Wear resistance of Si_3N_4–graphene and Si_3N_4–CNT composites was compared. GCMC were more wear resistant in comparison to CNT-based composites for the same filler loading.
3. The tribological behavior of Si_3N_4–graphene composites at moderate temperatures of 300°C, 500°C, and 700°C were studied. The authors reported increasing proportional relationship between coefficient of friction and wear rate with increasing temperature.

On the other hand, Belmonte et al. investigated the effect of different loads (50, 100, and 200 N) on the tribological behavior of Si_3N_4–graphene composites using the ball-on-disc method [63]. The authors reported that both coefficient of friction and wear rates are inversely proportional to load. Similarly, a study was performed by Li et al. in 2014 for zirconia–graphene composites [73]. According to the authors, the coefficient of friction and wear rate decreased by 29% and 50%, respectively, with the increase in load applied.

In 2015, Yazdani et al. for the first time measured the tribological behavior of Al_2O_3 reinforced with graphene and CNT using the common ball-on-disc method [59]. The authors reported that the hybrid addition of two different fillers (0.3 wt.% GNP and 1 wt.% CNT) showed significant reduction of 80% in wear rate and 20% in coefficient of friction, respectively. The reasons for the excellent wear resistance of the fabricated composite is twofold: (i) graphene played an essential role in the formation of tribofilm on worn surface by exfoliation and (ii) CNT improved fracture toughness and prevented grain from being pulled out during the tribological test.

5.4 Application of GCMC

5.4.1 Anode Materials for Li-Ion Batteries

SnO_2 is one of the best candidates of anode materials for Li-ion batteries due to its high theoretical reversible Li+ storage capacity of 782 mAh g^{-1}. This is greater than the storage capacity of graphite (372 mAh g^{-1}) currently used. Besides, SnO_2 is widely considered as a potential alternative material in Li-ion battery since it is inexpensive, abundant, and nontoxic. However, similar to many alloy-type anode materials, the main factor impeding the application of SnO_2 is its poor cyclic performance due to agglomeration, pulverization, and electrical disconnection of electrode. This is caused by the huge volume change of approximately 300% during insertion and extraction of Li+. To overcome these problems, carbonaceous materials can be incorporated within SnO_2 to buffer the volume change of SnO_2 and improve the electrical conductivity of electrodes. Furthermore, the SnO_2–carbon composite has the potential to improve cyclability due to the cushion effect of carbon fillers. Examples of carbon fillers include graphite, mesoporous and macroporous carbons, CNTs, and graphene. Among these carbon fillers, graphene has attracted great interest as a filler material in SnO_2 owing to its unique properties of high surface area and excellent electrical conductivity. However, it should be noted that reversible capacities and cycle stabilities of this composite are dependent on the dispersion and particle size, where small size and homogenous dispersion enable high capacities and high-rate performances.

For example, Zhang *et al.* synthesized 2–5 nm of SnO_2 onto the surface of graphene via a one-pot hydrothermal approach without the addition of any surfactant [74]. The excellent features of graphene allowed it to effectively buffer volume change of SnO_2 during the charge–discharge process. Moreover, the SnO_2–graphene composite facilitated fast diffusion of lithium ions in SnO_2 and electron transport in graphene when used as anode material for lithium storage, which resulted in a very high reversible capacity of 1037 mAh g^{-1} and a capacity retention of 90% over 150 cycles. A similar study was reported by Huang *et al.* where randomly connected SnO_2 rods like nanocrystals were laid uniformly on the surface of graphene sheet [75]. This morphology allows compact connections between SnO_2 particles and conductive graphene layers, whereas the void spaces between SnO_2 particles will provide room for buffering during the charge–discharge process. They reported a high reversible capacity of 838 mAh g^{-1} in the first cycle and improved cyclic performance. Besides, the charge capacity of the GCMC after 20 cycles remained 65.9% of the theoretical value. In another study by Ji *et al.*, graphene nanosheets (GNS) were inserted into silicon oxycarbide (SiOC) ceramics to develop electrochemically stable lithium batteries [37]. The GNS–SiOC composite shows a high initial discharge capacity of 1141 mAh g^{-1}, which decreases in the first eight cycles and remains at 364 mAh g^{-1} in the subsequent cycles. The reported value of discharging capacity is comparatively higher than graphite (328 mAh g^{-1}) or monolithic SiOC. More importantly, the authors suggested that the electrochemical performance of the composite material can be enhanced by increasing the amount of GNS in the ceramic. For example, the initial discharging capacity increased from 713 mAh g^{-1} to 1141 mAh g^{-1}, while reversible capacities increased from 173 mAh g^{-1} to 357 mAh g^{-1} when GNS composition was increased from 4 wt.% to 25 wt.%.

5.4.2 Supercapacitors

Another potential application of GCMC is to fabricate supercapacitors. For example, Chen *et al.* reported the possibility of using zinc oxide (ZnO)–graphene composites in supercapacitors [76]. ZnO was chosen since it has profound applications in optics, optoelectronics, sensors, and actuators owing to its semiconducting, piezoelectric, and pyroelectric properties. Besides, ZnO is eco-friendly and has facile growth on various substrates, which eases preparation of ZnO–graphene composites for supercapacitor applications. On the other hand, graphene is an excellent filler material since they overlap with each other and form a conducting three-dimensional network that eases electrolyte ions to access surface of graphene oxide. In the study by Chen *et al.*, the homogenous dispersion of reduced graphene oxide (rGO) showed 128% improvement in specific capacitance in comparison to pure ZnO samples. An important finding was reported by the group of authors where there was only 6.5% decay in the available capacity over 1500 cycles at a current density of 2 A g^{-1}. This finding is suggestive of using ZnO–graphene oxide composites as potential electrode material for supercapacitors.

5.4.3 Engine Components/Bearings/Cutting Tools

GCMCs have found some niche applications from engine components and bearings to high-speed cutting tools due to their high hardness, chemical inertness, and excellent

toughness due to fracture. Besides, studies have shown that addition of graphene platelets in the ceramic matrix enhances the wear resistance and reduces the friction coefficient of the composite material. These properties allow GCMC to be utilized in sliding contacts and as solid lubricants.

For example, Wang *et al.* used Al_2O_3–TiC–graphene platelets (ATG) composite to fabricate ceramic tool material by microwave sintering. While machining hard steel 40Cr (AISI 5140) at a cutting speed of 260 m/min, the lifetime of the ATG tool was approximately 15.4 min. The two factors investigated by the authors are as follows:

(i) Effects of sliding speed and normal load on the tribological properties (friction coefficient and wear rate) of ATG during sliding against GCr15 bearing steel.
(ii) Cutting performance of ATG tools for machining of hardened alloy 40Cr steel in comparison to commercial tools.

For (i), the authors reported that the friction coefficient of the tool material is proportional to the normal load while inversely proportional to the sliding speed. Besides, the friction coefficient and wear rate of the ATG ceramic tool are lower than those of the AT ceramic tool, where addition of graphene was proven to increase the wear resistance of the tool material. The major wear mechanism of ATG ceramic tool material was adhesive wear where, in the wear process, graphene disperses to form a lubricating film on the material surface. Thus, addition of graphene platelets in ceramics reduces the degree of adhesive wear in tool materials. For (ii), the cutting of hardened tool steel 40Cr (50 ± 2 HRC), the lifetime of the ATG tool is 125% and 174% higher in comparison to commercial ceramic tool LT55 and cemented carbide tool YT15, respectively. Furthermore, it was shown that addition of graphene platelets improved the breakage resistance and cutting depth of the tool.

In another study, Rutkowski *et al.* reported that Si_3N_4-GNP-based cutting tools were used in dry turning of NC6 steel (52 HRC) at a low cutting speed of 75 m/min with a lifetime of 9 min [62]. Jaroslaw *et al.* reported the use of Al_2O_3–graphene oxide composites for machining of hardened 145 Cr6 steel (50 ± 2 HRC) at a cutting speed of 370 m/min and reported a lifetime of 9.2 ± 2.4 min [77].

5.5 Conclusion

In conclusion, this chapter has successfully reported the progress and findings of graphene-based ceramic composites. The excellent mechanical and electrical properties of graphene render a huge potential for structural and functional applications of GCMC such as anode materials for Li-ion batteries, supercapacitors, and engine components. In the present chapter, we have provided an overview of research in GCMC with emphases on technical ceramics, processing of GCMC, properties of GCMC, and application perspective of GCMC. It should be noted that the homogenous distribution of graphene in ceramic matrix is essential since this factor, among the other processing parameters, governs the resulting properties of GCMC. Selection of one method or a combination of these methods should be based on the desired end properties since incorrect selection of the method may result in mechanical damage to graphene.

References

1. Markandan, K., Chin, J.K., Tan, M.T.T., Recent progress in graphene based ceramic composites: A review. *J. Mater. Res.*, 1–23, 2016.
2. Porwal, H., Grasso, S., Reece, M.J., Review of graphene–ceramic matrix composites, *Adv. Appl. Ceram.*, 112, 443–454, 2013.
3. Bekyarova, E., Itkis, M.E., Ramesh, P., Berger, C., Sprinkle, M., de Heer, W.A., Haddon, R.C., Chemical modification of epitaxial graphene: Spontaneous grafting of aryl groups. *J. Am. Chem. Soc.*, 131, 1336–7, 2009.
4. Jiang, J., Wang, J., Li, B., Young's modulus of Graphene: A molecular dynamics study. *Sci. York.*, 2, 3–6, 2009.
5. Stankovich, S., Dikin, D.A., Dommett, G.H.B., Kohlhaas, K.M., Zimney, E.J., Stach, E.A., Piner, R.D., Nguyen, S.T., Ruoff, R.S., Graphene-based composite materials. *Nature*, 442, 282–6, 2006.
6. Steurer, P., Wissert, R., Thomann, R., Mülhaupt, R., Functionalized graphenes and thermoplastic nanocomposites based upon expanded graphite oxide. *Macromol. Rapid Commun.*, 30, 316–327, 2009.
7. Balandin, A.A., Ghosh, S., Bao, W., Calizo, I., Teweldebrhan, D., Miao, F., Lau, C.N., Superior thermal conductivity of single-layer graphene. *Nano Lett.*, 8, 902–7, 2008.
8. Singh, V., Joung, D., Zhai, L., Das, S., Khondaker, S.I., Seal, S., Graphene based materials: Past, present and future. *Prog. Mater. Sci.*, 56, 1178–1271, 2011.
9. Hwang, J., Yoon, T., Jin, S.H., Lee, J., Kim, T.-S., Hong, S.H., Jeon, S., Enhanced mechanical properties of graphene/copper nanocomposites using a molecular-level mixing process. *Adv. Mater.*, 25, 6724–9, 2013.
10. Zhang, Y. and Pan, C., Measurements of mechanical properties and number of layers of graphene from nano-indentation. *Diam. Relat. Mater.*, 24, 1–5, 2012.
11. Li, H.Q., Wang, Y.G., Wang, C.X., Xia, Y.Y., A competitive candidate material for aqueous supercapacitors: High surface-area graphite. *J. Power Sources*, 185, 1557–1562, 2008.
12. Fugallo, G., Cepellotti, A., Paulatto, L., Lazzeri, M., Marzari, N., Mauri, F., Thermal conductivity of graphene and graphite: Collective excitations and mean free paths. *Nano Lett.*, 14, 6109–6114, 2014.
13. Brazhkin, V.V., Solozhenko, V.L., Bugakov, V.I., Dub, S.N., Kurakevych, O.O., Kondrin, M.V., Lyapin, A.G., Bulk nanostructured carbon phases prepared from C60: Approaching the 'ideal' hardness. *J. Phys. Condens. Matter.*, 19, 236209, 2007.
14. Peigney, A., Laurent, C.H., Flahaut, E., Rousset, A., Carbon nanotubes in novel ceramic matrix nanocomposites. *Ceram. Int.*, 26, 677–683, 2000.
15. Ruoff, R.S. and Lorents, D.C., Mechanical and thermal properties of carbon nanotubes. *Carbon N. Y.*, 33, 925–930, 1995.
16. Ruoff, R.S., Qian, D., Liu, W.K., Mechanical properties of carbon nanotubes: Theoretical predictions and experimental measurements. *C. R. Phys.*, 4, 993–1008, 2003.
17. Patterson, J.R., Vohra, Y.K., Weir, S.T., Akella, J., Single-wall carbon nanotubes under high pressures to 62 GPa studied using designer diamond anvils. *J. Nanosci. Nanotechnol.*, 1, 143–147, 2001.
18. Montes-navajas, P., Asenjo, N.G., Corma, A., Surface area measurement of graphene oxide in aqueous solutions. *Langmuir*, 29, 13443–13448, 2013.
19. Mahanta, N.K., Abramson, A.R., Thermal Conductivity of Graphene and Graphene Oxide Nanoplatelets. In *InterSociety Conference on Thermal and Thermomechanical Phenomena in Electronic Systems, ITHERM*; IEEE, pp 1–6; 2012.
20. Suk, J., Piner, R., An, J., Ruoff, R., Mechanical properties of monolayer graphene oxide. *ACS Nano*, 4, 6557–6564, 2010.

21. Alazmi, A., El Tall, O., Rasul, S., Hedhili, M.N., Patole, S.P., Costa, P.M.F.J., A process to enhance the specific surface area and capacitance of hydrothermally reduced graphene oxide. *Nanoscale*, 8, 17782–17787, 2016.
22. Sonifier products, Branson Ultrason. Corp. https://www.emerson.com/en-us/automation/branson.
23. Kun, P., Tapasztó, O., Wéber, F., Balázsi, C., Determination of structural and mechanical properties of multilayer graphene added silicon nitride-based composites. *Ceram. Int.*, 38, 211–216, 2012.
24. Michálková, M., Kašiarová, M., Tatarko, P., Dusza, J., Šajgalík, P., Effect of homogenization treatment on the fracture behaviour of silicon nitride/graphene nanoplatelets composites. *J. Eur. Ceram. Soc.*, 34, 3291–3299, 2014.
25. Centeno, A., Rocha, V.G., Alonso, B., Fernández, A., Gutierrez-Gonzalez, C.F., Torrecillas, R., Zurutuza, A., Graphene for tough and electroconductive alumina ceramics. *J. Eur. Ceram. Soc.*, 33, 3201–3210, 2013.
26. Yadhukulakrishnan, G.B., Karumuri, S., Rahman, A., Singh, R.P., Kaan Kalkan, A., Harimkar, S.P., Spark plasma sintering of graphene reinforced zirconium diboride ultra-high temperature ceramic composites. *Ceram. Int.*, 39, 6637–6646, 2013.
27. Walker, L.S., Marotto, V.R., Rafiee, M.A., Koratkar, N., Corral, E.L., Toughening in graphene ceramic composites. *ACS Nano*, 5, 3182–90, 2011.
28. Rincón, A., Moreno, R., Chinelatto, A.S.A., Gutierrez, C.F., Rayón, E., Salvador, M.D., Borrell, A., Al2O3-3YTZP-Graphene multilayers produced by tape casting and spark plasma sintering. *J. Eur. Ceram. Soc.*, 34, 2427–2434, 2014.
29. Fan, Y., Jiang, W., Kawasaki, A., Highly conductive few-layer graphene/Al2O3 nanocomposites with tunable charge carrier type. *Adv. Funct. Mater.*, 22, 3882–3889, 2012.
30. Wang, K., Wang, Y., Fan, Z., Yan, J., Wei, T., Preparation of graphene nanosheet/alumina composites by spark plasma sintering. *Mater. Res. Bull.*, 46, 315–318, 2011.
31. Palmero, P., Montanaro, L., Reveron, H., Chevalier, J., Surface coating of oxide powders: A new synthesis method to process biomedical grade nano-composites. *Materials (Basel)*, 7, 5012–5037, 2014.
32. Markandan, K., Tan, M.T.T., Chin, J., Lim, S.S., A novel synthesis route and mechanical properties of Si–O–C cured Yytria stabilised zirconia (YSZ)–graphene composite. *Ceram. Int.*, 41, 3518–3525, 2015.
33. Zheng, C., Feng, M., Zhen, X., Huang, J., Zhan, H., Materials investigation of multi-walled carbon nanotubes doped silica gel glass composites. *J. Non-Crystalline*, 354, 1327–1330, 2008.
34. Colombo, P., Mera, G., Riedel, R., Sorarù, G.D., Polymer-derived ceramics: 40 years of research and innovation in advanced ceramics. *J. Am. Ceram. Soc.*, 18371, 2010.
35. Riedel, R., Mera, G., Hauser, R., Klonczynski, A., Silicon-based polymer-derived ceramics: Synthesis properties and applications-A review. *J. Ceram. Soc. Jpn.*, 114, 425–444, 2006.
36. Ionescu, E., Francis, A., Riedel, R., Dispersion assessment and studies on AC percolative conductivity in polymer-derived Si–C–N/CNT ceramic nanocomposites. *J. Mater. Sci.*, 44, 2055–2062, 2009.
37. Ji, F., Li, Y.-L., Feng, J.-M., Su, D., Wen, Y.-Y., Feng, Y., Hou, F., Electrochemical performance of graphene nanosheets and ceramic composites as anodes for lithium batteries. *J. Mater. Chem.*, 19, 9063, 2009.
38. Cheah, K.H. and Chin, J.K., Fabrication of embedded microstructures via lamination of thick gel-casted ceramic layers. *Int. J. Appl. Ceram. Technol.*, 11, n/a–n/a, 2013.
39. Sarin, V., *Comprehensive Hard Materials*, Elsevier, 2014.
40. Lee, B., Koo, M.Y., Jin, S.H., Kim, K.T., Hong, S.H., Simultaneous strengthening and toughening of reduced graphene oxide/alumina composites fabricated by molecular-level mixing process. *Carbon N. Y.*, 78, 212–219, 2014.

41. Dimaio, J., Rhyne, S., Yang, Z., Fu, K., Czerw, R., Xu, J., Webster, S., Sun, Y., Carroll, D.L., Ballato, J., Transparent silica glasses containing single walled carbon nanotubes. Proc. SPIE 4452, *Inorganic Optical Materials* III, 47–53, 2001.
42. Jeong, H., Lee, Y.P., Jin, M.H., Kim, E.S., Bae, J.J., Lee, Y.H., Thermal stability of graphite oxide. *Chem. Phys. Lett.*, 470, 255–258, 2009.
43. Inam, F., Yan, H., Reece, M., Peijs, T., Structural and chemical stability of multiwall carbon nanotubes in sintered ceramic nanocomposite. *Appl. Ceram.*, 109, 240–247, 2010.
44. Munir, Z.A., Anselmi-Tamburini, U., Ohyanagi, M., The effect of electric field and pressure on the synthesis and consolidation of materials: A review of the spark plasma sintering method. *J. Mater. Sci.*, 41, 763–777, 2006.
45. Garay, J.E., Current-activated, pressure-assisted densification of materials. *Annu. Rev. Mater. Res.*, 40, 445–468, 2010.
46. Hulbert, D.M., Jiang, D., Dudina, D.V., Mukherjee, A.K., The synthesis and consolidation of hard materials by spark plasma sintering. *Int. J. Refract. Met. Hard Mater.*, 27, 367–375, 2009.
47. Milsom, B., Viola, G., Gao, Z., Inam, F., Peijs, T., Reece, M.J., The effect of carbon nanotubes on the sintering behaviour of zirconia. *J. Eur. Ceram. Soc.*, 32, 4149–4156, 2012.
48. Kwon, S.-M., Lee, S.-J., Shon, I.-J., Enhanced properties of nanostructured ZrO2–graphene composites rapidly sintered via high-frequency induction heating. *Ceram. Int.*, 41, 835–842, 2015.
49. Ahmad, I., Islam, M., Abdo, H.S., Subhani, T., Khalil, K.A., Almajid, A.A., Yazdani, B., Zhu, Y., Toughening mechanisms and mechanical properties of graphene nanosheet-reinforced alumina. *Mater. Des.*, 88, 1234–1243, 2015.
50. Todd, R.I., Zapata-Solvas, E., Bonilla, R.S., Sneddon, T., Wilshaw, P.R., Electrical characteristics of flash sintering: Thermal runaway of Joule heating. *J. Eur. Ceram. Soc.*, 35, 1865–1877, 2015.
51. Cologna, M., Rashkova, B., Raj, R., Flash sintering of nanograin zirconia in <5 s at 850°C. *J. Am. Ceram. Soc.*, 93, 3556–3559, 2010.
52. Grasso, S., Yoshida, H., Porwal, H., Sakka, Y., Reece, M., Highly transparent α-alumina obtained by low cost high pressure SPS. *Ceram. Int.*, 39, 3243–3248, 2013.
53. Dusza, J., Morgiel, J., Duszová, A., Kvetková, L., Nosko, M., Microstructure and fracture toughness of Si3N4 + graphene platelet composites. *J. Eur. Ceram. Soc.*, 32, 3389–3397, 2012.
54. Ramirez, C. and Osendi, M.I., Toughening in ceramics containing graphene fillers. *Ceram. Int.*, 40, 11187–11192, 2014.
55. Liu, J., Yan, H., Jiang, K., Mechanical properties of graphene platelet-reinforced alumina ceramic composites. *Ceram. Int.*, 39, 6215–6221, 2013.
56. Kvetková, L., Duszová, A., Hvizdoš, P., Dusza, J., Kun, P., Balázsi, C., Fracture toughness and toughening mechanisms in graphene platelet reinforced Si3N4 composites. *Scr. Mater.*, 66, 793–796, 2012.
57. Chen, Y.-F., Bi, J.-Q., Yin, C.-L., You, G.-L., Microstructure and fracture toughness of graphene nanosheets/alumina composites. *Ceram. Int.*, 40, 13883–13889, 2014.
58. Porwal, H., Tatarko, P., Grasso, S., Khaliq, J., Dlouhý, I., Reece, M.J., Graphene reinforced alumina nano-composites. *Carbon N. Y.*, 64, 359–369, 2013.
59. Yazdani, B., Xia, Y., Ahmad, I., Zhu, Y., Graphene and carbon nanotube (GNT)-reinforced alumina nanocomposites. *J. Eur. Ceram. Soc.*, 35, 179–186, 2015.
60. Kim, H., Lee, S., Oh, Y., Yang, Y., Lim, Y., Unoxidized graphene/alumina nanocomposite: Fracture-and wear-resistance effects of graphene on alumina matrix. *Sci. Rep.*, 4, 1–9, 2014.
61. Dusza, J., Blugan, G., Morgiel, J., Kuebler, J., Inam, F., Peijs, T., Reece, M.J., Puchy, V., Hot pressed and spark plasma sintered zirconia/carbon nanofiber composites. *J. Eur. Ceram. Soc.*, 29, 3177–3184, 2009.

62. Rutkowski, P., Stobierski, L., Zientara, D., Jaworska, L., Klimczyk, P., Urbanik, M., The influence of the graphene additive on mechanical properties and wear of hot-pressed Si3N4 matrix composites. *J. Eur. Ceram. Soc.*, 35, 87–94, 2015.
63. Belmonte, M., Ramírez, C., González-Julián, J., Schneider, J., Miranzo, P., Osendi, M.I., The beneficial effect of graphene nanofillers on the tribological performance of ceramics. *Carbon N. Y.*, 61, 431–435, 2013.
64. Hvizdoš, P., Dusza, J., Balázsi, C., Tribological properties of Si3N4–graphene nanocomposites. *J. Eur. Ceram. Soc.*, 33, 2359–2364, 2013.
65. Liu, J., Yan, H., Reece, M.J., Jiang, K., Toughening of zirconia/alumina composites by the addition of graphene platelets. *J. Eur. Ceram. Soc.*, 32, 4185–4193, 2012.
66. Shin, J.-H. and Hong, S.-H., Fabrication and properties of reduced graphene oxide reinforced yttria-stabilized zirconia composite ceramics. *J. Eur. Ceram. Soc.*, 34, 1297–1302, 2014.
67. Markandan, K., Chin, J.K., Tan, M.T.T., Enhancing electroconductivity of Yytria-stabilised zirconia ceramic using graphene platlets. *Key Eng. Mater.*, 690, 1–5, 2016.
68. Fan, Y., Wang, L., Li, J., Li, J., Sun, S., Chen, F., Chen, L., Jiang, W., Preparation and electrical properties of graphene nanosheet/Al2O3 composites. *Carbon N. Y.*, 48, 1743–1749, 2010.
69. Çelik, Y., Çelik, A., Flahaut, E., Suvaci, E., Anisotropic mechanical and functional properties of graphene-based alumina matrix nanocomposites. *J. Eur. Ceram. Soc.*, 36, 2075–2086, 2016.
70. Ramirez, C., Garzón, L., Miranzo, P., Osendi, M.I., Ocal, C., Electrical conductivity maps in graphene nanoplatelet/silicon nitride composites using conducting scanning force microscopy. *Carbon N. Y.*, 49, 3873–3880, 2011.
71. Ramirez, C., Figueiredo, F.M., Miranzo, P., Poza, P., Osendi, M.I., Graphene nanoplatelet/silicon nitride composites with high electrical conductivity. *Carbon N. Y.*, 50, 3607–3615, 2012.
72. Dong, H.S. and Qi, S.J., Realising the potential of graphene-based materials for biosurfaces – A future perspective. *Biosurf. Biotribol.*, 1, 229–248, 2015.
73. Li, H., Xie, Y., Li, K., Huang, L., Huang, S., Zhao, B., Zheng, X., Microstructure and wear behavior of graphene nanosheets-reinforced zirconia coating. *Ceram. Int.*, 40, 12821–12829, 2014.
74. Zhang, H., Gao, L., Yang, S., Ultrafine SnO2 nanoparticles decorated onto graphene for high performance lithium storage. *RSC Adv.*, 5, 43798–43804, 2015.
75. Huang, X., Zhou, X., Zhou, L., Qian, K., Wang, Y., Liu, Z., Yu, C., A facile one-step solvothermal synthesis of SnO2/graphene nanocomposite and its application as an anode material for lithium-ion batteries. *Chem. Phys. Chem.*, 12, 278–281, 2011.
76. Chen, Y.L., Hu, Z.A., Chang, Y.Q., Wang, H.W., Zhang, Z.Y., Yang, Y.Y., Wu, H.Y., Zinc oxide/reduced graphene oxide composites and electrochemical capacitance enhanced by homogeneous incorporation of reduced graphene oxide sheets in zinc oxide matrix. *J. Phys. Chem. C*, 115, 2563–2571, 2011.
77. Woźniak, J., Broniszewski, K., Kostecki, M., Czechowski, K., Jaworska, L., Olszyna, A., Cutting performance of alumina-graphene oxide composites. *Mechanik*, 88, 357–364, 2015.

6
Ab Initio Design of 2D and 3D Graphene-Based Nanostructure

Andrei Timoshevskii[1], Sergiy Kotrechko[1]*, Yuriy Matviychuk[1] and Eugene Kolyvoshko[2]

[1]*G.V. Kurdyumov Institute for Metal Physics, Kyiv, Ukraine*
[2]*Taras Shevchenko Kyiv National University, Kyiv, Ukraine*

Abstract

Based on the findings of *ab initio* computations, the ability to create 2D and 3D nanostructures made of carbyne chains and graphene sheets is substantiated. The possibility of creation of three previously unknown allotropic forms of carbon is predicted, namely, structurally modified graphene (M-graphene), hexagonal phase of carbon based on graphene (3D graphene), and a 3D crystal consisting of graphene sheets connected by carbyne chains (carbynophene). The results of *ab initio* simulation of the atomic and electronic structures of these phases are presented, their basic mechanical properties are predicted, and the enthalpies of the formation of these structural forms of carbon are also given.

The key regularities of instability and break of interatomic bonds in such structures are ascertained. It is indicated that strength and stability of such structures is predetermined by strength and stability of the contact bonds that are formed by interaction of carbyne with graphene. Within the framework of the proposed fluctuation model, the main factors of atomic interaction that determine stability of carbyne–graphene nanostructures over a wide range of temperatures and mechanical loads are established. The regularities of the effect of temperature and mechanical load on the lifetime of 2D carbyne–graphene nanoelements have been found. The temperature ranges are determined at which service life values of such nanoelements are sufficient for their practical use.

Keywords: Carbyne, graphene, nanostructure, stability, lifetime, thermal fluctuation, *ab initio* design

6.1 Introduction

Over the past three decades, qualitative changes have occurred in the theory of solids. Revolutionary changes in the field of computer hardware led to the development of computational methods based on quantum mechanics and molecular dynamics. The role of calculations in the field of solid state physics has increased especially. Gradually, calculations from the first principles (*ab initio*) displace semiempirical methods containing a large number of fitting parameters. The complexity of *ab initio* modeling is the need to have a high accuracy of the calculation result. Therefore, such calculations always have the nature of computer experiment, and stability, accuracy, and efficiency of the computational algorithm

---

Corresponding author: serkotr@gmail.com

are incredibly important. Recently, there has been a tendency to expand the scope of application of accurate, quantum-mechanical methods for calculating the electronic structure and physical properties of solids. First-principles methods are increasingly being used not only to solve fundamental problems (interpretation of numerous experimental evidence) but also to develop physical principles for creating new materials and predicting their properties. The present work is devoted to both *ab initio* modeling of the atomic structure of carbon nanostructures based on graphene and carbyne chains and calculating the formation enthalpies as well as predicting their strength and lifetime.

One-dimensional chains of carbon of finite length (carbynes) are of interest to many researchers, since they may have a set of remarkable physical properties. At the present time, *ab initio* computations are a main tool for examination of the carbyne structure and properties. The results of such calculations, presented in this paper, predict exceptional strength properties of carbynes, which depend on whether the number of atoms in the chain is even or odd. This effect is due to different atomic interactions in "even" and "odd" chains, which gives rise to significant differences in their atomic structure. Carbyne chains and graphene sheets are considered in this work as "building blocks" for design of 2D and 3D nanostructures. The maximum keeping of the exceptional properties of carbyne chains in these structures is the main task of such design. This requires solving the problem of interaction between chains and graphene sheets. At the first stage, we solved the problem of creating and analyzing the model 2D structure consisting of two graphene sheets connected by a carbyne chain. This paper presents the results of *ab initio* computation of the atomic structure and strength of such a nanoobject. At the next stage of the study, the more complex task was solved—creation of 3D structures consisting of a set of parallel graphene sheets connected by carbyne chains. Modeling of the atomic structure of such a material has shown that a structural modification of graphene (M-graphene) may exist. 3D structures were obtained in which the extremely high levels of strength of carbynes are realized to the maximum extent.

Stability and service time of nanoelements are the second issue considered in this chapter. This is due to the fact that these properties determine the possibility of their practical use in various nanodevices, in particular, all-carbon-based nanoelectronics. A specific feature of such nanodevices is that break of only one atomic bond in a carbyne–graphene nanoelement can entail a failure of the functional properties of the entire device. This requires the development of new methods of diagnostics and prediction of the lifetime of such devices.

6.2 The Subject and the Methods of Simulation

Calculations were performed by pseudopotential technique, realized as the code PWSCF, using the Quantum-ESPRESSO (QE) program package [1]. The ultrasoft pseudopotential for carbon, generated according to Vanderbilt scheme (code version 7.3.4) [2] with exchange-correlation potential PBE [3], was utilized. Monkhorst-Pack mesh [4] in the Brillouin zone was employed. The lattice parameters and the positions of atoms in the cell were calculated using the Broyden–Fletcher–Goldfarb–Shanno (BFGS) algorithm [5–7]. The value of kinetic energy cutoff, E_{cutoff}, amounted to 820 eV. Methfessel–Paxton [8] smearing of 0.6 eV for partial occupancies was used. The precision of calculations of the forces acting on the atoms was 0.03 eV/Å. Total energies of the model structures were estimated with an accuracy of 1 meV. Calculations were carried out without accounting for zero oscillations of nuclei.

6.2.1 1D Modeling

At modeling of an atomic structure and mechanical properties of carbynes (CACs), the elementary cell "molecule in a box" was utilized; the size of the box was 9 × 9 × 30 (Å); thus, actually, it allowed to eliminate interactions between chains. The unit cell size did not change when calculating the strain diagrams of chains. The number of atoms in the chain was varied from 2 to 21. When calculating the strain diagrams, one of the edge atoms was fixed, and the second was displaced. For each given displacement, the equilibrium positions of the remaining atoms were found. The maximum increase in chain length at each strain step did not exceed 2% of its initial equilibrium length. The set of points in the Brillouin zone was determined by a mesh 1 × 1 × 100. The value of kinetic energy cutoff, E_{cutoff}, amounted to 450 eV.

6.2.2 2D Modeling

2D superstructures were simulated by two infinite half-planes of graphene, connected by periodically repeating chains of carbyne. The unit cell contained 48 graphene atoms and one carbyne chain. The number of atoms in the chain was varied from 3 to 10. Therefore, the unit cell parameter c changed within the interval c = 20.96–29.98 Å. The cell parameter a (the distance between planes) was 10 Å. The cell parameter b was chosen to be 7.41 Å (the distance between chains), which eliminated interactions between chains. The set of points in the Brillouin zone was determined by a mesh 4 × 1 × 4. Tension of such a structure was simulated by increasing the lattice parameter c; this corresponded to uniaxial uniform strain of both graphene sheets and chains along the Z axis.

6.2.3 3D Modeling

Model 3D structures based on carbyne and graphene had lattice parameters $a = b = 3 \times a_0$ (3 × 3), where a_0 is the graphene lattice constant. The set of points in the Brillouin zone was determined by a mesh 4 × 1 × 4. Stress–strain diagrams of these structures were calculated by increasing the lattice parameter c; this corresponded to uniaxial uniform deformation of structures along the Z axis. A smearing width of 0.6 eV for partial occupancies was used. The sets of points in the Brillouin zone was determined by a mesh 8 × 8 × 1 for M-graphene, 8 × 8 × 8 for 3D graphene, and 8 × 8 × 4 for carbynophene. The strain diagram of carbynophene was computed by increasing the lattice parameter c.

Based on the results of modeling the tension of the structures studied, the diagrams "force F and/or stress σ vs. the total strain e of the structural element" were plotted, as well as "force F and/or stress σ vs. strain of the contact bond ε". Forces and stresses were calculated as:

$$F = \frac{dE}{dc} \quad (6.1)$$

$$\sigma = \frac{F}{S} \quad (6.2)$$

where E is the total energy of the system, c is the length of the chain in the case of individual carbines and the cell parameter c in all other cases, and S is the effective cross-sectional area. The strength $F_C(R_C)$ was estimated as the maximum value of the force (stress) on the curve "force (stress) vs. strain".

The total strain e and strain of the contact bond ε were calculated as follows:

$$e = \int_{c_0}^{c} \frac{dc}{c} = \ln \frac{c}{c_0} \tag{6.3}$$

$$\varepsilon = \int_{a_0}^{a} \frac{da}{a} = \ln \frac{a}{a_0} \tag{6.4}$$

where c_0 and c are the chain length in the case of carbynes or the cell parameter c in the equilibrium and strained states, respectively; a_0 and a are the contact bond length in the equilibrium and strained states, respectively.

The values of the elasticity (stiffness) coefficient k_Y and the elasticity modulus Y were calculated as:

$$k_Y = \left. \frac{dF}{de} \right|_{e=0} \tag{6.5}$$

$$Y = \frac{k_Y}{S} \tag{6.6}$$

6.3 *Ab Initio* Modeling of the Atomic Structure and Mechanical Properties

6.3.1 Atomic Structure and Strength of Carbynes

Recently, a lot of studies on the strength and stability of monatomic carbon chains have appeared [9–12]. The remarkable physical properties of carbynes [13–15] and their promising applications [16–21] attract the attention of an increasing number of researchers. Interest in these objects is also due to the possibility of real obtaining carbynes by unraveling out of nanotubes or graphene sheets [16, 17, 22, 23]. Several years ago, experimental findings of tests on determination of tensile strength of carbon atomic chains were published in [24, 25]. An extremely high level of strength of these chains was ascertained, which exceeds 270 GPa [25]. However, at present, there are no targeted investigations of the dependence of the strength of chains on their atomic structure. Data on the strength of monatomic carbon chains in various works differ by an order of magnitude [26]. In most cases, the properties of chains of infinite length were investigated. However, the structure and properties of a carbyne chain containing a finite number of atoms are significantly different from those of an infinite chain [27, 28]. Therefore, *ab initio* calculations of carbyne chains in the initial and deformed states were performed to ascertain regularities of the effect of the number of atoms in carbyne on both its atomic structure and mechanical properties such as strength and elasticity.

It is well known that the atomic structures of carbyne and infinite carbon chain are essentially different. The results of calculations show that in carbynes, the lengths of interatomic bonds depend on the position of atoms in the chain, as well as on the length of the chain itself. This is the fundamental difference between the atomic structure of carbyne and infinite carbon chains. Figure 6.1 shows the lengths of bonds between two nearest neighboring atoms in carbon chains of different lengths. The maximal was the length of the bond a_{12} between the first and second atoms. The length of the bond between the third and fourth atoms turned out to be the minimal. In chains with a number of atoms less than 16, the distance a_{12} depends on the total number of atoms in the chain and also on whether this number is odd or even. Thus, in the atomic structure of carbines, "scale" and "even–odd" effects are observed. Calculations have shown that in carbynes with more than 10 atoms, the internal structure of the chain is cumulene. This is a feature of the atomic structure of carbyne. Presence of edge atoms is the reason for stability of the cumulene structure in the central part of chains of finite length. In chains of infinite length, the structure of cumulene is unstable, and the structure of polyyne is more favorable [29].

Significant changes in the interatomic distances along the chain (Figure 6.1) exhibit that the energy of atomic interaction should also vary depending on the position of atom in the chain. Therefore, to describe the atomic interaction in carbyne, the binding energy $E_i^b(N)$

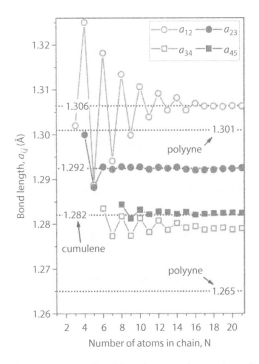

Figure 6.1 The dependence of the interatomic bond length, a_{ij}, on the number of atoms in the chain N.

of each atom with the rest of the chain was estimated. The total energy of a finite chain consisting of N atoms, $E(N)$, may be represented as:

$$E(N) = \sum_{i}^{n} E_i^b(N) + NE_{at} \tag{6.7}$$

where E_{at} is the energy of a free carbon atom. In the carbyne structure with the number of atoms $N \geq 6$, three types of carbon atoms with different binding energies can be identified. Analysis of the interatomic distances (Figure 6.1) shows that the first (from the edge) atom should have the lowest binding energy, since the distances a_{12} between the first and second atoms are the largest in carbyne. On the contrary, it is expected that atoms in the central part of carbyne will have the highest binding energies, since the distances between these atoms are approximately the same and close to the length of the bond in cumulene. The second atoms from the edge are expected to have intermediate values of the binding energy. This enables to represent the total energy (Equation 6.7) as follows:

$$E(N) = NE + 2\left[E_1^b(N) + E_2^b(N)\right] + (N-4)E_{cum}^b \tag{6.8}$$

where $E_1^b(N)$ and $E_2^b(N)$ are the binding energies of the first and second atoms from the edge of the chain, and $E_{cum}^b(N) = -7.71$ eV is the binding energy of the carbon atom in the cumulene structure. Calculations of the total energies of finite chains with different numbers of atoms made it possible to estimate the binding energy of carbon atoms inside the chain. The results of calculations are shown in Figure 6.2. In the first approximation, the value $E_1^b(N)$ in chains with $N \geq 16$ is equal to the average binding energy of carbon atom in the chain of three atoms $E_1^b(N) = -5.80$ eV (Figure 6.2). Such an approximation is based on the fact that interatomic distances in the chain are close to the value a_{12} in the chain

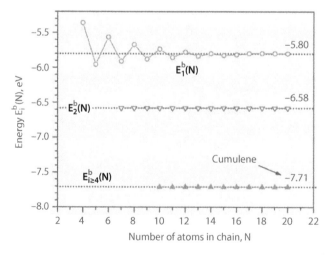

Figure 6.2 The dependence of binding energies of both two-edge atoms of the chain, $E_1^b(N)$, $E_2^b(N)$, and atoms located in its central part, $E_i^b(N)$, on the number of atoms in the chain.

with the number of atoms $N \geq 16$ (Figure 6.1). The binding energy of the atom closest to the edge $[E_2^b(N) = -6.58$ eV$]$ was derived from Equation 6.8 for the total energy of a chain containing 16 atoms, using $E_1^b(N) = -5.80$ eV. It is interesting that the average energy of an atom in a chain of five atoms is 6.55 eV and is close to the value $E_2^b(N)$. This can be explained by the fact that the interatomic distances in a chain of five atoms are the same and close to the value a_{23} in chains with $N \geq 16$ (see Figure 6.1). According to our calculations, the value $E_1^b(N)$ depends both on the total number of atoms in carbyne and on the parity of this number (see Figure 6.2). In chains with an odd number of atoms, the values $E_1^b(N)$ are greater in magnitude and decrease with increasing number of atoms in the chain, approaching a value of $E_1^b(N) = -5.80$ eV. The situation is different for chains with an even number of atoms, where the binding energy of the edge atoms increases with N growth (see Figure 6.2). It is interesting to note the fact that stronger carbynes with odd number of atoms are "insulators", and even ones are conductive systems. This is due to the fact that in chains with an odd number of atoms, all electronic levels are filled, and in chains with an even number, the last π-level is half-filled [30]. The total number of valence electrons in the chain C_n is equal to $4n$. The energy range of $2s$ states contains $(n-1)$ σ-orbital with the number of electrons $2(n-1)$. There are $2(n+1)$ electrons in the energy interval of $2p$ states. Four electrons of them are two lone pairs, which occupy two orbitals—σ_u and σ_g. Thus, the π orbitals are filled with $2(n-1)$ electrons. Since the π orbitals are doubly degenerated (they are filled with 4 electrons), the number of occupied π orbitals is equal to $(n-1)/2$, and in all odd chains, all π orbitals are always filled. This effect is demonstrated in Figure 6.3, which shows the energy levels of valence electrons for carbynes C_5 and C_6.

The data presented in Figure 6.2 exhibit that interatomic bonds of the edge atoms are the weakest. This results in the fact that the edge atoms break away when the load is applied. To determine the mechanical properties of carbynes, a tension of chains of different lengths was simulated. Based on the computation results, the values of the elasticity coefficient, k_Y, and the elasticity modulus, Y, were estimated, as well as the maximum value of force, F_{un}, at which the instability of interatomic interaction occurs; the corresponding critical strains

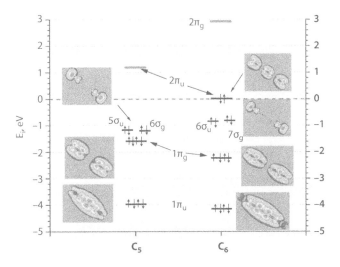

Figure 6.3 The energy levels of the valence electrons for carbynes C_5 and C_6.

for the entire chain, e_{un}, and the edge bond, ε_{un}, were also determined. The value F_{un} may be used as a measure of the strength of carbyne chain at absolute zero. Both a detailed description of the technique of modeling the chain tension and a detailed analysis of the findings obtained are given in [31].

The results of *ab initio* calculations have shown that carbyne strength is predetermined by strength of bond of the edge atom. The magnitude of this strength depends on the total number of atoms in chain and also whether this number is odd or even. Herewith, the strength of carbynes with an odd number of atoms is higher in comparison with the strength of "even" carbynes. Carbyne consisting of five atoms has maximum strength $F_c = 13.1$ nN. Figure 6.4 indicates that the difference between the strength of even and odd chains decreases with increasing number of atoms in the chain, and at $N \geq 12$, it actually disappears. Increasing the number of atoms in the chain gives rise to an increase in it stiffness, k_Y. From Figure 6.4, it follows that disappearance of both the "scale effect" and "even–odd" effect for strength arises with a smaller number of atoms in the chain than for the energy of atomic interaction (12 atoms instead of 16 atoms). For the elasticity coefficient, k_Y, the difference between "even" and "odd" chains disappears, starting at $N \geq 5$ atoms (Figure 6.4).

Typically, the strength of interatomic bonds is estimated employing the binding energy value of atom. A comparison was made between the binding energy of the edge atom, $E_1^b(N)$, and the value of the critical strength of instability of the edge bond, which predetermines strength of the entire chain. As shown in Figure 6.5, for chains containing more than three atoms, an increase in the binding energy of the edge atom, $E_1^b(N)$, is accompanied by an increase in strength of the chain. The strength of carbynes lies within the range from 11.3 nN to 13.1 nN (Figure 6.5). The difference in the mechanical properties of chains with an even and odd number of atoms is due to the difference in their electronic structure (Figure 6.3).

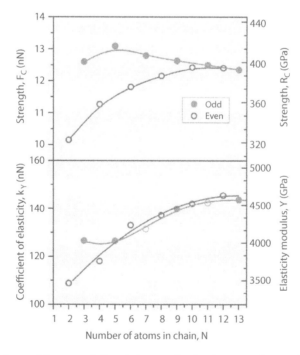

Figure 6.4 The dependence of the strength F_c and elasticity k_Y of carbynes on the number of atoms in them.

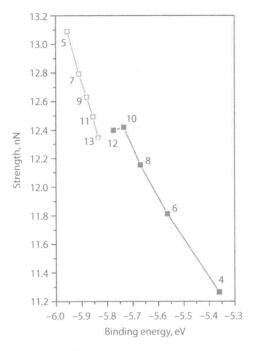

Figure 6.5 The dependence of strength F_c of even and odd chains on binding energies $E_1^b(N)$ of edge atoms of the chain.

These properties characterize a separate chain of carbyne. For applications, it is of interest how these properties are realized in 2D and 3D carbyne–graphene structures.

6.3.2 Atomics of Instability and Break of a Contact Bond in 2D Structures

A nanoelement consisting of two graphene sheets connected by a carbyne chain is a typical 2D structure based on carbyne and graphene [32]. Figure 6.6 demonstrates the nanoelement studied in this work, containing 10-atomic carbyne chain. DFT calculation of an atomic and electronic structure of the investigated carbyne–graphene nanoelement ascertains its metallic conductivity. In this case, 10-atomic chain is polyyne. The calculation results for the tension of the studied nanodevice are exhibited in Figures 6.7 and 6.8. The key points ("A", "B", and "C") that characterize the behavior of the considered nanodevice during its

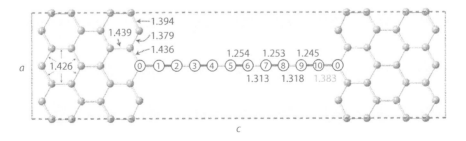

Figure 6.6 Schematic representation of the unit cell used in DFT calculations.

180 Handbook of Graphene: Volume 4

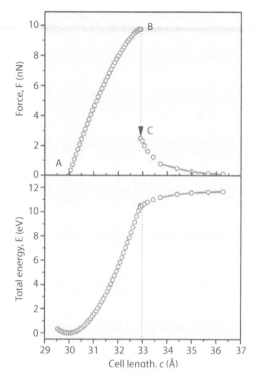

Figure 6.7 The dependences of total energy, E, and force, F, on a cell length c.

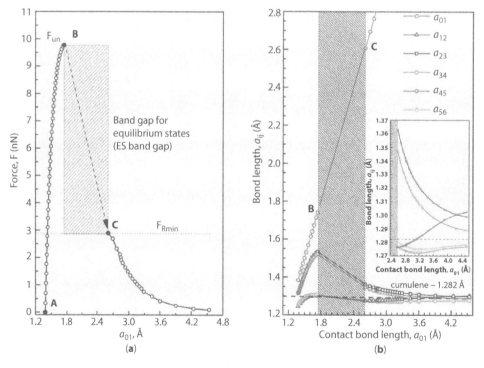

Figure 6.8 The dependence of force, F, (a) and bonds length a_{ij} inside the carbyne chain (b) on the contact bond length a_{01}: F_{un} is the force of instability; $F_{R\min}$ is the lower boundary of ES bandgap.

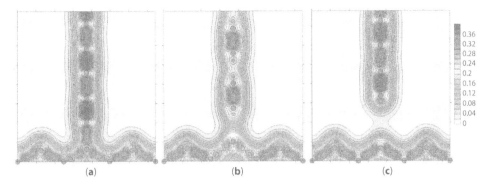

Figure 6.9 Distribution of the electron density at various stages of tension of nano-elements (points A, B, C on Figure 6.8, respectively): (a) is the initial state; (b) is the moment of instability; (c) is the moment after restoration of equilibrium.

tension are indicated in these figures. Figure 6.9 illustrates the distribution of the electron density and atomic structure of the system in the initial state and in key stages of its tension.

As is shown in Figure 6.8b, the contact bond instability (point "B") appears in a steep increase in its length. This is accompanied by the stress relaxation, which manifests itself in an abrupt decrease in the interatomic distances in the chain. Elastic strain energy released at that is spent to perform the work of internal forces for moving the atom of contact bond from a point "B" to a point "C". However, in this case, the magnitude of this energy is not enough to complete the chain break, so, at point "C", equilibrium between the applied force and the force acting in the contact bond was restored; i.e., at this point, the integrity of a nanodevice is maintained. Quantitatively, this is demonstrated in the diagram of dependence of the force on the contact bond length (Figure 6.8). Spatial distribution of the electron density in point "C" is given in Figure 6.9c. It is clearly seen that the edge atom of chain still interacts with graphene. Comparison of the data in Figure 6.9b and Figure 6.9c clearly illustrates the effect of stress relaxation in the chain after instability of a contact bond (point "C"). It should be emphasized that after a complete contact bond break, a change occurs in the structure of the carbyne chain. A resulting chain has a cumulene structure in its central part (Figure 6.8b). This agrees well with the data on isolated chains obtained earlier [31].

Presence of the region in which the equilibrium positions of atoms is impossible is a specific feature of the failure kinetics of the nano-object under consideration (Figure 6.8). In other words, between points "B" and "C", there is *bandgap for equilibrium states of atoms* (ES bandgap). The width of the ES bandgap is determined by the position of its lower boundary F_R. DFT calculations enable to find the minimum values of F_{Rmin} (Figure 6.8). These values are higher for "even" chains, and they have the tendency to decrease with an increase in the number of atoms in chain (Figure 6.10a). It is necessary to emphasize the opposite tendencies of the effect of even and odd number of atoms on the contact bond strength, F_{un}, and the lower limit of ES bandgap, F_{Rmin} (Figure 6.10a). The contact bond strength F_{un} in "odd" nanodevices is higher, but they have lower values of F_{Rmin}. As a result, the maximum width of ES bandgap, ($\Delta_{max} = F_{un} - F_{Rmin}$) for "odd" nanodevices is greater than that for "even" ones.

DFT calculations enable one to determine the maximum width of ES bandgap. In the general case, the width of ES bandgap decreases with decreasing magnitude of the applied force F_f. Its value changes from $\Delta_{min} = 0 nN$ for the unloaded state ($F_f = 0$) to $\Delta_{max} = F_{un} - F_{Rmin}$

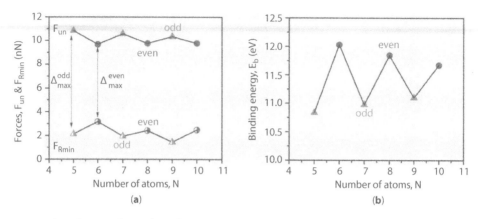

Figure 6.10 The influence of a number of atoms on the contact bond strength F_{un}, level of the lower boundary of ES bandgap F_{Rmin} (a) and the value of binding energy E_b (b): Δ_{max}^{odd} and Δ_{max}^{odd} are the maximum values of ES bandgap for nanodevices containing "odd" and "even" number of atoms.

at the maximum load ($F_{fmax} = F_{un}$). In a first approximation, the dependence for F_R can be determined as [33]:

$$F_R \approx \sqrt{F_{un}^2 - \alpha F_f^2} \quad (6.9)$$

where α is the coefficient characterizing sensitivity of the ES bandgap width to the magnitude of tensile force. At $\alpha = 0$, ES bandgap does not appear, and at $\alpha = 1$, maximum sensitivity is reached. In the latter case, the value of F_{Rmin} equals zero ($F_{Rmin} = 0$), and the descending branch of the deformation curve disappears. In general, the value of α is predetermined by the initial electronic structure and the features of its change in the process of chain tension.

Numerically, the value of α is determined as:

$$\alpha = 1 - \left(\frac{F_{Rmin}}{F_{un}}\right)^2 \quad (6.10)$$

According to the data in Figure 6.11, unlike "even" chains, the values of α for "odd" chains are very close to unity. This means that in "odd" chains, the ES bandgap formation effect is more pronounced. As the applied force increases, the width of the ES bandgap increases, which is manifested in a decrease of F_R [dependence (Equation 6.9)]. In addition, the relationship between the value F_f and the boundary of the ES bandgap, F_R, has a crucial effect on the regularities of the contact bond break and the waiting time for this break.

It should be emphasized that ES bandgap is also observed in separate chains. In [31], the descending branch of the deformation curve of the edge bond was not analyzed in detail; therefore, the effect considered in this article was not noticed. The detailed calculations performed in this work show that it takes place in separate chains with a number of atoms in the chain greater than 3 (Figure 6.12). Herewith, the ES bandgap width grows with the

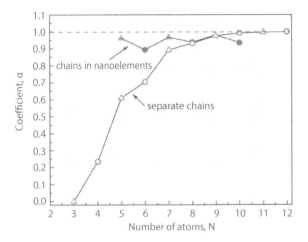

Figure 6.11 Dependence of the value of a coefficient α on the number of atoms in the separate chains and the chains in carbyne–graphene nanoelements.

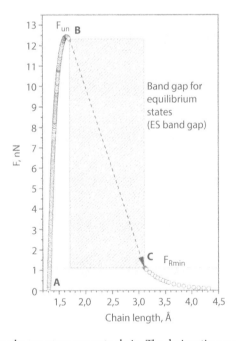

Figure 6.12 Strain diagram for the ten-atom separate chain. The designations are the same as Figure 6.8.

number of atoms in the chain, reaching saturation at the number of atoms that equals 12. It is convenient to demonstrate this by the example of dependence of the value of coefficient α on N (Figure 6.11). In contrast to nanoelement, for individual chains of carbyne, there is a pronounced tendency to α growth with increasing number of atoms in the chain. The effect of even–odd number of atoms in the chain on the value of α, on the contrary, is weakly expressed.

6.4 Modeling of 3D Crystal Structures

Employment of carbyne and graphene as building blocks for creating three-dimensional structures is of interest for both nanophysics and applications. Such 3D structure may be composed of graphene sheets connected by carbyne chains, which are located perpendicular to the graphene sheets. The simulation findings showed that the enthalpy of formation of three-dimensional structures consisting of sheets of graphene connected by carbine chains depends on the type of contact bond and location of carbynes on the graphene sheet. There are two main types of contact bonds of the edge atom of the chain with graphene sheet: involving only one atom of graphene ["atom"–"atom" type (Figure 6.13a)] and with two atoms ("bridge" type) (Figure 6.13b). Herewith, two fundamentally different ways of joining graphene sheets with carbynes may be realized. In the *first* way, chains are attached to a graphene sheet from two sides in different places. In the *second* way, the carbynes are attached to the graphene sheet from two sides in one place. In the latter case, graphene sheets are penetrated by quasi-infinite chains, as shown in Figure 6.14.

Analysis of different contact bonds of the first way of attachment to the graphene sheet allowed us to choose three, the most interesting model structures. The structures No. 1 and

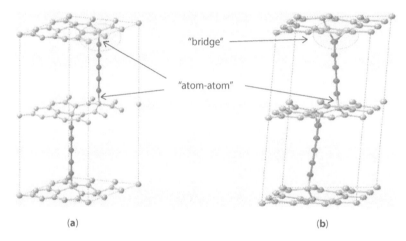

Figure 6.13 Structures with two types of contact bonds: (a) "atom–atom" only; (b) "bridge" and "atom–atom" in the same structure.

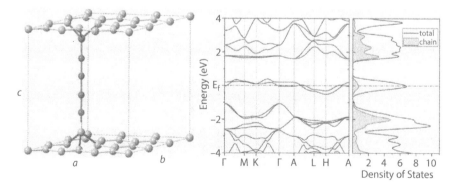

Figure 6.14 Unit cell, band structure, and density of electronic states for structure No. 3.

No. 2 consist of two five-atom chains each (Figure 6.13). The unit cell of these structures contains 46 atoms each. In structure No. 1, the bond of type "atom–atom" is realized. Structure No. 2 contains the combined "atom–atom" and "bridge" contact bonds of carbynes with graphene (Figure 6.13b). Structures with two "bridge" bonds proposed in [23] were not considered in this work because they immediately turn into structures with the "atom–atom" bond type, as soon as the tensile force is applied. An example of the second way of fixing, when carbynes are attached to a graphene sheet from two sides in one place, is the model structure No. 3 (Figure 6.14). In this structure, the same type of contact bond is realized, as in structure No. 1. The unit cell contains 23 atoms ($a_0 = 0.7375$ nm; $c_0 = 1.708$ nm, symmetry group # 187 P-6m2). According to the results of simulation, the strength of the proposed structure is influenced by two main factors: strength of the contact bond and concentration of chains on the graphene sheet. Among the structures under consideration, structure No. 3 has the maximum contact bond strength. Strength of the contact bond in this structure is equal to 8.77 nN (Figure 6.15a). It is 67% of the maximum attainable strength of a single five-atom carbyne chain ($F_{un} = 13.09$ nN) [31]. Therefore, utilizing this type of bond, nearly 70% of carbyne strength can be realized in carbynophene. In general, the contact bond strength for considered types of model structures changes in a sufficiently wide range: from 6.34 to 8.77 nN (Figure 6.15a). It may be explained by the difference in the lengths of contact bond. The maximum strength of structure No. 3 is due to the shortest contact bond length (0.1396 nm) among the structures considered.

Figure 6.15a shows the strain diagrams of the carbynophenes examined. When modeling tension of this structure, there is a monotonic increase in the distance between the sheets of graphene (parameter c) until force reaches its critical value, F_{un}. A further increase in the distance c gives rise to instability of the contact bond with its subsequent break. After the contact

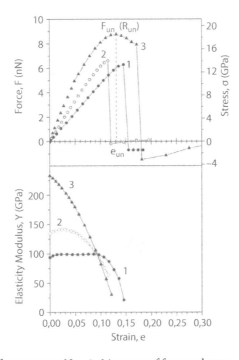

Figure 6.15 Strain diagram for structures Nos. 1–3 in terms of force and stress: F_{un}/R_{un} is the value of strength; e_{un} is the critical strain.

bond break, the chain becomes five-atom one again. Thus, the chain remains to be anchored to the sheet of graphene with only one end by the type of "atom–atom". A further change in the parameter c will only lead to an increase in the distance between the graphene sheets.

The enthalpies of formation of these structures from free carbon atoms may be obtained using *ab initio* calculations:

$$\Delta H = \frac{\left(E_{total}(N) - N \times E_{atom}\right)}{N} \qquad (6.11)$$

where $E_{total}(N)$ is the total energy of the model structure containing N atoms; E_{atom} is the energy of a free carbon atom. This value was almost the same for all three structures ≈ −8.30 eV. Therefore, it is more logical to use a value characterizing the interaction of graphene sheets with carbynes:

$$\Delta H^* = E_{grf+chain} - E_{grf} - E_{chain} \qquad (6.12)$$

where $E_{grf+chain}$ is the energy of model structure; E_{grf} is the graphene energy; E_{chain} is the carbyne energy.

For structures No. 2 and No. 3, the values of ΔH^* are equal to −1.17 and −1.21 eV, respectively. For structure No. 1, this value is almost two times less (−0.66 eV). The value ΔH^* is more sensitive to the arrangement of the chains and to the type of their anchoring to graphene, and it characterizes the binding energy of a carbyne with a graphene sheet.

A concentration of chains on a graphene sheet is the second factor governing the strength of carbynophene. Four types of model structures considered above predict relatively low strength of carbynophenes in comparison with an isolated chain. This is due to the fact that there is only one carbon chain per nine graphene cells in these model systems.

Combination of high strengths with relatively low elastic moduli is a specific feature of the investigated type of 3D carbyne-based structures. It should be noted that it is possible to change the values of elastic moduli over a wide range: from $Y \approx 95$ GPa for model structure No. 1 to $Y \approx 236$ GPa for structure No. 3 (Figure 6.15b). This is due to the fact that the elastic moduli of these structures are determined not only by the elastic properties of the contact bond but also by its location on the graphene sheet. This can be clearly demonstrated by the example of structures No. 1 and No. 3. The more than twofold decrease in the modulus of elasticity at transition from structure No. 3 to structure No. 1 is due to the possibility of bending deformation of the graphene sheet under tension of carbynophene No. 1 since, in the latter case, the chain is fixed to the opposite vertices of the graphene cell (Figures 6.13 and 6.14).

In general, a combination of extremely high strengths (of the order of tens of GPa) with such small values of the elastic moduli and possibility of their variations by the order of magnitude is a specific feature of carbynophenes. These interesting properties may be effectively used in different technological applications.

The most interesting is the model structure No. 3. In this case, the atomic structure of graphene significantly changes at the places where the chains are anchored. This leads to a characteristic feature in the electronic structure—the appearance of localized zones (bands) near the Fermi level (Figure 6.14). Calculations showed that the main contribution to DOS at the Fermi level is given not by chain atoms, but by graphene atoms. From Figure 6.14, it follows that when two contact bonds are formed, two outer atoms of chains and three atoms in the graphene sheet have a coordination number of 4. Five-atom cluster is formed with the characteristic sp^3 bond. Thus, the carbyne chains structurally modify the graphene sheet. The chain "gives" graphene one added atom to form the contact bond of the rest of the chain with the graphene sheet. The simulation results indicated that in the graphene sheet, the existence of structural defect "pair", shown in Figure 6.16a, is energetically favorable. In fact, the graphene sheet thus becomes "corrugated". To characterize the structural defect, it is convenient to use the value E_f—the energy of formation of the added atom configuration on the graphene sheet:

$$E_f = E_{gr+atom} - E_{gr} - E_{atom} \tag{6.13}$$

where $E_{gr+atom}$ is the energy of the "graphene + added atom" system; E_{gr} is the energy of graphene. The energy of formation of the configuration "pair" is equal to $E_b = -2.20$ eV. This atomic configuration is by 0.47 eV less favorable than the configuration "bridge" $E_f = -2.67$ eV (Figure 6.16b). This fact was noted earlier in publications, for example, in (such as) [34].

Graphene lattice may be considered as two interpenetrating hexagonal sublattices of carbon atoms (α and β). When a defect of the "pair" type is created in the α-sublattice, three atoms with sp^3 bonds (red, Figure 6.16a) are in the β-sublattice. An orderly arrangement of such defects is possible. A variant of this ordering is shown in Figure 6.17. In this figure, a unit cell of structurally modified graphene (M-graphene) is schematically indicated. The unit cell contains seven carbon atoms (Figure 6.18). Figure 6.18 shows the band structure and DOS of M-graphene. The results are obtained in the unpolarized approximation. The ground state is nonmagnetic. The electronic states of atoms with sp^3 and sp^2 bonds are strongly hybridized throughout the valence band. These atoms and their partial DOS are shown in Figure 6.18 with red and blue colors, respectively. Electronic states of atoms (marked in blue), which ensure the corrugation of graphene, are highly localized (Figure 6.18). The unit cell contains one such atom. Two electrons of this atom occupy a flat zone in the region −2.5 eV (Figure 6.18). Accordingly, at this energy, a narrow peak in DOS is observed. It follows from Figure 6.18 that the main contribution to DOS at the Fermi level is provided by electronic states of atoms from the α-sublattice (marked in gray). These states are hybridized with states of atoms from the β-sublattice (marked in red). The electronic structure of M-graphene is of undoubted interest and requires detailed researches, which is beyond the scope of this chapter. The enthalpy of formation of M-graphene from free carbon atoms is −8.23 eV, which is comparable with the enthalpy of other structural forms of carbon (Table 6.1).

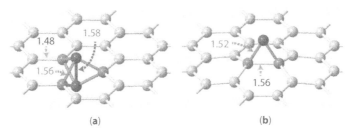

Figure 6.16 The "pair" and "bridge" adatom configurations on a graphene sheet.

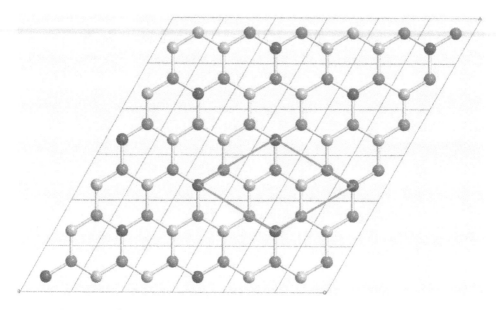

Figure 6.17 The pattern of ordered distribution of "pair" adatom configuration (blue); the M-graphene unit cell is marked with red solid lines.

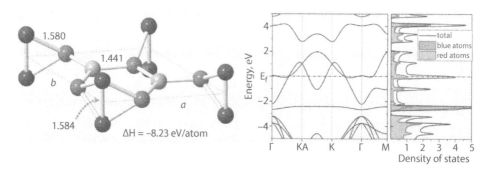

Figure 6.18 Unit cell and electronic structure of M-graphene.

Table 6.1 Structural and energy characteristics of different polymorphic forms of carbon.

Carbon allotropes	Space-group symmetry	Number of atoms in cell	Cell parameters, Å			Density, g/cm³	ΔH, eV/atom
			a	b	c		
Graphene	–	2	2.47	2.47	11.0	–	−9.16
M-graphene	–	7	4.25	4.25	11.0	–	−8.23
3D graphene	189 P$\bar{6}$2m	7	4.22	4.22	3.35	2.70	−8.41
Carbynophene-5		12	4.22	4.22	9.83	1.58	−8.24
Diamond	227 Fd$\bar{3}$m	8	3.57	3.57	3.57	3.52	−9.03

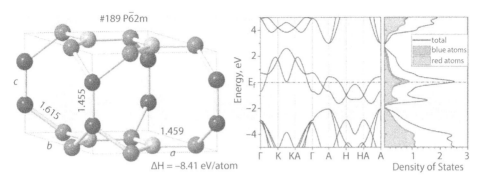

Figure 6.19 Unit cell and electronic structure of 3D graphene.

The possibility of existence of such material as M-graphene has two important consequences:

First, the calculations showed that combinations of M-graphene sheets are energetically favorable to form a 3D graphene crystal. The sheets are connected by sp^3 bonds. The enthalpy of crystal formation from M-graphene sheets is −1.22 eV. The enthalpy of formation from free carbon atoms is −8.41 eV, which is slightly higher than that for M-graphene, but ≈0.6 eV less than that for diamond (Table 6.1). Elementary cell contains seven carbon atoms of three types (Figure 6.19). The symmetry group of such a crystal is # 189 $P\bar{6}2m$. Thus, our calculations predict that the existence of a hexagonal phase of carbon is possible. Figure 6.19 shows the band structure and DOS of a 3D graphene crystal. The ground state is nonmagnetic. At the Fermi level, an intense peak is observed, the main contribution to which is made by electron states of two types of atoms—of both atoms connecting graphene sheets to each other (colored blue in Figure 6.19) and nearest atoms from the graphene plane (colored red). These states are strongly hybridized (see Figure 6.19). We also calculated the elastic constants, and the elastic moduli are given in Voigt-Reuss-Hill approximation for the 3D graphene crystal. The results are shown in Table 6.2.

Table 6.2 Elastic constants of various forms of carbon and elastic moduli (GPa).

##	Diamond		3D graphene
	Present	Exp. [33]	
C_{11}	1046.2	1079	911.8
C_{12}	122.7	124	133.7
C_{13}	–	–	84.6
C_{33}	–	–	338.7
C_{44}	563.6	578	77.0
B	430.5	442	276.5
G	520.4	535	184.7
E	1112.8	1050	451.6

Elastic moduli are given in Voigt-Reuss-Hill approximation.

As expected, 3D graphene has a pronounced anisotropy of elastic properties. Herewith, this structure has much lower (smaller) values of resistance to shear deformations. In [35], it was shown that the ratio between the values of Young's modulus and shear modulus for defect-free crystals may characterize the level of their brittleness. From this point of view, 3D graphene nanocrystal should be more ductile than diamond nanocrystal.

Second, the possibility of existence of M-graphene gives rise to a solution of the problem of anchoring carbyne chains to graphene. Of course, we are talking only about the modeling of the atomic structure of the "graphene + carbine" material (carbynophene), which is schematically shown in Figure 6.20. It is energetically favorable for the carbyne chains to be anchored to those atoms on the graphene sheet that provide graphene corrugation. The elementary cell (Figure 6.21) contains 12 carbon atoms, 5 of which belong to the carbyne chain ($\approx$42%). The symmetry group of such a crystal is # 189 $P\bar{6}2m$. The enthalpy of the formation of such carbynophene from the sheets of M-graphene and carbyne is −6.0 eV. The enthalpy of formation from free carbon atoms is −8.24 eV, which is ~0.8 eV less than that for diamond (Table 6.1). Figure 6.21 exhibits the band structure and DOS of carbynophene. It should be noted that the electron states of atoms in chains and states of graphene atoms are hybridized over the entire width of the valence. The deformation curve of carbynophene-5 is shown in Figure 6.22.

Transition from 2D to 3D structures gives rise to a change in the contact bond properties. This is clearly shown in Figure 6.23. This figure demonstrates dependences of the force acting in the contact bond on the increment of its length. The solid line is the strain curve for carbynophene-5; the dotted line is the strain diagram for the contact bond in the previously considered 2D carbyne–graphene nanoelement containing a five-atom carbyne chain. As follows from these data, the ascending branches (AB) of the deformation

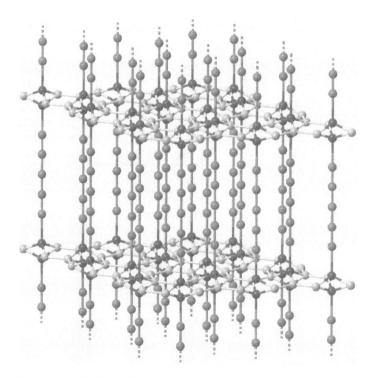

Figure 6.20 Atomic structure of carbynophene-5.

AB INITIO DESIGN OF GRAPHENE NANOSTRUCTURE 191

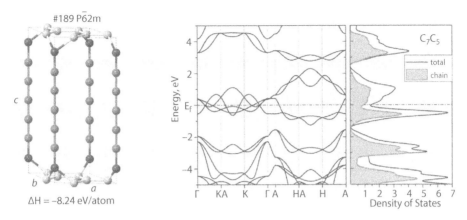

Figure 6.21 Unit cell and electronic structure of carbynophene-5.

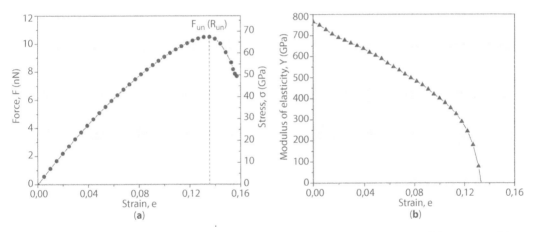

Figure 6.22 The strain curve of carbynophene-5 (a) and the dependence of its elasticity modulus on strain (b).

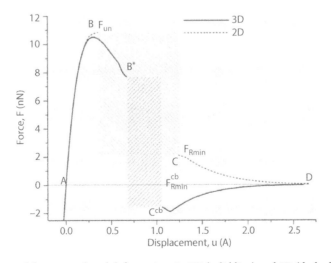

Figure 6.23 Diagrams of the contact bond deformations in 2D (solid line) and 3D (dashed line) structures; F_{un}—critical force of the contact bond stability loss (strength); F_{Rmin}—lower boundary of ES bandgap; u—increase of the contact bond length (ES bandgap region is shown with shaded areas).

curves of contact bonds in 2D and 3D nanocrystals actually coincide; that is, the elasticity coefficients of the contact bonds are equal, and the values of their strengths, F_{un}, are virtually identical. The descending branches BD have a cardinal difference. The contact bond in the 2D-nanoelement becomes unstable when $F \geq F_{un}$, but at $F_{Rmin} \approx 2nN$, the interaction between the chain and the graphene sheet is restored and the contact bond continues to resist tension. In carbynophene, the situation is fundamentally different. After passing through a maximum, a contact bond is able to resist tension to the point B^*. With larger displacements, there are no equilibrium positions for atoms in the chain; accordingly, the contact bond becomes unable to offer resistance to tension. Moreover, with displacements of the contact bond atom of the order of $B^* \approx 1\text{Å}$ and more, *repulsion* of the chain atom from the graphene sheet atom is observed. This means that after point B^*, an accelerated break of the chain occurs. This is due to differences in the rearrangement of the electronic structure in 2D and 3D crystals after passing through a maximum of the tensile strength (point "B").

These differences in the atomic interaction in the descending branches of the tension curves, i.e., for large deviations of the atoms from their equilibrium position, play a decisive role when the contact bond break is initiated by thermal fluctuations of the atoms. This issue is discussed below.

6.5 Thermomechanical Stability

The lifetime of nanoelements containing monatomic chains is predetermined by the waiting time of break of a contact bond. As noted above, a classical approach to this problem consists in using the Arrhenius equation, or its later modifications within the framework of the theory of reactions. In this case, the probability of atomic bond break is equivalent to probability of appearance of fluctuation of the atom kinetic energy, sufficient to overcome the energy barrier whose magnitude is equal to the binding energy E_0. This takes place in the case of a mechanically unloaded crystal.

The attempt to take into account the force field effect within the framework of these approaches meets considerable difficulties. Usually, in this case, the linear law of decreasing the height of energy barrier with increasing stress is postulated [36, 37]. However, the results of MD calculations show that this dependence is nonlinear [36]. This means that the magnitude of activation volume used in these dependences is not a material constant, since it depends on the level of stresses acting in the crystal. Even in the case of a one-dimensional atomic chain, the height of energy barrier decreases nonlinearly with increasing tensile force [38].

Besides, these models do not account for the specific feature of kinetics of instability and bond breaking due to the existence of the ES bandgap.

6.5.1 Fluctuation Model

For the general case of a mechanically loaded crystal, the probability of an atomic bond break can be formulated as follows:

$$P(\delta \geq \delta_c) = P_c \tag{6.14}$$

where P_c is the probability of failure and δ_c is the critical value of bond length fluctuation δ.

As shown in Figure 6.23, the magnitude of critical strain δ_c is predetermined by the level of applied force F_f. This is due to the fact that thermal fluctuations cause only short-term instability of atomic interaction. To break the bond, it is necessary for these fluctuations to be "picked up" by the applied force (Figure 6.24).

Accounting for Equation 6.14, the expression for the probability of a bond breaking, i.e., of realization of critical fluctuation, δ_c, takes the form:

$$P(\delta > \delta_c) = \frac{\int_{\delta_c}^{\delta_{br}} \exp\left[\frac{-\varepsilon(\delta)}{k_B T}\right] d\delta}{\int_0^{\delta_{br}} \exp\left[\frac{-\varepsilon(\delta)}{k_B T}\right] d\delta} \qquad (6.15)$$

where T is the temperature, k_B is the Boltzmann constant, δ_{br} is the fluctuation corresponding to the maximum displacement u_{br} (Figure 6.24), and $\varepsilon(\delta)$ is the energy fluctuation:

$$\varepsilon(\delta) = E(u_f + \delta) - E(u_f) \qquad (6.16)$$

where u_f is the displacement of atom due to the applied force F_f.

The value of the critical fluctuation, δ_c, is determined from the condition

$$\delta_c = u_c - u_f \qquad (6.17)$$

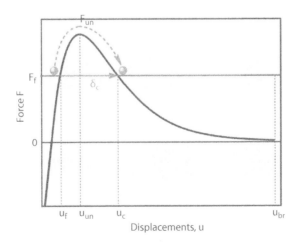

Figure 6.24 Strain diagram for contact bond (scheme): F_f is the value of "applied" force; u_f is the deviation of an atom from the equilibrium position due to mechanical load; F_{un} and u_{un} are the force and displacement of a contact bond instability, respectively; u_c and δ_c are the critical displacement and critical fluctuation, respectively; u_{br} is the displacement of bond break.

where the magnitude of the critical displacement of the atom u_c is determined by solving the nonlinear equation:

$$F_f = F_{DFT}(u_c) \tag{6.18}$$

relatively to displacement u_c at a given value of the applied load F_f, where $F_{DFT}(u_c)$ is the DFT dependence of force on the contact bond length.

Figure. 6.24 clearly demonstrates the essence of the procedure for determining the value u_c. In Equation 6.17, δ_c is the lower limit of integration and has a significant effect on the value of bond break probability. This means that the nature of the descending branch of dependence "force vs. interatomic distance" [the region between u_{un} and u_{br} (Figure 6.24)] should play a key role in the fluctuation-induced atomic bond break. At the same time, at an athermal break of the bond, its strength is determined only by the magnitude of F_{un}.

The average time before the bond break, τ, at a constant value of the applied force F_f is estimated as:

$$\tau = \frac{\tau_0}{P(\delta > \delta_c)} \tag{6.19}$$

where τ_0 is the average oscillation period.

The value of τ_0 can be found by solving the Lagrange equation for the motion of an atom in the potential field $E(u)$:

$$\delta t = \sqrt{2m} \int \frac{du}{\sqrt{U - E(u)}} \tag{6.20}$$

where m is the mass of an atom, δt is the time, and U is the total (kinetic and potential) energy of an atom.

Averaging over the ensemble of states gives:

$$\tau_0 = \frac{\int_0^\infty \delta t \exp\left(-\frac{U}{k_B T}\right) dU}{\int_0^\infty \exp\left(-\frac{U}{k_B T}\right) dU} \tag{6.21}$$

As calculations show, τ_0 values vary in a quite narrow range. For the investigated range of temperatures (600K–1500K) this interval for τ_0 is 0.032–0.037ps. Reducing the rigidity of the atomic bond (the *anharmonism phenomena*) within the region of high loads gives rise to a certain increase in τ_0. Thus, for example, at $F_f = 0.85F_{un}$, the period of vibration of the contact bond atom increases by approximately 7%, and when $F_f = 0.99F_{un}$ is reached, it increases by 20%.

As noted above, to break the bond, it is necessary for these fluctuations to be "picked up" by the applied force F_f; therefore, in a rigorous formulation, this model cannot be employed to predict the probability of the contact bond break at $F_f = 0$. However, in calculations, this difficulty can be overcome if the calculations are performed for sufficiently small but non-zero values of the applied force. In this case, the minimum value was assumed to be $F_{f\min} = 0.065$ nN. Accordingly, the value of the fracture probability at zero load was determined by extrapolation for the value $F_f = 0$.

6.5.2 Lifetime Prediction

As mentioned above, the ratio between F_f and the boundary of the ES bandgap, F_R, has a crucial effect on the regularities of the contact bond break and the waiting time for this break. The proposed fluctuation model enables both to describe quantitatively effects due to the ES bandgap and to predict the effect of temperature and mechanical loading on the lifetime of nanoelements.

As follows from this model, depending on the ratio between F_f and F_R, two different mechanisms of bond break are possible. When the magnitude of the applied force F_f is less than F_R, the short-term bond instability due to fluctuation ($u \geq u_{un}$) cannot cause bond break. This requires the realization of a larger fluctuation δ_c, at which the atom will be "picked up" by the applied force (Figure 6.25a) (mechanism "I"). Another mechanism will be observed when the magnitude of the applied force F_f exceeds the level of F_R (Figure 6.25b). In this case, the fluctuation-induced instability of the contact bond is sufficient for its break; i.e., the bond instability ($u \geq u_{un}$) becomes both a necessary and sufficient condition for such break (mechanism "II"). Accordingly, the value of critical fluctuation required for bond break, $\delta_c = \delta_{un}$, is significantly reduced.

These differences in the mechanisms of fluctuation-induced bond break have a significant influence on both the behavior of the time dependence of strength and the absolute lifetime of a carbyne-based nanoelement. Accordingly, it is possible to distinguish two regions on the dependence of the lifetime on the magnitude of applied force; these regions differ both in the nature of time variation before the contact bond break and in the absolute

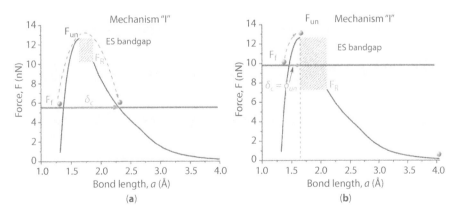

Figure 6.25 Mechanisms of the contact bond break [(a) and (b)]: at $F_f < F_R$, instability of the contact bond is a necessary but insufficient condition for its break (mechanism "I") (a); at $F_f \geq F_R$, instability of the contact bond results in its break (mechanism "II") (b) ($T = 600$K; the 2D nanostructure contains a 10-atomic chain).

value of nanoelement lifetime (Figure 6.26). In this case, the lifetime of a nanoelement ($N = 10$, $T = 600K$) can reach millions of years when the first bond break mechanism is realized (region "I" in Figure 6.26), and in the second case (region "II", Figure 6.26), lifetime does not exceed 0.1 s. From an applied point of view, this means that at realization of the first mechanism, a long-term operation of a nanoelement is possible. The second mechanism should result in a rapid failure of the nanoelement. It is important that a rapid break of the nanoelement will occur at loads less than the strength of the contact bond.

From Equation 6.10, it follows that the critical value of the load, F_f^*, starting from which the bond instability due to thermal fluctuations results in its break (mechanism "II"), is defined as:

$$\frac{F_f^*}{F_{un}} = \frac{1}{\sqrt{1+\alpha}} \tag{6.22}$$

In accordance with the data for α shown in Figure 6.11, for a carbyne–graphene nanoelement, the relative value of critical load, which delimitates the "I" and "II" regions (Figure 6.25), depends weakly on the chain parameters and lies within a rather narrow range $F_f^* = F_{un} \times (0.71-0.74)$. This means that actually attainable strength of such nanostructure is always 30% lower than the contact bond strength; i.e., in such systems, the strength of the contact bond cannot be reached. Existence of ES bandgap is the reason for this effect.

The regularities of change in the lifetime of a carbyne–graphene nanoelement over a wide temperature range are shown in Figure 6.27. As it follows from the data, the lifetime of carbyne-based nanodevice is sufficient for application at temperatures not higher than 800K

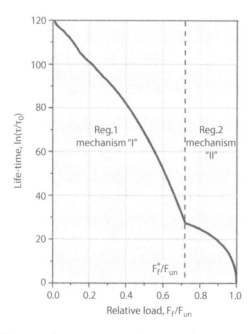

Figure 6.26 Regularities of change in the average time to breaking of the contact bond when the first and second mechanisms of a fluctuation-induced break of the interatomic bond are realized ($T = 600K$; the nanodevice contains a 10-atomic chain).

AB INITIO DESIGN OF GRAPHENE NANOSTRUCTURE 197

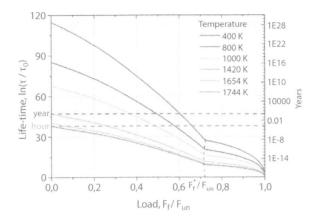

Figure 6.27 Lifetime dependence on load and temperatures: $\tau_0 = 0.042$ ps is the average period of atom vibration; F_f is the value of applied load; F_{un} is the tensile strength of a contact bond; F_f^* is the critical force of change in the mechanism of contact bond breaking.

and loads not exceeding half of tensile strength of the contact bond. At mechanical loads not exceeding 30% of the ultimate one, the considered nanodevice can operate also at a temperature of 1000K. In the unloaded state, such nanodevice has a level of thermal stability sufficient for application up to temperatures not greater than 1500K. It should be emphasized that these results were obtained for a nanoelement containing a carbyne chain of 10 atoms. As calculations show, the expected time to the contact bond break is significantly dependent on whether the number of atoms in the carbyne chain is even or odd (Figure 6.28).

Thus, two opposite regularities will be observed. Without a mechanical load $F = 0$, "even" chains have a longer lifetime, and at loads greater than the critical one $F_f \geq F_f^*$, an opposite regularity is observed. This is due to realization of two different mechanisms of thermofluctuation bond breaking considered above (Figure 6.25). In the first case ($F_f = 0$), the width of ES bandgap equals zero, and the value of work of failure is the main factor that determines the expected time to the contact bond breaking. According to the DFT calculations, the value of E_b for "even" chains is higher (Figure 6.10a). In the region of loads greater than the critical one, $F_f \geq F_f^*$, an opposite regularity is observed. "Odd" chains have a longer waiting time for contact bond break. As was shown above, in this case, the contact bond instability is both a necessary and sufficient condition for its break; i.e., the second mechanism is realized (Figure 6.25b). Accordingly, the instability force, F_{un}, has a decisive influence on the lifetime. For "odd" chains, F_{un} is higher (Figure 6.10a). With increasing load from $F_f = 0$ to $F_f \geq F_f^*$, there is a transition from the first mechanism to the second one. As a result, more reliable "even" chains become less reliable. In the transition region, the "even–odd" effect disappears. This occurs when the load is close to $F_f \approx 0.3 F_{un}$.

Thus, depending on the mechanism of contact bond break, the expected time of this break is governed either by the contact bond strength or by the value of binding energy. In turn, the value of these characteristics depends on whether the number of atoms in the chain is even or odd. The ascertained regularities of the effect of temperature and mechanical load on the value of lifetime exhibit the fact that the *descending* branch of dependence "the force vs. the displacement of atom" plays a decisive role in the fluctuation-induced break of interatomic bonds. Only in the region of high mechanical loads ($F_f \geq 0.7 F_{un}$),

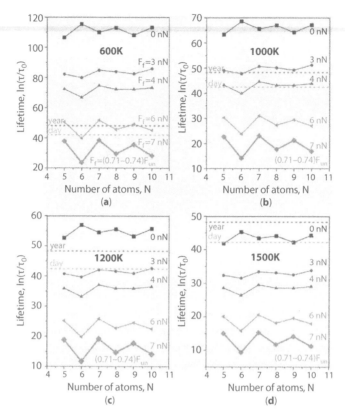

Figure 6.28 Dependence of the lifetime on the number of atoms in the chain at fixed temperatures T and mechanical load F_f. The bottom curve in each graph corresponds to a critical load.

where the magnitude of critical fluctuation does not exceed $\delta_c \leq u_{un} - u_f$ (Figure 6.25b), is lifetime predetermined by the characteristics of the ascending branch on this dependence.

$$F_f^* = (0.71 - 0.74) \times F_{un}.$$

The data shown in Figure 6.28 indicate that from a practical point of view, when using carbyne-based nanoelement, mechanical loads exceeding the critical one $F_f^* = (0.71 - 0.74) \times F_{un}$ cannot be tolerated. In an unloaded state, nanoelements containing six-atom carbyne chains should have the longest lifetime. In the loaded state ($F_f^* > F_f \geq 0.3 F_{un}$), devices with "odd" carbyne chains become more preferable. From an applied point of view, it means that these nanoelements should be used in systems in which they are strained during operation.

6.6 Conclusions

1. *Ab initio* calculations demonstrate that the formation of three-dimensional structures, where graphene sheets are connected by carbyne chains, is

energetically favorable. This fact paves the way to possible synthesis of a new type of 3D carbon structure—"carbynophenes".

2. Two main types of contact bonds connecting carbyne chain with graphene sheet were identified: "atom–atom" and "bridge" ones. The combination of these bond types enables one to obtain a great number of carbynophene architectures, which are characterized by a wide range of formation enthalpy, and vary in strength and elastic properties. The possibility of the existence of a structurally modified graphene (M-graphene) is the key to the creation of carbynophenes, as it enables to solve the problem of anchoring carbyne chains to graphene sheets.

3. The strength of contact bonds and their concentration on the graphene sheet are two main factors governing the strength of carbynophenes. The highest bond strength is realized in carbynophene, obtained on the basis of M-modified graphene. It is equal to 10.5 nN, which is 80% of the strength of the five-atom carbyne chain. In this case, one chain exists per three M-graphene cells, which makes it possible to obtain the high level of strength of the carbynophene in a direction perpendicular to the plane of M-graphene sheets. It is 68 GPa at a modulus of elasticity in this direction, equal to 766 GPa.

4. The ability to change the elastic moduli over a wide range of values (from 95 GPa to 766 GPa) while retaining high values of their strengths is one of the specific features of carbynophenes, which is very attractive for possible nano-devices and medical applications.

5. The lifetime of carbyne–graphene nanostructures is governed by the phenomenon of fluctuation-induced contact bond break. Short-term fluctuation in the contact bond length can result in its break only under the condition if its magnitude reaches a critical value, which is determined by the magnitude of the mechanical load and the nature of the descending branch of the interatomic interaction dependence.

6. Existence of a bandgap for equilibrium states of atoms (ES bandgap) is a key feature of the behavior of carbyne chains and 2D–3D carbyne–graphene nanostructures under mechanical stresses. This has a crucial effect on the lifetime of such nanoelements and is the reason for the existence of two mechanisms for the contact bond break. These mechanisms are distinguished by a different "response" of the interatomic bond to the fluctuation of its length. One of them is manifested in the fact that short-term bond instability of interatomic bond is a necessary but insufficient condition for its break. This means that to break the bond, the fluctuation of the interatomic distance must reach a critical value at which it can be "picked up" by the applied force. The peculiarity of the second mechanism is that the fluctuation-induced instability of the contact bond is a sufficient condition for its break. In this case, considerably smaller fluctuations lead to the contact bond break. As a result, lifetime decreases by many orders of magnitude compared to the first case. For the carbyne–graphene nanostructures considered, the transition from the first to the second mechanism occurs when the value of the applied force exceeds 71%–74% of the contact bond strength.

7. The lifetime of 2D carbyne–graphene nanoelements depends on whether the number of atoms in carbyne chain is even or odd ("even–odd" effect). This is a consequence of the existence of an "even–odd" effect for strength (value of instability force) and binding energy of a contact bond. The kind of mechanism of the interatomic bond break influences the regularity of manifestation of the "even–odd" effect. When realizing the first mechanism, nanoelements with an "even" chain have a greater value of the lifetime. When realizing the second mechanism, on the contrary, nanoelements with "odd" carbyne chains become more stable. From an applied point of view, it means that this type of nanoelements is preferable for systems in which carbyne chains are strained during operation. In the absence of deformation, it is advisable to use carbyne-based nanoelements with even chains.

Funding

This work was supported by the National Academy of Sciences of Ukraine (grant numbers #011U002131, #011U006351 and #0116U003051).

References

1. Giannozzi, P. *et al.*, QUANTUM ESPRESSO: A modular and open-source software project for quantum simulations of materials. *J. Phys. Condens. Matter.*, 21, 395502, 2009.
2. Original QE PP library http://www.quantum-espresso.org/pseudopotentials/
3. Perdew, J.P., Burke, K., Ernzerhof, M., Generalized gradient approximation made simple. *Phys. Rev. Lett.*, 77, 3865–3868, 1996.
4. Monkhorst, H.J. and Pack, J.D., Special points for brillouin-zone integrations. *Phys. Rev. B*, 13, 5188–5192, 1976.
5. Fletcher, R., *Practical Methods of Optimization*, Wiley, New York, 1987.
6. Billeter, S.R., Turner, A.J., Thiel, W., Linear scaling geometry optimisation and transition state search in hybrid delocalised internal coordinates. *Phys. Chem. Chem. Phys.*, 2, 2177, 2000.
7. Billeter S., R. and Curioni A. and Andreoni, W., Efficient linear scaling geometry optimization and transition-state search for direct wavefunction optimization schemes in density functional theory using a plane-wave basis. *Comput. Mater. Sci.*, 27, 437, 2003.
8. Methfessel, M. and Paxton, A.T., High-precision sampling for Brillouin-zone integration in metals. *Phys. Rev. B*, 40, 3616, 1989.
9. Kavan, L., Hlavat´y, J., Kastner, J., Kuzmany, H., Electrochemical carbyne from perfluorinated hydrocarbons: Synthesis and stability studied by Raman scattering. *Carbon*, 33, 1321, 1995.
10. Casari, C.S., Li Bassi, A., Ravagnan, L., Siviero, F., Lenardi, C., Piseri, P., Bongiorno, G., Bottani, C.E., Milani, P., Chemical and thermal stability of carbyne-like structures in cluster-assembled carbon films. *Phys. Rev. B*, 69, 075422, 2004.
11. Jin, C., Lan, H., Peng, L., Suenaga, K., Iijima, S., Deriving carbon atomic chains from graphene. *Phys. Rev. Lett.*, 102, 205501, 2009.
12. Liu, M., Artyukhov, V.I., Lee, H., Xu, F., Yakobson, B.I., Carbyne from first principles: Chain of C atoms, a nanorod or a nanorope. *ACS Nano*, 7, 11, 10075, 2013.

13. Banhart, F., Chains of carbon atoms: A vision or a new nanomaterial? *Beilstein J. Nanotechnol.*, 6, 559, 2015.
14. Ravagnan, L., Manini, N., Cinquanta, E., Onida, G., Sangalli, D., Motta, C., Devetta, M., Bordoni, A., Piseri, P., and Milani, P., Effect of axial torsion on sp carbon atomic wires. *Phys. Rev. Lett.*, 102, 245502, 2009.
15. Cinquanta, E., Ravagnan, L., Castelli, I.E., Cataldo, F., Manini, N., Onida, G., Milani, P., Vibrational characterization of dinaphthylpolyynes: A model system for the study of end-capped sp carbon chains. *J. Chem. Phys.*, 135, 194501, 2011.
16. Durgun, E., Senger, R.T., Mehrez, H., Dag, S., Ciraci, S., Nanospintronic properties of carbon-cobalt atomic chains. *Europhys. Lett.*, 73, 642, 2006.
17. Wang, Y., Ning, X.-J., Lin, Z.-Z., Li, P., Zhuang, J., Preparation of long monatomic carbon chains: Molecular dynamics studies. *Phys. Rev. B*, 76, 165423, 2007.
18. Erdogan, E., Popov, I., Rocha, C.G., Cuniberti, G., Roche, S., Seifert, G., Engineering carbon chains from mechanically stretched graphene-based materials. *Phys. Rev. B*, 83, 041401(R), 2011.
19. Rinzler, G., Hafner, J., Nikolaev, P., Nordlander, P., Colbert, D.T., Smalley, R.E., Lou, L., Kim, S.G., Tomanek, D., Unraveling nanotubes: Field emission from an atomic wire. *Science*, 269, 1550, 1995.
20. Lang, N.D. and Avouris, P., Oscillatory conductance of carbon-atom wires. *Phys. Rev. Lett.*, 81, 3515, 1998.
21. Yazdani, A., Eigler, D.M., Lang, N.D., Off-resonance conduction through atomic wires. *Science*, 272, 1921, 1996.
22. Ragab, T. and Basaran, C., The unravelling of open-ended single walled carbon nanotubes using molecular dynamics simulations. *J. Electron. Packag.*, 133, 020903, 2011.
23. Ataca, C. and Ciraci, S., Perpendicular growth of carbon chains on graphene from first-principles. *Phys. Rev. B*, 83, 235417, 2011.
24. Kotrechko, S., Mazilov, A.A., Mazilova, T.I., Sadanov, E.V., Mikhailovskij, I.M., Experimental determination of the mechanical strength of monatomic carbon chains. *Tech. Phys. Lett.*, 38, 132, 2012.
25. Mikhailovskij, I.M., Sadanov, E.V., Kotrechko, S., Ksenofontov, V.A., Mazilova, T.I., Measurement of the inherent strength of carbon atomic chains. *Phys. Rev. B*, 87, 045410, 2013.
26. Huang, Y., Wu, J., Hwang, K.C., Thickness of graphene and single-wall carbon nanotubes. *Phys. Rev. B*, 74, 245413, 2006.
27. Fan, X.F., Liu, L., Lin, J., Shen, Z.X., Kuo, J.-L., Density functional theory study of finite carbon chains. *ACS Nano*, 3, 3788, 2009.
28. Cahangirov, S., Topsakal, M., Ciraci, S., Long-range interactions in carbon atomic chains. *Phys. Rev. B*, 82, 195444, 2010.
29. Peierls, R.E., *Quantum Theory of Solids*, p. 108, Oxford University Press, New York, 1955.
30. Pitzer, K.S. and Clementi, E., Large molecules in carbon vapor. *J. Am. Chem. Soc.*, 81, 4477, 1959.
31. Timoshevskii, A., Kotrechko, S., Matviychuk, Yu., Atomic structure and mechanical properties of carbyne. *Phys. Rev. B*, 91, 245434, 2015.
32. Lin, Z.Z., Yu, W.F., Wang, Y., Ning, X.J., Predicting the stability of nanodevices. *EPL*, 94, 40002, 2011.
33. Kotrechko, S., Timoshevskii, A., Kolyvoshko, E., Yu. Matviychuk, N., Stetsenko, Thermo-mechanical stability of carbyne-based nanodevices. *Nanoscale Res. Lett.*, 12, 327, 2017.
34. Ataca, C., Aktürk, E., Şahin, H., Ciraci, S., Adsorption of carbon adatoms to graphene and its nanoribbons. *J. Appl. Phys.*, 109, 013704, 2011.

35. Krenn, C.R., Roundy, D., Morris, J.W., Jr., Marvin, L., Cohen, ideal strengths of bcc metals. *Mater. Sci. Eng. A*, 319–321, 111–114, 2001.
36. Zhu, T., Li, J., Samanta, A., Leach, A., Gall, K., Temperature and strain-rate dependence of surface dislocation nucleation. *Phys. Rev. Lett.*, 100, 025502, 2008.
37. Zhao, H. and Aluru, N.R., Temperature and strain-rate dependent fracture strength of graphene. *J. Appl. Phys.*, 108, 064321, 2010.
38. Regel, V.R., Slutsker, A.Zh., Tomashevskiy, E.E., Kineticheskaya priroda prochnosti tverdyh tel. *Uspehi fizicheskih nauk*, 106, 2, 193–223, 1972 (in Russian).

7

Graphene-Based Composite Nanostructures: Synthesis, Properties, and Applications

Mashkoor Ahmad[1]* and Saira Naz[1,2]

[1]Nanomaterials Research Group (NRG), Physics Division, PINSTECH, P.O. Nilore, Islamabad, Pakistan
[2]Institute of Chemical Sciences, University of Peshawar, Pakistan

Abstract

Graphene has recently attracted great attention because of its unique properties such as giant electron mobility, extremely high thermal conductivity, extraordinary elasticity and stiffness, and ultralarge specific surface area. In addition, graphene has attracted much attention because of its wide potential applications in nanoelectronics, energy storage and conversion, chemical and biological sensors, composite materials, and biotechnology. This chapter provides a comprehensive investigation on the current research activities that focus on the synthesis, properties, and applications of graphene-based composite nanostructures. We briefly describe the most commonly applied methodologies for the synthesis of graphene-based composite nanostructures. A range of remarkable properties is then presented. Finally, we include a brief analysis on the potential applications of graphene-based composite nanostructures in various fields including energy storage, medicine, catalysis, and development of biosensors. These studies constitute the basis for developing versatile applications of graphene-based composite nanostructures.

Keywords: Nanocomposite, graphene oxide, electrochemical sensor, lithium ion batteries, supercapacitor, catalysis, exfoliation, antibacterial activity, chemical vapor deposition, reduced graphene oxide, hybrid materials, photocatalytic activity

7.1 Introduction

Until the mid-1980s, pure solid carbon was thought to exist in only two physical forms, diamond and graphite. Diamond and graphite have different physical structures and properties; however, their atoms are arranged in covalently bonded networks. These two different physical forms of carbon atoms are called allotropes.

Graphite, usually known as the main ingredient of lead pencil, has several large-scale industrial applications, namely, battery electrodes and industrial-grade lubricants. Mostly, graphite is an active precursor material to engineer various types of carbon-based nanomaterials including single- or multiwall nanotubes, fullerenes, and graphene [1, 2]. Carbon nanotubes have a cylindrical carbon structure and possess a wide range of electrical and

---

Corresponding author: mashkoorahmad2003@yahoo.com; mashkooreml@gmail.com

optical properties not only because of their extended sp² carbon but also because of their tunable physical properties (e.g., diameter, length, single-walled vs. multi-walled, chirality, and surface functionalization). The physical properties of carbon nanotubes such as mechanical strength, electrical conductivity, and optical properties could be of great value for creating advanced carbon-based biomaterials. The electrical properties of CNTs rely on the fabrication of electronic devices owing to the high length of carbon nanotubes ranging from 100 nm to bigger than several hundred micrometers. Fullerene, commonly known as the buckyball, is a spherical closed cage structure made up of 60 sp² hybridized carbon with highly symmetrical electronic structure that has somewhat lost its popularity in recent years with the rise of more scalable and useful carbon-based materials such as carbon nanotubes and graphene. Graphene is a soft membrane with high Young's modulus. Single-layered graphene is transparent with good thermal and electric conductivities and specific surface of approximately 2600 m²/g. Graphene enriched with oxygen containing functional groups (graphene oxide) have great affinity for nanoparticle growth [3–7].

7.2 Carbon Nanomaterials

Materials derived from carbon including graphite, diamond, fullerenes, nanotubes, nanowires, and nanoribbons have been used for innumerable applications such as electronics, optics, optoelectronics, biomedical engineering, tissue engineering, medical implants, medical devices, and sensors [8]. In graphite, every single carbon atom is attached to other carbon atoms through strong covalent bonds in one plane. However, the interlayer binding through weak van der Waals forces is responsible for its softness, in contrast to diamond. Likewise, carbon nanotubes and fullerenes are other forms of carbon, having tubular and spherical arrangements (Figure 7.1). Carbon nanofibers are sp²-bonded linear filaments (diameter of 100 nm) known for their flexibility. Fibrous materials are of great importance owing to their significant high specific area in combination with flexibility and high mechanical strength, allowing its use in daily life. Conventional carbon fibres have several micrometer-sized diameters and are different from carbon nanotubes. Carbon nanofibers grow by passing carbon feedstock over nanosized metal particles at

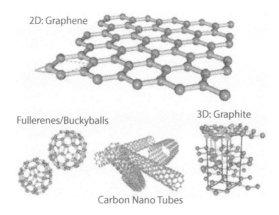

Figure 7.1 Carbon structures representing diamond, graphite, fullerene, and carbon nanotube.

elevated temperatures very similar to the growth condition of carbon nanotubes. Still, carbon nanotubes are different in geometry, having an entire hollow core along the filament length. Graphene is a novel addition to carbon family materials with its unique and versatile properties. With its strong carbon–carbon in-plane bonding, aromatic structure, freely motile π electrons, and surface-active sites, graphene is a unique material with extraordinary mechanical, thermal, electronic, optical, and biomedical properties.

7.3 Graphene

Graphene recently emerged as an attractive and alternative energy storage material with superior and distinctive properties like chemically inert, low weight, and low price. Graphene is a large monolayer sp^2-bonded carbon sheet with unique optical, electrical, mechanical, and electrochemical properties. The surface area of graphene is 2630 $m^2\ g^{-1}$, which is massively favorable for numerous applications. Graphene is conductive and easy to functionalize with other molecules. A family of graphene-related materials, comprising structural or chemical derivatives of graphene, is called "graphenes" by the research community. These include double- and few (3 to 9)-layer graphene and graphene restricted along a plane (resembling a polyaromatic molecule) called a graphene nanoribbon (single, double, few, or multilayer). The most important chemically derived graphene is graphene oxide (single layer of graphite oxide), usually synthesized from graphite by oxidization to graphite oxide and consequent exfoliation to graphene oxide. Graphene nanomaterials are classified based on either number of layers in the sheet or their chemical modification, comprising single-layer graphene, bi-layer graphene, multilayer graphene, graphene oxide (GO), and reduced graphene oxide (rGO). Each member fluctuates from the other in terms of number of layers, surface chemistry, purity, lateral dimensions, defect density and composition. Single-layer graphene is an isolated single layer of carbon atoms bonded together in a planar 2D structure. Graphene oxide (GO) is a highly oxidized form of chemically modified graphene that consists of a single-atom-thick layer of graphene sheets with carboxylic acid, epoxide, and hydroxyl groups in the plane (Figure 7.2). The peripheral carboxylate group provides colloidal stability and pH-dependent negative surface charge. Epoxide (-O-) and hydroxyl (-OH) groups present on the basal plane are uncharged but polar, allowing weak interactions, hydrogen bonding, and other surface reactions [9, 10]. The basal plane also contains free surface π electrons from unmodified areas of graphene, which are hydrophobic and capable of π–π interactions [11]. Three-dimensional (3D) graphene-based frameworks such as aerogels, foams, and sponges are an important class of new-generation porous carbon materials, exhibiting interconnected macroporous structures, low mass density, large surface area, and high electrical conductivity. These materials can serve as a robust matrix for accommodating metal, metal oxide, and electrochemically active polymers for various applications in capacitors, batteries, and catalysis [12].

7.3.1 Graphene Structure

Graphene is a two-dimensional (2D) sp^2-bonded carbon sheet, arranged in a hexagonal honeycomb lattice. From a fundamental point of view, graphene is nothing but a single layer of graphite, which is an infinite three-dimensional (3D) material made up of stacked

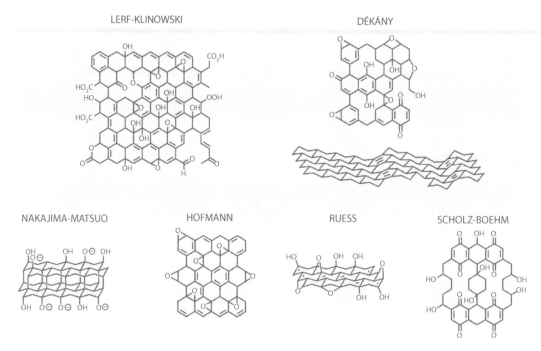

Figure 7.2 Summary of proposed structural models of graphene [17].

layers of graphene. The layers in graphite interact weakly through van der Waals forces. From a condensed matter viewpoint, graphene is constructed of sp²-bonded carbon atoms *via* hybridization of s, p_x, and p_y atomic orbitals, forming three strong σ bonds with three adjacent atoms. The remaining p_z orbital on each carbon overlaps with those from neighboring atoms, establishing a filled band of π orbitals (valence band) and an empty band of π* orbitals (conduction band).

In terms of properties, graphene is unique; it has a soft membrane and, at the same time, possesses a high Young's modulus and good thermal and electrical conductivities. In addition, a single-layer graphene is a zero-band gap material, is highly transparent, and exhibits optical transmittance of 97.7%. With its high theoretical specific surface area of ~2600 m²/g, graphene provides a rich platform for surface chemistry. The combined extraordinary physical and chemical properties of graphene, in turn, has ignited extensive research in nanoelectronics, supercapacitors, fuel cells, batteries, photovoltaics, catalysis, gas sorption, separation and storage, and sensing.

Graphene is not soluble in most solvents. To date, graphene is only soluble in solvents exhibiting surface tension close to 40–50 mJ m⁻², like benzyl benzoate, *N,N*-dimethylacetamide (DMA), g-butyrolactone, or 1,3-dimethyl-2-imidazolidinone. Frequent problems of graphene sheets forming irreversible agglomerates or restacking to form graphite via p–p stacking and van der Waals interactions are imminent concerns as well [13, 14]. The chemical structure of graphene oxide intrinsically originates from graphite oxide. Several structures have been proposed including Hofmann, Ruess, Scholz-Boehm, Nakajima-Matsuo, Lerf-Klinowski, and Dékány models. Among them, the Lerf-Klinowski model is currently the most widely accepted configuration (Figure 7.2) [12].

This model consists of unoxidized aromatic regions and aliphatic six-membered rings containing OH and epoxide, whereas the edges are terminated with OH and COOH groups [15]. The additional presence of five- and six-membered-ring lactols decorating along the peripheral edges of GO as well as esters of tertiary alcohols on the surface has been reported recently [16]. The type of oxygen functionalities and their relative proportion and coverage density on graphene oxide depend on synthetic methods and graphite sources used.

7.3.2 Graphene Synthesis

Numerous methods have been developed to synthesize graphene; the raw material used to prepare graphene such as natural graphite, carbons, polymers, and biomass waste is abundantly available. The availability of effective methods of synthesis and a high range of precursors as well as unique properties of graphene make it a very promising candidate for large-scale production and commercialization. Graphene synthesis can be categorized into two methodologies: top-down and bottom-up; the top-down technique involves (i) isolating graphene from the stacked parent materials by solid-phase, liquid-phase, or electrochemical exfoliation of pristine graphite and graphite intercalated compounds, and (ii) exfoliating graphite oxide into graphene oxide (GO) followed by chemical, thermal, and electrochemical reduction. The bottom-up approach involves building up graphene from molecular precursors, typically including chemical vapor deposition (CVD) and epitaxial growth [18].

7.3.2.1 *Exfoliation of Graphite*

The simplest technique is the "scotch tape method" used for freeing graphene layers from graphite (Figure 7.3). In 2004, Geim and Novosolev isolated one-layer-thick graphene by the Scotch tape method, also known as "the micro-mechanical exfoliation method" [20]. Efficient exfoliation of graphite can therefore be achieved by providing an external force above van der Waals forces by increasing the interlayer spacing in the solid and liquid states. A graphene monolayer sheet with a thickness of about 0.4 nm and a lateral size up to microns can be isolated by this process. The method is quite simple yet highly reliable to achieve the best samples in terms of purity, defects, charge mobility, and optoelectronic properties but is not applicable on a large scale. Liquid-phase exfoliation of graphite involves dispersion followed by sonication-induced exfoliation in suitable solvents in the absence/presence of surfactants. Solvent molecules by themselves cannot inherently dissolve graphene, and solvent–graphene interactions thus need to balance intersheet attractions of graphene after exfoliation to avoid their restacking. The exfoliated graphene sheets consist of 28% monolayer and nearly 100% few-layer (up to 5) pristine graphene. The yields of monolayer graphene can significantly be enhanced by increasing sonication time, repeated exfoliation, and subjecting it to solvothermal and supercritical treatments. The addition of surfactants, organic molecules, and polymers to organic solvents enhances exfoliation of graphite; it also stabilizes graphene suspensions by molecular adsorption onto the basal planes and edges of exfoliated sheets [21–24]. Furthermore, they tune the water surface tension to an appropriate level for aqueous exfoliation of graphite. Nonionic surfactants, attached to both sides of graphene through

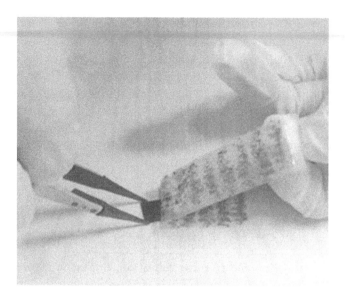

Figure 7.3 Scotch tape method for graphene synthesis [19].

hydrophobic interaction, produce steric repulsion to separate graphene sheets. Beside conventional surfactants, pyrene and perylene-containing molecules, with hydrophobic aromatic rings and hydrophilic functional groups, also act as surfactants to assist in the exfoliation of graphite in aqueous solutions [25–28]. Since sonication weakens van der Waals attractive forces in between layers, these conjugated molecules intercalate into the interlayers and adsorb *in situ* onto the graphene surface through hydrophobic and p–p interactions between layers [29]. Stable liquid-exfoliated graphene dispersions can function as conducting inks and hence enable top-down approaches to print electronics and can also be processed into flexible, transparent, conducting, and freestanding films for cell electrodes [30–32]. Electrochemical exfoliation of graphite into graphene involves utilizing graphite rods or foils as electrodes (mostly anode) in an electrolytic cell and then collecting the exfoliated graphene from the electrolyte solution. Various aqueous and nonaqueous electrolyte solutions have been developed. The aqueous electrolytes of surfactants and polymers are responsible for the electrolytic exfoliation of graphite into graphene due to their hydrophobic aromatic rings interacting with the p-orbitals of graphene [33, 34]. However, the adsorbed surfactants and polymers cannot be fully removed, resulting in interference with the electrical and electrochemical properties of graphene. Protonic acids, such as sulfuric acid (H_2SO_4) and phosphoric acid (H_3PO_4), are found to be good electrolytes for the exfoliation of graphite due to the intercalation of electrolyte anions, radicals, and their solvated complexes between the layers [35–39].

7.3.2.2 CVD Synthesis

Chemical vapor deposition has emerged as an important method for scalable production of high-quality graphene films [40]. This technique involves the pyrolysis of hydrocarbon compounds on the surface of transition metal catalysts. The quality of graphene is mainly determined by processing parameters such as catalysts, precursors, gas flow rate, temperature, pressure, and time [41, 42]. CVD is the most promising route toward the synthesis of large-area graphene required for electronic and optoelectronic applications [43]. In general, the CVD process includes four steps: (i) adsorption and catalytic decomposition of precursors (gas phase), (ii) diffusion and dissolution of decomposed carbon species into bulk (iii), segregation of dissolved carbon atoms onto the surface of metals, and (iv) surface nucleation and graphene growth [44]. Thus, graphene film grows on a catalytic metallic (like copper) layer coated on a substrate prior to graphene growth, forming carbon species when the substrate is exposed to precursor molecular flux. Other metals like nickel, silver, gold, platinum, and cobalt can be used as the catalytic layer. Both low- and high-temperature CVD is used for graphene growth, producing graphene with large surface area; however, its efficiency depends on the quality of polycrystalline metallic film (catalyst) and it requires multiple processing steps to obtain transferable sheets [45].

With the chemical vapor deposition (CVD) method, graphene with well-defined basal plane enriched and edge plane enriched have been fabricated on a Cu and Ni substrate, respectively. Moreover, the efficacious syntheses of large-area mono- and multilayer graphene and the feasibility to transfer onto any substrate give an opportunity to explore numerous essential science issues. Cu- and Ni-based CVD graphene has received huge attention, and other transition metals, such as Fe, Ru, Co, Rh, Ir, Pd, Pt, Au, and alloys such as Co–Ni, Au–Ni, and Ni–Mo are able to support the growth of graphene (Figure 7.4) [46–48]. By tuning CVD parameters and composition of catalysts and precursors, graphene with desired layer number, grain size, band gap, and doping effect can be achieved [49]. However, CVD is usually limited to the use of gas precursors. In addition, the current methods for transferring graphene is to etch away the metal substrate with etchants, which leads to higher cost, toxic wastes, and structural damage to graphene. Recently, vertically oriented graphene (VG) nanosheets have been grown on various substrates (e.g., planar or cylindrical metals, and carbon nanotubes) through plasma-enhanced chemical vapor deposition (PECVD).

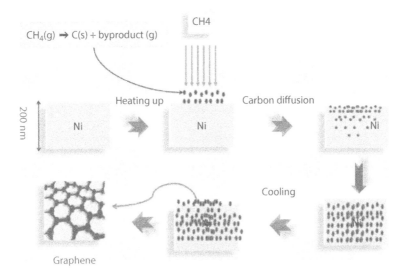

Figure 7.4 CVD graphene growth mechanism on nickel substrate [50].

As graphene shows a higher in-plane than out-of-plane electrical conductivity, this vertical graphene serves as an ideal electrical "bridge" linking the current collector and active materials.

7.3.2.3 Epitaxial Growth

Thermal decomposition of SiC to produce monolayer graphene has been demonstrated through the graphitization of SiC by Si sublimation during high-temperature vacuum annealing. The advantage of this approach is that insulating SiC substrates are used so that transfer to another insulator is not required. Unfortunately, thermal annealing under vacuum often yields graphene layers with small graphene domains (30–200 nm). Thermal decomposition of SiC is also not a self-limiting process and hence graphene regions with different thicknesses often coexist. The presence of disilane during SiC decomposition was found to reduce the Si sublimation rate, thus enabling the formation of high-quality graphene [51, 52]. Epitaxial graphene seems to be suitable for wafer-based electronic and component applications; however, commercial SiC is still expensive, particularly for large-area films. Moreover, for epitaxial graphene, high temperature (>1000°C) is usually required, and this is not compatible with current silicon electronics technology [53, 54].

7.3.2.4 Chemical Method

Graphene oxide is typically synthesized by Hummer's method (Figure 7.5) [55, 56]. It requires graphite flakes, sodium nitrate, concentrated acid (like sulfuric acid), permanganate, and deionized water. The components are mixed under stirring conditions in an ice bath to quench

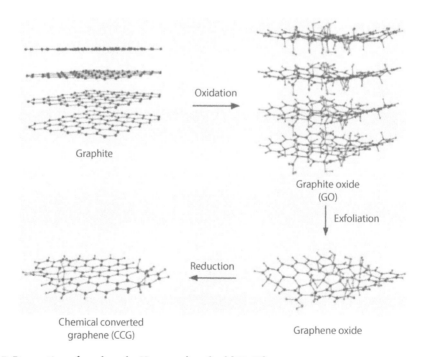

Figure 7.5 Preparation of graphene by Hummers' method [57, 58].

the reaction heat. This mixture is then treated with hydrogen peroxide for an optimized time. Afterward, the mixture is cleaned with deionized water by repeated centrifugation followed by filtration. The resulting wet powders of graphene oxide are vacuum dried. Multilayered GO is produced by coarse oxidation of crystalline graphite followed by dispersion in aqueous medium through sonication or other processes. However, repeated treatment, centrifugation, and severe situation lead to production of monolayer oxidized graphene. Reduced graphene oxide (rGO) can be obtained by thermal, chemical, and UV treatment of GO under reducing conditions with hydrazine or other reducing agents. Reduced GO is mainly produced to restore the electrical conductivity and optical absorbance in GO while reducing the oxygen content, surface charge, and hydrophilicity. Functionalized graphene is modification of any graphene family member by polymers, small molecules, nanoparticles, etc. to enhance or alter the properties required for definite application.

7.3.3 Graphene Properties

Graphene is an indefinitely extended two-dimensional (2D) carbon crystal, in which carbon atoms are packed in a hexagonal lattice resembling a honeycomb. Significantly, a lot of fascinating properties of graphene have been reported, including high specific surface area, excellent mechanical strength and flexibility, unparalleled thermal and electrical conductivity, as well as superior electronic properties. Graphene can be considered either as a metal with vanishing Fermi surface or a semiconductor with zero band gap. The lack of intrinsic band gap greatly limits the applications of pristine graphene in such areas as nanoelectronics, energy storage, and electro catalysis; hence, it is appealing to induce a band gap in graphene. Since 2004, many interesting properties of graphene have been discovered, which include high thermal conductivity, ultra-high charge carrier mobility, large theoretical specific surface area, and extraordinary mechanical properties. Extremely high intrinsic thermal conductivity, the thermal conductivity of a single-layer graphene sheet, is measured to be as high as ~5000 W mK^{-1} at room temperature. Chemically organized graphene exhibits divergent electrochemical properties compared to graphite, attributed to the presence of some residual oxygen groups.

7.3.3.1 *Physicochemical Properties*

The honeycomb lattice structure of a single atomic layer of graphene comprises two equivalent sublattices connected through σ bonds with each carbon atom having free π electrons contributing toward a delocalized electronic system. The free π electrons provide high electron density above and below the 2D plane of graphene. These free electrons interact freely with the boundary molecular orbitals of many organic compounds by electrophilic substitution compared to nucleophilic substitution. The planar structure of graphene also enables it to participate in several reactions like click reactions, cyclo-additions, and carbine insertion reactions. This transforms the sp^2 system to an sp^3 arrangement leading to formation of topological faults (pentagon, heptagon, or their combinations) [59]. The chemical reactivity of geometrically trained areas and zigzag edges of graphene is found greater than unstrained areas or armchair edges due to the ease of electron displacement from the upper plane of the aromatic ring. Zigzag edges are distorted by the aromatic sextet that causes thermodynamic instability and makes them more reactive than armchair edges [59].

Thus, geometric strains or defects may be deliberately imparted to graphene for applications requiring higher chemical reactivity. Pristine graphene is hydrophobic in nature (water contact angle in the range of 95–100°) [60, 61]. Due to slight dispersion in water, a surfactant or another stabilizing agent is added to achieve suspension and prevent agglomeration. However, graphene oxide (water contact angle of 30.7°) [61] forms hydrogen bonds and metal ion complexes because of the polar basal plane and negative charges (having carboxylate groups on the edge site). Reduced graphene oxide has basal vacancy defects formed during deoxygenation, making it less hydrophobic than graphene, and exhibits less basal reactivity than graphene oxide [62, 63]. Physicochemical properties like the unique planar 2D structure, high specific surface area, and availability of free π electrons make graphene a good candidate for interaction with organic molecules.

7.3.3.2 Thermal and Electrical Properties

One-atom-thick carbon membranes turned out to have the highest known electrical and thermal conductivity with low coefficient of thermal expansion and low defect density in the crystal lattice as well as the highest stiffness and strength. The thermal conductivity of single-layer, defect-free graphene is ~4500 to 5200 W/mK, notably higher than graphene oxide (~2000 W/mK) [64], multiwall carbon nanotubes (~3000 W/mK), and single-wall carbon nanotubes (~3500 W/mK) [65, 66]. The electrical conductivity of defect-free single-layer graphene is 10^4 S/cm and that of graphene oxide is 10^{-1} S/cm at room temperature. During chemical modification or processing, defects arise, which disturb the flow of electrons and heat, thereby reducing conductivity. For instance, the thermal conductivity of supported graphene (graphene on silicon carbide substrate) is significantly lower (~600 W/mK) than that of pure graphene. Other phenomena such as defect edge scattering and isotopic doping due to scattering or localization of phonons at the defect sites widely affect thermal properties [67, 68]. Electron mobility of suspended graphene is greatly affected by impurities on graphene surface and those trapped between the substrate and graphene as well [69]. The outstanding thermal and electrical conductivity of graphene is useful not only in electronic devices but also in biomedical devices for measuring cell potential and biosensors.

7.3.3.3 Optical Properties

Graphene has gained a lot of curiosity due to its superb electric charge transport and optical properties. Single-layer graphene transmits 97.7% of the total incident light over a broad range of wavelengths. Light absorption and optical image contrast increase with increase in the number of layers of graphene [70]. Graphene-based optoelectronic devices can also be developed as tunable IR detectors, modulators, and emitters by electrical gating and charge injection. Depending on the density of electrons and holes, electron–hole pairs generated upon light absorption on graphene surface can recombine rapidly (picoseconds), but they can be separated by applying an external or internal field formed near the electron graphene interface to generate photo current [71]. This ability to control the recombination and separation of surface electrons can be exploited in developing bioimaging applications.

Graphene can be made luminescent by cutting into nanoribbons and quantum dots to induce a suitable band gap or by physicochemical treatment using various gases to trim down the π electron network [72, 73]. Recombination of electron–hole pairs also contributes to the photoluminescence of graphene. Eminent light transmittance, photoluminescence, and outstanding charge mobility make graphene a significant material for applications in magnetic resonance imaging (MRI) and biomedical imaging.

7.3.3.4 Mechanical Properties

The breaking strength of single-layer defect-free graphene is approximately 200 times higher than steel, making it one of the strongest materials tested [74]. Young's modulus, Poisson's ratio, and fracture strength for defect-free graphene are 1 TPa, 0.149 GPa, and 130 GPa, respectively [75]. Methods like numerical simulations (e.g., molecular dynamics) force displacement, force volume, and nano-indentation atomic force microscopy (AFM) are used for mechanical strength of graphene. GO have significantly lower mechanical strength than pure graphene (Young's modulus in the range of 0.15–0.35 TPa) [76]. GO platelets (paper-like layer) exhibit an elastic modulus of 32 GPa and a fracture strength of 120 MPa [77]. Because of its outstanding mechanical strength, graphene has been discovered for enhancing mechanical properties of polymeric materials and significantly increased the modulus and hardness of the composites for biological applications [78]. When graphene is used with other carbon materials like carbon nanotubes (CNTs), the mechanical strength of polymer composites increased up to 400% due to synergistic effect [79, 80]. High strength and capability of tuning the mechanical properties using various functionalization approaches imply the potential of graphene as fillers or reinforcements in medical implants, hydrogels, and scaffolds used in tissue engineering.

7.3.3.5 Biological Properties

Graphene nanomaterials with different physicochemical properties exhibit unique modes of interaction with biomolecules, cells, and tissues based on number of layers, dimensions, and hydrophilicity. It is important to understand such interactions from two points of view, one for biomedical applications and another for their toxicity and biocompatibility. A comprehensive discussion on biologically relevant properties of graphene nanomaterials and their toxicity has been studied so far [81, 82]. Graphene-based materials show unique interactions with DNA and RNA, which make them attractive in DNA or RNA sensing and delivery. GO shows preferential adsorption of single-stranded DNA over double-stranded DNA and protects the adsorbed nucleotides from attack by nuclease enzymes [83–85]. Due to interactions of negative charges on DNA with graphene, adsorption of small oligomers was enhanced in high ionic strength solution at low pH. In contrast to interaction with DNA and RNA, graphene interacts less with proteins and lipids. Graphene forms stable and functional hybrid structures with lipids [86]. Graphene and other carbon-based materials are nonbiodegradable, causing environmental hazards. The high surface area of graphene promotes its cellular interactions although precise uptake mechanism is not studied until recently. Different forms of graphene interact differently with the cell membranes and also differ in different cell types. Graphene sheets (10 μm thick) can enter the cells by edge-first or corner-first penetration of the cell membranes and are completely engulfed by epithelial cells in lungs. Plate-like graphene microsheets physically disrupted the cytoskeletal organization. However, cell attachment decreased significantly with oxygen content in few-layer rGO as reduced few-layer graphene enhanced cell adhesion due to increased extracellular matrix protein adsorption whereas highly reduced few-layer graphene did not support cell adhesion. Graphene-based materials are being explored for antimicrobial activity. Many studies report the antibacterial activity of CNTs, graphene, GO, and rGO against *Escherichia coli* and *Staphylococcus aureus* bacteria, with rGO having the strongest antibacterial effectiveness [87–89]. On the contrary, the *Shewanella* family of bacteria with the ability to reduce metals have been shown to reduce GO in suspension cultures with no inhibition of bacterial growth [87]. The antibacterial activity of graphene-based materials can be exploited in various wound-healing applications or external injuries to prevent infections. The shape, size, and chemistry of graphene play important role in determining its interaction with cell membrane, intracellular uptake, and fate.

7.4 Carbon-Based Nanocomposites

Nanocomposites are multiphase material made from two or more constituent materials with significantly altered physical or chemical properties that, when combined, produce a material with enhanced properties because of the high surface area of building material and the extremely reactive surface of metal nanoparticles. Literature studies show that the properties of carbon-based nanomaterials (CNTs and graphene) can be made more versatile by incorporating other active materials like metal, metal oxide, and noble metals into the matrix to form a hybrid system. The large demand of carbon nanomaterial-based composites and hybrids can never be underestimated because of the unique properties they have and thus have attracted the attention of researchers all over the world in the area of material sciences [3, 90–92]. To date, majority of metals and metal oxides were decorated onto carbon-based materials such as graphene, carbon nanotubes, and carbon nanowires to create a new advanced class of carbon-based nanocomposites [93, 94]. Well-known binary oxides like SnO_2, TiO_2, MnO_2, ZnO, NiO, WO_3, CuO, Co_3O_4, Fe_2O_3, Fe_3O_4, CuO_2, etc. were reported to form a composite with carbon-based nanomaterials having excellent characteristics to be used in different fields such as energy harvesting, conversion and storage devices, photovoltaic devices, sensing technology, and photocatalysis [57, 95–98]. The integration of carbonaceous nanomaterials with graphene is highly conducive to enlarging the interlayer spacing and preventing the restacking of graphene sheets during fabrication and cycling operation of the electrode. Carbon allotropes such as fullerenes, CNTs, CNFs, and graphene were found to combine well with graphene for fabricating carbonaceous hybrid electrodes [99, 100].

7.4.1 Graphene-Based Composites

In graphene-based composites, graphene acts either as a functional component or as a substrate for immobilizing the other components. The large surface areas and the conductive robust structure of graphene often facilitate charge transfer and redox reaction as well as enforce the mechanical strengths of resulting composites. Therefore, anchoring metal oxides on graphene will boost the efficiency of various catalytic and storage reactions in energy conversion applications.

Graphene-based materials have generated tremendous interest in a wide range of research activities. A wide variety of graphene-related materials have been synthesized for potential applications in electronics, energy storage, catalysis, gas sorption, storage, separation, and sensing. A graphene layer decorated with nanoparticles (NPs) leads to a well-demarcated, innovative graphene with exceptional properties. These NPs act as a stabilizer against the aggregation of discrete graphene sheets, which is generally caused by a strong van der Waals interaction between graphene layers. Modifications of graphene decorated with metal oxide NPs have been reported. The incorporation of nanomaterials on graphene surface is highly desirable for tuning surface morphology, electronic structure, and following intrinsic properties of graphene. The hybrid structures that combine graphene with other functional materials such as metal oxides or organic molecules have shown better performance compared to the pristine graphene due to the synergetic effect between them. Doping of heteroatoms that introduces more surface defects and improves electrical conductivity of pure graphene also gives superior enhancement in performance. Over the past decades, extensive efforts have been devoted into increasing the capacity and energy density of existing cathode materials as well as exploring their possible alternatives to satisfy future demands in the electronics market. Graphene and its derivatives have been extensively introduced into the cathode system to compensate for some deficiencies suffered by common cathode materials in LIBs, such as the poor electrical conductivity, sluggish kinetics of electron and Li-ion transportation, low specific capacities and particle agglomeration generated from their nanostructures. Meanwhile the integration of inorganic nanostructures with the graphene layers may reduce the restacking of graphene sheets and consequently maintain the high surface area. Besides inorganic/rGO composites, preparation of organic molecular/rGO composites as anode materials has also been reported. In addition, a series of graphene–polymer composites have been prepared. For example, graphite oxide has shown efficient heterogeneous catalytic activity for the polymerization of various olefin monomers.

Graphene nanocomposites are the latest additions to the wonderful applications of graphene. One of the promising applications of the graphene-oxide nanocomposites is chemical sensing, which is useful for monitoring the toxicity, inflammability, and explosive nature of chemicals. Well-known binary oxides like ZnO, TiO_2, SnO_2, WO_3, and CuO, when combined with graphene in the form of nanocomposites, have excellent potential for detecting trace amounts of hazardous gases and chemicals. Graphene and graphene-related materials are mostly conductors or insulators. Hence, an uphill task of the graphene research community is to produce semiconducting graphene material for sensor and other electronic applications. A major contribution in this direction has been achieved through chemical modifications of graphene molecules, mostly by composite formation. Graphene–metal oxide hybrid composite is one such example, since graphene–metal oxide is semiconducting; therefore, all the three aspects of electrical conductivity, e.g., conducting, semiconducting, and insulating characteristics, are available in the carbon family, which offers great compatibility for electronic applications.

7.4.2 Graphene-Based Composite Synthesis

Nanomaterials are added to the graphene framework by a number of effective and prospective methods. The aim is to combine metal or metal oxides and transition metals in addition to noble metals with graphene to form graphene-based composites because the composite has a synergistic effect of all the components of the composite combined, delivering outstanding efficiency in a variety of applications. The most effective strategies for synthesizing high quality graphene-metal oxide composites are discussed underneath.

7.4.2.1 Solution Mixing Method

Solution mixing is an efficient and direct method. It has been widely used to prepare graphene–metal oxide composites. A solution of suspended graphene acts as a precursor for an integrated support network for discrete metal nanoparticles. First, graphene dispersed in aqueous or organic solvents by electrostatic stabilization and chemical functionalization. The presence of hydrophilic oxygen-containing functional groups such as epoxides, hydroxides, and carboxylic groups on the surface enables graphene to be well dispersed. Such dispersions are a good template suspension for chemical reaction with metal ions from the precursors of inorganic and organic metal salts, which undergo hydrolysis or *in situ* reactions to anchor them on the surface of graphene with rich functionalities followed by annealing (Figure 7.6).

7.4.2.2 Sol–Gel Method

The sol–gel process is a popular approach for the preparation of metal oxide structures and film coatings, with the metal alkoxides or chlorides as precursors that undergo a series of hydrolysis and polycondensation reactions. The key advantage of the *in situ* sol–gel process lies in the fact that the functional groups on GO/RGO (reduced graphene oxide) provide reactive and anchoring sites for nucleation and growth of NPs, so that the resulting metal oxide nanostructures are chemically bonded to the GO/RGO surfaces [101, 102].

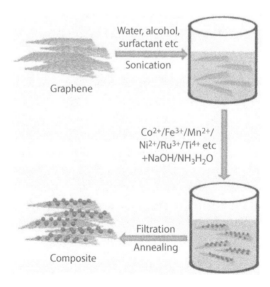

Figure 7.6 A general wet-chemistry strategy to fabricate graphene/metal oxide composites [91].

7.4.2.3 Hydrothermal/Solvothermal Method

Hydrothermal/solvothermal is a powerful tool for the synthesis of inorganic nanocrystals, which operates at an elevated temperature in a confined volume to generate high pressure. The one-pot hydrothermal/solvothermal process can give rise to nanostructures with high crystallinity without postsynthetic annealing or calcination and at the same time reduce GO to RGO. The whole process is simple, scalable, and industrially compatible [103, 104].

The hydrothermal technique is convenient for graphene-oxide nanocomposite synthesis. Graphene/graphene oxide or reduced graphene oxide and nanoparticle oxides are synthesized separately and dispersed in aqueous solution. Then, the aqueous dispersions are sonicated, mixed, and heat treated in a closed ambient. The heat treatment is slow and requires long-time period. After that, the composite is washed and dried at relatively low temperatures for 12–24 h. The last drying step can be modified by freezing the yield and then drying, which is termed as freeze-drying. Controlling the parameters like concentrations of precursor solutions and reaction time, graphene composites with controlled crystal facets can be easily obtained by using the hydrothermal method. The porosity, composition, grain size, and surface area of nanocomposites were tunable through controlled annealing processes, e.g., under different gas environments and different temperatures. These graphene/metal oxide hybrids showed structural-process-dependent performances as anode materials for Li-ion batteries [105, 106].

7.4.2.4 Self-Assembly

Self-assembly is an efficient process to assemble micro-objects into ordered macroscopic structures. It is utilized to produce functional materials like photonic crystals, composites, and ordered DNA structures. To obtain a layered structure of nanocomposites, a novel method has been used to prepare the ordered graphene–metal oxide hybrids through a surfactant-aided ternary self-assembly process [107]. Anionic surfactant modified reduced graphene oxide as starting material. Surfactant assisted the dispersal of graphene sheets and loading of metal cations. After converting the metal cations to oxides at graphene, graphene–metal oxide composites were obtained with layered structure. Due to the negatively charged state of graphene nanosheets, another feasible and low-cost assembly process based on the negative–positive electrostatic attraction has also been widely used to construct graphene-based nanocomposites.

7.4.2.5 Other Methods

Microwave irradiation is a facile method to provide energy for chemical reactions. Microwave irradiation has been used to prepare graphene/metal oxide hybrids [108]. Direct electrochemical deposition of inorganic crystals on graphene substrates, without the requirement for postsynthetic transfer of the composite materials, is an attractive approach for thin-film-based applications. Nanostructures have been successfully deposited on reduced graphene oxide or CVD–graphene films [109].

7.4.3 Graphene-Based Composite Properties

Graphene serves as a 2D support for uniformly anchoring or dispersing metal oxides with well-defined sizes, shapes, and crystallinity and the metal oxides suppressing the re-stacking of graphene. Graphene acts as a two-dimensional conductive template or builds a 3D conductive porous network for improving the electrical properties and charge transfer pathways of pure oxides. Graphene also suppresses the volume change and agglomeration of metal oxides. Moreover, oxygen-containing groups on graphene ensure good bonding, interfacial interactions, and electrical contacts between graphene and metal oxides.

The main problem associated with nanoparticle is that they agglomerate; graphene in a nanocomposite suppresses the agglomeration of metal oxide nanoparticles. In graphene-oxide composite, metal oxides have interfacial interactions with functional groups (such as HO–C=O and –OH) by chemisorption, which bridge metal centers with carboxyl or hydroxyl groups at oxygen-defect sites as well as through van der Waals interactions between the pristine region of graphene and metal oxides. Reduced graphene oxide usually suffers from serious agglomeration and restacking due to the van der Waals interactions between adjacent sheets, leading to a great loss of effective surface area and electrochemical properties. Thus, keeping graphene from restacking is a key role in improving the electrochemical performance of graphene-based materials in batteries and electrochemical devices. The loading of metal oxide particles can inhibit or decrease agglomeration and restacking of graphene and increase the available electrochemically active surface area of graphene. Due to the synergistic effect between graphene and metal oxides, metal oxide nanoparticles supported on both side of graphene can serve as a nano spacer to separate the adjacent graphene sheets (Figure 7.7).

Graphene is not only a good electrically conductive carbon material but also an electrochemically active material. Therefore, graphene as a conductive carbon material in metal oxide electrodes is expected to construct a 3D conductive network among metal oxide particles [91, 110]. Moreover, graphene can host the nanostructured electrode materials by providing a support for anchoring nanoparticles and work as an excellent conductive matrix for better contact between electrode and current collector. More importantly, graphene layers can prevent the volume expansion/contraction and the aggregation of nanoparticles effectively during the charge and discharge process. The introduction of graphene into the cathode system mostly improves the electrical conductivity or serves as a protection barrier for the dissolution of conventional cathode materials.

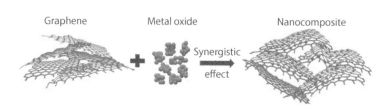

Figure 7.7 Schematic of the preparation of graphene/metal oxide composites with synergistic effects between graphene and metal oxides [91].

7.5 Applications

Carbon and carbon-based materials attracted great attention around the world because of their potential for a variety of applications ranging from electronics, optoelectronics, gas sensing, energy storage, and photocatalysis. The large specific surface area of these materials and versatile modification make them excellent adsorbents for diverse applications. For graphene, a wide range of applications include the computer industry to replace silicon in new-generation processors. Other possible electronic applications of graphene/graphene-based materials are touch screens, nanotransistors, microprocessors, and LED including so-called flexible electronics. Besides the electronic applications of graphene, other uses include energy storage (supercapacitors, batteries, fuel, and solar cells) [111–113], filters, adsorbents, and detectors. Due to its high aspect ratio and unique physicochemical properties, graphene is an extremely valuable component for the development of various composite materials [114, 115]. Also, it can be used in medicine, e.g., anticancer therapy, and as a drug carrier or bactericidal material [116–118].

7.5.1 Gas Sorption and Storage

Carbon-based materials have been considered promising for gas sorption, storage, and separation because of the abundance, robust pore structure, tunable porosity and surface area, lightweight, high thermal and chemical stability, and easy synthesis in industrial scale. There is a considerable amount of interest in graphene-related materials for gas sorption, storage, and separation. The high adsorption capacities of graphene-based materials are mainly determined by their unique nanostructures, high specific surface areas, and tailorable surface properties, which make them suitable for storage or capture of various molecules relevant for environmental and energy-related applications (Figure 7.8).

Gas sorption, storage, and separation in carbon materials are mainly based on physisorption on the surfaces and particularly depend on the electrostatic and dispersion interactions. The strength of the interaction is determined by the surface characteristics of the

Figure 7.8 A 3D graphene membrane for gas separation [127].

adsorbent and the properties of the targeted adsorbate molecule, including the size and shape of the adsorbate molecule along with its polarizability, magnetic susceptibility, permanent dipole moment, and quadrupole moment. Normally, the binding or adsorption strength with a carbon nanostructure is relatively low for H_2 and N_2, moderate for CO, CH_4 and CO_2, and relatively high for H_2S, NH_3, and H_2O. Thus, surface modifications, such as doping, functionalization, and improving the pore structure and specific surface area of nanocarbons, are important to enhance gas adsorption.

The adsorption properties of various solids including graphene-based materials in relation to H_2 storage have been broadly studied. The volumetric density is not well defined for a single graphene layer; thus, the estimation of hydrogen storage capabilities should be considered for multilayer or three-dimensional graphene structures. The maximum gravimetric density achievable by chemisorption in a graphene layer is calculated to be 8.3%, corresponding to a completely saturated graphene sheet with one hydrogen atom per carbon atom (graphene). Reduced graphene oxide is potentially one of the promising candidates for the development of sorbents for efficient gas storage applications [119–126].

7.5.2 Hydrogen Storage

Molecular hydrogen has very advantageous properties, which make it an important fuel. The hydrogen molecule, the lightest known element, has high combustion heat. Although hydrogen storage technologies for transport applications are already available, they require high density of stored fuel at the operating temperatures from 0°C to 150°C and respectively rapid load/unload of the storage systems. Hydrogen storage in solid materials has the potential to surpass the densities of compressed hydrogen. Physical H_2 adsorption, which takes place on nanoporous carbon materials, is advantageous due to reversibility of the process and good adsorption kinetics. A disadvantage of H_2 physisorption is that low temperatures (−196°C) and high pressures are usually required to store sufficiently high quantities of this gas. Therefore, to achieve a reasonable storage of H_2 by physisorption, the conditions are required. Chemisorption in turn relies on the chemical binding, which seems to be more suitable for the storage and transport of hydrogen for a longer time. In this case, higher temperatures are required as compared to physisorption. The environmental-friendly methods suitable for large-scale production of graphene-based materials with desired adsorption properties are needed for the effective storage of energy-relevant gases. The gas uptakes obtained for unmodified graphene oxides (GO or rGO) alone are rather low; however, significant progress has been made in recent years in this area, especially in the development of new graphene-based composite materials toward improving their adsorption properties. Graphene oxides can be doped with heteroatoms (e.g., B, N, S) and decorated with polymers and nanoparticles (e.g., Fe, Pd, Fe_3O_4, V_2O_5). These two-dimensional nanosheets can be used to design and prepare 3D structures with large surface area and well-developed porosity. In the case of some metal or metal-oxide-decorated graphene materials, an enhanced chemisorption and specific gas adsorption may take place in addition to physisorption resulting in an enhanced sorption capacity. A comparison of various materials ranging from graphene, GO, and activated rGO to graphene-based composites with metal and metal oxide nanoparticles shows that the best H_2 capacities were reported for activated rGO samples and rGO composites with incorporated Pd nanoparticles. Transition metal oxide nanoparticles wrapped with single- or few-layered graphene oxide nanosheets are better H_2 adsorbents than pristine transition metal oxide or graphene oxide alone [123, 128–132].

7.5.3 Energy Storage Devices

A major advantage of graphene over other carbon materials such as graphite and CNTs is the presence of many oxygen-containing functional groups on the edges and surface of GO and reduced GO. Graphene can host the nanostructured electrode materials by providing a support for anchoring nanoparticles and work as a highly conductive matrix for good contact between electrode and current collector. More importantly, graphene layers can prevent the volume expansion/contraction and the aggregation of nanoparticles effectively during charge and discharge process. The functional groups on graphene strongly influence the size, shape, and distribution of metal oxide particles. Nanomaterials generally undergo severe structural and volume changes during lithium insertion and removal, leading to the pulverization of their electrodes and consequently fast capacity loss. Graphene-based 3D structures, such as metal oxide anchoring on graphene, graphene-wrapped metal oxide, and graphene-encapsulated metal oxide, were reported, in which metal oxides are uniformly anchored onto the surface of graphene, or wrapped between graphene layers, or encapsulated by individual graphene sheets (Figure 7.9) [133–137]. Graphite is the most commonly used anode material in Li-ion batteries and has a specific capacity of 372 mAh/g by forming LiC_6 upon Li intercalation between the stacked layers [96, 138]. It has been proposed that graphene can accommodate Li ions through an adsorption mechanism on both sides to form Li_2C_6 with a theoretical capacity of 744 mAh/g, which is twice that of graphite and other carbonaceous materials such as CNTs [139, 140]. Substituting graphene for graphite has been explored to increase the lithiation sites and storage capacity. Previously, graphene materials are the best choice as anode for Li-ion batteries because the three-dimensional graphene-based structures have large spaces for the accommodating volume expansion/contraction of metal oxide

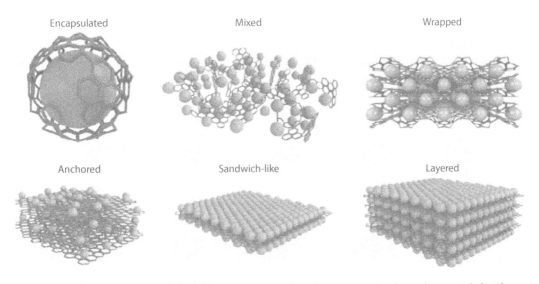

Figure 7.9 Schematic diagram of the different structures of graphene composite electrode materials [142].

during the Li insertion and extraction process [140, 141], which also prevents the aggregation and cracking of electrode material upon cycling, retaining the large capacity, good cycling performance, and high rate capability.

Carbon materials including graphene usually provide a low voltage window (below 1.5 V vs. Li/Li$^+$) and were widely used as anodes compared to cathode material in LIBs [143–145]. The reaction between Li ions and carbonyl/carboxylic acid groups occurs at a higher voltage of up to 3 V vs. Li/Li$^+$ [146]. The presence of hydroxyl, carbonyl, and carboxylic groups on reduced graphene oxide was identified as active sites for lithiation. Pristine graphene cannot be used directly to increase the storage capacity of Li ions. It provides mechanical support to cathode in addition to high electrical conductivity and large surface to anchor and separate metal oxides. The cathode composites can also form wrapped, anchored, encapsulated, layered, mixed, and sandwich-like architectures as that of anode [147–154]. The introduction of graphene into the cathode system not only improves the electrical conductivity but also serves as a protective barrier for the dissolution of conventional cathode materials. Due to these synergistic effects, integration of metal oxides and graphene in a composite fully uses each active component and consequently achieves excellent electrochemical performance in Li-ion batteries and electrochemical capacitors through materials design and fabrication.

7.5.4 Antibacterial Activity

Graphene and graphene-based nanocomposites have also been used in bacteria detection and antibacterial applications. The unique synergistic antibacterial effects may be attributed to the existence of graphene sheet in the nanocomposite. As graphene nanostructures have been found to exhibit low toxicity toward eukaryotic cells, the graphene derivatives for biological applications have been attracting significant attention. Graphene with hydrophobic domains in the nanocomposites may allow effective attachment of nanomaterials on the surface of bacteria. Sharp edges of graphene nanosheets damage cell membranes, causing membrane stress, thereby contributing to the loss of bacterial membrane integrity and the leakage of RNA. Different types of nanomaterials like copper and its oxide, zinc oxide, titanium oxide, magnesium, gold, and silver have been investigated. It is found that silver nanoparticles show the most effective inhibitory and bactericidal properties. Particularly, silver nanoparticles form stable nanocomposites with graphene by electrostatic interactions. Graphene acts as a support and prevents the aggregation of silver nanoparticles. Therefore, graphene-based silver composites were widely studied for antibacterial effectiveness due to their synergistic effect (Figure 7.10) [155–158].

7.5.5 Bioimaging

Optical imaging, a noninvasive technique, uses visible light and the special properties of photons to obtain detailed images of organs and tissues as well as smaller structures including cells and even molecules. It has advantages over other imaging modalities including relatively low cost, high sensitivity, nonionizing radiation, real-time imaging, short acquisition time, and multiplexing capability. Graphene-based nanomaterials were actively explored for optical imaging, mainly including fluorescence imaging, two-photon

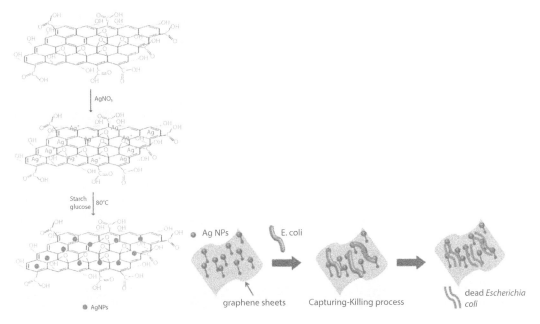

Figure 7.10 Schematic of the procedure for preparing GO–Ag nanocomposite for antibacterial activity [159] and capture killing mechanism of the enhanced antibacterial property of the Ag@rGO nanocomposites [160].

fluorescence imaging, Raman imaging, and so on. The dye functionalized GO/rGO have been widely investigated for fluorescence imaging. Two-photon fluorescence imaging has attracted much attention for its promising applications in both basic research and biomedical diagnostics, owing to the minor autofluorescence background, larger imaging depth, reduced photo-bleaching, and photo-toxicity compared to single photon fluorescence imaging. Carbon-based nanocomposites, including carbon dots, graphene quantum dots, and GO, are attracting considerable interest in the field of two-photon fluorescence imaging (Figure 7.11) [161–165].

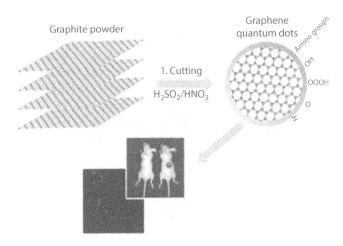

Figure 7.11 Synthetic scheme of graphene QDs and *in vivo* imaging in visible light [165].

7.5.6 Biosensing

Biosensor is an analytical device utilizing sensing elements to detect/sense an analyte or family of analytes. As such, biosensors are applicable to the biomedical field, with applications ranging from medical diagnostics to drug discovery, environmental monitoring, food and defense. Fundamentally, biosensors are composed of two elements: a receptor and a transducer. The receptor consists of either organic or inorganic materials that interact with the target analyte or a family of analytes. On the other hand, the transducer converts the recognition incident that occurs between the analyte and the receptor into a measurable signal in the form of electronic, electrochemical, and optical signals. Graphene–nanoparticle hybrids are particularly well suited for biosensing applications. There has been significant effort for utilizing graphene for biosensing. Moreover, nanoparticles have also been widely investigated in this regard due to the exquisite sensitivity required for this type of application. By combining these two excellent and unique modalities as graphene nanoparticle hybrids, many advantageous properties are attained for biosensing applications. Graphene can increase the available surface area for analyte binding as well as improve their electrical conductivity and electron mobility, thus enhancing the achievable sensitivity and selectivity. Graphene–nanoparticle hybrid materials for biosensing applications can be generally divided into three classes based on the underlying mechanism of detection. These classes include (1) electronic, (2) electrochemical, and (3) optical sensors [166, 167]. Because of the large specific surface area, graphene–nanoparticle hybrids are advantageous for the immobilization of biomolecules. Moreover, the excellent electrical properties of graphene significantly improve the electronic and ionic transport capacity of the resulting hybrid electrochemical sensor, thereby enhancing achievable sensitivities and measurement ranges. There have been numerous reports demonstrating the graphene–nanoparticle hybrid-based immobilization of enzymes for electrochemical detection. Specifically, the mechanism of action is based on utilizing enzymes that catalyze redox reactions. In this case, when the immobilized enzyme catalyzes a redox reaction of the target analyte, a direct electron transfer from the enzyme to the electrode occurs, which provides an amperometric signal that is proportional to the concentration of analyte [168, 169].

7.5.7 Photocatalysis

Recently, semiconductor-based photocatalysis has gained worldwide attention for its rule in energy applications [170]. However, the recombination of photo-generated electron–hole pairs within photocatalysts results in low efficiency, thus limiting its practical applications. The suppression of this recombination is key for the improvement of photocatalysis of semiconductor photocatalysts. Carbon/semiconductor hybrid materials are emerging as a new class of photocatalysts and have attracted a lot of attention [171, 172]. Composites, combining carbon and semiconductor photocatalysts, offer better efficiency for separating electron–hole pairs. In this regard, graphene has been examined in combination with a semiconductor photocatalyst, resulting in improved photocatalytic activity. These composites possess high dye adsorption capacity, extended light absorption range, and enhanced charge separation and transportation properties. Therefore, graphene-based semiconductor photocatalysts have been extensively applied to photocatalytic degradation of organic compounds [173, 174].

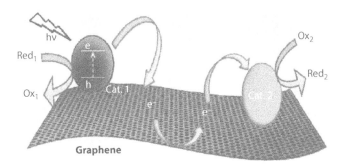

Figure 7.12 Schematic illustration of selective catalysis at different sites on graphene used as a conducting support [175].

Water splitting into hydrogen and oxygen using semiconductor photocatalysts by photocatalysis has promising approach to produce hydrogen energy. A number of semiconductor photocatalysts have been studied for the evolution of hydrogen from water by photocatalysis. The superior electron mobility and high specific surface area of graphene can be used as an efficient electron acceptor to increase the photo-induced charge transfer and the backward reaction by separating the evolution sites of hydrogen and oxygen (Figure 7.12) [175].

Acknowledgments

The authors are thankful to the Pakistan Science Foundation and TWAS for financial support through projects PSF/Res/C-PINSTECH/Phys (172) and 13-319RG/MSN/AS-C-UNESCO FR:3240279202, respectively.

References

1. Cha, C. et al., Carbon-based nanomaterials: Multifunctional materials for biomedical engineering. *ACS Nano*, 7, 4, 2891–2897, 2013.
2. Reddy, A.L.M. et al., Hybrid nanostructures for energy storage applications. *Adv. Mater.*, 24, 37, 5045–5064, 2012.
3. Hazra, S.K. and Basu, S., Graphene-oxide nanocomposites for chemical sensor applications. *C*, 2, 2, 12, 2016.
4. Li, Y., Wu, J., Chopra, N., Nano-carbon-based hybrids and heterostructures: Progress in growth and application for lithium-ion batteries. *J. Mater. Sci.*, 50, 24, 7843–7865, 2015.
5. Goriparti, S. et al., Review on recent progress of nanostructured anode materials for Li-ion batteries. *J. Power Sources*, 257, 421–443, 2014.
6. Sehrawat, P., Julien, C., Islam, S., Carbon nanotubes in Li-ion batteries: A review. *Mater. Sci. Eng., B*, 213, 12–40, 2016.
7. Wang, H. and Dai, H., Strongly coupled inorganic–nano-carbon hybrid materials for energy storage. *Chem. Soc. Rev.*, 42, 7, 3088–3113, 2013.
8. Goenka, S., Sant, V., Sant, S., Graphene-based nanomaterials for drug delivery and tissue engineering. *J. Controlled Release*, 173, 75–88, 2014.
9. Kim, J. et al., Graphene oxide sheets at interfaces. *J. Am. Chem. Soc.*, 132, 23, 8180–8186, 2010.

10. Kim, F., Cote, L.J., Huang, J., Graphene oxide: Surface activity and two-dimensional assembly. *Adv. Mater.*, 22, 17, 1954–1958, 2010.
11. Guo, F. et al., Hydration-responsive folding and unfolding in graphene oxide liquid crystal phases. *ACS Nano*, 5, 10, 8019–8025, 2011.
12. He, H. et al., A new structural model for graphite oxide. *Chem. Phys. Lett.*, 287, 1, 53–56, 1998.
13. Hernandez, Y. et al., High-yield production of graphene by liquid-phase exfoliation of graphite. *Nat. Nanotechnol.*, 3, 9, 563–568, 2008.
14. Chua, C. and Pumera, M., Covalent chemistry on graphene. *Chem. Soc. Rev.*, 42, 8, 3222–3233, 2013.
15. Casabianca, L.B. et al., NMR-based structural modeling of graphite oxide using multidimensional 13C solid-state NMR and *ab initio* chemical shift calculations. *J. Am. Chem. Soc.*, 132, 16, 5672–5676, 2010.
16. Gao, W. et al., New insights into the structure and reduction of graphite oxide. *Nat. Chem.*, 1, 5, 403–408, 2009.
17. Dreyer, D.R., Todd, A.D., Bielawski, C.W., Harnessing the chemistry of graphene oxide. *Chem. Soc. Rev.*, 43, 15, 5288–5301, 2014.
18. Chen, L. et al., From nanographene and graphene nanoribbons to graphene sheets: Chemical synthesis. *Angew. Chem. Int. Ed.*, 51, 31, 7640–7654, 2012.
19. Van Noorden, R., Production: Beyond sticky tape. *Nature*, 483, 7389, S32–S33, 2012.
20. Novoselov, K.S. et al., Electric field effect in atomically thin carbon films. *Science*, 306, 5696, 666–669, 2004.
21. Vadukumpully, S., Paul, J., Valiyaveettil, S., Cationic surfactant mediated exfoliation of graphite into graphene flakes. *Carbon*, 47, 14, 3288–3294, 2009.
22. Geng, J. et al., Preparation of graphene relying on porphyrin exfoliation of graphite. *Chem. Commun.*, 46, 28, 5091–5093, 2010.
23. Park, J.S. et al., Liquid-phase exfoliation of expanded graphites into graphene nanoplatelets using amphiphilic organic molecules. *J. Colloid Interface Sci.*, 417, 379–384, 2014.
24. Skaltsas, T. et al., Graphene exfoliation in organic solvents and switching solubility in aqueous media with the aid of amphiphilic block copolymers. *J. Mater. Chem.*, 22, 40, 21507–21512, 2012.
25. Parviz, D. et al., Dispersions of non-covalently functionalized graphene with minimal stabilizer. *ACS Nano*, 6, 10, 8857–8867, 2012.
26. Yang, H. et al., A simple method for graphene production based on exfoliation of graphite in water using 1-pyrenesulfonic acid sodium salt. *Carbon*, 53, 357–365, 2013.
27. Schlierf, A. et al., Nanoscale insight into the exfoliation mechanism of graphene with organic dyes: Effect of charge, dipole and molecular structure. *Nanoscale*, 5, 10, 4205–4216, 2013.
28. Sampath, S. et al., Direct exfoliation of graphite to graphene in aqueous media with diazaperopyrenium dications. *Adv. Mater.*, 25, 19, 2740–2745, 2013.
29. Bjoürk, J. et al., Adsorption of aromatic and anti-aromatic systems on graphene through π–π stacking. *J. Phys. Chem. Lett.*, 1, 23, 3407–3412, 2010.
30. Torrisi, F. et al., Inkjet-printed graphene electronics. *ACS Nano*, 6, 4, 2992–3006, 2012.
31. Shin, K.Y., Hong, J.Y., Jang, J., Micropatterning of graphene sheets by inkjet printing and its wideband dipole-antenna application. *Adv. Mater.*, 23, 18, 2113–2118, 2011.
32. De, S. et al., Flexible, transparent, conducting films of randomly stacked graphene from surfactant-stabilized, oxide-free graphene dispersions. *Small*, 6, 3, 458–464, 2010.
33. Kakaei, K., One-pot electrochemical synthesis of graphene by the exfoliation of graphite powder in sodium dodecyl sulfate and its decoration with platinum nanoparticles for methanol oxidation. *Carbon*, 51, 195–201, 2013.

34. Lee, S.-H. et al., A graphite foil electrode covered with electrochemically exfoliated graphene nanosheets. *Electrochem. Commun.*, 12, 10, 1419–1422, 2010.
35. Su, C.-Y. et al., High-quality thin graphene films from fast electrochemical exfoliation. *ACS Nano*, 5, 3, 2332–2339, 2011.
36. Xia, Z.Y. et al., The exfoliation of graphene in liquids by electrochemical, chemical, and sonication-assisted techniques: A nanoscale study. *Adv. Funct. Mater.*, 23, 37, 4684–4693, 2013.
37. Liu, J. et al., Improved synthesis of graphene flakes from the multiple electrochemical exfoliation of graphite rod. *Nano Energy*, 2, 3, 377–386, 2013.
38. Wu, L. et al., Powder, paper and foam of few-layer graphene prepared in high yield by electrochemical intercalation exfoliation of expanded graphite. *Small*, 10, 7, 1421–1429, 2014.
39. Cui, X. et al., Liquid-phase exfoliation, functionalization and applications of graphene. *Nanoscale*, 3, 5, 2118–2126, 2011.
40. Yan, Z., Peng, Z., Tour, J.M., Chemical vapor deposition of graphene single crystals. *Acc. Chem. Res.*, 47, 4, 1327–1337, 2014.
41. Chen, S. et al., Millimeter-size single-crystal graphene by suppressing evaporative loss of Cu during low pressure chemical vapor deposition. *Adv. Mater.*, 25, 14, 2062–2065, 2013.
42. Li, X. et al., Large-area synthesis of high-quality and uniform graphene films on copper foils. *Science*, 324, 5932, 1312–1314, 2009.
43. Ning, J. et al., Review on mechanism of directly fabricating wafer-scale graphene on dielectric substrates by chemical vapor deposition. *Nanotechnology*, 2017.
44. Yan, K. et al., Designed CVD growth of graphene via process engineering. *Acc. Chem. Res.*, 46, 10, 2263–2274, 2013.
45. Tatarova, E. et al., Towards large-scale in free-standing graphene and N-graphene sheets.
46. Mattevi, C., Kim, H., Chhowalla, M., A review of chemical vapour deposition of graphene on copper. *J. Mater. Chem.*, 21, 10, 3324–3334, 2011.
47. Muñoz, R. and Gómez-Aleixandre, C., Review of CVD synthesis of graphene. *Chem. Vap. Deposition*, 19, 10-11-12, 297–322, 2013.
48. Zhang, Y., Zhang, L., Zhou, C., Review of chemical vapor deposition of graphene and related applications. *Acc. Chem. Res.*, 46, 10, 2329–2339, 2013.
49. Wei, D. et al., Controllable chemical vapor deposition growth of few layer graphene for electronic devices. *Acc. Chem. Res.*, 46, 1, 106–115, 2012.
50. Al-Shurman, K. and Naseem, H., CVD graphene growth mechanism on nickel thin films, in: *Proceedings of the 2014 COMSOL Conference in Boston*, 2014.
51. Sutter, P., Epitaxial graphene: How silicon leaves the scene. *Nat. Mater.*, 8, 3, 171–172, 2009.
52. Tromp, R. and Hannon, J., Thermodynamics and kinetics of graphene growth on SiC (0001). *Phys. Rev. Lett.*, 102, 10, 106104, 2009.
53. Kunc, J. et al., Planar edge Schottky barrier-tunneling transistors using epitaxial graphene/SiC junctions. *Nano Lett.*, 14, 9, 5170–5175, 2014.
54. Kim, J. et al., Principle of direct van der Waals epitaxy of single-crystalline films on epitaxial graphene. *Nat. Commun.*, 5, 4836, 2014.
55. Lian, P. et al., Porous SnO_2@C/graphene nanocomposite with 3D carbon conductive network as a superior anode material for lithium-ion batteries. *Electrochim. Acta*, 116, 103–110, 2014.
56. Hummers, W.S., Jr. and Offeman, R.E., Preparation of graphitic oxide. *J. Am. Chem. Soc.*, 80, 6, 1339–1339, 1958.
57. Hu, C. et al., A brief review of graphene–metal oxide composites synthesis and applications in photocatalysis. *J. Chin. Adv. Mater. Soc.*, 1, 1, 21–39, 2013.
58. El-Maghrabi, H.H. et al., Magnetic graphene based nanocomposite for uranium scavenging. *J. Hazard. Mater.*, 322, 370–379, 2017.
59. Loh, K.P. et al., The chemistry of graphene. *J. Mater. Chem.*, 20, 12, 2277–2289, 2010.

60. Taherian, F. et al., What is the contact angle of water on graphene? *Langmuir*, 29, 5, 1457–1465, 2013.
61. Xue, Y. et al., Functionalization of graphene oxide with polyhedral oligomeric silsesquioxane (POSS) for multifunctional applications. *J. Phys. Chem. Lett.*, 3, 12, 1607–1612, 2012.
62. Hasan, S.A. et al., Transferable graphene oxide films with tunable microstructures. *ACS Nano*, 4, 12, 7367–7372, 2010.
63. Hsieh, C.-T. and Chen, W.-Y., Water/oil repellency and work of adhesion of liquid droplets on graphene oxide and graphene surfaces. *Surf. Coat. Technol.*, 205, 19, 4554–4561, 2011.
64. Mahanta, N.K. and Abramson, A.R., Thermal conductivity of graphene and graphene oxide nanoplatelets, in: *Thermal and Thermomechanical Phenomena in Electronic Systems (ITherm)*, 2012 13th IEEE Intersociety Conference, 2012.
65. Kuila, T. et al., Chemical functionalization of graphene and its applications. *Prog. Mater. Sci.*, 57, 7, 1061–1105, 2012.
66. Afanasov, I. et al., Preparation, electrical and thermal properties of new exfoliated graphite-based composites. *Carbon*, 47, 1, 263–270, 2009.
67. Nika, D. et al., Phonon thermal conduction in graphene: Role of Umklapp and edge roughness scattering. *Phys. Rev. B*, 79, 15, 155413, 2009.
68. Jiang, J.-W. et al., Isotopic effects on the thermal conductivity of graphene nanoribbons: Localization mechanism. *J. Appl. Phys*, 107, 5, 054314, 2010.
69. Bolotin, K.I. et al., Ultrahigh electron mobility in suspended graphene. *Solid State Commun.*, 146, 9, 351–355, 2008.
70. Kravets, V. et al., Spectroscopic ellipsometry of graphene and an exciton-shifted van Hove peak in absorption. *Phys. Rev. B*, 81, 15, 155413, 2010.
71. Rana, F. et al., Carrier recombination and generation rates for intravalley and intervalley phonon scattering in graphene. *Phys. Rev. B*, 79, 11, 115447, 2009.
72. Elias, D.C. et al., Control of graphene's properties by reversible hydrogenation: Evidence for graphane. *Science*, 323, 5914, 610–613, 2009.
73. Avouris, P. and Freitag, M., Graphene photonics, plasmonics, and optoelectronics. *IEEE J. Sel. Top. Quantum Electron.*, 20, 1, 72–83, 2014.
74. Suk, J.W. et al., Mechanical properties of monolayer graphene oxide. *ACS Nano*, 4, 11, 6557–6564, 2010.
75. Li, J.-L. et al., Oxygen-driven unzipping of graphitic materials. *Phys. Rev. Lett.*, 96, 17, 176101, 2006.
76. Gómez-Navarro, C., Burghard, M., Kern, K., Elastic properties of chemically derived single graphene sheets. *Nano Lett.*, 8, 7, 2045–2049, 2008.
77. Dikin, D.A. et al., Preparation and characterization of graphene oxide paper. *Nature*, 448, 7152, 457–460, 2007.
78. Das, B. et al., Nano-indentation studies on polymer matrix composites reinforced by few-layer graphene. *Nanotechnology*, 20, 12, 125705, 2009.
79. Prasad, K.E. et al., Extraordinary synergy in the mechanical properties of polymer matrix composites reinforced with 2 nanocarbons. *Proc. Natl. Acad. Sci.*, 106, 32, 13186–13189, 2009.
80. Rao, C. et al., Some novel attributes of graphene. *J. Phys. Chem. Lett.*, 1, 2, 572–580, 2010.
81. Sanchez, V.C. et al., Biological interactions of graphene-family nanomaterials: An interdisciplinary review. *Chem. Res. Toxicol.*, 25, 1, 15–34, 2011.
82. Bianco, A., Graphene: Safe or toxic? The two faces of the medal. *Angew. Chem. Int. Ed.*, 52, 19, 4986–4997, 2013.
83. Ren, H. et al., DNA cleavage system of nanosized graphene oxide sheets and copper ions. *ACS Nano*, 4, 12, 7169–7174, 2010.

84. Lu, C.-H. *et al.*, Using graphene to protect DNA from cleavage during cellular delivery. *Chem. Commun.*, 46, 18, 3116–3118, 2010.
85. Xu, Y. *et al.*, Three-dimensional self-assembly of graphene oxide and DNA into multifunctional hydrogels. *ACS Nano*, 4, 12, 7358–7362, 2010.
86. Titov, A.V., Král, P., Pearson, R., Sandwiched graphene–membrane superstructures. *ACS Nano*, 4, 1, 229–234, 2009.
87. Wang, G. *et al.*, Microbial reduction of graphene oxide by *Shewanella*. *Nano Res.*, 4, 6, 563–570, 2011.
88. Akhavan, O. and Ghaderi, E., Toxicity of graphene and graphene oxide nanowalls against bacteria. *ACS Nano*, 4, 10, 5731–5736, 2010.
89. Shi, X. *et al.*, Regulating cellular behavior on few-layer reduced graphene oxide films with well-controlled reduction states. *Adv. Funct. Mater.*, 22, 4, 751–759, 2012.
90. Fan, X., Chen, X., Dai, L., 3D graphene based materials for energy storage. *Curr. Opin. Colloid Interface Sci.*, 20, 5, 429–438, 2015.
91. Wu, Z.-S. *et al.*, Graphene/metal oxide composite electrode materials for energy storage. *Nano Energy*, 1, 1, 107–131, 2012.
92. Leung, K.C.-F. *et al.*, Gold and iron oxide hybrid nanocomposite materials. *Chem. Soc. Rev.*, 41, 5, 1911–1928, 2012.
93. Akbulut, H. *et al.*, Co-deposition of Cu/WC/graphene hybrid nanocomposites produced by electrophoretic deposition. *Surf. Coat. Technol.*, 284, 344–352, 2015.
94. Liu, T., Fan, W., Zhang, C., Carbon nanotube-based hybrid materials and their polymer composites. *Polymer Nanotube Nanocomposites: Synthesis, Properties, and Applications, Second Edition*, pp. 239–277, 2014.
95. Dong, X.-C. *et al.*, 3D graphene–cobalt oxide electrode for high-performance supercapacitor and enzymeless glucose detection. *ACS Nano*, 6, 4, 3206–3213, 2012.
96. Wang, X. *et al.*, Constructing aligned γ-Fe_2O_3 nanorods with internal void space anchored on reduced graphene oxide nanosheets for excellent lithium storage. *RSC Adv.*, 5, 111, 91574–91580, 2015.
97. Jimenez-Villacorta, F. *et al.*, Graphene–ultrasmall silver nanoparticle interactions and their effect on electronic transport and Raman enhancement. *Carbon*, 101, 305–314, 2016.
98. Bonaccorso, F. *et al.*, Graphene, related two-dimensional crystals, and hybrid systems for energy conversion and storage. *Science*, 347, 6217, 1246501, 2015.
99. Chen, T. *et al.*, Microwave-assisted synthesis of reduced graphene oxide–carbon nanotube composites as negative electrode materials for lithium ion batteries. *Solid State Ionics*, 229, 9–13, 2012.
100. Hu, Y. *et al.*, Free-standing graphene–carbon nanotube hybrid papers used as current collector and binder free anodes for lithium ion batteries. *J. Power Sources*, 237, 41–46, 2013.
101. Azarang, M. *et al.*, One-pot sol–gel synthesis of reduced graphene oxide uniformly decorated zinc oxide nanoparticles in starch environment for highly efficient photodegradation of methylene blue. *RSC Adv.*, 5, 28, 21888–21896, 2015.
102. Li, H. *et al.*, *In situ* sol-gel synthesis of ultrafine ZnO nanocrystals anchored on graphene as anode material for lithium-ion batteries. *Ceram. Int.*, 42, 10, 12371–12377, 2016.
103. Dong, X. *et al.*, One-step growth of graphene–carbon nanotube hybrid materials by chemical vapor deposition. *Carbon*, 49, 9, 2944–2949, 2011.
104. Li, Q. *et al.*, Graphene and its composites with nanoparticles for electrochemical energy applications. *Nano Today*, 9, 5, 668–683, 2014.
105. Park, S.-K. *et al.*, *In situ* hydrothermal synthesis of Mn_3O_4 nanoparticles on nitrogen-doped graphene as high-performance anode materials for lithium ion batteries. *Electrochim. Acta*, 120, 452–459, 2014.

106. Gao, Y. et al., Novel $NiCo_2S_4$/graphene composites synthesized via a one-step *in-situ* hydrothermal route for energy storage. *J. Alloys Compd.*, 704, 70–78, 2017.

107. Wang, D. et al., Ternary self-assembly of ordered metal oxide–graphene nanocomposites for electrochemical energy storage. *ACS Nano*, 4, 3, 1587–1595, 2010.

108. Yan, J. et al., Fast and reversible surface redox reaction of graphene–MnO_2 composites as supercapacitor electrodes. *Carbon*, 48, 13, 3825–3833, 2010.

109. Wu, S. et al., Electrochemical deposition of semiconductor oxides on reduced graphene oxide-based flexible, transparent, and conductive electrodes. *J. Phys. Chem. C*, 114, 27, 11816–11821, 2010.

110. Liu, Y. et al., Mesoporous Co_3O_4 sheets/3D graphene networks nanohybrids for high-performance sodium-ion battery anode. *J. Power Sources*, 273, 878–884, 2015.

111. Wang, T. et al., Interaction between nitrogen and sulfur in co-doped graphene and synergetic effect in supercapacitor. *Sci. Rep.*, 5, 2015.

112. Xu, Y. and Liu, J., Graphene as transparent electrodes: Fabrication and new emerging applications. *Small*, 12, 11, 1400–1419, 2016.

113. Ma, X. et al., Phosphorus and nitrogen dual-doped few-layered porous graphene: A high-performance anode material for lithium-ion batteries. *ACS Appl. Mater. Interfaces*, 6, 16, 14415–14422, 2014.

114. Du, J. and Cheng, H.M., The fabrication, properties, and uses of graphene/polymer composites. *Macromol. Chem. Phys.*, 213, 10–11, 1060–1077, 2012.

115. Eda, G. and Chhowalla, M., Graphene-based composite thin films for electronics. *Nano Lett.*, 9, 2, 814–818, 2009.

116. Das, M.R. et al., Synthesis of silver nanoparticles in an aqueous suspension of graphene oxide sheets and its antimicrobial activity. *Colloids Surf., B*, 83, 1, 16–22, 2011.

117. de Faria, A.F. et al., Cellulose acetate membrane embedded with graphene oxide-silver nanocomposites and its ability to suppress microbial proliferation. *Cellulose*, 24, 2, 781–796, 2017.

118. Ran, X. et al., Hyaluronic acid-templated Ag nanoparticles/graphene oxide composites for synergistic therapy of bacteria infection. *ACS Appl. Mater. Interfaces*, 2017.

119. Tozzini, V. and Pellegrini, V., Prospects for hydrogen storage in graphene. *Phys. Chem. Chem. Phys.*, 15, 1, 80–89, 2013.

120. Bénard, P. et al., Comparison of hydrogen adsorption on nanoporous materials. *J. Alloys Compd.*, 446, 380–384, 2007.

121. Ghosh, A. et al., Uptake of H_2 and CO_2 by graphene. *J. Phys. Chem. C*, 112, 40, 15704–15707, 2008.

122. Choma, J. et al., Highly microporous polymer-based carbons for CO_2 and H_2 adsorption. *RSC Adv.*, 4, 28, 14795–14802, 2014.

123. Hong, W.G. et al., Agent-free synthesis of graphene oxide/transition metal oxide composites and its application for hydrogen storage. *Int. J. Hydrogen Energy*, 37, 9, 7594–7599, 2012.

124. Divya, P. and Ramaprabhu, S., Hydrogen storage in platinum decorated hydrogen exfoliated graphene sheets by spillover mechanism. *Phys. Chem. Chem. Phys.*, 16, 48, 26725–26729, 2014.

125. Moradi, S.E., Enhanced hydrogen adsorption by Fe_3O_4–graphene oxide materials. *Appl. Phys. A*, 119, 1, 179–184, 2015.

126. Kostoglou, N. et al., Few-layer graphene-like flakes derived by plasma treatment: A potential material for hydrogen adsorption and storage. *Microporous Mesoporous Mater.*, 225, 482–487, 2016.

127. Wesołowski, R.P. and Terzyk, A.P., Pillared graphene as a gas separation membrane. *Phys. Chem. Chem. Phys.*, 13, 38, 17027–17029, 2011.

128. Yuan, W., Li, B., Li, L., A green synthetic approach to graphene nanosheets for hydrogen adsorption. *Appl. Surf. Sci.*, 257, 23, 10183–10187, 2011.

129. Zhou, C. and Szpunar, J.A., Hydrogen storage performance in Pd/graphene nanocomposites. *ACS Appl. Mater. Interfaces*, 8, 39, 25933–25940, 2016.

130. Zhou, C., Szpunar, J.A., Cui, X., Synthesis of Ni/graphene nanocomposite for hydrogen storage. *ACS Appl. Mater. Interfaces*, 8, 24, 15232–15241, 2016.
131. Ismail, N., Madian, M., El-Shall, M.S., Reduced graphene oxide doped with Ni/Pd nanoparticles for hydrogen storage application. *J. Ind. Eng. Chem.*, 30, 328–335, 2015.
132. Burress, J. et al., Gas adsorption properties of graphene-oxide-frameworks and nanoporous benzene–boronic acid polymers, in: *APS Meeting Abstracts*, 2010.
133. Wu, W.-M., Zhang, C.-S., Yang, S.-B., Controllable synthesis of sandwich-like graphene-supported structures for energy storage and conversion. *New Carbon Mater.*, 32, 1, 1–14, 2017.
134. Wang, H. et al., Rechargeable $Li-O_2$ batteries with a covalently coupled $MnCo_2O_4$–graphene hybrid as an oxygen cathode catalyst. *Energy Environm. Sci.*, 5, 7, 7931–7935, 2012.
135. Zhou, W. et al., Fabrication of Co_3O_4-reduced graphene oxide scrolls for high-performance supercapacitor electrodes. *Phys. Chem. Chem. Phys.*, 13, 32, 14462–14465, 2011.
136. Wang, T. et al., Graphene–Fe_3O_4 nanohybrids: Synthesis and excellent electromagnetic absorption properties. *J. Appl. Phys*, 113, 2, 024314, 2013.
137. Wei, W., *Controllable Assembly of Graphene Hybrid Materials and Their Application in Energy Storage and Conversion*, Universitätsbibliothek Mainz, 2015.
138. Winter, M. et al., Insertion electrode materials for rechargeable lithium batteries. *Adv. Mater.*, 10, 10, 725–763, 1998.
139. Kaskhedikar, N.A. and Maier, J., Lithium storage in carbon nanostructures. *Adv. Mater.*, 21, 25–26, 2664–2680, 2009.
140. Dahn, J.R. et al., Mechanisms for lithium insertion in carbonaceous materials. *Science*, 590, 1995.
141. Sun, H. et al., Mesoporous Co_3O_4 nanosheets-3D graphene networks hybrid materials for high-performance lithium ion batteries. *Electrochim. Acta*, 118, 1–9, 2014.
142. Raccichini, R. et al., The role of graphene for electrochemical energy storage. *Nat. Mater.*, 14, 3, nmat4170, 2014.
143. Wang, Z.-L. et al., In situ fabrication of porous graphene electrodes for high-performance energy storage. *ACS Nano*, 7, 3, 2422–2430, 2013.
144. Bhardwaj, T. et al., Enhanced electrochemical lithium storage by graphene nanoribbons. *J. Am. Chem. Soc.*, 132, 36, 12556–12558, 2010.
145. Liu, F. et al., Folded structured graphene paper for high performance electrode materials. *Adv. Mater.*, 24, 8, 1089–1094, 2012.
146. Lee, S.W. et al., High-power lithium batteries from functionalized carbon-nanotube electrodes. *Nat. Nanotechnol.*, 5, 7, 531–537, 2010.
147. Jiang, K.-C. et al., Superior hybrid cathode material containing lithium-excess layered material and graphene for lithium-ion batteries. *ACS Appl. Mater. Interfaces*, 4, 9, 4858–4863, 2012.
148. Zhu, K. et al., Synthesis of $H_2V_3O_8$/reduced graphene oxide composite as a promising cathode material for lithium-ion batteries. *ChemPlusChem*, 79, 3, 447–453, 2014.
149. Han, S. et al., Graphene aerogel supported $Fe_5(PO_4)_4(OH)_3 \cdot 2H_2O$ microspheres as high performance cathode for lithium ion batteries. *J. Mat. Chem. A*, 2, 17, 6174–6179, 2014.
150. Li, B. et al., An in situ ionic-liquid-assisted synthetic approach to iron fluoride/graphene hybrid nanostructures as superior cathode materials for lithium ion batteries. *ACS Appl. Mater. Interfaces*, 5, 11, 5057–5063, 2013.
151. Fei, H. et al., $LiFePO_4$ nanoparticles encapsulated in graphene nanoshells for high-performance lithium-ion battery cathodes. *Chem. Commun.*, 50, 54, 7117–7119, 2014.
152. Hu, J. et al., Alternating assembly of Ni–Al layered double hydroxide and graphene for high-rate alkaline battery cathode. *Chem. Commun.*, 51, 49, 9983–9986, 2015.
153. Ma, R. et al., Fabrication of LiF/Fe/Graphene nanocomposites as cathode material for lithium-ion batteries. *ACS Appl. Mater. Interfaces*, 5, 3, 892–897, 2013.

154. Kim, W. et al., Fabrication of graphene embedded LiFePO4 using a catalyst assisted self assembly method as a cathode material for high power lithium-ion batteries. *ACS Appl. Mater. Interfaces*, 6, 7, 4731–4736, 2014.

155. de Faria, A.F. et al., Eco-friendly decoration of graphene oxide with biogenic silver nanoparticles: Antibacterial and antibiofilm activity. *J. Nanopart. Res.*, 16, 2, 2110, 2014.

156. Zhu, Z. et al., Preparation of graphene oxide–silver nanoparticle nanohybrids with highly antibacterial capability. *Talanta*, 117, 449–455, 2013.

157. Li, S.-K. et al., Bio-inspired *in situ* growth of monolayer silver nanoparticles on graphene oxide paper as multifunctional substrate. *Nanoscale*, 5, 24, 12616–12623, 2013.

158. He, G. et al., Photosynthesis of multiple valence silver nanoparticles on reduced graphene oxide sheets with enhanced antibacterial activity. *Synth. React. Inorg. Met.-Org., Nano-Met. Chem.*, 43, 4, 440–445, 2013.

159. Shao, W. et al., Preparation, characterization, and antibacterial activity of silver nanoparticle-decorated graphene oxide nanocomposite. *ACS Appl. Mater. Interfaces*, 7, 12, 6966–6973, 2015.

160. Xu, W.-P. et al., Facile synthesis of silver@ graphene oxide nanocomposites and their enhanced antibacterial properties. *J. Mater. Chem.*, 21, 12, 4593–4597, 2011.

161. Yoo, J.M., Kang, J.H., Hong, B.H., Graphene-based nanomaterials for versatile imaging studies. *Chem. Soc. Rev.*, 44, 14, 4835–4852, 2015.

162. Janib, S.M., Moses, A.S., MacKay, J.A., Imaging and drug delivery using theranostic nanoparticles. *Adv. Drug Delivery Rev.*, 62, 11, 1052–1063, 2010.

163. Wang, J. et al., Imaging-guided delivery of RNAi for anticancer treatment. *Adv. Drug Delivery Rev.*, 104, 44–60, 2016.

164. Lin, J., Chen, X., Huang, P., Graphene-based nanomaterials for bioimaging. *Adv. Drug Delivery Rev.*, 105, 242–254, 2016.

165. Zhu, S. et al., Photoluminescent graphene quantum dots for *in vitro* and *in vivo* bioimaging using long wavelength emission. *RSC Adv.*, 5, 49, 39399–39403, 2015.

166. Shao, Y. et al., Graphene based electrochemical sensors and biosensors: A review. *Electroanalysis*, 22, 10, 1027–1036, 2010.

167. Holzinger, M., Le Goff, A., Cosnier, S., Nanomaterials for biosensing applications: A review. *Front. Chem.*, 2, 63–63, 2014.

168. Park, S., Boo, H., Chung, T.D., Electrochemical non-enzymatic glucose sensors. *Anal. Chim. Acta*, 556, 1, 46–57, 2006.

169. Yin, P.T. et al., Design, synthesis, and characterization of graphene–nanoparticle hybrid materials for bioapplications. *Chem. Rev.*, 115, 7, 2483–2531, 2015.

170. Li, C., Wang, F., Jimmy, C.Y., Semiconductor/biomolecular composites for solar energy applications. *Energy Environm. Sci.*, 4, 1, 100–113, 2011.

171. Yu, J., Fan, J., Cheng, B., Dye-sensitized solar cells based on anatase TiO_2 hollow spheres/carbon nanotube composite films. *J. Power Sources*, 196, 18, 7891–7898, 2011.

172. Yu, J. et al., Enhanced photocatalytic activity of bimodal mesoporous titania powders by C 60 modification. *Dalton Trans.*, 40, 25, 6635–6644, 2011.

173. Yoo, D.-H. et al., Enhanced photocatalytic activity of graphene oxide decorated on TiO_2 films under UV and visible irradiation. *Curr. Appl. Phys.*, 11, 3, 805–808, 2011.

174. Liu, J. et al., Gram-scale production of graphene oxide–TiO 2 nanorod composites: Towards high-activity photocatalytic materials. *Appl. Catal., B*, 106, 1, 76–82, 2011.

175. Lightcap, I.V., Kosel, T.H., Kamat, P.V., Anchoring semiconductor and metal nanoparticles on a two-dimensional catalyst mat. Storing and shuttling electrons with reduced graphene oxide. *Nano Lett.*, 10, 2, 577–583, 2010.

8

Graphene-Based Composites with Shape Memory Effect—Properties, Applications, and Future Perspectives

André Espinha[1], Ana Domínguez-Bajo[2], Ankor González-Mayorga[3] and María Concepción Serrano[2]*

[1]*UAM-IFIMAC-Condensed Matter Physics Center, Departamento de Física de Materiales, Universidad Autónoma de Madrid, Madrid, Spain*
[2]*Group of Materials for Health, Instituto de Ciencia de Materiales de Madrid, Consejo Superior de Investigaciones Científicas, Madrid, Spain*
[3]*Hospital Nacional de Parapléjicos, Servicio de Salud de Castilla-La Mancha, Toledo, Spain*

Abstract

Graphene, a honeycomb lattice of carbon atoms, is being explored as an attractive material for diverse applications including electronics, sensors, and biomaterials. It is characterized by an extraordinary surface area, high charge carrier mobility, and Young modulus, among others. Shape memory effect concerns the property that some materials possess for recovering a permanent shape after being programmed in a temporary shape, triggered by an external stimulus. Shape memory polymers (SMPs) stand out due to advantages related to their broad chemical modification capabilities, high throughput, and lower cost. SMPs are attractive for applications such as biomedicine, textile, packaging, aerospace, and actuators. A convenient way to add value to SMPs is by preparing composite materials with fillers of specific functionalities. In this scenario, graphene-derived materials have already demonstrated a great interest as SMP fillers, leading to composites with advanced properties (mechanical reinforcement, electrical conductivity, surface wettability). From a chemical point of view, materials as diverse as polyurethanes, polylactic acid, polyacrylamide, poly(propylene carbonate), epoxy resins, and chitosan have been used as polymeric matrices. To date, an extensive work on the fabrication and characterization of these composites has been done. Nevertheless, industrial applications remain challenging. Some potential uses include the domain of electronics, sensors, actuators, and biomedical devices. In this book chapter, major advances in graphene-based composites displaying shape memory properties are presented. A final discussion on the applicability and future perspectives of these advanced composites is also included.

Keywords: Actuators, composites, graphene, responsive materials, shape memory effect

Corresponding author: mc.terradas@csic.es

List of Abbreviations

CNTs	Carbon nanotubes
DMA	Dynamic mechanical analysis
GO	Graphene oxide
IR	Infrared light
NIR	Near-infrared light
PAAm	Polyacrylamide
PCL	Poly(ε-caprolactone)
PLA	Poly(lactic acid)
PNIPAM	Poly(N-isopropylacrylamide)
PPC	Poly(propylene carbonate)
PU	Polyurethane
PVA	Poly(vinyl alcohol)
PVAc	Poly(vinyl acetate)
R_f	Strain fixity rate
R_r	Strain recovery rate
RGO	Reduced graphene oxide
SEM	Scanning electron microscopy
SERS	Surface enhanced Raman spectroscopy
SME	Shape memory effect
SMP	Shape memory polymer
T_g	Glass transition temperature
T_m	Melting transition temperature
T_{max}	Maximum temperature
T_{trans}	Transition temperature
TME	Temperature memory effect

8.1 Introduction

Graphene-based composites displaying shape memory are relatively novel materials whose research is still in an early stage, with the most significant development carried out in the last 5 years. Although promising, marketable applications are still under way. The incorporation of graphene into the so-called shape memory polymers (SMPs) has enabled the improvement of their mechanical properties, as well as provided them with electrical conductivity or light absorption features, which have expanded their functionality. These materials have significant technological interest especially for the fabrication of actuators, useful for areas such as aerospace or soft robotics. In the next sections, we start by presenting the two most important actors for the following discussion, namely, graphene and SMPs. Afterwards, a brief introduction to SMP-based composites is exposed, followed by the main focus of this book chapter: graphene-doped shape memory composites. This section is divided into fabrication methods, enhancement of mechanical properties, electrical and thermal conductivities, and characterization of shape memory properties. A final discussion of current applications and future perspectives is also included.

8.1.1 Graphene

Graphene was isolated for the first time in 2004 by Novoselov and Geim [1], a discovery that granted them the Nobel Prize in Physics in 2010. It is a two-dimensional allotrope of carbon that consists of a single layer of atoms arranged in a honeycomb structure (Figure 8.1a). In this configuration, the sp^2 orbitals of carbon atoms separated by 1.42 Å hybridize, leading to the formation of a molecular σ bound between them (Figure 8.1b). On their turn, the p orbitals orienting perpendicular to the basal plane form a π band (Figure 8.1b) [2]. The long-range conjugation of these leads to the extraordinary mechanical, electrical, and thermal properties of graphene. It presents Young modulus values in the range of 0.5–1 TPa, electrical conductivity in the order of 10^6 S cm^{-1}, charge carrier mobility of 2×10^5 cm^2 V^{-1} s^{-1} at electron densities of 2×10^{11} cm^{-2}, thermal conductivity of 5×10^3 Wm^{-1} K^{-1}, low optical absorbance in the visible range (~2.3%), and a large specific surface area (>2.5×10^3 m^2 g^{-1}). Another useful feature is its relatively easy chemical functionalization [3, 4].

The most straightforward method to prepare graphene, and the one initially proposed by Novoselov and Geim, is by micromechanical cleavage of graphite. In fact, graphene layers in that material are weakly bounded through van der Waals forces and are easily separable. Using scotch tape, individual monolayers are exfoliated and then transferred to a suitable substrate. Although very convenient, this method presents drawbacks such as difficult scalability and poor reproducibility. Nowadays, techniques in use try to circumvent these problems. Some worth noting are epitaxial growth, chemical vapor deposition, arc discharge of graphite, and preparation in colloidal suspensions using selected solvents by solution phase exfoliation of graphite [5].

Since its isolation in 2004, the research in graphene has expanded at a rhythm hardly observed in any other area. Remarkably, this is due to its unusual properties and the promise for new and exciting applications and devices. Graphene has an enormous potential for implementation in applications ranging from chemical sensors, flexible electronics, optoelectronic, biomedical devices, and gas storage, to cite a few. For a comprehensive review on the subject, readers are referred elsewhere [4]. Among other remarkable uses, graphene and its derivatives (mainly, graphene oxide, GO, and reduced graphene oxide, RGO) are especially suited as a filler for the production of high-performance nanocomposites.

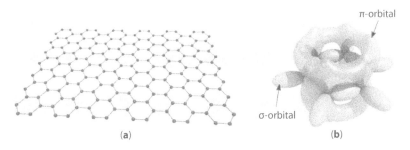

Figure 8.1 Scheme illustrating the structure of graphene, where carbon atoms organize into a honeycomb lattice (a). Artistic representation of the orbitals hybridization in graphene atoms showing the σ bounds in blue and the π bounds in light red (b).

8.1.2 Shape Memory Polymers

The shape memory effect (SME) was first discovered in a gold-cadmium alloy [6]. It concerns the property that some materials present for recovering a permanent set shape after being fixed into a temporary one, by applying some mechanical programming, and later exposed to an external stimulus responsible for triggering the recovery back. The effect was later discovered in polymers [7], the so-called shape memory polymers (SMPs) [8], with some important advantages such as higher strains before rupture, higher recoverable strains, lower densities, corrosion resistance, lower production costs, and more flexibility in their properties design. Also, they are easier to implement with high throughput and additional features aiming at a multifunctional performance [9].

In a typical programming, cross-linked slabs of a SMP are configured into a permanent shape. After a heating/deformation/cooling cycle, the samples are programmed into a deformed temporary state. In polydiolcitrate-based polymers [10], for example, the SME is driven by a melting transition (one of the possible SME mechanisms in polymers) at a temperature T_{trans}, which is close to the human body temperature. Above T_{trans}, the material becomes rubbery-like and moldable. During the cooling stage of the cycle, the temperature drops below T_{trans} and the polymeric blocks regain their stiffness. If the deformation forces are maintained during this step, the polymer adopts and retains the new temporary shape, even after stress release. Finally, after reheating the polymer, the initial permanent shapes are restored.

A specific molecular architecture is required for an elastomer to present SME. It must contain some sort of netpoints responsible for fixing the permanent shape and some molecular switches that are sensitive to the external stimulus and thus responsible for the temporary state. Netpoints can have origin for example in covalent bonds established during the polymer synthesis and cross-linking. In cases where the polymer consists of segregated domains, those associated to the highest transition temperature are responsible for the netpoints while the ones associated to the subsequent transition act as the switching mechanism. The most common switching transitions, occurring at a temperature T_{trans}, are melting transitions (where T_{trans} is denoted as T_m) in which the temporary shape is fixed by solidification of the switching domains—as in the case described in the last paragraph—or glass/rubber transitions (where T_{trans} is labeled as T_g) where the temporary shape is fixed by vitrification.

In a broad variety of SMPs, the SME is directly triggered by heat due to a thermal transition, such as the ones described above. There are cases however where the triggering is sensitive to other external stimulus such as electrical current [11] or alternating magnetic fields [12]. These mechanisms may be useful, for example, for remote actuation of the SME. Nevertheless, in this scenario, the trigger responsible for the shape restoration is still temperature as the elastomers are indirectly heated by those stimuli. Interestingly, other alternative options for SME activation in isothermal conditions such as light irradiation [13], water induction [14], or pH variation [15] are also under development. To accomplish these effects, many of these strategies rely on fabricating smart composites or nanocomposites.

The shape memory properties in SMPs are typically quantified by cyclic mechanical tests, the so-called thermomechanical programming tests. These measurements are done in a tensile tester equipped with a thermochamber. In general, each cycle consists of programming the elastomeric sample in a temporary shape and then recovering its permanent shape. One of such cycles is illustrated in a strain/temperature/stress diagram as the one exemplified in Figure 8.2. In a conventional experiment, the sample at the original shape of the corresponding cycle (ε_p) is heated to a maximum temperature (T_{max}) above T_{trans}, then deformed to the maximum strain (ε_m), and cooled down. Once cooled, the stress constraints are removed in order to configure the temporary shape. In nonideal samples, some relaxation occurs; therefore, the final strain is not the programmed one (ε_m) but becomes ε_u. At this point, the original shape can be restored by increasing the temperature (in the absence of stress) or reprogrammed by deforming the sample again.

The important magnitudes required to characterize SMPs are the switching temperature T_{trans} (considering thermally induced SMPs), the elastic modulus (determined from the slope of the stress–strain curve at the initial linear part of the curve), and two rates denoted as the strain recovery rate (R_r) and the strain fixity rate (R_f) at a specific iteration number N. The strain recovery rate evaluates the capacity of a material to memorize its permanent shape and is calculated according to Equation 8.1.

$$R_r(N) = \frac{\varepsilon_m - \varepsilon_p(N)}{\varepsilon_m - \varepsilon_p(N-1)} \tag{8.1}$$

The strain fixity rate compares the programmed strain after stress release and relaxation with the maximum imposed strain according to Equation 8.2.

$$R_f(N) = \frac{\varepsilon_u(N)}{\varepsilon_m} \tag{8.2}$$

SMPs have a tremendous technological interest in many areas such as automobile engineering, aerospace industry, photonics, construction, household products, and biomedicine, with exemplary applications including microfluidic devices, self-foldable packaging, smart textile, erasable Braille, smart adhesives, and vascular stents [16].

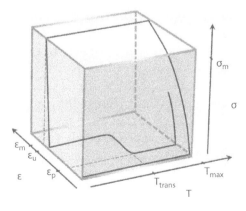

Figure 8.2 Three-dimensional strain–temperature–stress diagram illustrating a thermomechanical programing cycle used in a typical experiment for the characterization of the SME in polymers.

8.1.3 Shape Memory Polymer Composites

As previously mentioned, SMPs are extraordinary materials with notable properties and a marked multifunctional character, due to their intrinsic responsiveness. It is frequently convenient to prepare SMP-based composites by incorporating specific fillers into the polymeric matrix to take advantage of the synergy between both components. In this way, the functionality of the initial SMPs may be enormously augmented, boosting their performance [17]. One common reason for producing such composites is mechanical reinforcement of the original SMP. Other motives include the integration of additional functionalities such as, for example, electrical conductivity, the development of different stimuli-triggering mechanisms, and the exploration of different memory effects like multiple shape memory, spatially localized SME, and two-way shape memory.

Common fillers used in the fabrication of SMP composites include microfibers made of cellulose, carbon, glass, or Kevlar, which help to increase the elastic modulus of the resulting materials and enable them to bear higher mechanical loads [17]. Carbon nanotubes (CNTs) are also being used aiming at the development of electroactive SME derived from Joule heating [18]. Different kinds of particles ranging from metallic, dielectric, and ferromagnetic are also under investigation, targeting at novel properties such as SERS substrates [19], diffusive photonic gain media [20], or magnetic fields excitation [21], respectively. In the next section, composites based on graphene-doped SMPs are presented and their major features are discussed.

8.2 Graphene-Doped SMP Composites

The incorporation of graphene into SMPs helps to improve their mechanical properties, shape memory performance, and electrical conductivity and accelerates electrically driven shape restoration due to resistive heating. It may be also responsible for improved thermal stability of the composites [22]. Interestingly, as pure graphene films are nearly impermeable to water, the preparation of polymeric composites with designed graphene fillers may improve the hydrophilicity of the whole system, enabling water diffusion into the material [23].

To the best of our knowledge, the first reference to graphene-doped SMPs was published by Liang and coworkers in 2009 [24]. These researchers incorporated up to 1.0 wt.% of sulfonated functionalized graphene sheets into a thermoplastic polyurethane (PU). This material is essentially transparent. By doping it with graphene, light absorption was enabled, particularly at the infrared (IR) region of the spectrum. As so, graphene played the role of the energy transfer unit, allowing the temperature of the composite to increase when illuminated and, consequently, the SME prompted. In fact, when programmed in a deformed state, and after exposure to IR light, nanocomposites with graphene loadings as low as 0.1 wt.% contracted almost completely to the original shape, while nondoped PUs (used as control) remained unaffected. In this sense, the samples performed as light-triggered actuators.

Solution mixing processes are probably the most used methods for the production of graphene-based shape memory composites [22, 24, 25]. Generally, a solution of graphene (typically functionalized, so that its solubility is enhanced) is prepared in a selected solvent using sonication. The SMP, also in solution (ideally in the same solvent), is added to the first one by stirring for several hours. This method could be classified as a physical one where a dispersion of the graphene flakes is added to a solution of the already synthetized SMP or vice versa. Alternatively, the SMP is polymerized in the presence of the graphene dispersion—*in situ* processing. Naturally, in this case, the specificities of the chemical syntheses depend on the selected SMPs and the graphene surface treatment. The *in situ* method has been reported, for example, by Choi and coworkers for the preparation of shape memory PU nanocomposites, starting from a mixture of RGO and poly(ε-caprolactone) diol (PCL) [26]. An alternative approach was pursued for *in situ* cationic polymerization of linseed oil [27]. Specifically, GO nanoplatelets modified by using cetyltrimethyl ammonium bromide were dispersed into the solution containing the polymer precursors (i.e., styrene, divinylbenzene, and linseed oil). As a result, shape memory actuation in these composites could be triggered by thermal heating, sunlight, and microwaves. In general, for both physical mixture and *in situ* processing, the final step consists of casting the mixture in some sort of container and letting the solvent to evaporate or vacuum dry, yielding the final composites. Typical doping concentrations used for the preparation of composites range from approximately 0.1 wt.% to values as high as 20 wt.%.

Another aspect of fundamental importance concerns the functionalization of the graphene surface. It plays an essential role in both the dispersibility of the fillers within the polymer and the properties of the final composite, such as density, mechanical performance, conductivity, and chemical stability. These derive from both the inherent properties of graphene and the interfacial interactions between the filler and the polymeric matrix. Chemical groups may be attached to the graphene surface by either covalent bonding or physical adsorption through van der Waals forces or π–π interactions. In this regard, covalent functionalization is frequently more suitable for the fabrication of composites, especially in applications where a good transfer of mechanical load between the polymer and the filler is required. On their turn, the functional groups attached to the graphene surface can also establish covalent bonds with the SMP matrix leading to higher performance. Literature reports on graphene surface functionalization include chemical groups such as isocyanate, sulfonate [24], hydroxyl, alkoxy [28], epoxide, amine, imide [29], carbonyl, and carboxylic [30]. To this respect, GO and RGO, whose surfaces contain several of those groups, are frequently preferable to pure graphene for the preparation of advanced nanocomposites. Nevertheless, one significant drawback is the damaging of the sp^2 carbon network and the introduction of defects, which drives to an inherent cost in performance [24]. Examples include the use of GO sheets covalently functionalized with hyperbranched PU chains serving to the preparation of high-performance nanocomposites by *in situ* polymerization [31]. In a different approach, GO sheets were modified with small amounts of octadecylamine (0.25–1 wt.%) and incorporated into a 3-amino-1,2,4-triazole cross-linked maleated polyethylene–octene elastomer via melt blending [32].

It is interesting to note that the mass density of the composites may become smaller than that of the original polymers [26], and further, the density may decrease as the graphene content increases, pointing out that additional free volume is created in the composites. This effect is attributed not only to the lower packing density of the polymer at the interface between the graphene filler and the matrix, in comparison to bulk packing, but also to the constrained rearrangement of the polymer chains due to the presence of graphene when the temperature is lowered below the T_{trans} of the SMP.

Table 8.1 summarizes some of the most outstanding graphene-doped SMP composites reported to date. It sums up major details on their chemical composition and properties, namely, the T_{trans} (for thermally activated SMPs), Young modulus, electrical conductivity, and recovery ratios. As can be appreciated, the most explored polymeric matrices for the preparation of graphene-based SMP composites are PUs due to their good melt processability, flame retardancy, high recovery and fixity ratios, and short recovery times [25, 33, 34]. For instance, PUs made from PCL and isophorone diisocyanate have been used for the fabrication of hybrid composites with aluminum hydroxide and RGO displaying shape memory properties triggered by heat (60°C), microwaves (300 W), and sunlight (10^5 lux) [35]. The mechanical properties of the system significantly improved by doping the polymer matrix, reaching values of 350% for tensile strength, 292% for elongation at break, and 441% for toughness. As another example, acrylate-terminated PUs have been also doped with allyl isocyanate-modified GO [36]. In a more recent work by these authors, electroactive graphene-doped shape memory PUs were synthesized from poly(tetramethylene ether) glycol, 4,4-methylenebis(phenyl isocyanate), and 1,3-butandiol [37].

Besides PUs, other chemical polymer compositions may also be extremely favorable, enabling new phenomena. For example, Qi et al. developed a GO and polyvinyl alcohol (PVA) composite in which the SME was triggered by water [30]. The same group described the fabrication of poly(propylene carbonate) (PPC)/GO composites with tunable shape memory properties by controlling the filler content. Specifically, dual SME was obtained for those composites with GO contents below 10 wt.%, while triple SME was achieved in those doped with GO percentages higher than 10 wt.% [38]. Noteworthy, in the first case, one single T_g corresponding to PPC was identified (i.e., slightly confined system). In the second one, two distinctive T_g were observed (i.e., highly confined system). More recently, another alternative composition was reported using microcrystalline cellulose nanofibers extracted from *Agave sisalana* as the matrix and doped with GO to produce nanocomposite papers [39]. In a different work, Tang et al. explored the incorporation of RGO into composites of polyester and vapor-grown carbon nanofibers [40]. Sabzi et al. doped poly(vinyl acetate) (PVAc) with graphene nanoplatelets, improving the glassy and rubbery moduli of the composite [41]. Interestingly, these hybrid composites also showed scratch self-healing capability. Other authors have also reported the improvement of shape memory properties of poly(L-lactide)-based SMPs by incorporating graphene nanoplatelets [42, 43]. In these polymers, the SME could be triggered by both thermal heating and IR light irradiation. Last, it is worth to note that hybrids of graphene and carbon nanotubes dispersed in SMPs are also an interesting possibility, conjugating features from both CNTs and graphene [44]. In the fabrication of graphene-doped SMPs, graphene has been typically used as a dopant within the SMP matrix, as described in the above examples. Alternatively, although rarely explored, graphene can also be the major component of the material, being its amount comparable to or even higher than the matrix, as in the composites fabricated with amyloid fibrils by Li and coworkers [23].

In an attempt to design biodegradable and more environmentally friendly SMP materials, biopolymers have also been selected to fabricate composites with graphene displaying SME. For instance, recent work by Zhang et al. described layered chitosan–GO nanocomposite hydrogels with a nacre-like brick-and-mortar structure [45]. These hybrid materials displayed pH-triggered SME due to the reversible physical cross-linking of chitosan chains by hydrogen bonding and hydrophobic interactions. They nearly reached 100% of shape recovery after 9 min of immersion in aqueous solutions at pH = 3 and 25°C, while chitosan hydrogels

Table 8.1 Summary of reference values for some outstanding graphene-doped SMP composites.

SMP material	T_{trans} (°C)	Nondoped E (MPa)	Doped E (MPa)	Electrical conductivity (S cm^{-1})	Pristine recovery rate (%)	Doped recovery rate (%)	Ref.
Thermoplastic polyurethane	–	10	14–22	–	–	70–90	[24]
Polyurethane	24–43	–	–	8.1×10^{-3}	63.2	98.6	[26]
Polyurethane block copolymer	–	5	62.9	1.6×10^{-3}	–	83–90	[22]
Polyimide resin	230	–	–	–	89	95	[29]
Polyurethane	–	5.5	6.5–10.5	–	95–96.5	97–99	[25]
Polyurethane	36–46	–	–	1.7×10^{-3}	–	–	[28]
Polyurethane block copolymer	36	10	30	1.7×10^{-3}	89	97	[54]
Hyperbranched polyurethane	50	2.8	4.2–6.6	–	88	95–99	[48]
Epoxy	75	1.2×10^3	$(0.9–1.5) \times 10^3$	–	100	100	[47]
Microcrystalline cellulose nanofibers	–	5.8×10^3	9.9×10^3	–	–	–	[39]
Poly(caprolactone)/polyurethane blend	–(41–27)	–	–	$10^{-12}–10^{-1}$	100	100	[51]
Chitosan	–	2.51	3.0–8.3	–	72	98	[45]

attained only 74% under those conditions. *N*-succinyl chitosan was also used to prepare similar films containing GO via zinc ion cross-linking and evaporation process [46]. The resulting hybrid matrices displayed remarkable mechanical properties and antibacterial capacity against *Escherichia coli* and *Staphylococcus aureus*. Shape memory properties were driven by swelling/evaporation after exposure to alcohols due to the ability of chitosan to absorb/desorb water.

Finally, it is interesting to mention that apart from the traditional scheme of SME, the shape of graphene-vitrimer composites could be randomly changed by means of a dynamic covalent transesterification reaction (dual-triggered shape recovery and shape reconfigurable shape memory composites) [47]. These composites were fabricated by dispersing graphene nanosheets (0.1, 0.5, 1, and 3 wt.%) into DGEBA E51 epoxy resin and using sebacic acid as the curing agent.

8.2.1 Morphological Properties

Morphological characterization of the composites is typically performed by scanning electron microscopy (SEM) observation of cryofractured surfaces of the samples. The low temperature is an important requisite; otherwise, high-temperature cleavage may introduce plastic deformations, altering their morphology. Additionally, SEM analysis might aid to evaluate the dispersion homogeneity of the graphene filler within the polymeric matrix. Results show that, as a general trend, the surface of nondoped SMPs presents a rather smooth aspect, independently of the chemical composition used. However, the surface roughness increases with the amount of graphene incorporated into the composite [25]. Changing the graphene concentration is therefore an effective method to control the morphology of the final material.

As an example, Figure 8.3 shows a sequence of SEM images of composites made of PPC doped with GO at concentrations varying from 0 wt.% to 20 wt.% [38]. As can be observed, the morphology of the material changes significantly when increasing graphene doping. With 1 wt.% of GO, the composite exhibited a dotted structure where graphene flakes are quite separated from each other. As GO content increases, graphene sheets acquire a more connected morphology, thus becoming a compacted layered structure at high loadings. Using the Halpin–Tsai model, Thakur and coworkers concluded that at low dopant concentrations in hyperbranched PUs, there is no preferential orientation of the graphene sheets within the matrix [48]. On the contrary, due to the lower accessible space to the fillers at higher concentrations, a deviation from the three-dimensional random model emerges. Interactions between graphene flakes start to become important and some orientation builds up.

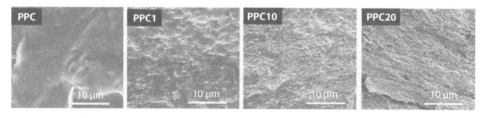

Figure 8.3 SEM images of cryofractured surfaces of pure PPC and PPC/GO nanocomposites. Graphene content increases from left to right as follows: 0 wt.% (PPC), 1 wt.% (PPC1), 10 wt.% (PPC10), and 20 wt.% (PPC20) of GO content, respectively. Adapted with permission from Ref. [38]. Copyright 2016 American Chemical Society.

As exemplified above for PPC, the most common aspect for graphene–SMP composites is a dense and compact morphology. However, lighter porous materials also offer interesting opportunities as advanced materials. Figure 8.4 shows an example of a nontrivial morphology—an interconnected network of PU nanofibers. These were produced by electrospinning [49]. Tan *et al.* also described the preparation of GO-doped PU nanofibers made of PCL by electrospinning [50]. The nanofiber diameters varied from 437 to 543 nm depending on the GO content. Other examples include, for instance, layered structures with a preferential orientation obtained by a solvent casting method on PVA/GO composites [30]. Furthermore, disordered rough morphologies composed of a ternary mixture of PCL/PU and graphene nanoplatelets were prepared by a three-step technique (master batch preparation via solution compounding, composite preparation by diluting the master batch via melt compounding, and subsequent compression molding) [51]. Nacre-like structures based on chitosan were prepared via evaporation technique by Liu and coworkers [46]. Additionally, highly porous foams with a honeycomb-like morphology were attained in a polyacrylamide (PAAm) system by vacuum, air, or freeze drying [52]. In this particular system, the improved packing organization of RGO layers assisted by PAAm during freeze-casting was demonstrated to have a pivotal role for the enhancement of the mechanical properties of the resulting foams. In a different approach, aerogels prepared by a freeze-drying/annealing method were incorporated into a shape memory epoxy resin via a vacuum infusion methodology [53].

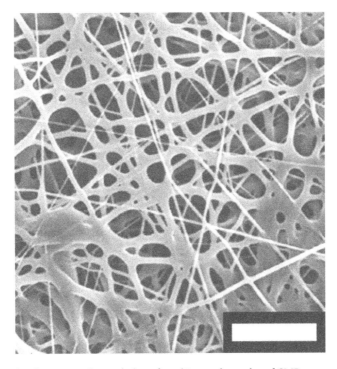

Figure 8.4 An example of a nontrivial morphology found in graphene-doped SMP composites, containing some degree of porosity. Scale bar corresponds to 20 μm. Adapted with permission from Ref. [49]. Copyright 2014 American Chemical Society.

8.2.2 Optical Properties

The optical properties of the resulting doped composites are typically quite distinct from the ones of the initial polymers. In fact, SMPs are usually either transparent or whitish and opaque due to their low light absorption and (where applicable) high scattering. On the contrary, with the incorporation of graphene, absorption is enhanced and the typical aspect of the composites is dark. Although absorption of a single graphene monolayer is extremely low (~2.3%) and therefore it is hardly observable, when light interacts with many graphene layers in the composite, the transparency is severely reduced, the reason why they become black. For example, nondoped thermoplastic PU shows a transmittance of approximately 43% (Figure 8.5a). After doping with sulfonated graphene (1 wt.%), transmittance dropped to about 0.3% at a wavelength of 850 nm [24]. The authors assigned the high absorbance of the composites to the good preservation of the sp^2 network of graphene sheets and to their homogeneous dispersion within the polymeric matrix. The functionalization of graphene is therefore also critical in this matter. As may be appreciated in Figure 8.5b, the absorbance highly depends on the presence of sulfonated graphene, which is the one presenting higher absorbance for the same doping quantity.

8.2.3 Mechanical Properties

As previously mentioned, doping SMPs with graphene typically produces a significant mechanical reinforcement, a feature widely pursued in most of the applications of this kind of materials. Mechanical properties are evaluated measuring stress vs. strain curves in a conventional tensile testing equipment. From these studies, important information such as the Young modulus (also known as elastic modulus), the tensile strength, and the elongation at break is extracted. Moreover, dynamic mechanical analysis (DMA) is also a useful technique for studying the complex modulus of these materials as a function of temperature (storage modulus and loss modulus), thus enabling the determination of the transitions at T_{trans} (i.e., T_g, T_m; for further details, please refer to Section 8.1.2) or other molecular mechanisms.

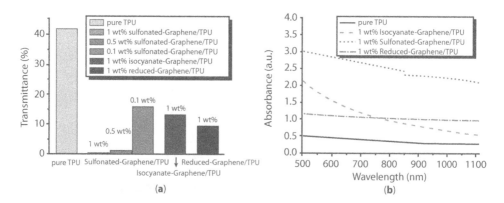

Figure 8.5 Dependence of the optical transmittance of PU composites on the graphene content at a wavelength of 850 nm (a). Comparison of the absorbance spectra of the thermoplastics for different kinds of graphene functionalization (b). Adapted with permission from Ref. [24]. Copyright 2009 American Chemical Society.

For example, the incorporation of sulfonated graphene into thermoplastic PUs enables the composites to become stronger and stiffer (Figure 8.6a). An increase of the tensile stress by 75% at the strain of 100% has been observed for this system, and the elastic modulus was shown to increase 120% at 1 wt.% graphene loading [24]. Furthermore, a monotonous increasing trend of the Young modulus with increasing graphene content was observed (Figure 8.6b). A reinforcement effect can be achieved with even smaller amounts of the graphene filler. Specifically, Jung and coworkers observed an increase of both Young modulus and tensile strength from 5 to 62.9 MPa (~1200%) and from 15.6 to 37.4 MPa (140%), respectively, with just 0.1 wt.% doping [22]. A decrease in the elongation at break, of the order of 7.5% for the doped sample, could also be appreciated from the mechanical characterization. Rana et al. reported similar effects reaching ~16% elongations at break [54]. This trend for the tensile strength and the elongation at break, although widely reported, is not universal for all these polymer-based materials. For example, the incorporation of GO into hyperbranced PUs led to a systematic increase of both the tensile strength and the elongation at break when increasing GO content, with remarkable values as high as 800% [48]. Interestingly, the tensile stress and Young modulus in epoxy-based composites showed a minimum for a graphene amount of 0.5 wt.%. The authors assigned this effect to the presence of an epoxy toughener used for dispersing graphene, which has a larger effect than the reinforcement due to graphene at such low concentrations. For higher concentrations, the typical trend previously described is obtained [55].

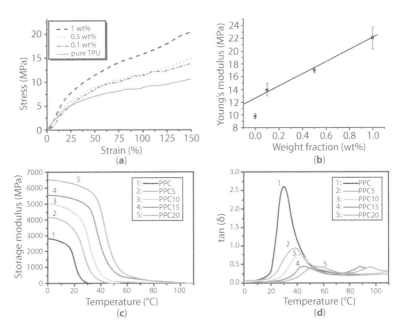

Figure 8.6 Mechanical characterization of graphene-doped SMP composites by stress/strain curves (a) and dependence of the Young modulus on graphene concentration (b) for a thermoplastic PU doped with sulfonated graphene. Adapted with permission from Ref. [24]. Copyright 2009 American Chemical Society. Storage modulus/temperature (c) and tan δ/temperature (d) plots for an exemplary PPC/GO composite. Adapted with permission from Ref. [38]. Copyright 2016 American Chemical Society.

As exemplified in Figure 8.6c, the incorporation of graphene leads to an increase of the storage modulus, which typically increases with the graphene content [29, 38, 51]. This increase is explained by the higher cross-link density when graphene concentration increases, thus leading to enhanced mechanical reinforcement and network formation [25]. It is also apparent that the area of the maximum loss tangent peak (tan δ) was decreased (Figure 8.6d), possibly due to constrained relaxation mechanisms due to the graphene presence. Importantly, the incorporation of graphene into SMPs may also be responsible for changes in the T_m of the composite. As appreciated in Figure 8.6d, the relaxation peak temperature was shifted to higher values. This result implies a restriction effect from graphene on the mobility of the macromolecular chains [51]. More interesting is the appearance of new transitions, observed as new peaks in the tan δ curve. These results have been reported by Qi et al. and pointed out the appearance of new mechanisms responsible for the chain confinement [38]. DMA may also provide useful information about rheological properties of the composites, enabling one to identify characteristic transitions. For example, Zhang and coworkers observed a change from liquid-like to solid-like viscoelastic behavior, implying the formation of a percolated graphene network structure at about 2 wt.% loading [51].

The mechanical properties of the resulting composites also depend on other parameters such as the relative humidity [23] or the graphene functionalization. For instance, the storage moduli of surface-imidized graphene nanocomposites were 25%–30% higher than those containing only RGO [29]. Reported work using nanofibers compared three types of graphene incorporated to PCL-based PUs: GO, RGO, and PCL-functionalized GO [49]. Interestingly, the largest improvement in mechanical properties was found in those composite nanofibers containing PCL-functionalized GO due to the better interaction between the GO fillers and the polymer chains.

8.2.4 Electrical Properties

The majority of pure SMPs reported in the literature are electrical insulators and consequently cannot by stimulated by electricity. Nevertheless, conductivity is an essential requisite for the realization of electroactive shape recovery. This feature is extremely useful, for example, for remote actuation of devices. As we shall see in what follows, the incorporation of graphene or graphene derivatives is a very attractive strategy for providing SMPs with conductivity in a wide range and, thus, to design their electrical properties with high flexibility. In line with electrical conductivity boosting, graphene-doped SMP composites also experience dramatic changes in their thermal conductivity, with reported enhancements in the range of 100% to 200% [22]. This enables a fast transfer of the induced Joule heating to the material and a subsequent acceleration of the shape recovery process.

As expected, the electrical conductivity of composites has a strong dependence on the concentration of the filler added to the polymeric matrix. For example, as the amount of graphene nanoplatelets increased from 0 to 4.5 wt.% in PVAc, the conductivity augmented from 10^{-11} to 24.7 S m^{-1}, a remarkable change by ~13 orders of magnitude [41]. The same authors reported a change of ca. 14 orders of magnitude (from 3×10^{-13} to 31.3 S m^{-1}) in a blend of PVAc/poly(lactic acid) (PLA) when the graphene amount was varied in a similar interval (Figure 8.7) [58]. The enormous enhancement in conductivity was assigned

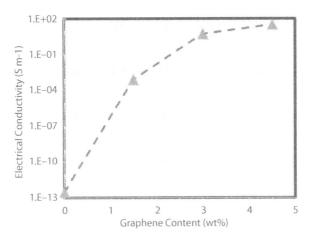

Figure 8.7 Dependence of the electrical conductivity of a blend of PLA/PVAc SMP on the amount of graphene filler. Reprinted with permission from Ref. [58]. Copyright 2017 American Chemical Society.

to the good quality of the exfoliated graphene flakes, with a high aspect ratio and surface area, being able to form a dense network of electron conductive paths. Studying doped polyesters, Zhang and coworkers observed a significant conductivity jump around 2 wt.% [51]. They applied a classical percolation scaling law to their results, describing the conductivity σ as

$$\sigma(p) = \sigma_0 (p - p_c)^\xi \tag{8.3}$$

where σ_0 is the scaling factor, p is the content of graphene nanoplatelets, p_c is the percolation threshold, and ξ is the percolation exponent. With this approach, they determined the percolation threshold of the system at 1.62 wt.% of graphene content.

The electrical conductivity of the composites depends not only on the doping concentration as described above but also on the graphene functionalization. For instance, Choi and coworkers reported composite conductivities in the range of 1.20×10^{-10} to 8.14×10^{-3} S cm^{-1} depending on the oxidation degree of graphene and its concentration [26]. These values represent an increase of the conductivity of at least two orders of magnitude in comparison to the nondoped PUs (1.96×10^{-12} S cm^{-1}). In the case of the most conductive samples, the enhancement corresponds to nine orders of magnitude. Naturally, the more oxidized the graphene state is, the worse the conductivity of the composite result. For electrically triggering the SME, authors applied voltages in the range of 100 to 200 V.

The electroactive SME is triggered by Joule heating generated by the electrical current passing through the sample (Figure 8.8) [58]. The amount of Joule heating (Q) depends on the heating power and the time that the current flows through the material (t). It can be calculated as a function of the electrical resistance of the composite sample (R) and the externally applied voltage (U) as $Q = U^2 t/R$ [51]. Accordingly, more heat is generated for higher applied voltages, if the stimulus is kept for a large duration, or alternatively maintaining U and t constant for more conductive samples. This effect leads to a faster temperature increase for more doped samples and, in the case of a concrete graphene concentration, for higher applied voltages.

Figure 8.8 Electroactive shape recovery in graphene-doped poly(vinyl acetate)/poly(lactic acid) induced by resistive heating. The samples undergo the transition within less than 3 s due to the temperature increase, with an applied voltage of 70 V. Reprinted with permission from Ref. [58]. Copyright 2017 American Chemical Society.

8.2.5 Shape Memory Characterization

The fabrication of graphene-based SMP composites may solve several major challenges that still hinder the broad use of SMPs in advanced applications nowadays. Some of their important limits are low recovery forces, low recovery speeds due to poor thermal conductivities, low stiffness and tensile strength, and inertness to electromagnetic stimuli. The introduction of graphene leads to mechanical reinforcement and to an increase of electrical and thermal conductivities, as previously described in Sections 8.2.3 and 8.2.4. In what follows, we shall discuss the effect of graphene incorporation into the shape memory properties of composites, with a special focus on new effects.

As a general trend, the recovery rate increases by doping SMPs with graphene fillers. This is illustrated in Table 8.1, in which a comparison between R_r of undoped SMPs and graphene SMP composites is presented. Enhancements as high as 60% have been reported in champion PU composites [26]. Liang *et al.* reported a monotonous increase of R_r with the content of sulfonated graphene on thermoplastic PUs [24], which is also quite reproducible when increasing the number of actuating cycles, even if a decrease of performance is expected with cycling [44]. Han *et al.* observed R_r improvements for PUs doped with RGO and functionalized graphene in ranges 0.5–1.0 wt.%, although smaller R_r than that for the pristine PU was obtained with a 2.0 wt.% [68]. Dong *et al.* also showed a dramatic enhancement of R_r in composites of poly(acrylamide-co-acrylic acid) grafted on graphene [62]. They reported poor SME for the bare polymer ($R_r \sim 20\%$), while shape recovery was complete ($R_r \sim 100\%$) for composites with 10 wt.% of graphene. Regarding shape fixity, results seem to be more dependent on the specific material. For example, both R_r and R_f improved due to the presence of graphene in SMPs based on polyimide [29] and PU [54]. Nevertheless, Ponnamma and coworkers observed an increase of R_r and a noticeable decrease of R_f in PU/GO composites when increasing graphene content [25]. A similar behavior was reported by Thakur *et al.* [48]. In most of the cases, the improvements in R_r are assigned to an increase in the stored elastic strain energy driven by graphene (which also leads to an increased recovery stress) resulting from the augmented cross-link density for higher graphene contents. On its turn, R_f is highly dependent on the crystalline phase, and therefore, the presence of graphene may decrease the amount of crystalline phases available.

These results must be interpreted accordingly to the molecular mechanism of the particular system under study. In fact, the shape memory properties of pure SMPs are typically dependent on the crystalline nature of the polymer (determining hard and soft segments), dipole–dipole interactions, hydrogen bonds, and molecular entanglements. For composites incorporating graphene, more complex mechanisms emerge. The restricted mobility of the polymeric chains, cross-link density, glass transitions, and interfacial interactions between graphene and the polymeric matrices are some additional factors that have to be taken into account to understand changes in the resulting SME [25]. Demonstrating the importance of these interactions, Zhang et al. functionalized graphene to enhance its dispersion into a PU/epoxy resin with shape memory [56]. The functionalization consisted in a treatment of GO nanosheets with (3-aminopropyl)triethoxysilane (KH-550), followed by reduction with hydrazine hydrate. The resulting SMP composites displayed thermo-electrical dual-responsive shape memory properties, with shape fixity ratios of 96% and shape recovery values of 94% for those composites containing 1 wt.% of GO (maximum improvement of the SME).

Beyond changes in R_r and R_f, another important SME parameter to study is the shape restoration time. Epoxy-based systems might be interesting in this regard due to the very high recovery ratios, virtually 100% [55]. In this work, the graphene amount seemed not to have an influence on the recovery ratio, but it played a role on the recovering time (i.e., composites recovering their shape faster than the pristine epoxy). In another example, RGO paper was incorporated into the surface of epoxy-based SMP sheets via resin transfer molding to confer electrically driven shape memory properties to the resulting composites [57]. Shape recovery was triggered by heating indirectly produced by electrically resistive heating of the RGO component. Under the application of a voltage of 6 V, the composite was able to reach 100% of shape recovery in only 5 s. Particularly, the rapidity for shape recovery was directly proportional to the applied voltage. Similarly, Zhang and coworkers demonstrated that the incorporation of graphene nanoplatelets into a polyester blend induced a faster electrically stimulated shape recovery (from about 296 s in the binary polymer blend to 36 s in composites containing 5 wt.% of graphene) [51]. Nonetheless, this concentration-dependent effect on R_r reached a minimum value at 5 wt.% of graphene, increasing at higher graphene contents. This effect was likely due to the balance between heat conduction and molecular chain restriction, as hypothesized by the authors. Furthermore, R_r gradually increased with the conduction time and the application of higher voltages required shorter times for attaining similar R_r values.

Another relevant rationale for preparing graphene-doped composites is the search for new, unexpected features in novel materials. One of them is the achievement of electrically triggered dual shape memory, as previously mentioned in this book chapter. In this case, the material can switch between two predefined shapes. Indeed, multishape SME materials, which are able to memorize more than two shapes, are attracting a great deal of attention in the area of responsive materials. In this regard, recent work by Sabzi et al. has revealed the capacity of graphene nanoplatelets to confer triple shape memory properties to blends of PLA and PVAc (30:70) [58]. As demonstrated by atomic force microscopy and thermal analyses, the graphene nanofillers (3 and 4.5 wt.%) induced phase separation and the subsequent appearance of two distinctive T_{trans} (25–26°C and 45–47°C), corresponding to the respective glass transitions of PVAc- and PLA-rich domains (Figure 8.9). In this context, temporary shape C could be achieved at temperatures below T_g of PLA and PVAc, while temporary shape B was attained at temperatures between both T_g values. The permanent shape A was only reached at temperatures above both glass transitions. In these composites, the amorphous PVAc and PLA segments acted as reversible phases, while physical netpoints were attained by physical entanglement of amorphous PVAc and the amorphous phase of PLA and crystallites of PLA (fixed phase). As the concentration of graphene nanoplatelets

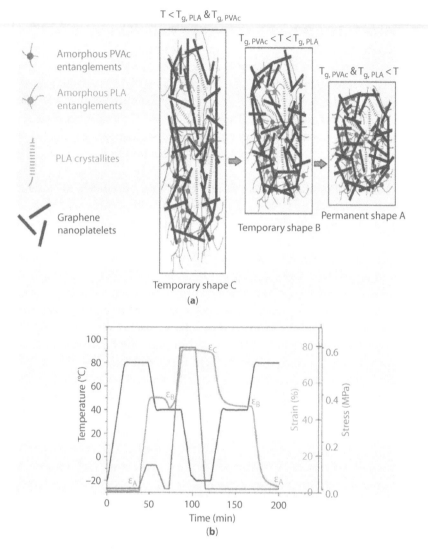

Figure 8.9 Example of multishape effect in mixtures of PLA and PVAc. Schematic illustration of the triple shape memory mechanism (a) and strain/temperature/stress vs time diagram recorded for composites with 4.5 wt.% of graphene (b). Reprinted with permission from Ref. [58]. Copyright 2017 American Chemical Society.

was larger than required for the percolation threshold, the stiff and elastic network that they formed acted as second physical netpoints. Both thermal and electrical actuation of the triple SME in these composites were demonstrated, along with their remarkable dual shape memory behavior. In the case of electrically driven SME, temporary shapes in each region of the composite could be either individually or simultaneously recovered by varying the voltage applied. Triple shape memory behavior has also been described for blends of PLA and PCL (30:70, 50:50, 70:30) doped with either oxidized or functionalized graphene nanoplatelets (0.75 and 1.50 wt.%) [59]. The resulting composite blends displayed two distinctive T_m values corresponding to those of PCL and PLA (51°C and 160°C, respectively) and responsible for the thermally driven triple shape transitions. DMA studies demonstrated enhancement of both the fixity and recovery ratios due to an increase in blend crystallinity and thermal conductivity.

Another fascinating effect allowed by the fabrication of graphene-doped SMP composites includes triggering the shape recovery by immersion in water under isothermal conditions. Qi and coworkers demonstrated that temporarily programmed samples doped with graphene could fully regain their original shape after water immersion, while pristine PVA samples (SMP selected for this study) could not do so [30]. Moreover, the water-induced shape restoration was faster for composites loaded with 1.0 wt.% of graphene (~14 s) than for those having only 0.5 wt.% (~60 s). The strong hydrogen bonding between PVA and GO might be responsible for additional physically cross-linking points that improved the shape memory properties in comparison to pure PVA. Due to the plasticizing effect of water on PVA and the weakened hydrogen bonding between GO and PVA in the presence of water, the SME was triggered by immersion in water.

In a different approach, work by Zhang et al. reported pH-responsive SME of composite hydrogels made of GO and chitosan [45]. The sample was programmed by immersion in an aqueous solution at pH = 3, where it became soft, thus allowing for its bending and then storing the temporary shape by immersion in an aqueous solution at pH = 12. The original shape was restored after reimmersion in the acidic solution (pH = 3). The authors also observed better shape memory properties in graphene-doped samples than pure chitosan control ones. The mechanism of pH-driven SME of the hydrogel was ascribed to the reversible transition of partial physically cross-links corresponding to hydrogen bonding and hydrophobic interactions between chitosan polymeric chains and hydroxyl and carbonyl groups in GO.

Qi et al. [38] observed that high loadings of graphene in PPC (15 and 20 wt.%; highly confined system) led to the achievement of temperature memory effect (TME), which is the capacity of some materials to remember the temperature at which they were mechanically programmed. Graphene loadings in that system originated the appearance of a second transition at higher temperatures. For those particular composites, the two transitions became very broad and superposed in the range 40°C–100°C. In this way, free-strain recovery curves of the samples were shifted to higher temperatures when increasing the programming temperature. The switching temperature was observed to coincide with the programming one in a good degree (linear correlation). Slightly confined systems (with graphene loadings < 15 wt.%) also exhibited TME with a linear correlation between those temperatures. Nevertheless, the switching temperature was lower than the programming one in this particular case.

In summary, graphene is generally responsible for the improvement of the shape memory properties of the resulting composites, mainly exemplified by the recovery and fixity rates. However, care must be taken with the specificities of the polymeric matrix under study. In fact, due to particular interactions of graphene and the polymer, confinement effects, and so on, improvements can be observed only to certain extents. For example, too high graphene contents may severely increase the brittleness of the material and also lead to the deterioration of the shape memory properties. Therefore, depending on the final pursued application, a precise control of graphene loading is necessary for the actual optimization of the material properties. Furthermore, graphene doping of SMPs seems to enable a faster shape recovery in comparison with pristine SMPs. Finally, a notorious advantage of producing graphene-based composites is the potential emergence of new properties and shape memory features, as those described above including electrically and IR light triggered SME, solvent and pH-sensitive shape restoration, multishape memory effects, and TME. Having thoroughly discussed the most relevant composite properties, we now turn to the applications for which these advanced materials are being investigated.

8.3 Applications

Although still in their infancy, as revealed by the limited number of publications found in the literature, applications based on graphene-doped SMP composites hold the promise for the development of multifunctional and responsive devices for impacting areas as diverse as biomedicine, chemical and biological testing, smart coatings, aerospace industry, and soft robotics, among many others. In what follows, we highlight four different promising uses of these composites: light-triggered actuators, self-healing surfaces, surfaces with tunable wettability, and biosensors for enzymatic activity.

One of the most explored applications for these materials is undoubtedly related to IR light-triggered systems. Chen *et al.* developed bilayer actuators made of GO and poly(N-isopropylacrylamide) (PNIPAM) [60]. These devices displayed bending/unbending behavior in response to near-IR (NIR) light, as a result of the combination of the NIR absorbance and the photothermal conversion capability of GO and thermo-responsiveness of PNIPAM. Specifically, the NIR light absorbed by GO sheets was converted into heat, which acted as a trigger of the SME in the PNIPAM film and caused bending. Interestingly, repeated NIR irradiations induced shrink/relax cycles that led to reversible bending/unbending of the actuator and complex deformations such as twisting. These features enabled the demonstration of this hybrid actuator as a smart NIR-driven "forklift truck" able to lift goods by exposure to NIR light. In another exemplary study, Zhang and colleagues combined a PCL/PU (50/50) polymer blend with graphene nanoplatelets to attain electrical/IR actuation [51]. To illustrate the versatile applicability of these actuators, the authors designed a smart switch with the capacity to turn on/off a current flow through an electrical circuit under IR irradiation. Thus, the on/off switch of a bulb lamp was achieved by varying IR illumination, thus supporting the potential utility of these devices in smart switches, robot hands, and biomedical devices.

An additional application in which these composites may have a substantial impact is in self-healing surfaces, which may be used, for example, in advanced smart coatings. Self-healing capabilities could bring great benefits such as the increased lifetime of the coatings and protection capacity of what is underneath. By the addition of nanolayered graphene (less than 10 stacked single sheets) fillers into an epoxy SMP, a considerable improvement of scratch resistance and thermal healing capability, in comparison to the pure matrix, could be obtained [61]. After scratching tests by an indenter and subsequent heating of the samples above their T_g, scratches on the unfilled polymer surface largely disappeared but the permanent damage remained. On the contrary, scratches on the graphene-doped composites recovered almost completely because of their resilience to crack formation. This was attributed to a better dissipation of the mechanical energy by sufficient interfacial adhesion between the filler and the polymer matrix. Moreover, this enhancement of properties was also assigned to the exceptional in-plane fracture strength of the individual graphene sheets. Self-healing features were also explored in a system based on poly(acrylamide-co-acrylic acid), exploring the "zipper effect" (reversible hydrogen-bond cross-linking) of the copolymer [62]. A cut was made on a strip containing 10 wt.% of graphene and then the two pieces were pressed together and heated at 37°C for 20 min. After the healing completion, the cut wound disappeared and the surface morphology was similar to that of the original samples.

Although rarely explored, some reports exist on grafting graphene-based materials with SMPs in an attempt to confer the resulting scaffolds with improved properties including SME. For instance, inspired by the *Nepenthes* pitcher plant, Wang *et al.* developed a lubricant-infused RGO/SMP film displaying shape memory properties and electrothermal tunable wettability [63]. Specifically, a porous 3D RGO sponge was prepared from a GO solution containing acrylamide to aid GO gelation, methylene-bis-acrylamide as an acrylamide cross-linker, potassium peroxydisulfate as an initiator of the reaction, and ascorbic acid to reduce GO. The resulting foam was then coated with *trans*-1,4-polyisoprene, a SMP triggered by a T_m, which reduced the size of the remaining pores at the sponge surface. Compressive tests revealed no significant alterations of the porous structure of the hybrid foam for 10 cycles of loading and unloading, as well as excellent resilience and cyclic reproducibility above T_m. The authors pointed to polymerization of acrylamide chains and RGO sheets as a major responsible element for the excellent mechanical properties and cyclability of the films attained. Krytox 103 was added to the substrate to gain liquid lubrication of the substrate. Interestingly, the final surface properties of the hybrid film were controlled through the application of electric fields (constant DC voltage of 6 V). Particularly, the dynamic interface of the hybrid sponge reversibly varied from a lubricant-coated surface (in which liquids could be repelled and easily slip down) to a surface with a textured roughness (where liquid droplets are pinned) in response to electrical stimulation. These features allowed on demand tunable repellency, paving the way for liquid transport and manipulation. The applicability of this novel composite was demonstrated by preparing arrays with electrothermally controlled surface wettability in each independent unit for accurately pipetting of different samples into different wells of multiwell microplates (Figure 8.10). Concentration gradients of solutions could be even achieved. Interestingly, the shape memory properties of the hybrid films allowed for recycling by compression, thus reducing the associated costs. In summary, these novel arrays could simplify the liquid handling process in comparison with the use of multichannel pipettes or

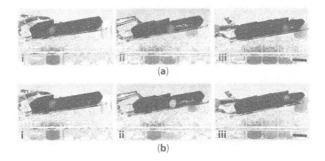

Figure 8.10 Demonstration of liquid handling capabilities derived from electrothermal tunable wettability of a hybrid film composed of graphene and SMPs. Droplets of the same liquid sample could be separated in different containers (a) and droplets of different liquids could be also separated in different containers (b). Adapted from Ref. [63]. © The Authors, some rights reserved; exclusive licensee American Association for the Advancement of Science. Distributed under a Creative Commons Attribution NonCommercial License 4.0 (CC BY-NC) http://creativecommons.org/licenses/by-nc/4.0/.

automatic microarrays, in which the original samples need to be prepared one by one, and a great amount of pipette tips are consumed to avoid sample pollution.

Unconventional applications of graphene/amyloid composites include biosensors for enzymatic activity. Using a vacuum filtration method, Li et al. obtained rigid composite films formed by organized layered structures of alternating graphene and amyloid layers [23]. These materials exhibited a reversible water-tunable SME, became biodegradable, and displayed enzyme activity sensing features. In fact, the composites entirely degraded and dispersed back into a stable colloid after being treated with a water-pepsin solution for 2 days and ultrasounds (1 min). On the contrary, similar treatments in simple water resulted in the preservation of the structural integrity of the films. The enzymatic biosensing activity was accomplished by measuring the time evolution of the electric resistance of a film and, based on it, calculating the cumulative pepsin activity. The sensor response was demonstrated by exposing the nanocomposites to two different pepsin solutions: one where pepsin was folded in the native state and the other where it was unfolded and denatured. The cumulative activity was then monitored for a period of 16 h in both cases. Results showed a remarkable activity difference between both scenarios, with a fast increase of the cumulative activity for the folded enzyme. Activity for the denatured pepsin solution remained one order of magnitude lower. These findings demonstrated an excellent hint of the activity loss when enzyme is denatured (unfolded), thus demonstrating the utility of this system as a biosensor with applicability in biomedical devices.

8.4 Future Perspectives

As discussed along this book chapter, although promising, graphene-doped SMP composites have been barely explored to date. Even though their potential and interest have been experimentally proven on the bench, further investigation is required to bring them closer to the design of marketable materials. In this sense, boosting the applicability of these materials might entail the development of more efficient methodologies for the commercial production of large, planar and crystalline graphene, in the first place. Improvement might be also necessary for the production of more homogeneous graphene dispersions into polymeric matrices, in order to avoid undesired agglomeration that could drive to heterogeneous and unpredictable properties in the resulting composites. Last but not least, strategies to better control the interface between graphene sheets and polymer chains (for instance, involving alternative surface functionalization of graphene) might benefit strain transfer from the polymer to the graphene dopants.

The attainment of more complex actuation and movement schemes might be achieved by advancing on the design of graphene-doped SMP composites. In this sense, interesting progress already reported in different polymeric systems could be potentially transferred to graphene-doped SMPs. For instance, recent findings describing novel bilayer substrates made of poly(dimethylsiloxane), thermally expanding microspheres, and RGO with IR light-driven properties for autonomic origami assembly of 3D structures might be a source of inspiration [64]. Additionally, advances from research on the applicability of CNT-based SMPs, as that described for the development of nanocomposites with caterpillar-like locomotion [65], can be eventually applied to graphene-based systems.

The improvement of graphene-based SMP composites in combination with the incorporation of other kind of fillers can also lead to systems with more complex functionalities and responsiveness. In these lines, Lee *et al.* first described the fabrication of a 3D networked GO-ferromagnetic hybrid for ultrafast magneto-responsiveness. These doped PU SMPs displayed enhanced mechanical stiffness and thermal conductivity in 3D space [66]. In this approach, magnetic nanoparticles were used to decorate graphene flakes surface and acted as remotely heat-transferring materials under alternating magnetic fields. These results could inspire the invention of technologically advanced composites able to perform multifunctionally depending on the stimuli supplied.

Finally, unprecedented discoveries on the own nature of graphene behavior might bring new paths for the achievement of both known and unknown properties and functionalities. Interesting theoretical calculations have revealed the co-existence of two stable phases in a coherent crystal lattice of 2D atomically thin GO crystal with ordered epoxy groups (C_8O), thus paving the way to the potential fabrication of multiple temporary shapes in a single graphene-based material [67]. This work illustrates the possibility of designing novel materials with properties as advanced as the ones discussed in this book chapter by simply relying in a material such as graphene itself. Such progress would represent significant savings in both fabrication costs and materials.

Acknowledgments

This work was supported by the *Ministerio de Economía y Competitividad* and the *Fondo Europeo de Desarrollo Regional* (MAT2016-78857-R, MINECO/FEDER, UE).

References

1. Novoselov, K.S., Geim, A.K., Morozov, S.V., Jiang, D., Zhang, Y., Dubonos, S.V., Grigorieva, I.V., Firsov, A.A., Electric field effect in atomically thin carbon films. *Science*, 306, 666, 2004.
2. Allen, M.J., Tung, V.C., Kaner, R.B., Honeycomb carbon: A review of graphene. *Chem. Rev.*, 110, 132, 2010.
3. Dreyer, D.R., Ruoff, R.S., Bielawski, C.W., From conception to realization: An historical account of graphene and some perspectives for its future. *Angew. Chem. Int. Ed.*, 49, 9336, 2010.
4. Novoselov, K.S., Fal'ko, V.I., Colombo, L., Gellert, P.R., Schwab, M.G., Kim, K., A roadmap for graphene. *Nature*, 490, 192, 2012.
5. Rao, C.N.R., Sood, A.K., Subrahmanyam, K.S., Govindaraj, A., Graphene: The new two-dimensional nanomaterial. *Angew. Chem. Int. Ed.*, 48, 7752, 2009.
6. Chang, L.C. and Read, T.A., Plastic deformation and diffusionless phase changes in metals—The gold-cadmium beta-phase. *Trans. AIME*, 189, 47, 1951.
7. Osada, Y. and Matsuda, A., Shape-memory in hydrogels. *Nature*, 376, 219, 1995.
8. Lendlein, A. and Kelch, S., Shape-memory polymers. *Angew. Chem. Int. Ed.*, 41, 2034, 2002.
9. Espinha, A., Serrano, M.C., Blanco, A., López, C., Thermoresponsive shape-memory photonic nanostructures. *Adv. Opt. Mater.*, 2, 516, 2014.
10. Serrano, M.C., Carbajal, L., Ameer, G.A., Novel biodegradable shape-memory elastomers with drug-releasing capabilities. *Adv. Mater.*, 23, 2211, 2011.

11. Xiao, Y., Zhou, S.B., Wang, L., Gong, T., Electro-active shape memory properties of poly-(epsilon-caprolactone)/functionalized multiwalled carbon nanotube nanocomposite. *ACS Appl. Mater. Interfaces*, 2, 3506, 2010.
12. Kumar, U.N., Kratz, K., Wagermaier, W., Behl, M., Lendlein, A., Non-contact actuation of triple-shape effect in multiphase polymer network nanocomposites in alternating magnetic field. *J. Mater. Chem.*, 20, 3404, 2010.
13. Lendlein, A., Jiang, H.Y., Junger, O., Langer, R., Light-induced shape-memory polymers. *Nature*, 434, 879, 2005.
14. Mendez, J., Annamalai, P.K., Eichhorn, S.J., Rusli, R., Rowan, S.J., Foster, E.J., Weder, C., Bioinspired mechanically adaptive polymer nanocomposites with water-activated shape-memory effect. *Macromolecules*, 44, 6827, 2011.
15. Chen, H.M., Li, Y., Liu, Y., Gong, T., Wang, L., Zhou, S.B., Highly pH-sensitive polyurethane exhibiting shape memory and drug release. *Polym. Chem.*, 5, 5168, 2014.
16. Serrano, M.C. and Ameer, G.A., Recent insights into the biomedical applications of shape-memory polymers. *Macromol. Biosci.*, 12, 1156, 2012.
17. Meng, H. and Li, G., A review of stimuli-responsive shape memory polymer composites. *Polymer*, 54, 2199, 2013.
18. Jung, Y.C., Yoo, H.J., Kim, Y.A., Cho, J.W., Endo, M., Electroactive shape memory performance of polyurethane composite having homogeneously dispersed and covalently cross-linked carbon nanotubes. *Carbon*, 48, 1598, 2010.
19. Mengesha, Z.T. and Yang, J., Silver nanoparticle-decorated shape-memory polystyrene sheets as highly sensitive surface-enhanced raman scattering substrates with a thermally inducible hot spot effect. *Anal. Chem.*, 88, 10908, 2016.
20. Espinha, A., Serrano, M.C., Blanco, A., López, C., Random lasing in novel dye-doped white paints with shape memory. *Adv. Opt. Mater.*, 3, 1080, 2015.
21. Cuevas, J.M., Alonso, J., German, L., Iturrondobeitia, M., Laza, J.M., Vilas, J.L., León, L.M., Magneto-active shape memory composites by incorporating ferromagnetic microparticles in a thermo-responsive polyalkenamer. *Smart. Mater. Struct.*, 18, 075003, 2009.
22. Jung, Y.C., Kim, J.H., Hayashi, T., Kim, Y.A., Endo, M., Terrones, M., Dresselhaus, M.S., Fabrication of transparent, tough, and conductive shape-memory polyurethane films by incorporating a small amount of high-quality graphene. *Macromol. Rapid Commun.*, 33, 628, 2012.
23. Li, C.X., Adamcik, J., Mezzenga, R., Biodegradable nanocomposites of amyloid fibrils and graphene with shape-memory and enzyme-sensing properties. *Nat. Nanotechnol.*, 7, 421, 2012.
24. Liang, J.J., Xu, Y.F., Huang, Y., Zhang, L., Wang, Y., Ma, Y.F., Li, F.F., Guo, T.Y., Chen, Y.S., Infrared-triggered actuators from graphene-based nanocomposites. *J. Phys. Chem. C*, 113, 9921, 2009.
25. Ponnamma, D., Sadasivuni, K.K., Strankowski, M., Moldenaers, P., Thomas, S., Grohens, Y., Interrelated shape memory and Payne effect in polyurethane/graphene oxide nanocomposites. *RSC Adv.*, 3, 16068, 2013.
26. Choi, J.T., Dao, T.D., Oh, K.M., Lee, H.I., Jeong, H.M., Kim, B.K., Shape memory polyurethane nanocomposites with functionalized graphene. *Smart. Mater. Struct.*, 21, 075017, 2012.
27. Das, R., Banerjee, S.L., Kundu, P.P., Fabrication and characterization of *in situ* graphene oxide reinforced high-performance shape memory polymeric nanocomposites from vegetable oil. *RSC Adv.*, 6, 27648, 2016.
28. Oh, S.M., Oh, K.M., Dao, T.D., Lee, H.I., Jeong, H.M., Kim, B.K., The modification of graphene with alcohols and its use in shape memory polyurethane composites. *Polym. Int.*, 62, 54, 2013.
29. Yoonessi, M., Shi, Y., Scheiman, D.A., Lebron-Colon, M., Tigelaar, D.M., Weiss, R.A., Meador, M.A., Graphene polyimide nanocomposites; thermal, mechanical, and high-temperature shape memory effects. *ACS Nano*, 6, 7644, 2012.

30. Qi, X.D., Yao, X.L., Deng, S., Zhou, T.N., Fu, Q., Water-induced shape memory effect of graphene oxide reinforced polyvinyl alcohol nanocomposites. *J. Mater. Chem. A*, 2, 2240, 2014.
31. Mahapatra, S.S., Ramasamy, M.S., Yoo, H.J., Cho, J.W., A reactive graphene sheet *in situ* functionalized hyperbranched polyurethane for high performance shape memory material. *RSC Adv.*, 4, 15146, 2014.
32. Kashif, M. and Chang, Y.W., Supramolecular hydrogen-bonded polyolefin elastomer/modified graphene nanocomposites with near infrared responsive shape memory and healing properties. *Eur. Polym. J.*, 66, 273, 2015.
33. Li, Y.T., Lian, H.Q., Hu, Y.N., Chang, W., Cui, X.G., Liu, Y., Enhancement in mechanical and shape memory properties for liquid crystalline polyurethane strengthened by graphene oxide. *Polymers*, 8, 236, 2016.
34. Jiu, H.F., Jiao, H.Q., Zhang, L.X., Zhang, S.M., Zhao, Y.A., Graphene-cross-linked two-way reversible shape memory polyurethane nanocomposites with enhanced mechanical and electrical properties. *J. Mater. Sci. Mater. Electron.*, 27, 10720, 2016.
35. Bayan, R. and Karak, N., Renewable resource derived aliphatic hyperbranched polyurethane/aluminium hydroxide-reduced graphene oxide nanocomposites as robust, thermostable material with multi-stimuli responsive shape memory features. *New J. Chem.*, 41, 8781, 2017.
36. Kim, J.T., Kim, B.K., Kim, E.Y., Park, H.C., Jeong, H.M., Synthesis and shape memory performance of polyurethane/graphene nanocomposites. *React. Funct. Polym.*, 74, 16, 2014.
37. Kim, J.T., Jeong, H.J., Park, H.C., Jeong, H.M., Bae, S.Y., Kim, B.K., Electroactive shape memory performance of polyurethane/graphene nanocomposites. *React. Funct. Polym.*, 88, 1, 2015.
38. Qi, X.D., Guo, Y.L., Wei, Y., Dong, P., Fu, Q., Multishape and temperature memory effects by strong physical confinement in poly(propylene carbonate)/graphene oxide nanocomposites. *J. Phys. Chem. B*, 120, 11064, 2016.
39. Song, L.F., Li, Y.Q., Xiong, Z.Q., Pan, L.L., Luo, Q.Y., Xu, X., Lu, S.R., Water-induced shape memory effect of nanocellulose papers from sisal cellulose nanofibers with graphene oxide. *Carbohyd. Polym.*, 179, 110, 2018.
40. Tang, Z.H., Kang, H.L., Wei, Q.Y., Guo, B.C., Zhang, L.Q., Jia, D.M., Incorporation of graphene into polyester/carbon nanofibers composites for better multi-stimuli responsive shape memory performances. *Carbon*, 64, 487, 2013.
41. Sabzi, M., Babaahmadi, M., Samadi, N., Mahdavinia, G.R., Keramati, M., Nikfarjam, N., Graphene network enabled high speed electrical actuation of shape memory nanocomposite based on poly(vinyl acetate). *Polym. Int.*, 66, 665, 2017.
42. Lashgari, S., Karrabi, M., Ghasemi, I., Azizi, H., Messori, M., Paderni, K., Shape memory nanocomposite of poly(L-lactic acid)/graphene nanoplatelets triggered by infrared light and thermal heating. *Exp. Polym. Lett.*, 10, 349, 2016.
43. Keramati, M., Ghasemi, I., Karrabi, M., Azizi, H., Sabzi, M., Incorporation of surface modified graphene nanoplatelets for development of shape memory PLA nanocomposite. *Fiber. Polym.*, 17, 1062, 2016.
44. Feng, Y.Y., Qin, M.M., Guo, H.Q., Yoshino, K., Feng, W., Infrared-actuated recovery of polyurethane filled by reduced graphene oxide/carbon nanotube hybrids with high energy density. *ACS Appl. Mater. Interfaces*, 5, 10882, 2013.
45. Zhang, Y.Q., Zhang, M., Jiang, H.Y., Shi, J.L., Li, F.B., Xia, Y.H., Zhang, G.Z., Li, H.J., Bio-inspired layered chitosan/graphene oxide nanocomposite hydrogels with high strength and pH-driven shape memory effect. *Carbohyd. Polym.*, 177, 116, 2017.
46. Liu, S.L., Yao, F., Oderinde, O., Li, K.W., Wang, H.J., Zhang, Z.H., Fu, G.D., Zinc ions enhanced nacre-like chitosan/graphene oxide composite film with superior mechanical and shape memory properties. *Chem. Eng. J.*, 321, 502, 2017.

47. Yang, Z.H., Wang, Q.H., Wang, T.M., Dual-triggered and thermally reconfigurable shape memory graphene-vitrimer composites. *ACS Appl. Mater. Interfaces*, 8, 21691, 2016.
48. Thakur, S. and Karak, N., Bio-based tough hyperbranched polyurethane-graphene oxide nanocomposites as advanced shape memory materials. *RSC Adv.*, 3, 9476, 2013.
49. Yoo, H.J., Mahapatra, S.S., Cho, J.W., High-speed actuation and mechanical properties of graphene-incorporated shape memory polyurethane nanofibers. *J. Phys. Chem. C*, 118, 10408, 2014.
50. Tan, L., Gan, L., Hu, J.L., Zhu, Y., Han, J.P., Functional shape memory composite nanofibers with graphene oxide filler. *Compos. Part A – Appl. S*, 76, 115, 2015.
51. Zhang, Z.X., Dou, J.X., He, J.H., Xiao, C.X., Shen, L.Y., Yang, J.H., Wang, Y., Zhou, Z.W., Electrically/infrared actuated shape memory composites based on a bio-based polyester blend and graphene nanoplatelets and their excellent self-driven ability. *J. Mater. Chem. C*, 5, 4145, 2017.
52. Li, C.W., Qiu, L., Zhang, B.Q., Li, D., Liu, C.Y., Robust vacuum-/air-dried graphene aerogels and fast recoverable shape-memory hybrid foams. *Adv. Mater.*, 28, 1510, 2016.
53. Liu, X.F., Li, H., Zeng, Q.P., Zhang, Y.Y., Kang, H.M., Duan, H.A., Guo, Y.P., Liu, H.Z., Electroactive shape memory composites enhanced by flexible carbon nanotube/graphene aerogels. *J. Mater. Chem. A*, 3, 11641, 2015.
54. Rana, S., Cho, J.W., Tan, L.P., Graphene-cross-linked polyurethane block copolymer nanocomposites with enhanced mechanical, electrical, and shape memory properties. *RSC Adv.*, 3, 13796, 2013.
55. Zhao, L.M., Feng, X., Li, Y.F., Mi, X.J., Shape memory effect and mechanical properties of graphene/epoxy composites. *Polym. Sci. Ser. A*, 56, 640, 2014.
56. Zhang, L.X., Jiao, H.Q., Jiu, H.F., Chang, J.X., Zhang, S.M., Zhao, Y.A., Thermal, mechanical and electrical properties of polyurethane/(3-aminopropyl) triethoxysilane functionalized graphene/epoxy resin interpenetrating shape memory polymer composites. *Compos. Part A – Appl. S*, 90, 286, 2016.
57. Wang, W.X., Liu, D.Y., Liu, Y.J., Leng, J.S., Bhattacharyya, D., Electrical actuation properties of reduced graphene oxide paper/epoxy-based shape memory composites. *Compos. Sci. Technol.*, 106, 20, 2015.
58. Sabzi, M., Babaahmadi, M., Rahnama, M., Thermally and electrically triggered triple-shape memory behavior of poly(vinyl acetate)/poly(lactic acid) due to graphene-induced phase separation. *ACS Appl. Mater. Interfaces*, 9, 24061, 2017.
59. Molavi, F.K., Ghasemi, I., Messori, M., Esfandeh, M., Nanocomposites based on poly(L-lactide)/poly(epsilon-caprolactone) blends with triple-shape memory behavior: Effect of the incorporation of graphene nanoplatelets (GNps). *Compos. Sci. Technol.*, 151, 219, 2017.
60. Chen, Z., Cao, R., Ye, S.J., Ge, Y.H., Tu, Y.F., Yang, X.M., Graphene oxide/poly(N-isopropylacrylamide) hybrid film-based near-infrared light-driven bilayer actuators with shape memory effect. *Sensor. Actuat. B - Chem.*, 255, 2971, 2018.
61. Xiao, X.C., Xie, T., Cheng, Y.T., Self-healable graphene polymer composites. *J. Mater. Chem.*, 20, 3508, 2010.
62. Dong, J., Ding, J.B., Weng, J., Dai, L.Z., Graphene enhances the shape memory of poly(acrylamide-co-acrylic acid) grafted on graphene. *Macromol. Rapid Commun.*, 34, 659, 2013.
63. Wang, J., Sun, L.Y., Zou, M.H., Gao, W., Liu, C.H., Shang, L.R., Gu, Z.Z., Zhao, Y.J., Bioinspired shape-memory graphene film with tunable wettability. *Sci. Adv.*, 3, e1700004, 2017.
64. Tang, Z.H., Gao, Z.W., Jia, S.H., Wang, F., Wang, Y.L., Graphene-based polymer bilayers with superior light-driven properties for remote construction of 3d structures. *Adv. Sci.*, 4, 1600437, 2017.

65. Peng, Q.Y., Wei, H.Q., Qin, Y.Y., Lin, Z.S., Zhao, X., Xu, F., Leng, J.S., He, X.D., Cao, A.Y., Li, Y.B., Shape-memory polymer nanocomposites with a 3D conductive network for bidirectional actuation and locomotion application. *Nanoscale*, 8, 18042, 2016.
66. Lee, S.H., Jung, J.H., Oh, I.K., 3D networked graphene-ferromagnetic hybrids for fast shape memory polymers with enhanced mechanical stiffness and thermal conductivity. *Small*, 10, 3880, 2014.
67. Chang, Z.Y., Deng, J.K., Chandrakumara, G.G., Yan, W.Y., Liu, J.Z., Two-dimensional shape memory graphene oxide. *Nat. Commun.*, 7, 11972, 2016.
68. Han, S. and Chun, B.C., Preparation of polyurethane nanocomposites via covalent incoporation of functionalized graphene and its shape memory effect. *Composites: Part A*, 58, 65, 2014.

9

Graphene-Based Scroll Structures: Optical Characterization and Its Application in Resistive Switching Memory Devices

Janardhanan R. Rani* and Jae-Hyung Jang[†]

School of Electrical Engineering and Computer Science, Gwangju Institute of Science and Technology, Oryong-dong, Buk-gu, Gwangju, South Korea

Abstract

Graphene, an emerging two-dimensional (2D) material, with a single-atom-thick sheet of hexagonally arrayed sp^2 bonded carbon atoms, has received significant attention due to its unique electronic, mechanical, and thermal properties. Graphene-based scrolls (GS) are new members of the graphene family, formed through the rolling of graphene layers in one or more directions. GS is considered as interesting carbon materials since their interlayer distance can be easily adjusted and they are proposed to synthesize novel graphene-based materials/composites for a wide variety of applications. GS can encapsulate other nanomaterials into the interior cavities and can exhibit various important applications including resistive switching. The electronic structure of graphene can be modified through encapsulating nanomaterials using graphene sheets. Such modification occurs because of the disorientation between the individual layers of graphene scrolls. Thus, GS can exhibit outstanding optical properties due to the π–π interaction between the inner and outer surfaces of scrolled graphene. Since pristine graphene has no bandgap, no photoluminescence would be expected. However, photoluminescence from graphene-based scroll structures can realize various graphene-based optoelectronic devices such as optical modulators, light-emitting diodes, ultrafast lasers, etc. In particular, the reduced graphene oxide scrolls can significantly enhance photoluminescence emission. Graphene-based scrolls can be used as two-terminal nonvolatile memory devices that have important application in the next-generation information technology industry. Graphene-based scrolls can be effectively used as active layers in resistive switching devices. The resistive switching device can be electrically switched between the "ON" and "OFF" states. The electrical behavior of these devices can be affected by the materials in the active layer and the electrodes. In this chapter, we first briefly overview the preparation of graphene-based scrolls. Then, we systematically discuss the optical, especially the photoluminescence properties of graphene-based scrolls. Next, the mechanism of resistive switching and applications of these scrolls in various memory devices have been summarized. The chapter is finalized with concluding remarks and a perspective for future study. The book chapter also gathers new progress on research on graphene scrolls and their important applications on graphene-based resistive switching memory. Also, the chapter will provide a comprehensive state-of-the-art overview of the optical properties of graphene.

Keywords: Graphene, scroll structures, nanomaterials, photoluminescence, resistive switching, nonvolatile memory

---

*Corresponding author: ranijnair@gmail.com
[†]Corresponding author: jjang@gist.ac.kr

9.1 Graphene-Based Scroll Structures

9.1.1 Introduction

Graphene has received significant attention due to its unique electronic, mechanical, and thermal properties, and has been among the leading topics of scientific research in optics, electronics, and materials science [1–5]. Among carbon-based materials, the high conductivity and remarkable optical properties of graphene have attracted attention toward a number of potential applications such as touch screens [6], liquid crystal displays (LCDs) [6], resistive random access memory (RRAM) based on the resistive switching (RS) [7, 8], etc. Recently, new graphene architecture resulting from the scrolling of graphene into spiral-wound structures in one or more directions has been reported and is known as graphene scrolls (GNS) [9, 10]. GNS have attracted much attention due to their novel physical and chemical properties. In GNS, the interlayer distance can be easily adjusted [11]. The increasing interest in synthesizing GNS is because of their promising physicochemical properties and they are proposed for a wide variety of applications. GNS are different from carbon nanotubes due to their open cap configuration and hollow tubular structures. GNS has exposed an interlayer in which different materials can be intercalated in it and its adjustable interlayer distances, and flexible interior volume made them feasible for ion transport [11], hydrogen storage [12], supercapacitors [13], batteries [14], and nanodevices [15]. Free edges and large out-of-plane thermal fluctuations in graphene make it susceptible to edge defects, which have important limitations for its practical applications. However, the closed edges in GNS provide much higher conductivity than open edges and the curvature-induced disorientation between the individual layers affects the electronic, transport, and optical properties of GNS due to the π–π interaction between the inner and outer surfaces of scrolled graphene. Thus, it exhibits outstanding properties such as quantum electronic transport, variable electronic structure, large thermal conductivity, and high elasticity [16–20].

Recently, many methods have been reported for the fabrication of graphene/reduced graphene oxide (rGO)-based scrolls (GONS), such as compression of graphene oxide (GO) nanosheets with liquid nitrogen [21], scrolling of the mechanically exfoliated graphene sheets [22], sublimation-induced scrolling of GO nanosheets [23], Langmuir–Blodgett (LB) compression methods [24], sonication of graphite intercalation compounds [25], lyophilization of GO [25], etc. Langmuir–Blodgett (LB) nanosheets can produce large-scale loosely scrolled structures but the lyophilization method can produce only thin GONSs. A graphene nanoribbon can spontaneously scroll up into a GNS via scrolling of graphene enabled with single-sided hydrogenation. Chemical methods are simple and easy ways to produce graphene-oxide-based scroll structures. The physical and chemical properties of GONS/GNS can be easily tuned by tailoring the structure and modifying the edges.

9.1.2 Reduced Graphene-Oxide-Based Scroll Fabrication: Iron Oxide Intercalation with rGO Powder

Fe_2O_3 powders (Sigma-Aldrich) are used for the scroll preparation. GO (0.3 g) was dispersed in 30 mL of water and 0.1 g of Fe_2O_3 was added to the dispersion, and this mixture was suspended in water and was centrifuged at 10,000 rpm for 10 min and the experiment was repeated multiple times. The resulting solution was dried at 90°C overnight in order to obtain the powder and the powder was further mixed in an agate mortar for 30 min, and it was then annealed at 400°C in a nitrogen ambient for 2 h.

Figure 9.1a shows the SEM image of Fe_2O_3-doped planar rGO structure in which the powder samples are annealed at low temperature [13]. Figure 9.1b to d shows the SEM images of Fe_2O_3-doped rGO scroll structures in which the powder samples are annealed at high temperature. These SEM images confirm the scroll formation and Fe_2O_3 are encapsulated in rGO scrolls. In the initial stage, Fe_2O_3 molecules are attached to flat GO layers, and when the temperature increases during annealing, the scroll structures are formed. When the temperature increases, the rGO layer bends around Fe_2O_3 molecules to minimize the total surface energy of the hybrid system. The HRTEM images also clearly confirm the scroll formation (Figure 9.1f to i).

9.1.3 Reduced Graphene-Oxide-Based Scroll Fabrication: Scrolls Formed due to Phosphor Intercalation

The GO was synthesized by a modified Hummer's method and phosphor material ($SrBaSi_2O_2N_2:Eu^{2+}$) is doped to GO solution by chemical method. The GO solution with concentration 0.3 mg/mL and the phosphor colloid solution of concentration 0.1 mg/mL were separately sonicated in deionized water and mixed together, and the resulting GO–phosphor

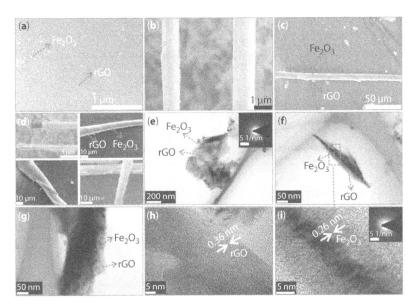

Figure 9.1 Scanning electron microscopy (SEM) images of (a) Fe_2O_3-doped planar rGO and (b–d) Fe_2O_3-doped rGO scroll structures. Magnified images of the scrolls are presented in (d). HRTEM images of (e) Fe_2O_3-doped planar rGO and (f–i) Fe_2O_3-doped rGO scroll structures. The SAED patterns of the Fe_2O_3-doped planar rGO and Fe_2O_3-doped rGO scroll structures are depicted in the insets of (e) and (i), respectively. rGO and Fe_2O_3 are marked in the figures. (Reprinted with permission from Rani, J. R., Thangavel, R., Oh, S.-I., Woo, J. M., Chandra Das, N., Kim, S.-Y., Lee, Y.-S., Jang, J.-H., High Volumetric Energy Density Hybrid Supercapacitors Based on Reduced Graphene Oxide Scrolls. *ACS Appl. Mater. Interfaces*, 9, 22398–22407, 2017. Copyright [2017] American Chemical Society.)

hybrid solution was spin-coated on a Si substrate with multiple rotation speed. The films were named GO70 (GO film annealed at 70°C), GO180 (GO film annealed at 180°C), GP70 (rGO–phosphor hybrid film annealed at 70°C), and GP180 (GO–phosphor hybrid film annealed at 180°C). The schematics of the scroll formation are shown in Figure 9.2.

Figure 9.3a shows the SEM image of undoped rGO in which scroll structures are not observed. The morphology of the scroll structure (Figure 9.3b and c) shows that the edges of the GO layer folded back, resulting in a tubular scroll structure with an average length of ~3 μm. The phosphor particles are attached to the GO film during sonication of the GO–phosphor hybrids. During annealing, the attached phosphor particles make GO roll up and form entangled structures. More discussions about scrolling are provided in [27]. A size distribution analysis of more than 100 scrolls reveals that their diameters are typically 1–2 μm (average diameter, 1.4 ± 0.4 μm), and their lengths are 10–20 μm (average length: 15 ± 7 μm). Upon sonication of the GO dopant in water, the dopants are attached to the rGO film and spontaneously bends the rGO flakes. The scrolling can be attributed to the fact that the high aspect ratio of the material and the effect of dopant atoms make it extremely unfavorable to maintain the planar structure. Van der Waals attraction and the π–π stacking effect between graphene layers play an important role in scrolling, and thus, due to scrolling, the graphene layer can overcome the energy barrier and fully wrap onto the dopants. The spontaneous wrapping of the rGO layers is affected by the dopant nature and the large graphene sheets rapidly rolled up to form an entangled structure due to their high inability to self-maintain a high aspect ratio 2D structure due to doping. The sizes including the widths and lengths of the rGO can be controlled by varying the deposition temperature. Previous reports show that the driving force of graphene scrolling is due to the energy difference between the total surface energy of the system and the elastic energy associated with graphene bending [27]. The fact that scroll shapes are formed in the presence of

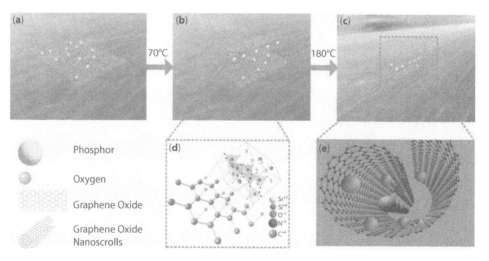

Figure 9.2 Schematic diagram showing (a) phosphor particles attached to GO sheets, (b) the GO sheets just beginning to scroll, (c) GO sheets completely scrolled at 180°C, (d) the bonding between GO and phosphor, and (e) three-dimensional view of a single GO–phosphor scroll. (Reprinted with permission from Rani, J. R., Oh, S.-I., Woo, J. M., Tarwal, N. L., Kim, H.-W., Mun, B. S., Lee, S., Kim, K.-J., Jang, J.-H., Graphene Oxide–Phosphor Hybrid Nanoscrolls with High Luminescent Quantum Yield: Synthesis, Structural, and X-ray Absorption Studies. *ACS Appl. Mater. Interfaces*, 7, 5693–5700, 2015. Copyright [2015] American Chemical Society.)

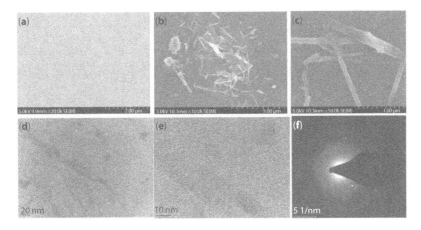

Figure 9.3 (a) SEM image of rGO, (b and c) SEM images of rGO–phosphor hybrid scrolls, (d and e) high-resolution transmission electron microscope (HRTEM) images of rGO–phosphor hybrid scrolls, (f) selected area electron diffraction (SAED) patterns of rGO–phosphor hybrid scrolls.

temperature suggests that temperature initiates scrolling. Scrolls are produced in two stages: At first the dopant atoms are attached to flat rGO layers and rGO layer bends around the dopants because of the temperature increase. When the temperature increases the surface energy of the dopant-rGO system also increases and in order to minimize the total surface energy of the dopant–rGO system the bending of rGO layer occurs. The bending of the rGO layer reduces the exposed surfaces of both rGO and dopants resulting in the minimization of the total surface energy. Second, after the dopant is enwrapped by rGO, subsequent rolling of rGO is driven by the reduction of the total area of the exposed graphene surface, and, consequently, the total surface energy of the system is minimized by tightly wrapping adjacent layers of the scroll together and spontaneously roll into full or partial scrolls [26].

9.1.4 Optical Properties of rGO–Phosphor Hybrid Scrolls

The rGO–phosphor hybrid scroll structure shows different optical properties including high quantum yield photoluminescence. The quantum yield of photoluminescence from graphene lies in a range from 1×10^{-12} to 1×10^{-9}, which is too low for practical applications. Different research groups report different methods and materials to be doped in graphene to enhance the quantum yield of photoluminescence from graphene, via the introduction of defects in the carbon network [28–31]. Converting graphene into rGO is one of the methods for the generation of luminescence in graphene; rGO possesses a finite electronic bandgap generated by the disruption of π-networks. But even with disruption of sp^2 groups, rGO films usually exhibit weak luminescence and low quantum yield (0.02 to 0.2%), which limit their application in optoelectronic devices [31–33]. But the phosphor-doped rGO hybrid scrolls exhibited enhanced luminescence with high quantum yield (48 times higher than that of pristine rGO films), which is highly favorable for optoelectronic device applications. The HOMO and LUMO levels are changed due to the phosphor doping as well as due to the scroll structure. UPS analysis shows that HOMO shifted by 1.65, 0.9, 2.1, and 1.9 eV downward with respect to EF for the GO70, GO180, GP70, and GP180 films, respectively, which resulted in the change in the band gap of the GO–phosphor hybrid films [27].

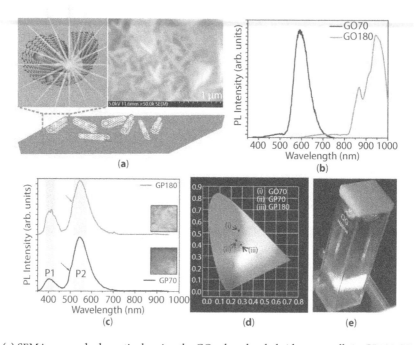

Figure 9.4 (a) SEM image and schematic showing the GO–phosphor hybrid nanoscrolls in GP180, PL emission spectra of (b) GO70 and GO180 films, (c) GP70 and GP180 films measured with an excitation wavelength of 280 nm. (d) CIE chromaticity coordinates of the PL emission from the synthesized thin films and (e) digital image of the strong PL emission from the GO–phosphor hybrid solution. (Reprinted with permission from Rani, J. R., Oh, S.-I., Woo, J. M., Tarwal, N. L., Kim, H.-W., Mun, B. S., Lee, S., Kim, K.-J., Jang, J.-H., Graphene Oxide–Phosphor Hybrid Nanoscrolls with High Luminescent Quantum Yield: Synthesis, Structural, and X-ray Absorption Studies. *ACS Appl. Mater. Interfaces*, 7, 5693–5700, 2015. Copyright [2015] American Chemical Society.)

The curvature-induced scroll structure and the bonding between the π and oxygen states result in the shift of HOMO levels. The annealing temperature also affects the shift of HOMO values. The bandgap values of these hybrids films were measured using UPS spectra and cyclic voltammetry measurements [27]. The band gap values of the GO–phosphor hybrid scrolls vary with annealing temperature; the values are found to be 2.05, 1.3, 2.56, and 2.43 eV for different GO70, GO180, GP70, and GP180 films, respectively. Emission at ~400 nm is due to the C–N bonding in GO–phosphor hybrids (Figure 9.4c). C–N bonding resulted in the formation of the lone pair electrons in the π valence band and is responsible for emission at ~400 nm. The quantum yields of the emissions from the GP70 and GP180 films were 7.0% and 9.6%, respectively, and the quantum yields of the GP180 films are 48 times higher than those of the GO180 films, respectively [27]. The scrolling and the phosphor embedding in the rGO sheets enhanced their quantum yields.

Figure 9.5 schematically represents the E_{HOMO}, E_{LUMO}, calculated band gap values, and schematic of PL emission of the films.

9.1.5 Raman Spectra of the Scrolls

The GO–phosphor hybrid scrolls are produced under different conditions. The GO solution (concentration 0.3 mg/mL) and the phosphor colloid solution (various concentrations of

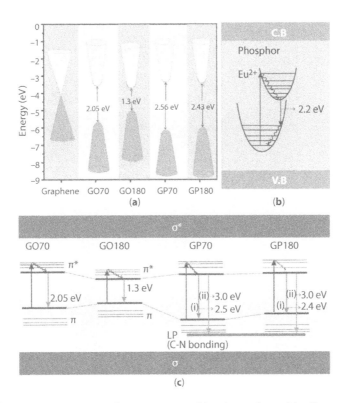

Figure 9.5 (a) Schematic representation of E_{HOMO}, E_{LUMO}, and bandgap values of the films measured through CV and UPS spectra, (b) schematic of PL emission from the phosphor, and (c) schematic of PL emission from GO70, GO180, GP70, and GP180 films. (Reprinted with permission from Rani, J. R., Oh, S.-I., Woo, J. M., Tarwal, N. L., Kim, H.-W., Mun, B. S., Lee, S., Kim, K.-J., Jang, J.-H., Graphene Oxide–Phosphor Hybrid Nanoscrolls with High Luminescent Quantum Yield: Synthesis, Structural, and X-ray Absorption Studies. *ACS Appl. Mater. Interfaces*, 7, 5693–5700, 2015. Copyright [2015] American Chemical Society.)

0.05, 0.07, 0.1, 0.125, and 0.15 mg/mL) were used. The GO–phosphor hybrid solutions were spin-casted onto standard Si/SiO$_2$ substrates at 500, 800, and 1600 rpm for 30 s and were then annealed at 160°C for 5 min. The films were termed GO (undoped), GNP1, GNP2, GNP3, GNP4, and GNP5 (phosphor-attached GO film with concentrated solutions at 0.05, 0.07, 0.1, 0.125, and 0.15 mg/mL, respectively). SEM image of GNP1 film is shown in Figure 9.6.

Figure 9.7 shows that Raman spectra of scroll structures and planar rGO look different. The spectral features of rGO correspond to peaks at 1590 cm^{-1} (G peak), 1350 cm^{-1} (D peak), 2697 cm^{-1} (2D peak), and ~2940 cm^{-1} (G + D peak) [34–36]. The E2g vibrational modes within aromatic carbon rings result in the formation of the G band, while the out-of-plane vibrational modes within aromatic carbon rings result in the formation of the 2D bands. The D band requires scattering at defect sites in order to conserve momentum. The D band is found to be broadened in scrolled GO. This is due to the conservation of the momentum of the intra-valley electrons scattered via iTO phonons (at the K-point) by curvature-induced defect scattering [37]. Also, the curvature in the scrolled structure resulted in the formation of low-frequency radial breathing-like (RBLM) modes, located between 75 and 300 cm^{-1}. Usually, the RBLM modes are observed in CNT-like structures [37] and correspond to a simple translation of the honeycomb network. The curvature in the scroll structures

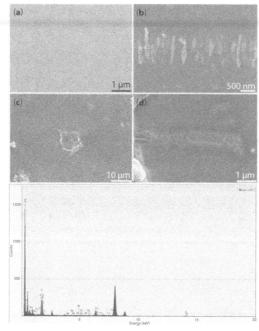

Figure 9.6 SEM images of (a) GO, (b and c) various regions of the GNP1 film, (d) an expanded view of the single scroll shown in the red dotted square in (c), and (e) the EDAX spectra from the scroll [33].

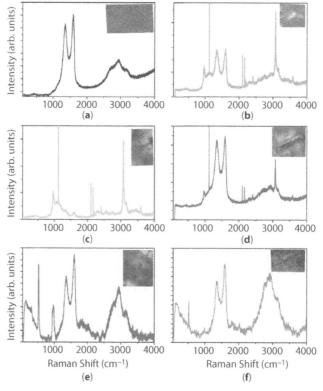

Figure 9.7 Raman spectra of (a) GO and (b–f) different regions of the GNP1 film. The corresponding optical images are shown in the inset of each figure [33].

transforms this translation into a phonon mode and results in the formation of RBLM modes located between 75 and 300 cm^{-1}. Figure 9.7b to f show RBM modes around 140 cm^{-1} and Figure 9.7e and f show additional radial breathing mode (RBM) features around 290 cm^{-1}. The intensity of the modes varies in different type of scrolls. The RBLM modes are found to be less intense where no perfect scroll is formed and is found to be absent in the Raman spectrum of the flat GO. These modes provide a strong evidence for the formation of scroll structures. Bands in the range of 910–1050 cm^{-1} correspond to the iTA phonons that are previously reported for graphite whiskers and CNTs [37]. This band is due to the scrolling of the rGO layers and is found to be absent in pure GO. Out-of-plane phonons around the Γ point result in the formation of bands around 1800 cm^{-1} and 2000 cm^{-1}, which are due to the combination of LA and iTA modes [38]. The emergence of these modes is due to the intervalley scattering process. The combination of iTA and LO phonons results in the peak at 1800 cm^{-1}. The combination of the oTO (out-of-plane tangential optical) and LO phonon mode around the K point in the rGO Brillouin zone presents a peak around 2000 cm^{-1} [38]. For planar GO, the oTO phonon at the G point is not a Raman active mode due to the odd symmetry for the mirror operation on a GO plane. However, the Raman band of the overtone of the oTO phonon is observed in single-walled carbon nanotubes (SWNTs) and is known as the M band. M bands that are activated in SWCNT are due to their cylindrical shape, and for a perfectly scrolled tubular structure, M bands are active due to the similarity of the scroll structure with that of CNTs. Bonding between the carbon and nitrogen in the GO–phosphor hybrid scrolls results in the formation of stretching modes around 2200 cm^{-1}. The combination of the iTO and iTA phonons results in the formation of mode observed around the K point at ~2280 cm^{-1} [37], and the combination of the zone boundary in-plane longitudinal acoustic (iLA) phonon and the in-plane transverse optical (iTO) phonon modes results in a G* band around 2450 cm^{-1} [38].

The PL emission spectra of films show different emission peaks with different phosphor concentrations (Figure 9.8). Figure 9.8b shows the 1931 CIE chromaticity coordinates of

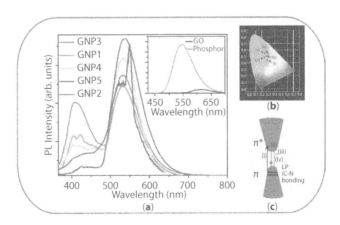

Figure 9.8 PL emission spectra of (a) the GNP1–5 films at various phosphor concentrations as measured with an excitation wavelength of 280 nm. The inset shows the PL emission from pure GO and the reference phosphor; (b) CIE chromaticity coordinates of the PL emission from the synthesized thin films; (i), (ii), (iii), (iv), and (v) correspond to GNP3, GNP1, GNP4, GNP5, and GNP2, respectively; and (c) presents a schematic of the PL emission from the films [33].

the PL emission from the synthesized thin films. The photoluminescence emission spectrum of films consists of peaks around 400 nm, 537 nm, and weak emission around 620 nm [33]. The emission around 537 nm is the typical photoluminescence emission from GO and is due to the emission from the $\pi-\pi^*$ energy gap, which is formed due to the disruption of π networks due to the attachment of oxygen functional groups onto the basal plane of rGO [38]. Previously, many research groups report about the emission from GO in the UV-visible region [31, 39, 40] and these rGO–phosphor hybrid scrolls show blue emissions centered around 390–440 nm. This blue PL emission is due to the radiative recombination of electron–hole pairs generated within localized states [34]. Since the bandgap emission from GO is very weak, merging takes place with the intense peak around 537 nm. The additional broad emission observed around 400 nm is due to the C–N bonding [41]. Lone pair electrons of nitride are not hybridized with the carbon, and it is located in the sp^2 C–N valence band. The transitions between the lone pair electrons in the valence band and the π^* conduction band result in the emission around the 400-nm region [34]. It is observed from the photoluminescence emission spectra that the PL emission changes with the concentration of the phosphor. The intensity of the emission around 400 nm varies with the phosphor concentration. First, as the concentration increases, the probability of a transition between the π^* band and the LP electron level increases, resulting in an increase in the emission intensity of this peak. However, a further increase in the phosphor concentration results in the re-absorption of the emitted photons, which reduces the emission intensity of the 400-nm light [33].

9.2 Reduced Graphene-Oxide-Based Resistive Switching Devices

One of the major challenges that we face is that the current memory technologies are pacing toward the end of the road. More components such as transistors are deliberately packed onto a single chip to achieve higher data storage density and faster access to the information [42]. The feature size has shrunk from 130 nm in 2000 to 16 nm in 2018. Therefore, from both technological and economic points of view, novel information storage materials and devices should be developed to fulfill the current requirements of the electronics industry. Due to the reduction of size, additional issues such as higher power consumption and unwanted heat generation also need to be addressed. In summary, in the near future, flash memories are approaching its physical and technical limitations. The conventional memory technology timeline is shown in Figure 9.9.

As an alternative, resistive memory devices have been intensively studied due to their potential for the replacement of flash memories. Owing to their simple structure, high density, excellent scalability, low power consumption, fast switching speed, easy operation, flexibility, and compatibility with conventional complementary metal-oxide semiconductor (CMOS) technology, resistive memories have become an attractive platform for memory devices [43–46]. Unlike transistor and capacitor memories, ReRAM-based devices do not require a specific cell structure or do not need to be integrated with the CMOS (complementary metal-oxide semiconductor) technology because memories store data as low/high resistance states. Moreover, it is possible to achieve huge storage density because a memory cell can be fabricated using just one device. ReRAMs are also capable of replacing both DRAM and hard drives. As compared to transistors, reduction in size will not give rise

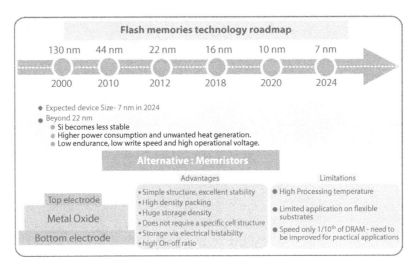

Figure 9.9 Conventional memory technology timeline.

to higher heat generation and quicker boot up is possible and could enable more energy-efficient high-density circuits.

Resistive random access memory (ReRAM) has attracted great attention for next-generation memory applications due to its simple structure, facile processing, high density, high operation speed, long retention time, and low power consumption [47, 48]. Detailed studies on ReRAM devices are important for next-generation technology because of their high speed, low operation voltage, high packing density device fabrication, etc. Recently, carbon-based materials such as graphene/graphene oxide [48], fullerenes [49], carbon nanotubes [50], graphene-like conductive carbon [51], and amorphous carbon [51, 52] have also been considered as a potential element for nonvolatile memory application resistance-change materials. Graphene-based electronic devices have attracted great interest for next-generation optoelectronics devices due to their flexibility, low cost, and compatibility with the existing silicon-based technology. Due to its highly conductive nature, the read/write process can be done in nano/microseconds, capable of very high data storage density, as compared to currently available memory devices. However, the gapless nature of graphene results in a small ON/OFF ratio and in turn a large read power consumption. But creating an appropriate energy gap in graphene for this purpose without any defect formation is a herculean task; researchers focus on graphene-oxide-based resistive switching devices. The possible alternative is graphene oxide memristor in which the on and off currents are determined by the sp^2/sp^3 fractions on the graphene oxide layers. Its solution process compatibility and high thermal conductivity enable manufacturing of low-cost and flexible memory cells. The high thermal conductivity results in effective heat dissipation, which is crucial to avoid device malfunction or failure. But currently there is a sizable performance gap between the current results and realistic memory. For practical applications, factors such as low energy consumption, low power dissipation, mechanical flexibility, non-destructive readout, long retention time, etc. are of equal importance when designing and fabricating new memory devices.

Graphene oxide (GO) can be widely used for nonvolatile switching memory applications because of its large surface area, excellent scalability, retention, and endurance properties

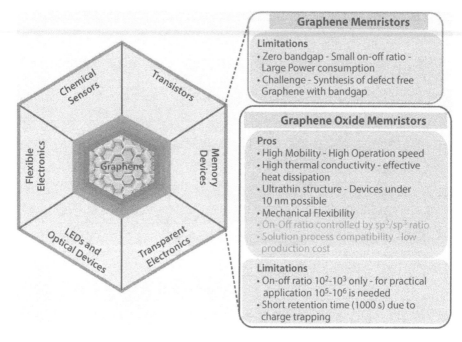

Figure 9.10 Schematic diagram shows the advantages and limitations of graphene oxide memristors.

[8, 52–55] (Figure 9.10). Graphene oxide can be uniformly deposited onto the form of thin films using drop casting [56], spin coating [57], Langmuir–Blodgett deposition [58], and vacuum filtration due to its high solubility in water and other solvents, which makes it capable of large-scale device fabrication. Most of the RRAM devices reported to date use various insulating, semiconducting, and binary transition metal oxides such as ZnO [58], NiO [58], HfO [59], TiO_2 [59], Ta_2O_5 [59], perovskite oxides [60], solid electrolytes [60], organic materials [61], etc.

There are certain disadvantages in the present GO-based memory, and immense research is needed to improve the practicability of the GO-based memristors. The major limitation of GO for practical application is average ON/OFF ratio (~10–10^2) [62, 63]. Also, for practical applications, switching speed (currently >100 ns) and SET–REST voltage must be improved. For practical applications, the ON/OFF value should be increased up to the range 10^5–10^6. If we could enhance the switching speed and ON/OFF ratio and reduce the threshold voltage, then the GO-based memristors will revolutionize the next-generation memory devices.

9.2.1 Resistive Switching in GO–Phosphor Hybrid Scrolls

The GO–phosphor hybrid solution was prepared as discussed in Section 9.1.3. In Figure 9.11, ITO is used as a bottom electrode and Pt is used as a bottom electrode, deposited onto a substrate via electron beam evaporation. For the film deposition, the substrate was completely covered with sufficient amount of the phosphor–GO hybrid solution, allowed to stand for 60 s and spin coated at 500, 800, and 1600 rpm for 30 s each, followed by annealing at 90°C for 7 min. Finally, a patterned top metal electrode will be deposited via electron beam evaporation using a shadow mask. The device *I–V* measurements and endurance and

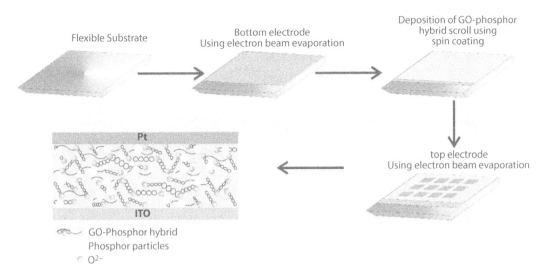

Figure 9.11 Resistive switching device fabrication using GO–phosphor hybrid solution.

retention measurements were performed by using a semiconductor parameter analyzer under ambient conditions.

The rGO–phosphor hybrid solution can be used effectively in memristor-based memory devices, where the mechanism of charge storage is entirely different from conventional memristors. $I-V$ characteristics, retention, and endurance behavior of the device measured at room temperature are shown in Figure 9.12. The device shows an ON/OFF ratio of around 2×10^2.

The rGO–phosphor hybrid layer can be considered as a graphene sheet with oxygen functional attached to both sides and has unique properties of graphene. In contrast to pristine graphene, in which all atoms are sp^2 hybridized, the GO–phosphor hybrid layer consists of sp^3 carbon atoms covalently bonded to oxygen-bearing functional groups [63]. The sp^2-hybridized state is highly conducting due to the presence of delocalized π electrons, while in sp^3 hybridization, these available π electrons are bonded with functional groups [64]. Thus, controlling the ratio of sp^2/sp^3 fractions opens up possibilities of adjusting resistance and an enhanced ON/OFF ratio. Chemical phase change between sp^2 and sp^3 carbon obtained from electrical redox processes can result in a distinct alternation of sheet conductivity in rGO.

The working principle of an rGO–phosphor hybrid-based resistive switching device is shown in Figure 9.13. Resistive switching is possible due to the desorption/absorption of oxygen-related groups on the rGO. When there are epoxide, hydroxyl, and carboxyl groups on the top GO layer, the conductance of the device is assumed to be low due to the sp^3 bonding feature [65]. As a negative bias (V_{SET}) is applied on the top electrode, oxygen-related functional groups in the rGO–phosphor hybrid scroll layer close to the top electrode diffuse down to the bottom ITO layer [65], leaving oxygen vacancies in the rGO–phosphor hybrid film, resulting in an increased amount of sp^2 bonds in the rGO–phosphor hybrid layer. Thus, the interlayer electrons forming metal-like conducting paths result in the LRS or ON state [65]. During V_{RESET}, the oxygen functional groups diffuse toward the top rGO–phosphor hybrid layer, which becomes sp^3 hybridized again, and this decreases the conductivity.

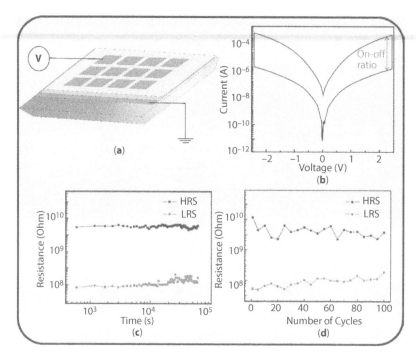

Figure 9.12 (a) The fabricated device using GO–phosphor hybrid solution, (b) I–V characteristics, (c) retention, and (d) endurance behavior of the device measured at room temperature.

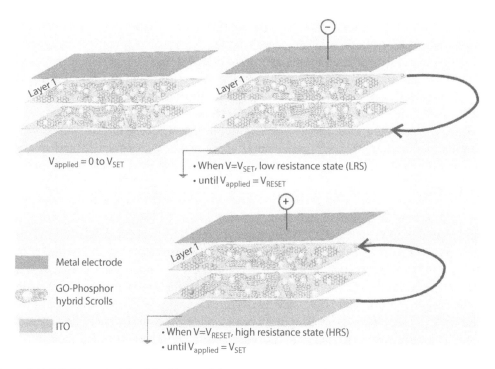

Figure 9.13 Working principle of the fabricated device using an rGO–phosphor hybrid solution.

Thus, the memory cell returns to the high resistance (HRS) or OFF state. Binary digital data can be recorded in these resistance states, for example, LRS for logic "0" and HRS for logic "1". Pristine device state is defined as the OFF state or high resistance state (HRS). By increasing the negative voltages imposed on the device, the current increases abruptly at V_{SET}, indicating that the device switches from an HRS to a low resistance state (LRS or ON state). The switching from HRS to LRS is called the SET or "Write" process. By sweeping the voltage from V_{SET} to V_{RESET}, the device holds on the LRS, suggesting nonvolatile memory characteristics. Once the voltage exceeds V_{RESET}, the device switches from the LRS to HRS, which is called the RESET or "Erase" process. The HRS can be retained in the next Read process. An rGO-based memory can be easily fabricated using a room-temperature spin-casting method on flexible substrates and has reliable memory performance in terms of retention and endurance.

It is reported that the electronic properties of GNSs are related to the number of overlapping layers and the diameter of the scroll [15]. In the GNS, the current is carried through the whole scroll structure, and the different layers can contribute to the overall current conduction and thus GNS can exhibit higher ON current [15]. Xie et al. report that π electrons can tunnel through different layers in the scrolls and the interaction between the inner and outer surfaces of the scrolls results in excellent electrical conductivity [22]. In GNS, however, current flows through the whole scrolled graphene layer and can increase the ON state conduction in switching devices. Also below the threshold voltage, the off current will be low due to the decreased π electron migration between layers. In the HRS state, sp^3 states result in the low conductivity through scrolls and results in low OFF current. The high thermal conductivity of the scrolls makes the scrolls a better candidate with higher heat dissipation capacity compared to that of other graphene oxide derivatives. The high ON current in rGO scrolls is shown schematically in Figure 9.14.

9.2.2 Resistive Switching in Graphene-Oxide–Iron Oxide Hybrid Thin Films

Iron oxide–graphene oxide hybrids were prepared by solution method and the mixture was suspended in water by sonication for 30 min, using an ultrasonic reactor operated at a frequency of 33 kHz.

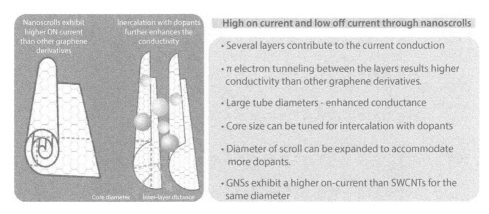

Figure 9.14 Schematic diagram showing the properties of rGO/phosphor hybrid scrolls.

The iron oxide nanoparticles are formed according to the following reaction mechanism, shown in Equations 9.1 to 9.3 [66].

$$FeCl_3 + 6H_2O \rightarrow [Fe(H_2O)_6]^{3+} + 3Cl^- \quad \quad (9.1)$$

$$[Fe(H_2O)_6]^{3+} + H_2O \rightarrow [Fe(H_2O)_5OH]^{2+} + H_3O^+ \quad \quad (9.2)$$

$$[Fe(H_2O)_6]^{3+} + 3OH^- \rightarrow FeOOH + 7H_2O \quad \quad (9.3)$$

The solution was spin-cast on the substrate and annealed at 90°C (GF90) and 180°C (GF180). For comparison, GO90 and GO180 were also prepared. Pt is used as top electrode coating. SEM image of GF 90 film is shown in Figure 9.15.

The bipolar resistive switching behavior of the iron oxide–graphene oxide hybrid-based devices is shown in Figure 9.16. During the negative bias voltage, the device obtained a SET condition in which there is an abrupt change in the resistance [67]. The resistance was changed from a high resistance state (HRS) to a low resistance state (LRS). This is the SET process. During a particular positive bias voltage V_{RESET}, the current abruptly decreases and the device switches from the LRS to the HRS state. This is the "RESET" process.

In GO-based devices, resistive switching is due to the desorption/adsorption of oxygen-related functional groups, and the oxygen vacancies result in the formation of electron trap states [68]. These electron traps result in the resistance switching via forming electron hopping paths.

During V_{SET}, the device switches to LRS or ON state and the electric field repels the oxygen ions in GO to the bottom ITO electrode, creating metal-like conducting paths [68]. During the RESET process, the oxygen ions migrate back from the bottom ITO electrode into the GO layer and the device is said to be HRS or OFF state [68]. The SET/RESET voltages changes with nature of doping and the annealing temperature.

Figure 9.15 SEM image of the GF90 film.

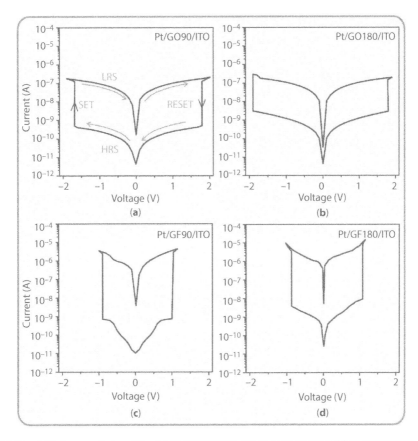

Figure 9.16 The I–V characteristics of the (a) Pt/GO90/ITO, (b) Pt/GO180/ITO, (c) Pt/GF90/ITO, and (d) Pt/GF180/ITO devices measured at room temperature. The voltage bias was applied to the top electrode (anode) while the bottom electrode (cathode) was grounded. (Reprinted from *Carbon*, 94, Rani, J. R., Oh, S.-I., Woo, J. M., Jang, J.-H., Low voltage resistive memory devices based on graphene oxide–iron oxide hybrid, 362–368. Copyright (2015), with permission from Elsevier.)

The Pt/GF90/ITO device shows the highest ON/OFF current ratio (5×10^3) while those of the Pt/GF180/ITO, Pt/GO100/ITO, and Pt/GO180/ITO devices are 900, 600, and 100, respectively. The higher ON current in the Pt/GF90/ITO device, which is due to the reduced interface resistance, results in the enhancement of the ON/OFF current ratio. At higher temperature annealing, the OFF current also increased due to the lower density of oxygen functional groups in the films annealed at a higher temperature (180°C) than those annealed at a lower temperature (90°C). The conduction mechanism is different for iron oxide–graphene oxide hybrid films compared to that of graphene oxide films. In iron oxide–graphene oxide hybrid films, the conducting Fe_3O_4 phase is mixed with the nonconducting γ-Fe_2O_3 phase at the interface between the top electrode and GO [69]; when the negative bias is applied, O^{2-} ions from nonconducting γ-Fe_2O_3 move downward toward the bottom ITO electrode and are transformed into the conducting Fe_3O_4 phase. The resistance switching behavior can be expressed in terms of the redox reaction between γ-Fe_2O_3 and Fe_3O_4 as shown in Equation 9.4 [69].

$$2Fe_3O_4 + O^{2-} \leftrightarrow 3\gamma\text{-}Fe_2O_3 + 2\overline{e} \tag{9.4}$$

Also, an additional conduction path is formed due to the d–π orbital interaction between carbon and iron, which makes a charge transfer channel between the C2p and Fe3d states [111].

Due to these additional current paths, devices with GF films exhibit a higher ON current and a lower resistance than the devices with GO films at the LRS state. During the positive bias, the oxygen ions migrate from the bottom ITO electrode back into the GO and to the Fe_3O_4 layer and the interface becomes nonconducting γ-Fe_2O_3 and thereby ruptures the conducting channels [67, 69]. Also in this process, GO becomes more sp^3 hybridized, and the conducting channels formed at the interface layer and inside the GO layer are locally disconnected and the resistance dramatically increases.

This bistable resistive switching can be repeated through the reversible redox reaction between Fe_3O_4 and γ-Fe_2O_3 near the interface between the top electrode and GO layer. A schematic view of the resistive switching mechanism in the Pt/GF/ITO device is shown in Figure 9.17.

I–V characteristics (positive voltage sweep region) were plotted on a log–log scale as shown in Figure 9.18a to d. Ohmic conduction dominates in both the LRS and HRS for the Pt/GO90/ITO device (slope of around 1). The LRS plot of the Pt/GF90/ITO device is not entirely linear and can be divided into R-1 and R-2 as shown in Figure 9.18. The Pt/GF90/ITO device exhibits a linear relationship of log vs $V^{0.5}$ at the high-voltage LRS region, which is a characteristic of the Schottky conduction mechanism. In the HRS region of the Pt/GF90/ITO device, the log I–log V curve can be divided into three regions, and the current

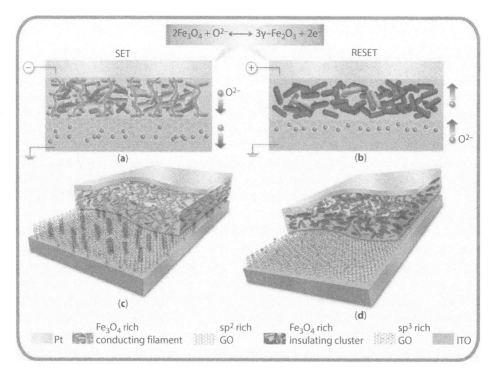

Figure 9.17 Schematic representation of the resistive switching mechanism in the Pt/GF/ITO device. (Reprinted from *Carbon*, 94, Rani, J. R., Oh, S.-I., Woo, J. M., Jang, J.-H., Low voltage resistive memory devices based on graphene oxide–iron oxide hybrid, 362–368. Copyright (2015), with permission from Elsevier.)

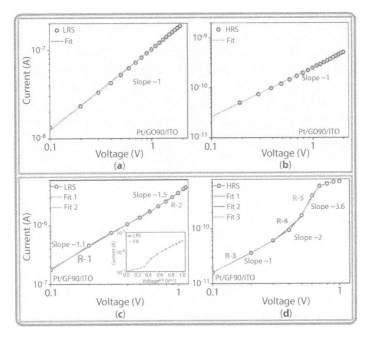

Figure 9.18 The log I– log V plot of the LRS and HRS region of Pt/GO90/ITO and Pt/GF90/ITO devices. Inset of (c) shows the log I–$V^{0.5}$ plot of the Pt/GO90/ITO device. (Reprinted from *Carbon*, 94, Rani, J. R., Oh, S.-I., Woo, J. M., Jang, J.-H., Low voltage resistive memory devices based on graphene oxide–iron oxide hybrid, 362–368. Copyright (2015), with permission from Elsevier.)

conduction mechanism changes from ohmic conduction in R-3 (slope ~1) to the space charge limited current (SCLC) conduction in R-4 (slope ~2), followed by a trap-charge limited current (TCLC) conduction in R-5 (slope ~3.6) [69].

I–V characteristics at the high-voltage regime exhibit a nonlinear relationship of $I \sim V^n$ ($n \geq 2$). According to the SCLC model, the factor n of 2 implies that the space charges are present only in the conduction band, while the factor $n > 2$ implies that space charges

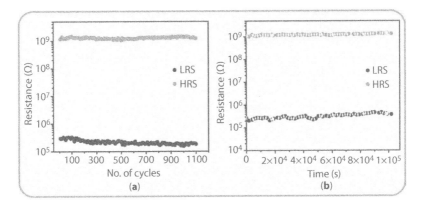

Figure 9.19 (a) Endurance and (b) retention behavior of the Pt/GF90/ITO memory device [67]. (Reprinted from *Carbon*, 94, Rani, J. R., Oh, S.-I., Woo, J. M., Jang, J.-H., Low voltage resistive memory devices based on graphene oxide–iron oxide hybrid, 362–368. Copyright (2015), with permission from Elsevier.)

are distributed in the trap sites [69]. Ohmic conduction is predominant in the low-voltage region (R-3), in which the intrinsic carrier density is higher than the injected carrier density. As the applied voltage increases, the applied electric field and injected charge density are high enough to fill all the trap states; the conduction mechanism becomes trap-charge limited current conduction (TCLC in the R5 region) [69]. Figure 9.18a and b show the respective endurance and retention behavior of the Pt/GF90/ITO memory device [67].

Resistive switching in the iron oxide–GO hybrid is due to the formation/rupture of metallic filaments due to the redox reaction in Fe_2O_3/Fe_3O_4 states, and the device showed a high ON/OFF current ratio as high as 5×10^3, with a retention time 10^5 s and an endurance cycle of more than 1100 (Figure 9.19).

References

1. Geim, A.K. and Novoselov, K.S., The rise of graphene. *Nat. Mater.*, 6, 183–191, 2007.
2. Novoselov, K.S., Geim, A.K., Morozov, S.V., Jiang, D., Katsnelson, M.I., Grigorieva, I.V., Dubonos, S.V., Firsov, A.A., Two-dimensional gas of massless Dirac fermions in graphene. *Nature*, 438, 197–200, 2005.
3. Wang, X., Zhi, L., Müllen, K., Transparent, conductive graphene electrodes for dye-sensitized solar cells. *Nano Lett.*, 8, 323–327, 2007.
4. Elias, D.C., Nair, R.R., Mohiuddin, T.M.G., Morozov, S.V., Blake, P., Halsall, M.P., Ferrari, A.C., Boukhvalov, D.W., Katsnelson, M.I., Geim, A.K., Novoselov, K.S., Control of graphene's properties by reversible hydrogenation: Evidence for graphane. *Science*, 323, 610–613, 2009.
5. Yan, M., Wang, F., Han, C., Ma, X., Xu, X., An, Q., Xu, L., Niu, C., Zhao, Y., Tian, X., Hu, P., Wu, H., Mai, L., Nanowire templated semihollow bicontinuous graphene scrolls: Designed construction, mechanism, and enhanced energy storage performance. *J. Am. Chem. Soc.*, 135, 18176–18182, 2013.
6. Wang, J., Liang, M., Fang, Y., Qiu, T., Zhang, J., Zhi, L., Rod-coating: Towards large-area fabrication of uniform reduced graphene oxide films for flexible touch screens. *Adv. Mater.*, 24, 2874–2878, 2012.
7. Ho, N.T., Senthilkumar, V., Kim, Y.S., Impedance spectroscopy analysis of the switching mechanism of reduced graphene oxide resistive switching memory. *Solid-State Electron.*, 94, 61–65, 2014.
8. Gwang Hyuk, S., Choong-Ki, K., Gyeong Sook, B., Jong Yun, K., Byung Chul, J., Beom Jun, K., Myung Hun, W., Yang-Kyu, C., Sung-Yool, C., Multilevel resistive switching nonvolatile memory based on MoS_2 nanosheet-embedded graphene oxide. *2D Mater.*, 3, 034002, 2016.
9. Mpourmpakis, G., Tylianakis, E., Froudakis, G.E., Carbon nanoscrolls: A promising material for hydrogen storage. *Nano Lett.*, 7, 1893–1897, 2007.
10. Wu, J., Yang, J., Huang, Y., Li, H., Fan, Z., Liu, J., Cao, X., Huang, X., Huang, W., Zhang, H., Graphene oxide scroll meshes prepared by molecular combing for transparent and flexible electrodes. *Adv. Mater. Technol.*, 2, 1600231, 2017.
11. Braga, S.F., Coluci, V.R., Legoas, S.B., Giro, R., Galvão, D.S., Baughman, R.H., Structure and dynamics of carbon nanoscrolls. *Nano Lett.*, 4, 881–884, 2004.
12. Coluci, V.R., Braga, S.F., Baughman, R.H., Galvão, D.S., Prediction of the hydrogen storage capacity of carbon nanoscrolls. *Phys. Rev. B*, 75, 125404, 2007.
13. Rani, J.R., Thangavel, R., Oh, S.-I., Woo, J.M., Chandra Das, N., Kim, S.-Y., Lee, Y.-S., Jang, J.-H., High volumetric energy density hybrid supercapacitors based on reduced graphene oxide scrolls. *ACS Appl. Mater. Interfaces*, 9, 22398–22407, 2017.

14. Pei, L., Zhao, Q., Chen, C., Liang, J., Chen, J., Phosphorus nanoparticles encapsulated in graphene scrolls as a high-performance anode for sodium-ion batteries. *Chem. Electro. Chem.*, 2, 1652–1655, 2015.
15. Karimi, H., Ahmadi, M.T., Khosrowabadi, E., Rahmani, R., Saeidimanesh, M., Ismail, R., Naghib, S.D., Akbari, E., Analytical prediction of liquid-gated graphene nanoscroll biosensor performance. *RSC Adv.*, 4, 16153–16162, 2014.
16. Liu, Y., Wang, L., Zhang, H., Ran, F., Yang, P., Li, H., Graphene oxide scroll meshes encapsulated Ag nanoparticles for humidity sensing. *RSC Adv.*, 7, 40119–40123, 2017.
17. Wang, L., Yang, P., Liu, Y., Fang, X., Shi, X., Wu, S., Huang, L., Li, H., Huang, X., Huang, W., Scrolling up graphene oxide nanosheets assisted by self-assembled monolayers of alkanethiols. *Nanoscale*, 9, 9997–10001, 2017.
18. Yu, Y., Li, G., Zhou, S., Chen, X., Lee, H.-W., Yang, W., Self–adaptive Si/reduced graphene oxide scrolls for high–performance Li–ion battery anodes. *Carbon*, 120, 397–404, 2017.
19. Li, H., Wu, J., Qi, X., He, Q., Liusman, C., Lu, G., Zhou, X., Zhang, H., Graphene oxide scrolls on hydrophobic substrates fabricated by molecular combing and their application in gas sensing. *Small*, 9, 382–386, 2013.
20. Zhou, W., Liu, J., Chen, T., Tan, K.S., Jia, X., Luo, Z., Cong, C., Yang, H., Li, C.M., Yu, T., Fabrication of Co_3O_4-reduced graphene oxide scrolls for high-performance supercapacitor electrodes. *Phys. Chem. Chem. Phys.*, 13, 14462–14465, 2011.
21. Zheng, Q., Shi, L., Ma, P.-C., Xue, Q., Li, J., Tang, Z., Yang, J., Structure control of ultra-large graphene oxide sheets by the Langmuir–Blodgett method. *RSC Adv.*, 3, 4680–4691, 2013.
22. Xie, X., Ju, L., Feng, X., Sun, Y., Zhou, R., Liu, K., Fan, S., Li, Q., Jiang, K., Controlled fabrication of high-quality carbon nanoscrolls from monolayer graphene. *Nano Lett.*, 9, 2565–2570, 2009.
23. Xu, Z., Zheng, B., Chen, J., Gao, C., Highly efficient synthesis of neat graphene nanoscrolls from graphene oxide by well-controlled lyophilization. *Chem. Mater.*, 26, 6811–6818, 2014.
24. Gao, Y., Chen, X., Xu, H., Zou, Y., Gu, R., Xu, M., Jen, A.K.Y., Chen, H., Highly-efficient fabrication of nanoscrolls from functionalized graphene oxide by Langmuir–Blodgett method. *Carbon*, 48, 4475–4482, 2010.
25. Savoskin, M.V., Mochalin, V.N., Yaroshenko, A.P., Lazareva, N.I., Konstantinova, T.E., Barsukov, I.V., Prokofiev, I.G., Carbon nanoscrolls produced from acceptor-type graphite intercalation compounds. *Carbon*, 45, 2797–2800, 2007.
26. Sharifi, T., Gracia-Espino, E., Reza Barzegar, H., Jia, X., Nitze, F., Hu, G., Nordblad, P., Tai, C.-W., Wågberg, T., Formation of nitrogen-doped graphene nanoscrolls by adsorption of magnetic γ-Fe_2O_3 nanoparticles. *Nat. Commun.*, 4, 2319, 2013.
27. Rani, J.R., Oh, S.-I., Woo, J.M., Tarwal, N.L., Kim, H.-W., Mun, B.S., Lee, S., Kim, K.-J., Jang, J.-H., Graphene oxide–phosphor hybrid nanoscrolls with high luminescent quantum yield: Synthesis, structural, and x-ray absorption studies. *ACS Appl. Mater. Interfaces*, 7, 5693–5700, 2015.
28. Baskey, M. and Saha, S.K., A graphite-like zero gap semiconductor with an interlayer separation of 2.8 Å. *Adv. Mater.*, 24, 1589–1593, 2012.
29. Dikin, D.A., Stankovich, S., Zimney, E.J., Piner, R.D., Dommett, G.H.B., Evmenenko, G., Nguyen, S.T., Ruoff, R.S., Preparation and characterization of graphene oxide paper. *Nature*, 448, 457, 2007.
30. Eda, G., Lin, Y.-Y., Mattevi, C., Yamaguchi, H., Chen, H.-A., Chen, I.S., Chen, C.-W., Chhowalla, M., Blue photoluminescence from chemically derived graphene oxide. *Adv. Mater.*, 22, 505–509, 2010.
31. Mei, Q., Zhang, K., Guan, G., Liu, B., Wang, S., Zhang, Z., Highly efficient photoluminescent graphene oxide with tunable surface properties. *Chem. Commun.*, 46, 7319–7321, 2010.

32. Loh, K.P., Bao, Q., Eda, G., Chhowalla, M., Graphene oxide as a chemically tunable platform for optical applications. *Nat. Chem.*, 2, 1015, 2010.
33. Chien, C.-T., Li, S.-S., Lai, W.-J., Yeh, Y.-C., Chen, H.-A., Chen, I.S., Chen, L.-C., Chen, K.-H., Nemoto, T., Isoda, S., Chen, M., Fujita, T., Eda, G., Yamaguchi, H., Chhowalla, M., Chen, C.-W., Tunable photoluminescence from graphene oxide. *Angew. Chem. Int. Ed.*, 51, 6662–6666, 2012.
34. Graf, D., Molitor, F., Ensslin, K., Stampfer, C., Jungen, A., Hierold, C., Wirtz, L., Spatially resolved Raman spectroscopy of single- and few-layer graphene. *Nano Lett.*, 7, 238–242, 2007.
35. Ferrari, A.C. and Robertson, J., Interpretation of Raman spectra of disordered and amorphous carbon. *Phys. Rev. B*, 61, 14095–14107, 2000.
36. Dresselhaus, M.S., Dresselhaus, G., Saito, R., Jorio, A., Raman spectroscopy of carbon nanotubes. *Phys. Rep.*, 409, 47–99, 2005.
37. Rao, R., Podila, R., Tsuchikawa, R., Katoch, J., Tishler, D., Rao, A.M., Ishigami, M., Effects of layer stacking on the combination Raman modes in graphene. *ACS Nano*, 5, 1594–1599, 2011.
38. Cong, C., Yu, T., Saito, R., Dresselhaus, G.F., Dresselhaus, M.S., Second-order overtone and combination Raman modes of graphene layers in the range of 1690–2150 cm^{-1}. *ACS Nano*, 5, 1600–1605, 2011.
39. Pan, D., Zhang, J., Li, Z., Wu, M., Hydrothermal route for cutting graphene sheets into blue-luminescent graphene quantum dots. *Adv. Mater.*, 22, 734–738, 2010.
40. Gan, Z.X., Xiong, S.J., Wu, X.L., He, C.Y., Shen, J.C., Chu, P.K., Mn^{2+}-bonded reduced graphene oxide with strong radiative recombination in broad visible range caused by resonant energy transfer. *Nano Lett.*, 11, 3951–3956, 2011.
41. Zhang, Y., Pan, Q., Chai, G., Liang, M., Dong, G., Zhang, Q., Qiu, J., Synthesis and luminescence mechanism of multicolor-emitting g-C_3N_4 nanopowders by low temperature thermal condensation of melamine. *Sci. Rep.*, 3, 1943, 2013.
42. Chen, Y., Zhang, B., Liu, G., Zhuang, X., Kang, E.-T., Graphene and its derivatives: Switching ON and OFF. *Chem. Soc. Rev.*, 41, 4688–4707, 2012.
43. Sawa, A., Resistive switching in transition metal oxides. *Mater. Today*, 11, 28–36, 2008.
44. Lin, W.-P., Liu, S.-J., Gong, T., Zhao, Q., Huang, W., Polymer-based resistive memory materials and devices. *Adv. Mater.*, 26, 570–606, 2014.
45. Lu, W. and Lieber, C.M., Nanoelectronics from the bottom up. *Nat. Mater.*, 6, 841, 2007.
46. Yang, Y. and Lu, W., Nanoscale resistive switching devices: Mechanisms and modeling. *Nanoscale*, 5, 10076–10092, 2013.
47. Zhuge, F., Hu, B., He, C., Zhou, X., Liu, Z., Li, R.-W., Mechanism of nonvolatile resistive switching in graphene oxide thin films. *Carbon*, 49, 3796–3802, 2011.
48. Khurana, G., Misra, P., Katiyar, R.S., Forming free resistive switching in graphene oxide thin film for thermally stable nonvolatile memory applications. *J. Appl. Phys.*, 114, 124508, 2013.
49. Jo, H., Ko, J., Lim, J.A., Chang, H.J., Kim, Y.S., Organic nonvolatile resistive switching memory based on molecularly entrapped fullerene derivative within a diblock copolymer nanostructure. *Macromol. Rapid Commun.*, 34, 355–361, 2013.
50. Cava, C.E., Persson, C., Zarbin, A.J.G., Roman, L.S., Resistive switching in iron-oxide-filled carbon nanotubes. *Nanoscale*, 6, 378–384, 2014.
51. Choi, H., Pyun, M., Kim, T.W., Hasan, M., Dong, R., Lee, J., Park, J.B., Yoon, J., Seong, D., Lee, T., Hwang, H., Nanoscale resistive switching of a copper–carbon-mixed layer for nonvolatile memory applications. *IEEE Electron Device Lett.*, 30, 302–304, 2009.
52. Shumkin, G.N., Zipoli, F., Popov, A.M., Curioni, A., Multiscale quantum simulation of resistance switching in amorphous carbon. *Procedia Comput. Sci.*, 9, 641–650, 2012.
53. Pradhan, S.K., Xiao, B., Mishra, S., Killam, A., Pradhan, A.K., Resistive switching behavior of reduced graphene oxide memory cells for low power nonvolatile device application. *Sci. Rep.*, 6, 26763, 2016.

54. Chengbin, P., Enrique, M., Marco, A.V., Na, X., Xu, J., Xiaoming, X., Tianru, W., Fei, H., Yuanyuan, S., Mario, L., Model for multi-filamentary conduction in graphene/hexagonal-boron-nitride/graphene based resistive switching devices. *2D Mater.*, 4, 025099, 2017.
55. Kim, S.K., Kim, J.Y., Choi, S.-Y., Lee, J.Y., Jeong, H.Y., Direct observation of conducting nano-filaments in graphene-oxide-resistive switching memory. *Adv. Funct. Mater.*, 25, 6710–6715, 2015.
56. Schniepp, H.C., Li, J.-L., McAllister, M.J., Sai, H., Herrera-Alonso, M., Adamson, D.H., Prud'homme, R.K., Car, R., Saville, D.A., Aksay, I.A., Functionalized single graphene sheets derived from splitting graphite oxide. *J. Phys. Chem. B*, 110, 8535–8539, 2006.
57. Robinson, J.T., Zalalutdinov, M., Baldwin, J.W., Snow, E.S., Wei, Z., Sheehan, P., Houston, B.H., Wafer-scale reduced graphene oxide films for nanomechanical devices. *Nano Lett.*, 8, 3441–3445, 2008.
58. Cote, L.J., Kim, F., Huang, J., Langmuir–Blodgett assembly of graphite oxide single layers. *J. Am. Chem. Soc.*, 131, 1043–1049, 2009.
59. Huang, C.-H., Chang, W.-C., Huang, J.-S., Lin, S.-M., Chueh, Y.-L., Resistive switching of Sn-doped In_2O_3/HfO_2 core-shell nanowire: Geometry architecture engineering for nonvolatile memory. *Nanoscale*, 9, 6920–6928, 2017.
60. Choi, J., Le, Q.V., Hong, K., Moon, C.W., Han, J.S., Kwon, K.C., Cha, P.-R., Kwon, Y., Kim, S.Y., Jang, H.W., Enhanced endurance organolead halide perovskite resistive switching memories operable under an extremely low bending radius. *ACS Appl. Mater. Interfaces*, 9, 30764–30771, 2017.
61. Ling, H., Yi, M., Nagai, M., Xie, L., Wang, L., Hu, B., Huang, W., Controllable organic resistive switching achieved by one-step integration of cone-shaped contact. *Adv. Mater.*, 29, 1701333, 2017.
62. Jeong, H.Y., Kim, J.Y., Kim, J.W., Hwang, J.O., Kim, J.-E., Lee, J.Y., Yoon, T.H., Cho, B.J., Kim, S.O., Ruoff, R.S., Choi, S.-Y., Graphene oxide thin films for flexible nonvolatile memory applications. *Nano Lett.*, 10, 4381–4386, 2010.
63. Jilani, S.M., Gamot, T.D., Banerji, P., Chakraborty, S., Studies on resistive switching characteristics of aluminum/graphene oxide/semiconductor nonvolatile memory cells. *Carbon*, 64, 187–196, 2013.
64. Lee, D.W. and Seo, J.W., sp2/sp3 carbon ratio in graphite oxide with different preparation times. *J. Phys. Chem. C*, 115, 2705–2708, 2011.
65. Singh, M., Yadav, A., Kumar, S., Agarwal, P., Annealing induced electrical conduction and band gap variation in thermally reduced graphene oxide films with different sp2/sp3 fraction. *Appl. Surf. Sci.*, 326, 236–242, 2015.
66. Kyzas, G., Travlou, N., Kalogirou, O., Deliyanni, E., Magnetic graphene oxide: Effect of preparation route on reactive black 5 adsorption. *Materials*, 6, 1360, 2013.
67. Rani, J.R., Oh, S.-I., Woo, J.M., Jang, J.-H., Low voltage resistive memory devices based on graphene oxide–iron oxide hybrid. *Carbon*, 94, 362–368, 2015.
68. He, C.L., Zhuge, F., Zhou, X.F., Li, M., Zhou, G.C., Liu, Y.W., Wang, J.Z., Chen, B., Su, W.J., Liu, Z.P., Wu, Y.H., Cui, P., Li, R.-W., Nonvolatile resistive switching in graphene oxide thin films. *Appl. Phys. Lett.*, 95, 232101, 2009.
69. Verwey, E.J.W. and Haayman, P.W., Electronic conductivity and transition point of magnetite ("Fe_3O_4"). *Physica*, 8, 979–987, 1941.

10

Fabrication and Properties of Copper–Graphene Composites

Vladimir G. Konakov[1,2*], Ivan Yu. Archakov[2,3] and Olga Yu. Kurapova[1,2]

[1]St. Petersburg State University, St. Petersburg, Russia
[2]Peter the Great St. Petersburg Polytechnic University, St. Petersburg, Russia
[3]Institute for Problems of Mechanical Engineering, Russian Academy of Sciences, St. Petersburg, Russia

Abstract

The present chapter summarizes the recent advances in bulk and foil copper–graphene (Cu–Gr) composites fabrication. Graphene introduction into copper results in significant Cu matrix reinforcement with no loss in material conductivity. The most effective techniques for copper matrix composites fabrication are discussed. The benefits and limitations of each technique are pointed out depending on the target application of the composite material. The experimental data on metal–graphene composites and theoretical predictions are critically reviewed. The emphasis is made on nanotwinned copper–graphene composite foils obtained by the electrochemical deposition as very promising reinforced materials. Mechanical properties and thermal and electrical conductivities of Cu–Gr composites are regarded with respect to their manufacturing method, type, and concentration of the graphene additive. The suggested mechanisms of conductivity and reinforcement in Cu–Gr composites are discussed.

Keywords: Graphene, copper matrix composites, reinforcement, mechanical properties, composites fabrication, bulk composites, composite foils

10.1 Introduction

Copper (Cu) is a well-known metallic material that has been widely used due to its high electrical and thermal conductivities, as well as low expansion coefficient. Pure copper is applied as an electrical contact in electronics, machinery, civil engineering, automobile industry, etc. [1–3]. However, its application is rather limited due to its low hardness coupled with high plasticity. Moreover, copper oxidation takes place at elevated temperatures, which drastically decreases the mechanical properties of the material. In recent years, there is a demand for both high strength and high conductive material at a wide operating temperature range. So, the search for the ways to improve copper mechanical properties with no loss in electrical conductivity is a great challenge for modern materials science.

---

*Corresponding author: glasscer@yandex.ru

The required balance in conductivity and mechanical properties can be achieved through Cu-based composites development via the addition of a rather small fraction of graphene [4–6]. Graphene (Gr) is a 2D material that has electrical conductivity of the same order as pure copper [7]. Carrier mobility of graphene is 2×10^5 cm^2/(V · s) [8]. At the same time, its thermal conductivity is higher than the one for copper, i.e., 5000 J/(m · K · s) [9, 10]. Besides, Gr possesses superior mechanical properties. Its hardness is ~110–120 GPa, tensile strength is ~125 GPa, and Young's modulus is about 1100 GPa [11]. It makes Gr a very promising reinforcement additive for soft copper matrix. It should be noted that the data on conductivity and mechanical properties listed correspond to single-layer graphene. Practically, various graphene derivatives having different number of layers, structure, and deficiency [i.e., bilayered graphene, few-layered graphene, graphene oxide (GO), reduced graphene oxide (rGO), graphene nanoplatelets (GNP), etc.] are introduced into the copper matrix depending on target composite material properties [12]. Thus, the above listed data for single-layered graphene should be used with regard to the real graphene derivative introduced in the material. It has been proven that the introduction of certain amount of Gr to copper results in very promising reinforcement of metallic matrix along with the conductivity retained [12–15].

In recent years, an impressive variety of approaches have been developed for the synthesis of copper–graphene (Cu–Gr) composites. In case of bulk copper-graphene composites, powder metallurgy, microwave sintering, hot pressing, spark plasma sintering, cold pressing, hot isostatic pressing, etc. should be mentioned. Lately, copper-graphene films were produced by electrochemical deposition, chemical vapor deposition (CVD), and molecular-level mixing (MLM). The properties of Cu–Gr composites highly depend on their manufacturing method, concentration, form, and microstructure of the graphene derivative chosen as a dopant. Each method has its advantages and limitations depending on the target application of a composite material. General problems of copper–graphene composites should be mentioned: the difficulty of homogeneous graphene distribution in a metallic matrix, very low bonding between Gr and most metallic matrices, including copper, and the problem of full densification for bulk copper matrix composites. The present chapter summarizes the recent progress in copper–graphene composites fabrication. The properties of Cu–Gr materials are regarded with respect to their manufacturing method, type, and concentration of graphene additive content. Emphasis is placed on electrochemical deposition as a very promising technique for reinforced materials fabrication. The suggested conductivity and reinforcement mechanisms are discussed.

10.2 Powder Metallurgy Technique

The powder metallurgy (PM) technique is a common approach, providing a rather simple pathway for bulk metal–matrix composites (MMc) production. In modern materials science, it is generally used for high-volume (low-cost) nickel–graphene and aluminum–graphene composites development [1]. The method includes a powder mixing step that is usually realized as continuous grinding of a premade graphene–metal powder mixture in a ball mill with the chemically inert balls (e.g., made from tungsten carbide). A large number of defects, i.e., lattice distortion, dislocations, etc., are introduced during the process, so that the interdiffusion is strengthened and the activation energy is lowered.

In the case when powders are reactive to air (e.g., nanosized nickel powder), ball milling takes place under argon or nitrogen atmosphere. Homogeneity of a graphene additive in a mixture is reached by the optimization of the treatment conditions, i.e., milling time, milling rate (rpm), and number of reverses. High-energy milling and milling in a planetary ball mill can be regarded as the most efficient variation of the PM technique. It results in mechanical activation of the mixture and provides higher homogenization level.

Along with other MMc, copper–graphene composites can be also obtained via the PM technique. The schematic illustration of the ball milling process of the copper and graphene mixture is shown in Figure 10.1.

As seen from Figure 10.1, graphene coating on the surface of copper powder particle is formed during the ball milling process [1]. The thickness of this graphene layer depends on the ratio between the specific areas of Cu powder particles and graphene flakes. When these values are close to each other, the monolayer of Gr on the surface of each particle can be formed. Thus, the optimal amount of graphene addition should be chosen with respect to the initial Cu powder dispersity.

Summarizing the literature data on copper–graphene composites obtained by PM, the maxima of mechanical properties and electrical and heat conductivities are observed when ~1–3 wt.% of graphene is introduced into commercial microsized copper powders [1, 12, 15–17]. It should be noted that copper is a much softer material compared to nickel and is close to aluminum. Indeed, hardness of Cu is ~46 HV [18]. For Ni and Al, these values are ~80 and ~60 HV [18]. According to the reference data, high milling rate is necessary to provide proper Cu–Gr mixing (1400–1700 rpm) [19]. In contrast to Al particles, which are covered by alumina layer, the agglomeration of copper particles cannot be avoided during milling [19, 20]. Since Cu possesses high ductility and possible dynamic recovery occurs during milling, the particles of Cu powder are first crushed by plastic deformation under the impact of the balls and then subsequently cold-welded; as a result, large particles with rough surface are formed [19, 20]. Hence, copper powders become coarser (agglomerated) and irregular in shape after the ball milling process. Note that bonding of graphene and the copper matrix is very low. Obviously, Cu–Gr materials produced within the frames of the PM method followed by conventional compacting and sintering suffer from inevitable porosity, i.e., decreased mechanical and electrical properties. Indeed, the typical porosity level of copper–graphene composites, obtained by ball milling and cold-pressing compaction, is higher than 10–12% [16, 21]. The use of ultrafine-grained and nanosized Cu powders allows the increase in contact surface between metallic and graphene particles, but the intensive nanosized copper oxidation makes this approach rather complicated. As a result,

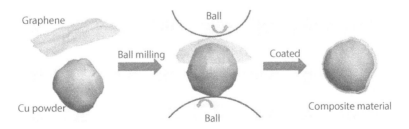

Figure 10.1 Schematic illustration of the fabrication process for the copper–graphene composite by ball milling. Reprinted with permission [1].

one can expect some decrease in electro- and heat conductivities of the composite as compared to metallurgical and electrodeposited copper. On the other hand, i.e., modification of Cu surface, the introduction of active sites on Cu surface for graphene bonding or the use of advanced sintering methods (hot pressing, spark plasma sintering) makes it possible to develop bulk Cu–Gr composites with enhanced mechanical, thermal, and electrical properties. Let us consider the most effective techniques.

10.2.1 Hot Pressing Technique

Significant reinforcement of Cu–Gr composites is achieved by the combination of PM and hot pressing technique [15] due to copper and carbon filler adhesion enhancement. Compared to pure Cu, the copper–graphene composites show a remarkable increase in yield strength and Young's modulus [19]. The extraordinary reinforcement is attributed by many researchers to the homogeneous dispersion of graphene in copper matrix and overall microstructure refinement (i.e., enhanced densification parameter and copper grain refinement). Indeed, hot pressing provides fast sintering of pre-compacted powder under elevated pressure (i.e., 40–60 MPa) at the high heating rate (~50°C/min) [19]. This approach reduces pores (voids) created on the compaction step in the green sample due to the high plasticity of copper. The relative density of 96.4%–99.6% can be reached after hot pressing. Besides, short sintering time allows the retention of the homogeneous distribution of graphene filler achieved on the ball milling step [19]. Copper grain refinement is observed in the composites with moderate graphene content. It is attributed to the fact that a continuous interphase boundary is formed between two-dimensional graphene platelets and the copper grain surface. Since graphene has negative thermal expansion coefficient (CTE) [22], greater compression stress is applied on copper grain, limiting its growth on the initial heating stage. Similar to carbon nanotubes–metal composites [23], it is thought that grain size refinement in graphene–Cu composites is attributed to the blocking effect of the nanosized graphene plates on the grain boundaries, where the dislocation motion could be blocked at the sites of GNPs. Consequently, the dislocation accumulation on the grain boundary eliminates the growth of the recrystallized grains during the processing. That, no doubt, contributes to the high strength of Cu matrix composites.

One more important factor affecting the mechanical properties of Cu–Gr materials is graphene nanoplatelet (GNP) thickness. Recent investigations of hot pressed composites (GNP amount 1 and 2 wt.%) manufactured from ball milled powders showed that the addition of graphene with a platelet thickness of 2–4 nm leads to about 50% higher hardness and about 30% lower electrical resistivity than the use of graphene with a platelet thickness of 10–20 nm [12]. In the case of few-layered GNPs, a more homogeneous composite microstructure is obtained. For instance, it is found that 8 vol.% GNPs with a typical thickness ~3.5 nm added to the copper matrix results in a remarkable increase in yield strength and Young's modulus to 114 GPa and 37%, respectively. A more evident difference in properties is observed when comparing to copper–matrix composites containing 0.5 wt.% of few-layered graphene and graphite consolidated by hot pressing at 600°C under a pressure of 30 MPa (see Figure 10.2) and consequent annealing at different temperatures [15].

The use of graphene as reinforcing additive resulted in ~1.5 times decrease in the average Cu grain size. The copper grains in the Cu–Gr composite are typically smaller than 10 μm,

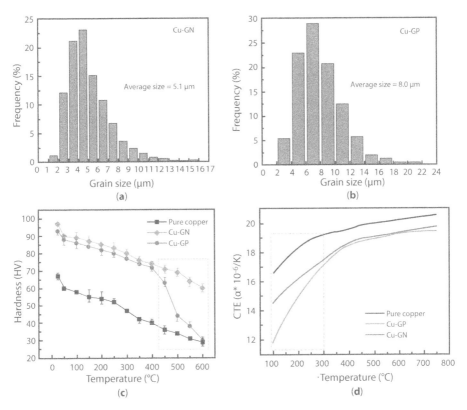

Figure 10.2 Comparison of copper–graphene and copper–graphite composites. Reprinted with permission from [15].

while more than 25% of grains in the copper–graphite composite are characterized by grain sizes over 10 μm (see Figure 10.2a and b). The temperature dependence of Vickers hardness (HV) is demonstrated in Figure 10.2c. One can see that carbon additions resulted in 40%–50% HV increase at a temperature up to 450°C. Similar tendency is observed for pure Cu and for both composites. The hardness of the Cu–Gr composite is slightly higher than that for the copper–graphite material. This excess can be estimated as 5%–7%. However, the hardness behavior of carbon-containing composites drastically differs at higher temperatures (450–600°C). The decrease of Vickers hardness, HV, is about 15% for the Cu–Gr composite, while copper–graphite composite hardness decreases nearly twice. Note that the hardness tests were carried out under neutral atmosphere (Ar), so the above effect was not due to any kind of oxidation. Located at the grain boundaries, 2D graphene likely prevents grain boundary failure at high temperatures. The mechanism similar to that discussed above for low temperatures (<250°C) is suggested: due to negative CTE, graphene located at the grain boundaries hinders the material expansion.

The data on electrical conductivity of copper-matrix composites doped by graphene and graphite differ as well. Upon graphene addition, the conductivity of composite slightly decreases compared to pure annealed copper. It can be estimated as 0.95 IACS (International Annealed Copper Standard, 100% IACS at 20°C = 5.8001×10^7 S/m). In turn, graphite addition results in a 25% decrease in the electroconductivity down to 0.75 IACS.

10.2.2 Microwave Heating

Along with hot pressing, microwave heating is a rapid sintering technique for the consolidation of various materials. The main advantage of microwave sintering is that it provides rapid heating resulting in microstructure refinement [24–26]. Even though the heating rates are high, the homogeneity of material is preserved. Short sintering times make it somewhat similar to hot pressing. During microwave processing, the composite material is heated up as a result of the reaction taking place under the electromagnetic wave. In addition to this, the grain boundary diffusion is promoted by the decrease in the activation energy for sintering [24, 25]. It was admitted that microwave sintering could result in higher density and smaller matrix grain size [27].

The data on the porosity of copper–graphene composites obtained by microwave sintering (2.45 GHz, 10 kW, heating rate of 20°C/min, 900°C in 95% N_2–5% H_2 for 1 h, cooling with furnace) and conventional heating (tubular furnace, heating rate of 5°C/min, 900°C in 95% N_2–5% H_2 for 1 h, cooling with furnace) is presented in Table 10.1. Let us note that coarse-grained copper powder with the typical linear size of the spherical particles <42.5 μm and graphene nanoplatelets with thickness 50–100 nm were used to manufacture powder mixtures containing 0.9, 1.8, 2.7, and 3.6 vol.% of graphene.

As seen from Table 10.1, microwave heating provides enhanced densification level of composites. The difference in porosity for both pure copper and Cu–Gr composites fabricated by microwave and conventional heating is ~20%. Graphene additive also affects sample porosity. Minimal Gr addition (0.9 vol.%) results in the material porosity decrease. The increase in the Gr content results in an increase in the composite porosity. It becomes equal to the porosity of the pure bulk copper at 2–2.5 vol.% of graphene. Higher Gr content results in further porosity increase.

Table 10.1 Comparison of the properties of Cu–Gr composites produced by powder metallurgy with those of reference samples manufactured within the same approach, data from [27].

Gr content in the initial powder mixture, vol.%	Reference sample – pure copper	0.9	1.6	2.7	3.6
Porosity, %					
Conventional heating	14	12	12.6	15	15.6
Microwave heating	11	8	10	11	12
Typical grain size, μm					
Conventional heating	50.3	47.7	44.9	53.1	46.2
Microwave heating	43.9	42.8	42.3	48.4	40
Electrical conductivity, % IACS					
Conventional heating	89	92	91	88	84
Microwave heating	92	94	92	89	86

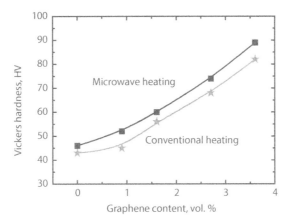

Figure 10.3 Vickers hardness of the samples synthesized by powder metallurgy. Reprinted with permission from [27].

The values of electric conductivity of copper–graphene composites obtained by microwave synthesis (Table 10.1) generally agree with the data for copper–graphene composites, obtained by hot pressing [15]. Indeed, small addition of graphene results in a slight decrease in the composite's conductivity: 92%–94% for microwave sintering against 95% for hot pressing. Further increase of Gr content leads to ~15% decrease in electric conductivity.

Results on the Vickers hardness are demonstrated in Figure 10.3 [27]. As seen from the figure, Gr addition results in a nearly twice increase in the Vickers hardness of the material; i.e., the application of microwave heat treatment leads to some additional hardness improvement. Note that hardness increase is accompanied by evident decrease in friction coefficient (from 0.5–0.6 to 0.22–0.27 at different measuring regimes), while the decrease in the wear rate registered by pin-on-disc wear testing machine is extremely high (abound 20 times). Graphene here acts as a solid lubricant minimizing the wear of the composite. Obviously, these results are very promising.

10.2.3 Spark Plasma Synthesis

Spark plasma sintering (SPS) is one of the most promising techniques used to produce bulk copper–graphene composite materials [28]. Here, the consolidation of powder mixture takes place under the passage of electric current. The properties of the composites depend on current characteristics, temperature, and pressure applied during sintering. The main advantage of SPS is the possibility of performing a short-time sintering process of various powders. Another particularly important feature of this method is that electric current passing through the ball milled powder does not cause drastic grain growth and agglomeration. Figure 10.4 shows the microstructure of the copper–graphene composite obtained by SPS sintering of ball milled powder (sintering temperature 950°C, heating rate 100°C/min, holding time 10 min, pressure 50 MPa). Commercial graphene powder with 1- to 5-nm platelet thickness was used as the filler to coarse-grained copper matrix.

As seen from Figure 10.4, graphene addition is distributed homogeneously in the grain boundaries of composites after SPS. Note that the overall microstructure differs from that usually obtained by other sintering techniques. Grain boundary thickness significantly

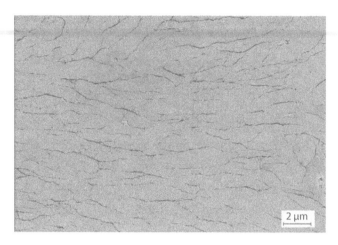

Figure 10.4 The microstructure of 97Cu-3Gr composite (vol.%). Reprinted with permission from [28].

increases with the increase of the graphene amount in the composite. For instance, when the composite contains 3 vol.% addition of graphene, the grain boundary thickness is lower than 15 nm [28]. Upon graphene addition increases to 10 (vol.%), the maximum thickness value reaches several hundred nanometers. No aggregation of graphene to the separate region is observed.

Composites sintered via SPS are characterized by the high relative density of the materials (>93% of theoretical density). It may be due to elevated pressure applied during the sintering that brings the individual graphene flakes and powder particles closer, i.e., eliminates voids between them. The other reason could be the arching appearing in pores during SPS and thus intensifying the sintering processes. The relative density decreases when the amount of graphene in the composite is increased.

Even a small amount of graphene in the copper matrix results in a remarkable increase in the composite material hardness versus pure copper (88.5 HV for 97Cu-3Gr (vol.%) composite and 44.6 HV for pure copper). When this amount increases to 10 vol.%, the composite hardness decreases due to the growing number of pores in the composite bulk. Graphene addition affects the wear resistance of Cu–graphene composite materials. During the friction contact, graphene acts as a solid lubricant minimizing the wear of both the composite and the steel ball. The wear of the steel ball decreases as the concentration of graphene in the composite exceeds 5 vol.%; these results agree with the data reported by Ayyappadas *et al.* [27].

The electrical conductivity of the copper–graphene composites obtained via SPS (650°C and 60 MPa for 5 min, 1 wt.% of high-quality graphene, HQG) is 8% higher than that measured for pure copper samples manufactured within the same technique [1]. The addition of 5 wt.% HQG to the copper matrix results in the same conductivity as in pure copper. Higher concentrations of HQG further decrease the conductivity of the composite. It is most likely due to excess of carbon additive and inhomogeneous mixing during ball milling. For copper-reduced graphene oxide composites obtained by SPS, the electrical conductivity also shows the increase in 0.3%. Vickers hardness of the copper–HQG composite shows an increase of 13% compared to pure copper (0.5 wt.% HQG sample). In the work by W. Li *et al.*, factors related to the enhancement of the conductivity of the copper–HQG composite were suggested [1]: (i) the electron mobility of the HQG is higher than that of regular reduced graphene oxide, produced by a modified Hummer's approach; (ii) the formation of an interconnected conductive network of the HQG in the copper matrix. In other words, the electrical conductivity at low HQG content in a composite is decreased since the system cannot construct a conduction network structure in the composite bulk. When the HQG content is too high, a large amount of voids (pores) is formed in the microstructure and the scattering of carriers is increased, resulting in the decrease of the electrical conductivity.

Summarizing the discussion above, the use of the PM technique even in combination with novel sintering approaches makes it complicated to produce fully dense Cu–Gr composites with enhanced properties; this is due to low copper–graphene bonding. The special role of the Cu–C interface is especially emphasized when it comes to thermal and electrical properties of composites. Indeed, despite the outstanding thermal properties of graphene, the absence of bonding makes it difficult to achieve the heat flux in the composite [29, 30]. Among the other 3d-transition metals, copper has the lowest affinity to carbon. It does not form any carbide phases. Also, compared to Co and Ni, Cu is characterized by very low carbon solubility (0.001–0.008 wt.% at 1084°C) [31, 32]. The low reactivity with carbon could also be due to the fact that copper has the most stable electron configuration with a filled d-electron shell. Thus, the use of alternative approaches is needed to provide sufficient copper–graphene bonding in composite materials. Novel synthesis techniques such as electrochemical deposition, electroless deposition, molecular-level mixing (MLM), chemical vapor deposition (CVD), etc. can provide the necessary chemical bonding between carbon additive (graphene) and copper. Let us discuss the principles of these techniques as well as the microstructure and properties of thus-synthesized Cu–Gr composites in more detail.

10.3 Electrochemical Deposition

Electrochemical deposition is a technique widely used for copper foil (ribbon) production; the details of the process can be found in a lot of handbooks and patents (see, e.g., [33]). Generally, the process can be described as follows. Copper anode serving as a source of copper ions is placed into a deposition cell filled with an electrolyte (copper sulfate is widely used for this task). Some substrates for copper foil (film) deposition (e.g., from stainless steel) are used as a cathode. External electric field is applied, resulting in copper Cu^{2+} ions migration to the cathode; they are discharging there, forming the metal layer. At the same time, copper from the anode replaces the metal deposited on the cathode, maintaining electrical neutrality. Obviously, this description (see Figure 10.5) is done in a most generalized form; the real mechanisms (metal ions hydration, convection/diffusion/migration under electric current competition, ion dehydration and neutralization, electron transfer, etc.) and technical application of the technique (deposition regimes, necessity to control pH level, etc.) are much more complex.

Mechanical properties of the copper manufactured by electrochemical deposition as well as its electro- and heat conductivities could be very close to the properties typical for metallurgical copper; this is the indisputable advantage of the process. Another advantage is its relative simplicity and cheapness. On the other hand, the significant limitation of the

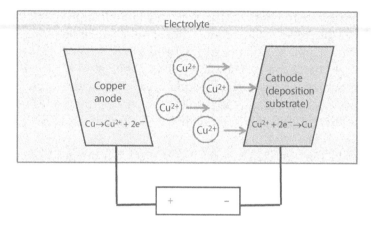

Figure 10.5 General scheme of the electrodeposition cell for copper film production.

technique is the thickness of the produced copper film—it depends on the deposition time; usually, it is tens of micrometers.

This section discusses the application of the electrochemical deposition for copper–graphene film production.

10.3.1 Deposition in the Direct Current Regime

A paper by Kasichainula Jagannadham [34] reports synthesis and investigation of copper and copper-graphene composites deposited on a copper foil. Referring to [4, 35], the author claims that graphite additions should reduce copper electroconductivity. It was expected that graphene possessing thermal conductivity higher than and electroconductivity close to those of copper should provide the required electro- and heat conductivities of the composite.

Thin copper foil (135 mm thickness, two 1 × 5 cm plates connected by 1 mm ⌀ copper wire) was used as a substrate for electrochemical deposition; this substrate was preliminarily polished and chemically cleaned by a 50/50 nitric and sulfuric acid mixture. The copper plate (3 × 5 cm with 1 mm thickness) was used as an anode of the deposition cell. Hence, the anode-to-cathode surface ratio here was 1.5. In order to prepare electrolyte for pure copper deposition, technical grade copper sulfate was dissolved in distilled water up to 0.2 M concentration, the required concentration value was maintained via permanent electrolyte stirring by a magnetic stirrer.

Direct current regime at a low current density of ~1.75 A/cm² was used to perform copper deposition. The obtained deposition rates of copper films were relatively low (2–3 μm per hour), providing high film quality: smooth films with rather high grain sizes.

Cu–Gr foils were deposited in a similar cell design using the same deposition regime; the only difference was in the electrolyte composition. GrO suspension was added to the sulfate solution in order to provide the graphene source in the electrolyte. This suspension was prepared as follows (see also [36]). At the first step, microcrystalline graphite was oxidized in a boiling acid mixture ($HNO_3 + H_2SO_4$). It was considered that the change of the graphite powder color to brown indicated the completeness of the oxidation. The oxidized graphite was cleaned from acids, washed by distilled water, and filtered.

At the second preparation step, the graphite oxide suspension was prepared. The powder obtained at the first step was placed into the distilled water, mixed, and exfoliated by ultrasound treatment. Note the extremely long duration of such sonic treatment: several days were necessary to provide proper mixing and graphite exfoliation. The suspension was filtered and then dried (150°C for several hours); the obtained powder was mixed with isopropyl alcohol and subjected to ultrasound treatment for several hours. Following [37, 38], the authors consider such an approach as an effective route to exfoliate graphite oxide and to convert it into "exfoliated graphene oxide".

The GrO suspension manufactured according to the above-described procedure was used as a Gr source in the electrolyte; it was added to sulfur electrolyte for Cu–Gr foil deposition. The pH of the final mixture was maintained at ~6 (sulfuric acid was used to control this level). It was stated that such a pH level decreases GrO agglomeration due to high GrO particle hydrophobicity.

GrO-to-Gr reduction in the grown Cu–Gr composites was performed by sample heat treatment in the hydrogen flow (20 Torr H_2 at 400°C for 3 h); heating and cooling rates were slow to eliminate possible film cracking. It was expected that GrO at elevated temperatures will reduce to graphene forming CuO and Cu_2O oxides with their further reduction to pure copper due to oxygen reaction with hydrogen. Resistivity measurements were used to prove the complete graphene oxide reduction: the ~15% decrease in the film's resistivity was shown during the first 3 h of the procedure, while further heat treatment in hydrogen did not affect this characteristic.

Cu–Gr foils were deposited on both sides of the copper substrate; the results obtained for these composites were compared with those for reference samples—pure copper foils deposited on a similar substrate. Let us discuss the results of sample examination in more detail. First, the results of XRD analysis should be mentioned. GrO oxide reflexes were detected for the samples studied before reduction in hydrogen atmosphere, while no peaks that can be attributed to carbon allotropes were found. High-temperature hydrogen treatment made the intensity of GrO reflexes negligible, proving the reduction of graphene oxide. However, peaks typical for any carbon allotropes were still under the sensitivity limit of the method. SEM and optical microscopy results indicated the presence of both copper and some additives in the copper matrix. The continuous dark regions were treated as some impurities, while the separated regions with a typical grain size up to 10 μm were attributed to exfoliated Gr or residual GrO platelets. The secondary electron analysis also showed the presence of twinned copper; these regions had similar typical linear dimensions: from 5 to 10 μm. Figure 10.6 [34] demonstrates typical SEM images of Cu–Gr composites produced by the above-discussed technology. The continuous dark regions in Figure 10.6a were attributed to contaminations; the region marked by an arrow appears to be separated and was treated as graphene platelets. Figure 10.6b presents the image taken in a secondary electron mode; the region marked by an arrow is considered as being the twinned copper.

The estimates of the graphene volume fraction in the deposited composites were carried out. It was shown that the graphene contents in the composites can be estimated as 0.08–0.11 (volume fraction). The difference in these values obtained for samples deposited under the same conditions was attributed to the electrolyte depletion in graphene source during 50- and 60-h process.

Analysis of the Cu–Gr composites' electrical resistivity was performed. It was shown that the results for graphene-containing composites (1.87–2.03) ±0.003 mΩ cm were quite close

Figure 10.6 Typical SEM images of Cu–Gr composites: (a) backscattered electrons; (b) secondary electrons. Reprinted with permission from [34].

to the values determined for the reference samples of pure deposited copper (1.97–2.05) ±0.005 mΩ cm. However, both Cu–Gr and reference pure Cu samples demonstrated a resistivity increase of ~20% as compared with that of commercial copper foil. Moreover, the above data refer to composites/substrate samples; the values measured for composite films separated from the substrate are generally in the range ~2.0–2.2 mΩ cm. The author of the work considers this resistivity increase as being due to contaminations during the deposition process. Similar behavior was also shown for the temperature coefficient of resistance: it was (3.03–3.57) × 10^{-3} K^{-1} for composites on Cu substrate, (3.61–3.76) × 10^{-3} K^{-1} for the reference samples (pure copper deposited on copper substrate), and 4.04 × 10^{-3} K^{-1} for commercial copper foil; the experimental error for all measured values here was less than 0.003.

The special work by K. Jagannadham [39] is devoted to thermal conductivity of Cu–Gr composites produced by the above-described approach. Some important points dealing with samples characterization should be mentioned here. XRD patterns registered before the high-temperature treatment showed a reflex at θ = 11.8° attributed to the graphene oxide, while the peak at θ = 26.5° typical for carbon allotropes was not found. Thus, it was concluded that all the carbon of the sample is present in the graphene form. Since only θ = 26.5° reflex was registered in the patterns after the high-temperature hydrogen treatment, it was assumed that the GrO contents in the samples are below the XRD sensitivity limit (~5 wt.%); in addition, all carbon in the sample is still graphene. However, EDS analysis showed the presence of some oxygen in the material; it can be due to residual graphene oxide or due to hydroxyl groups (see theoretical calculations in [40]). A special discussion is devoted to the estimation of the Gr contents in the composite; the estimates of 8–11 vol.% from [34] were confirmed by theoretical calculations based on electro- and thermal conductivity values. However, metallography data provided a much higher value of 19–25 vol.%.

As it follows from the results presented in [34], the electrical conductivity of Cu–Gr composites was slightly lower than that of metallurgical copper and comparable to that of the electrodeposited copper. However, thermal conductivity behavior differs from the

Table 10.2 Thermal conductivity of Cu–Gr composites compared with metallurgical and electrodeposited copper (data from [39]).

Sample	Thermal conductivity, W/mK, ±10, at different temperatures		
	250 K	300 K	350 K
Metallurgical Cu	420	390	380
Electrodeposited Cu	400	380	370
Cu–Gr composite	510	460	440

above tendency (see results in Table 10.2). As seen from the table, metallurgical and electrodeposited copper showed similar thermal conductivity in the temperature range from 250 to 350 K; some decrease is observed with the temperature increase. In contrast to electrical conductivity, the thermal conductivity of Cu–Gr composites is higher than that for pure copper; the difference is quite evident: it decreases from 25% at 250 K to 17% at 350 K.

The authors of [41] used GrO as a graphene source in the deposition process; graphene oxide here was prepared by Hummer's method from graphite powder. An aqueous solution containing copper sulfate and sulfuric acid (250 and 130 g/L, respectively) was prepared; GrO contents of 0.2, 0.5, and 0.8 g/L were tested. Polyacrilic acid (PAA 5000) was added to the electrolyte in order to avoid graphene oxide agglomeration and precipitation, the amount of surfactant was $50 \times N$ ppm, where N is the GrO concentration in g/L. A two-electrode electrodeposition cell with the anode made from pure copper (6 × 10 cm) and the cathode made from titanium (3 × 10 cm) was used; the electrolyte stirring by air bubbles was used as an additional factor preventing GrO agglomeration/precipitation. The effect of the electrolyte temperature (25, 40, and 55°C) and electric current density (0.5, 10, and 20 A/dm^2) was investigated; the typical film thickness produced in all deposition regimes was ~20 μm.

The deposited films were separated from Ti substrate after the deposition. It was shown that the inner film surface, i.e., surface adjusting to substrate, was very smooth, while the outer surface was rather rough. Analyzing the results of XPS and Raman analysis, the authors claimed partial GrO-to-Gr conversion as a result of electrochemical deposition. The optimal deposition conditions were chosen as 0.5 g/L GrO contents at room temperature and an electric current density of 10–20 A/dm^2. The following mechanical properties were reported for the samples tested: hardness from 2.7 to 4.0 GPa, elastic modulus from 136 to 192 GPa, and tensile strength from 353 to 452 MPa (data obtained for a bath electrolyte temperature of 60°C were excluded).

10.3.2 Deposition of Cu–Gr Composites in a Pulse Regime

Attempts to apply the electrochemical deposition approach for copper–graphene film production were made in [42]. A number of procedures were tested. Conventional electrolyte, widely used for pure copper film deposition (copper sulfate with sulfuric acid), was mixed with pure Gr to prepare the so-called "composite electrolyte", i.e., graphene-containing electrolyte. Discussing the use of pure graphene instead of graphene oxide/reduced graphene oxide, the authors claimed the following advantages. First, the GrO reduction process makes the complete elimination of hydroxyl groups in the final composite impossible; their amount in it cannot be lower than 6 vol.% (see theoretical calculations in [40]). Second, high-temperature composite treatment in the hydrogen ambient is undesirable in a number of electronic applications of the material, e.g., in Micro-Electro-Mechanical Systems (MEMS) production. Thus, pure graphene produced by CVD in the form of powder was added to the electrolyte, and a wide range of Gr content in the electrolyte was tested (0.5, 2, 5, 50, 100, and 300 mg/L). In order to maintain the constant Gr level in the electrolyte during the long-term deposition, polymeric surfactants polyvinylpyrrolidone, sodium dodecyl sulfate, and polyacrilic acid (PVP, SDS, and PAA3000, respectively) were added to the electrolyte.

Electrochemical deposition was performed in both Direct Current (DC) and pulse regimes. Both cathode and anode were made from copper; polished 3×3 cm copper plate with a thickness of 135 μm was used as an anode, while copper foil was used as a cathode. Experiments were carried out in the different electrodeposition cell geometry: a horizontal one with the cathode placed below the anode (i.e., with ion movement along the gravity field lines) and a vertical one (movement in the direction perpendicular to the gravity field). The electric current density was maintained at 10 mA/cm^2 during the 1-h deposition; the average deposition rate obtained was ~10 μm/h. A longer deposition process (20 h) was tested for 2 mg/L Gr-containing electrolyte in order to clarify the possibility of achieving some extra graphene incorporation into the composite matrix. The deposition process here was performed under permanent electrolyte stirring using the magnetic stirrer.

The pulse regime was used to produce Cu–Gr composites from the electrolytes with high Gr content (over 50 mg/L). Since the DC deposition in these conditions provides a very rough composite surface (a large number of bulges were observed at this surface), it was expected that pulse regime will improve the surface quality. The following electric current parameters were tested: on-time = 0.1 ms, off-time = 0.9 ms, and current density = 2 mA/cm^2. Ultrasonic stirring was used in the pulse regime experiments to maintain the constant graphene contents in the electrolyte.

Analyzing the results reported in [42], one could state the following important conclusions. First, the attempts to increase the Gr distribution uniformity in the composite via surfactant application failed. Regarding the results of the 6-h-long deposition, it can be concluded that there is no evident difference between the quality of the films deposited from electrolytes containing PVP, SDS, and PAA3000 surfactants and from the electrolyte without any surfactant. Further increase in the deposition duration from 2 to 5 h resulted in lower uniformity; electrolyte free of surfactants and PVP-containing electrolyte provided a better composite quality here. Evident nonuniformity was shown for samples manufactured at a process duration longer than 5 h. Thus, the authors claimed that the use of surfactants gives no advantages for the graphene distribution in the electrolyte/final composite.

The second important statement of the authors is that the DC regime in the described cell geometry and deposition parameters is not effective for Cu–Gr composite production. Indeed, Raman analysis indicated the presence of peaks attributed to graphene for the extra-long deposition (>20 h); shorter processes resulted in the graphene peak intensity below the sensitivity limit of the method. Attempts to increase the graphene contents in the electrolyte led to the significant surface roughening (bulge formation). Pulse regime provided better results: it made the use of electrolytes with high graphene contents possible; samples produced were characterized by smooth and bright surfaces; graphene incorporation into the metal matrix of the composite was enough to register Raman spectra identical to that of pure graphene (see Figure 10.7).

Analyzing the results obtained in [42], the authors recommended an optimal deposition regime (see [43]). The pulse regime has the following parameters: electric current density = 10 mA/cm^2; on- an off-times of 0.2 and 0.4 ms, respectively; the Gr content in the electrolyte of 300 mg/L was chosen to produce samples for sample fabrication. Both magnetic and ultrasonic stirring were tested; it was shown that ultrasonic stirring provides better composite surface quality. In particular, surface roughness was more than 150 times lower in the case of ultrasound application (0.054 µm vs. 9.5 µm). SEM characterization demonstrated that the average linear dimensions of the graphene flakes incorporated into the copper matrix (they can be estimated as 2–5 µm) were 5–15 times higher than those typical for copper grains. The specific form of the Raman spectrum (see Figure 10.8) should be mentioned.

As seen from this figure, D and 2D peak intensities are negligible in the case of Cu–Gr films. The authors assumed the absence of the D peak to the low defects level in the graphene incorporated into the composite due to the low defect level in the initial graphene produced by the CVD method and due to the specific features of the deposition process. The absence of the 2D peak was considered as resulting from the multilayered graphene flakes. In addition, it is stated that the 2D peak is an overtone of the D peak; hence, the elimination of the initial D peak should eliminate the 2D peak also. Surely, these results require further discussion.

Electric and thermal conductivity tests of the samples produced showed promising results. In contrast to results reported in [34], the electric resistivity of the Cu–Gr composite was better than that obtained for pure electrodeposited copper (1.66 × 10^{-8} Wm vs. 1.78 × 10^{-8} Wm, respectively); moreover, this value was quite close to that reported in [34] for metallurgical copper foil. The same dependence was shown for the thermal conductivity:

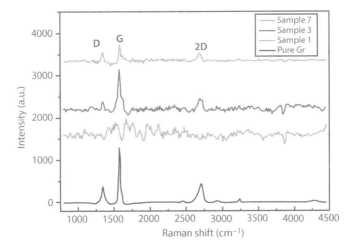

Figure 10.7 Raman spectra of Cu–Gr samples synthesized in [42] in a pulse regime; the spectrum for Sample 7 (the upper one) is similar to that of pure graphene (the bottom one). Reprinted with permission from [42].

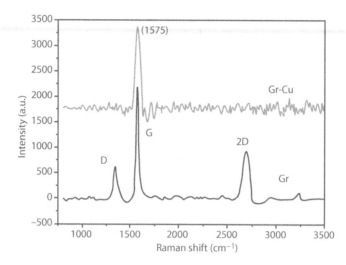

Figure 10.8 Comparison of the Cu–Gr Raman spectrum of the sample produced in pulse electrodeposition regime in [43] with that of pure graphene. Reprinted with permission from [43].

the values measured for the Cu–Gr composite were 5% higher than that for pure electrodeposited copper (300.5 W/mK vs. 286.5 W/mK, respectively). However, the thermal conductivity of the metallurgical copper foil was reported to be much higher—400 W/mK.

The mechanical properties of the composites were tested using a dynamic mechanical analyzer under tension mode. It was proven that the Gr addition significantly increases the mechanical properties of the material: ~15% increase in the Young modulus, ~40% increase in the yield strength, and ~17% increase in the breaking stress were reported. Surely, the downside of these improvements was the decrease in the material elongation. A summary of the mechanical test results is presented in Table 10.3.

The reverse pulse electrodeposition method was used in [44]; the procedure was as follows. GrO was prepared by electrochemical graphite exfoliation in a nitric-acid-based electrolyte; the obtained GrO platelets were washed using ultrasound, centrifugated, and dried at 60°C. The Cu–GrO composite was produced in the electrodeposition cell with a Cu_2SO_4-based electrolyte; pH regulation up to ~1 was done using sulfuric acid. A number of surfactants were tested in order to avoid GrO agglomeration and precipitation during

Table 10.3 The increase in the mechanical properties of the Cu–Gr composite vs. pure electrodeposited Cu (data from [43]).

Sample	Property (averaged values, estimated uncertainty ≤3%)			
	Young modulus, GPa	Yield strength, MPa	Breaking strength, MPa	Elongation, %
Pure electrodeposited Cu	70.4	174.1	319.2	14.4
Cu–Gr composite deposited under the same conditions	82.5	242.2	386.7	2.0

the deposition process. It was stated that the use of sodium dodecyl sulfate (SDS) and cetrimonium bromide (CTAB) was not effective; moreover, their addition resulted in GrO precipitation in powder form. In contrast, the addition of polyacrilic acid (PAA, 25 ppm per 0.5 g/L GrO) showed a positive result. 0.1–1 g/L range of GrO content in the electrolyte was tested and 0.5 g/L of GrO was shown to be an optimal concentration. Copper anode and Ti cathode were used. At the first deposition step, a pure copper layer of 2 μm thickness was grown on the cathode in order to simplify further Cu–CrO film separation from the titanium substrate. GrO was placed in the electrolyte and treated by ultrasound for 3 h. This procedure provided the uniformity of GrO distribution in the electrolyte. However, the treatment time seems to be quite insufficient to provide a proper exfoliation level; note that the duration of a similar process in [34] was tens of hours. DC deposition regime was carried out at 0.025 A/cm² electric current density. The parameters of the reversed pulse deposition are summarized in Table 10.4. The duration of both regimes was chosen in order to produce ~30 μm film thickness. The deposited films (both pure Cu and Cu–GrO) were heat treated in a neutral (Ar) atmosphere at 400°C for 30 min.

The experimental data on as-synthesized pure Cu and Cu–GrO foils can be summarized as follows. Analyzing TEM images, the authors concluded that the exfoliating procedure for graphene oxide resulted in GO flakes of one to five layers with typical linear dimensions 0.5–1 μm. SEM images of the polished Cu–Gr surface demonstrated the uniformity of the Gr distribution into the metal matrix of the composite; the average distance between them was ~0.8–1.2 μm. Note that this value is close to the typical grain size in the material (1.3 ± 0.3 μm and 1.2 ± 0.4 μm for pure Cu and Cu–GrO composite, respectively), so one can expect graphene distribution along the grain boundaries; this fact can be very interesting for the development of the materials with required properties. XRD analysis did not prove the presence of Gr/GrO; the authors attributed this fact to the total carbon phase content below the sensitivity limit of the method. Raman spectra obtained are characterized by rather high D peak intensity (Figure 10.9); the authors assumed it as a result of stress/strain

Table 10.4 Parameters of the reversed pulse deposition regime in [44].

	Electric current density, A/cm²	**On-time, ms**	**Off-time, ms**
Forward pulse	0.05–0.2	15–50	50–100
Reverse pulse	0.005–0.15	1–10	1–10

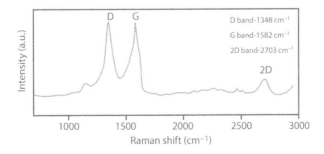

Figure 10.9 Raman spectra of the Cu–Gr composite. Adapted from [44].

at the Cu–grain/Gr–GrO interface; such an assumption correlates with the general tendency of the D-peak increase as a result of defect formation.

It was stated that annealing in a neutral (Ar) atmosphere should reduce graphene oxide to individual graphene. The following important result should be mentioned here: annealing at elevated temperature resulted in the significant grain growth in the case of pure copper films (up to 10 μm). However, the presence of graphene at the grain boundaries in the case of the Cu–Gr composite suppressed grain growth; their typical dimensions were similar to those determined before annealing.

Hardness and elastic modulus of pure Cu films were compared with those of Cu–Gr films electrodeposited in DC and pulse regimes; a nanoindentation approach was used for this task (see results summarized in Table 10.5). As seen from the table, annealing generally decreases both hardness and elastic modulus; this tendency was shown for Cu films deposited in the DC regime and for all graphene-containing composites. Surprisingly, the opposite hardness behavior was observed for copper films deposited in a pulse regime: a slight increase of ~5% was registered. The deposition regime seems to provide similar results for both pure Cu and Cu–GrO films before annealing; the data on sample hardness and elasticity modulus are quite similar. In contrast, the samples deposited in a pulse regime demonstrated higher mechanical properties after annealing.

Thus, the significant increase in hardness (>90%) and elasticity (> 3%) resulted from graphene reinforcement of the material.

As for the electric resistivity, the data obtained by Pavithra *et al.* appeared to be somewhat higher than those reported in the above papers by K. Jagannadham and G. Huang: 3.4 mΩ cm and 2.3 mΩ cm for the initial Cu–GrO–Gr foil and final Cu–Gr composites, respectively [36, 39, 42–44].

A number of attempts to manufacture Cu–GrO–Gr composites using electrochemical deposition were also undertaken in [45]. Commercial graphene oxide was used as a GrO source in the electrodeposition process in [45]; its aqueous solution (0.1 and 0.5 mg/L) was added to a copper sulfate solution (0.005–0.5 M) with a volume ratio of 1:1, and the final

Table 10.5 Mechanical properties of Cu–Gr composites in comparison with the data for pure Cu (results from [44]).

Sample	Deposition regime	Hardness, GPa	Elastic modulus, GPa
Pure Cu (before annealing)	DC	1.53	115
	Pulse	1.5	117
Pure Cu (after annealing)	DC	1.4	93
	Pulse	1.58	110
Cu–Gr–GrO composites (before annealing)	DC	2.3	127
	Pulse	2.35	132
Cu–Gr composites (after annealing)	DC	2.0	100
	Pulse	2.12	125

pH value of the electrolytes varied from 4 to 5.5. Copper foil (ground, rinsed, and dried) was used as a working electrode in a three-electrode electrochemical cell (Ag/AgCl as a reference electrode and Pt mesh as a counting one). Some reduction of the graphene oxide during the deposition process can be seen from the comparison of Raman spectra of pure graphene oxide and reduced GrO in the composite (see Figure 10.10). Indeed, the ratio of D and G peak intensities significantly changed after the deposition process. Surely, this can be assumed as a result of the graphene oxide reduction during the electrodeposition; however, the alternative idea of the D peak intensity increase following the increase in a number of stress/strain/defects at the grain interfaces in the multiphase material should also be considered. Discussing the results of high-resolution XPS data, the authors estimated the content of carbon in its reduced form (C–C bond) at ~85 at.%. The results of the electrical resistivity measurements are presented in Table 10.6; one can see some evident difference with the results reported in the above-described works.

AFM analysis showed the presence of relatively small Cu particles (~80 nm in diameter) uniformly distributed between the micron-sized regions (with the typical linear size in the range 0.4–1.3 μm); the authors reported that both Cu and reduced GrO areas could be considered as having these typical dimensions.

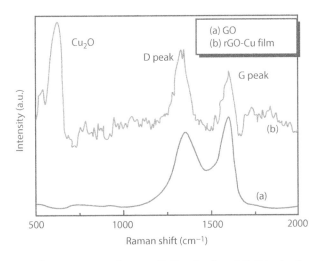

Figure 10.10 Comparison of Raman spectra for pure GrO and reduced GrO in the deposited Cu–GrO-Gr composite. Reprinted with permission from [45].

Table 10.6 Electrical resistivity results from [45].

Sample	Electrical resistivity, $m\Omega \cdot m$
Commercial copper foil	35.4 ± 1.9
Electrodeposited copper film	38.4 ± 1.3
Electrodeposited Cu–GrO–Gr composites	30.15–34.8*

*Data for different deposition conditions.

10.3.3 Electrochemical Deposition of Nanotwinned Copper–Graphene Composites

Papers of V. Konakov et al. [46–52] report electrochemical deposition of nanotwinned copper–graphene composites. The following advantages of these composites were mentioned. First, nanotwinned metals including copper demonstrated high mechanic properties, see, e.g., a review [53] and experimental works of L. Lu [54–57]. Second, the use of nanosized grains in a metal matrix provides an increase in the grain boundaries' surface, giving the opportunity to increase graphene incorporation in the form of single few-layered flakes and eliminate graphene–graphite regions in the manufactured composite.

The detailed results of these studies can be found in a review [52]; they could be briefly described as follows. Both direct current and pulse electrodeposition regimes were tested; DC was finally chosen for sample production. Copper sulfate ($CuSO_4$) ethanol aqueous solution was used as sulfur electrolyte (1 M $CuSO_4$*$6H_2O$, acidified by H_2SO_4 up to pH 1; ethanol content 37.5 mL/L); a two-electrode electrochemical deposition cell (coplanar electrodes 25 × 20 × 5 mm X10CrNi18-8, i.e., SAE grade 301 stainless steel cathode and 100 × 80 × 5 mm copper anode with 30 mm distance between them) was taken as a basic experimental setup. All composites produced were washed in distilled water, dried by ethanol, and then removed from the substrate. A number of graphene sources were tested. As it is known, micromechanical splitting of exfoliated graphite [58] is an effective pathway for graphite-to-graphene conversion. Thus, the first Gr source was a commercial exfoliated graphite subjected to micromechanical splitting by ball milling with mechanical activation (Pulversette 6 FRITSCH planetary ball mill, 400 rpm). Another graphene source used was the graphene–graphite mixture purchased from Active-NanoCo. (Russia). Graphene source was mixed with distilled water up to suspension state; this suspension was added to a copper electrolyte in the required amount.

A set of preliminary experiments was performed; they manifested two general problems of the experimental approach. The first one was the stability of the graphene content during the long-term deposition procedure; this problem was previously mentioned in [45]. Indeed, graphene precipitation and agglomeration occur in the deposition cell; as was shown in a number of previous works, these processes cannot be completely eliminated by mechanical or ultrasound stirring. Some recent patents suggested the use of neutral gas bubbling [58] in addition to traditional stirring. Graphene precipitation causes some deviations in graphene content from the initial one, resulting in nonuniformity of Cu–Gr composition, while graphene agglomeration gives rise to the formation of additional growth centers on the substrate surface; these centers differ in dimensions from those typical to pure copper deposition. As a result, the structure of the composite metal matrix is disturbed and the required level of twinning cannot be reached. In addition, graphene agglomerates also incorporate into the metal matrix, forming significant carbon regions. The authors of [46–52] considered the use of nonionic surfactants (Pluronic F127 and polyacrilic acid) as an effective way to overcome this problem. These surfactants were used to prepare graphene suspensions, and a number of concentrations (25, 50, and 100 ppm in the final electrolyte) were tested. It was shown that the use of surfactants provides graphene precipitation and agglomeration in the case of deposition processes with some hours' duration, resulting in a good level of material uniformity. In addition, varying the combination of the electrolyte type and graphene-to-surfactant ratio, it was possible to control the grain size distribution in the metal matrix and the level of its twinning.

The second problem was to maintain the proper nanotwinning of the composite metal matrix. The deposition conditions chosen in [46] provided the required nanotwinning in the case of pure copper; however, the presence of additional carbon growth centers at the very first step of composite deposition made the task of nanotwinned metal matrix production quite complicated. Hence, the use of two-step deposition was suggested. At the first step, a relatively thin layer of nanotwinned copper was deposited on the cathode (a similar idea is widely used in CVD techniques, e.g., for AlGaN material production); the graphene-free electrolyte was used at this step. At the second step, the deposition was performed using a graphene-containing electrolyte and the composite with required thickness was produced. The results reported in [50–52] proved the positive effect of this procedure application.

Let us discuss the principal results of [46–52] in more detail.

The study performed using Raman spectroscopy (see Figure 10.11) proved the carbon phase incorporation into the metal matrix in the form of graphene.

Following the recommendations of [59], the number of layers in the graphene flakes could be estimated from the ratio of G and 2D peak intensities (1550 and 2880 cm^{-1}, respectively); the authors of [50–52] concluded that this number was four to six in the samples studied. Note that this result was typical for samples produced using both split exfoliated graphite and commercial graphene–graphite mixture. Some data obtained for materials deposited from electrolytes prepared with Pluronic F127 surfactant showed that the increase in the graphene and surfactant content could increase the relative number of mono- and bilayered graphene flakes in the resulting foils.

Figure 10.12 compares the electron back-scattering diffraction data (EBSD) obtained for pure nanotwinned copper and for nanotwinned Cu–Gr composites. As seen from the figure, the fraction of 60° misorientation angles (usually treated as twinned boundary type indication) in the case of pure nanotwinned copper was ~50%. The use of the approach suggested in [46–52] resulted in the possibility of controlling the nanotwinning level depending on the graphene-to-surfactant ratio and the type of surfactant. A similar conclusion could be made from the analysis of grain size distributions (see [49–52]).

Some results on the mechanical properties of nanotwinned copper–graphene composites are shown in Figure 10.13. As seen from this figure, the results obtained for composite microhardness are generally similar to those reported for pure nanotwinned copper (see [54–57]). However, some compositions demonstrated the unique microhardness of ~3 GPa (data taken for the layers adjacent to the cathode substrate).

Summarizing the results reported in [46–52], one can conclude that the application of the suggested technique (two-step deposition with buffer layer coupled with the use of nonionic surfactants to maintain the graphene contents uniformity) gives the opportunity to increase the quality of copper–graphene composites. Deposition of the thin buffer layer from a graphene-free electrolyte led to the absence of additional growth centers on the deposition surface; it provided nanotwinned copper growth that is further inherited by a composite metal matrix. Surfactants stabilize graphene concentration in the electrolyte during long-term deposition; it was shown that variations in graphene-to-surfactant ratio and the type of surfactant provide the possibility of controlling composite microstructure and their mechanical properties.

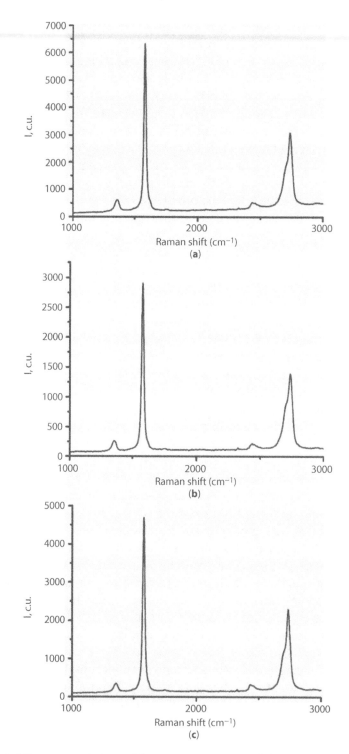

Figure 10.11 Typical Raman spectra obtained for Cu–Gr composites in [52]. Reprinted with permission from [52].

FABRICATION OF CU-GR COMPOSITES 307

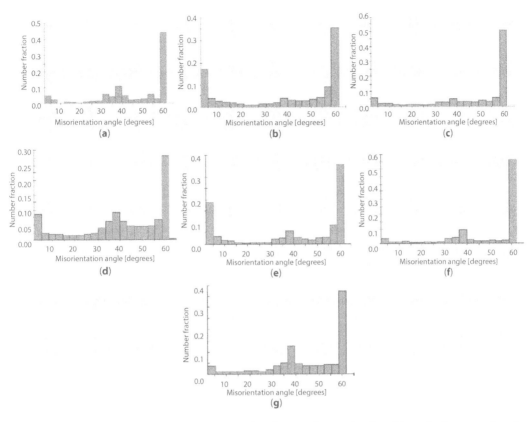

Figure 10.12 Comparison of misorientation angle distributions from EBSD data: (a) results for nanotwinned Cu, (b–g) nanotwinned copper–graphene composites. Reprinted with permission from [49].

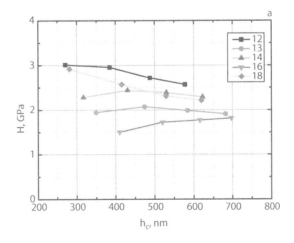

Figure 10.13 Microhardness test results obtained for the inner surface of nanotwinned copper–graphene films. Reprinted with permission from [52].

10.4 Electroless Deposition

Electroless deposition (or electroless plating) is a versatile platform for the synthesis of metal-based composites [60, 61]. The technique is based on the difference in the redox potential between metallic substrate (Cu or Zn foil) and the metal ions in the corresponding precursor (M^{m+}/M), as well as the high conductivity of graphene platelets. Since the reducing electrons are derived from the metallic substrate, the synthesis is usually carried out in the absence of an external reducing agent. The schematic illustration of metal–graphene composite fabrication via electroless deposition is presented in Figure 10.14.

Metal foil (Cu or Zn, with typical dimensions 1 × 2 cm) is immersed into the solution, containing the graphene platelets. Then, the foil is slowly removed from the suspension. After drying under Ar atmosphere, the foil containing Gr (substrate) is immersed into the corresponding precursor (metal ion solution, such as H_2PtCl_4, $HAuCl_4$, H_2PdCl_4, and $AgNO_3$), having a higher redox potential than that of Cu or Zn, for different periods of time. Finally, metal decorated by graphene is prepared. Layered metal–graphene composites can be obtained by repeating the processes of graphene coating and metal depositing in a similar way. The produced metal nanoparticles are characterized by clean surfaces and large surface area. This type of composites is promising for catalysis in chemical reaction. As it is mentioned in [60], the size and density of the metal nanoparticles on the graphene surface can be controlled by optimization of the experimental conditions. When it comes to copper–graphene composite fabrication, Zn foil should be used as a substrate to ensure the necessary redox potential. If the Zn foil is used and the reaction time is 10 s, Cu nanoparticles with a diameter of ~50 nm are deposited homogeneously on the surface of graphene. The use of graphene oxide is unfavorable for electroless deposition. It can result in lower yield of metal nanoparticles and their inhomogeneous distribution on substrate surface.

The described strategy has been extended to manufacture the hybrid metal–graphene composite powders. For instance, electroless silver coating on graphene and copper powder was used to enhance the interface bonding between graphene and copper matrix [62]. In order to improve the adhesion between the graphene and silver, the graphene surface was modified and sensitized in aqueous $SnCl_2$ solution and then in $PdCl_2$ solution (see sample "after treatment" in Figure 10.15). Then, silver-coated graphene and silver-coated copper powders were prepared successfully by electroless deposition [62]. The Gr–Cu composites were prepared by ball milling and the cold compacting sintering process (see Section 10.1 for details). Brinell hardness of the obtained composites is presented in Figure 10.15.

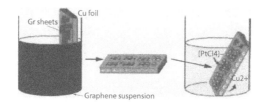

Figure 10.14 Schematic illustration of electroless deposition process powders. Reprinted with permission from [60].

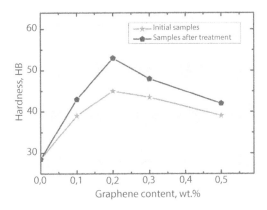

Figure 10.15 Brinell hardness of Cu–Gr composites, obtained through electroless deposition. Adapted from [62].

As it is seen from the figure, graphene efficiently reinforces copper matrix. The maximum value of hardness is observed at 0.2 wt.% graphene content. It is higher in the case of the treated sample, i.e., 53.4 HB. Thus, the addition of silver-plated graphene and silver-plated copper powder is effective for additional composite reinforcement. Further increase of Gr content results in the decrease of composite hardness. The electrical conductivity of composites, produced from treated and untreated powders, decreases continuously to 63%–67% IACS at 0.5 wt.% graphene content. However, the electrical properties of treated composites are better than those for samples produced without application of the electroless deposition approach. The same tendency is observed in the case of relative density, antioxidant, and arc-erosion properties. The correlation of all the properties suggests stronger interface bonding between the graphene and the copper matrix introduced via electroless deposition. However, the application of such composites seems to be limited due to the fact that the third component, i.e., silver, is easily oxidized by air oxygen.

10.5 Molecular-Level Mixing (MLM) Technique

Molecular-level mixing (MLM) is a potential method to provide sufficient interfacial bonding between carbon additive and copper. The molecular-level mixing process consists of attaching functional groups onto carbon additives in the aqueous solution of GO, RGO, or Gr and Cu acetate. This approach provides chemical bonding between the carbon phase and the composite matrix. Copper matrix composites reinforced by CNT, RGO, and graphene are successfully fabricated via MLM [12, 14, 63]. A schematic diagram of the fabrication process is shown in Figure 10.16 for RGO–Cu nanocomposite powder. First, graphene oxide (GO) and soluble copper salt (typically copper acetate) are homogeneously mixed in deionized water (Figure 10.16, steps b and c); at this step, the negatively charged RGO surface could attract Cu^{2+} in the solution. Thus, chemical bonds are formed between the functional groups of the GO flakes and the Cu ions. Then, the mixture of GO and Cu ions is oxidized to GO–CuO nanocomposite powder by adding NaOH solution. It prevents GO reduction before forming chemical bonds with Cu^{2+} (see Figure 10.16, step d). However, NaOH might rapidly reduce both Cu^{2+} ions and GO upon heating. This may create an effect opposite to chemical bonding [64].

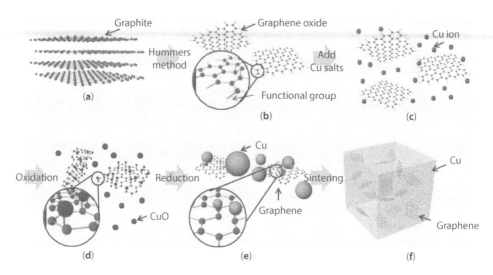

Figure 10.16 Schematic illustration of the fabrication process of the RGO/Cu nanocomposite. (a) Pristine graphite. (b) Graphene oxide fabricated by the Hummer's method. (c) Dispersion of Cu salt in graphene oxide solution. (d) Oxidation of Cu ions to Cu-oxide on graphene oxide. (e) Reduction of Cu-oxide and graphene oxide. (f) Sintered RGO/Cu nanocomposite powders. Reprinted with permission from [63].

Thermal reduction of powder mixture by H_2 results in RGO flakes decorated with metallic copper particles (see Figure 10.16, step e). On the final step, the composite powders are sintered and densified (Figure 10.16, step f). The described technique allows homogeneous dispersion of graphene in the Cu matrix. Since no high temperatures or milling is implied during the synthesis of the graphene/metal nanocomposite, there is no thermal or mechanical damage of graphene flakes. Once metal particles are attached to the carbon filler, no further agglomeration of individual graphene, reduced graphene oxide flakes, or CNTs is possible.

The stress–strain curves for bulk RGO/Cu nanocomposites with different RGO content fabricated by MLM and spark plasma sintering in [63] and pure copper are presented in Figure 10.17.

As it is seen, the tensile strength of the 2.5 vol.% RGO/Cu nanocomposite (≈335 MPa) is about 30% higher than that of pure Cu (≈255 MPa). The elastic modulus and yield strength

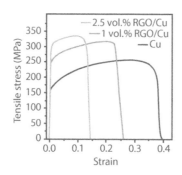

Figure 10.17 Stress–strain curves for bulk RGO/Cu nanocomposites and pure copper. Reprinted with permission from [63].

of the composite increased by ≈30% (from 102 to 131 GPa) and by ≈80% (from 160 to 284 MPa), respectively, compared to bulk Cu. According to [63], remarkable mechanical properties of the composite can be explained by the high load-transfer efficiency of RGO in the copper matrix, i.e., strong bonding between Cu and reduced graphene oxide mediated by oxygen. Indeed, adhesion energy between graphene and Cu in the sintered composite measured by double cantilever beam (DCB) test (164 J m^{-2}) is much stronger than the adhesion energy of 0.72 J m^{-2} for as-grown graphene on a Cu substrate. It is believed that the successful application of the molecular-level mixing process results in the strong adhesion energy between graphene (~200 times enhancement of adhesion energy). Even though the measured adhesion energy does not exceed Cu/Cu adhesion energy, the combined effects of dislocation blocking and pinning even by a single layer of graphene could explain the strengthening effects of graphene in the Cu matrix [63, 65]. The electrical and thermal conductivity of RGO–Cu composite, produced by MLM, are similar to the properties of pure Cu [63]. The introduction of high-shear mixing in the MLM process improves the homogenization level of graphene flakes in the mixture. It results in a compressive strength of 2.4 vol.% RGO–Cu composite increment to 501 MPa [66].

The graphene–copper composite with a micro-layered structure and excellent tensile properties was fabricated in [67] via the MLM method at 45°C with rotor–stator mixing. Here, Cu(OH)$_2$ nanorods were *in situ* produced on the surface of graphene oxide (GO) sheets, which serve as excellent supporters. A roughly plane structure was formed. Dehydrated by drying at 110°C, CuO nanorods could form a composite with graphene oxide (GO) sheets with a layered structure. Van der Waals forces and hydrogen bonding bring the composite sheets together into a self-assembled micro-layered structure on the mixing step. Further, the micro-layered structure is retained during the reduction process. The tensile strength and compressive strength of micro-layered RGO–Cu composites are shown in Figure 10.18.

As seen from the figure, the ultimate tensile strength of 2.5 vol.% RGO–Cu and 5 vol.% bulk composites are 524 and 608 MPa, respectively, which is more than two times higher compared to pure Cu (255 MPa [63]). Gradual transition between elastic and plastic deformation is observed in Figure 10.18a, which suggests strain hardening occurring at the initial stage of plastic deformation. The obvious strain hardening in composites may be interpreted in terms of glide dislocation interaction with the interface between graphene and the Cu matrix. Specifically, dislocations are generated in the Cu matrix and they glide

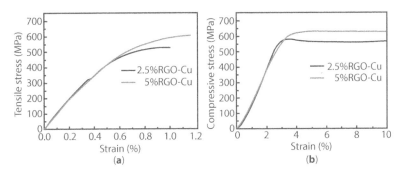

Figure 10.18 (a) Tensile and (b) compressive strength of RGO–Cu composites. Reprinted with permission from [67].

to the interface between graphene and the Cu matrix. The reinforcement mechanism is due to the fact that graphene efficiently blocks the dislocation motion further due to its high strength and elastic modulus, which leads to dislocation accumulation at the interface. The compressive strength of 2.5 vol.% RGO–Cu and 5 vol.% RGO–Cu composites is 576 and 630 MPa, respectively. Note that the tensile strength and the compressive strength of RGO–Cu composites, measured in the same direction of anisotropic material, are almost equal. Since the tensile strength is sensitive to macro defects, it can be concluded that the composites have the uniform structure of layers with less macro defects such as cracks and holds.

Single-layered graphene and Gr derivatives (GNPs, GO, RGO, etc.) show a different strengthening effect on the Cu matrix due to their different structure. Indeed, reduced graphene oxide is characterized by a significant number of structural defects and by the presence of residual groups on its surface; thus, the intrinsic strength of the graphene layer should be decreased [14, 68]. The structure of GNPs exhibits few defects and larger thickness; hence, the thermal stability of GNPs is higher.

The mechanical properties of GNPs and RGO-doped Cu matrix composites, synthesized by the modified MLM process [14], are shown in Figure 10.19.

As seen from Figure 10.19, the strength of GNP–Cu and RGO–Cu composites show a different tendency toward strengthening. In the case of GNP–Cu composites, the local maximum is observed at 0.1 vol.% of GNPs and then the composite strength decreases with a further increase in GNPs amount to 1.0 vol.%. At the same time, the strength of RGO–Cu composites increases gradually with an increase in RGO from 0.05 to 1.0 vol.%. According to [14], the observed difference might be attributed to more homogeneous distribution, higher structural integrity, and interfacial bonding of small GNP amounts as compared to the same amount of RGO addition. Upon the increase of the additive volume fraction, agglomerates of graphene nanoplatelets are formed in the composite. The aggregation causes higher porosity and, hence, the decrease of the mechanical performance of the composites. Compared with GNPs, RGO disperses better at high reinforcing additive content ($\geq$ 0.5 vol.%). It is mainly due to many hydrophilic functional groups on its surface. Interfacial bonding and adhesion energy between RGO and Cu in the composites prepared by MLM are stronger than those of GNP–Cu composites [14, 63].

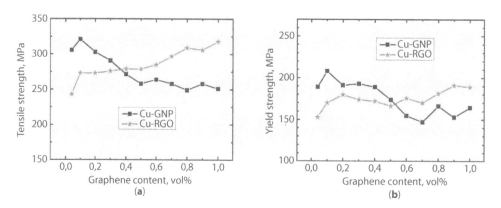

Figure 10.19 (a) Tensile and (b) yield strength of Cu matrix composites doped by RGO and GNPs. Reprinted with permission from [14].

10.6 Chemical Vapor Deposition (CVD) Technique

Chemical vapor deposition (CVD) is a reliable technological process for fabricating large-area graphene layers on transition metals. Copper and nickel are the most common substrates used for that. Comparing experimental results on graphene growth on different metals with computed data, M. Losurdo *et al.* showed that the Cu-catalyzed process differs from growth on other metals [69]. Hydrogen also plays an important role in graphene growth kinetics, slowing down the deposition kinetics of graphene on Cu due to site blocking on the Cu surface. Since the CVD growth process is mostly surface controlled, the thickness and defectiveness of graphene layer can be well adjusted. The use of a polycrystalline Cu substrate can allow even a monolayer of graphene. In terms of interfacial bonding, the use of monolayer is also important to measure the interfacial bond strength between graphene and Cu, instead of the bond strength between graphene layers [63].

Besides a layer of Gr grown on a substrate, reinforced bulk copper–graphene composites with high thermal conductivity and reinforced multilayered copper–graphene composite films consisting of alternating layers of copper and graphene monolayer can be successfully fabricated through the CVD technique [13, 30]. The use of CVD graphene enables precise control of graphene layer thickness, which is not achievable by other techniques. The multilayered copper–graphene composite fabrication is rather close to the original technique developed for CVD synthesis of graphene [13]. As seen from the schematic illustration shown in Figure 10.20, graphene is first grown using CVD and then transferred to the metal thin film substrate via the support layer. The layer is then removed, and the next metal thin film layer is deposited. Cu–graphene nanolayered composites with different repeated metal

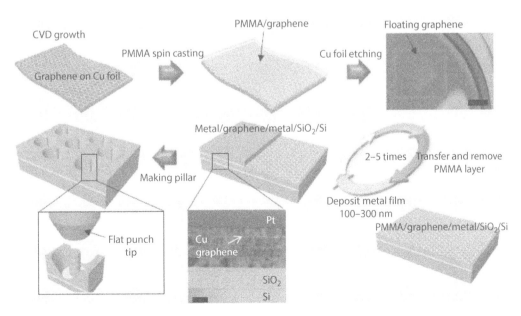

Figure 10.20 Schematic of metal–graphene multilayer system synthesis. The scale bar for the floating graphene is 10 μm and that for the TEM is 200 nm. Reprinted with permission from [13].

thicknesses of 70, 125, and 200 nm were synthesized in [13] using this approach, i.e., repeating the metal deposition and graphene transfer processes.

The highest strength observed for the composites with a layered design was 1.5 GPa at the smallest Cu layer spacing of 70 nm. This value is several times more than those for bulk single crystal of copper (580 MPa [13, 70]). Nanolayered copper–graphene composites demonstrated a Hall–Petch-like behavior at a length scale greater than 100 nm [70–72]. In other words, the strengthening mechanism suggested by Kim *et al.* is as follows: multiple dislocations pile up at the interface and eventually propagate through the interface when a critical shear stress is applied. The critical event in the case of metal–graphene nanolayered composite would be the activation of complex slip systems at high stresses and/or the piled-up dislocations escaping through the free surface due to interfacial shear because of the extreme difficulty in shearing through the graphene layer. Comparison of mechanical properties with pure Cu thin film synthesized using the same procedure as that used for the Cu–graphene multilayered composite with a 100-nm thickness confirms the mechanism above. The flow stress at 5% plastic strain of the pure Cu foil was 600 MPa against 1.5 GPa for the Cu–graphene composite.

The oxidation stability of copper and copper–graphene composite foils obtained by CVD (graphene deposited on high-quality copper surface) after the exposure in the air ambient from 1 month to 1.5 years at low and high humidity differs significantly [32]. Figure 10.21 presents XPS spectra of graphene–copper composites after exposure in oxygen at low- and high-humidity environments.

As seen from Figure 10.21, graphene coating protects Cu surface from oxidation under long-term exposure. The XPS spectra of the initial sample of pure Cu metal and the spectra of graphene–Cu composites are quite similar. At the same time, the XPS spectrum of Cu foil after the exposure in the air ambient at room temperature appears to be closer to that of Cu_2O. Note that the higher-energy fine structure of Cu_2O and CuO (features c, d, and e) is not seen in the spectra of graphene–Cu composites. Thus, one can suggest that the Cu substrate is not oxidized. For evaluation of composite stability, the experimental results in [32]

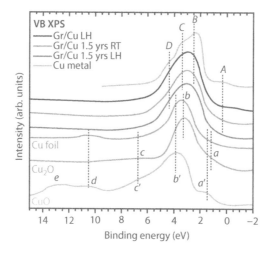

Figure 10.21 XPS valence band spectra of the copper-graphene composite and the reference samples (Cu foil after exposure to ambient air, Cu metal, CuO, and Cu_2O). Reprinted with permission from [32].

were compared with the density functional theory (DFT) calculations. A two-step model was proposed. The first step is the oxidation of the graphene, and the second is the perforation of graphene with the removal of carbon atoms as a part of the carbon dioxide molecules. Modeling results coupled with the experimental measurements provided evidence that defect-free graphene grown on high-quality copper substrate is rather stable in time (1.5 years). However, the stability of this interface depends on the number of defects in the graphene and substrate. The coating of graphene monolayer is unsuitable for application in industry as an anti-corrosion cover of metallic surfaces because only a high-quality defect-free graphene cover provides sufficient protection of the metallic substrate from oxidation [5]. The presence of defects and impurities, which is unavoidable for large-scale industrial production, significantly decreases the protective properties of graphene. According to [32], bi- and multilayered graphene can be used for that.

The fabrication of reinforced bulk copper–graphene composites by CVD involves the additional steps of compacting and sintering. At the first step, copper powders are covered by a graphene layer via CVD. Then, powders are mixed, compacted, and sintered. Here, the thickness and quality of graphene layer on the copper powder, as well as the choice of sintering method, are critical for dense composite manufacturing. The effect of sintering technique and type of carbon additive on the heat conductivity of bulk copper matrix composites was examined in [30]. Cold pressing (pressure of 15 MPa, sintering in an atmosphere of dry hydrogen at 1030°C), hot isostatic pressing (pressure of 30 MPa, temperature of 1000°C, time 30 min, in an argon atmosphere), and spark plasma synthesis (pressure of 50 MPa, a vacuum of 10^{-4} hPa, temperature of 950°C, time 15 min) techniques were examined. The composites synthesized by the above-discussed SPS approach demonstrated the highest material density of 99.8% which is comparable with the best results ever obtained. Thermal conductivity values higher than the ones for pure copper were obtained for graphene-coated Cu powders obtained by CVD and sintered by SPS (see G for graphene in Figure 10.22).

As seen from Figure 10.22, thermal conductivity values of copper matrix composites produced from the CVD graphene-coated powders and sintered by SPS are about 10% higher than the values registered for pure copper in all investigated temperature range. According

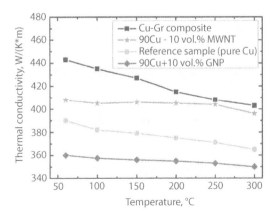

Figure 10.22 Thermal conductivity of copper matrix composite materials, obtained by CVD and SPS. G corresponds to graphene additive, MWNT denotes multiwalled nanotubes, GNP indicates graphene nanopowder. Adapted from [30].

to theoretical estimation performed in [8, 30], 1 vol.% of graphene should improve the thermal conductivity of the composites by 10% relative to pure copper. However, the thickness of graphene coating and the microstructural features in the composite manufactured from copper coated by CVD graphene were not revealed in the research. Thus, the amount of graphene introduced into the copper matrix is not known. High thermal conductivity values of composites can be understood from the positions of low porosity, high-quality structure of graphene obtained by CVD (i.e., less number of defects in the Gr coating), and the possibility of obtaining a monolayer of Gr [30]. Unfortunately, few works on thermal conductivity or mechanical properties of bulk CVD graphene-doped Cu matrix composites are known to date, which makes comprehensive analysis of the graphene effect on the thermal conductivity rather complicated. However, it can be assumed that graphene monolayer coating with the lowest amount of defects could show the highest result in terms of composite properties.

10.7 Functionalization of Copper Powder Surface

Functionalization of the surface of Cu powder can be regarded as an alternative way to enhance copper–graphene bonding. One way of keeping copper from oxidizing is the use of 2,2′-bithiophene to form self-assembled monolayers (SAMs). SAMs not only act as an antioxidizing agent but also provide active sites for preparing covalent bond with functional groups of reduced GO (rGO). A simple method to attach SAMs of 2,2-bithiophene on Cu particle surface has been recently suggested by Hwang et al. [73]. First, Cu powders are immersed in a concentrated HCl solution (5 M) to remove oxide layer. Then, powders are filtered and subsequently rinsed by acetone and ethanol prior to chemical modification. These Cu powders were then treated in an ethanol-based 2,2-bithiophene (0.25 M) solution for 5 days at 50°C under a nitrogen atmosphere with permanent stirring to create Cu–S chemical bonding. In order to fabricate thermal and electrical interface adhesive, rGO (0.1 wt.%) obtained by chemical reduction of GO by hydrazine was added to the binder containing F–Cu (80 wt.%) and milled by 3 roll mill. The Raman and FT-IR spectra shown in Figure 10.23 confirm that the covalent bonding and functional groups were successfully attached to GO and rGO.

The attachment of rGO through self-assembled monolayers on Cu surface (the F–Cu + rGO composite) significantly prevents oxidation and improves the electrical properties of material. Despite the originality of the technique, it has very limited application as a composite copper matrix filler for polymeric compositions.

10.8 Conclusions

Although copper has been one of the most important metallic materials since the Chalcolithic period and the Bronze Age, the development of graphene-reinforced metal matrix composites opened a bright prospect for the novel structural applications of copper. The extraordinary reinforcement of copper matrix is observed for both bulk Cu–Gr composites and composite foils. For instance, the hardness value of composite copper–graphene foils reaches 4.0 GPa, its elastic modulus is up to 192 GPa, and tensile strength is up to 452 MPa. The

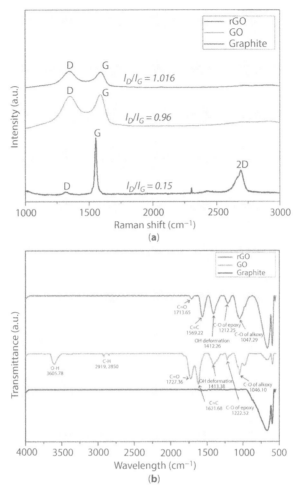

Figure 10.23 (a) Raman spectra and (b) Fourier transform infrared spectroscopy spectra for rGO, GO, and graphite. Reprinted with permission from [73].

hardness of layered Cu–Gr composites obtained using the MLM technique is ~1.5 GPa. It is three to eight times higher than that for pure Cu foil (~600 MPa). The electrical conductivity of composites, produced by various techniques, can be retained as 85%–95% IACS along with enhanced mechanical properties. In contrast to electrical conductivity, the thermal conductivity of Cu–Gr composites is ~30% higher than that for pure copper. The overall increase in mechanical and thermal properties is attributed to the homogeneous dispersion of graphene on the grain boundaries of copper matrix, grain refinement, and enhanced densification parameter upon graphene derivative addition. The strengthening effect is attributed by many researchers to the dislocation motion blocking by Gr due to its high strength and elastic modulus, which leads to dislocation accumulation at the interface. The presence of nanotwinned copper crystals provides additional strengthening. In the case of bulk copper–graphene composites, reinforcement is not as impressive as it is for foil composites; this is due to very low copper–graphene bonding. Indeed, the copper–graphene interface plays a key role when it comes to the thermal and electrical properties of composites.

At the same time, the field of copper matrix composites is very young and is developing fast. Thus, the literature data on mechanical, thermal, and electrical properties of composites are rather random. Sometimes, it is difficult to compare the results presented in literature as, often, the individual property is investigated depending on the target application of the material. An additional problem is the correct estimation of the graphene content in the materials studied since volumetric-to-mass fraction conversion is rather complicated due to (i) the different densities of the graphene-containing powders and (ii) the different types of graphene derivatives (GO, rGO, NGP, graphene–graphite mixtures, etc.). There is an obvious need for more detailed understanding of the optimal amount of graphene addition to ensure the necessary balance of composite properties.

In this way, it seems that the future, at least for the short and middle term, gravitates toward approaches that provide sufficient copper–graphene bonding in the copper matrix composite, i.e., electrochemical deposition, CVD, and MLM techniques coupled with modern sintering procedures. Despite their originality, some techniques, such as electroless deposition and functionalization of copper surface, seem to have very limited industrial applications. A complementary study of the synthesis technique effect and graphene derivative type on thermal, electrical, and mechanical properties could open the pathway to high-performance applications of copper–graphene composites.

References

1. Li, W., Li, D., Fu, Q., Pan, C., Conductive enhancement of copper/graphene composites based on high-quality graphene. *RSC Adv.*, 98, 80428, 2015.
2. Ibrahim, I.A., Mohamed, F.A., Lavernia, E.J., Particulate reinforced metal matrix composites—A review. *J. Mater. Sci.*, 26, 1137, 1991.
3. Lee, Y., Choi, J.R., Lee, K.J., Stott, N.E., Kim, D., Large-scale synthesis of copper nanoparticles by chemically controlled reduction for applications of inkjet-printed electronics. *Nanotechnology*, 41, 415604, 2008.
4. Stankovich, S., Dikin, D.A., Dommett, G.H.B., Kohlhaas, K.M., Zimney, E.J., Stach, E.A., Piner, R.D., Nguyen, S.T., Ruoff, R.S., Graphene-based composite materials. *Nature*, 442, 282, 2006.
5. Uddin, S.M., Mahmud, T., Wolf, C., Effect of size and shape of metal particles to improve hardness and electrical properties of carbon nanotube reinforced copper and copper alloy composites. *Compos. Sci. Technol.*, 70, 2253, 2010.
6. Zhang, L., Duan, Z., Zhu, H., Yin, K., Advances in synthesizing copper/graphene composite material. *Mater. Manuf. Processes*, 32, 475, 2017.
7. Geim, A.K. and Novoselov, K.S., The rise of graphene. *Nat. Mater.*, 6, 183, 2007.
8. Stoller, M.D., Park, S., Zhu, Y., An, J., Ruoff, R.S., Graphene-based ultracapacitors. *Nano Lett.*, 10, 3498–3502, 2008.
9. Balandin, A.A., Ghosh, S., Bao, W., Calizo, I., Teweldebrhan, D., Miao, F., Lau, C.N., Superior thermal conductivity of single-layer graphene. *Nano Lett.*, 3, 902, 2008.
10. Chang, S.W., Nair, A.K., Buehler, M.J., Geometry and temperature effects of the interfacial thermal conductance in copper- and nickel-graphene nanocomposites. *J. Phys.: Condens. Matter*, 24, 245301, 2012.
11. Lee, C., Wei, X., Kysar, J.W., Hone, J., Measurement of the elastic properties and intrinsic strength of monolayer graphene. *Science*, 321, 385, 2008.

12. Dutkiewicz, J., Ozga, P., Maziarz, W., Pstruś, J., Kania, B., Bobrowski, P., Stolarska, J., Microstructure and properties of bulk copper matrix composites strengthened with various kinds of graphene nanoplatelets. *Mater. Sci. Eng., A*, 628, 124, 2015.
13. Kim, Y., Lee, J., Yeom, M.S., Shin, J.W., Kim, H., Cui, Y., Han, S.M., Strengthening effect of single-atomic-layer graphene in metal–graphene nanolayered composites. *Nat. Commun.*, 4, 2013.
14. Zhang, D. and Zhan, Z., Strengthening effect of graphene derivatives in copper matrix composites. *J. Alloys Compd.*, 654, 226, 2016.
15. Wang, X., Li, J., Wang, Y., Improved high temperature strength of copper–graphene composite material. *Mater. Lett.*, 181, 309, 2016.
16. Manjunath, S., Manjunatha, L.H., Kumar, V., Development and characterization of copper metal matrix composite by powder metallurgy technique. *Int. J. Adv. Sci. Res. Eng.*, 3, 147, 2017.
17. Varol, T. and Canakci, A., Microstructure, electrical conductivity and hardness of multilayer graphene/copper nanocomposites synthesized by flake powder metallurgy. *Met. Mater. Int.*, 21, 2015.
18. Chawla, K.K., *Metal Matrix Composites*, Wiley-VCH Verlag GmbH & Co. KGaA, 2006.
19. Chu, K. and Jia, C., Enhanced strength in bulk graphene–copper composites. *Phys. Status Solidi A*, 211, 184, 2014.
20. Wang, L., Choi, H., Myoung, J.M., Lee, W., Mechanical alloying of multi-walled carbon nanotubes and aluminium powders for the preparation of carbon/metal composites. *Carbon*, 47, 3427, 2009.
21. Varol, T. and Canakci, A., Microstructure, electrical conductivity and hardness of multilayer graphene/copper nanocomposites synthesized by flake powder metallurgy. *Met. Mater. Int.*, 21, 704, 2015.
22. Yoon, D., Son, Y.W., Cheong, H., Negative thermal expansion coefficient of graphene measured by Raman spectroscopy. *Nano Lett.*, 11, 3227, 2011.
23. Chokshi, A.H., Rosen, A., Karch, J., Gleiter, H., On the validity of the Hall–Petch relationship in nanocrystalline materials. *Scr. Metall.*, 23, 1679, 1989.
24. Roy, R., Agrawal, D., Cheng, J., Gedevanishvili, S., Full sintering of powdered-metal bodies in a microwave field. *Nature*, 6737, 668, 1999.
25. Cheng, Y., Zhang, Y., Wan, T., Yin, Z., Wang, J., Mechanical properties and toughening mechanisms of graphene platelets reinforced Al_2O_3/TiC composite ceramic tool materials by microwave sintering. *Mater. Sci. Eng., A*, 680, 190, 2017.
26. Nawathe, S., Wong, W.L.E., Gupta, M., Using microwaves to synthesize pure aluminum and metastable Al/Cu nanocomposites with superior properties. *J. Mater. Process. Technol.*, 209, 4890, 2009.
27. Ayyappadas, C., Muthuchamy, A., Annamalai, A.R., Agrawal, D.K., An investigation on the effect of sintering mode on various properties of copper–graphene metal matrix composite. *Adv. Powder Technol.*, 28, 1760, 2017.
28. Chmielewski, M., Michalczewski, R., Piekoszewski, W., Kalbarczyk, M., Tribological behaviour of copper–graphene composite materials. *Key Eng. Mater.*, 674, 219, 2016.
29. Park, M., Kim, B.H., Kim, S., Han, D.S., Kim, G., Lee, K.R., Improved binding between copper and carbon nanotubes in a composite using oxygen-containing functional groups. *Carbon*, 49, 811, 2011.
30. Pietrzak, K., Gładki, A., Frydman, K., Wójcik-Grzybek, D., Strojny-Nędza, A., Wejrzanowski, T., Copper–carbon nanoforms composites—Processing, microstructure and thermal properties. *Arch. Metall. Mater.*, 62, 1307, 2017.
31. Zhang, L., Pollak, E., Wang, W.C., Jiang, P., Glans, P.A., Zhang, Y., Cabana, J., Kostecki, R., Chang, C., Salmeron, M., Guo, J., Zhu, J., Electronic structure study of ordering and interfacial interaction in graphene/Cu composites. *Carbon*, 50, 5316, 2012.

32. Boukhvalov, D.W., Bazylewski, P.F., Kukharenko, A.I., Zhidkov, I.S., Ponosov, Y.S., Kurmaev, E.Z., Chang, G.S., Atomic and electronic structure of a copper/graphene interface as prepared and 1.5 years after. *Appl. Surf. Sci.*, 426, 1167, 2017.
33. Durney, L.J., *Electroplating Engineering Handbook*, 4th edition, Van Nostrand Reinhold Company, 1984.
34. Jagannadham, K., Electrical conductivity of copper–graphene composite films synthesized by electrochemical deposition with exfoliated graphene platelets. *J. Vac. Sci. Technol., B*, 30, 03D109, 2012.
35. Kováčik, J. and Bielek, J., Electrical conductivity of Cu/graphite composite material as a function of structural characteristics. *Scr. Mater.*, 35, 151, 1996.
36. Sruti, A.N. and Jagannadham, K., Electrical conductivity of graphene composites with In and In-Ga alloy. *J. Electron. Mater.*, 39, 1268, 2010.
37. Stankovich, S., Dikin, D.A., Piner, R.D., Kohlhaas, K.A., Kleinhammes, A., Jia, Y., Ruoff, R.S., Synthesis of graphene-based nanosheets via chemical reduction of exfoliated graphite oxide. *Carbon*, 45, 1558, 2007.
38. Celzard, A., Mareche, J.F., Furdin, G., Modelling of exfoliated graphite. *Prog. Mater. Sci.*, 50, 93, 2005.
39. Jagannadham, K., Thermal conductivity of copper-graphene composite films synthesized by electrochemical deposition with exfoliated graphene platelets. *Metall. Mater. Trans. B*, 43, 316, 2012.
40. Park, S. and Ruoff, R.S., Chemical methods for the production of graphenes. *Nat. Nanotechnol.*, 4, 217, 2009.
41. Song, G., Yang, Y., Fu, Q., Pan, C., Preparation of Cu-graphene composite thin foils via DC electro-deposition and its optimal conditions for highest properties. *J. Electrochem. Soc.*, 164, D652, 2017.
42. Huang, G., Cheng, P., Wang, H., Ding, G., Optimizing electrodeposition process for preparing copper–graphene composite film. *Advanced Material Engineering: Proceedings of the 2015 International Conference on Advanced Material Engineering.* 497, 2016. https://doi.org/10.1142/9789814696029_0058.
43. Huang, G., Wang, H., Cheng, P., Wang, H., Sun, B., Sun, S., Ding, G., Preparation and characterization of the graphene–Cu composite film by electrodeposition process. *Microelectron. Eng.*, 157, 7–12, 2016.
44. Pavithra, C.L.P., Sarada, B.V., Rajulapati, K.V., Rao, T.N., Sundararajan, G., A new electrochemical approach for the synthesis of copper-graphene nanocomposite foils with high hardness. *Sci. Rep.*, 4, 1, 4049, 2014.
45. Xie, G., Forslund, M., Pan, J., Direct electrochemical synthesis of reduced graphene oxide (rGO)/copper composite films and their electrical/electroactive properties. *ACS Appl. Mater. Interfaces*, 6, 7444, 2014.
46. Konakov, V.G., Kurapova, O.Yu., Novik, N.N., Golubev, S.N., Osipov, A.V., Graschenko, A.S., Zhilyaev, A.P., Sergeev, S.N., Archakov, I.Yu., Optimized approach for synthesis of nanotwinned copper with enhanced hardness. *Rev. Adv. Mater. Sci.*, 39.
47. Konakov, V.G., Kurapova, O.Yu., Novik, N.N., Graschenko, A.S., Osipov, A.V., Archakov, I.Yu., Approach for electrochemical deposition of copper-graphite films. *Mater. Phys. Mech.*, 24, 61, 2015.
48. Konakov, V.G., Kurapova, O.Yu., Novik, N.N., Golubev, S.N., Microstructure of copper-graphene composites manufactured by electrochemical deposition using graphene suspensions stabilized by non-ionic surfactants. *Mater. Phys. Mech.*, 24, 382, 2015.
49. Konakov, V.G., Kurapova, O.Yu., Novik, N.N., Golubev, S.N., Zhilyaev, A.P., Sergeev, S.N., Archakov, I.Yu., Ovid'ko, I.A., Nanotwinned copper–graphene composite: Synthesis and microstructure. *Rev. Adv. Mater. Sci.*, 45, 1, 2016.

50. Kurapova, O.Yu., Konakov, V.G., Grashchenko, A.S., Novik, N.N., Golubev, S.N., Ovid'ko, I.A., Nanotwinned copper–graphene composites with high hardness. *Rev. Adv. Mater. Sci.*, 48, 71, 2016.
51. Kurapova, O.Yu., Konakov, V.G., Grashchenko, A.S., Novik, N.N., Golubev, S.N., Orlov, A.V., Ovid'ko, I.A., Structure and microhardness of two-layer foils of nanotwinned copper with graphene nanoinclusions. *Mater. Phys. Mech.*, 32, 58, 2017.
52. Konakov, V.G., Kurapova, O.Yu., Grashchenko, A.S., Golubev, S.N., Solovyeva, E.N., Archakov, I.Yu., Nanotwinned copper–graphene foils—A brief review. *Rev. Adv. Mater. Sci.*, 51, 160, 2017.
53. Ovid'ko, I.A. and Sheinerman, A.G., Mechanical properties of nanotwinned metals: A review. *Rev. Adv. Mater. Sci.*, 44, 1, 2016.
54. Lu, L., Shen, Y., Chen, X., Qian, L., Lu, K., Ultrahigh strength and high electrical conductivity in copper. *Science*, 304, 422, 2004.
55. Lu, L., Chen, X., Huang, X., Lu, K., Revealing the maximum strength in nanotwinned copper. *Science*, 323, 607, 2009.
56. You, Z.S., Lu, L., Lu, K., Temperature effect on rolling behavior of nano-twinned copper. *Scr. Mater.*, 62, 415, 2010.
57. You, Z.S., Lu, L., Lu, K., Tensile behavior of columnar grained Cu with preferentially oriented nanoscale twins. *Acta Mater.*, 59, 6927, 2011.
58. Method for preparing nano-copper/graphene composite particles under assistance of ultrasonic wave, patent CN 103769602 A, 2014.
59. Shakalova, V. and Kaiser, A.B., *Woodhead Publishing Series in Electronic and Optical Materials*, vol. 57, p. 401, 2014.
60. Liu, X.W., Mao, J.J., Liu, P.D., Wei, X.W., Fabrication of metal–graphene hybrid materials by electroless deposition. *Carbon*, 49, 477, 2011.
61. Qu, L. and Dai, L., Substrate-enhanced electroless deposition of metal nanoparticles on carbon nanotubes. *J. Am. Chem. Soc.*, 127, 10806, 2005.
62. Liu, H., Teng, X., Wu, W., Wu, X., Leng, J., Geng, H., Effect of graphene addition on properties of Cu-based composites for electrical contacts. *Mater. Res. Express*, 4, 066506, 2017.
63. Hwang, J., Yoon, T., Jin, S.H., Lee, J., Kim, T.S., Hong, S.H., Jeon, S., Enhanced mechanical properties of graphene/copper nanocomposites using a molecular-level mixing process. *Adv. Mater.*, 25, 6724, 2013.
64. Fan, X., Peng, W., Li, Y., Li, X., Wang, S., Zhang, G., Zhang, F., Deoxygenation of exfoliated graphite oxide under alkaline conditions: A green route to graphene preparation. *Adv. Mater.*, 20, 4490, 2008.
65. Yoon, T., Shin, W.C., Kim, T.Y., Mun, J.H., Kim, T.S., Cho, B.J., Direct measurement of adhesion energy of monolayer graphene as-grown on copper and its application to renewable transfer process. *Nano Lett.*, 12, 1448, 2012.
66. Wang, L., Cui, Y., Li, B., Yang, S., Li, R., Liu, Z., Fei, W., High apparent strengthening efficiency for reduced graphene oxide in copper matrix composites produced by molecule-lever mixing and high-shear mixing. *RSC Adv.*, 5, 51193, 2015.
67. Wang, L., Yang, Z., Cui, Y., Wei, B., Xu, S., Sheng, J., Wang, M., Zhu, Yu., Fei, W., Graphene–copper composite with micro-layered grains and ultrahigh strength. *Sci. Rep.*, 7, 41896, 2017.
68. Dreyer, D.R., Ruoff, R.S., Bielawski, C.W., From conception to realization: An historial account of graphene and some perspectives for its future. *Angew. Chem. Int. Ed.*, 49, 9336, 2010.
69. Losurdo, M., Giangregorio, M.M., Capezzuto, P., Bruno, G., Graphene CVD growth on copper and nickel: Role of hydrogen in kinetics and structure. *Phys. Chem. Chem. Phys.*, 13, 20836, 2011.
70. Jennings, A.T., Burek, M.J., Greer, J.R., Microstructure versus size: Mechanical properties of electroplated single crystalline Cu nanopillars. *Phys. Rev. Lett.*, 104, 135503, 2010.

71. Misra, A., Hirth, J.P., Kung, H., Single-dislocation-based strengthening mechanisms in nanoscale metallic multilayers. *Philos. Mag. A*, 82, 2932, 2002.
72. Misra, A., Hirth, J.P., Hoagland, R.G., Length-scale-dependent deformation mechanisms in incoherent metallic multilayered composites. *Acta Mater.*, 53, 4817, 2005.
73. Hwang, J., Park, M., Jang, S., Choi, H., Jang, J., Yoo, Y., Jeon, M., Copper–graphene composite materials as a conductive filler for thermal and electrical interface adhesive. *J. Nanosci. Nanotechnol.*, 17, 3487, 2017.

11

Graphene–Metal Oxide Composite as Anode Material in Li-Ion Batteries

Sanjaya Brahma[1], Shao-Chieh Weng[1] and Jow-Lay Huang[1,2]*

[1]*Department of Materials Science and Engineering, National Cheng Kung University, Tainan, Taiwan (R.O.C.)*
[2]*Center for Micro/Nano Science and Technology, National Cheng Kung University, Tainan, Taiwan (R.O.C.)*
[3]*Hierarchical Green-Energy Materials (Hi-GEM) Research Center, National Cheng Kung University, Tainan, Taiwan (R.O.C.)*

Abstract

Lithium ion batteries (LIBs) have drawn considerable attention in physics/materials science, chemistry/chemical engineering, and computational chemistry, as well as in industries. Owing to its outstanding properties such as superior energy density, good cycle life, high operating voltage, and wide working temperature range, LIBs have applications in consumer electronic devices and electric/hybrid vehicles. The low theoretical capacity (372 mAh g^{-1}) of the currently used graphite anode limits its applications in higher-capacity devices. Metal oxide (MO) has been pursued previously to overcome the issues related to the specific capacity, but the low conductivity and capacity fading due to severe volume expansion are some of the major limitations for future applications in LIBs. Recently, graphene–metal oxide (MO) composite has attracted huge attention among researchers because of its synergistic effects, where graphene/reduced graphene oxide (rGO) can work as an excellent conducting layer for better charge transport and strong adhesion of the MO with the oxygen functional groups of graphene/rGO. Although graphene has high surface area with excellent electrical conductivity, graphene sheets usually aggregate, reducing the overall surface area as well as the properties. MO nanoparticles/nanostructures grow on/attach to the graphene/rGO sheets, preventing the aggregation of the graphene sheets to improve capacity, cyclic stability, and rate capability of the anode materials. Porous graphene and the MO composite are new concepts, and several researches have also been carried out to enhance the capacity of the anode materials. In this book chapter, we describe different methods for the synthesis of graphene–MO/porous graphene–MO composites, their microstructure, bond vibrations/binding energies, thermal studies, and electrochemical properties, and we will also compare all the available data with the results obtained by our graphene/rGO–MO composite. In this book chapter, we first introduce the basics of LIBs followed by the use of metal oxide and the graphene–metal oxide composite as anode in LIBs.

Keywords: Lithium-ion battery, graphene, reduced graphene oxide, Raman studies, anode materials, electrochemical properties

*Corresponding author: jlh888@mail.ncku.edu.tw

11.1 Introduction

The day-to-day inventions in science and technology lead to the production of a large number of digital electronic devices that currently demand huge amounts of energy, and this demand will surge over time. Even commonly used household consumer electronic devices such as electric bulbs, fans, televisions, laptops, mobile phones, refrigerators, and water heaters need energy for their regular operation, let alone high-end applications such as electric hybrid vehicles/medical devices. Naturally available energy resources such as fossil fuels and natural gas are heavily consumed to generate energy to meet the essential needs and the fast depletion of these sources; these along with the increase in greenhouse gases are a major concern around the globe. In order to reduce global warming due to the greenhouse emissions and to overcome the energy crisis, it has become a challenge to harvest energy from various renewable sources, such as wind/solar energy, which are clean and environmentally friendly. Another way to overcome this crisis is to produce energy conversion and storage devices such as rechargeable batteries that can store electrical energy in the form of chemical energy.

Lithium-ion batteries (LIBs), a kind of electrochemical energy conversion and storage device, have been widely used energy storage systems, with several excellent features such as high energy density, long cycle life, high operating voltage, wide working temperature range, no memory effect, low maintenance, and low self-discharge [1–9]. The first lithium-ion battery is commercialized by Sony (Japan) in 1991, and it is estimated that the lithium-ion battery market will reach US$77 billion by 2024, with the consumer electronics market taking up the majority of this growth. Because of its wide range of applications from commonly used mobile phones to high-end electric hybrid vehicles, it has attracted enormous attention among researchers from the fields of physics/materials science, chemistry, chemical engineering, and computational chemistry. However, the growing requirements for better LIBs require constant innovation, in terms of improved safety, longer lifetime, smaller size, lighter weight, and lower cost.

In lithium ion batteries, lithium ions move from the positive electrode to the negative electrode through the electrolyte during charging, and the electrons flow from the positive electrode to the negative electrode through the external circuit. The battery is said to be fully charged if there are no more ions that flow, and then the battery is ready to use. The motion of the ions (electrons) is reversed during discharge. Figure 11.1 shows a schematic diagram of the operation of LIBs. The key to improved LIB performance lies in the electrode materials, and a significant amount of research has been devoted to improve the capacity as well as the stability of anode materials in LIBs. The most commonly used cathode materials are $LiCoO_2$, $LiMn_2O_4$, and $LiFePO_4$ (lithium intercalation compounds). Graphite (theoretical capacity of 372 mAh g^{-1}) is a widely used anode material in commercial LIBs [10] because of its high coulombic efficiency and better cycle performance, and the anode based on graphite carbon stores one Li$^+$ for every six carbon atoms between its graphene layers. Traditional intercalation-type graphite materials show low Li storage capacity due to limited Li-ion storage sites within an sp^2 hexagonal carbon structure [11]. The low specific capacity limits the Li storage capacity due to limited Li-ion storage sites and consequently hinders its applications in higher-capacity devices. Therefore, the development of alternative electrodes with high capacity has been one of the most urgent tasks

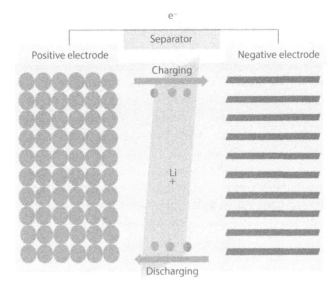

Figure 11.1 Schematic diagram of operation of LIBs.

in advancing the overall performance of LIBs. Anode materials such as metal oxide, metal sulfide, metal oxide, carbon/graphene composite, and nonmetal with larger specific capacity are getting a lot of attention.

11.2 Type of Anode Materials

Anode materials can be divided into three different types depending on the type of the mechanism for storage of lithium ions. First, for intercalation/de-intercalation materials, lithium ions can reversibly intercalate into the materials without destroying their structure. This kind of material, such as graphite and titanium oxide, shows good stability and great safety. However, in this mechanism, anode materials cannot react with large amounts of Li^+, resulting in low device performance. In the case of alloy/de-alloy materials, for example, Si or tin, lithium ions react with elements or metals to form alloys or intermetallic compounds. In alloy reaction, it can supply deliver capacities, but it is often accompanied by high volume expansion of the particles leading to stress and even pulverization problems. The last type is the conversion mechanism that is common in many transition metal oxides. The conversion materials store lithium ions by converting Li^+ into LiO_2. During the charging (lithiation) process, the conversion materials are reduced and the lithium ions are oxidized into LiO_2, and then during discharging, the materials are re-oxidated and LiO_2 releases Li^+ back to cathode.

11.3 Metal Oxides as Anode Materials in Lithium Ion Battery

Nanoscale transition metal oxides (TMO_S) [12, 13] have been applied as anode materials in LIBs to deal with the mechanical strains produced by the volume changes caused by Li-ion intercalation/deintercalation and suppress the corrosion of electrodes. Poizot et al. [14] first introduced nanosized TMOs (M = Co, Ni, Cu, or Fe) for LIB applications due to their characteristics, such as high energy density, high theoretical capacity, long cycle life, low cost, low toxicity, and natural abundance, and their unique conversion reactions [15, 16]. The electrodes fabricated with these TMO nanoparticles demonstrate electrochemical capacities of 700 mAh g^{-1}, with 100% capacity retention for up to 100 cycles and high recharging rates. The mechanism of Li reactivity involves the formation and decomposition of Li$_2$O, through the reduction and oxidation of metal nanoparticles (1±5 nm).

Electrochemically active transition metal oxides (TMOs), such as CuO [17], Co$_3$O$_4$ [18], SnO$_2$ [19, 20], manganese oxide (MnO) [21–24], and Fe$_2$O$_3$ [25], have been intensively exploited as potential anode materials because of their high theoretical capacity and improved safety. Among them, tin dioxide (SnO$_2$) is regarded as one of the promising anode materials for the next-generation LIBs due to its high theoretical capacity (782 mAh g^{-1}), low cost, wide availability, and nontoxicity [26]. However, poor electrical conductivity and the rapid capacity fading due to the large volume change during the Li$^+$ insertion/extraction process is a major challenge for its industrial applications [26]. Several synthesis methods are followed to produce SnO$_2$ nanostructures with varying shape and size, such as nanoparticles, nanorods [27], nanowires [28], nanotubes [29], and hollow structures [13, 30], to reduce the volume expansion. Similarly, TiO$_2$ is another type of anode material that delivers a relatively low capacity (e.g., ~170 mAh g^{-1} for anatase) at a higher potential of about 1.7 V vs. Li+/Li, but the low volume expansion upon lithiation, good stability, and lack of lithium plating enhance its potential to be charged/discharged at high current rates for many cycles. Low electronic and ionic conductivities still present serious problems [31, 32], and it is expected that the nanostructures of TiO$_2$ would overcome these disadvantages [33–35].

Iron oxide, as an important member of these transition metal oxides, is regarded as a potential electrode material to replace graphite owing to its much higher theoretical capacity (~1005 mAh g^{-1}) than most other oxides, natural abundance, and nontoxicity [36, 37]. However, the poor electrical conductivity and large volume change during the lithiation/delithiation processes and subsequent pulverization of particles result in the rapid drop in cycling performance, which severely encumbers its practical application [38]. α-Fe$_2$O$_3$ nanoflakes [39] prepared by a simple hotplate technique over copper substrates have also been used as anodes in LIBs, and these nanostructures show high capacity (680 ± 20) mAh g^{-1}, with negligible capacity fading up to 80 cycles. Similarly, α-Fe$_2$O$_3$ nanorods (~40 nm in diameter and ~400 nm in length) synthesized by the hydrothermal method [40] are investigated as anode material for Li-ion batteries that exhibit relatively high capacities of 908 mAh g^{-1} at 0.2 C rate and 837 mAh g^{-1} at 0.5 C rate, and these capacities were completely retained after 100 cycles. These nanorods provide a short path for lithium-ion diffusion and effective accommodation of the strain generated from volume expansion during the lithiation/delithiation process. α-Fe$_2$O$_3$ nanorods show much better capacitance as compared with submicron and micron-sized α-Fe$_2$O$_3$ particles. A typical one-dimensional core–shell Fe/Fe$_2$O$_3$ nanowire prepared in a facile aqueous synthesis process [41] under ambient conditions with subsequent annealing maintains an excellent reversible capacity of over 767 mAh g^{-1} (538 mAh g^{-1}) at 500 mA g^{-1} (2000 mA g^{-1}) after 200 cycles with a high average coulombic efficiency of 98.6%. When this hybrid is used as an anode material for LIBs, the outer Fe$_2$O$_3$ shell can act as an electrochemically active material to store and release lithium ions, whereas the highly conductive and inactive

Fe core acts as an electrical conducting pathway and a good buffer to tolerate volume changes of the electrode materials during the insertion and extraction of lithium ions. The mesoporous α-Fe_2O_3/cyclized-polyacrylonitrile (C-PAN) composite synthesized by a rapid two-step method [42] exhibits high reversible capacity (996 mAh g^{-1} after 100 cycles at 0.2 C, 773 mAh g^{-1} at 1 C, and 655 mAh g^{-1} at 2 C), enhanced cycling performance, and superior rate capacity. 3D dendritic Fe_2O_3 nanoparticles [43] wrapped with carbon (3DD-Fe_2O_3@C) exhibit an excellent discharge/charge capacity (982 mAh g^{-1} and 971 mAh g^{-1} after 100 cycles at a rate of 100 mA g^{-1}) and good cyclic stability and rate capability. The nanocomposite of Fe_2O_3 nanoparticle (~4.0 nm) MoS_2 nanosheets synthesized by facile method [44] is used for high-performance anode material for Li-ion battery, and it shows a very stable, high reversible capacity (829 mAh g^{-1} to 864 mAh g^{-1} at a high current density of 2 Ag^{-1} up to 140 cycles. The capacity is mainly attributed to the Fe_2O_3 nanoparticles, whereas the MoS_2 nanosheets act as scaffolds and accommodate the large volume change of Fe_2O_3 during the charge/discharge process.

Cobalt oxide (Co_3O_4) nanostructures (nanoparticles, nanobelts, microspheres, nanoplatelets, nanocubes, nanodiscs, and nanocomposite) have been studied extensively for their applications as anode material in Li-ion batteries. Co_3O_4 has attracted much more attention because of its high theoretical capacity (890 mAh g^{-1}) according to the electrochemical reaction $Co_3O_4 + 8Li^+ + 8e^- \leftrightarrow 3Co + 4Li_2O$, and the capacity is more than two times higher than graphite. However, this material usually suffers from poor capacity retention while cycling and poor rate capability, which still remain major challenges when used in practical cells. Co_3O_4 nanoparticles prepared by the thermal decomposition of nanoparticles of cobalt-based Prussian blue at different temperatures (450, 550, 650, 750, and 850°C) showed a high discharge capacity of 970 mAh g^{-1} (for samples annealed at 550°C) after 30 cycles at a current density of 50 mA g^{-1} [45]. This high performance is attributed to the size of the grain, the porous structure (with the pore size distribution centered at 3 and 9 nm) with improved crystallinity, and the large specific surface area. A detailed analysis of the microstructure with the property reveals that the nanoparticle having an optimum structure including crystallinity, morphology, inner structure, and chemical composition is more important for the device performance rather than the optimum size. Leng et al. [46] have synthesized ultrafine Co_3O_4 nanoparticles homogeneously embedded in ultrathin porous graphitic carbon and achieved a very high reversible capacity of up to 1413 mAh g^{-1} (0.1 A g^{-1}) after 100 cycles with a high rate capability (845, 560, 461, and 345 mAh g^{-1} at 5, 10, 15 and 20 C, respectively, 1 C = 1 A g^{-1}) and a supreme cycling performance (760 mAh g^{-1} at 5 C after 1000 cycles). Porous Co_3O_4 materials [47] prepared by a nanocasting route using mesoporous silicas as templates and $Co(NO_3)_2 \cdot 6H_2O$ as precursor have achieved high reversible capacities around 1141 mAh g^{-1}, and this relatively high capacity is attributed to the lithium storage in the interconnected mesopores via an electric double-layer capacitive mechanism. Similarly, Co_3O_4 hierarchical nanostructures [48] have been prepared by multiple process steps, which includes hydrothermal synthesis (95°C for 8 h) followed by annealing (450°C in air for 2 h) and reduction in 1 M $NaBH_4$. Electrochemical analysis reveals a high reversible capacity of 1053.1 mAh g^{-1} after 50 cycles at a current density of 0.2 C (1 C = 890 mA g^{-1}), good cycling stability, and rate capability. Similarly, a variety of other Co_3O_4 nanostructures along with composites with carbon such as microspheres (550.2 mAh g^{-1}) [49], nanocapsules (1026.9 mAh g^{-1} after 50 cycles) [50], nanodiscs (1161 mAh g^{-1} after 100 cycles) [51], nanoplatelets, nanosheets (970 mAh g^{-1} at

1 A g^{-1} after 500 cycles) [52], snowflake-shaped crystals (977 mAh g^{-1} at 3000 mA g^{-1}) [53], nanobelts (857 mAh g^{-1} after 60 cycles) [54], and hollow cubes (1032 mAh g^{-1} at 1000 mA g^{-1} after 150 cycles) [55] are synthesized, and electrochemical properties reveal good capacitance, high cyclic stability, and high rate capability, which are considered for potential anode materials for LIBs.

11.4 Graphene/Graphene–Metal Oxide as Anode in Li-Ion Battery

11.4.1 Graphene as Anode Materials in Lithium Ion Battery

Graphene, a monolayer of graphite, the basic unit of all graphitic materials such as carbon nanotube (CNT) and fullerene, is a two-dimensional sheet of sp^2-bonded carbon atoms arranged in a honeycomb-like crystal lattice structure and is recognized as an important material of the future [56–60]. Graphene, a zero gap semiconductor, which can be synthesized by a low-cost easy chemical treatment, is gifted with many superior properties such as very high surface area (~1500 m^2/g), high room-temperature electron mobility (2.53105 cm^2 V^{-1} s^{-1}), superb electrical conductivity (2000 S cm^{-1}), 97.7% optical transmittance, excellent thermal conductivity (4840–5300 Wm^{-1} K^{-1}), high hardness, Young's modulus of 1 TPa, good chemical stability, nontoxic, lightweight, and an intrinsic strength of 130 GPa. The above outstanding properties of this miracle material attract enormous attention among researchers in a variety of research fields including academics as well as industry, and extensive research has been carried out in nanoelectronics (flexible electronics, transistors), photonics (photodetectors, optical modulators), energy generation and storage devices (supercapacitors, Li-ion batteries, solar cells), gas sensors, and biological applications.

It was Eizenberg and his team [61] who observed monolayer carbon condensed over Ni(111) surface for the first time while investigating the surface properties of carbon-doped Ni sheets. The doping was carried out at high temperature by heating the single crystal nickel sheet covered with graphite powder in an evacuated quartz capsule. Monolayer graphite was also obtained over the TaC(111) surface when exposed to ethylene at 800–1200°C [62]. However, Lu et al. [63] for the first time introduced the word graphene while investigating the peeled outer layer of graphite islands obtained by patterning the highly oriented pyrolytic graphite (HOPG). Similarly, mechanical exfoliation [64], chemical vapor deposition (CVD) [65], wet chemical synthesis by reduction of graphene oxide [66], and solvothermal synthesis [67] were used extensively for the synthesis of graphene. Mechanical exfoliation and CVD yield single- or few-layer graphene for applications in electronics and gas sensors, but mass production and cost are underlying issues. Large-scale production of graphene is the major advantage of the wet chemical method, but the defects generated during the synthesis process are the main disadvantage.

As far as the application of graphene in LIBs is concerned, it is expected that the synergistic effect of high electrical conductivity that helps the rapid transport of electrons to and from the active material intercalation sites and huge surface-to-volume ratio would enhance the specific capacity (500–1100 mAh g^{-1}) of LIBs through high Li storage [68, 69]. Furthermore, graphene can act as a binder and replace the poly(vinylidene fluoride) that works as a binding polymer material [70, 71].

Yoo et al. [72] have used graphene nanosheets (GNS) as anode materials in the Li-ion battery, and they have obtained relatively high reversible capacity (540 mAh g^{-1}) at a current density of 0.05 A g^{-1} and a retention of 54% (290 mAh g^{-1}) after 20 cycles. A comparative study of GNS composites with carbon nanotubes (CNTs) and fullerenes (C_{60}) reveals relatively high capacity (784 mAh g^{-1}) with 77% retention (600 mAh g^{-1}) after 20 cycles, which is much higher than that of graphite. The enhanced lithium storage capacity is attributed not only to the formation of the LiC_6 compound but also to the electronic structure of the GNS and the expansion in the *d*-spacing of the graphene layers, which may create additional sites for accommodation of lithium ions. Similarly, GNS and doped GNS [73–82] prepared by different methods have also been used as anode materials in LIBs, and Table 11.1 summarizes the preparation methods, microstructure, and the electrode performance as anode in LIBs. Although the capacity in the first cycle can reach more than 1000 mAh g^{-1}, the fading is quite significant and the final capacity can be within 600–700 mAh g^{-1} after 150–500 cycles.

Even though graphene has many advantages, aggregation remains the major problem that hinders its wider application. Graphene-based materials can be used as a 2D buffer layer for the anisotropic growth of various metals (M)/metal oxide (MO) nanoparticles (NPs), which not only effectively prevent the aggregation and volume expansion of these M/MO NPs but also enhance the capacity. Due to the presence of oxygen-containing functional groups on the surface of the GOs, the steric effect facilitates better dispersion of the GOs in solvents and thus extends their range of application. In addition, the oxygen-containing functional groups of GOs can act as active sites to react with transition metal ions for the formation of transition metal oxide (TMO) nanostructures having a uniform distribution over the surface of the GOs. The nanocomposites of graphene–MnO_2 and graphene–SnO_2 are most widely investigated as the anode in Li-ion battery because of the high theoretical specific capacity of both metal oxides.

11.4.2 Graphene–MnO_2 as Anode in Li-Ion Battery

The strategy to combine MnO_2 and graphene/reduced graphene oxide (rGO) aims to obtain the synergistic effects of their respective advantages and to improve the performance of the anode materials for high-power LIB applications. The graphene–MnO_2 composite [83–89] has been used extensively as anode materials in Li-ion battery, and Table 11.2 summarizes the performance of all these composites. MnO_2 nanoparticles, nanosheets, nanowires, nanoneedles, and nanotubes are incorporated within the graphene layers. One of these reports shows high (1215 mAh g^{-1}) initial reversible capacity and good retention (1100 mAh g^{-1}) after 100 cycles. However, majority of the reports show capacity below 1100 mAh g^{-1}, and this may be due to stacking of the graphene layers and large volume change during Li intercalation/deintercalation. Although there are many reports about MnO_2/carbon nanocomposites, the synthesis methods are complicated, are time-consuming, and are carried out at high temperature (>100°C). Recently, we have prepared MnO_2 nanoneedles and the MnO_2/rGO nanocomposite at 83°C by a simple chemical method, and they were used successfully as the anode material in LIBs [90]. The synthesis procedure includes the preparation of graphene oxide (oxidation of graphite powder) by the modified Hummer's method and dispersion of graphene oxide (0.066 g), $MnCl_2 \cdot 4H_2O$ (0.27 g) in isopropyl alcohol (50 ml) under ultrasonication. The mixture was heated to 83°C in a three-neck flask under

Table 11.1 Physical properties and electrochemical Li cycling data of graphene and graphene–carbon composite.

Authors	Morphology	Graphene morphology/size	Synthesis method	Current rate	Reversible capacity of 1st cycle (mAh g^{-1})	Voltage range	Capacity retention after n cycles	Ref.
Yoo EJ et al.	Graphene nanosheets (GNS)	Curled morphology consisting of a thin wrinkled paper-like structure, platelet thickness: 2–5 nm, (6–15 layers)	Exfoliation	0.05 A g^{-1}	540 mAh g^{-1}	0.01 to 3.0 V	290 mAh g^{-1} $n = 20$	[72]
	GNS+CNT			0.05 A g^{-1}	730 mAh g^{-1}	0.01 to 3.0 V	480 mAh g^{-1} $n = 20$	
	GNS+C60			0.05 A g^{-1}	784 mAh g^{-1}	0.01 to 3.0 V	600 mAh g^{-1} $n = 20$	
Guo P et al.	GNS	Crumpled paper, thickness: 7–10 nm, 20–30 layers	Oxidation, heat treatment, sonication	0.2 mA/cm^2	1233 mAh g^{-1}	0.01 to 3.0 V	502 mAh g^{-1} $n = 30$	[73]
Wang G et al.	GNS	Flower-like nanosheets (2–3 layers)	Oxidation, reduction, reflux	1 C	945 mAh g^{-1}	0.01 to 3.0 V	460 mAh g^{-1}, $n = 100$	[74]

(Continued)

Table 11.1 Physical properties and electrochemical Li cycling data of graphene and graphene–carbon composite. (*Continued*)

Authors	Morphology	Graphene morphology/size	Synthesis method	Current rate	Reversible capacity of 1st cycle (mAh g^{-1})	Voltage range	Capacity retention after n cycles	Ref.
Lian P *et al.*	GNS	Curled morphology, thin wrinkled paper-like structure, thickness: 2.1 nm (~4 layers), surface area: 492.5 m^2 g^{-1}	Oxidation, rapid heating	100 mA g^{-1}	2035 mAh g^{-1}	0.01 to 3.0 V	848 mAh g^{-1}, $n = 20$ at low current density of 50 mA/g^{-1}	[75]
Wu Z.S. *et al.*, Graphene, doped graphene sheets	Graphene	Scrolled sheets		50 mA g^{-1}	955 mAh g^{-1}	0.01 to 3.0 V	638 mAh g^{-1}, $n = 30$	[76]
	N-doped graphene	2D ultrathin flexible structure, corrugations/ scrolling. Surface area: 290 m^2 g^{-1}	Chemical exfoliation, heat treatment in a gas mixture of NH$_3$ and Ar.	50 mA g^{-1}	1043 Ah g^{-1}	0.01 to 3.0 V	872 mAh g^{-1}, $n = 30$	
	B-doped graphene	Same morphology, surface area: 256 m^2 g^{-1}.	Chemical exfoliation, heat treatment in a mixture of BCl$_3$, Ar.	50 mA g^{-1}	1549 mAh g^{-1}	0.01 to 3.0 V	1227 mAh g^{-1}, $n = 30$	

(*Continued*)

Table 11.1 Physical properties and electrochemical Li cycling data of graphene and graphene–carbon composite. (*Continued*)

Authors	Morphology	Graphene morphology/size	Synthesis method	Current rate	Reversible capacity of 1st cycle (mAh g^{-1})	Voltage range	Capacity retention after n cycles	Ref.
Li X et al.	N-doped graphene	Worm-like appearance, surface area: 599 m^2 g^{-1}	Oxidation, heat treatment in N atmosphere		454 mAh g^{-1}	0.01 to 3.0 V	684 mAh g^{-1}, $n = 501$	[77]
Hassoun J et al.	Graphene	Flake-like morphology, ~30–100 nm,	Chemical wet dispersion	1C (170 mA g^{-1} vs. LiFePO$_4$)	~7500 mAh g^{-1} at 700 mA g^{-1}	0.01 to 3.0 V	~650 mAh g^{-1}, $n = 150$	[80]
Fu C et al.	N-doped graphene	Corrugated sheet-like morphology	Oxidation to form GO, solvothermal method at 180°C for 2 h.	100 mA g^{-1}	Charge (discharge) capacity are 732.5 mAh g^{-1} (1245 mAh g^{-1})	0.01 to 3.0 V	332 mAh g^{-1} at 0.5 A·g^{-1}, $n = 600$	[81]
Cheng Q et al.	Graphene-like graphite (GLG)	Flake-type (~5 μm), surface area of GLG was 31.3 m^2/g,	Oxidation and heat treatment at different temperatures		Charge (discharge) capacity 1033 and (608) mAh/g	0.01 to 3.0 V	90% retention, $n = 100$	[82]

Table 11.2 Physical properties and electrochemical Li cycling data of MnO$_2$/graphene nanocomposites.

Authors	Morphology	MnO$_2$ morphology/size	Current rate	Reversible capacity of 1st cycle (mAh g^{-1})	Voltage range	Capacity retention after n cycles (cycling range)	Ref.
Yu A et al.	Graphene–MnO$_2$ nanotube (NT) thin-film composites	MnO$_2$ nanotubes; diameters: 70 to 80 nm; lengths: 1 μm and	100 mA g^{-1}	686 mAh g^{-1}	0.01 V to 3.0 V	495 mAh g^{-1} (n = 2–40)	[83]
Xing L et al.	α-MnO$_2$/graphene nanocomposites	α-MnO$_2$ nanosheets	0.1 C	726.5 mAh g^{-1}	0.01 to 3.0 V	575 mAh g^{-1} (n = 2–20)	[84]
Zhang Y et al.	Graphene/α-MnO$_2$ nanocomposites	α-MnO$_2$ nanowire; diameter: 40–50 nm; length: 5–10 μm	60 mA g^{-1}	~1150 mAh g^{-1}	0.01 to 3.0 V	998 mAh g^{-1} (n = 2–30)	[85]
Chen J et al.	MnO$_2$–GNRs (MG)	MnO$_2$ nanorod	100 mA g^{-1}	753 mAh g^{-1}	0.01 to 3.0 V	470 mAh g^{-1} (n = 2–5)	[86]
Kim SJ, et al.	MnO$_2$/rGO nanocomposites	Diameters: 10 to 30 nm	123 mA g^{-1}	1215 mAh g^{-1}	0.01 to 3.0 V	1100 mAh g^{-1} (n = 2–100)	[87]
Wen K et al.	MnO$_2$–graphene composite	MnO$_2$ nanoparticles	100 mA g^{-1}	746 mAh g^{-1}	0.01 to 3.0 V	752 mAh g^{-1} (n = 2–65)	[88]
Jiang Y et al.	MnO$_2$–nanorods/ rGO nanocomposites	RGO supported MnO$_2$ nanorods	1.0 A g^{-1}	1945.8 mAh g^{-1}	0.01 to 3.0 V	1635.3 mAh g^{-1} (n = 2–450)	[89]
Our work	α-MnO$_2$/rGO nanocomposites	α-MnO$_2$ nanoneedeles; diameter: 15–20 nm; length: 450–550 nm	123 mA g^{-1}	855.2 mAh g^{-1}	0.002 to 3.0 V	660.9 mAh g^{-1} (n = 2–50)	[90]

vigorous stirring. Then, 0.15 g of $KMnO_4$ was added in 5 ml of deionized water (DI), and the solution was added quickly into the above solution mixture. The mixture was allowed to be at the same temperature under a reflux condition for 4.5 h, which was then cooled to room temperature naturally. The precipitate was cleaned by repeated washing, collected by centrifugation at 6000 rpm, and finally dried at 55°C in air for 24 h.

The structure/surface morphology of the MnO_2/rGO nanocomposite was studied by X-ray diffraction and field emission scanning electron microscopy. The microstructure and the phases are studied by using a transmission electron microscope. The thermogravimetric analysis was carried out in N_2 atmosphere from 20°C to 800°C at a heating rate of 15°C/min to measure the percentage of graphene as well as the metal oxide. CR2032-type coin cells are used to measure electrochemical properties at room temperature. Four different powders such as active material 80 wt.%, 10 wt.% of Super P as a conductive additive, 5 wt. % of LiOH, and 5 wt.% of polyacrylic acid (PAA) as a binder are used to prepare the working electrode. All these four components were mixed in deionized water to form a slurry that was loaded on a copper foil as a current collector. Circular electrodes are prepared and the cells were assembled in a glove box (Ar-filled) with lithium foil as the counter electrode and a solution of 1.0 M $LiPF_6$ dissolved in 1:1 (v/v) EC/DEC as the electrolyte. Galvanostatic Li^+ charge/discharge analysis was carried out using a Wonatech WBCS3000 automatic battery cycler. All electrochemical measurements were conducted at a potential range of 0.002 V to 3 V (vs. Li^+/Li).

XRD patterns (Figure 11.2) of the as-prepared MnO_2 nanoneedle and the MnO_2/rGO nanocomposite match well with standard XRD of α-MnO_2 (JCPDS, card NO. 44-0141) that crystallizes to the pure tetragonal phase [space group = I4/m (87)]. No other peaks either from the starting material or from the impurity are observed, confirming that the α-type MnO_2 could be prepared by an easy chemical method.

The microstructure and crystallinity of the MnO_2 nanoneedle and the MnO_2/rGO nanocomposite are investigated by FESEM and FETEM. The FESEM image (Figure 11.3a) shows that the nanoneedles are attached to each other and shows a kind of agglomeration. A single MnO_2 nanoneedle is observed in the FETEM image (Figure 11.3b) and the high-resolution image (Figure 11.3c) shows clear distinct atomic planes that signify that these nanostructures are highly crystalline. The typical length and diameter of the as-prepared nanoneedles are measured to be 480±40 nm and 20±2 nm, respectively. The MnO_2/rGO nanocomposite shows fiber-like morphology (Figure 11.3d), and the shape of the MnO_2 nanoneedles in the composite seems similar to that with as-prepared MnO_2 nanoneedles (Figure 11.3a). Figure 11.3d shows TEM images of the MnO_2/rGO nanocomposite, which shows a clear distinction between the rGO and MnO nanoneedles attached to the rGO sheets. Each MnO_2 nanoneedle is a single crystal and the MnO_2 nanoneedles are uniformly distributed on the surface of the reduced graphene oxide sheet (Figure 11.3e). The size of the rGO sheet is estimated to be 3–4 μm^2. The SAED pattern of rGO is shown in Figure 11.3f, which confirms the crystallinity of the rGO. The d-spacing 2.39 Å and 6.79 Å correspond to the (211) and (110) planes [90], and this agrees well with the interplanar spacing (2.395 Å and 6.919 Å) obtained from the standard XRD pattern.

Thermal analysis was carried out in nitrogen atmosphere (from room temperature to 800°C) to confirm the ratio of the composites having carbon and metal oxide. Figure 11.4 shows the mass loss of MnO_2 (black curve) and the graphene–MnO_2 composite (red curve), which shows loss of water molecules (room temperature to 180°C), thermolysis of graphene

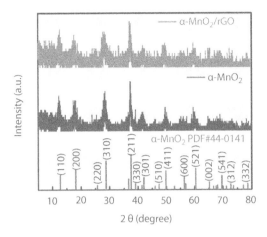

Figure 11.2 XRD patterns of (a) MnO_2/rGO, (b) MnO_2, and (c) α-MnO_2 PDF #44–0141.

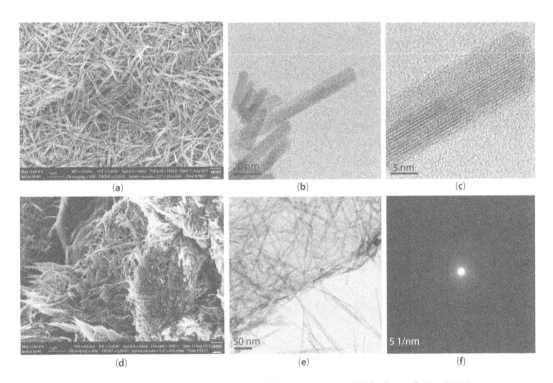

Figure 11.3 MnO_2 nanoneedles: (a) FESEM image, (b) FETEM image, (c) high-resolution TEM image; graphene–MnO_2 nanocomposite: (d) FESEM image, (e) FETEM image, (f) SAED pattern of graphene.

oxide to generate carbon monoxide (CO) or carbon dioxide (CO_2) within 210–460°C, and loss of oxygen (460–580°C), respectively. After 600°C, the weight loss is negligible and finally the ratio of MnO_2 to graphene oxide is evaluated as 7:3.

A half-cell is fabricated to investigate the electrochemical performance of the MnO_2 nanoneedle and the MnO_2-rGO nanocomposite by using metallic lithium film as the counter/reference electrode. The galvanostatic charge–discharge analysis

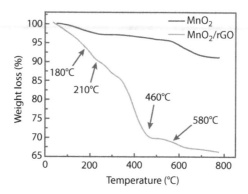

Figure 11.4 TGA analysis of α-MnO$_2$ (black line) and the α-MnO$_2$/rGO nanocomposite (red line) [90].

(Figure 11.5) at a current density of 123 mA g^{-1} shows initial charge (discharge) capacities of 688.4 mAh g^{-1} (665.5 mAh g^{-1}) for MnO$_2$ nanoneedle and 1100.4 mAh g^{-1} (855.2 mAh g^{-1}) for MnO$_2$/rGO nanocomposite electrodes. The discharge capacity of the MnO$_2$/rGO nanocomposite is higher than that of MnO$_2$ and is attributed to the presence of reduced graphene oxide. The charge/discharge curve shows plateaus at ~2.0, ~1.25, ~0.75, and ~0.4 V for both MnO$_2$ and the composite, which are attributed to reactions between Li ions and MnO$_2$ to form Li$_x$MnO$_2$ (~2.0 V and ~1.25 V), the decomposition of electrolyte and the deposition of the SEI layer (~0.75 V), and conversion reactions of the MnO$_2$ nanoneedles to Mn metal with Li$_2$O formation (0.4 V). The conversion reactions between the MnO$_2$ nanoneedle, the MnO$_2$/rGO nanocomposite, and Li ions can be expressed by the following four equations:

$$MnO_2 + Li^+ + e^- \leftrightarrow LiMnO_2 \tag{11.1}$$

$$LiMnO_2 + Li^+ + e^- \rightarrow Li_2MnO_2 \tag{11.2}$$

$$Li_2MnO_2 + 2Li + 2e^- \rightarrow Mn + Li_2O \tag{11.3}$$

$$Mn + xLi_2O \leftrightarrow MnO_x \ (0 < x < 1) + 2xLi^+ + 2xe^- \tag{11.4}$$

The coulombic efficiency of the as-prepared MnO$_2$ nanoneedle electrodes in the first charge/discharge cycle are as high as 96.7% and that of MnO$_2$/rGO nanocomposite electrodes is just ~77.7%. The capacity fading is almost negligible after the 20th cycle, which indicates that a combination of MnO$_2$ with reduced graphene oxide is an effective way to enhance the capacity and reduce its fading. The cyclic stability experiment shows a relatively good retention (547.8 mAh g^{-1} for MnO$_2$ and 660.9 mAh g^{-1} for the MnO$_2$/rGO nanocomposite) after 50 cycles. Therefore, the combination of MnO$_2$ nanoneedles and rGO not only maintains the structure and accommodates the volume change upon lithiation/delithiation but also enhances the cyclic stability and performance of LIBs.

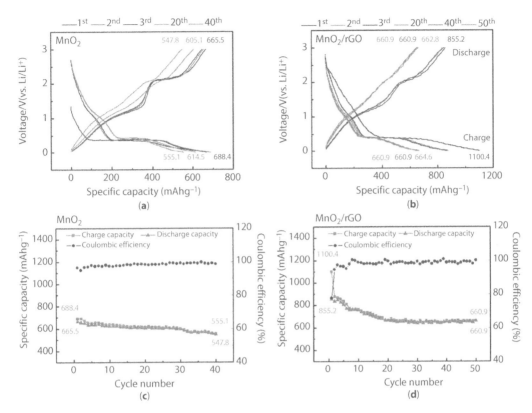

Figure 11.5 Charge/discharge curve of (a) α-MnO$_2$ and (b) the α-MnO$_2$/rGO nanocomposite. Capacity vs. cycle number plots of (c) α-MnO$_2$ and (d) the α-MnO$_2$/rGO nanocomposite [90].

To summarize, our research work on the α-MnO$_2$/rGO nanocomposite is an easy process for the synthesis of the material, and the nanocomposite not only maintained a reversible capacity of 660.9 mAh g^{-1} after 50 cycles at a current density of 123 mA g^{-1} but also had steady cyclic performance. The enhancement in the electrochemical properties of the α-MnO$_2$/rGO nanocomposite is attributed to the synergistic effects of α-MnO$_2$ with rGO. Our results indicate that the α-MnO$_2$/rGO nanocomposite has the potential to work as an anode material in next-generation LIBs.

11.4.3 Graphene–SnO$_2$ as Anode in Li-Ion Battery

SnO$_2$ has been investigated extensively as an anode material in LIBs due to its superior theoretical capacity (782 mAh g^{-1}), low cost, wide availability, and nontoxicity. However, the major issues lie with low electrical conductivity, large volume expansion, and quick capacity fading, which limit its applications. In order to overcome these issues, efforts have been taken to produce nanostructures (nanoparticles, nanorods, nanowires) of SnO$_2$ and combine them with carbon nanomaterials [carbon nanotubes (CNTs), graphene, and graphene oxides] to enhance the stability of SnO$_2$-based LIB anode. However, the production of the SnO$_2$–graphene composite is energy- and time-consuming, involving a complex synthesis procedure at high temperatures followed by annealing for a longer duration at elevated temperatures. A variety of different starting materials, such as chloride ($SnCl_2 \cdot 2H_2O$, $SnCl_4$) and sulfate ($SnSO_4$), are used for the synthesis of SnO_2. A low-temperature (60°C) synthesis process has also been used for the production of SnO_2-reduced graphene oxide (RGO) composite by using $Sn(BF_4)_2$ as the starting precursor material. A detailed synthesis of SnO_2-reduced graphene oxide is reported elsewhere [91]. In brief, graphene oxide (GO) is first prepared by Hummer's method and a specific concentration of GO powder is added to the precursor solution [$Sn(BF_4)_2$ in deionized water with HBF_4] and the mixture was stirred at 60°C for 30 min. Varying concentrations of $Na_2S_2O_4$ with HBF_4 were added into the solution to produce an RGO–SnO_2 nanocomposite. The RGO–SnO_2 composites are annealed at 500°C under an argon (Ar) atmosphere for 2 h. Table 11.3 shows the different amounts RGO/SnO_2 in the chemical treatment.

The structure/microstructure/composition of GO, RGO, and the RGO–SnO_2 nanocomposite is investigated by X-ray diffraction, high-resolution field emission scanning electron microscopy (HR-FESEM), field emission transmission electron microscopy (FEG-TEM), and energy-dispersive X-ray spectroscopy (EDS). Fourier transform infrared spectrometer is used to study the oxygen-containing functional group. Electron spectroscopy for chemical analysis is used to study the variation in the binding energy of carbon and tin of the RGO–SnO_2 nanocomposite. Thermogravimetric analysis (TGA) is done from 20 to 800°C (heating rate 15°C/min) in N_2 atmosphere to determine the carbon content in the composite. Electrochemical analysis such as charge discharge testing was done at room temperature within 0.2–3.0 V range on a coin-type cell fabricated in a glove box. Details about the fabrication of the cell are published in our earlier reports [91].

XRD spectra in Figure 11.6 shows the details of the diffraction peaks obtained from GO and the RGO–SnO_2 composite and at varying levels of reductant concentration. The presence of a strong 001 peak at 10.98° (d = 0.80 nm) for GO (Figure 11.6a) confirms the complete oxidation of graphite. The reduction is done by adding the tin precursor and the reducing agent that led to the reduction of GO as well as the formation of SnO_2. This reduction leads to the disappearance of the peak at 10.98° and a broad peak at 24.50° (002 peak, d-spacing = 0.36 nm) with a shoulder (d-spacing = 0.4 nm) emerging (Figure 11.6b) after adding the tin precursor in the GO solution (without any reducing agent), which indicates the removal of oxygen-containing functional groups after sufficient reduction. It is interesting to note that SnO_2 does not from at this synthesis condition, which is confirmed

Table 11.3 The different reductant concentrations for GO and RGO–SnO_2 synthesis.

Species SnO_2/GO	GO (g)	HBF_4 (mol)	DI (g)	$Sn(BF_4)_2$ (mol)	$Na_2S_2O_4$ (mol)
0.075 mol of reductant (31.8 wt.% of SnO_2)	6	0.5	300	0.279	0.075
0.05 mol of reductant (30.2 wt.% of SnO_2)	6	0.5	300	0.279	0.05
0.025 mol of reductant (27.7 wt.% of SnO_2)	6	0.5	300	0.279	0.025
0 mol of reductant (6.2 wt.% of SnO_2)	6	0.5	300	0.279	0

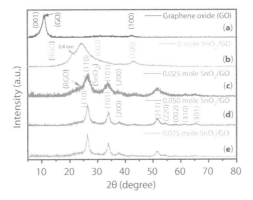

Figure 11.6 X-ray diffraction pattern of (a) GO, RGO–SnO$_2$ composite with different reductant concentrations (b) 0 mol, (c) 0.025 mol, (d) 0.050 mol, and (e) 0.075 mol [91].

by the absence of any peak from SnO$_2$. However, the addition of the reducing agent helps the crystallization of SnO$_2$ along with the reduction of GO as shown by the XRD spectra (Figure 11.6c–e) of the RGO–SnO$_2$ nanocomposite produced after the addition of different concentrations (0.025 mol, 0.050 mol, and 0.075 mol) of reductants.

The FESEM images (Figure 11.7a–d) reveal layer-like morphology for GO sheets (Figure 11.7a), and the attachment of the SnO$_2$ nanoparticles on rGO sheets is very clear after adding a high concentration of the reductant (0.050 mol), as shown in Figure 11.7d. The FETEM analysis provides detailed information about GO and the RGO–SnO$_2$ nanocomposite, as shown in Figure 11.8. The flat GO sheets (length ~ 5–7 μm, width ~ 3–5 nm) (Figure 11.8a) have many active sites at the edges for the growth of the SnO$_2$ nanoparticles. Addition of the tin precursor only facilitates the reduction of GO to RGO and the formation of fine SnO$_2$ nanoparticles (15–20 nm) over RGO (Figure 11.8b). However, the presence of the reducing agent along with the tin precursor helps

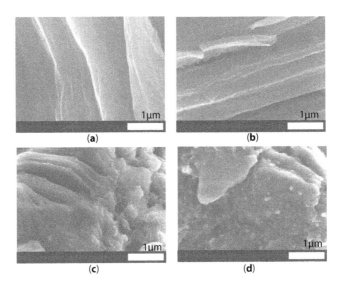

Figure 11.7 FESEM image of (a) GO, RGO–SnO$_2$ composite with different reductant concentrations: (b) 0 mol, (c) 0.025 mol, and (d) 0.050 mol [91].

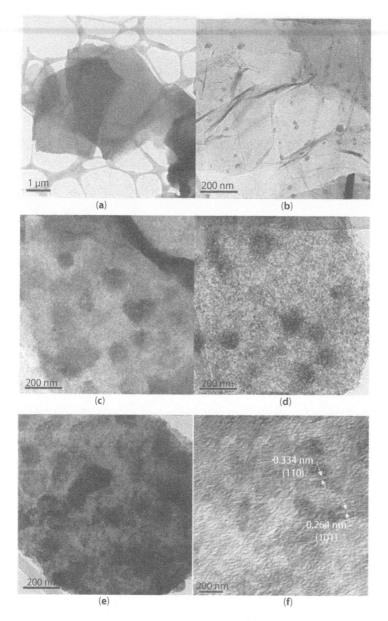

Figure 11.8 FESEM image of (a) GO, RGO–SnO$_2$ composite with different reductant concentrations: (b) 0 mol, (c) 0.025 mol, (d) 0.050 mol, and (e) 0.075 mol, (f) HRTEM image of SnO$_2$ nanoparticles [92].

in the production of high concentration of crystalline SnO$_2$ nanoparticles that agglomerate and cover the RGO sheets (Figure 11.8c–e). The HR TEM image shows clear distinct atomic planes with a d-spacing of 3.34 Å and 2.64 Å, corresponding to the (110) and (101) plane of SnO$_2$.

Fourier transform infrared (FTIR) analysis (Figure 11.9a) is able to distinguish different bond vibrations of GO such as C=O (1732 cm^{-1}), C-O-C (1252 cm^{-1}), and C-O stretching vibrations (1058 cm^{-1}) and confirms the presence of large concentrations of

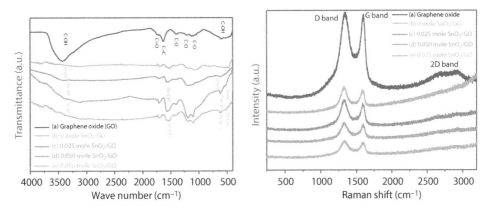

Figure 11.9 (a) FT-IR spectrum of GO, RGO–SnO$_2$ composite, (b) Raman spectra of GO, RGO–SnO$_2$ composite [91].

the oxygen functional group, which are beneficial for the growth of any metal oxide (e.g., SnO$_2$). After the addition of the tin precursor, the intensity of the absorption peaks becomes weak and the oxygen-containing functional groups are completely removed after the addition of the reducing agent, and Sn–O bond vibrations are clearly observed at 614 cm^{-1}. Similarly, Raman analysis, as shown in Figure 11.9b, clearly distinguishes different bands at 1342 cm^{-1} (D band, defect peak in the carbon), 1594.8 cm^{-1} (G band), and 2700 cm^{-1} (2D band). The 2D band, which is the characteristic signature of graphene, disappears for all RGO–SnO$_2$ composites. The intensity ratio (I_D/I_G) increases with the addition of the tin precursor in the absence of the reductant and remains the same (~1.06 ± 0.01) after using the reducing agent.

The thermogravimetric analysis in Figure 11.10 shows the mass loss as a function of temperature for GO and RGO–SnO$_2$ composites from room temperature to 800°C in air atmosphere. The mass loss at different temperatures within 100°C–550°C, is due to

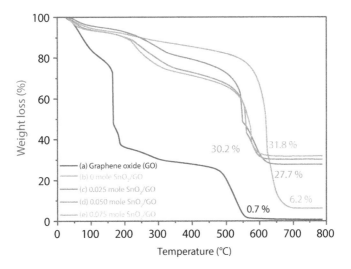

Figure 11.10 TGA spectra of GO and RGO–SnO$_2$ composites [91].

the loss of water molecules and oxygen functional groups. After 600°C, the mass loss is almost constant that the remnant corresponds to the amount of SnO_2 in the composite and it varies from 6.2% for 0 mol of reducing agent to 27.7%, 30.2%, and 31.8% for 0.025 mol, 0.050 mol, and 0.075 mol of reducing agent, respectively. This analysis confirms the increase of SnO_2 loading on RGO with the increase of the concentration of the reducing agent.

The electrochemical performance (Figure 11.11a) shows a relatively high discharge capacity of 498.7 mAh g^{-1} for the SnO_2 composite prepared without using any reducing agent and which varies to 635.3 mAh g^{-1}, 593 mAh g^{-1}, and 404 mAh g^{-1} for the RGO–SnO_2 composite prepared by using 0.025 mol, 0.075 mol, and 0.05 mol of reductant, respectively. This indicates that the highest capacity could be achieved for a relatively low level of reduction of GO. The cyclic voltammograms (CVs) for the same sample (0.025 mol) show several peaks in the reduction cycle: 0.12 V, 0.35 V due to the formation of Li and Sn alloys, 0.95 V,

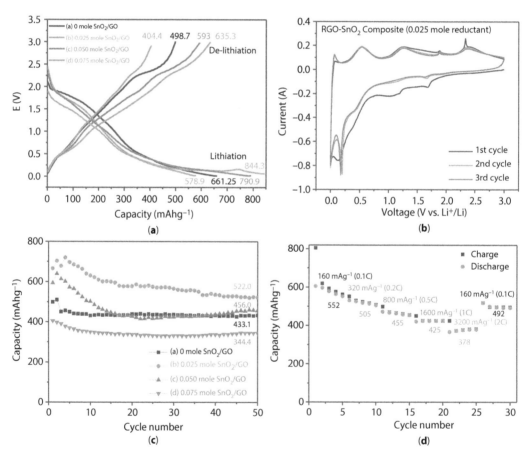

Figure 11.11 (a) Lithiation and delithiation of the RGO–SnO_2 composite, (b) cyclic voltammetry (CV) of the RGO–SnO_2 composite (0.025 mole reductant), (c) discharge capacity versus cycle number of the RGO–SnO_2 composite, (d) capacity versus cycle number of the 0.025 mole SnO_2/RGO nanocomposite. The cell was first cycled at 160 mA g^{-1} for 5 cycles with the voltage ranging between 0.2 and 3.0 V, after which the rate was increased in stages to 320 mA g^{-1} for 5 cycles, 800 mA g^{-1} for 5 cycles, 1600 mA g^{-1} for 5 cycles, and 3200 mA g^{-1} and 160 mA g^{-1} for a final 5 cycles [91].

1.17 V due to the formation of Sn and Li_2O during the reduction reaction of SnO_2 with Li^+ and the solid electrolyte interface (SEI) layer; 1.7 V is due to the conversion of Sn(IV) oxide to Sn(II) oxide. Similarly, the peaks 0.53 V, 1.24 V, and 1.88 V in the oxidation cycle corresponds to the de-alloying of Li-Sn, the partial conversion of Sn into SnO_2, and the conversion of Sn into SnO. The cyclic stability experiment (Figure 11.11c) for the same sample shows relatively high capacity (522 mAh g^{-1}) at 50 cycles, which is much better than earlier published reports on SnO_2 nanostructures as well as SnO_2-carbon composites. The uniform distribution of SnO_2 over RGO and the effective reaction between the tin oxide and the Li ion are the main reasons for this high capacity. The RGO–SnO_2 composite having high SnO_2 loading suffers from large capacity fading due to the increase in volume expansion, which leads to lower capacity (344 mAh g^{-1}).

The RGO–SnO_2 composites prepared by using high concentrations of the reducing agent show lower capacity but they show better cyclic stability, which may be ascribed to the increase in the electrical conductivity of the composite after reduction. The variation in the capacitance with the charge–discharge rate shows 552 mAh g^{-1} at 160 mA g^{-1} (0.1 C), which reduces to 378 mAh g^{-1} at 3200 mA g^{-1} (2 C) after 25 cycles. This capacity is still higher than the conventional graphite (372 mAh g^{-1}). The capacity is finally retained at 492 mAh g^{-1} at a current rate of 160 mA g^{-1} (0.1 C) after 30 cycles. This indicates that the composite is quite stable after 30 cycles, and it retains its structure after being tested with a high current charge–discharge test.

11.4.4 Graphene–Co_3O_4 as Anode in Li-Ion Battery

Owing to its high theoretical specific capacity (890 mAh g^{-1}), Co_3O_4 has the potential to meet the future requirements of energy storage devices. However, capacity fading and poor stability due to the severe volume expansion during the Li^+ insertion/extraction process damage the electrode and result in loss of contact within particles. Several efforts have been taken to overcome this problem by creating unique nanostructures of Co_3O_4 and by combining them with carbon-based materials. However, major challenges of obtaining high coulombic efficiency with good cycle life at high rate capability persist. Graphene-based Co_3O_4 composites are considered one of the alternatives to achieve the above goals because of the high surface area, excellent conductivity, flexibility, and chemical stability of graphene. Graphene layers can also provide support to the nanoparticles that help to avoid the restacking of graphene sheets during the synthesis process and keep the active surface area relatively high to facilitate lithium storage capacity as well as cyclic performance [92–96].

Synthesis of the Co_3O_4–graphene composite follows a typical procedure [92] where a $Co(OH)_2$/graphene composite precursor is first prepared by a chemical method and is then calcined at 450°C in air for 2 h to prepare the Co_3O_4/graphene composite. $Co(OH)_2$/graphene composite is prepared by adding $Co(NO_3)_2 \cdot 6H_2O$ and ammonia solution to graphene (synthesized by chemical exfoliation) dispersed in alcohol–water solution. A comparative study about electrochemical performance is carried out by using the as-prepared Co_3O_4/graphene composite, graphene, and Co_3O_4. The initial discharge capacity is 2179 mAh g^{-1}, 1105 mAh g^{-1}, and 1097 mAh g^{-1} for graphene, Co_3O_4, and Co_3O_4/graphene composite electrodes. After 30 cycles, the reversible capacity of the composite reaches ~935 mAh g^{-1}. The high capacity of the composite is due to the high surface area of graphene as well as the grain boundary area of Co_3O_4 nanoparticles. Furthermore, the rate capability of the composite is also good, i.e., 800 mAh g^{-1} (@50 mA g^{-1}) after the 10th cycle, 715 mAh g^{-1} (@150 mA g^{-1}) after the 20th cycle, and again comes back to 767 mAh g^{-1} after the 50th cycle when the rate returns to the initial 50 mA g^{-1}. Therefore, the Co_3O_4/graphene composite is a good candidate as anode materials in LIBs. The high level of performance of the composite is attributed to the flexibility of the ultrathin graphene sheets to hold Co_3O_4 nanoparticles, the good electrical conductivity of graphene that facilitates easy charge transfer between nanoparticles, the high surface area of the composite due to the very fine size of the particles and ultrathinness of the graphene, and the uniform distribution of the Co_3O_4 nanoparticles that prevents the restacking of the graphene layers.

Microwave has also been used [93] to exfoliate GO prepared by Hummer's method, which is then annealed at 850°C in NH_3 and Ar atmosphere to prepare nitrogen-modified exfoliated graphene (NMEG). The composite (Co_3O_4/NMEG) is prepared by dispersing NMEG in water, followed by the addition of different concentrations of $CoCl_2 \cdot 6H_2O$ and urea, and the resultant solution is refluxed at 90°C with magnetic stirring to form $Co(OH)_2$/NMEG. The precipitate is collected by filtration, dried overnight at 60°C, and annealed at 300°C in N_2 atmosphere for 4 h. The as-prepared Co_3O_4 shows a flower-like structure composed of Co_3O_4 nanoneedles. Similarly, the NMEG shows corrugated morphology. However, the morphology of the Co_3O_4/NMEG composite is completely different where fine Co_3O_4 nanoparticles (~5 nm) are seen uniformly distributed over the NMEG sheets. Electrochemical performance is investigated with different concentrations of Co_3O_4 loading over the NMEG sheets. Cyclic performance has shown better capacity with high stability for 70% loaded Co_3O_4 in NMEG. This also shows large initial capacity and large retention after 100 cycles (>900 mAh g^{-1}).

Atomically thin mesoporous Co_3O_4 nanosheets/graphene composite (*ATMCNs/GE*) [94] is prepared by a simple chemical procedure where previously prepared ultrathin Co_3O_4 nanosheets and mildly oxidized graphene are dispersed in an ethanol/water mixture, stirred overnight at 70°C, followed by the addition of L-ascorbic acid to reduce mildly oxidized graphene at room temperature for 24 h. The composite is centrifuged, washed with water and organic solvents for several times, and dried at 60°C.

The composite has shown remarkably high discharge capacities (2014.7 mAh g^{-1} at 0.11 C), very good rate capability, and 92.1% capacity retention after 2000 cycles over all the reports on Co_3O_4 and the composites. This high performance is ascribed to the atomic thinness and porosity of Co_3O_4 sheets, the high conductivity and flexibility of the graphene sheets, high structural stability, huge surface area, and unique layer-by-layer morphology.

Microwave has also been used [95] to prepare reduced graphene oxide (RGO) and Co_3O_4 nanocomposite where graphene oxide (GO) is prepared by Hummer's method, Co_3O_4 is prepared by microwave irradiation of $Co(acac)_3$ and urea dissolved in water and ethanol, RGO–Co_3O_4 is prepared by similar microwave irradiation of the mixture of Co(acac)3, urea, and GO dissolved/dispersed in water and ethanol. In another interesting report, the Co_3O_4 rose spheres are prepared by a hydrothermal method, and the surface of these spheres is positively charged by 3-aminopropyltrimethoxysilane (APS) under reflux condition in Ar atmosphere. The graphene and Co_3O_4 (G–Co_3O_4) composite was prepared by a self-assembly process where positively charged Co_3O_4 is attached to negatively charged GO. The composite is fabricated into LIBs, which show very high capacity

(1300 mAh g^{-1}) at a high rate of 1 C (1 C = 890 mA g^{-1}) and long cycle life over 600 cycles. The good performance may be attributed to the high surface area (222 m^{2}g^{-1}) and wide pore distribution (1.4–300 nm).

11.4.5 Graphene–Fe$_2$O$_3$ as Anode in Li-Ion Battery

Fe$_2$O$_3$ is an environmentally friendly material having several interesting properties such as high theoretical capacity (1007 mAh g^{-1}), high electrical conductivity (~2 × 10^{-4} S m^{-1}), economic, easy availability, and ease of synthesis process and is considered as one of the potential candidates for application as anode in LIBs. However, severe aggregation, large volume expansion, huge capacity fading, and simultaneously the device suffer from poor stability/rate capability over the cycles of operation during the lithium insertion/extraction process. Several innovative methods have been explored to overcome the above problems, either by fabrication of nanostructures (nanoparticles, nanoribbons, nanorods, hollow structures) of Fe$_2$O$_3$ or by doping carbon materials with Fe$_2$O$_3$. Other possible ways are to prepare a composite of Fe$_2$O$_3$ with carbon host matrix (e.g., graphene) to act as a mechanical buffer to accommodate the volume expansion during Li diffusion and improve the electrochemical property [97–101].

A variety of methods have been undertaken for the synthesis of the graphene–Fe$_2$O$_3$ composite. In one type of procedure, graphene oxide (GO) is prepared by Hummer's method and Fe$_2$O$_3$ is grown over GO. GO is reduced by the direct synthesis procedure without using a reducing agent, or a reducing agent is added in the same step or in the following step. In a typical procedure [97], aqueous solution of FeCl$_3$ and urea is slowly added to the graphene oxide dispersion, and the mixture is heated to 90°C for 1.5 h and then allowed to cool to room temperature. Then, hydrazine is added to the mixture, which is then irradiated with microwave for 2 min. The black precipitate is collected by filtration, washed with DI water to remove hydrazine as well as other impurities, and dried at 80°C for 24 h under vacuum to obtain the RG-O/Fe$_2$O$_3$ composite. The composite shows curl-like morphology, and Fe$_2$O$_3$ nanoparticles (60 nm) are uniformly distributed over the RGO sheets. The electrochemical property has also shown a significantly high capacity ~800 mAh g^{-1} at a very high current density of 800 mA g^{-1}. The discharge capacities of the RGO-Fe$_2$O$_3$ composite vary from 1693, 1142, 1120, 1098, to 1027 mAh g^{-1} for the 1st, 10th, 20th, 30th, and 50th cycles, respectively. This very good performance of this composite is ascribed to the very good interaction of Fe$_2$O$_3$ nanoparticles and RG-O platelets, the creation of space by the nanoparticles for efficient Li-ion insertion/extraction, and the high conductivity of RGO. The Li-ion diffusion mechanism with Fe$_2$O$_3$ in Li-ion batteries is described as Fe$_2$O$_3$+6Li$^+$+6e = 2Fe+$_3$Li$_2$O.

A comparative study with Fe$_2$O$_3$ only synthesized separately shows very low capacity, 130 mAh g^{-1}, after 30 cycles after achieving a high discharge capacity of 1542 mAh g^{-1} in the first cycle. Therefore, the RGO-Fe$_2$O$_3$ composite is considered as a potential candidate for anode material in LIBs.

The hydrothermal method has also been used for the synthesis of the graphene–Fe$_2$O$_3$ composite [98]. In a typical procedure, FeCl$_3$·6H$_2$O, ascorbic acid, PEG, and urea are mixed in a desired concentration in deionized (DI) water followed by dispersion of graphene sheets, and the resultant solution mixture is subjected to hydrothermal treatment at 120°C for 12 h and then cooled to room temperature naturally. The resulting black precipitate of graphene–Fe$_2$O$_3$ is collected by centrifugation, washed, and dried at 80°C in vacuum.

The graphene–Fe_2O_3 composite obtained by the above hydrothermal method has graphene platelets (3–5 nm thickness, 9–15 layers) and microsize Fe_2O_3 particles. The composite is used as the anode material in LIBs showed very high discharge capacity (660 mAh g^{-1} at a current density of 160 mA g^{-1}) up to 100 cycles. Discharge capacity as obtained in the first cycle is nearly 1800 mAh g^{-1}.

Highly conducting graphene nanoribbons (GNRs) and the Fe_2O_3 (~10 nm) nanocomposite have also been synthesized [99] and used as anode in LIBs that showed very high capacity (>910 mAh g^{-1} at a current density of 200 mA g^{-1}) over 130 cycles. The very high electrochemical property of the as-prepared nanocomposite is due to the distinct structure of the Fe_2O_3 NPs uniformly coated over GNRs. The synthesis of the GNR/Fe_2O_3 composite follows two steps where the GNR/Fe composite is prepared in the first step, oxidized to form GNR/Fe_2O_3, and annealed in air at 250 or 300°C for 14 h. GNR has a width of 200 nm and a length up to 100 μm. Iron oxide (Fe_2O_3) fiber/reduced graphene oxide (rGO) composites synthesized [100] by electrospinning followed by annealing and reduction by far-infrared radiation reveal good capacity (1085.2 mAh g^{-1} at 0.1 A g^{-1}), excellent cycle life (407.8 mAh g^{-1} at 5 A g^{-1} for 1500 cycles), high coloumbic efficiency with very good rate capability. In a very easy synthesis procedure, the composite is prepared by mixing the previously prepared GO and Fe_2O_3 in appropriate ratio in alcohol and the mixture is irradiated with far-infrared light.

Electron beam irradiation has also been used for the synthesis of the Fe_2O_3/rGO nanocomposite where narrow-size (~2 nm), amorphous Fe_2O_3 nanoparticles are anchored on RGO sheets. Commercially obtained GO dispersed in DI water is added to aqueous solution $Fe(NH_4)_2(-SO_4)_2 6H_2O$, and the resultant solution is packaged well before irradiating with a 1-MeV electron beam accelerator. The as-obtained product is washed several times in DI water and heated to 300°C for 3 h under vacuum. The composite shows good capacity of 1064 mAh g^{-1} (current density = 200 mA g^{-1}) after 100 cycles with very good retention (88%) and very high rate capability (580 mAh g^{-1} at 5000 mA g^{-1}). The superior electrochemical performance is due to the fine size of the nanoparticle, amorphous structure, high surface area (236 m^2 g^{-1}), and RGO sheets having very high electrical conductivity.

11.5 Conclusion

The graphene–metal oxide composite is considered a suitable alternative to metal oxide as anode material in LIBs. Facile, low-temperature chemical processes are usually used for the production of these composites. The nanoparticles of metal oxide can be grown directly (*in situ*) during the chemical reaction or it may also be possible to incorporate the metal oxide nanoparticles in graphene by soft chemical processes. It is also possible to reduce graphene oxide by the *in situ* chemical process or by using a reducing agent (e.g., hydrazine). The graphene–metal oxide composite can provide high reversible capacity for longer cycles (>100 cycles) with good coloumbic efficiency, high stability, and high rate capability. The superior electrochemical properties are generally attributed to the high conductivity of graphene/reduced graphene oxide, the narrow size of the nanoparticles, the uniform distribution of the metal oxide nanostructure over the graphene sheets, the high surface area of both metal oxide and graphene, and space creation by the metal oxide within graphene sheets.

Acknowledgment

This work was financially supported by the Hierarchical Green-Energy Materials (Hi-GEM) Research Center, from The Featured Areas Research Center Program within the framework of the Higher Education Sprout Project by the Ministry of Education (MOE) and the Ministry of Science and Technology (MOST 107-3017-F-006-003) in Taiwan.

References

1. Etacheri, V., Marom, R., Elazari, R., Salitra, G., Aurbach, D., Challenges in the development of advanced Li-ion batteries: A review. *Energy Environ. Sci.*, 4, 3243, 2011.
2. Wu, H.B., Chen, J.S., Hng, H.H., (David) Lou, X.W., Nanostructured metal oxide-based materials as advanced anodes for lithium-ion batteries. *Nanoscale*, 4, 2526, 2012.
3. Chen, J., Recent progress in advanced materials for lithium ion batteries. *Materials*, 6, 156, 2013.
4. Reddy, M.V., Subba Rao, G.V., Chowdari, B.V.R., Metal oxides and oxysalts as anode materials for Li ion batteries. *Chem. Rev.*, 113, 5364, 2013.
5. Goriparti, S., Miele, E., Angelis, F.D., Fabrizio, E.D., Zaccaria, R.P., Capiglia, C., Review on recent progress of nanostructured anode materials for Li-ion batteries. *J. Power Sources*, 257, 421, 2014.
6. Deng, D., Li-ion batteries: Basics, progress, and challenges. *Energy Sci. Eng.*, 3, 5, 385, 2015.
7. Nitta, N., Wu, F., Lee, J.T., Yushin, G., Li-ion battery materials: Present and future. *Mater. Today*, 18, 5, 252, 2015.
8. Wu, S., Xu, R., Lu, M., Ge, R., Iocozzia, J., Han, C., Jiang, B., Lin, Z., Graphene-containing nanomaterials for lithium-ion batteries. *Adv. Energy Mater.*, 1, 1500400, 2015.
9. Zhao, Y., Li, X., Yan, B., Xiong, D., Li, D., Lawes, S., Sun, X., Recent developments and understanding of novel mixed transition-metal oxides as anodes in lithium ion batteries. *Adv. Energy Mater.*, 6, 1502175, 2016.
10. Tarascon, J.M. and Armand, M., Issues and challenges facing rechargeable lithium batteries. *Nature*, 414, 359, 2001.
11. Sato, K., Noguchi, M., Demachi, A., Oki, N., Endo, M., A mechanism of lithium storage in disordered carbons. *Science*, 264, 556, 1994.
12. Etacheri, V., Marom, R., Elazari, R., Salitra, G., Aurbach, D., Challenges in the development of advanced Li-ion batteries: A review. *Energy Environ. Sci.*, 4, 3243–3262, 2011.
13. Lee, K.T. and Cho, J., Roles of nanosize in lithium reactive nanomaterials for lithium ion batteries. *Nano Today*, 6, 28, 2011.
14. Poizot, P., Laruelle, S., Grugeon, S., Dupont, L., Tarascon, J.M., Nano-sized transition-metal oxides as negative-electrode materials for lithium-ion batteries. *Nature*, 407, 496, 2000.
15. Chen, J., Wang, H, Y.X., Xu, S., Fang, M., Zhao, X., Shang, Y., Electrochemical properties of MnO_2 nanorods as anode materials for lithium ion batteries. *Electrochim. Acta*, 142, 152, 2014.
16. Jin-Yun Liao, D.H., Lui, G., Chabot, V., Xiao, X., Chen, Z., Multifunctional TiO_2-C/MnO_2 core-double-shell nanowire arrays as high-performance 3D electrodes for lithium ion batteries. *Nano Lett.*, 13, 5467, 2013.
17. Hu, Z. and Liu, H., Three-dimensional CuO microflowers as anode materials for Li-ion batteries. *Ceram. Int.*, 41, 8257, 2015.
18. Wang, D., Yu, Y., He, H., Wang, J., Zhou, W., Abru-na, H.D., Template-free synthesis of hollow-structured Co_3O_4 nanoparticles as high-performance anodes for lithium-ion batteries. *ACS Nano*, 9, 1775, 2015.

19. Yu, S.H., Lee, D.J., Park, M., Kwon, S.G., Lee, H.S., Jin, A., Lee, K.S., Lee, J.E., Oh, M.H., Kang, K., Sung, Y.E., Hyeon, T., Hybrid cellular nanosheets for high-performance lithium-ion battery anodes. *J Am. Chem. Soc.*, 137, 11954, 2015.
20. Li, Z., Tan, Y., Huang, X., Zhang, W., Gao, Y., Tang, B., Three-dimensionally ordered macroporous SnO_2 as anode materials for lithium ion batteries. *Ceram. Int.*, 2016.
21. Su, H., Xu, Y.F., Feng, S.C., Wu, Z.G., Sun, X.P., Shen, C.H., Wang, J.Q., Li, J.T., Huang, L., Sun, S.G., Hierarchical Mn(2)O(3) hollow microspheres as anode material of lithium ion battery and its conversion reaction mechanism investigated by XANES. *ACS Appl. Mater. Interfaces*, 7, 8488, 2015.
22. Wang, J.G., Jin, D., Zhou, R., Li, X., Liu, X.R., Shen, C., Xie, K., Li, B., Kang, F., Wei, B., Highly flexible graphene/MnO nanocomposite membrane as advanced anodes for Li-ion batteries. *ACS Nano*, 2016.
23. Zhang, Y., Luo, Z., Xiao, Q., Sun, T., Lei, G., Li, Z., Li, X., Freestanding manganese dioxide nanosheet network grown on nickel/polyvinylidene fluoride coaxial fiber membrane as anode materials for high performance lithium ion batteries. *J. Power Sources*, 297, 442, 2015.
24. Liu, H., Hu, Z., Tian, L., Su, Y., Ruan, H., Zhang, L., Hu, R., Reduced graphene oxide anchored with δ-MnO_2 nanoscrolls as anode materials for enhanced Li-ion storage. *Ceram. Int.*, 42, 13519, 2016.
25. Cho, J.S., Hong, Y.J., Kang, Y.C., Design and synthesis of bubble-nanorod-structured Fe_2O_3-carbon nanofibers as advanced anode material for Li-Ion batteries. *ACS Nano*, 9, 2015.
26. Wu, H.B., Chen, J.S., Hng, H.H., (Davod) Lou, X.W., Nanostructured metal oxide-based materials as advanced anodes for lithium-ion batteries. *Nanoscale*, 4, 2526, 2012.
27. Jiao, Z., Chen, D., Jiang, Y., Zhang, H., Ling, X., Zhuang, H., Su, L., Cao, H., Hou, M., Zhao, B., Synthesis of nanoparticles, nanorods, and mesoporous SnO_2 as anode materials for lithium-ion batteries. *J. Mater. Res.*, 29, 609, 2014.
28. Ying, Z., Wan, Q., Cao, H., Song, Z.T., Feng, S.L., Characterization of SnO_2 nanowires as an anode material for Li-ion batteries. *Appl. Phys. Lett.*, 87, 113108, 2005.
29. Wang, J., Du, N., Zhang, H., Yu, J., Yang, D., Large-scale synthesis of SnO_2 nanotube arrays as high-performance anode materials of Li-ion batteries. *J. Phys. Chem. C*, 115, 22, 11302, 2011.
30. Han, S., Jang, B., Kim, T., Oh, S.M., Hyeon, T., Simple synthesis of hollow tin dioxide microspheres and their application to lithium-ion battery anodes. *Adv. Funct. Mater.*, 15, 1845, 2005.
31. Cao, F.F., Guo, Y.G., Zheng, S.F., Wu, X.L., Jiang, L.Y., Bi, R.R., Wan, L.J., Maier, J., Symbiotic coaxial nanocables: Facile synthesis and an efficient and elegant morphological solution to the lithium storage problem. *Chem. Mater.*, 22, 1908, 2010.
32. Wang, D.H., Choi, D.W., Li, J., Yang, Z.G., Nie, Z.M., Kou, R., Hu, D.H., Wang, C.M., Saraf, L.V., Zhang, J.G., Aksay, I.A., Liu, J., Self-assembled TiO_2–graphene hybrid nanostructures for enhanced Li-ion insertion. *ACS Nano*, 3, 907, 2009.
33. Chen, J.S., Tan, Y.L., Li, C.M., Cheah, Y.L., Luan, D.Y., Madhavi, S., Boey, F.Y.C., Archer, L.A., Lou, X.W., Constructing hierarchical spheres from large ultrathin anatase TiO_2 nanosheets with nearly 100% exposed (001) facets for fast reversible lithium storage. *J. Am. Chem. Soc.*, 132, 6124, 2010.
34. Ding, S., Chen, J.S., Luan, D., Boey, F.Y.C., Madhavi, S., Lou, X.W., Graphene-supported anatase TiO_2 nanosheets for fast lithium storage. *Chem. Commun.*, 47, 5780, 2011.
35. Sun, C.H., Yang, X.H., Chen, J.S., Li, Z., Lou, X.W., Li, C., Smith, S.C., Lu, G.Q., Yang, H.G., Higher charge/discharge rates of lithium-ions across engineered TiO_2 surfaces leads to enhanced battery performance. *Chem. Commun.*, 46, 6129, 2010.
36. Liu, H., Wang, G., Park, J., Wang, J., Liu, H., Zhang, C., Electrochemical performance of -Fe_2O_3 nanorods as anode material for lithium-ion cells. *Electrochim. Acta*, 54, 1733, 2009.

37. Xiao, W., Wang, Z., Guo, H., Zhang, Y., Zhang, Q., Gan, L., A facile PVP-assisted hydrothermal fabrication of Fe_2O_3/graphene composite as high performance anode material for lithium ion batteries. *J. Alloys Compd.*, 560, 208, 2013.
38. Chan, C.K., Peng, H., Liu, G., McIlwrath, K., Zhang, X.F., Huggins, R.A., Cui, Y., High-performance lithium battery anodes using silicon nanowires. *Nat. Nanotechnol.*, 3, 31, 2008.
39. Reddy, M.V., Yu, T., Sow, C.-H., Shen, Z.X., Lim, C.T., Subba Rao, G.V., Chowdari, B.V.R., α-Fe_2O_3 nanoflakes as an anode material for Li-ion batteries. *Adv. Funct. Mater.*, 17, 2792, 2007.
40. Lin, Y.-M., Abel, P.R., Heller, A., Mullins, C.B., α-Fe_2O_3 nanorods as anode material for lithium ion batteries. *J. Phys. Chem. Lett.*, 2, 2885, 2011.
41. Na, Z., Huang, G., Liang, F., Yin, D., Wang, L., A core–shell Fe/Fe_2O_3 nanowire as a high-performance anode material for lithium-ion batteries. *Chem. Eur. J.*, 22, 12081, 2016.
42. Wang, D., Dong, H., Zhang, H., Zhang, Y., Xu, Y., Zhao, C., Sun, Y., Zhou, N., Enabling a high performance of mesoporous α-Fe_2O_3 anodes by building a conformal coating of cyclized-PAN network. *ACS Appl. Mater. Interfaces*, 8, 19524, 2016.
43. Zhang, X., Zhou, Z., Ning, J., Nigar, S., Zhao, T., Lub, X., Caog, H., 3D dendritic-Fe_2O_3@C nanoparticles as an anode material for lithium ion batteries. *RSC Adv.*, 7, 18508, 2017.
44. Qu, B., Sun, Y., Liu, L., Li, C., Yu, C., Zhang, X., Chen, Y., Ultra-small Fe_2O_3 nanoparticles/MoS_2 nanosheets composite as high-performance anode material for lithium ion batteries. *Sci. Rep.*, 7, 42772 (1–11).
45. Yan, N., Hu, L., Li, Y., Wang, Y., Zhong, H., Hu, X., Kong, X., Chen, Q., Co_3O_4 nanocages for high-performance anode material in lithium-ion batteries. *J. Phys. Chem. C*, 116, 7227, 2012.
46. Leng, X., Wei, S., Jiang, Z., Lian, J., Wang, G., Jiang, Q., Carbon-encapsulated Co_3O_4 nanoparticles as anode materials with super lithium storage performance. *Sci. Rep.*, 5, 16629.
47. Sun, S., Zhao, X., Yang, M., Wu, L., Wen, Z., Shen, X., Hierarchically ordered mesoporous Co_3O_4 materials for high performance Li-ion batteries. *Sci. Rep.*, 6, 19564.
48. Mujtaba, J., Sun, H., Huang, G., Mølhave, K., Liu, Y., Zhao, Y., Wang, X., Xu, Zhu, S.J., Nanoparticle decorated ultrathin porous nanosheets as hierarchical Co_3O_4 nanostructures for lithium ion battery anode materials. *Sci. Rep.*, 6, 20592.
49. Liu, Y., Mi, C., Su, L., Zhang, X., Hydrothermal synthesis of Co_3O_4 microspheres as anode material for lithium-ion batteries. *Electrochim. Acta*, 53, 2507, 2008.
50. Liu, X., Wing, S., Jin, C., Lv, Y., Li, W., Feng, C., Xiao, F., Sun, Y., Co_3O_4/C nanocapsules with onion-like carbon shells as anode material for lithium ion batteries. *Electrochim. Acta*, 100, 140, 2013.
51. Pan, A., Wang, Y., Xu, W., Nie, Z., Liang, S., Nie, Z., Wang, C., Cao, G., Zhang, J.-G., High-performance anode based on porous Co_3O_4 nanodiscs. *J. Power Sources*, 255, 125, 2014.
52. Wang, H., Mao, N., Shi, J., Wang, Q., Yu, W., Wang, X., Cobalt oxide-carbon nanosheet nanoarchitecture as an anode for high-performance lithium-ion battery. *ACS Appl. Mater. Interfaces*, 7, 2882, 2015.
53. Wang, B., Lu, X.-Y., Tang, Y., Synthesis of snowflake-shaped Co_3O_4 with a high aspect ratio as a high capacity anode material for lithium ion batteries. *J. Mater. Chem. A*, 3, 9689, 2015.
54. Zheng, F., Shi, K., Xu, S., Liang, X., Chena, Y., Zhang, Y., Facile fabrication of highly porous Co_3O_4 nanobelts as anode materials for lithium-ion batteries. *RSC Adv.*, 6, 9640, 2016.
55. Li, L., Zhang, Z., Ren, S., Zhang, B., Yang, S., Cao, B., Construction of hollow Co_3O_4 cubes as a high performance anode for lithium ion batteries. *New J. Chem.*, 41, 7960, 2017.
56. Yu, X., Cheng, H., Zhang, M., Zhao, Y., Qu, L., Shi, G., Graphene-based smart materials. *Nat. Rev.*, 2, 17046, 1–13, 2017.
57. Novoselov, K.S., Falko, V.I., Colombo, L., Gellert, P.R., Schwab, M.G., Kim, K., A roadmap for graphene. *Nature*, 490, 192, 2012.
58. Rao, C.N.R., Sood, A.K., Voggu, R., Subrahmanyam, K.S., Some novel attributes of graphene. *J. Phys. Chem. Lett.*, 1, 572, 2010.

59. Allen, M.J., Tung, V.C., Kaner, R.B., Honeycomb carbon: A review of graphene. *Chem. Rev.*, 110, 132, 2010.
60. Rao, C.N.R., Sood, A.K., Subrahmanyam, K.S., Govindaraj, A., Graphene: The new two-dimensional nanomaterial. *Angew. Chem. Int. Ed.*, 48, 7752, 2009.
61. Eizenberg, M. and Blakely, J.M., Carbon monolayer phase condensation on Ni (111). *Surf. Sci.*, 82, 228, 1979.
62. Aizawa, T., Souda, R., Otani, S., Ishizawa, Y., Oshima, C., Anomalous bond of monolayer graphite on transition-metal carbide surfaces. *Phys. Rev. Lett.*, 64, 7, 768, 1990.
63. Lu, X., Yu, M., Huang, H., Ruoff, R.S., Tailoring graphite with the goal of achieving single sheets. *Nanotechnology*, 10, 269, 1999.
64. Novoselov, K.S., Geim, A.K., Morozov, S.V., Jiang, D., Zhang, Y., Dubonos, S.V., Grigorieva, I.V., Firsov, A.A., Electric field effect in atomically thin carbon films. *Science*, 306, 666, 2004.
65. Wang, J.J., Zhu, M.Y., Outlaw, R.A., Zhao, X., Manos, D.M., Holloway, B.C., Mammana, V.P., Free-standing subnanometer graphite sheets. *Appl. Phys. Lett.*, 85, 1265, 2004.
66. Zheng, X., Peng, Y., Yang, Y., Chen, J., Tian, H., Cui, X., Zheng, W., Hydrothermal reduction of graphene oxide; effect on surface-enhanced Raman scattering. *J. Raman Spectrosc.*, 48, 97, 2017.
67. Choucair, M., Thordarson, P., Stride, J.A., Gram-scale production of graphene based on solvothermal synthesis and sonication. *Nat. Nanotech.*, 4, 30, 2009.
68. Yoo, E.J., Kim, J., Hosono, E., Zhou, H.S., Kudo, T., Honma, I., Large reversible Li storage of graphene nanosheet families for use in rechargeable lithium ion batteries. *Nano Lett.*, 8, 2277, 2008.
69. Wang, G., Wang, B., Wang, X., Park, J., Dou, S., Ahn, H., Kim, K., Sn/graphene nanocomposite with 3D architecture for enhanced reversible lithium storage in lithium ion batteries. *J. Mater. Chem.*, 19, 8378, 2009.
70. Abouimrane, A., Compton, O.C., Amine, K., Nguyen, S.T., Non-annealed graphene paper as a binder-free anode for lithium-ion batteries. *J. Phys. Chem. C*, 114, 12800, 2010.
71. Zhu, N., Liu, W., Xue, M.Q., Xie, Z.A., Zhao, D., Zhang, M.N., Chen, J.T., Cao, T.B., Graphene as a conductive additive to enhance the high-rate capabilities of electrospun $Li_4Ti_5O_{12}$ for lithium-ion batteries. *Electrochim. Acta*, 55, 5813, 2010.
72. Yoo, E.J., Kim, J., Hosono, E., Zhou, H.-S., Kudo, T., Honma, I., Large reversible Li storage of graphene nanosheet families for use in rechargeable lithium ion batteries. *Nano Lett.*, 8, 2277, 2008.
73. Guo, P., Song, H., Chen, X., Electrochemical performance of graphene nanosheets as anode material for lithium-ion batteries. *Electrochem. Commun.*, 11, 1320, 2009.
74. Wang, G., Shen, X., Yao, J., Park, J., Graphene nanosheets for enhanced lithium storage in lithium ion batteries. *Carbon*, 47, 2049, 2009.
75. Lian, P., Zhu, X., Liang, S., Li, Z., Yang, W., Wang, H., Large reversible capacity of high quality graphene sheets as an anode material for lithium-ion batteries. *Electrochim. Acta*, 55, 3909, 2010.
76. Wu, Z.-S., Ren, W., Xu, L., Li, F., Cheng, H.-M., Doped graphene sheets as anode materials with super high rate and large capacity for lithium ion batteries. *ACS Nano*, 5, 7, 5463, 2011.
77. Li, X., Geng, D., Zhang, Y., Meng, X., Li, R., Sun, X., Superior cycle stability of nitrogen-doped graphene nanosheets as anodes for lithium ion batteries. *Electrochem. Commun.*, 13, 822, 2011.
78. Li, N., Chen, Z., Ren, W., Li, F., Cheng, H.-M., Flexible graphene-based lithium ion batteries with ultrafast charge and discharge rates. *PNAS*, 109, 43, 17360, 2012.
79. Kheirabadi, N. and Shafiekhani, A., Graphene/Li-ion battery. *J. Appl. Phys.*, 112, 124323, 1–5, 2012.
80. Hassoun, J., Bonaccorso, F., Agostini, M., Angelucci, M., Grazia Betti, M., Cingolani, R., Gemmi, M., Mariani, C., Panero, S., Pellegrini, V., Scrosati, B., An advanced lithium-ion battery based on a graphene anode and a lithium iron phosphate cathode. *Nano Lett.*, 14, 4901, 2014.

81. Fu, C., Song, C., Liu, L., Xie, X., Zhao, W., Synthesis and properties of nitrogen-doped graphene as anode materials for lithium-ion batteries. *Int. J. Electrochem. Sci.*, 11, 3876–3886, 2016.
82. Cheng, Q., Okamoto, Y., Tamura, N., Tsuji, M., Maruyama, S., Matsuo, Y., Graphene-like-graphite as fast-chargeable and high-capacity anode materials for lithium ion batteries. *Sci. Rep.*, 7, 14782, 1–14, 2017.
83. Yu, A., Park, H.W., Davies, A., Higgins, D.C., Chen, Z., Xiao, X., Free-standing layer-by-layer hybrid thin film of graphene–MnO_2 nanotube as anode for lithium ion batteries. *J. Phys. Chem. Lett.*, 2, 1855, 2011.
84. Xing, L., Cui, C., Ma, C., Xue, X., Facile synthesis of α-MnO_2/graphene nanocomposites and their high performance as lithium-ion battery anode. *Mater. Lett.*, 65, 2104, 2011.
85. Zhang, Y., Liu, H., Zhu, Z., Wong, K.-W., Mi, R., Mei, J., Lau, W.-M., A green hydrothermal approach for the preparation of graphene/α-MnO_2 3D network as anode for lithium ion battery. *Electrochim. Acta*, 108, 465, 2013.
86. Chen, J., Wang, Y., He, X., Xu, S., Fang, M., Zhao, X., Shang, Y., Electrochemical properties of MnO_2 nanorods as anode materials for lithium ion batteries. *Electrochim. Acta*, 142, 152, 2014.
87. Kim, S.J., Yun, Y.J., Kim, K.W., Chae, C., Jeong, S., Kang, Y., Choi, S.Y., Lee, S.S., Choi, S., Superior lithium storage performance using sequentially stacked MnO_2/reduced graphene oxide composite electrodes. *Chem. Sus. Chem.*, 8, 1484, 2015.
88. Wen, K., Chen, G., Jiang, F., Zhou, X., Yang, J., A facile approach for preparing MnO_2–graphene composite as anode material for lithium-ion batteries. *Int. J. Electrochem. Sci.*, 10, 3859, 2015.
89. Jiang, Y., Jiang, Z.-J., Chen, B., Jiang, Z., Cheng, S., Rong, H., Huang, J., Liu, M., Morphology and crystal phase evolution induced performance enhancement of MnO_2 grown on reduced graphene oxide for lithium ion batteries. *J. Mater. Chem. A*, 4, 2643, 2016.
90. Weng, S.-C., Brahma, S., Chang, C.-C., Huang, J.-L., Synthesis of MnOx/reduced graphene oxide nanocomposite as an anode for lithium-ion battery. *Ceram. Int.*, 43, 50, 2017.
91. Hou, C.-C., Brahma, S., Weng, S.-C., Chang, C.-C., Huang, J.-L., Facile, low temperature synthesis of SnO_2–RGO nanocomposite as negative electrode materials for lithium-ion batteries. *Appl. Surf. Sci.*, 413, 160, 2017.
92. Wu, Z.-S., Ren, W., Wen, L., Gao, L., Zhao, J., Chen, Z., Zhou, G., Li, F., Cheng, H.-M., Graphene anchored with Co_3O_4 nanoparticles as anode of lithium ion batteries with enhanced reversible capacity and cyclic performance. *ACS Nano*, 4, 6, 3187, 2010.
93. Lai, L., Zhu, J., Li, Z., Yu, D.Y.W., Jiang, S., Cai, X., Yan, Q., Lam, Y.M., Shen, Z., Lin, J., Co_3O_4/nitrogen modified graphene electrode as Li-ion battery anode with high reversible capacity and improved initial cycle performance. *Nano Energy*, 3, 134, 2014.
94. Dou, Y., Xu, J., Ruan, B., Liu, Q., Pan, Y., Sun, Z., Dou, S.X., Atomic layer-by-layer Co_3O_4/graphene composite for high performance lithium-ion batteries. *Adv. Energy Mater.*, 6, 1501835, 2016.
95. He, J., Liu, Y., Meng, Y., Sun, X., Biswas, S., Shen, M., Luo, Z., Miao, R., Zhang, L., Mustain, W.E., Suib, S.L., High-rate and long-life of Li-ion batteries using reduced graphene oxide/Co_3O_4 as anode materials. *RSC Adv.*, 6, 24320, 2016.
96. Jing, M., Zhou, M., Li, G., Chen, Z., Xu, W., Chen, X., Hou, Z., Graphene-embedded Co_3O_4 rose-spheres for enhanced performance in lithium ion batteries. *ACS Appl. Mater. Interfaces*, 9, 9662, 2017.
97. Zhu, X., Zhu, Y., Murali, S., Stoller, M.D., Ruoff, R.S., Nanostructured reduced graphene oxide/Fe_2O_3 composite as a high-performance anode material for lithium ion batteries. *ACS Nano*, 5, 4, 3333, 2011.
98. Wang, G., Liu, T., Luo, Y., Zhao, Y., Ren, Z., Bai, J., Wang, H., Preparation of Fe_2O_3/graphene composite and its electrochemical performance as an anode material for lithium ion batteries. *J. Alloys Compd.*, 509, L216, 2011.

99. Lin, J., Raji, A.-R.O., Nan, K., Peng, Z., Yan, Z., Samuel, E.L.G., Natelson, D., Tour, J.M., Iron oxide nanoparticle and graphene nanoribbon composite as an anode material for high-performance Li-ion batteries. *Adv. Funct. Mater.*, 24, 2044, 2014.
100. Cai, J., Zhao, P., Li, Z., Li, W., Zhong, J., Yub, J., Yang, Z., A corn-inspired structure design for an iron oxide fiber/reduced graphene oxide composite as a high performance anode material for Li-ion batteries. *RSC Adv.*, 7, 44874, 2017.
101. Zhu, X., Jiang, X., Chen, X., Liu, X., Xiao, L., Cao, Y., Fe_2O_3 amorphous nanoparticles/graphene composite as high-performance anode materials for lithium-ion batteries. *J. Alloys Compd.*, 711, 15, 2017.

12

Graphene/TiO$_2$ Nanocomposites: Synthesis Routes, Characterization, and Solar Cell Applications

Chin Wei Lai*, Foo Wah Low, Siti Zubaidah Binti Mohamed Siddick and Joon Ching Juan

Nanotechnology & Catalysis Research Centre (NANOCAT), Level 3, Institute Graduate Studies (IGS) Building, University of Malaya (UM), Kuala Lumpur, Malaysia

Abstract

Renewable solar cell energy is a key target for sustainable energy development, which is inexhaustible and nonpolluting for our energy system. Today, nanomaterials are widely applied in solar-cell-related technologies, including photovoltaic as well as dye-sensitized solar cell (DSSC) systems. The commonly used nanomaterials are metal oxide, organic-based substances, and polymer-based materials. The utmost concern of these nanomaterials in practical application is the constraint of its high recombination losses, low photo-conversion efficiency, and toxicity matter. To bring more solar-related technologies to the point of commercial readiness and viability in terms of performance and cost, substantial research on the development of highly efficient renewable solar cell energy system is necessary. Recent studies have indicated that graphene (Gr) is a relatively novel material with unique properties that could be applied in photoanode/counter electrode components such as efficient electrodes. In fact, the atom-thick 2D structure of Gr provides an extraordinarily high conductivity, repeatability, productivity, and prolonged lifetime to the related solar cell applications. Continuous efforts have been exerted to further improve the Gr textural and electronic properties by loading an optimum content of photocatalyst for a high-efficiency renewable solar cell energy system. In the field of photocatalysis today, titanium dioxide (TiO$_2$) has emerged as an efficient photocatalyst in solar cell applications because of its unique characteristics, such as high stability against corrosion, nontoxicity, good photocatalytic property, and ready availability. However, the high efficiency of the Gr/TiO$_2$ nanocomposite (NC) as photoanode/counter electrode requires a suitable architecture that minimizes electron loss at nanostructure connections and maximizes photon absorption. Notably, Gr/TiO$_2$ NC-based photoanodes/counter electrodes will benefit photon absorption, charge separation, and charge carrier transport. In this chapter, different synthesis strategies and characterization analyses for Gr/TiO$_2$ NC as well as its prospects in solar-cell-related applications will be reviewed in detail. Indeed, innovative new approaches and synthesis of high-quality Gr/TiO$_2$ NC is crucial for determining the potential of the material as an efficient photoanode/counter electrode in solar-cell-related applications.

Keywords: Gr/TiO$_2$ nanocomposites, renewable solar cell energy, photoanode/counter electrode, photon absorption, photoconversion efficiency

---

*Corresponding author: cwlai@um.edu.my

12.1 Introduction

Nowadays, there has been an increasing energy demand for fossil fuels, and this can be seen by the growing trend of energy produced by fossil fuels. Indeed, the utilization of non-renewable resources mainly could bring out many environmental and public health risks associated with burning fossil fuels, the most serious in terms of its universal and potentially irreversible consequences is global warming. Thus, many scientists have been doing research to obtain the best solution to secure our future energy management and ensure the availability of the energy source to produce sufficient electricity [1]. It seems that homeowners around the world are finally waking up to the need for a mass switch to renewable energy or environmental friendly energy resources with less dependency toward fossil fuels [2, 3]. Among all the renewable power sources, solar energy is the most easily exploitable, inexhaustible, quiet, and adjustable to enormous applications [2]. In this case, sunlight is a potential energy that can be utilized to harness the sun's energy and make it useable in our photovoltaic technology industry for generating electrical energy. A typical photovoltaic system has attractive features for large-area applications including less secondary environment pollution contribution, no nuclear waste by-products, inexhaustibility, and no greenhouse by-product waste [3]. The discovery of the photovoltaic effect was made in 1839 by Antoine-Cesar Becquerel. He discovered that light falling upon a solid electrode in an electrolyte solution produced a phenomenon now known as the photoelectric effect, whereby electrons were released from the electrode surface [4]. Then, Albert Einstein clearly reported the photoelectric effect in which electrons are emitted from matter after the absorption of energy from electromagnetic radiation in early 1905 [5].

Bell Labs demonstrates the first practical silicon solar cell in 1954, known as the first generation of p–n junction photovoltaic solar cell. The two major types of photovoltaic cell materials used are monocrystalline silicon and polycrystalline silicon doped with other materials, which vary from each other in terms of light absorption efficiency. Monocrystalline solar panels have the highest efficiency rates since they are made out of the highest-grade silicon. The efficiency rates of monocrystalline solar panels are typically 15%–20%. Nevertheless, monocrystalline solar panels are the most expensive. From a financial standpoint, a solar panel that is made of polycrystalline silicon can be a better choice [5]. In 1954, Hoffman introduced a commercial photovoltaic cell using amorphorus polycrystalline compound semiconductors such as amorphous silicon (A-Si), cadmium telluride (CdTe), and copper indium gallium selenite (CIGS) to further improve its efficiency, which were known as the second generation of photovoltaic cell. The third generation of photovoltaic cell is the amorphous or thin-film solar cell, which consists of three types of film cell structure, namely, single junction, twin junction, and multiple junctions, wherein it was differentiated by the number of p–i–n junctions. In order to further increase the efficiency of the thin-film solar cell, the processing of potential materials into thin films allows easy integration and has been studied and explored extensively [8, 9]. In general, both first and second generation of solar cells are fundamentally derived from semiconductor materials. Meanwhile, the third generation of solar cells could potentially overcome the Shockley–Queisser limit of 31%–41% power efficiency for single band gap solar cells. The third generation of solar cells also covers expensive high-performance experimental multijunction solar cells that hold the world record in solar cell performance.

In fact, third-generation solar cells consist of dye-sensitized solar cells (DSSCs), heterojunction cells, polymer solar cells, and quantum dots. However, the new type of quantum dot applied to solar cells still at initial research stage. In this manner, DSSCs have emerged as so-called "simple and cheap solar cells" that have been universally promoted as an economically and environmentally viable renewable technology option to traditional solar cell technologies [6]. DSSCs were first developed by Michael Gratzel and Brian O'Regan using the combination of nanostructure electrodes and efficient charge transportation through commercial dye injection [7, 8]. The use of thin nanocrystalline mesoporous TiO_2 thin film and intensify photocatalytic activity under the light illumination was developed in this DSSC device.

Then, the conversion of sunlight into electrical energy occurred in the DSSC device [5, 9–11]. In general, DSSC device built up by five main components, where transparent conductive oxide (TCO), photoanode, dye, electrolyte, and counter electrode [12–16]. The DSSC device can operate accordingly at where photon absorption and selectable charge generation transports occur at the interface of TiO_2/dye/electrolyte through different components. In other words, dye molecules absorb the photon and then photo-induced charge carriers are generated. Meanwhile, TiO_2 acts as a transport path for photo-induced charge transport across the electrolyte [17]. The structure of the DSSC device is obstructed by a series of resistance, including the ionic diffusion resistance at the semiconductor/dye/electrolyte interface and between the counter electrode/electrolyte interface. In fact, the efficiency of the DSSC device can be improved by minimizing the resistance between the elements.

Indeed, the third generation of solar cell is focused on the clean, abundant, low-cost, and easy fabricated solar cell. Nevertheless, the main challenges are low efficiency, high production cost, and short lifetime of the solar cell. Theoretically, a successful DSSC device must fulfill several requirements, such as long-term stability, and the function must be retained after millions of times of turnover for the catalytic cycle, which is excitation, charge injection, and regeneration. Besides, material selection and structure of each layer within the DSSC device can affect their reliability and the efficiency of the solar cells significantly. In this manner, two-dimensional crystal thin films of graphene (Gr) appeared as a new novel material with many unique properties, including excellent electrical and thermal conductivity, high mechanical strength, large active surface area, and incredible high mobility of charge carriers. However, in practice, Gr thin films fabricated using solution processing (Hummer's method) will contain many lattice defects and grain boundaries that act as recombination centers and decrease the electrical conductivity of the material significantly. Besides, Gr thin films can only absorb 2.3% of visible light from solar illumination. Thus, continuous efforts have been exerted to further improve the Gr's textural and electronic properties by loading an optimum content of TiO_2 nanoparticles for high photoactive electrode for further improvement of photoconversion efficiency and their immigration of photo-induced charge carriers. In fact, two major essential parts in the DSSC device played an important role, at where the sensitizer absorbs light to excite electrons; the electrons would then migrate to an electrode that produces a current; the photoanode is the path for the electron transport [17]. Indeed, fabrication of photoanode is a relatively important aspect in the production of higher DSSC efficiency [18].

12.2 History of Solar Cells

Over centuries, humans have been trying to find an alternative way to produce electricity and ensure the availability of the source to producing the electricity [1]. Energy rules the economic growth as the demands increase each year, and energy experts predicted that, by 2050, the world needs 30 TW to maintain the stability of energy production [19]. The biggest challenge of replacing the current energy source in the sense that energy consumption is predominately rely on fossil fuel [20]. Fossil fuel energy faces a problem with the continuation of consumption causing the source to deplete and harmful effects to the environment, plus it is nonrenewable energy [21]. More alternative potential energy sources are being discovered and studied, such as nuclear energy, nuclear fission, and solar cell. The challenge is to find a sustainable energy that is an abundant, clean, and renewable raw material with low-cost solution [22].

Solar/sun energy is the primary energy that is abundantly available and holds a tremendous potential for the survival of the future generation, especially in Malaysia. Therefore, the photovoltaic cell is developed where it can harness power from the sun and produce electrical energy [22]. First, silicon solar cells were produced in 1954 by Bell Lab and were known as the first-generation solar cell, which refers to p–n junction photovoltaic. The photovoltaic (PV) was made from mono- and polycrystalline silicon doped with other materials. The monocrystalline silicon recorded the highest efficiency for the first-generation solar cell, but this cell is not consumer efficient as it has high fabrication cost and composition [5]. In 1954, Hoffman develops a method to increase the PV cell efficiency by using amorphous polycrystalline compound semiconductors such as amorphous silicon (A-Si), cadmium telluride (CdTe), and copper indium gallium selenide (CIGS), which were known as second-generation PV devices [5]. A thin-film PV cell consists of three types of film cell structure (single junction, twin junction, and multiple junctions) where it is differentiated by the number of p–i–n junctions. To increase the efficiency of thin solar cell, several processes are involved in depositing thin-film material, which will make the fabrication and device production cost more expensive [5, 12]. CdTe thin film PV is known as the most expensive thin-film candidate. Both the first- and second-generation solar cell were fundamentally derived from semiconductor materials. Then, the third-generation solar cell was introduced with the aim to optimize device efficiency and particularly to reduce the device production cost.

The first solar cell was created in 1839 by Antoine–Cesar Becquerel [4] via the photovoltaic effect of a solid electrode in the electrolyte solution when he observed a voltage developing when light strikes the electrode [4]. A photovoltaic cell holds attractive features such as it does not contribute secondary environment pollution, it does not contribute to nuclear waste by-products, it is inexhaustible, and it has no greenhouse by-product waste [3]. Albert Einstein won a Nobel prize in 1921 by a reported photon absorption generating a photoelectric effect in early 1905 [5]. The first generation solar cell has been assembled so-called clean and abundant energy source. The first-generation solar cell is made and based on silicon material [20]. Furthermore, the second generation of the solar cell is made of thin-film material; such an example is cadmium telluride and copper indium selenide [20]. Basically, DSSC is a third-generation solar cell that is more focused on the environmental comparison of electricity generation from the DSSC system [20]. The generation of the solar cell is classified in Figure 12.1.

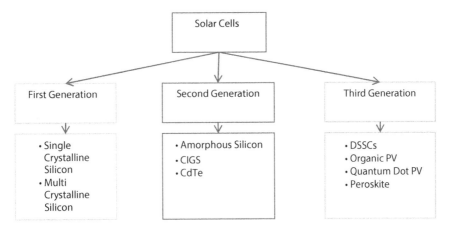

Figure 12.1 Solar cell history and their generations.

DSSCs were first developed by Michael Gratzel and Brian O'Regan in 1991 [8, 23] by the combination of nanostructure electrode and efficient charge inject dyes [8]. By using thin nanocrystalline mesoporous TiO_2 film and intensifying the light absorbed by the sponge-like structure, it can increase the intensity of the light absorption by 11% efficiency. The TiO_2 was soaked with dye-sensitizer and found that the conversion of sunlight towards the chemical reaction of ion dye-sensitizer in DSSCs resemble the photosynthesis process and known as artificial photosynthesis. The conversion of sunlight towards energy brings about the photoelectrochemical principle [23]. The discovery of sunlight energy being captured and converted into electric power brought out lots of great ideas for scientists and researchers, who seek alternative energy sources.

Renewable solar cell energy source, which is considered to be practically inexhaustible and nonpolluting, is a key component in sustainable energy development. Taking this fact into consideration, plenty of research works have been done to generate green and renewable energy from natural resources with the aim of creating a sustainable environment that is in harmony with nature and enhancing the quality of life. Nevertheless, turning sunlight into electricity as a renewable source, which is a more controllable and useful energy form while keeping the cost low, remains one of the biggest challenges.

In general, the dye sensitizer and the photoanode are two major essential parts of DSSCs, in which the sensitizer absorbs light to excite electrons; the electrons would then migrate to an electrode that produces a current; the photoanode is the path for the electron transport. In recent years, natural resources have been extensively studied as possible inexpensive and eco-friendly alternatives to conventional materials. The dye-sensitized and photoanode structure is one of the most crucial tasks to be considered as it contributes to the efficiency of the device. In fact, the selection of photoanode material is necessary to ensure the good crystallinity and semiconductor properties for DSSSC device [19].

Lately, rGO, which is inexpensive and abundant, has attracted tremendous research interest in the DSSC field and will be applied as a photoanode enhancer in this study. In order to bring the renewable solar energy to the point of commercial readiness, substantial research efforts toward the development of a hybrid semiconductor/photoelectrode for highly efficient solar cells, particularly DSSCs, have been widely made in recent years.

Basically, in the present study, the N719 commercial dye, red anthocyanin dye, and green chlorophyll dye were used as photosensitizers for DSSCs. On top of that, dyes derived from natural sources such as the anthocyanin and chlorophyll photosensitizer for DSSCs have been extensively studied in association with their large absorption coefficient, high light-harvesting efficiency, low cost, and environmental friendliness [18, 24, 25]. This breakthrough has triggered subsequent interests in modifying semiconductor oxide research on rGO-TiO$_2$ by scientists and researchers from all over the world and made photoanode an important component in DSSC applications.

12.3 DSSC Structure and Working Operation

DSSCs are a photovoltaic device that the performances depending to its physical and chemical characteristics which is slightly different as compared with other photovoltaic cell. It combines the operation of liquid and solid phase material to produce current–voltage density. In real life, the DSSC working principle is analogous to photosynthesis where it absorbed light to gain energy and excite electrons. DSSCs consist of a layer component that is sandwiched together by two conductive transparent glass (mainly ITO/FTO glass) [26]. As shown in Figure 12.2, a typical DSSC consists of a transparent cathode (e.g., FTO), a highly porous semiconductor (Gr composited with TiO$_2$ nanocrystals) layer with a soaked layer of dye (e.g., ruthenium polypyridine dye/organic dye), an electrolyte solution containing redox pairs (e.g., iodine/triiodide), and a counter electrode (e.g., platinum sheet) [12, 26]. For the semiconductor, *TiO$_2$* acts as the electron acceptor; the electrolyte process of redox reaction (electron donor and oxidation) of iodide/triiodide (I^-/I_3^-) resembles the water and oxygen in photosynthesis. At the same time, the multilayer structure DSSCs function together to enhance the light absorption and electron collection efficiency, which is the same as the thylakoid membrane in photosynthesis [8]. The operational processes of DSSCs are divided into a charge separation process and charge collection when the device was introduced to the photo from the sunlight/illumination.

A transparent conductive film (TCF) is a transparent electrode that is used in DSSCs to maximize the transparency of the DSSC device and protect the inner layer of material against harsh chemical and thermal treatments. Normally, the industry standard TCF is

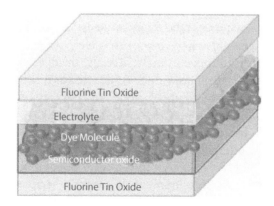

Figure 12.2 DSSC structure.

the indium tin oxide (ITO). The ITO glass increase R_{sh} is 5 Ω/sq. However, indium is a rare earth metal that is not cost-effective and ITO glass is not compatible with a strong acid, not stable at high temperature, and mechanically brittle [27, 28]. Fluorine tin oxide (FTO) glass is used as an alternative due to its properties to overcome the harsh chemical and higher thermal treatments almost at ~700°C. In terms of cost, FTO glass is more cost-effective and durable than ITO [12, 16, 28].

The working principle of DSSCs is illustrated in Figure 12.3: (1) First, the dye molecules will harvest the high-energy photons from solar irradiation and then the electrons will be released into the conduction band of the Gr composites with TiO_2 nanocrystals. (2) The injected photo-induced electrons will then move to the transparent anode and through the external circuit before transporting to the cathode.

Meanwhile, the dye molecules strip one electron from iodine in the electrolyte by oxidizing it to triiodide (redox reaction) [8, 14]. (3) The triiodide then recovers its missing electron from one circle of electron migration through the external circuit by diffusing to the counter electrode (i.e., cathode) [13]. (4) The counter electrode of a DSSC catalyzes the reduction of redox pairs after electron injection. (5) Voltage generated by the cell depends on the illumination and shows the difference in the Fermi level of the electron and the redox potential in the electrolyte [8, 14]. (6) By completing this cycle, the device is generating power from the light without undergoing any permanent chemical transformation [8, 14]. This proposed project will demonstrate a novel approach to enhance DSSC photovoltaic performance by applying modified photoanode of TiO_2 nanocomposite rGO nanocrystals. From the study, photovoltaic performance in DSSCs will be addressed with physical and chemical properties of TiO_2 nanocrystals on rGO. Furthermore, the detailed kinetic mechanism regarding TiO_2 nanocomposite rGO nanocrystals will be established in this proposed project.

Semiconductor oxide has the ability to absorb dye molecules due to the indispensable large surface area. The dye molecules contain electrons to effectively absorb photons. The device absorbs photon that strikes into the photoanode, and the process leads to the excitation of dye to an electronically excited state (S*) and make the electron lie energetically above the conduction band of the semiconductor oxide. The above state of semiconductor oxide is known as lower unoccupied molecular orbital (LUMO). In the excited state, the

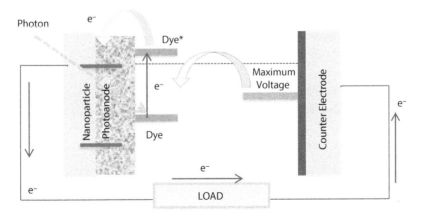

Figure 12.3 Working operation of DSSCs [29].

dye-sensitizer collects photons and produces excited electron (S*) from HOMO to LUMO, and the energy difference between the HOMO level and the LUMO level of the photosensitizer generates the amount of photocurrent in DSSCs.

When the excited electron injects to the conduction band, it leaves hole/oxidation of photosensitizer (Table 12.1). The dye-sensitizer LUMO needs to exceed the conduction band of semiconductor oxide for the excited electron to be kinetically favorable to the conduction band of semiconductor oxide where the LUMO energy needs to be sufficiently negative. To ensure the effectiveness of electron injection, the substantial electronic coupling of the conduction band of semiconductor oxide level and LUMO of dye-sensitized level needs to have a strong electronic interaction with a special anchoring group. The morphology of the semiconductor oxide plays an important role as it smoothens the electron transfer in the device and the connectivity of the particles. One of the important steps is the regeneration of the iodide (I^-) to replace the oxide dye that excites the LUMO. For the continuation of the current generation, the dye that is excited to the conduction band of the semiconductor will be reduced back to the ground state when it loses the energy to the surroundings. Electrolyte acts as the bridge to connect the counter electrode and semiconductor oxide photoelectrode. The I ion redox aims replace the lost electron in the oxidized dye.

In DSSCs, the charge transfers have a large contribution where it occurs in (1) a nanocrystalline structure and (2) a hole transfer in the electrolyte. The charge transfer in the electrolyte was represented by the redox reaction by tri-iodide and iodide. When the electrons travel from the conducting oxide layer into external load and the passing through counter electrode, the electron from the counter electrode will flow to the electrolyte where the I_3^- ions float around until the electrolyte compensates for the missing electrons. By the regeneration process, I_3^- will reduce back into I^- by the migration of electrons from the catalysis in the counter electrode through the external load. The finalization process is the generation of voltage by the illumination corresponding to the Fermi level of the electron in semiconductor oxide and the electrolyte redox potential [2, 12, 29, 30].

Electron recombination is the main issue in any semiconductor device including DSSCs. There is the probability of recombination process or back reaction in the DSSC reaction mainly in the charge separation and electron injection process. The back reaction may occur in the excitation of dye which from HOMO level to the LUMO level where the dye will decay or the energy loss of dye will be formed. Next, the probability of recombination occurs in the electron injection from a dye excited state to the conduction band and thus electrolyte is needed to overcome this problem. The speed of dye regeneration needs to be maintained in *ns* to lower down the recombination process of the DSSCs.

Table 12.1 Detailed equations of the working principle of DSSCs [2, 12, 15, 30].

S + Photon (hv) → S*	(Light absorption)
S* → S$^+$ + e$^-$ TiO$_2$	(Electron injection)
S$^+$ + I^- → S	(Dye regeneration)
I_3^- + 2e$^-$ (counter electrode) → 3I^-	(Redox mediator reduction)

12.3.1 Transparent Conductive Films

A transparent conductive film (TCF) is a transparent electrode used in DSSCs to maximize the transparency of the DSSC device and protect the inner layer of material against harsh chemical and thermal treatments. Normally, the industry standard TCF is the ITO. The ITO glass increase R_{sh} is 5 Ω/sq. However, indium is a rare earth metal that is not cost-effective and ITO glass is not compatible with a strong acid, not stable at high temperature, and mechanically brittle [27, 28]. FTO glass is used as an alternative due to its properties to overcome the harsh chemical and higher thermal treatments almost at ~700°C. In terms of cost, FTO glass is more cost-effective and durable than ITO [12, 16, 28].

12.3.2 Semiconductor Film Electrodes

Taking into account the processes involved in the DSSCs on particulate photocatalysts under solar irradiation, the materials used as semiconductor film electrode must satisfy several functional requirements with respect to DSSC properties as shown below: (1) Band gap: The electronic band gap should be low for most of the solar light spectrum so that it can be used for photoexcitation. (2) Transportation of charge carriers: Charge carriers should be transported with minimal losses from the bulk oxide material to the counter electrode for a highly efficient photovoltaic characteristic of DSSCs. (3) Stability: The photocatalyst must be stable against photocorrosion in the electrolyte.

In this research, rGO-TiO_2 is a potential material as the photoanode for DSSCs and the composition of this material will addressed to a great exploration in terms of enhance the conductivity, shorten the transfer path, increasing the active area for the attachments of the dye-sensitizer and reduce the interface resistance. A typical photoanode film is made by a glass sheet with transparent glass that has a side of the conductive oxide and known as transparent conductive oxide (TCO). The glass used in DSSCs is mainly FTO glass or ITO glass. The characteristic of the substrate is needed to allow the light to enter the solar cell. The semiconductor oxide is deposited to the conductive surface to allow the electron travelling from the film to the external load and entering the counter electrode. The ideal semiconductor oxide must satisfy these characteristics to improve the performance of DSSCs: (i) transparency to increase the light absorption of dye, (ii) high surface area with uniform nanostructured mesoscopic film to maximize the dye absorption, (iii) porous surface for electrolyte accessible, and (iv) fast electron transport against the semiconductor grain [31].

12.3.3 TiO_2

The semiconductor material is the main component that acts as a photoanode in DSSCs. Mainly, a mesoporous oxide film is used to activate the photoanode characteristic in order to facilitate dye-sensitized attach to the mesoporous semiconductor surface. The mesoporous material as the semiconductor is the highlight of semiconductor oxide materials. Basically, the material is a nanometer crystal array with mesoscopic pores that act as dye attachment site (active area) and electron transfer passage through the DSSCs. Examples of materials used as semiconductor oxide are metal oxides such as TiO_2, zinc oxide (ZnO), or tin oxide (SnO_2); inorganic materials such as carbon nanotube, Gr, or graphite [12, 15, 28, 32]. In 1990, M. Gratzel and coworkers had a large exploration in DSSC history where they successfully combined TiO_2 nanoparticles, charge injection dye, and electrodes to produce the third-generation solar cell. In order to bring renewable solar energy to the point of commercial readiness, substantial research on the development of hybrid semiconductor/photoelectrode for the highly efficient solar cell has been developed lately, and it shows a positive impact toward the performances of DSSCs.

A regards the nanoparticle semiconductor small size, it has the ability to provide a large surface area and a relatively high porosity for the semiconductor [33]. Figure 12.4 provides the FESEM image of a TiO_2 nanoparticle film semiconductor, which has a surface 10 μm thick, a porosity about 50%, and an area surface that is extremely good for dye absorption. Semiconductor plays an important role in DSSC system especially reduce in electron recombination rate. The semiconductor material also needs to easily access the dye and the electrolyte redox couple for the closed circuit system of the DSSCs [34]. Through the study, nanoparticle DSSCs depends on the network of the crystallography of the crystalline and lattice where usually the pattern of the crystal is random plus it would affect or sway the electron or light scattering [33]. The crystal structure can cause limitation of the electron transport where it will affect the rate (becomes slower), especially in the bright light that has longer wavelengths. Thus, the effect of the recombination process will resulting electron transporting take a longer time (~ms) to reach contact area [35].

The nanocrystals of TiO_2 are divided into few types of crystal phase: anatase, rutile, and brookite. For the DSSCs, both anatase and rutile are the most common types used as semiconductor oxide. The anatase crystalline form at low temperatures has a pyramid-like crystal and is stable. Meanwhile, the rutile crystal has a needle-like shape, and the crystal only forms in the high-temperature process [12]. The band gap for anatase is slightly higher than that in rutile, but in terms of recombination rate, anatase has a lower recombination rate compared to rutile phase [2]. In DSSCs, the anatase TiO_2 polymorph is more efficient as semiconductor oxide to cater the charge transport and charge separation process compared to rutile. It is proven that the TiO_2 anatase phase is a remarkable material for DSSCs, as it has a higher electric conductivity that is favorable to transport electron for the production of energy [2]. So far, anatase phase of TiO_2 is widely used as photoanode in DSSC rather than rutile phase due to the lower photocatalytic degradation of organic compound under UV radiation while rutile phase of TiO_2 only absorb 4% of incident light in near UV range and band gap excitation generated holes reduced the stability of DSSCs which is not favorable [30].

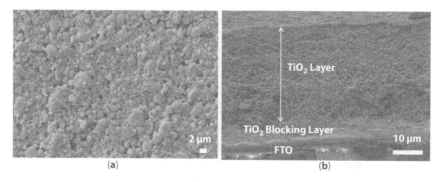

Figure 12.4 FESEM image of TiO_2 nanoparticle deposited on FTO glass. (a) TiO_2 morphology (low magnification). (b) TiO_2 nanoparticle on FTO cross-section [36].

TiO$_2$ was chosen as one of the best photoanodes for the sensitized photoelectrochemistry due to its properties. One of the advantages of TiO$_2$ is having a stable photoelectrode that has good chemical stability under irradiation, is eco-friendly, is cheap, and is widely available [12]. Since it is a high dielectric constant ($c = 80$ for anatase phase), so it has capable to reduce the recombination process when the electron excited that are injected from dye by providing electrostatic shielding. The high refractive index of TiO$_2$ (anatase reflective index = 2.5) helps in the efficient diffuse scattering of sunlight in the semiconductor. The porosity of the semiconductor TiO$_2$ is a crucial characteristic whereby it holds the dye molecules that will make the semiconductor surface electron-rich. The dye molecule is used as sensitizer and is coated to nanocrystalline TiO$_2$ film that will convert the photons into excited electrons and produce current flow in the device.

Normally, a DSSC photoanode consists of thick TiO$_2$ nanoparticles (~10–15 μm) that form a mesoporous network. The thick mesoporous structure (Figure 12.4) provides a large surface area that acts as an anchoring site for dye molecules to absorb in the photoanode. In past research, TiO$_2$ alone has recorded a remarkable photocurrent–density efficiency for DSSCs, but due to its large band gap, it leads to a fast recombination rate. The morphology (particle size, porosity, pore size, and nanostructure) of photoanode plays significant roles to modulate the photovoltaic characteristics. The physical properties of each photoanode are determined by the nature of coating such as nature of binder, solvent, viscosity, etc. of the respective paste.

The recombination occurs due to the injected electron travelling in the random colloidal particle matrix and grain boundaries in TiO$_2$, thus creating a random transfer route followed by a trap-limited diffusion process [37]. When photogenerated electrons travel in the random transfer path, it increases carrier recombination, which contributes to a reduction of photocurrent efficiency in DSSCs. It is favorable to have a highly conductive material as the composite to the semiconductor oxide to enhance the low photocurrent–density voltage. Therefore, rGO, which is low cost and available in abundance, has attracted tremendous research interest. In the DSSCs, the photoanode is an essential part in DSSCs.

12.3.4 rGO

Over the past years, the study of DSSCs has becomes popular for enhancing the excellent material and structure to boost the efficiency and stability of the DSSCs. The novelty of DSSCs is that it is a molecular device that transitions from microelectronic technology to nanotechnology [8]. Each layer has been manipulated to find the suitable material and design for the DSSCs. In order to reduce the cost of solar cell devices, low-cost and effective materials are investigated intensively. Therefore, rGO, which is low cost and available in abundance, has attracted tremendous research interest [38–41]. Dr. Andre Geim and Konstantin Novoselov [42] first discovered Gr in 2004 as a result of the duplicate properties of carbon nanotubes in flat sheets of carbon lattice [43].

A tremendous material such as rGO has been known as the thinnest and strongest material with a single-layer structure of graphite and one-atom-thick honeycomb-shaped two-dimensional crystal structure as shown in Figure 12.5 [43]. rGO shows that it is a 2D carbon-based material that has a flat single layer of carbon atoms, making it a simple nanostructure material for nanotechnology application [15, 27, 44, 45]. rGO has drawn much attention due to its unique properties, which show promise in various applications such

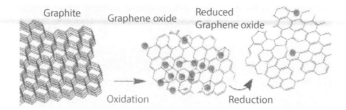

Figure 12.5 Reduction process of graphite to rGO.

as sensors, photovoltaics, nanoelectronics, and supercapacitors [46]. In fact, rGO exhibits unique electrical [47], electrochemical [47], and optical properties [48], as well as incredibly high mobility of charge carriers. Furthermore, rGO possesses high quality of easy accessibility, good flexibility and transparency property addressed to utilize as photoelectrode [15]. The unique combination of high electrical conductivity and optical transparency of rGO has made it as the leading candidate in photovoltaic solar cell applications [40, 41, 49].

Nowadays, the production of graphene oxide (GO) from graphite in bulk is being developed (Figure 12.5). In this approach for GO production, graphite is oxidized with strong oxidants and intercalating compounds (e.g., $KMnO$, H_2SO_4, HNO_3, $NaClO_2$). Some researchers use exfoliation technique through mechanical ultrasonic agitation, producing stable suspensions of GO demonstrated by Tanaka *et al.* [50, 51]. Surface functionalization for GO is important for controlling surface behavior in order for it to be used in a different application. Functionalization of GO is needed because GO utilizes weak interaction, e.g., the π–π interaction and van der Waals interaction between GO and molecules. By chemically reducing GO, the rGO is more stable when it is composite and is producing functional composite materials [27, 40]. The changes from GO to rGO provides structural changes such as electrical conductivity, carrier mobility, optical band gap, and thermal stability, which are beneficial for Gr-based solar cells [15].

The rGO electrical and optical properties depend on the spatial distribution of the functional groups and structural defects. Many methods were applied for the production of rGO such as electrochemical, thermal, or chemical synthesis. A minimal procedure is carried out for the reduction of GO to rGO such as chemically rGO. The most frequent method is using hydrazine vapor on GO as a reduction agent. The process path for producing the rGO is crucial to maintain a simple methodology. However, in practice, rGO films produced via solution processing will contain lattice defects and grain boundaries that act as recombination centers and decrease the electrical conductivity of the material significantly [52, 53]. Besides, rGO only can absorb 2.3% of visible light from solar illumination [41, 49]. Thus, continuous efforts have been exerted to further improve rGO textural and electronic properties by loading an optimum content of metal oxide photocatalyst for the high photoactive electrode [40, 52].

12.3.5 rGO-TiO$_2$ NC

Design and development of nanostructure of rGO composited TiO_2 assemblies has gained significant scientific interest and become the most studied material as it exhibits excellent properties. Among the vast number of different metal oxide photocatalysts, TiO_2 is one of the most capable candidate to be coupled with Gr for enhancement in numerous diverse applications, such as solar cell, hydrogen conversion catalyst, water treatment, and many more. The reasons mainly attribute to TiO_2 has a stable photocatalyst with large energy band gap, random porosity structure, low cost, nontoxicity, ready availability, strong photocatalytic activity, stability against photo-corrosion, high trap state which favors recombination rate of electron and retard the electron transfer [12, 54–55].

One of the important reasons for combining TiO_2 with rGO is that TiO_2 alone has a high diffusion coefficient where the structure of TiO_2 consists of random Ti grain that has significant neck size of porosity. When the electron travels on the Ti grain, it might lose its "power" due to the random and uncoordinated travel site. Whereby rGO is known as a flat 2D structure with a highly conductive surface that favors the ultrafast electron to transfer. However, the ability of rGO to trap and absorb light is limited due to its low diffusion coefficient, which leads to low absorption of the dye. By combining TiO_2 and rGO, the material becomes stable and has tremendous properties. However, TiO_2 will provide defects on the rGO surface that will increase the material porosity and placeholder for the dye absorption. This will contribute to the enhancement of DSSC J_{SC} and creates travel site for the photogenerated electron [19, 54].

By chemically controlling the reduction process of graphite to GO and rGO, the chemical properties of rGO are improved and create remarkable properties due to its high aspect ratio; hence, it provides a low percolation threshold [28]. Gr is one of the carbonaceous materials that have been widely used in solar photovoltaic parts, and in this research, the implication of rGO into TiO_2 as photoanodes in DSSCs is determined [47, 49]. It is noteworthy that the rGO has a matching conduction band with TiO_2, and thereby, a charge transfer could be formed between the rGO and TiO_2 surface. Also, the photo-induced electron can move through the rGO bridge where it shuttles electron to the current collector rather than to the TiO_2–TiO_2 grain boundary as a result of the diminished charge recombination [28, 56]. In view of these facts, the combination of TiO_2 and rGO to produce rGO–TiO_2 composites is an alternative method to improve the conduction pathways and photocurrent–voltage density of DSSCs from the point of photoinduced electrons at photoanode to the charge collector electrode. To extend the performance of rGO in the photoanode, the rGO material should be used, which has minimal defects to effectively coat the dense TiO_2 particle. Many previous studies recorded improvements in terms of increasing photocurrent and elucidated enhanced mechanism. Table 12.2 shows past research on rGO-TiO_2 DSSC performances as a reference for tailoring a suitable and simple method to elevate DSSC performances using rGO. The high efficiency of rGO composited with TiO_2 as a photoanode in a DSSC requires a suitable architecture that minimizes electron loss at nanostructure connections and maximizes photon absorption [55, 57]. In order to further improve the immigration of photo-induced charge carriers, considerable effort has to be exerted to increase the photoconversion efficiency of DSSCs under visible illumination.

From all the previous studies in Table 12.2, rGO plays a positive role in accelerating the electron transfer in the DSSC device where most of the studies show improvement in J_{sc} performances compared to pure TiO_2. From Figure 12.6, rGO was reported to be a bridging agent as the incorporation of rGO enhances the conductivity, which accelerates the electron transfer from TiO_2 to the FTO glass, which will reduce electron–hole recombination [63, 64]. The incorporation of rGO in TiO_2 will enhance the performance of DSSCs; thus, a simplified and stable sol method is developed to fulfill the requirement of coating TiO_2 to the rGO sheet efficiently to improve the morphology of the photoanode surface [67].

Table 12.2 Past reference of rGO–titania composite DSSC performances.

Author/Ref.	Gr DSSCs	Preparation method	Reference cell	Gr DSSCs J_{sc} (mA/cm^{-2})	η (%)	Reference cell J_{sc} (mA/cm^{-2})	η (%)
Routh et al. [58]	PHET with grafted rGO	Molecular grafting	TiO$_2$	7.50	3.06	5.6	2.66
Sharma et al. [59]	rGO-TiO$_2$ layer	Hydrothermal method and spin coating	TiO$_2$ layer	10.95	5.33	9.97	4.18
Bonaccorso et al. [60]	P25-Gr	Heterogeneous coagulation	P25 electrode	8.38	4.28	5.04	2.70
Kazmi et al. [61]	Gr-TiO$_2$(rGO)	Sonication	Pure TiO$_2$	9.80	–	–	–
Zhang et al. [62]	TiO$_2$-G	Synthesis	TiO$_2$	7.80	1.50	4.06	0.89
Wang et al. [63]	TrGO scaffold layer	Ultrasonication	TiO$_2$	7.60	2.8	5.0	1.8
Kim et al. [64]	Underlayer T-CrGO	Solvothermal	TiO$_2$	12.90	6.1	5.0	4.4

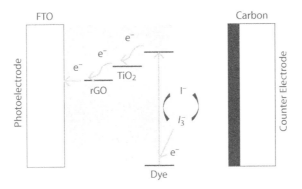

Figure 12.6 Electron flow in the photoanode contains rGO [63].

12.3.6 Dye Sensitizer

In general, dye sensitizer is a major essential part in DSSCs, in which the sensitizer absorbs light to excite electrons; the electrons would then migrate to an electrode that produces a current. In this research, two different type of dye is used in DSSCs are being studied with attachment of rGO-TiO_2 film. The two highly used dyes in DSSCs are metal complex dye and dye from natural derivatives. The charge transfer sensitizers that are employed are Ru–bipyridyl complexes, crude chlorophyll, and crude anthocyanin. In traditional DSSCs, the standard dye was tris (2,2'-bipyridyl-4,4'-carboxylate) ruthenium(II) (N3 dye). Ruthenium complexes (Ru-complexes) are known as the most effective sensitizers due to their high efficiency, where we could test the maximum performances of the composite, good photoelectrochemical properties, and intense charge transfer in the wide visible range [5]. However, Ru-complexes are known to not be nonenvironmental friendly because they contain a heavy metal that is harmful to the environment [5, 65]. Furthermore, with the presence of water, Ru-complexes tend to degrade. Ru-complexes have a general formula of $RuL_xL_y'SCN_z$, where the L and L' are polypyridyl ligands and Figure 12.6 shows the N719 Ru metal-based complexes with ligand attachments. The common alteration of the absorption spectrum of a complex π^* level tuning (ligands) where the π^* level energy and methyl/phenyl group tends to increase the absorption of the metal to ligand charge transfer (MCLT) [17].

On top of that, dyes derived from natural sources such as the photosensitizer for DSSCs have been extensively studied in association with their large absorption coefficient, high light-harvesting efficiency, low cost, and environmental friendliness [18, 24, 25]. Natural resources were extensively studied to find the possible substitute for expensive and non-environmental friendly dye compounds that were previously used in the present study. Many extraction methods and sources of natural dye pigments or color can be obtained from natural sources such as flowers, leaves, and bacteria [30]. The use of synthetic dyes as dye sensitizer in DSSCs tend to produce higher efficiency, but they suffer from certain limitation such as a tendency to undergo degradation, higher cost, and usage of toxic materials. Due to this limitation of synthetic dye, an alternative sensitizer, natural sensitizer, which is biocompatible, is being introduced. Natural dye has advantages to DSSC application such as large absorption coefficient in the visible region, easily obtainable, abundant, environmentally friendly, and easy preparation [30]. Plant pigments have the ability to exhibit electronic structures where they can interact with sunlight and alter the wavelength that is transmitted or reflected from the plant tissue. Table 12.3 classifies pigments into four different types.

Accordingly, natural dyes especially cyanine, chlorophyll, anthocyanin, carotene, and flavonoid have been extensively studied as sensitizers in DSSCs. Natural dye is extracted from plant pigments as shown in Figure 12.6 and each pigment holds a different molecular structure and absorption spectrum. Most importantly, the functional group of pigments is necessary to interact with the photoanode surface. In the present study, green chlorophyll extracted from pandan leaves (*Pandanus amaryllifolius*) was extracted as the green chlorophyll dye pigment and mulberry (*Morus alba*) was extracted as red anthocyanin dye pigment as in Figure 12.7. Specifically, chlorophyll could be classified as a unique pigment ascribed to its ability to conduct photosynthesis, converting light energy to transduction energy in the plant. Additionally, chlorophyll (a mixture of two pigment complexes, namely, chlorophyll a and b) is an attractive candidate as a sensitizer in DSSCs because of its tendency to absorb blue and red light. Also, many research works have focused on the preparation of a porphyrin-type organic dye from chlorophyll due to its low loss, ease of preparation, and eco-friendliness (Figure 12.8).

Table 12.3 Type of plant pigments.

Pigments	Common types	Occurrence
Betalains	Betacyanins Betaxanthins	Caryophyllales and some fungi
Carotenoids	Carotenes Xanthophylls	Photosynthetic plants and bacteria Retain from diet by some birds, fish, and crustaceans
Chlorophyll	Chlorophyll	All photosynthetic plant
Flavonoids	Anthocyanins Aurones Chalcones Flavonols Proanthocyanidins	Widespread and common plant including angiosperms, gymnosperms, ferns, and bryophytes

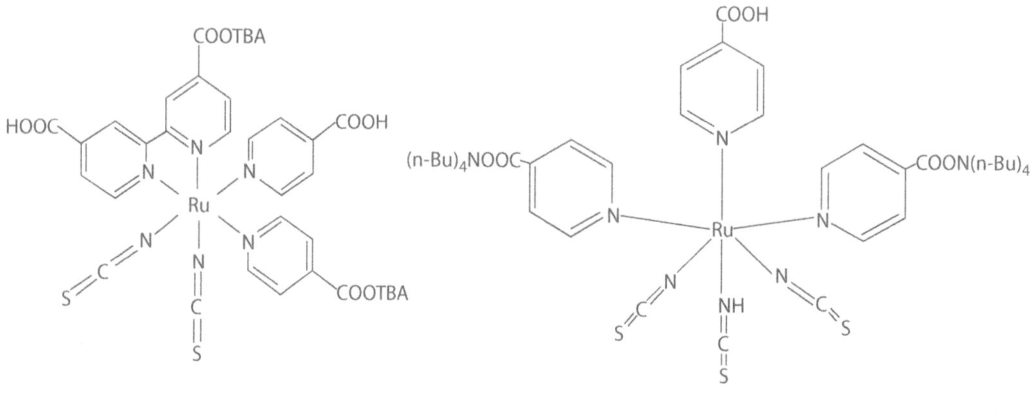

Figure 12.7 Ruthenium metal-based complexes N719 [5, 17].

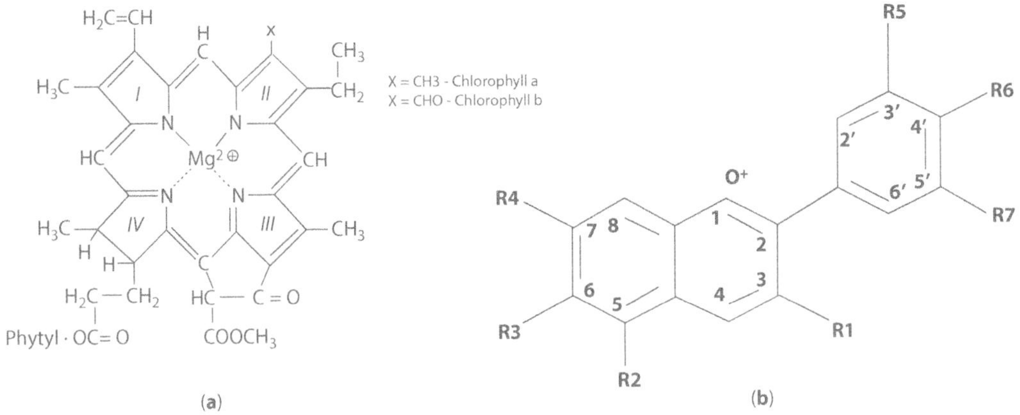

Figure 12.8 (a) Chemical structure of chlorophyll a and b, and (b) chemical structure of anthocyanin [30].

Anthocyanin is a red-blue plant pigment, which is an abundant type of plant with a production of ~10^9 tons/year [66]. From the literature, 17 types of anthocyanin have been reported so far, and it is classified per number of sugar molecule such as bioside, monoside, trioside etc. Anthocyanin exists in the plant at flowers, leaves, and fruits, and in some types of mosses or ferns, and the attractive color of anthocyanin ranges from scarlet to blue [30]. The anthocyanin counts in a plant determine the modified quantity and quality of light that is incident in the chloroplast. Common anthocyanidins found in a flower are the pelargonidin (orange), cyaniding (orange-red), delphinidin (blue-red), petunidin (blue-red), and malvidin (blue-red). One more advantage of anthocyanin in DSSCs is it contains a hydroxyl and carbonyl group that can bind to the surface of the semiconductor film, which will alleviate the excitation and transfer of the electron from the anthocyanin molecule to the semiconductor oxide conduction band. The binding makes the electron transport from the anthocyanin molecule to the conduction band of TiO_2. Anthocyanin molecules could help in organic solar cells due to the ability of light absorbing and able convert it into the excited electron.

The natural dye data that are used so far in DSSCs are shown in Table 12.4. The data are obtained by the different research on DSSC condition, and the natural dyes were extracted with a different method. Chlorophyll and anthocyanin as natural photosensitizers are sustainable and are available in large amount. In order for commercial readiness, fast extraction methods for purification are necessary for an efficient sensitizer. The exploration and modification of natural dye pigments hope to bring about new findings to the DSSC community.

12.3.7 Liquid Electrolyte

The heart of DSSCs is the junction that formed between liquid electrolytes for interaction in between the semiconductor electrode and a counter electrode. The catalytic activity in the counter electrode needs to be efficient in reducing triiodide and continuously helps in regenerating dye molecules [3, 15]. The electrolyte generates the regeneration process in DSSCs where it fills in the hole (oxidized state) in the dye sensitizer by donating ground state electrons from the redox mediator in an electrolyte to act as a media between counter

Table 12.4 Natural dyes used in DSSCs.

Plant source	Structure	Photoanode area (cm^2)	J_{sc} (mA/cm^{-2})	V_{oc} (mV)	η/FF	Extract method
Rosa xanthina	Anthocyanin	1	0.637	492	–/0.52	Fractionated extract
Black rice	Anthocyanin	1	1.142	551	–/0.52	Fractionated extract
Kelp	Chlorophyll	1	0.433	441	–/0.63	Fractionated extract
Mangosteen pericarp	–	0.2	2.69	686	1.17/0.63	Extract
Spinach	Modified chlorophyll/ neoxanthin	–	11.8	550	3.9/0.60	Isolated compounds
Spinach	Modified chlorophyll/ β-carotene	–	13.7	530	4.2/0.58	Isolated compounds
Mulberry	–	–	0.86	422	Na/0.61	Extract
Tradescantia zebrina	Anthocyanin	–	0.63	350	0.23/0.55	Ethanolic extract
Ixora sp.	Anthocyanin	–	6.26	351	0.96/0.44	Ethanolic extract
C. odontophyllum + Ixora sp.	Anthocyanin	–	6.26	384	1.13/0.47	Mixed ethanolic extract
C. odontophyllum + Ixora sp.	Anthocyanin	–	9.80	343	1.55/0.46	Ethanolic extract in consecutive layers

electrode and metal oxide in order to reduce the oxidized dye and fasten diffusion of charge carrier to sustain energy conversion under light illumination. To maintain the highly conducting surface area in any operating conditions, the ionic shielding by cations from a conducting salt that is dissolved in liquid phase increases the surface area in the nonporous structure of DSSCs. At very high interfacial contact, the charge carrier in the same phase is rapidly separated into different phases [3].

The liquid electrolyte is chosen because it offers an effective solution to the problem of electroneutrality in DSSC heterogeneous converters. The evolution of liquid electrolyte in DSSCs started as metal salt-based liquid electrolyte, and then it evolves to ionic liquid and lastly to the combination of metal salts and ionic liquid used as redox couple in DSSCs as iodide/triiodide (I^-/I_3^-). The triiodide possesses remarkable characteristics in DSSCs as a support system to each component in the device system such as (i) the regeneration of dye-sensitized hole with unity efficiency by I^- odide, (ii) the back reaction in the TiO_2 to I_3^- is slowed by complicated multielectron transfer mechanism by the liquid electrolyte, (iii) I^-/I_3^- has a high diffusion coefficient and is highly soluble in any solvents due to its small molecule to allow optimization of concentration to reach the solubility or diffusion limit, (iv) it has low light absorbance to reduce competition with dye, and (v) the I^-/I_3^- redox couple is stable and will not decompose under operation conditions [3].

From Figure 12.9 it shows the kinetics of the electron the DSSCs where the electron injection in the TiO_2 conduction band occurs in femtosecond which is faster than electron recombination with I_3^- and the injected electron combining with oxidized dye from the semiconductor react with I^-. The diffusion coefficient of I_3^- ions = 7.6×10^{-6} cm²/s on the porous semiconductor TiO_2 oxide [29]. The difference between quasi-Fermi levels of TiO_2 and redox potential in electrolyte determines the maximum voltage generated in DSSCs. Corrosion limits the DSSCs to obtain higher open circuit voltage; thus, additives are added to alter the concentration of iodide introduced such as 4-tert-butylpyridine (4TBP) (used in the research), quanidiumthiocyanate, and methyl benzimidazole (MBI) [29]. Many research has been dedicated to enhance the properties and effectiveness of redox potential such as matching the oxidation potential of dye sensitization with redox potential to minimize the energy loss in the dye regeneration in fact that it can strikingly high open-circuit voltage up to 1V [29].

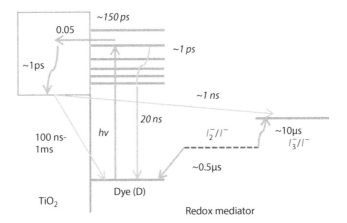

Figure 12.9 Kinetics of the electron in the DSSCs with the I_3^-/I^- redox mediator [29].

12.3.8 Cathode Electrodes

The main role of the cathode electrode in DSSCs is to catalyze the I^- regeneration from I_3^- in the redox couple to help the regeneration of dye. Nanostructure plays a crucial part in cathode electrode, especially the morphological part where it determines the performance of the DSSC device. The counter electrode also carries the photocurrent over the width of each solar cell device. Therefore, the counter electrode must be well conducting and exhibit low overvoltage. The most common type of cathode electrode is platinum (Pt) where it (i) acts as collection center of the electron from the external load to the electrolyte and (ii) acts as a catalyst that enhances the regeneration process of redox mediator from I^- to I_3^-.

Pt is commonly used due to its good photocatalytic activity and excellent stability toward the I^-/I_3^- electrolyte. By finely dispersing the expensive platinum on a conducting substrate, such as ITO-, FTO-, and SnO_2-coated glass, the required amount has been kept low only about < $0.1g/m^2$ Usually, the platinum-loaded FTO glass serves as the counter electrode for DSSCs. However, platinum is a rare and high-cost metal on earth and some of the researchers report that the corrosion of Pt would occur by reacting with triiodide containing the electrolyte form PtI_4 [68]. The catalytic activity of platinum was found to diminish on exposure of the dye solution, probably due to the blocking of its surface by the adsorbed dye. There is some concern regarding the small amount of platinum that might be dissolved in the electrolyte by oxidation and complex formation with iodide, I^-/triiodide, I_3^-, with H_2PtI_6 as an example [69]. However, if a tiny amount of platinum dissolves in the electrolyte, it will slowly redeposit on the TiO_2 layer and short circuit would occur by catalyzing triiodide reduction on the photoelectrode.

Carbon is a low cost material and widely used as cathode/counter electrode instead of platinum because of the combination of sufficient conductivity and heat resistance as well as corrosion resistance and electrocatalytic activity for triiodide reduction. Carbonaceous materials contain significant features such as high electronic conductivity, corrosion resistance toward triiodide reduction, and low cost, which are quite attractive to replace platinum. In 1996, Kay *et al.* reported that the use of carbon black as a counter electrode that shows the conversion efficiency of 6.7% [69]. Since then, carbonaceous materials such as carbon black, graphite, carbon nanotubes, and activated carbon have been alternative candidates for counter electrode.

Conductive carbon paste (CC) is an electrically conductive printing ink, made from nonmetal conductive carbon particles and thermoplastic resins. After heat curing, the thin film does not easily oxidize and possesses good corrosion resistance to acid, alkaline, and solvent [70]. Counter electrodes' catalytic activity of triiodide reduction as well as the conductivity were considerably enhanced by adding about 20% of carbon black [69]. Catalytic activity is increased according to the high surface area of carbon black, while the improvement of conductivity results from the partial filling of large pores between the graphite flakes with smaller carbon black aggregates. The excellent performance and low cost of conductive carbon has been widely applied in printed circuit boards and also in membrane switches as a burgeoning electronic paste. Carbon black is cheap in terms of industrial mass production and is widely used in printing toners, and it can be easily sprayed onto FTO substrates, but the conductivity is lower compared to highly oriented carbon materials such as graphite and carbon nanotubes. A comparison of conversion efficiency for carbon black, carbon nanotube, and platinum is listed in Table 12.5 based on Chen *et al.* [68].

Table 12.5 Parameters of DSSCs for carbon black counter electrode and carbon black nanotube counter electrodes [7].

Electrodes	V_{oc}	J_{sc} (mA cm^{-2})	FF	η(%)
Carbon black	0.71	9.44	0.57	3.97
Carbon black nanotube	0.72	12.69	0.61	5.57
Platinum	0.73	12.63	0.67	6.13

From Table 12.5, carbon black shows a performance comparable to Pt with 9.44 mA cm^{-2} current density for carbon black and carbon black nanotube with 12.63 mA cm^{-2} current density. This phenomenon was attributed to the addition of carbon nanotubes whose electric conductivity and surface area are large that they not only reduce the electrical resistance and facilitate the electron transfer but also increase the activity of the catalytic site. Despite the superior characteristics of carbon nanotubes in DSSC devices, it was obviously found that the conversion efficiency of CBNT-CE was low (5.57%) compared to the pure platinum (6.13%). Therefore, carbon is one of the potential substitutes compared to Pt to achieve low-cost and environmental friendly DSSCs for future development in solar cell industry.

12.4 rGO-TiO$_2$ NC Properties

Based on literature review, the rGO film is a potential candidate to improve the PCE of DSSCs, but it was normally studied and applied as a counter electrode [13]. Then, TiO$_2$ nanomaterials with superior photocatalytic activity have attracted great attention for use in DSSCs. However, photocatalysts suffer from drawbacks such as high electron–hole pair recombination resulting in a low PCE. Considering this fact, hybridization of rGO-TiO$_2$ NC could enhance photocatalyst activity by increasing the electron mobility and consequently reducing the charge recombination of the electron and hole [39]. On top of that, agglomeration of TiO$_2$ can be overcome since the free electrons trapped in the active area are fully occupied by the C–C bonding of rGO. This provides electron–hole separation and facilitates the interfacial electron transfer [13]. In this case, a hybrid of rGO-TiO$_2$ NC has gained much attention and has been intensively studied because of the unique features of enhancement in photocatalyst activity and accelerated electron mobility to suppress the charge recombination. Among the vast number of different dopants, TiO$_2$ is one of the most capable candidate to be coupled with rGO for enhancement in numerous diverse applications, such as DSSC photovoltaics. Several researchers have reported that the band gap of TiO$_2$ decreases with the tunable amount of rGO dopants in NC as shown in Table 12.6. This is due to the formation of the Ti–O–C bond and the hybridization of C $2p^2$ orbitals and O $2p^4$ orbitals to form new valence bands [13, 71, 72].

According to the electrical properties of rGO-TiO$_2$ NC, Zhang and co-researchers clarified that the photocatalytic performance can be improved with enhancement of carrier concentration and mobility between the rGO and TiO$_2$ materials [41, 71, 78–80]. To enhance the photocatalytic activity of rGO-TiO$_2$ NC, Khalid and co-researchers have shown that the function of TiO$_2$ can be easily enhanced in photocatalytic activity properties under visible light irradiation in terms of great absorptivity of dyes, extended light absorption range, and

Table 12.6 Band gap energy values of rGO-TiO$_2$ NC.

Methods	Results (eV)	Reference
Thermal	Pure TiO$_2$ = 3.10 rGO-TiO$_2$ = 2.95	[73]
Hydrothermal	Pure TiO$_2$ = 3.20 1 wt.% rGO-TiO$_2$ = 3.16 2 wt.% rGO-TiO$_2$ = 3.13 5 wt.% rGO-TiO$_2$ = 3.04 10 wt.% rGO-TiO$_2$ = 3.00	[74]
Solvothermal	Pure TiO$_2$ = 3.28 rGO-TiO$_2$ = 2.72	[75]
Hydrothermal	Pure TiO$_2$ = 3.03 rGO-TiO$_2$ = 2.78	[76]
Sonication	0.01 wt.% rGO-TiO$_2$ = 2.95	[77]

efficient charge separation with rGO [74, 81]. Khalid and co-researchers demonstrated that the band gap energy is decreased from 3.20 eV for TiO$_2$ to 3.00 eV when incorporated with rGO; it indicates the influence of rGO on the optical properties where an increase in rGO amount will result in the light absorption of TiO$_2$ [74]. Moreover, Khalid and co-researchers claimed that the presence of rGO in the TiO$_2$ composite could reduce the emission intensity in photoluminescence characterization and lead the enhancement of electron–hole pair separation efficiency [74].

12.4.1 Mechanism of rGO-TiO$_2$ NC

Zhang and co-researchers formed rGO-TiO$_2$ NC using a simple liquid phase deposition method by utilizing titanium tetrafluoride (TiF$_4$) and electron beam (EB) irradiation-pretreated rGO [82]. He discovered that the preparation condition had a significant effect on the structure and properties of rGO-TiO$_2$ NC. Through this method, it can be synthesized more uniform, smaller in size of TiO$_2$ nanoparticles and exhibited higher photocatalytic activities. Figure 12.10 shows the mechanism of rGO-TiO$_2$ NC that underwent a simple liquid phase deposition method.

12.4.2 Mechanism of rGO-TiO$_2$ NC in DSSCs

Figure 12.11 illustrates the electron flow when the rGO is loaded in between the TiO$_2$ molecules. The electron flow will be further enhanced if the rGO is well connected with TiO$_2$. This phenomenon is caused by the suppression of back-transport electron from the photoanode of the FTO/ITO electrode to the I_3^- ions, which subsequently increases the dye adsorption. Sung and co-researchers have mentioned that the presence of rGO oxide will reduce the back-transport in DSSCs and also assist in UV reduction in TiO$_2$ [74, 83].

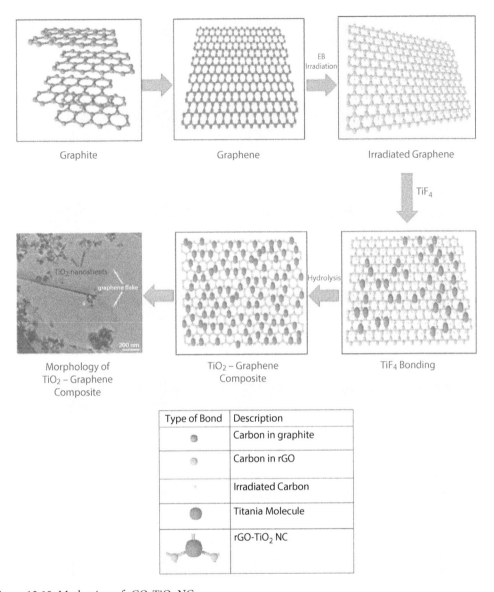

Figure 12.10 Mechanism of rGO-TiO$_2$ NC.

12.5 rGO-TiO$_2$ NC Synthesis

In this thin-film photovoltaic cell technology, second-generation solar cells are derived from the first-generation solar cell by depositing one or more thin layers of semiconductor materials on the specified substrate such as metal, glass, or silicon wafer. According to Thien and co-researchers, a higher photocurrent density is attributed to a delayed recombination rate and longer electron lifetime [84]. The photocurrent response of a solar cell is defined as the photo-generated electron–hole pair interaction between the photoanode and photocathode

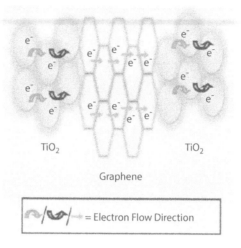

Figure 12.11 rGO-TiO$_2$ NC bonding mechanism.

electrode [21, 85–90]. The charge separation efficiency is increased due to the electronic interaction between rGO and the photo-induced electrons of TiO$_2$ in the NC [91–97].

On top of low cost and high reproducibility, rGO-TiO$_2$ NC also shows high interfacial contact and potential to enhance the photocatalytic activities of TiO$_2$. In the last two decades, there are a variety of techniques used to synthesize the rGO-TiO$_2$ NC-based materials to advance photovoltaic technology especially in DSSC application. For rGO-TiO$_2$ NC, rGO could be easily synthesized from the graphite flakes through the intermediate product of GO [98]. This technique was beneficial to form the TiO$_2$ nanocrystals during the synthesis of rGO-TiO$_2$ NC via the oxygenation of the functional groups from a GO or rGO product [55]. Kim and co-researchers reported that GO could be reduced via the UV-assisted photocatalytic reduction process using the 450-W xenon arc lamp forming the rGO-TiO$_2$ NC with low surface roughness and good adhesion at the photoanode element [83]. Dubey and co-researchers also reported that GO could be reduced by the UV radiation in the presence of ethanol solvent and TiO$_2$ nanoparticles to form the rGO-TiO$_2$ NC [99]. Another efficient technique to prepare the rGO-TiO$_2$ NC is the direct growth process to enhance the photocatalytic activity. Recently, Xu and co-researchers reported that rGO quantum dots could directly grow on 3D micropillar/microwave arrays of rutile TiO$_2$ nanorods forming the rGO-TiO$_2$ NC [100]. Additionally, the pathway for the large-scale production of the rGO-TiO$_2$ NC is the self-assembly approach of the *in situ* grown nanocrystalline TiO$_2$ with stabilization of rGO in the aqueous solutions by the anionic sulfate surfactants [101]. Furthermore, Liu and co-researchers reported an accessible synthetic route of solvothermal approach to form the rGO-TiO$_2$ NC with a better adsorption–photocatalytic activity than that of the pure TiO$_2$ [102].

12.5.1 Sol–Gel Synthesis

The sol-gel technique is widely used in the synthesis of rGO-based semiconductor composites. This method depends on the phase transformation of a sol obtained from metallic alkoxides or organometallic precursors. For instance, tetrabutyl titanate dispersed in rGO-containing absolute ethanol solution will gradually form a sol with continuous magnetic stirring and eventually change into rGO-TiO$_2$ NC after drying and after heat treatment [103, 104].

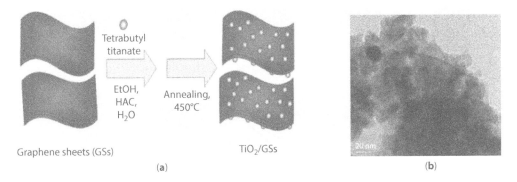

Figure 12.12 Schematic synthesis procedure (a) and typical TEM image of the rGO-TiO$_2$ NC (b) [104].

The synthesis process is illustrated in Figure 12.12a [104]. The resulting TiO$_2$ nanoparticles closely dispersed on the surface of 2D rGO NS (Figure 12.12b) [104]. Wojtoniszak and co-researchers used a similar strategy to prepare the rGO-TiO$_2$ NC via the hydrolysis of titanium (IV) butoxide in GO-containing ethanol solution [105]. The reduction of GO to rGO was performed in the post-heat treatment process. Meanwhile, Farhangi and co-researchers prepared Fe-doped TiO$_2$ nanowire arrays on the surface of functionalized rGO sheets using a sol-gel method in the green solvent of supercritical carbon dioxide [106]. During the preparation, the rGO NS acts as a template for nanowire growth through surface -COOH functionalities.

12.5.2 Solution Mixing Synthesis

Solution mixing is a simple method used to fabricate rGO/semiconductor composite photocatalysts. The uniform distribution of photocatalysts is facilitated by the oxygenated functional groups on GO under vigorous stirring or ultrasonic agitation [107]. The rGO-based composites can be obtained after the reduction of GO in the composite. Bell and co-researchers fabricated rGO-TiO$_2$ NC by ultrasonically mixing TiO$_2$ nanoparticles and GO colloids together, followed by ultraviolet (UV)-assisted photocatalytic reduction of GO to rGO [94]. Similarly, GO dispersion and N-doped Sr$_2$Ta$_2$O$_7$ have been mixed together, followed by reduction of GO to yield Sr$_2$Ta$_2$O$_{7-x}$N$_x$-rGO composites under xenon lamp irradiation [108]. Paek and co-researchers have prepared the SnO$_2$ sol by hydrolysis of SnCl$_4$ with NaOH and then mixed with the prepared rGO dispersion in ethylene glycol to form the SnO$_2$-rGO composite [109]. On the other hand, Geng co-researchers have synthesized the CdSe-rGO quantum dot composites [110]. In their work, pyridine-modified CdSe nanoparticles were mixed with GO sheets, where pyridine ligands provide π–π interactions for the assembly of CdSe nanoparticles on GO sheets.

12.5.3 *In Situ* Growth Synthesis

The *in situ* growth strategy provides efficient electron transfer between rGO and semiconductor nanoparticles through their intimate contact. The functional GO and metal salts are commonly used as precursors. The presence of epoxy and hydroxyl functional groups on rGO can act as heterogeneous nucleation sites and anchor semiconductor nanoparticles avoiding the agglomeration of the small particles [111]. Lambert and co-researchers have reported the *in situ* synthesis of petal-like TiO_2-GO by the hydrolysis of TiF_4 in the presence of aqueous dispersions of GO, followed by post-thermal treatment to produce rGO-TiO_2 NC [112]. With a high concentration of GO and stirring, long-range ordered assemblies of TiO_2-GO sheets were self-assembled. Besides that, Guo and co-researchers synthesized rGO-TiO_2 NC sonochemically from $TiCl_4$ and GO in an ethanol–water system, followed by a hydrazine treatment to reduce GO into rGO [113]. The average size of the TiO_2 nanoparticles was controlled at around 4–5 nm on the sheets, which is attributed to the pyrolysis and condensation of the dissolved $TiCl_4$ into TiO_2 by ultrasonic waves. Lastly, rGO-TiO_2 were synthesized with various method not only apply in photovoltaic application but also useful in other applications and summarized in Table 12.7.

12.6 Fabrication Technique of rGO-TiO_2 NC-Based Photoanode in DSSC Application

In this particular section, different preparations and various deposition technologies for the fabrication of rGO-TiO_2 NC as the photoanode in DSSC application will be highlighted and emphasized. Moreover, the schematic diagram/mechanism and PCE of DSSCs will be discussed as well. On top of that, different depositions of the rGO-TiO_2 NC will be reviewed, owing to the very limited studies being reported on the physical methods. The physical method is defined as the physical spectacle for the preparation and deposition on the materials. Generally, there are two major sources/mediums that have been applied using a variety of physical depositions such as liquid phase and gas phase. In this section, the PVD approach based liquid phase processes for rGO-TiO_2 NC preparation such as spin coating, doctor blade printing, and eletrohydrodynamic deposition. Nonetheless, gas phase processes like thermal evaporation, electron beam evaporation, sputtering, pulsed DC sputtering, and DC magnetron sputtering & radio frequency magnetron sputtering will be briefly explained as well. Lastly, recent studies for both liquid- and gas-phase processes as photoanode in DSSC-based materials instead of rGO-TiO_2 NC are summarized in Tables 12.8 and 12.9, respectively.

Table 12.7 Summary of rGO-TiO_2 synthesis in various applications.

Synthesis method	Materials	Application	Reference
Sol-gel	Ce-rGO-TiO_2	Photoelectrocatalytic	[99, 114]
Sol-gel	Anatase TiO_2-rGO	Photoelectrochemical water splitting	[115]
Solution mixing	rGO-TiO_2	Photocatalytic selectivity	[116]
Solution mixing	rGO-TiO_2	Hydrogen production	[117]
In situ growth	rGO-TiO_2	Sodium/lithium ion batteries	[118]
In situ growth	rGO-TiO_2	Photocatalytic activity	[119]

Table 12.8 List of deposition techniques of liquid-phase processes in DSSC application.

Deposition method	Photoanode materials	PCE, η (%)	Reference/Year
Spin-coating	Luminescent species-TiO$_2$	5.02	[130]/2015
Spin-coating	Ga-doped ZnO seed	1.23	[131]/2015
Spin-coating	Li-doped ZnO and SnO$_2$ NC	2.06	[132]/2016
Spin-coating	TiO$_2$	2.00	[133]/2017
Doctor blade printing	Mesoporous TiO$_2$	4.20	[134]/2016
Doctor blade printing	TiO$_2$	2.56	[135]/2016
Doctor blade printing	TiO$_2$	1.14	[136]/2016
Doctor blade printing	ZnO NS	2.00	[137]/2017
Electrospray deposition	TiO$_2$ NPs	1.674	[138]/2015

Table 12.9 List of deposition techniques based on gas-phase processes in DSSC application.

Deposition method	Photoanode materials	PCE, η (%)	Reference/Year
PVD	GO-TiO$_2$	4.65	[140]/2015
PVD	Ag-TiO$_2$	4.80	[141]/2016
PVD	Mg^{2+}-TiO$_2$	5.90	[142]/2016
Thermal evaporation	rGO-TiO$_2$-P3HT-PC$_{61}$BM/PEDOT/PSS/Ag	2.32	[143]/2015
Thermal evaporation	GO-ZnO	4.52	[144]/2016
EBE	Au-TiO$_2$	–	[145]/2016
Pulsed DC sputtering	NiO$_x$-TiO$_2$	2.79	[146]/2013
DC sputtering	TiO$_2$	4.00	[147]/2007
DC sputtering	AZO/Ag/AZO	0.60	[148]/2010
DC sputtering	TiO$_2$	2.07	[149]/2011
RF sputtering	TiO$_2$-rGO-TiO$_2$	3.93	[121]/2014
RF sputtering	Gr/ZnO	3.98	[150]/2014
RF sputtering	AZO/TiO$_2$/TiO$_2$ porous layer	5.69	[151]/2015

12.6.1 PVD Methods—rGO-TiO$_2$ NC (Liquid-Phase Processes)

12.6.1.1 Spin-Coating Technique

The spin-coating method is a technique where the specified chemical/solvent/polymer drops on the center of substrate during the spinning at vigorous stirring. A uniform thin film will be formed and deposited on the substrate with low surface roughness. A simple schematic of spin coating including the deposition, spin up, spin off, and evaporation is illustrated in Figure 12.13. In other words, the loaded solvent will be deposited uniformly on the substrate under high-speed rotation and coating by centrifugal force. The rGO-TiO$_2$ NC prepared via the spin-coating method was widely utilized as a photoanode assembly in DSSC application. Tsai and co-researchers demonstrated that the rGO-TiO$_2$ NC could be deposited on ITO substrates by the spin-coating method and used as an efficient electrode in DSSCs [120]. Accordingly, the presence of optimum rGO content (1 wt.%) in TiO$_2$ could give the highest PCE of 6.86% under a light illumination of 100 mW cm^{-2}. This implies the reduction in photogenerated electron loss and electron–hole pair recombination. Chen and co-researchers also investigated the TiO$_2$-rGO-TiO$_2$ sandwich structure as a working electrode via the spin-coating method. The ideal PCE of DSSCs was found to be 3.93% [121]. By using the spin-coating method, Lee and co-researchers reported that the rGO quantum dots incorporated with the TiO$_2$ working electrode could give a PCE of 7.95% [122]. Lately, Yao and co-researchers have established the hierarchical structures of rGO-TiO$_2$ seed layer on the FTO substrate using the spin-coating process, where the TiO$_2$ layer contains Er^{3+} and Yb^{3+} ions [123]. The PCE of the TiO$_2$:rGO-TiO$_2$:Er^{3+}, Yb^{3+} nanorod array was reported as 4.58%, as compared to that of 3.38% for TiO$_2$ nanorods. The modification of TiO$_2$ with TiO$_2$:Er^{3+}, Yb^{3+} and Al$_2$O$_3$:Eu^{3+} represented the up conversion (UC) and down conversion (DC) materials, respectively. Also, the light scattering capabilities of DSSCs could be improved via an increase in light absorption, shorter charge transportation, and also faster charge carrier mobility when incorporated with rGO material. Apart from that, another advantage of utilizing the TiO$_2$ nanorod arrays is its one-dimensional nanostructure, which can offer a direct pathway for photogenerated electrons.

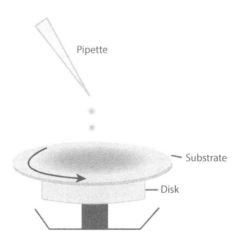

Figure 12.13 Spin-coating diagram.

12.6.1.2 Doctor Blade Printing Technique

Generally, the doctor blade method is one of the alternative approaches to produce a large-area thin film. Howatt and co-researchers are the first group to report on the tape casting process producing thin sheets of ceramic capacitors [124]. The doctor blade device to move plaster batts using aqueous and nonaqueous slurries was reported thereafter [125]. Accordingly, the doctor blade or tape casting method are conducted by three simple steps, in which (i) rGO-TiO_2 solution is applied along the glass substrate, (ii) a blade moves with a constant relative movement (pull and push) to spread the rGO-TiO_2 along the surface active area, and (iii) an rGO-TiO_2 thin film is formed uniformly in a gel layer upon drying/annealing processes, as illustrated in Figure 12.14, respectively. In DSSC application, rGO quantum dot-decorated TiO_2 nanofibers were coated onto FTO substrate by the doctor blade technique. The thickness was found to be approximately 10–12 μm, and a high PCE of about 6.22% could be obtained [126]. The strong interaction between rGO quantum dots and 1D TiO_2 nanofibers (without affecting the integrity) has consequently accelerated the photogenerated electron scan. Akbar and co-researchers have explored a one-step process where the TiO_2 and rGO sheets were mixed to form a rGO-TiO_2 paste. The paste was then deposited onto the FTO substrate using the doctor blade method [127]. However, a lower PCE of 0.7% was attained even if 1 wt.% of rGO content was used.

12.6.1.3 Electrohydrodynamic Deposition Technique

The liquid phase of electrohydrodynamic deposition is emphasized on electrospray deposition (ESD) (Figure 12.15). The ESD method is mainly used for the fabrication of MEMS and NEMS in order to obtain thin films (< 10 μm). Normally, the nanoparticle source (liquid phase) will be converted into droplets formed across the nozzle and spray to form a thin film on the FTO/ITO substrate. Among liquid phase deposition, ESD method has attracted the most attention and has the lowest cost merit for the large-area production [128]. The rGO-TiO_2 NC thin films will be deposited uniformly by evaporation or by heating the solvent on the FTO/ITO surface by sintering. Accordingly, the polymer-rGO-TiO_2 composite can be formed by using the ESD method in which the functionalized rGO was dissolved in N,N-dimethyl acetamide with polyvinyl acetate (PVAc) and a titanium precursor. The PVAc-rGO-TiO_2

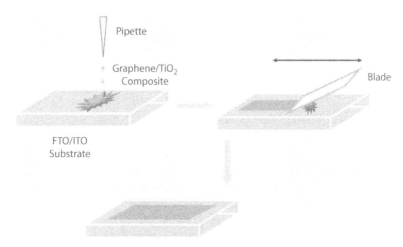

Figure 12.14 Doctor blade technique for rGO-TiO_2 as photoanode in DSSCs.

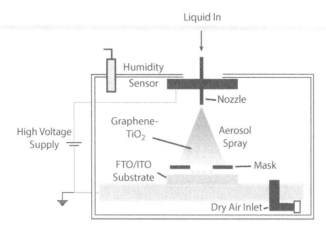

Figure 12.15 Schematic of the electrospray deposition technique.

composite fibers were formed as photoanode and potentially enhanced the PCE of DSSCs [129]. Recently, Liu and co-researchers studied the deposition time and the number of rGO layer deposition used as photoanode in DSSCs using the ESD method [128]. The deposition duration for a single Gr layer was 1 min, and the rGO-TiO_2 NC of three layers could achieve the optimum PCE of 7.8% and 8.9%, respectively.

12.6.2 PVD Methods—rGO-TiO_2 NC (Gas-Phase Processes)

12.6.2.1 Physical Vapor Deposition (PVD) Technique

Generally, the PVD technique is the deposition of thin film under evaporation condition and sputtering using vacuum chamber technology. Considering that particles tend to escape from the surface, TiO_2 particles with rGO material should be coated in a closed environment. The particles moved in direct motion heading to the substrate when the TiO_2 was being heated (Figure 12.16). TiO_2 is physically coated onto the rGO surface forming rGO-TiO_2 NC thin films on the FTO/ITO substrate within a short deposition duration under closed chamber conditions.

12.6.2.2 Thermal Evaporation Technique

Several deposition techniques are under evaporation where the source particle material (rGO-TiO_2 NC) needs to travel to the FTO/ITO substrate and to be sintered into the thin film as photoanode. The evaporation techniques include thermal evaporation, electron beam evaporation, sputtering, DC magnetron sputtering, and radio frequency magnetron sputtering. This is a low-cost and easy method to deposit a thin film. Upon source melting on a resistive heated boat via electrical heating, the source material is evaporated in the vacuum chamber and allows the vaporized particles to directly transfer to the substrate (Figure 12.17). The high vacuum pressure is able to prevent the particles from scattering, minimizing residual gas impurities. However, an adhesive layer is needed to strengthen the thin film due to its poor adhesion properties. So far, no research work has been conducted on rGO-TiO_2 thin film using this thermal evaporation method.

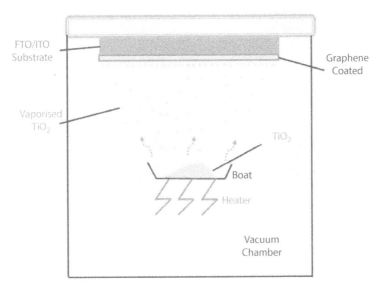

Figure 12.16 Schematic of the physical vapor deposition method.

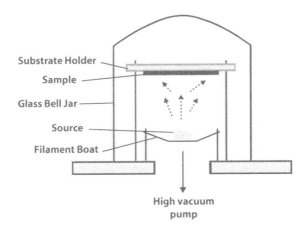

Figure 12.17 Schematic of the thermal evaporation process.

12.6.2.3 *Electron Beam Evaporation (EBE) Technique*

Electron beam evaporation (EBE) is one of the powerful deposition methods under PVD technique, applying high-speed electron to bombard the target source (Figure 12.18). The kinetic energy of electron beam is produced from the electron gun by using electric and magnetic fields to shoot the target and vaporize the surrounding vacuum area. Once the FTO/ITO substrate is heated by the radiation heating element, the surface atoms will have sufficient energy to leave the FTO/ITO substrate. At the same time, the FTO/ITO substrate will be coated when the thermal energy is less than 1 eV and the working distance is in the range of 300 mm to the 1 m. Jin and co-researchers studied the reduction of back transfer electrons with the direct contact between electrolyte and the FTO glass substrate-coated TiO_2 passivation layer [139]. The preparation of the TiO_2 passivation layer was performed using the EBE technique and has recorded a PCE of 4.93% due to the reduction in electron–hole pair charge recombination.

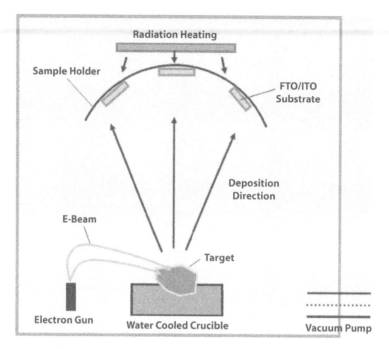

Figure 12.18 Schematic of the electron beam evaporation process.

12.6.2.4 Sputtering Technique

Sputtering is a useful deposition/modification technique whereby the accelerated ions are applied to expel the original particles on a target substrate via ion bombardment (Figure 12.19). In other words, the sputtering method is considered as the momentum transfer process of ions that accelerated from the source to the collision of substrate particles. Apart from that, the electrical potential will cause the ions to accelerate and the ions will be reflected or absorbed to the FTO/ITO substrate provided the kinetic energy is less than 5 eV. The substrate and lattice positions will be scratched once the kinetic energy is higher than that of the surface atom binding energy. Typically, there are two kinds of sputtering process utilizing the ions of an inert gas to eject atoms from the surface, such as (i) DC magnetron sputtering and (ii) radio frequency (RF) magnetron sputtering. The advantage of using DC magnetron sputtering is its ability to increase the deposition rate with a minimal damage to the FTO/ITO substrate while the RF magnetron sputtering provides a direct pathway deposition of insulators. The common use of deposition materials of DC magnetron sputtering and RF magnetron sputtering techniques are metal, alloy, and organic compound. In addition, there are also several sputtering with different deposition conditions like pulsed DC sputtering power (DCMSP), MF mid frequency AC sputtering power, high-power impulse magnetron sputtering (HIPIMS), etc.

12.6.2.5 Pulsed DC Sputtering Power Technique

Pulsed DC sputtering technique is generally used for metal deposition and dielectrics coating so that the insulator materials are able to receive donor charge (Figure 12.20). This coating technology is widely used in the industry sector such as semiconductors and

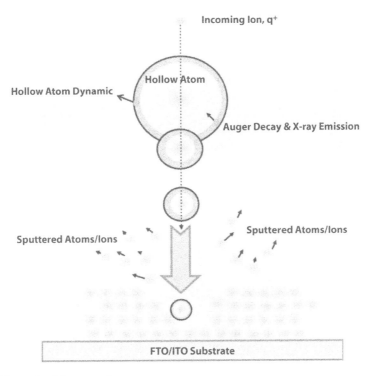

Figure 12.19 Schematic of the sputtering process.

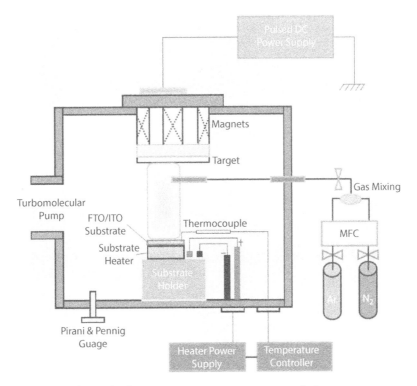

Figure 12.20 Schematic of the pulsed DC magnetron sputtering power technique.

optics for large area production. Also, this technique has found broad applications in reactive sputtering where the chemical reaction occurred in the plasma region between ionized gases and vaporized target materials. In reactive sputtering, the usage of oxygen (O_2) gas is to combine the target material within the plasma to form oxide molecules, while that of argon (Ar) gas is to convey kinetic energy upon impact with the target materials.

12.6.2.6 DC and RF Magnetron Sputtering Technique

The direct current (DC) magnetron sputtering concept is the oldest deposition technique among the magnetron sputtering field. Accordingly, this particular method would compromise the ionization of the mixture of argon (Ar) and nitrogen (N_2) gas while the positively charged sputtering gas could be easily accelerated toward the conductive target materials, causing the ejected target atoms to deposit on the ITO/FTO substrate easier (Figure 12.21). It is a controllable and low-cost sputter technique for relatively large substrate quantity and large-scale production. Typically, the working DC sputtering technique is under 1000–3000 V and the vacuum chamber is around 10^{-3} Pa with pressure around 0.075 to 0.12 torr. However, this technique can only be applicable to the conductive material but not to the dielectric target. This is mainly attributed to the termination of the discharge of insulator materials during the deposition process. In other words, the positively charged ions will be produced and accumulated on the surface of the dielectric or insulator films.

Radio frequency (RF) magnetron sputtering can be considered as an alternative way to overcome the DC magnetron sputtering by applying an alternating current (AC) power source. This approach is a suitable method for conductive and nonconductive target materials such as conductor, semiconductor, insulator, and dielectric films. Generally, the frequency used in this technique is an alternating voltage at a specific frequency of 13.56 MHz

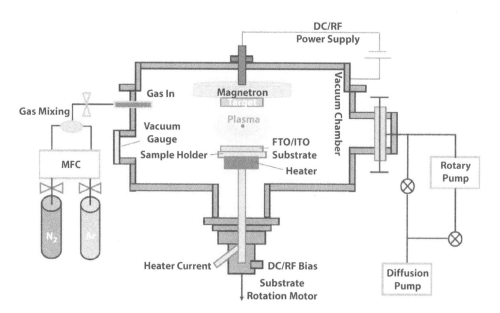

Figure 12.21 Schematic of the DC/RF sputtering deposition technique.

within the frequency range of 1 kHz to 103 MHz. During the positive electric field, the positively charged ions would be accelerated to the surface of the target and directly sputter the source on the FTO/ITO substrate. The positively charged ions on the surface of the target will be eliminated/neutralized by the electron bombardment force during the negative field moment. However, the high deposition rate of RF magnetron sputtering for FTO/ITO substrate deposition as compared to that of the DC sputtering magnetron sputtering has led to the mobility difference between the electrons and ions within the plasma region. Therefore, high heating temperature is required to accelerate the sputtering process. RF magnetron sputtering deposition is generally only limited to smaller substrate quantities and sizes due to the cost consideration of RF power supplies.

Currently, Chen and co-researchers have deposited the rGO film onto the ITO substrate for 2 min using the RF magnetron sputtering technique with the carbon target (99.99%) as the first photoanode layer under an input power of 90 W and an Ar gas flow rate of 90 sccm, with the TiO_2 film as the second photoanode layer with the spin-coating technique. The attained PCE was found to be 2.46% [131]. Among these deposition based gas phase processes approach, the RF magnetron sputtering method is aimed to utilize in this project due to the Ti^{3+} ions might simply accelerated with high energy under high applied power in a short period of time and implanted onto rGO surface or lattice.

In terms of deposition technique, to the best of the authors' knowledge, there is less literature that report on the gas-phase processes to incorporate TiO_2 with rGO material. Table 12.9 shows the photoanode materials prepared using a variety of deposition methods to enhance the PCE of DSSCs. These gas-phase processes included conductor and insulator materials coating other than that of the rGO-TiO_2 NC materials for DSSC application.

Acknowledgments

This research was supported by Universiti Malaya Prototype Grant (RU005G-2016), Transdisciplinary Research Grant Scheme (TRGS) (TR002A-2014B), and Global Collaborative Programme–SATU joint research scheme (ST007-2017) from the University of Malaya.

References

1. Grätzel, M., Solar energy conversion by dye-sensitized photovoltaic cells. *Inorg. Chem.*, 44, 6841–6851, 2005.
2. Lai, C.W. *et al.*, An overview: Recent development of titanium dioxide loaded graphene nanocomposite film for solar application. *Curr. Org. Chem.*, 19, 1882–1895, 2015.
3. Su'ait, M.S., Rahman, M.Y.A., Ahmad, A., Review on polymer electrolyte in dye-sensitized solar cells (DSSCs). *Solar Energy*, 115, 452–470, 2015.
4. Chandrasekaran, J. *et al.*, Hybrid solar cell based on blending of organic and inorganic materials—An overview. *Renewable Sustainable Energy Rev.*, 15, 1228–1238, 2011.
5. Ludin, N.A. *et al.*, Review on the development of natural dye photosensitizer for dye-sensitized solar cells. *Renewable Sustainable Energy Rev.*, 31, 386–396, 2014.
6. Low, F.W., Lai, C.W., Hamid, S.B.A., Surface modification of reduced graphene oxide film by Ti ion implantation technique for high dye-sensitized solar cells performance. *Ceram. Int.*, 43, 625–633, 2017.

7. De Souza, J.D.S., de Andrade, L.O.M., Müller, A.V., Polo, A.S., Nanomaterials for solar energy conversion: Dye-sensitized solar cells based on ruthenium (II) Tris-heteroleptic compounds or natural dyes. In *Nanoenergy*, pp. 69–106. Springer, 2018.
8. Narayan, M.R., Dye sensitized solar cells based on natural photosensitizers. *Renewable Sustainable Energy Rev.*, 16, 208–215, 2012.
9. Kay, A. and Graetzel, M., Artificial photosynthesis. 1. Photosensitization of titania solar cells with chlorophyll derivatives and related natural porphyrins. *J. Phys. Chem.*, 97, 6272–6277, 1993.
10. Kay, A., Humphry-Baker, R., Graetzel, M., Artificial photosynthesis. 2. Investigations on the mechanism of photosensitization of nanocrystalline TiO_2 solar cells by chlorophyll derivatives. *J. Phys. Chem.*, 98, 952–959, 1994.
11. Shanmugam, V. *et al.*, Green grasses as light harvesters in dye sensitized solar cells. *Spectrochim. Acta, Part A*, 135, 947–952, 2015.
12. Al-Alwani, M.A. *et al.*, Dye-sensitised solar cells: Development, structure, operation principles, electron kinetics, characterisation, synthesis materials and natural photosensitisers. *Renewable Sustainable Energy Rev.*, 65, 183–213, 2016.
13. O'regan, B. and Grätzel, M., A low-cost, high-efficiency solar cell based on dye-sensitized colloidal TiO_2 films. *Nature*, 353, 737, 1991.
14. Grätzel, M., Dye-sensitized solar cells. *J. Photochem. Photobiol., C*, 4, 145–153, 2003.
15. Singh, E. and Nalwa, H.S., Graphene-based dye-sensitized solar cells: A review. *Sci. Adv. Mater.*, 7, 1863–1912, 2015.
16. Sugathan, V., John, E., Sudhakar, K., Recent improvements in dye sensitized solar cells: A review. *Renewable Sustainable Energy Rev.*, 52, 54–64, 2015.
17. Basheer, B. *et al.*, An overview on the spectrum of sensitizers: The heart of dye sensitized solar cells. *Solar Energy*, 108, 479–507, 2014.
18. Ito, S. *et al.*, Fabrication of thin film dye sensitized solar cells with solar to electric power conversion efficiency over 10%. *Thin Solid Films*, 516, 4613–4619, 2008.
19. Asim, N. *et al.*, A review on the role of materials science in solar cells. *Renewable Sustainable Energy Rev.*, 16, 5834–5847, 2012.
20. Hemmatzadeh, R. and Mohammadi, A., Improving optical absorptivity of natural dyes for fabrication of efficient dye-sensitized solar cells. *J. Theor. Appl. Phys.*, 7, 57, 2013.
21. Anandan, S., Recent improvements and arising challenges in dye-sensitized solar cells. *Solar Energy Mater. Solar Cells*, 91, 843–846, 2007.
22. Nazeeruddin, M.K., Baranoff, E., Grätzel, M., Dye-sensitized solar cells: A brief overview. *Solar Energy*, 85, 1172–1178, 2011.
23. de Souza, J.D.S., de Andrade, L.O.M., Polo, A.S., Nanomaterials for solar energy conversion: Dye-sensitized solar cells based on ruthenium (II) Tris-heteroleptic compounds or natural dyes, in: *Nanoenergy*, pp. 49–80, Springer, 2013.
24. Ito, S. *et al.*, Fabrication of dye-sensitized solar cells using natural dye for food pigment: Monascus yellow. *Energy Environ. Sci.*, 3, 905–909, 2010.
25. Maurya, I.C., Srivastava, P., Bahadur, L., Dye-sensitized solar cell using extract from petals of male flowers *Luffa cylindrica L. as a natural sensitizer*. *Opt. Mater.*, 52, 150–156, 2016.
26. Wan, H.Y., *Dye Sensitized Solar Cells*, University of Alabama Department of Chemistry, p. 3, 2004.
27. Singh, V. *et al.*, Graphene based materials: Past, present and future. *Prog. Mater. Sci.*, 56, 1178–1271, 2011.
28. Roy-Mayhew, J.D. and Aksay, I.A., Graphene materials and their use in dye-sensitized solar cells. *Chem. Rev.*, 114, 6323–6348, 2014.
29. Gong, J., Liang, J., Sumathy, K., Review on dye-sensitized solar cell (DSSCs): Fundamental concepts and novel material. *Renewable and Sustainable Energy Reviews*, 16, 5848–5860, 2012.

30. Shalini, S. et al., Review on natural dye sensitized solar cells: Operation, materials and methods. *Renewable Sustainable Energy Rev.*, 51, 1306–1325, 2015.
31. Raj, C.C. and Prasanth, R., A critical review of recent developments in nanomaterials for photoelectrodes in dye sensitized solar cells. *J. Power Sources*, 317, 120–132, 2016.
32. Sengupta, D. et al., Effects of doping, morphology and film-thickness of photo-anode materials for dye sensitized solar cell application—A review. *Renewable Sustainable Energy Rev.*, 60, 356–376, 2016.
33. Qu, J. and Lai, C., One-dimensional TiO_2 nanostructures as photoanodes for dye-sensitized solar cells. *J. Nanomater.*, 2013, 2, 2013.
34. Chergui, Y., Nehaoua, N., Mekki, D.E., Comparative study of dye-sensitized solar cell based on ZnO and TiO_2 nanostructures. *Solar Cells-Dye-Sensitized Devices*. InTech, 2011.
35. Pagliaro, M. et al., Nanochemistry aspects of titania in dye-sensitized solar cells. *Energy Environ. Sci.*, 2, 838–844, 2009.
36. Tripathi, B. et al., Investigating the role of graphene in the photovoltaic performance improvement of dye-sensitized solar cell. *Mater. Sci. Eng., B*, 190, 111–118, 2014.
37. Tang, Y.-B. et al., Incorporation of graphenes in nanostructured TiO_2 films via molecular grafting for dye-sensitized solar cell application. *ACS Nano*, 4, 3482–3488, 2010.
38. Li, S. et al., Vertically aligned carbon nanotubes grown on graphene paper as electrodes in lithium-ion batteries and dye-sensitized solar cells. *Adv. Energy Mater.*, 1, 486–490, 2011.
39. Wu, J. et al., Dual functions of YF3: Eu3+ for improving photovoltaic performance of dye-sensitized solar cells. *Sci. Rep.*, 3, 2013.
40. Zhu, Y. et al., Graphene and graphene oxide: Synthesis, properties, and applications. *Adv. Mater.*, 22, 3906–3924, 2010.
41. Xiang, Q., Yu, J., Jaroniec, M., Graphene-based semiconductor photocatalysts. *Chem. Soc. Rev.*, 41, 782–796, 2012.
42. Geim, A.K., Graphene: Status and prospects. *Science*, 324, 1530–1534, 2009.
43. Bell, N., On the design and synthesis of titanium dioxide–graphene nanocomposites for enhanced photovoltaic and photocatalytic performance. *Thesis Citation*.
44. Wang, P. et al., Graphene oxide nanosheets as an effective template for the synthesis of porous TiO_2 film in dye-sensitized solar cells. *Appl. Surf. Sci.*, 358, Part A, 175–180, 2015.
45. Dey, A. et al., A graphene titanium dioxide nanocomposite (GTNC): One pot green synthesis and its application in a solid rocket propellant. *RSC Adv.*, 5, 63777–63785, 2015.
46. Shi, M. et al., Preparation of graphene–TiO_2 composite by hydrothermal method from peroxotitanium acid and its photocatalytic properties. *Colloids Surf. A*, 405, 30–37, 2012.
47. Huang, N.M. et al., Simple room-temperature preparation of high-yield large-area graphene oxide. *Int. J. Nanomed.*, 6, 3443, 2011.
48. Wu, T.-T. and Ting, J.-M., Preparation and characteristics of graphene oxide and its thin films. *Surf. Coat. Technol.*, 231, 487–491, 2013.
49. Huang, C., Li, C., Shi, G., Graphene based catalysts. *Energy Environ. Sci.*, 5, 8848–8868, 2012.
50. Shearer, C.J., Cherevan, A., Eder, D., Chapter 16–Application of functional hybrids incorporating carbon nanotubes or graphene A2–Tanaka, K, in: *Carbon Nanotubes and Graphene (Second Edition)*, S. Iijima (Ed.), pp. 387–433, Elsevier, Oxford, 2014.
51. Roy-Mayhew, J.D. et al., Functionalized graphene as a catalytic counter electrode in dye-sensitized solar cells. *ACS Nano*, 4, 6203–6211, 2010.
52. Yang, S. et al., Fabrication of graphene-encapsulated oxide nanoparticles: Towards high-performance anode materials for lithium storage. *Angew. Chem. Int. Ed.*, 49, 8408–8411, 2010.
53. Blanita, G. and Lazar, M.D., Review of graphene-supported metal nanoparticles as new and efficient heterogeneous catalysts. *Micro Nanosystems*, 5, 138–146, 2013.

54. Gong, J. et al., Review on dye-sensitized solar cells (DSSCs): Advanced techniques and research trends. *Renewable Sustainable Energy Rev.*, 68, Part 1, 234–246, 2017.
55. Liang, Y. et al., TiO_2 nanocrystals grown on graphene as advanced photocatalytic hybrid materials. *Nano Res.*, 3, 701–705, 2010.
56. Chen, L. et al., Enhanced photovoltaic performance of a dye-sensitized solar cell using graphene-TiO_2 photoanode prepared by a novel *in situ* simultaneous reduction-hydrolysis technique. *Nanoscale*, 5, 3481–3485, 2013.
57. Jo, W.-K. and Kang, H.-J., Titanium dioxide–graphene oxide composites with different ratios supported by Pyrex tube for photocatalysis of toxic aromatic vapors. *Powder Technol.*, 250, 115–121, 2013.
58. Routh, P. et al., Graphene quantum dots from a facile sono-fenton reaction and its hybrid with a polythiophene graft copolymer toward photovoltaic application. *ACS Appl. Mater. Interfaces*, 5, 12672–12680, 2013.
59. Sharma, P., Saikia, B.K., Das, M.R., Removal of methyl green dye molecule from aqueous system using reduced graphene oxide as an efficient adsorbent: Kinetics, isotherm and thermodynamic parameters. *Colloids Surf., A*, 457, 125–133, 2014.
60. Bonaccorso, F. et al., Graphene photonics and optoelectronics. *Nat. Photonics*, 4, 611–622, 2010.
61. Kazmi, S.A. et al., Electrical and optical properties of graphene–TiO_2 nanocomposite and its applications in dye sensitized solar cells (DSSC). *J. Alloys Compd.*, 691, 659–665, 2017.
62. Zhang, N. et al., Waltzing with the versatile platform of graphene to synthesize composite photocatalysts. *Chem. Rev.*, 115, 10307–10377, 2015.
63. Wang, H., Leonard, S.L., Hu, Y.H., Promoting effect of graphene on dye-sensitized solar cells. *Ind. Eng. Chem. Res.*, 51, 10613–10620, 2012.
64. Kim, A. et al., Photovoltaic efficiencies on dye-sensitized solar cells assembled with graphene-linked TiO_2 anode films. *Bull. Korean Chem. Soc.*, 33, 3355–3360, 2012.
65. Pablo, C.V. et al., Construction of dye-sensitized solar cells (DSSC) with natural pigments. *Mater. Today: Proc.*, 3, 194–200, 2016.
66. Calogero, G. et al., Absorption spectra and photovoltaic characterization of chlorophyllins as sensitizers for dye-sensitized solar cells. *Spectrochim. Acta, Part A*, 132, 477–484, 2014.
67. Bisquert, J. et. al., Electron lifetime in dye-sensitized solar cells: Theory and interpretation of measurements. *J. Phys. Chem. C*, 113, 17278–17290, 2009.
68. Chen, J.Z., Yan, Y.C., Lin, K.J., Effects of carbon nanotubes on dye-sensitized solar cells. *J. Chin. Chem. Soc.*, 57, 1180–1184, 2010.
69. Kay, A. and Grätzel, M., Low cost photovoltaic modules based on dye sensitized nanocrystalline titanium dioxide and carbon powder. *Solar Energy Mater. Solar Cells*, 44, 99–117, 1996.
70. Gao, Y. et al., Improvement of adhesion of Pt-free counter electrodes for low-cost dye-sensitized solar cells. *J. Photochem. Photobiol., A*, 245, 66–71, 2012.
71. Zhang, H. et al., P25-graphene composite as a high performance photocatalyst. *ACS Nano*, 4, 380–386, 2009.
72. Li, K. et al., Preparation of graphene/TiO_2 composites by nonionic surfactant strategy and their simulated sunlight and visible light photocatalytic activity towards representative aqueous POPs degradation. *J. Hazard. Mater.*, 250, 19–28, 2013.
73. Zhang, Y. and Pan, C., TiO_2/graphene composite from thermal reaction of graphene oxide and its photocatalytic activity in visible light. *J. Mater. Sci.*, 46, 2622–2626, 2011.
74. Khalid, N. et al., Enhanced photocatalytic activity of graphene–TiO_2 composite under visible light irradiation. *Curr. Appl. Phys.*, 13, 659–663, 2013.
75. Wang, Y. et al., Low-temperature solvothermal synthesis of graphene–TiO_2 nanocomposite and its photocatalytic activity for dye degradation. *Mater. Lett.*, 134, 115–118, 2014.

76. Kumar, R. et al., Hydrothermal synthesis of a uniformly dispersed hybrid graphene–TiO$_2$ nanostructure for optical and enhanced electrochemical applications. *RSC Adv.*, 5, 7112–7120, 2015.
77. Kanta, U.-A. et al., Preparations, characterizations, and a comparative study on photovoltaic performance of two different types of graphene/TiO$_2$ nanocomposites photoelectrodes. *J. Nanomater.*, 2017, 2017.
78. Zhang, Y. et al., Engineering the unique 2D mat of graphene to achieve graphene-TiO$_2$ nanocomposite for photocatalytic selective transformation: What advantage does graphene have over its forebear carbon nanotube? *ACS Nano*, 5, 7426–7435, 2011.
79. Jiang, B. et al., Enhanced photocatalytic activity and electron transfer mechanisms of graphene/TiO$_2$ with exposed {001} facets. *J. Phys. Chem. C*, 115, 23718–23725, 2011.
80. Zhang, Y. et al., Improving the photocatalytic performance of graphene–TiO$_2$ nanocomposites via a combined strategy of decreasing defects of graphene and increasing interfacial contact. *Phys. Chem. Chem. Phys.*, 14, 9167–9175, 2012.
81. Geng, D., Wang, H., Yu, G., Graphene single crystals: Size and morphology engineering. *Adv. Mater.*, 27, 2821–2837, 2015.
82. Zhang, H. et al., A facile one-step synthesis of TiO$_2$/graphene composites for photodegradation of methyl orange. *Nano Res.*, 4, 274–283, 2011.
83. Kim, S.R., Parvez, M.K., Chhowalla, M., UV-reduction of graphene oxide and its application as an interfacial layer to reduce the back-transport reactions in dye-sensitized solar cells. *Chem. Phys. Lett.*, 483, 124–127, 2009.
84. Thien, G.S. et al., Improved synthesis of reduced graphene oxide-titanium dioxide composite with highly exposed 001 facets and its photoelectrochemical response. *Int. J. Photoenergy*, 2014, 2014.
85. Woan, K., Pyrgiotakis, G., Sigmund, W., Photocatalytic carbon-nanotube–TiO$_2$ composites. *Adv. Mater.*, 21, 2233–2239, 2009.
86. Khan, S.U., Al-Shahry, M., Ingler, W.B., Efficient photochemical water splitting by a chemically modified n-TiO$_2$. *Science*, 297, 2243–2245, 2002.
87. Park, J.H., Kim, S., Bard, A.J., Novel carbon-doped TiO$_2$ nanotube arrays with high aspect ratios for efficient solar water splitting. *Nano Lett.*, 6, 24–28, 2006.
88. Sellappan, R. et al., Influence of graphene synthesizing techniques on the photocatalytic performance of graphene–TiO$_2$ nanocomposites. *Phys. Chem. Chem. Phys.*, 15, 15528–15537, 2013.
89. Tryba, B., Morawski, A., Inagaki, M., Application of TiO$_2$-mounted activated carbon to the removal of phenol from water. *Appl. Catal., B*, 41, 427–433, 2003.
90. Wang, H. et al., Photoelectrocatalytic oxidation of aqueous ammonia using TiO$_2$ nanotube arrays. *Appl. Surf. Sci.*, 311, 851–857, 2014.
91. Wang, P. et al., Enhanced photoelectrocatalytic activity for dye degradation by graphene-titania composite film electrodes. *J. Hazard. Mater.*, 223, 79–83, 2012.
92. Min, Y. et al., Enhanced chemical interaction between TiO$_2$ and graphene oxide for photocatalytic decolorization of methylene blue. *Chem. Eng. J.*, 193, 203–210, 2012.
93. Lee, J.S., You, K.H., Park, C.B., Highly photoactive, low bandgap TiO$_2$ nanoparticles wrapped by graphene. *Adv. Mater.*, 24, 1084–1088, 2012.
94. Bell, N.J. et al., Understanding the enhancement in photoelectrochemical properties of photocatalytically prepared TiO$_2$-reduced graphene oxide composite. *J. Phys. Chem. C*, 115, 6004–6009, 2011.
95. Ng, Y.H. et al., Reducing graphene oxide on a visible-light BiVO4 photocatalyst for an enhanced photoelectrochemical water splitting. *J. Phys. Chem. Lett.*, 1, 2607–2612, 2010.
96. Liang, Y.T. et al., Effect of dimensionality on the photocatalytic behavior of carbon–titania nanosheet composites: Charge transfer at nanomaterial interfaces. *J. Phys. Chem. Lett.*, 3, 1760–1765, 2012.

97. Fan, W. et al., Nanocomposites of TiO_2 and reduced graphene oxide as efficient photocatalysts for hydrogen evolution. *J. Phys. Chem. C*, 115, 10694–10701, 2011.
98. Marcano, D.C. et al., Improved synthesis of graphene oxide. *ACS Nano*, 4, 4806–4814, 2010.
99. Dubey, P.K. et al., Synthesis of reduced graphene oxide–TiO_2 nanoparticle composite systems and its application in hydrogen production. *Int. J. Hydrogen Energy*, 39, 16282–16292, 2014.
100. Xu, Z. et al., 3D periodic multiscale TiO_2 architecture: A platform decorated with graphene quantum dots for enhanced photoelectrochemical water splitting. *Nanotechnology*, 27, 115401, 2016.
101. Wang, D. et al., Self-assembled TiO_2–graphene hybrid nanostructures for enhanced Li-ion insertion. *ACS Nano*, 3, 907–914, 2009.
102. Liu, X.-W., Shen, L.-Y., Hu, Y.-H., Preparation of TiO_2–graphene composite by a two-step solvothermal method and its adsorption-photocatalysis property. *Water Air Soil Pollut.*, 227, 1–12, 2016.
103. Zhang, X., Cui, X., Graphene/semiconductor nanocomposites: Preparation and application for photocatalytic hydrogen evolution. *Nanocomposites-New Trends and Developments*. InTech, 2012.
104. Zhang, X.Y., Li, H.P., Cui, X.L., Lin, Y., Graphene/TiO 2 nanocomposites: Synthesis, characterization and application in hydrogen evolution from water photocatalytic splitting. *J. Mat. Chem*, 20, 14, 2801–2806, 2010.
105. Wojtoniszak, M. et al., Synthesis and photocatalytic performance of TiO_2 nanospheres–graphene nanocomposite under visible and UV light irradiation. *J. Mater. Sci.*, 47, 3185–3190, 2012.
106. Farhangi, N. et al., Visible light active Fe doped TiO_2 nanowires grown on graphene using supercritical CO_2. *Appl. Catal., B*, 110, 25–32, 2011.
107. Zhang, Q. et al., Structure and photocatalytic properties of TiO_2–graphene oxide intercalated composite. *Chin. Sci. Bull.*, 56, 331–339, 2011.
108. Mukherji, A. et al., Nitrogen doped $Sr_2Ta_2O_7$ coupled with graphene sheets as photocatalysts for increased photocatalytic hydrogen production. *ACS Nano*, 5, 3483–3492, 2011.
109. Paek, S.-M., Yoo, E., Honma, I., Enhanced cyclic performance and lithium storage capacity of SnO_2/graphene nanoporous electrodes with three-dimensionally delaminated flexible structure. *Nano Lett.*, 9, 72–75, 2008.
110. Geng, X. et al., Aqueous-processable noncovalent chemically converted graphene–quantum dot composites for flexible and transparent optoelectronic films. *Adv. Mater.*, 22, 638–642, 2010.
111. Li, N. et al., Battery performance and photocatalytic activity of mesoporous anatase TiO_2 nanospheres/graphene composites by template-free self-assembly. *Adv. Funct. Mater.*, 21, 1717–1722, 2011.
112. Lambert, T.N. et al., Synthesis and characterization of titania–graphene nanocomposites. *J. Phys. Chem. C*, 113, 19812–19823, 2009.
113. Guo, J. et al., Sonochemical synthesis of TiO_2 nanoparticles on graphene for use as photocatalyst. *Ultrason. Sonochem.*, 18, 1082–1090, 2011.
114. Hasan, M.R. et al., Effect of Ce doping on RGO-TiO_2 nanocomposite for high photoelectrocatalytic behavior. *Int. J. Photoenergy*, 2014, 2014.
115. Morais, A. et al., Nanocrystalline anatase TiO_2/reduced graphene oxide composite films as photoanodes for photoelectrochemical water splitting studies: The role of reduced graphene oxide. *Phys. Chem. Chem. Phys.*, 18, 2608–2616, 2016.
116. Yu, H. et al., Phenylamine-functionalized rGO/TiO_2 photocatalysts: Spatially separated adsorption sites and tunable photocatalytic selectivity. *ACS Appl. Mater. Interfaces*, 8, 29470–29477, 2016.

117. Chen, D. et al., Nanospherical like reduced graphene oxide decorated TiO$_2$ nanoparticles: An advanced catalyst for the hydrogen evolution reaction. *Sci. Rep.*, 6, 2016.
118. Liu, H. et al., Ultrasmall TiO$_2$ nanoparticles in situ growth on graphene hybrid as superior anode material for sodium/lithium ion batteries. *ACS Appl. Mater. Interfaces*, 7, 11239–11245, 2015.
119. Xing, H., Wen, W., Wu, J.-M., One-pot low-temperature synthesis of TiO$_2$ nanowire/rGO composites with enhanced photocatalytic activity. *RSC Adv.*, 6, 94092–94097, 2016.
120. Tsai, T.-H., Chiou, S.-C., Chen, S.-M., Enhancement of dye-sensitized solar cells by using graphene–TiO$_2$ composites as photoelectrochemical working electrode. *Int. J. Electrochem. Sci.*, 6, 3333–3343, 2011.
121. Chen, L.-C. et al., Improving the performance of dye-sensitized solar cells with TiO$_2$/graphene/TiO$_2$ sandwich structure. *Nanoscale Res. Lett.*, 9, 1–7, 2014.
122. Lee, E., Ryu, J., Jang, J., Fabrication of graphene quantum dots via size-selective precipitation and their application in upconversion-based DSSCs. *Chem. Commun.*, 49, 9995–9997, 2013.
123. Yao, N. et al., Improving the photovoltaic performance of dye sensitized solar cells based on a hierarchical structure with up/down converters. *RSC Adv.*, 6, 11880–11887, 2016.
124. Howatt, G., Breckenridge, R., Brownlow, J., Fabrication of thin ceramic sheets for capacitors. *J. Am. Ceram. Soc.*, 30, 237–242, 1947.
125. *Method of producing high dielectric high insulation ceramic plates*, 1952, Google Patents.
126. Salam, Z. et al., Graphene quantum dots decorated electrospun TiO$_2$ nanofibers as an effective photoanode for dye sensitized solar cells. *Solar Energy Mater. Solar Cells*, 143, 250–259, 2015.
127. Eshaghi, A. and Aghaei, A.A., Effect of TiO$_2$–graphene nanocomposite photoanode on dye-sensitized solar cell performance. *Bull. Mater. Sci.*, 38, 1177–1182, 2015.
128. Liu, J. et al., Stacked graphene–TiO$_2$ photoanode via electrospray deposition for highly efficient dye-sensitized solar cells. *Org. Electron.*, 23, 158–163, 2015.
129. Zhu, P. et al., Facile fabrication of TiO$_2$–graphene composite with enhanced photovoltaic and photocatalytic properties by electrospinning. *ACS Appl. Mater. Interfaces*, 4, 581–585, 2012.
130. Bella, F. et al., Performance and stability improvements for dye-sensitized solar cells in the presence of luminescent coatings. *J. Power Sources*, 283, 195–203, 2015.
131. Dou, Y. et al., Enhanced photovoltaic performance of ZnO nanorod-based dye-sensitized solar cells by using Ga doped ZnO seed layer. *J. Alloys Compd.*, 633, 408–414, 2015.
132. Hung, I. and Bhattacharjee, R., Effect of photoanode design on the photoelectrochemical performance of dye-sensitized solar cells based on SnO$_2$ nanocomposite. *Energies*, 9, 641, 2016.
133. Ghann, W. et al., Fabrication, optimization and characterization of natural dye sensitized solar cell. *Sci. Rep.*, 7, 2017.
134. Sahu, S. et al., Fabrication and characterization of nanoporous TiO$_2$ layer on photoanode by using Doctor Blade method for dye-sensitized solar cells, in: *International Conference on Fibre Optics and Photonics*, Optical Society of America, 2016.
135. Bernacka-Wojcik, I. et al., Inkjet printed highly porous TiO$_2$ films for improved electrical properties of photoanode. *J. Colloid Interface Sci.*, 465, 208–214, 2016.
136. Kadachi, Z. et al., Effect of TiO$_2$ blocking layer synthesised by a sol–gel method in performances of fluorine-doped tin oxide/TiO$_2$/dyed-TiO$_2$/electrolyte/pt/fluorine-doped tin oxide solar cells based on natural mallow dye. *Micro Nano Lett.*, 11, 94–98, 2016.
137. Patil, S.A. et al., Photonic sintering of a ZnO nanosheet photoanode using flash white light combined with deep UV irradiation for dye-sensitized solar cells. *RSC Adv.*, 7, 6565–6573, 2017.
138. Tang, J. and Gomez, A., Control of the mesoporous structure of dye-sensitized solar cells with electrospray deposition. *J. Mater. Chem. A*, 3, 7830–7839, 2015.

139. Jin, Y.S. and Choi, H.W., Properties of dye-sensitized solar cells with TiO_2 passivating layers prepared by electron-beam evaporation. *J. Nanosci. Nanotechnol.*, 12, 662–667, 2012.
140. Agarwal, R. et al., Plasmon enhanced photovoltaic performance in TiO_2–graphene oxide composite based dye-sensitized solar cells. *ECS J. Solid State Sci. Technol.*, 4, M64–M68, 2015.
141. Noh, Y. et al., Properties of blocking layer with Ag nano powder in a dye sensitized solar cell. *J. Korean Ceram. Soc.*, 53, 105–109, 2016.
142. Cheng, G. et al., Nanoprecursor-mediated synthesis of Mg^{2+}-Doped TiO_2 nanoparticles and their application for dye-sensitized solar cells. *J. Nanosci. Nanotechnol.*, 16, 744–752, 2016.
143. Morais, A. et al., Enhanced photovoltaic performance of inverted hybrid bulk-heterojunction solar cells using TiO_2/reduced graphene oxide films as electron transport layers. *J. Photonics Energy*, 5, 057408–057408, 2015.
144. Ahmed, M.I. et al., Low resistivity ZnO-GO electron transport layer based $CH_3NH_3PbI_3$ solar cells. *AIP Adv.*, 6, 065303, 2016.
145. Lee, Y.K. et al., Hot carrier multiplication on graphene/TiO_2 Schottky nanodiodes. *Sci. Rep.*, 6, 2016.
146. Lin, Y.-C., Chen, Y.-T., Yao, P.-C., Effect of post-heat-treated NiO x overlayer on performance of nanocrystalline TiO_2 thin films for dye-sensitized solar cells. *J. Power Sources*, 240, 705–712, 2013.
147. Waita, S.M. et al., Electron transport and recombination in dye sensitized solar cells fabricated from obliquely sputter deposited and thermally annealed TiO_2 films. *J. Electroanal. Chem.*, 605, 151–156, 2007.
148. Sutthana, S., Hongsith, N., Choopun, S., AZO/Ag/AZO multilayer films prepared by DC magnetron sputtering for dye-sensitized solar cell application. *Curr. Appl. Phys.*, 10, 813–816, 2010.
149. Meng, L. and Li, C., Blocking layer effect on dye-sensitized solar cells assembled with TiO_2 nanorods prepared by DC reactive magnetron sputtering. *Nanosci. Nanotechnol. Lett.*, 3, 181–185, 2011.
150. Hsu, C.-H. et al., Enhanced performance of dye-sensitized solar cells with graphene/ZnO nanoparticles bilayer structure. *J. Nanomater.*, 2014, 4, 2014.
151. Huang, C., Chang, K., Hsu, C., TiO_2 compact layers prepared for high performance dye-sensitized solar cells. *Electrochim. Acta*, 170, 256–262, 2015.

13

Role of Reduced Graphene Oxide Nanosheet Composition with ZnO Nanostructures in Gas Sensing Properties

A.S.M. Iftekhar Uddin[1]* and Hyeon Cheol Kim[2]

[1]*Department of Electrical and Electronic Engineering, Metropolitan University, Sylhet, Bangladesh*
[2]*School of Electrical Engineering, University of Ulsan, Ulsan, Republic of Korea*

Abstract

Various metal-oxide semiconductor (MOS)-based gas sensors have been researched extensively for decades for the development of high-performance, accurate, and low-power-consuming gas sensors for the reliable detection and monitoring of various toxic and flammable pollutants in order to assure environmental and personal safety. With the continuation of such progress, to date, a number of promising and highly sensitive gas sensors have been reported in the literature. However, most of the reported works showed a high operating temperature requirement (nearly over 200°C) for such sensors, which are not desirable for those gases, which are highly flammable in nature and requires special safety measure and monitoring during operation. To overcome the abovementioned limitations and to fabricate high-performance sensors, two-dimensional (2-D) reduced graphene oxide (rGO) nanosheets are considered as a promising support catalyst candidate, which, as a template, can play a vital role in enhancing the sensor performances and help to reduce the operating temperature requirements. In the current contribution, we synthesized different dimensional ZnO nanostructures starting from 0-D to 3-D and rGO-loaded ZnO nanostructure hybrids, and investigated their morphological and compositional effects on acetylene (C_2H_2) sensing behaviors. We hope that this study will be beneficial for the readers to understand the role of rGO nanosheet composition with ZnO nanostructures in gas sensing properties and will help them to develop high-performance and low-temperature operable future generation metal oxide nanostructure-based gas sensors.

Keywords: ZnO nanostructures, reduced graphene oxide, hybrid, acetylene, gas sensor, composition effect, operating temperature, sensitivity

13.1 Introduction

In recent years, great efforts have been devoted in the development of high-performance, accurate, and low-power-consuming gas sensors for the reliable detection and monitoring of various toxic and flammable pollutants in order to assure environmental and personal safety. To meet the demand, during the last few decades, various metal-oxide semiconductor

---

*Corresponding author: iftekhar@metrouni.edu.bd; iftekhar_ece@yahoo.com

(MOS)-based gas sensors have been researched extensively, among which zinc oxide (ZnO) nanostructures (NSs) were considered as one of the promising candidates due to their numerous remarkable properties [1–3]. With the continued development, various ZnO NSs such as nanoparticles (NPs), nanograins (NGs), nanowires (NWs), nanorods (NRs), nanofibers (NFs), nanoflakes (NFls), microspheres (MSs), microdisks (MDs), etc. have shown promising advancements for ultrasensitive sensors due to their synthesis simplicity, low cost of preparation, and excellent chemical and thermal stability [4–10].

It is well known that the performance of ZnO-based sensors is significantly influenced by the synthesis process and the architecture of the sensing materials. Additionally, gas sensing behaviors of ZnO-based sensors significantly depend on the morphology, porosity, defects concentration, crystal orientation, and grain size of the sensing material [11]. Each dimensional ZnO NSs has its own unique property that modifies sensing performances such as the following: zero-dimensional (0-D) NSs exhibit higher active exposed surface area, smallest grain size, and uniform shape and size [12], which provide higher number of contact points; one-dimensional (1-D) NSs exhibit a high degree of crystallinity and more quantum effects compared to other dimensional NSs [13], which enables the production of sensors with good long-term stability; two-dimensional (2-D) NSs can provide additional support like base template; and three-dimensional (3-D) NSs exhibit unique and complex architectures including ordered crystalline properties, high degree of porosity, and excellent gas molecule adsorption capability [9].

Very recently, the development of high-performance acetylene (C_2H_2) sensors based on ZnO NSs has been focused intensively due to their wide applications in many chemical and mechanical industries. Acetylene (C_2H_2) is a colorless combustible hydrocarbon and the simplest alkyne with a distinctive odor, widely used as fuel and in many industrial applications, such as in the preparation of organic chemicals like 1,4-butanediol, in vitamins, in welding and metal cutting, in dry-cell batteries, etc. It is quite unstable in pure form and is usually handled in solution and becomes highly explosive when it is liquefied, compressed, heated, or mixed with air. For this reason, special safety measure is vital during its production and handling. At the same time, the range of interest for its detection is much wider, typically 100–100,000 ppm, allowing for early leakage warning and explosive indication. To date, numerous research results have already been reported in the literature [14–27]. However, the reported works showed high operating temperature requirement (nearly over 200°C) for such sensors, which are not desirable as C_2H_2 is highly flammable in nature and requires special safety measure and monitoring during its production and handling.

Dong et al. [14] reported arc plasma-assisted Ag/ZnO composites for C_2H_2 sensing, which had a maximum response of 42 to 5000 ppm C_2H_2 at 120°C. Tamaekong et al. [18] developed a C_2H_2 sensor based on Pt/ZnO thick films by using the flame spray pyrolysis method; their sensor showed a low detection limit of 50 ppm gas concentration at 300°C. Zhang et al. [19] hydrothermally synthesized a hierarchical nanoparticle-decorated ZnO microdisk as a sensing material. This material provided C_2H_2 sensing within the gas concentration range of 1–4000 ppm at 420°C, where 1 ppm was detected with a response of 7.9. Wang et al. [21] published their experimental results, showing effective improvement in the C_2H_2 detection ability of their ZnO sensor by doping 5 at% Ni in the ZnO nanofibers. Their sensor showed a maximum sensor response of ~17 to 2000 ppm C_2H_2 at 250°C. In addition, Sm_2O_3–SnO_2 [15], Au/MWCNT [16], Ag/Pd–SiO_2 [17], Pd–SnO_2 [20], SnO_2 NPs [27], etc. have been studied for C_2H_2 sensing.

To overcome the abovementioned limitations and to fabricate a high-performance C_2H_2 sensor, incorporation of noble metals or dimensional reduced graphene oxide (rGO) nanosheets on the surface of the base ZnO sensors can be a promising support catalyst candidate, which, as a template, can play a vital role in enhancing the sensor performances and help to reduce the operating temperature requirements [28–31]. It has been reported in the literature that the introduction of noble metals can produce some kind of synergistic effect, which influences the material's electronic and chemical distribution favorable to the adsorption of oxygen species, and results in high performance in metal-oxide-based sensors. Importantly, two-dimensional (2-D) carbon material such as reduced graphene oxide (rGO) nanosheets as a support catalyst acts as a bridge inside the sensing materials, which greatly enhances the charge transfer among them. It can also act as an electron acceptor to increase the depletion layer of metal-oxide sensor and helps to boost the sensing performance [31].

In the composition, rGO can act as an electron acceptor to increase the depletion layer of ZnO nanostructures and can activate the dissociation of molecular oxygen on the sensing surface. This phenomenon greatly increases both the quantity of oxygen to repopulate the oxygen vacancies on the ZnO surface and the rates of repopulation for more electron withdrawal from the ZnO/rGO composite than from the pristine ZnO at lower operating temperatures [22–24, 32, 33]. As a result, when C_2H_2 molecules react with the chemisorbed oxygen ions on the sensing surface to form CO_2 or H_2O, more electrons are transferred to the composite surface and a large change in resistance occurs, and therefore, ZnO/rGO shows better response than the pristine ZnO nanostructures.

In the current contribution, we synthesized different dimensional ZnO nanostructures starting from 0-D to 3-D and rGO-loaded ZnO nanostructure hybrids and investigated their morphological and compositional effects on C_2H_2 sensing behaviors. The C_2H_2 gas adsorption–desorption behavior of each pristine nanostructure and hybrid sample is analyzed extensively, and the results are discussed in detail. It is expected that this study will be supportive to the readers who are researching on the development of high-performance and low-temperature operable future generation metal oxide nanostructure-based gas sensors.

13.2 Experimental

All the chemicals used in the experiment were of analytical grade and obtained from Sigma-Aldrich, Dongwoo Fine-Chem., and Dae Jung Chem. & Inds. Co. Ltd., and were used without further purification. In the experimental process, the hydrothermal method was used entirely to synthesize the pristine ZnO NSs and chemical method for hybrid materials.

13.2.1 Synthesis of ZnO Nanostructures

13.2.1.1 Nanoparticles (NPs)

In a typical process, 4 M of zinc nitrate hexahydrate ($Zn(NO_3)_2 \cdot 6H_2O$) and 8 M of sodium hydroxide (NaOH) were dissolved in 40 mL of ethanol (C_2H_6O) using magnetic stirrer and continuously stirred for 1 h. The mixture was then transferred into a Teflon-lined stainless-steel autoclave, heated at 120°C for 8 h in a laboratory oven and naturally cooled to room temperature. Finally, the fine ZnO NPs powder was obtained using drying process.

13.2.1.2 Nanocapsules (NCs)

2 M of $Zn(NO_3)_2 \cdot 6H_2O$ and 2 M of NaOH were dissolved to 40 mL of ethanol using a magnetic stirrer and continuously stirred for 30 min. After 30 min, 3 mL of 0.1 M cetyltrimethyl ammonium bromide (CTAB) $(CH_3(CH_2)_{15}N(Br)(CH_3)_3)$ and 360 µL of ascorbic acid $(HC_6H_7O_6)$ were added to the as-prepared solution. The final solution was then transferred into a Teflon-lined stainless-steel autoclave, heated at 140°C for 8 h in a laboratory oven, and naturally cooled to room temperature. Finally, the fine ZnO NCs powder was obtained using drying process.

13.2.1.3 Nanograins (NGs)

2 M of $Zn(NO_3)_2 \cdot 6H_2O$ and 2 M of NaOH were dissolved to 40 mL of ethanol using a magnetic stirrer and continuously stirred for 30 min. After 30 min, 3 mL of 0.3 M CTAB and 360 µL of $HC_6H_7O_6$ were added to the as-prepared solution. The final solution was then transferred into a Teflon-lined stainless-steel autoclave, heated at 140°C for 8 h in a laboratory oven, and naturally cooled to room temperature. Finally, the fine ZnO NGs powder was obtained using drying process.

13.2.1.4 Nanorods (NRs)

3 M of $Zn(NO_3)_2 \cdot 6H_2O$, 16 M of NaOH, and 2 mL of ethylene glycol (EG) $(C_2H_6O_2)$ were dissolved to 40 mL of ethanol and vigorously stirred for 1 h using a magnetic stirrer. The suspension was then transferred into a Teflon-lined stainless-steel autoclave, heated at 200°C for 20 h in a laboratory oven, and naturally cooled to room temperature. Finally, the fine ZnO NRs powder was obtained using drying process.

13.2.1.5 Nanoflakes (NFls)

4 M of $Zn(CH_3COO)_2 \cdot 2H_2O$ and 2 M of NaOH were dissolved in DI water through vigorous stirring for 30 min. Subsequently, 600 µL of ammonium hydroxide (NH_4OH) were added dropwise to the as-prepared solution and stirring was continued for an additional 30 min. The suspension was then transferred into a Teflon-lined stainless-steel autoclave, heated at 170°C for 20 h in a laboratory oven, and naturally cooled to room temperature. Finally, the fine ZnO NFls powder was obtained using drying process.

13.2.1.6 Microflowers (MFs)

4 M of zinc acetate dehydrate $(Zn(CH_3COO)_2 \cdot 2H_2O)$ and 2 M of NaOH were dissolved in DI water through vigorous stirring for 30 min. Few drops of NH_4OH were then added very slowly dropwise to the solution as a capping agent to maintain the pH level of 9 (exactly). The suspension was then transferred into a Teflon-lined stainless-steel autoclave, heated at 200°C for 10 h in a laboratory oven, and naturally cooled to room temperature. Finally, the fine ZnO MFs powder was obtained using drying process.

13.2.1.7 Microurchins (MUs)

4 M of Zn(CH$_3$COO)$_2$·2H$_2$O and 2 M of NaOH were dissolved in DI water through vigorous stirring for 30 min. Few drops of NH$_4$OH were then added very slowly dropwise to the solution as a capping agent to maintain the pH level of 10.5 (exactly). The suspension was then transferred into a Teflon-lined stainless-steel autoclave, heated at 200°C for 10 h in a laboratory oven, and naturally cooled to room temperature. Finally, the fine ZnO MUs powder was obtained using drying process.

13.2.1.8 Microspheres (MSs)

4 M of Zn(CH$_3$COO)$_2$·2H$_2$O and 2 M of NaOH were dissolved in DI water through vigorous stirring for 30 min. Few drops of NH$_4$OH were then added very slowly dropwise to the solution as a capping agent to maintain the pH level of 11 (exactly). The suspension was then transferred into a Teflon-lined stainless-steel autoclave, heated at 200°C for 10 h in a laboratory oven, and naturally cooled to room temperature. Finally, the fine ZnO MSs powder was obtained using drying process.

The precipitation of white-colored ZnO suspension obtained from each synthesis process was washed several times using de-ionized (DI) water and subsequently dried at 60°C in the laboratory oven overnight to obtain the fine ZnO nanostructures powder.

13.2.2 Synthesis of Bare Reduced Graphene Oxide (rGO) Nanosheets

Graphene oxide (GO) was prepared from extra pure graphite powder (particle size <50 μm) by modified Hummer method [34]. In a typical process, 2 g of extra pure graphite powder (12.0 g/mol) was pre-oxidized by slowly adding it to a solution of 50 mL of sulfuric acid (H$_2$SO$_4$, 95%–97%) and 50 mL of nitric acid (HNO$_3$, 68%–70%) followed by stirring at 80°C for 4 h. After that, the mixture was cooled down to room temperature and then washed by DI water until the pH value was neutral, followed by drying at 40°C overnight. The resultant pre-oxidized graphite was dispersed into cold concentrated H$_2$SO$_4$ in a reaction vessel, which was kept in an ice bath and stirred followed by slow addition of 10 g of potassium permanganate (KMnO$_4$, 97%). The temperature was held below 10°C during the addition. Importantly, the mixture was stirred at 35°C for 2 h. During the mixing process, the solution thickened and turned into brownish gray in color. Afterward, 250 mL of DI water was added and the temperature was raised to 100°C and stirring was continued for 15 min. Subsequently, the addition of 700 mL of DI water with 30 mL of hydrogen peroxide (H$_2$O$_2$, 30%) followed by stirring for 1 h removed the Mn$^+$ ions, resulting in a yellow-brown colored solution. The solid products collected from the solution after 12 h were washed five times with 5% hydrochloric acid (HCl) to remove the impurities. Further centrifugation at 3000 rpm for 5 min was carried out to remove all visible particles from precipitates (unexfoliated GO). The supernatant was then subjected to high-speed centrifugation at 10,000 rpm for 10 min, and the resulting sediment was dried at 60°C in a vacuum oven to yield GO powder.

To synthesize the reduced graphene oxide (rGO), the aqueous suspension of GO (10 mg/mL) was diluted with N,N-dimethylformamide (DMF) through sonication treatment in an ultrasonic bath for 1 h to make a homogenous suspension of GO in DMF/water (80:20 v/v).

Then, 1 mL of hydrazine monohydrate was added to the solution as a reducing agent and stirred for 6 h at an elevated temperature of 80°C. The resulting rGO suspension was black in color and was preserved for further experimental use.

13.2.3 Synthesis of ZnO NSs–rGO Hybrids

ZnO nanostructures–reduced graphene oxide (ZnO NSs–rGO) hybrids were synthesized via a facile and rapid chemical route. In a typical process, 0.5 g of each pure ZnO NSs powder was dispersed into 50 mL of GO solution (0.5 mg/mL) using sonication treatment and continuous stirring for 1 h. Subsequently, 60 µL of hydrazine monohydrate was slowly added to the as-prepared mixers as an agent and heated at a temperature of 110°C for 8 h. The solutions were then cleaned by DI water using centrifugation and preserved in 20 mL of fresh DI water for further use.

13.2.4 Device Fabrication

For electrical and gas sensing measurements, sensor device was fabricated as follows: two electrodes (dimension: 2 mm × 4 mm, thickness: 100 nm) were fabricated by depositing gold (Au) on a 6 mm × 12 mm alumina (Al_2O_3) substrate by lift-off process. The spacing between two electrodes was measured to be nearly 210 µm. Figure 13.1 shows the schematic diagram of an as-fabricated sensor podium with patterned electrodes. Each ZnO NS, bare rGO, and ZnO NSs–rGO hybrid was then deposited on the center of the patterned electrodes via drop casting method. The device was then dried at 80°C on a hot plate until all solvent evaporates. This step was repeated two to three times. After deposition, the device was annealed at 300°C in air for 30 min for thermal stabilization.

13.2.5 Characterization and Sensor Test

Structural properties of the as-prepared sensing materials were investigated using an X-ray diffractometer (XRD) (Rigaku Ultima IV) with Cu Kα (λ = 0.154 nm) radiation with a 2θ scanning range of 10–80°. The surface morphology was examined by field emission scanning electron microscopy (FESEM; JEOL JSM-7600F) equipped with an energy-dispersive spectrometer (EDS) for compositional analysis with an accelerating voltage of 10 kV. The chemical compositions of GO and rGO were studied by the Fourier transform infrared

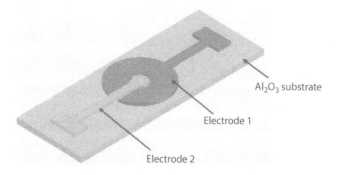

Figure 13.1 Schematic diagram of an as-fabricated sensor podium with patterned electrodes.

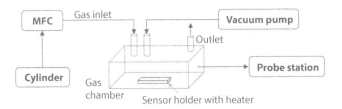

Figure 13.2 Block diagram of the overall construction of the chamber and the measurement setup.

spectroscopy (FTIR). FTIR was carried out using a Varian 2000 Scimitar spectrometer in the range of 4000 cm^{-1} to 350 cm^{-1}. Raman spectroscopy was carried out through WITec alpha300R in the range of 1800 cm^{-1} to 300 cm^{-1}.

Gas sensing measurements were conducted at atmospheric pressure within a temperature range of 25–450°C for various C_2H_2 concentrations in a fully sealed lab-made chamber using the flow-through technique. The gas chamber was composed of two inlets for gas supply and vacuum pressure, respectively, and one outlet for air passage. A vacuum pump (ULVAC, DAP-15, 39.9 kPa) and a programmable mass flow controller (MFC) were used to balance the atmosphere inside the chamber. A programmable heater integrated with the sensor holder in the chamber was used to adjust the temperature. The computerized mass flow controller (ATO-VAC, GMC 1200) system was used to vary the concentration of C_2H_2 in synthetic air. The gas mixture was delivered to the chamber at a constant flow rate of 50 sccm (standard cubic centimeters per minute) with different C_2H_2 concentrations. The gas chamber was purged with synthetic air between each C_2H_2 pulse to allow the surface of the sensor to return to atmospheric conditions. The overall construction of the chamber and the measurement setup is depicted in the block diagram as shown in Figure 13.2. The gas concentration was controlled and measured using the following equation:

$$Desired\ gas_{con.} = \{Flow\ rate_{gas}/(Flow\ rate_{gas} + Flow\ rate_{air})\} \times Supplied\ gas_{con.} \quad (13.1)$$

A Keithley probe station (SCS-4200) with a bias voltage fixed at 1 V was used for all measurements and data acquisition. The response magnitudes of the sensors were calculated using the following formula:

$$S = R_a/R_g \quad (13.2)$$

where S denotes the response of the sensor, R_a is the resistance of the sensor in the presence of synthetic air, and R_g is the resistance in the presence of C_2H_2 at certain concentrations. The response time and recovery time of the sensor are defined as the time to reach 90% of total resistance change.

13.3 Results and Discussions

To investigate the surface morphology, elemental analysis, and structural analysis, all the as-prepared samples (ZnO NSs, GO, rGO, and hybrid specimen) solution were drop casted on SiO_2/Si substrates. They were then dried on a hot plate at 80°C until all solvent evaporates and used for specific investigation.

13.3.1 Morphological Studies of the ZnO NSs

Figure 13.3 represents the surface morphologies of the as-synthesized different dimensional pristine ZnO NSs. Figure 13.3a to c show the successful synthesis of particle-like (Figure 13.3a), capsule-like (Figure 13.3b), and rice grain-like (Figure 13.3c) ZnO NSs with a mean diameter of 40 nm, 100 nm, and 120 nm, respectively. It is clearly observed that 0-D ZnO NPs and 1-D nanocapsules (NCs) and nanograins (NGs) are highly uniform in size and are well distinguished. ZnO NCs and NGs were prepared with the assistance of ascorbic acid (as a reductant) in the presence of CTAB (as a stabilizing agent). The synthesis of metal oxide NSs with a controlled size and shape is highly influenced by the concentration of precursor, surfactant/stabilizer, reducing agent, and reaction temperature [35]. In this case, the presence of CTAB played a vital role to obtain the desired shape of ZnO NSs, while ascorbic acid preferentially facilitated the nucleation process. The size of the ZnO NPs, NCs, and NGs can be controlled by varying the concentration of the $Zn(NO_3)_2 \cdot 6H_2O$ precursor and reaction time. In a similar way, ZnO NR formation as shown in Figure 13.3d, was controlled

Figure 13.3 FESEM micrographs of as-synthesized ZnO NSs: (a) nanoparticle (NP), (b) nanocapsule (NC), (c) nanograin (NG), (d) nanorod (NR), (e) nanoflake (NFl), (f) microflower (MF), (g) microurchin (MU), and (h) microsphere (MS). [Note: The entire scale bar is in 100 nm.]

by the amount of NaOH and the presence of ethylene glycol. It is observed that the as-grown ZnO NRs have a smooth surface throughout their lengths. The typical average diameter and length of the as-grown NRs were measured to be 40 ± 5 nm and 0.8 ± 0.05 μm, respectively.

Figure 13.3e shows the sheet-like 2-D ZnO NFls with an average flake thickness of 10 nm. Figure 13.3f to h depict the formation of 3-D microflowers (MFs) (Figure 13.3f), microurchins (MUs) (Figure 13.3g), and microspheres (MSs)-like (Figure 13.3h) hierarchical structures of ZnO with an average size of 1.5–2 μm. It can be seen that numerous ZnO nanopetals uniformly grew from the center and reduced in size in the c-axis direction to form the flower-like NSs, while randomly oriented ZnO nanoneedles and nanosheets were entangled to form the urchin-like and bunch of nanosheet-like ZnO NSs. The formation of the 3-D NSs might be attributed to the anisotropic growth of the ZnO, which was appreciably affected by the pH level. In the hydrothermal process, $[Zn(OH)_4]^{2-}$ were formed first through the decomposition of $Zn(OH)_2$ precipitates [9, 36]. With the continuation of the reaction process, these ions were aggregated to form ZnO nanopetals, nanoneedles, and nanosheets with the varied concentration of NH_4OH, and gradually assembled to form hierarchical 3-D ZnO MFs, MUs, and MSs, respectively.

13.3.2 Morphological and Elemental Studies of GO and rGO

Figure 13.4 shows the surface morphology of GO and rGO flakes on SiO_2 substrate (Figure 13.4a and b) and the corresponding EDS analysis (Figure 13.4c and d). Figure 13.4a shows several nanosheets of GO in an aggregated form. However, reduction of GO to rGO formed

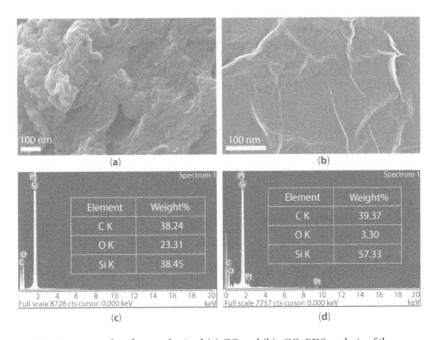

Figure 13.4 FESEM micrographs of as-synthesized (a) GO and (b) rGO. EDS analysis of the as-prepared (c) GO and (d) rGO.

few layers of nanosheets with many wrinkles and folding as shown in Figure 13.4b, revealing a very good 2-D structure of rGO. In addition, rGO nanosheets with broad wrinkles or folds indicate the presence of few-layer rGO. Importantly, reduction of the GO entirely reduced the presence of oxygen-containing functional groups while forming the rGO and the partial exfoliation of GO facilitated to structure the lower thickness of rGO with few layers of nanosheets. Additionally, the EDS results shown in Figure 13.4c and d confirm the presence of C and O in the as-prepared GO and rGO samples. It should be noted that platinum (Pt) and Si peaks in the Figure 13.4c and d, respectively, were originated from the metal coating and substrate used during the sample preparation. The EDS analysis indicates that the carbon (C) content in the GO and rGO was 38.24% and 39.37%, respectively. However, the oxygen (O) content of rGO was significantly reduced to 3.3% from 23.31% in the GO. The mass ratio of C/O was 1.6 in GO and increased to 11.9 in rGO. The O content in rGO was 3.3%, indicating that some oxygen-containing functional groups still present in the rGO sample.

13.3.3 Chemical Composition Studies of GO and rGO

In order to further confirm the reduction of GO to rGO, the chemical compositions were studied by the Fourier transform infrared spectroscopy (FTIR). Figure 13.5 shows the FTIR spectra of the as-prepared GO and rGO samples. The oxygen-containing functional groups of GO at 1095 cm^{-1}, 1404 cm^{-1}, and 1740 cm^{-1} correspond to the functional oxide after oxidation, epoxy, and carboxy groups, respectively. The broad absorption peak at 3400 cm^{-1} indicates the presence of O–H groups in GO [37]. It is observed that the oxygen-containing functional groups were almost entirely removed during the reduction process.

Raman spectra of GO and rGO are shown in Figure 13.6. A sharp, strong peak appeared at 521 cm^{-1} (discarded from the figure for better clarity) during Raman spectra acquisition, which was the characteristic peak related to the Si substrate (used as supporting wafer). The

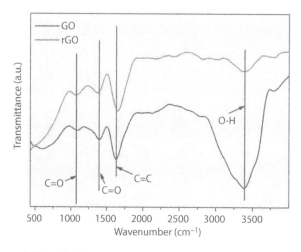

Figure 13.5 FTIR spectra of GO and rGO.

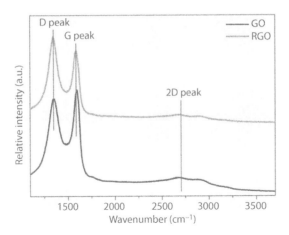

Figure 13.6 Raman spectra of GO and rGO.

characteristic peak at 1349 cm^{-1} and 1588 cm^{-1} corresponding to the D-band and G-band, respectively, in GO and rGO samples arose from the vibration of the sp^2-bonded carbon atoms and structural disorder at defect sites. 2-D characteristic peak relates to the graphitic chemical structure. The 2-D band shape, frequency, and intensity are strongly influenced by the perfection of graphitic chemical structure and the number of graphite layers in the graphite starting material. The D peak denotes the physical (hole, folding, or strain) and chemical defects (foreign or oxygen functionality) in carbon materials, and the G peak is related to the carbon material quality or graphitic domain [38]. The D/G intensity ratio I(D)/I(G) were 0.98 and 1.18 in GO and rGO, respectively. The result shows that the D/G intensity ratio of rGO is higher than that of GO, which indicates that the chemical reduction process induced defects in GO and caused a significant Raman intensity in the D and G band.

13.3.4 Morphological and Elemental Analysis of the ZnO NSs–rGO Hybrids

Representative FESEM micrographs of ZnO NSs–rGO hybrids are shown in Figure 13.7. Figure 13.7 demonstrates that the as-fabricated ZnO NSs were well mixed and closely attached to the 5- to 10-layer-thick rGO nanosheets. These results indicate that functional oxygen-containing groups of GO played a vital role in the synthesis process and facilitated the firm attachment of ZnO NSs in the rGO network, which ultimately endowed with excellent hybrid materials with less aggregation [39]. In the case of ZnO NSs–rGO hybrid, ZnO NPs, ZnO NCs, ZnO NGs, ZnO NRs, and ZnO NFls are closely affixed with the rGO nanosheets (Figure 13.7a to e), in which rGO acted as a template. No significant aggregation or deformation in these hybrids was observed, which might be attributed to the smaller and uniform sizes of NPs, NCs, NGs, NRs, and NFls. However, some aggregation and breakage were observed in the hybrids with the ZnO MFs, ZnO MUs, and ZnO MSs (Figure 13.7f to h), which might be attributed to the larger size and highly crowded ZnO NSs in the network.

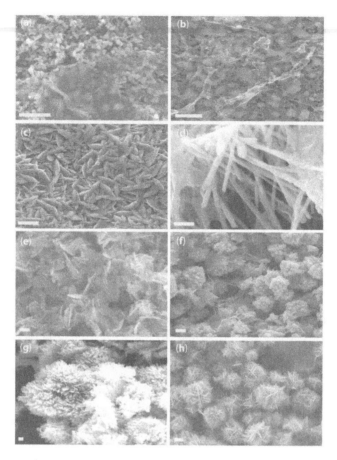

Figure 13.7 FESEM micrographs of as-synthesized ZnO NSs–rGO hybrids: (a) ZnO NPs–rGO, (b) ZnO NCs–rGO, (c) ZnO NGs–rGO, (d) ZnO NRs–rGO, (e) ZnO NFls–rGO, (f) ZnO MFs–rGO, (g) ZnO MUs–rGO, and (h) ZnO MSs–rGO. [Note: The entire scale bar is in 100 nm.]

Table 13.1 Summary of elemental composition within ZnO NSs–rGO hybrids.

Element	Weight (%)							
	NP	NC	NG	NR	NFl	MF	MU	MS
C K	2.60	2.82	2.52	2.59	3.10	1.67	1.88	2.09
O K	22.37	19.23	19.33	18.22	17.97	16.72	15.87	15.61
Zn k	13.09	15.28	16.32	16.29	16.47	18.97	19.55	20.23
Si k	61.24	61.97	61.13	62.20	61.76	62.64	62.00	61.37

Elemental compositions of the as-prepared ZnO NSs–rGO hybrids were investigated using EDS analysis and the results are summarized in Table 13.1. The presence of various well-defined elements such as Zn, oxygen (O), and carbon (C) confirmed the formation of high-purity ZnO NSs–rGO hybrids.

13.3.5 Structural Studies of GO, rGO, ZnO NSs, and ZnO NSs–rGO Hybrids

To investigate the phase purity and structural properties, XRD analysis of the as-fabricated GO, rGO, ZnO NSs, and ZnO NSs–rGO hybrids was carried out and the results are depicted in Figure 13.8. In Figure 13.8a, the characteristic reflection peak centered at 2θ values of 12.3° correspond to the (001) crystalline plane of GO with an interlayer spacing (*d*-spacing) of 0.7 nm, which is larger than the *d*-spacing (0.34 nm) of natural graphite. An intense peak centered at 2θ values of 24.6° can be indexed to the characteristic reflection peak (002) of rGO with interlayer distances of 0.40 nm. This peak is related to the exfoliation and reduction processes of GO after removing intercalated water molecules and the oxide groups [40]. Additionally, the appearance of this peak also suggest the restacking of carbon layers into an ordered crystalline structure of rGO [41]. Usually, diffraction peak appeared at 2θ = 26.5° indexed to the (002) plane of graphite's major peak, with an interlayer spacing of 0.33 nm. The shift in diffraction peak of the graphite from 26.5° to 24.6° was presumably induced due to the short-range order in stacked stacks and the partial reduction of residual oxygen-containing functional groups or other structural defects in GO [42]. Moreover, the absence of the (002) reflection peak at 2θ values of 43.3° (graphitic peak) for both GO and rGO samples suggests that the distance between the carbon sheets increased due to the insertion of the interplanar groups [41].

In Figure 13.8b, the appearance of characteristic diffraction peaks for pure ZnO NSs corresponds to (100), (002), (101), (102), (110), (103), (112), and (201) planes at 2θ = 31.8°, 34.4°, 36.3°, 47.5°, 56.6°, 62.9°, 68°, and 69.1°, respectively, and was in good agreement with the standard XRD peaks of crystalline ZnO with a hexagonal wurtzite structure (JCPDS card No. 36-1451). No characteristic peaks from the intermediates such as $Zn(OH)_2$ could be detected in the samples, which indicate the formation of high purity of the ZnO NSs.

In Figure 13.8c, the observed characteristic peaks of the hybrids exhibited a well-structured crystalline nature and mixed phases of ZnO NSs, and rGO. The diffraction peaks of ZnO (100), (002), (101), (102), (110), (103), (112), and (201) were similar to the standard ZnO hexagonal wurtzite structure (JCPDS card No. 36-1451). However, the broadened peaks of ZnO (100), (002), and (101) in the hybrid samples indicate small but finite degradation in the crystalline structure of ZnO NSs due to insertion of rGO nanosheets [31, 41]. In addition, the peak centered at 2θ values of 24.6° can be indexed to the characteristic reflection peak (002) of rGO. No other characteristic peaks from the intermediates and significant variations or shifts in the patterns were observed in the hybrid samples, indicating the formation of high-purity hybrid materials.

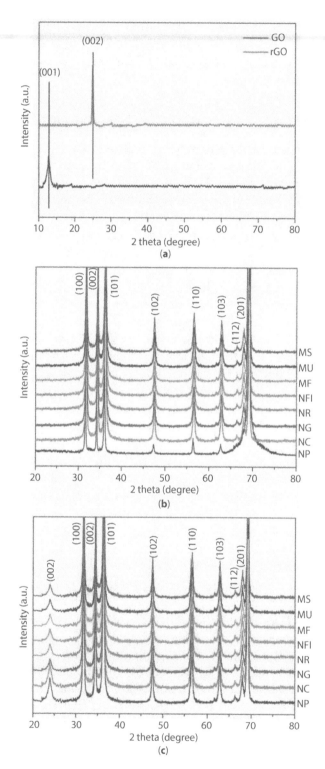

Figure 13.8 XRD patterns of (a) GO and rGO, (b) bare ZnO NSs, and (c) ZnO NSs–rGO hybrids.

13.3.6 Gas Sensing Mechanism

The overall sensing mechanism of the bare ZnO NSs and ZnO NSs–rGO sensor is depicted in Figure 13.9. The gas sensing mechanism of the ZnO-based sensor can be described in terms of oxygen adsorption on the surface of sensing materials and the reactions of gas molecules adsorbed on the material surface [43–47]. When a sensor is exposed in air, oxygen molecules in air can be chemisorbed on the sensing material's surface and form the chemisorbed oxygen anions $\left(O_{n\,ads}^{n-}\right)$ by capturing electrons from the oxide conduction band and the electron depletion layer at the surface of ZnO NSs. The reaction can be described as follows:

$$O_2 + 2e^- \leftrightarrow 2\,O_{n\,ads}^{n-} \qquad \text{Reaction 1}$$

This phenomenon causes a decrease in the concentration of free electrons, leading to an increase in resistance of the sensing surface. Takata et al. have found that the chemisorbed oxygen anions strongly depend on temperature, and the stable oxygen ions O_2^-, O^-, and O^{2-} operate below 100°C, within 100–300°C, and above 300°C, respectively [43]. Hence, the oxygen adsorption reaction can be represented as follows:

$$O_2(gas) + e^- \leftrightarrow O_2^-(adsorb)\,(low\ temperature) \qquad \text{Reaction 2}$$

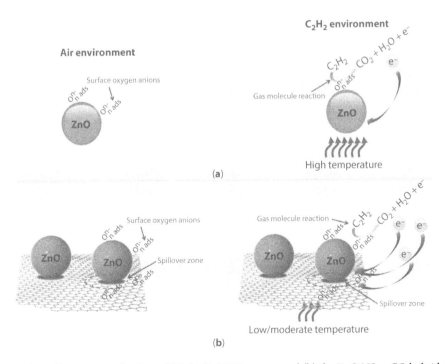

Figure 13.9 Overall sensing mechanism of (a) the ZnO NSs sensor and (b) the ZnO NSs–rGO hybrid sensor.

$$O_2(gas) + e^- \leftrightarrow O^-_{ads} \text{ (moderate temperature)} \qquad \text{Reaction 3}$$

$$O_2(gas) + e^- \leftrightarrow O^{2-}_{ads} \text{ (high temperature)} \qquad \text{Reaction 4}$$

When the ZnO NSs sensor is exposed to target gas (considering the target gas as C_2H_2) at a specific operating temperature, the adsorbed gas molecules then react with the chemisorbed oxygen ions of the ZnO NSs surface and hence the reaction can be described as follows:

$$C_2H_2 + 5O^{n-}_{n\,ads} \xrightarrow{K_{Eth}} 2CO_2 + H_2O + 5e^- \qquad \text{Reaction 5}$$

Electron emission in this reaction leads to an increase in concentration of free electrons and eventually decreases the resistance of the sensing surface of the sensor, which can be further used for the detection of target gas. Figure 13.9a shows the sensing mechanism of the bare ZnO-based sensor both in air and in a C_2H_2 environment.

In the ZnO NSs–rGO hybrid, rGO can be acted as an electron acceptor, which led to an increase in the depletion layer of ZnO NSs. Therefore, as compared with the pristine ZnO NSs, the change in resistance is larger, which led to an increase in response. Additionally, rGO can activate the dissociation of molecular oxygen [22, 23, 33], which can increase the quantity of oxygen to repopulate the oxygen vacancies on the ZnO NSs surface and the rate of repopulation for higher degree of electron withdrawal from the ZnO NSs–rGO hybrid than pristine ZnO NSs. From Reaction 5, the rate equation of electron density can be written as:

$$\frac{dn}{dt} = K_{Eth}(T)[O^{n-}_{n\,ads}][C_2H_2]$$

$$or, n = K_{Eth}(T)[O^{n-}_{n\,ads}][C_2H_2]t + n_o \qquad (13.3)$$

where $K_{Eth}(T)$ is the reaction rate constant or coefficient, n is the electron density under an acetylene atmosphere, and n_o is the electron density under an air atmosphere. Usually, the reaction rate coefficient and electron density have a strong relationship to operating temperature, and magnitude increases exponentially with rising temperatures. However, sensor response ($S = R_a/R_g$) is directly proportional to the reaction rate coefficient and is inversely proportional to electron density [48]. For C_2H_2 gas sensing, these two parameters compete with each other and can result in enhanced sensitivity of the ZnO NSs–rGO hybrid at a lower optimum operating temperature than that of the bare ZnO NSs sensor (as shown in Figure 13.9b).

13.3.7 Gas Sensor Studies

It is well known that sensing properties of the metal-oxide-based gas sensors are highly influenced by the structural defects of the sensing material, in which several factors such as grain size, pores, surface area, etc. play important roles. In order to determine the feasible

ZnO NS, highly sensitive to target gas, all the synthesized ZnO NSs-based sensors were exposed to C_2H_2 gas at different operating temperatures. Figure 13.10 shows the responses of different ZnO NSs toward 100 ppm C_2H_2 within the operating temperature between 25 and 450°C. Due to the experimental limitations, the maximum operating temperature was limited to 450°C. An increase in sensor response for all the samples was observed with increasing temperature. Generally, pristine ZnO NSs-based C_2H_2 sensors work at high temperatures (over 400°C) [19]. This phenomenon might be attributed to the potential barrier formed by the chemisorbed oxygen ions $\left(O_{n\,ads}^{n-}\right)$ on the ZnO surface, which prevents the C_2H_2 gas molecules from reacting at low temperature. However, it is observed that 0-D nanoparticles and 1-D nanorods exhibit maximum sensitivity in comparison to other nanostructures. This phenomenon might be attributed to the highly uniform and well-shaped nanoparticles with better crystallinity and smaller grain size. Additionally, these features can provide larger surface area for gas molecule exposure and ultimately can facilitate better sensing performance [49].

Importantly, sensor performance is also highly influenced by the addition of an effective and appropriate amount of additives [29]. In C_2H_2 sensing, rGO in the ZnO NSs–rGO hybrid sensor can facilitate the dissociation of molecular oxygen by enhancing the quantity of oxygen to repopulate the oxygen vacancies on the ZnO surface [22, 23]. In addition, the operating temperature has a significant influence on the fundamental sensing mechanism of metal-oxide-based and metal-oxide/additive-based gas sensors [22–25, 50]. Therefore, in order to investigate the effect of rGO composition with the ZnO NSs in sensing behaviors and to determine the effect of operating temperature on the hybrid samples, the as-fabricated ZnO NSs–rGO hybrid sensors were investigated within the temperature range between 25 and 350°C. Figure 13.11 represents the relationship between the response magnitudes of the hybrid sensors and the operating temperature at 100 ppm C_2H_2 gas concentrations. It was observed that due to the addition of rGO nanosheets to the ZnO NSs, the operating temperature of the hybrid samples considerably reduced and at the same time

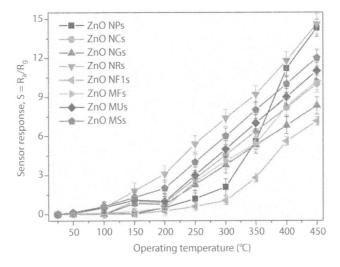

Figure 13.10 Response variations of bare ZnO NSs to 100 ppm C_2H_2 concentrations at different operating temperatures.

sensor response magnitude increased significantly. Notably, all the hybrid samples showed low response at lower operating temperatures, while with the increasing temperature, the response values gradually increased, reached the maximum at a certain operating temperature, and then decreased with further increase in temperature.

In Figure 13.11, it is observed that all the tested hybrid sensors have better functionality at 250°C. In ambient environment, at an elevated temperature, the hybrid materials modified the sensor surface by capturing the oxygen molecules and formed the chemisorbed oxygen ions $\left(O^-_{n\,ads}\right)$. In this situation, the sensors acquire sufficient energy to react with the C_2H_2 gas molecules. However, at higher temperatures (over 250°C), due to the excessive chemical activation, the adsorbed gas molecules might escape before their reaction and the dissociation of more oxygen molecules exceeds the percolation threshold, which ultimately obstructs the effective oxygen delivery system and reduces the probability of C_2H_2 adsorption and thus leads to a response drop. On the other hand, the formation of depletion zones around the ZnO NS due to the addition of rGO is attributed to the modulation of the nano-Schottky barriers and hence improves the surface reactivity. At the same time, the chemisorbed oxygen ions on the composite surface become more active with C_2H_2 gas molecules, which help to leave more electrons on the composite material surface than pure ZnO NS. Thus, an appropriate amount of rGO that increased these free electrons help to enhance the sensor response. Besides, additional active sensing sites can be created on the ZnO surface after adding rGO to pristine ZnO NS (due to the spillover effect), which increases the C_2H_2 gas molecule adsorption rate and decreases the desorption rate [22–24].

In Figure 13.11, it is also observed that among the 0-D, 1-D, 2-D, and 3-D ZnO nanostructures, the ZnO NPs–rGO hybrid exhibits the maximum response value in comparison to all other nanostructure-based hybrid sensors. In addition, the ZnO NRs–rGO hybrid exhibits a higher response value than the ZnO MSs–rGO hybrid. This phenomenon might be attributed to the following reasons: first, compared to long, smooth ZnO NRs and randomly oriented ZnO MSs in the hybrids, ZnO NPs exhibited highly uniform and

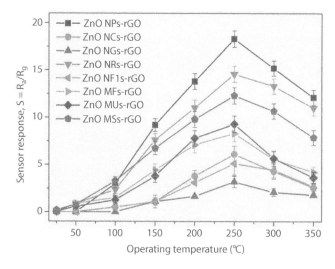

Figure 13.11 Response variations of ZnO NSs–rGO hybrid sensor to 100 ppm C_2H_2 concentrations at different operating temperatures.

well-shaped particles with better crystallinity and smaller grain size, which provided larger surface area and more contact points for the possible nanojunctions that are prominent for the enhancement of sensing performances [49]. Similarly, directional long ZnO NRs can trap more electrons on the ZnO surface due to their large surface area and pores, which leads to insufficient surface atomic coordination and high surface energy for enhanced oxygen adsorption [30]. Second, good contact among the ZnO NPs and the supporting rGO might facilitate efficient charge transfer, which helped the ZnO NPs–rGO hybrid to show better C_2H_2 sensing properties than ZnO NRs–rGO and ZnO MSs–rGO hybrids. A maximum response value of 18.3 was recorded for the ZnO NPs–rGO hybrid to a 100 ppm gas concentration, while ZnO NRs–rGO and ZnO MSs–rGO hybrids showed maximum responses of 14.6 and 12.3, respectively.

In other words, improved sensitivity of different ZnO NSs–rGO hybrid sensors toward C_2H_2 molecules can relate to the bond energy of H–C≡C–H (490 kJ mol^{-1}) [51–54]. The ZnO NSs–rGO hybrid sensor preferentially showed a higher response than the bare ZnO NSs at a lower operating temperature, because the reaction mechanism was accelerated due to the addition of the rGO. The C–H bond energy in C_2H_2 is quite large, because the carbon hybridization energy is shared by both the C–C and C–H bonds [53]. In the case of pristine ZnO NSs sensors, the required energy to break the C_2H_2 bond in the surface reaction between ZnO NSs and C_2H_2 molecules could not be supplied at lower operating temperatures (below 300°C). Additionally, decomposition of H_2 and CO on the ZnO NSs surface might require high operating temperatures (over 300°C). However, at an elevated temperature range of 200–250°C, the catalytic activity of the ZnO NSs–rGO toward C_2H_2 decomposition might be strongly facilitated. At this stage, the reaction between ZnO NSs–rGO and C_2H_2 molecules might be an exothermic and spontaneous process, which preferentially promoted the ZnO NSs–rGO hybrid sensor to achieve sufficient surface energy to deal with the bond energy of C_2H_2 and hence showed higher sensitivity.

Response–recovery time characteristics are an important parameter to evaluate the sensor performance. The response–recovery time characteristics of all the test samples are summarized in Table 13.2. It was observed that addition of rGO with the ZnO NSs also facilitated response–recovery time characteristics. Interfacial coordination of ZnO NSs and

Table 13.2 Summary of response–recovery time characteristics of pristine ZnO NSs sensors and ZnO NSs–rGO hybrid sensors.

Pristine ZnO NS sensor @ 250°C							
	NP	NC	NG	NR	NFl	MF	MU	MS
Response time (min)	>17	>17	>17	>4	>25	>5	>8	>4
Recovery time (min)	>22	>28	>30	>10	>38	>43	>41	>36
ZnO NS–rGO hybrid sensor @ 250°C							
	NP	NC	NG	NR	NFl	MF	MU	MS
Response time (s)	100	118	127	62	126	76	70	70
Recovery time (s)	24	86	106	46	90	115	110	94

rGO favorably promoted the adsorption and desorption of C_2H_2 during chemical reactions on the sensor surface and hence showed faster response–recovery time than that of the bare ZnO NSs sensor. On the other hand, rGO nanosheets as a template supported the ZnO NSs to form additional conducting channel for charge transfer within the sensing surface; thus, the surface reaction (adsorption and desorption of C_2H_2 molecules) was rapid. The resistance of the hybrid sensors quickly decreased when C_2H_2 was exposed and returned to its initial baseline value when the gas was stopped. This phenomenon might be attributed to the n-type behavior of the dominant ZnO NSs in the hybrid sensors and revealed good reversibility behavior of the as-synthesized hybrid sensing materials.

13.4 Conclusions

In summary, the role of reduced graphene oxide nanosheets with different dimensional ZnO NSs in gas sensing behaviors has been investigated. Different dimensional ZnO NSs (such as nanoparticles, nanocapsules, nanograins, nanorods, nanoflakes, microflowers, microurchins, and microspheres) and ZnO NSs–rGO hybrids have been synthesized through hydrothermal and chemical route, respectively, and investigated their sensing behaviors toward C_2H_2. Notably, structural defects of the ZnO NSs sensor such as grain size, pores, surface area, etc. favorably influence specific gas molecule detection. However, the high operating temperature requirement for such sensors limits their practical functionality in most cases. Carbon materials such as reduced graphene oxide and their composites can offer an attractive way to fabricate high functional sensors with the possibility of low energy consumption and ability to operate at lower operating temperatures. The unique properties of reduced graphene oxide such as high surface area, mechanical strength, and better electrical/temperature tolerance can also provide exceptional chemical and/or physical modification in the composition, which ultimately enhance the surface reaction with target gas molecules and show better sensing performances. For the experiment, it was observed that incorporation of rGO with the ZnO NSs significantly improved sensitivity at lower operating temperatures. The results demonstrate that the ZnO NSs–rGO hybrid exhibits remarkable potential for developing low-cost, simple, highly sensitive C_2H_2 sensors, which can be operable at lower operating temperatures.

References

1. Sun, Y.-F., Liu, S.-B., Meng, F.-L., Liu, J.-Y., Jin, Z., Kong, L.-T., Liu, J.-H., Metal oxide nanostructures and their gas sensing properties: A review. *Sensors*, 12, 2610, 2012.
2. Wei, A., Pan, L., Huang, W., Recent progress in the ZnO nanostructure-based sensors. *Mater. Sci. Eng. B*, 176, 1409, 2011.
3. Wang, Z.L., Zinc oxide nanostructures: Growth, properties and applications. *J. Phys. Condens. Matter.*, 16, 829, 2004.
4. Kolmakov, A. and Moskovits, M., Chemical sensing and catalysis by one-dimensional metal-oxide nanostructures. *Annu. Rev. Mater. Res.*, 34, 151, 2004.
5. Forleo, A., Francioso, L., Capone, S., Siciliano, P., Lommens, P., Hens, Z., Synthesis and gas sensing properties of ZnO quantum dots. *Sens. Actuators, B*, 146, 111, 2010.

6. Calestani, D., Zha, M., Mosca, R., Zappettini, A., Carotta, M., Natale, V., Zanotti, L., Growth of ZnO tetrapods for nanostructure-based gas sensors. *Sens. Actuators, B*, 144, 472, 2010.
7. Ra, Y.W., Choi, K.S., Kim, J.H., Hahn, Y.B., Im, Y.H., Fabrication of ZnO nanowires using nanoscale spacer lithography for gas sensors. *Small*, 4, 1105, 2008.
8. Gupta, S.K., Joshi, A., Kaur, M., Development of gas sensors using ZnO nanostructures. *J. Chem. Sci.*, 122, 57, 2010.
9. Guo, W., Liu, T., Zhang, H., Sun, R., Chen, Y., Zeng, W., Wang, Z.C., Gas-sensing performance enhancement in ZnO nanostructures by hierarchical morphology. *Sens. Actuators, B*, 166, 492, 2012.
10. Ozturk, S., Kilinc, N., Tasaltin, N., Ozturk, Z.Z., A comparative study on the NO_2 gas sensing properties of ZnO thin films, nanowires and nanorods. *Thin Solid Films*, 520, 932, 2011.
11. Wang, C., Yin, L., Zhang, L., Xiang, D., Gao, R., Metal oxide gas sensors: Sensitivity and influencing factors. *Sensors*, 10, 2088, 2010.
12. Eriksson, J., Khranovskyy, V., Soderlind, F., Kall, P.O., Yakimova, R., Spetz, A.L., ZnO nanoparticles or ZnO films: A comparison of the gas sensing capabilities. *Sens. Actuators, B*, 137, 94, 2009.
13. Sberveglieri, G., Barrato, C., Comini, E., Faglia, G., Ferroni, M., Ponzani, A., Vomiero, A., Synthesis and characterization of semiconducting nanowires for gas sensing. *Sens. Actuators, B*, 121, 208, 2007.
14. Dong, L.F., Cui, Z.L., Zhang, Z.K., Gas sensing properties of nano-ZnO prepared by arc plasma method. *Nanostruct. Mater.*, 8, 815, 1997.
15. Qi, Q., Zhang, T., Zheng, X., Fan, H., Liu, L., Wang, R., Zeng, Y., Electrical response of Sm_2O_3-doped SnO_2 to C_2H_2 and effect of humidity interference. *Sens. Actuators, B*, 134, 36, 2008.
16. Li, C., Su, Y., Lv, X., Xia, H., Wang, Y., Electrochemical acetylene sensor based on Au/MWCNTs. *Sens. Actuators, B*, 149, 427, 2010.
17. Miller, K.L., Morrison, E., Marshall, S.T., Medlin, J.W., Experimental and modeling studies of acetylene detection in hydrogen/acetylene mixtures on PdM bimetallic metal-insulator-semiconductor devices. *Sens. Actuators, B*, 156, 924, 2011.
18. Tamaekong, N., Liewhiran, C., Wisitsoraat, A., Phanichphant, S., Acetylene sensor based on Pt/ZnO thick films as prepared by flame spray pyrolysis. *Sens. Actuators, B*, 152, 155, 2011.
19. Zhang, L., Zhao, J., Zheng, J., Li, L., Zhu, Z., Hydrothermal synthesis of hierarchical nanoparticle-decorated ZnO microdisks and the structure-enhanced acetylene sensing properties at high temperatures. *Sens. Actuators, B*, 158, 144, 2011.
20. Chen, W., Zhou, Q., Gao, T., Su, X., Wan, F., Pd-doped SnO_2 based sensor detecting characteristic fault hydrocarbon gases in transformer oil. *J. Nanomater.*, 2013, 127345, 2013.
21. Wang, X., Zhao, M., Liu, F., Jia, J., Li, X., Cao, L., C_2H_2 gas sensor based on Ni-doped ZnO electrospun nanofibers. *Ceram. Int.*, 39, 2883, 2013.
22. Uddin, A.S.M.I. and Chung, G.-S., Synthesis of highly dispersed ZnO nanoparticles on graphene surface and their acetylene sensing properties. *Sens. Actuators, B*, 205, 338, 2014.
23. Uddin, A.S.M.I., Phan, D.-T., Chung, G.-S., Low temperature acetylene gas sensor based on Ag nanoparticles-loaded ZnO-reduced graphene oxide hybrid. *Sens. Actuators, B*, 207, 362, 2015.
24. Uddin, A.S.M.I., Lee, K.-W., Chung, G.-S., Acetylene gas sensing properties of an Ag-loaded hierarchical ZnO nanostructure-decorated reduced graphene oxide hybrid. *Sens. Actuators, B*, 216, 33, 2015.
25. Lee, K.-W., Uddin, A.S.M.I., Phan, D.-T., Chung, G.-S., Fabrication of low temperature acetylene gas sensor based on Ag nanoparticles-loaded hierarchical ZnO nanostructures. *Electron. Lett.*, 51, 572, 2015.
26. Uddin, A.S.M.I., Yaqoob, U., Phan, D.-T., Chung, G.-S., A novel flexible acetylene gas sensor based on PI/PTFE-supported Ag-loaded vertical ZnO nanorods array. *Sens. Actuators, B*, 222, 536, 2016.

27. Liewhiran, C., Tamaekong, N., Wisitsoraat, A., Phanichphant, S., Highly selective environmental sensors based on flame-spray-made SnO_2 nanoparticles. *Sens. Actuators, B*, 163, 51, 2012.
28. Rashid, T.R., Phan, D.-T., Chung, G.-S., A flexible hydrogen sensor based on Pd nanoparticles decorated ZnO nanorods grown on polyimide tape. *Sens. Actuators, B*, 185, 777, 2013.
29. Rai, P., Kim, Y.S., Song, H.M., Song, M.K., Yu, Y.T., The role of gold catalyst on the sensing behavior of ZnO nanorods for CO and NO_2 gases. *Sens. Actuators, B*, 165, 133, 2012.
30. Rashid, T.R., Phan, D.-T., Chung, G.-S., Effect of Ga-modified layer on flexible hydrogen sensor using ZnO nanorods decorated by Pd catalysts. *Sens. Actuators, B*, 193, 869, 2014.
31. Singh, G., Choudhary, A., Haranath, D., Joshi, A.G., Singh, N., Singh, S., Pasricha, R., ZnO decorated luminescent graphene as a potential gas sensor at room temperature. *Carbon*, 50, 385, 2012.
32. Yuan, W. and Shi, G., Graphene-based gas sensors. *J. Mater. Chem. A*, 1, 10078, 2013.
33. Basu, S. and Bhattacharyya, P., Recent developments on graphene and graphene oxide based solid state gas sensors. *Sens. Actuators, B*, 173, 1, 2012.
34. Phan, D.-T., Gupta, R.K., Chung, G.-S., Al-Ghamdi, A.A., Al-Hartomy, O.A., El-Tantawy, F., Yakuphanoglu, F., Photodiodes based on graphene oxide-silicon junctions. *Solar Energy*, 86, 2961, 2012.
35. Polsongkram, D., Chamninok, P., Pukird, S., Chow, L., Lupan, O., Chai, G., Khallaf, H., Park, S., Schulte, A., Effect of synthesis conditions on the growth of ZnO nanorods via hydrothermal method. *Physica B*, 403, 3713, 2008.
36. Zhang, H., Yang, D., Li, D., Ma, X., Li, S., Que, D., Controllable growth of ZnO microcrystals by a capping-molecule-assisted hydrothermal process. *Cryst. Growth Des.*, 5, 547, 2005.
37. Ju, H.M., Huh, S.H., Choi, S.H., Lee, H.L., Structures of thermally and chemically reduced graphene. *Mater. Lett.*, 64, 357, 2010.
38. Dang, T.T., Pham, V.H., Hur, S.H., Kim, E.J., Kong, B.-S., Chung, J.S., Superior dispersion of highly reduced graphene oxide in N,N-dimethylformamide. *J. Colloid Interface Sci.*, 376, 91, 2012.
39. Wu, J., Shen, X., Jiang, L., Wang, K., Chen, K., Solvothermal synthesis and characterization of sandwich-like graphene/ZnO nanocomposites. *Appl. Surf. Sci.*, 256, 2826, 2010.
40. Gao, W., Alemany, L.B., Ci, L., Ajayan, P.M., New insights into the structure and reduction of graphite oxide. *Nat. Chem.*, 1, 403, 2009.
41. Ullah, K., Zhu, L., Meng, Z.D., Ye, S., Sun, Q., Oh, W.C., A facile and fast synthesis of novel composite Pt-graphene/TiO_2 with enhanced photocatalytic activity under UV/Visible light. *Chem. Eng. J.*, 231, 76, 2013.
42. Park, S., An, J., Potts, J.R., Velamakanni, A., Murali, S., Ruoff, R.S., Hydrazine-reduction of graphite- and graphene oxide. *Carbon*, 49, 3019, 2011.
43. Takata, M., Tsubone, D., Yanagida, H., Dependence of electrical conductivity of ZnO on degree of sintering. *J. Am. Cer. Soc.*, 59, 4, 1976.
44. Bai, S.L., Chen, L.Y., Li, D.Q., Yang, W.H., Yang, P.C., Liu, Z.Y., Chen, A.F., Chung, C.L., Different morphologies of ZnO nanorods and their sensing property. *Sens. Actuators, B*, 146, 129, 2010.
45. Lim, S.K., Hwang, S.H., Kim, S., Park, H., Preparation of ZnO nanorods by microemulsion synthesis and their application as a CO gas sensor. *Sens. Actuators, B*, 160, 94, 2011.
46. Lin, Q., Li, Y., Yang, M., Tin oxide/graphene composite fabricated via a hydrothermal method for gas sensors working at room temperature. *Sens. Actuators, B*, 173, 139, 2012.
47. Afzal, A., Cioffi, N., Sabbatini, L., Torsi, L., NO_x sensors based on semiconducting metal oxide nanostructures: Progress and perspectives. *Sens. Actuators, B*, 171, 25, 2012.
48. Hongsith, N., Wongrat, E., Kerdcharoen, T., Choopun, S., Sensor response formula for sensor based on ZnO nanostructures. *Sens. Actuators, B*, 144, 67, 2010.

49. Mende, L.S. and Driscoll, J.L.M., ZnO—Nanostructures, defects, and devices. *Mater. Today*, 10, 40, 2007.
50. Dan, Y., Cao, Y., Mallouk, T.E., Evoy, S., Johnson, A.T.C., Gas sensing properties of single conducting polymer nanowires and the effect of temperature. *Nanotechnology*, 20, 434014, 2009.
51. Swaddle, T.W., *Inorganic Chemistry: An Industrial and Environmental Perspective*, Academic Press, USA, 1997.
52. Cottrell, T.L., *The Strengths of Chemical Bonds*, 2nd edition, Academic Press, New York, USA, 1961.
53. Bauschlicher, C.W., Jr. and Langhoff, S.R., Theoretical study of the C–H bond dissociation energies of CH_4, C_2H_2, C_2H_4, and H_2C_2O. *Chem. Phy. Lett.*, 177, 133, 1991.
54. Blanksby, S.J. and Ellison, G.B., Bond dissociation energies of organic molecules. *Acc. Chem. Res.*, 36, 255, 2003.

14
Functional Graphene Oxide/Epoxy Nanocomposite Coatings with Enhanced Protection Properties

H. Alhumade[3]*, R.P. Nogueira[2], A. Yu[1], L. Simon[1] and A. Elkamel[1,2]

[1]*Department of Chemical Engineering, University of Waterloo, Waterloo, Canada*
[2]*Department of Chemical Engineering, Khalifa University, The Petroleum Institute, Abu Dhabi, UAE*
[3]*Center of Research Excellence in Nanotechnology (CENT), King Fahd University of Petroleum and Minerals (KFUPM), Dhahran, Saudi Arabia*

Abstract

Functional graphene oxide (FGO) is synthesized by attaching the amino group to the surface of graphene oxide (GO) using a simple one-step preparation procedure. FGO is incorporated as a filler in epoxy (E) resin and the E/FGO composite is evaluated as a protective coating on cold rolled steel (CRS) metal substrates. The synthesized coatings are characterized using X-ray diffraction (XRD) and Fourier transform infrared (FTIR), while the dispersion of the fillers in the polymer resin is observed and evaluated using transmission electron microscopy (TEM). Corrosion resistance properties of the prepared coatings are examined in 3.5% NaCl solution using electrochemical impedance spectroscopy (EIS) and potentiodynamic polarization measurements. Long-term corrosion protection properties of the coatings are examined by conduction gravimetric analysis over 120 days in 3.5% NaCl solution. Interface adhesion between the composite coatings and the CRS metal substrates is inspected and evaluated according to the ASTM-D3359 standard. Thermal stability and the thermal behaviors properties of the prepared composites are evaluated using thermogravimetric analysis (TGA) and differential scanning calorimetry (DSC). Moreover, ASTM-D2794 and ASTM-D4587 standards are followed to examine the resistances of the prepared composites to impact deformation and UV degradation, respectively. Results demonstrate that the utilization of FGO as a filler in epoxy resin may deliver advanced protection properties over epoxy or E/GO composite coatings.

Keywords: Nanocomposites, graphene, corrosion, coatings, adhesion, UV degradation, impact

14.1 Introduction

Corrosion is the descent of metal substrates driven by interaction between metals and the environment, where electrons escape from the metal substrates to the surroundings, causing the metal to release ions that may react to form metal oxide. The rate of such electrochemical interactions between metal and the environment might be influenced by different factors including the nature of the metal and the surrounding conditions. For instance, misty surrounding conditions may accelerate the rate of corrosion and increase the severity of the damages to metal

---

Corresponding author: halhumade@uwaterloo.ca

substrates. Industry and economy are both facing serious threats that may result from the lack of mitigation of corrosion. Therefore, an increasing number of studies are devoted to investigate the possibilities of utilizing various techniques in order to prevent or mitigate the process of corrosion in various environments. Examples of such techniques include the utilization of anodic/cathodic protection, corrosion inhibitors, and protective coatings [1–3].

In particular, the use of protective coatings is one of the eminent approaches used in various fields of applications for corrosion mitigation purposes, and this can be attributed to easiness and the cost-effectiveness of the application of protective coatings. In the protective coating industry, the coatings act as a physical barrier between the surface of metal substrates and the surroundings, which may shield corrosive agents such as oxygen, chloride, and moisture from reaching the surface of the coated metal substrates. However, the curing chemical reaction process of protective coating may involve the production of side elements such as hydrogen gas or water, which can be trapped and form pores inside the protective coatings. The formation of these pores inside the protective coatings may attenuate the corrosion resistance performance of the coating. In addition, and depending on various factors including the nature of the protective coatings, curing process conditions, as well as the amount of the side products, these pores inside the protective coatings may network and form channels that allow the corrosive agents to migrate through the coatings and reach underneath metal substrates. The accumulation of corrosive agents at the interface between the protective coating and the coated metal substrate may accelerate the process of corrosion in addition to other effects such as blistering of the coating or loss of interface adhesion between the coating and metal substrate.

A growing number of studies have focused on the possibility of enhancing the corrosion resistance performance of protective coatings by the incorporation of fillers such as additives or corrosion resistance pigments in the polymeric matrix of the coating resin. For instance, Jiang et al. investigated the possibility of enhancing the corrosion protection as well as the interface adhesion properties of epoxy by the incorporation of active (amino-propyltrimethoxy) and non-active (bis-1,2-[triethoxysilyl]ethane)silane precursors [4, 5]. In another study, TiO_2-doped poly-pyrrole coating was utilized to enhance the corrosion resistance property of aluminum substrates [6], while hydroxyapatite and octacalcium phosphate coatings were utilized to extend the life span of magnesium alloy [7]. There are different types of fillers that have been explored as corrosion resistance pigments in protective coatings [8–10]. However, a growing number of studies have focused on the utilization of nanomaterials in order to further excel the corrosion resistance properties of protective coatings. For example, graphene and graphene derivatives have been widely utilized as fillers in different polymeric matrices to improve various properties of the polymer composites, including but not limited to corrosion resistance [11–15]. Moreover, various studies have investigated the deposition of graphene as protective coating on metal substrates using the chemical vapor deposition (CVD) method. The studies revealed that the graphene-deposited protective coating may extend the life span of the coated metal substrate by acting as a passive layer that extend pathways corrosive agents follow to reach underneath metal substrates and attenuate the transportation rates of electrons and ions between the coated metal substrates and the surroundings. The attention given to graphene and graphene derivative materials can be attributed to various factors including the unique properties of graphene such as higher surface area and aspect ratio in addition to the lower density compared to various fillers such as clay [16].

The investigation of graphene and graphene derivative materials has been extended beyond the use of pristine graphene as a filler in polymer composite protective coatings or the utilization of graphene as a barrier coating using the CVD techniques. For instance, a growing number of studies have examined the possibility of enhancing various properties of graphene-based composites including corrosion resistance by surface modification of graphene and graphene oxide. For example, a study has examined the surface modification of GO sheets by attaching titanium dioxide using 3-aminopropyltriethoxysilane as coupling agents before incorporating the filler in epoxy resin, and the study revealed that the corrosion resistance properties of epoxy as well as epoxy/GO composites can enhance this functionalization of GO sheets [17]. A different study examined the influences of the incorporation of fluorographene particles into polyvinyl butyral composites protective coatings for corrosion mitigation purposes [18]. The study demonstrates that such functionalization of graphene materials may deliver further enhancement in the corrosion resistance property of the coating by shielding the diffusion paths of corrosive agents and moisture.

In this study, the surface of GO sheets is chemically modified by attaching an amino functional group from silane material. The treated GO sheets are characterized using FTIR and XRD techniques in order to confirm the successful synthesis of FGO material. TEM is utilized to observe the dispersion of GO and FGO in the polymeric composite matrices. The study investigates the influences of the incorporation of FGO on the various properties of the hosting polymer resin as well as the impact of the surface modification of GO on the protection properties of E/GO composites. For example, the study demonstrates the influences of the utilization of FGO as filler on corrosion resistance, thermal stability, thermal behavior, impact resistance, and UV degradation properties of epoxy and E/GO composites. The long-term corrosion protection properties of the prepared composite coatings are examined by conducting gravimetric analysis over a 120-day exposure period in a temperature-controlled 3.5% NaCl solution. Moreover, different electrochemical testing techniques such as EIS and potentiodynamic measurements are carried out to examine and compare the corrosion mitigation properties of epoxy, E/GO, and E/FGO in a temperature-controlled 3.5% NaCl solution. TGA and DSC are used to evaluate the thermal stability properties and the thermal behaviors of the prepared composite coatings. Finally, the UV degradation properties and the resistance of the prepared composite coatings to sudden deformation are examined and evaluated according to ASTM-D4587 and ASTM-D2794 standards, respectively.

14.2 Experimental

14.2.1 Materials

Polished CRS sheet (McMASTER-CARR) was used as metal substrates, where the CRS substrates were polished with SIC 800 and then with 1200 grit discs, washed with acetone and then with DDI water, and finally cleaned with KIMTECH wipes before applying the coating. Bisphenol A diglycidyl ether (BADGE, Sigma-Aldrich) was used as epoxy resin, while poly(propylene glycol) bis(2-aminopropyl ether) (B230, Sigma-Aldrich) was used as hardener. GO sheets (ACS Material) were synthesized by the modified Hummer method and thermally treated to improve dispersion. GO sheets have average diameters of 1–5 μm and average thicknesses of 0.8–1.2 nm according to the supplier. (3-Aminopropyl)triethoxysilane (APS, Sigma-Aldrich) was utilized to functionalize the surface of GO. All materials were used as received.

14.2.2 Composite Synthesis

The synthesis procedures start with the functionalization of GO sheets, where 200-mg GO sheets were mixed with 2 g of APS in a mixture of 250 ml of DDI and 250 ml of ethanol. The GO suspension was stirred in a water bath at 70°C overnight. The GO suspension was collected under low pressure vacuum, and the collected materials were washed with DDI and then with ethanol three times before the FGO materials were allowed to dry under vacuum overnight in a vacuum oven at 90°C. 2.1 mg of FGO was dispersed in 1.5 g of BADGE under reflux for 4 h and bath sonicated for an additional 2 h. 0.5 g of the curing agent (B230) was added to the GO suspension in BADGE, and the final mixture was refluxed for 2 h, bath sonicated for 2 h, and finally homogenized (125, Fisher Scientific) for 1 h. CRS metal substrates were polished with different grids, washed with acetone and DDI water, and dried before the FGO suspension mixture was applied using a fine brush. Spin coater (SC 100, Smart Coater) was used to control the thickness of the composite coating on the CRS substrate, where the FGO suspension mixtures were rotated at 400 RPM for 1 min. Finally, the FGO/pre-polymer mixture was cured on the CRS substrate under vacuum at 50°C in a vacuum oven for 4 h to produce 123 ± 2 μm of the E/FGO-coated CRS substrate. Figure 14.1 depicts the functionalization process of GO and the preparation procedures of the E/FGO-coated CRS substrate. Similar procedures were followed to synthesis epoxy- and E/GO composite-coated CRS substrates, where 1.5 g of BADGE and 0.5 g of B230 were mixed with/without 2.1 mg of GO to produce 123 ± 2 μm of E/GO- and epoxy-coated CRS metal substrates.

14.2.3 Composites Characterization

The prepared protective polymer composite coatings were characterized using various techniques such as Fourier transfer infrared spectroscopy (FTIR) and X-ray diffraction (XRD). Furthermore, the dispersion of fillers in polymer composites was captured using scanning electron microscopy (SEM) and transmission electron microscopy (TEM). The objective here is to illustrate the sample preparations steps for each technique.

FTIR samples were prepared by scratching the polymer composites with a sharp fine knife in order to collect a small amount of the composites. The collected composite samples were then mixed with a certain amount of potassium bromide (KBr) in order to maintain the sample load to 2–5 wt.% of the mixture. The mixture was then compressed at 5000 lb for 2 min to form the FTIR disk sample. FTIR data were recorded from 400 to 4000 wavenumber at 4 cm^{-1} resolution and a scan time of 64 s. Unlike FTIR, the XRD samples did not require specific procedures to follow in sample preparation. However, the thicknesses of the prepared polymer composites were maintained below 20 μm in order to maximize the quality of diffraction peaks. XRD diffraction patterns for all prepared composites were recorded in the range of 2θ = 3–90° at a scan rate of 0.24°/s and 0.02° step size.

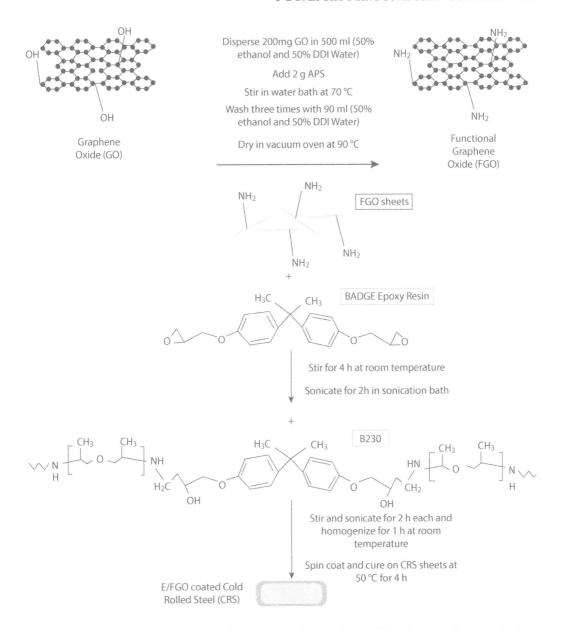

Figure 14.1 Functionalization process of GO and synthesis procedures of E/FGO composites using *in situ* polymerization.

The dispersion of fillers in the polymer composites was captured using TEM. TEM samples were collected by scraping the prepared polymer composites with a fine sharp knife, and the collected samples were dispersed in methanol. The dispersion was sonicated in a sonication bath for 30 min before the samples were fished with TEM copper grids. Finally, the collected samples in copper grids were allowed to dry under vacuum at room temperature overnight before capturing TEM imaging.

14.2.4 Adhesion

Interface adhesion between polymer composite coatings and the coated metal substrates was evaluated according to the ASTM-D3359 standard. An adhesion tape testing kit was used for this purpose with a standard 11-tooth and 1-mm spacing blade. The test was conducted by making parallel cuts on the coating before applying the adhesion tape on the cuts. After peeling off the tape from the surfaces of the composite coatings, the scanning electron microscopy (SEM) technique was used to observe and evaluate the adhesion property of the various prepared coatings, where samples were gold coated using the sputtering technique for 120 s. Moreover, the adhesion property of each coating was evaluated and rated according to the ASTM standard based on the amount of peeled materials from the coated samples after conducting the adhesion tape test.

14.2.5 Electrochemical Measurements

All electrochemical measurements were conducted in a 1-L double-jacked temperature-controlled corrosion cell using 3.5% NaCl solution as electrolyte at 25°C. A three-electrode configuration was used to conduct electrochemical measurements, where a silver/silver chloride (Ag/AgCl) electrode was used as a reference electrode (RE), a graphite rod was used as a counter electrode, and coated samples were used as the working electrode (WE). Coated samples were cleaned and dried and then mounted in a Teflon holder with 1 cm^2 exposed surface area, and the potentials of the testing samples were allowed to stabilize for at least 30 min before conducting electrochemical measurements. The potentials of the testing samples were recorded after stabilization as open circuit potentials (OCP). The electrochemical behaviors of coated samples were evaluated using electrochemical impedance spectroscopy (EIS) and potentiodynamic measurement techniques.

EIS measurements were carried out at a frequency range from 200 kHz to 100 mHz, and the collected raw impedance data were presented using Bode and Nyquist plots. Furthermore, equivalent circuits with a specific combination of elements were utilized for fitting raw impedance data, and the variations in the magnitude of the different elements of the circuits were used to evaluate the corrosion resistance properties of the prepared protective composite coatings.

Following the nondestructive EIS measurements, potentiodynamic measurements were also utilized to evaluate the corrosion protection properties of the prepared protective coatings using the similar testing setup. Here, the potentiodynamic measurements were carried out by scanning the potential of the testing sample from −0.5 V to 0.5 V around OCP at a rate of 0.02 V/min. The collected potentiodynamic measurements were used to generate Tafel plots in order to extract valuable corrosion parameters such as corrosion current (I_{corr}) and corrosion potential (E_{corr}). The variation of these parameters' corrosion was investigated to evaluate the corrosion resistance properties of the different protective coatings.

14.2.6 Gravimetric Analysis

The long-term corrosion resistance performances of the prepared protective coatings were examined by conducting weight loss measurements. A 500-ml temperature-controlled 3.5%

NaCl solution at 25°C was used as the corrosion medium. Testing samples were cleaned with acetone, dried with KIMTECH paper, and weighted before conducting the weight loss tests. After which, the samples were mounted in Teflon holders with 1 cm^2 exposed surface areas and immersed in the corrosive medium for 120 days. At the end of the exposure period and after removing the samples from the holders, samples were cleaned in order to remove the corrosive residues by washing the samples with distilled water and immersing the sample in bath sonication for 10 min. Samples were allowed to dry under vacuum overnight before the final weights were recorded. The corrosion protection properties of the different coatings were evaluated by comparing the weights of the samples before and after exposure to the corrosive medium. Furthermore, all weight loss measurement were conducted in triplicate in order to examine the reproducibility of the results.

14.2.7 Thermal Analysis and UV Degradation

Some of the prepared protective composite coatings are intended for utilization in an outdoor environment, and therefore, it was important to evaluate the thermal stability and UV degradation properties of the prepared coatings. The thermal stability property was evaluated using thermal gravimetric analysis (TGA) and differential scanning calorimetry (DSC) techniques. TGA was conducted over a temperature range of 25–800°C at a heating rate of 10°C/min; DSC analysis was conducted over a temperature range of 25–200°C at a heating rate of 10°C/min. Thermal analysis helped evaluate important thermal properties such glass transition temperature (T_g) and the onset temperature (T_{onset}), which is the temperature where the composites lose 5% of the original weight.

In addition to thermal behavior, it was important to evaluate UV degradation properties of coatings intended for utilization in outdoor applications. UV analysis was conducted and evaluated according to the ASTM standard D4587 using an accelerated weathering tester. In this test, the samples were continuously exposed to repeated cycles of UV light at 60 ± 2.5°C for 8 h, followed by water condensation at 50 ± 2.5°C for 4 h over 30 days. The surface morphology of the tested samples was examined by SEM, and here, too, the samples were gold coated using the sputtering technique for 120 s.

14.2.8 Impact Resistance

In addition to thermal stability and UV degradation, impact resistance was an important property to evaluate for coatings intended for usage in various environments, where the protective coatings might be exposed to impact deformation.

Resistance to impact deformation was assessed according to the ASTM standard D2794 using a universal impact tester with 2 Ib falling weight attached to a ball with 0.5-inch diameter. The test was conducted by raising the falling weight 1 inch above the surface of the coatings and releasing the falling weight to impact the coating. This process was repeated with 1-inch increment in the distance between the height of the falling weight and the surface of the coating until the coating cracks. The heights at which the coatings cracked were recorded and compared in order to examine the influences of the incorporated filler on the impact resistance properties of the composite coatings.

14.3 Results and Discussion

14.3.1 Composite Characterization

The functionalization process of GO sheets with the amino group from APS to prepare FGO sheets is confirmed using FTIR and XRD techniques. FTIR spectra depicted in Figure 14.2 show some typical characteristic peaks that correspond to a typical functional group attached to the surface of GO such as the peak at 1226 cm^{-1}, which corresponds to the epoxide group as well as the peaks at 1602 and 3410 cm^{-1}, which correspond to carboxyl and hydroxyl groups, respectively. Furthermore, the FTIR spectra for GO reflect the influences of the thermal reduction of GO sheets, which can be observed as an attenuation in the characteristic peak for O–H at 3410 cm^{-1}.

The chemical modification of GO sheets may take place as a replacement of the hydroxyl group attached to the surface of GO sheets with the APS particle. The formation of a C–C bond may occur between the carbon atoms on the basal planes of GO sheets, which give the –OH group with carbon atoms in the (H$_3$CO-SI) group on APS. The FTIR spectra of FGO

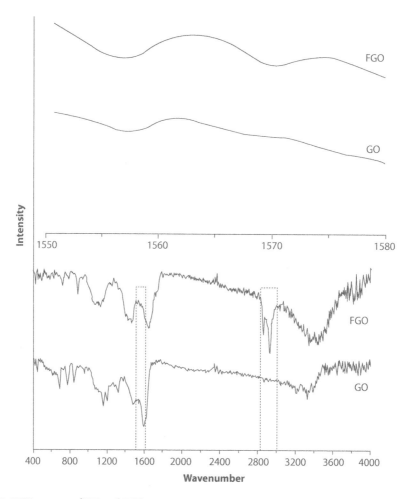

Figure 14.2 FTIR spectra of GO and FGO.

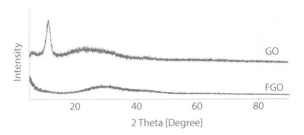

Figure 14.3 XRD patterns of GO and FGO.

confirm the attachment of APS to the GO sheets based on the appearance of some characteristic peak such as the strong absorption peak at 2800–3000 cm^{-1}, which corresponds to the C–H vibrations in the APS and the absorption peak at 1574 cm^{-1}, which corresponds to the NH_2 group attached to the APS coupling agent [19]. The XRD technique was also used to confirm the chemical modification of the surface of GO sheets as depicted in Figure 14.3. XRD patterns of GO and FGO reflect the influences of the surface functionalization of GO on the crystal structure as well as the d-spacing of GO sheets. From XRD patterns and Bragg's law, the d-spacing values were calculated as 7.96 Å and 29.4 Å for GO and FGO, respectively.

Both FTIR and XRD techniques were used to examine the completion of curing process and polymer chain linkage of the epoxy resin with the hardener in all composites. The FTIR spectra for epoxy, E/GO, and E/FGO are depicted in Figure 14.4, and all spectra indicate characteristic peaks that confirm the completion of the curing of epoxy. For instance, the characteristic peak at 3380 cm^{-1} represent -OH stretching, which is a result of the ring opening reaction of the epoxy resin with the amino group in the hardener. The FTIR spectra also show some common characteristic peaks of epoxy composites such as the peaks

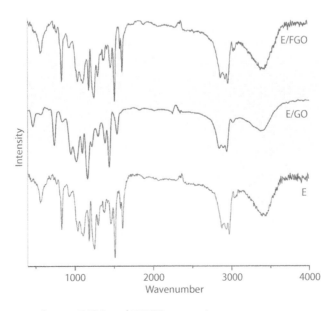

Figure 14.4 FTIR spectra of epoxy, E/GO, and E/FGO composites.

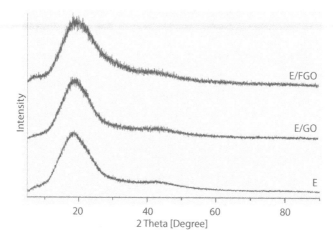

Figure 14.5 XRD patterns of epoxy, E/GO, and E/FGO composites.

at 1508 cm^{-1} and 1609 cm^{-1} (C–C skeletal stretching) and the peak at 915 cm^{-1} (epoxide ring). Furthermore, the XRD patterns depicted in Figure 14.5 for epoxy, E/GO, and E/FGO describe typical XRD patterns of epoxy composites. In these XRD patterns, broad diffraction peaks between 2θ values of 10°–30° represent the homogeneously amorphous phase of epoxy composites. In addition to the position of the XRD diffraction peaks, the reservation of the amplitudes of the diffraction peaks for all prepared composites indicates that the incorporation of GO and FGO as fillers in the E/GO and E/FGO composites did not affect the degree of crystallization of epoxy.

The degree of dispersion of the GO and FGO sheets in E/GO and E/FGO composites was investigated using TEM techniques, and the captured images are depicted in Figure 14.6. The TEM images for E/GO indicate that the GO sheets were agglomerated in the

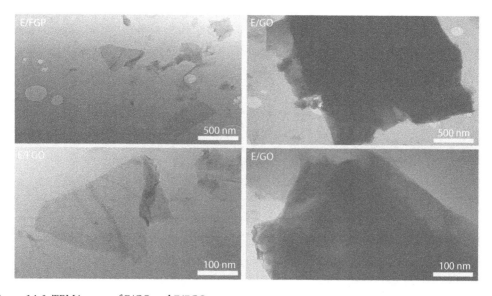

Figure 14.6 TEM images of E/GO and E/FGO.

composites and that the GO sheets were manifested in thick stacked layers. However, TEM images for E/FGO clearly illustrate the advantage of chemical modification of the functional group on the surface of GO sheets, where thin sheets of FGO can be observed in the E/FGO composite. The enhancement in the degree of dispersion of FGO sheets can be attributed to the interaction between the grafted amino functional group on the surface of FGO sheets and the epoxide group in epoxy resin.

14.3.2 Adhesion

In the coating industry and in particular corrosion protection applications, the main objective of coating metal substrates with protective coatings is to shield corrosive agents such as moisture, oxygen, and chloride ions from the coated metal substrates in order to extend the life span of the coated metal substrates. However, loss of interface adhesion between the protective coating and the underneath metal substrate in certain areas may result in void spaces at the interface. Corrosive agents may accumulate in the void spaces and accelerate the process of corrosion in these areas, causing pitting corrosion, which can be difficult to detect and repair before serious damages are encountered. Therefore, interface adhesion is one of the critical properties of protective coating that need to be carefully assessed and noble interface adhesion between the protective coating and the coated metal substrate is always a desire before conducting further analysis on the protective coating including corrosion resistance. In this study, the interface adhesion between the prepared protective coatings and the CRS metal substrates after exposing the coated CRS substrates to a temperature-controlled 3.5% NaCl solution for 120 days using adhesion tap test kit was determined in order to examine the long-term interface adhesion properties of the coatings.

The interface adhesion tests were conducted and evaluated according to the ASTM D3359 standard using a standard blade (11 teeth with teeth spacing of 1 mm), where perpendicular cuts were made on the surfaces of the coatings and the adhesion testing tape was applied on the surfaces and peeled after 2 min. Once the adhesion tape was peeled, the conditions of the surfaces of the coatings were observed using the SEM technique after coating the tested samples with gold using the sputtering technique for 120 s, and the adhesion test results were evaluated based on the amount of the peeled materials from the protective coatings. Figure 14.7 depicts the postadhesion test results for epoxy-, E/GO-, and E/FGO-coated CRS substrates. From the figure, no peelings were observed from any of the protective coatings, and hence, all prepared coatings were rated as 5B coatings (0% peeling) according to the ASTM standard.

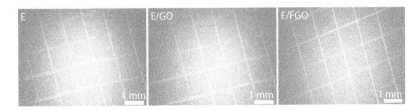

Figure 14.7 SEM images of post-adhesion tests of epoxy-, E/GO-, and E/FGO-coated CRS substrates.

14.3.3 Gravimetric Analysis

The long-term corrosion protection property of a protective composite coating can be evaluated by conducting a gravimetric analysis over a certain period of exposure time to a corrosive medium. Gravimetric analysis helps evaluate the long-term durability of a protective coating and confirms the possibility of utilizing the protective coating in various fields including those areas where the protective coating might be exposed to corrosive elements. Here, the long-term corrosion resistance properties of the prepared composite protective coatings were examined by conducting gravimetric analysis in a temperature-controlled 3.5% NaCl solution over a 120-day exposure period. In addition, all weight loss measurements were carried out in triplicate in order to examine the reproducibility of the corrosion mitigation performances of the coatings. The weight loss measurements were used to compute the corrosion rates (R_{corr}) for each sample by comparing the weight of the samples before and after the exposure period using Equation 14.1. Furthermore, the computed corrosion rates for the different testing samples were used to calculate the protection efficiency (P_{EF}) of the protective composite coatings using Equation 14.2.

$$R_{corr} = \frac{W_0 - W}{A \times t} \qquad (14.1)$$

$$P_{EF}[\%] = \left(1 - \frac{R_{corr}}{R_{corr}^\circ}\right) \times 100 \qquad (14.2)$$

In the above equations, A represents the exposed surface area of a testing sample (1 cm²), t is the time of exposure to the corrosive medium (120 days), R_{corr} and R_{corr}° are the average corrosion rates of coated and bare CRS substrates, respectively, while W and W_0 are the weight of the samples (mg) before and after exposure to the corrosive medium. Finally, statistical analysis was performed on the gravimetric analysis using the triplicate weight loss measurements, which allow the calculation of the average and the standard deviation of corrosion rates ($R_{corr,STD}$) for all samples. All weight loss measurements including the statistical analysis results are reported in Table 14.1.

Table 14.1 Weight loss measurements for bare CRS and epoxy-, E/GO-, and E/FGO-coated CRS substrates in a 3.5% NaCl solution.

Sample	W_0 Mg	W mg	R_{corr} mg cm^{-2} d^{-1}	$R_{corr,STD}$ mg cm^{-2} d^{-1}	P_{EF} %
CRS	120	82	0.42	0.03	–
Epoxy	140.5	136.8	0.04	0.05	90.3
E/GO	142.2	139.7	0.028	0.009	93.4
E/FGO	142	141.9	0.001	0.001	99.7

The observed weight loss measurement results reported in Table 14.1 demonstrate that the corrosion resistance property of CRS metal substrates can be improved by applying epoxy coating on CRS. Furthermore, the corrosion protection property of epoxy coating can be further excelled by the incorporation of GO sheets as a filler in polymer resin. However, the results reported on weight loss measurements clearly illustrate the advantage of the chemical modification of GO sheets before incorporation in the polymer resin. This can be evidently seen as further attenuation in corrosion rate and enhancement in protection efficiency for E/FGO over other protective coatings. In addition to the excelled corrosion resistance property of E/FGO, the statistical analysis of the gravimetric measurements explains the noble reproducibility of the corrosion protection performance of the E/FGO composite protective coating over other protective coatings and this can be observed as a lower magnitude of the standard deviation of the corrosion rate.

14.3.4 Impedance

In the field of corrosion studies and industry, there are various techniques that can be utilized to examine the corrosion behavior of metal substrates and evaluate the corrosion protection properties of protective coatings. For instance, electrochemical impedance spectroscopy is one of the electrochemical techniques that are widely used to evaluate the electrochemical behavior of bare and coated metal substrates and examine the corrosion mitigation performance of a protective coating. In impedance studies, an alternative current is passed through electrical circuits, which may consist of different elements including resistors, capacitors, and insulators in a certain order and combination. The observed outcome of passing the alternative current through the electrochemical circuit is a complex resistance known as impedance. In corrosion studies, the alternative current is passed through the bare and coated metal substrates over a certain range of frequencies, and the variation in the observed impedance results can be interpreted to explain the electrochemical behaviors of the bare and coated metal substrates and evaluate the corrosion protection properties of protective coatings. Moreover, equivalent circuits with particular combination of different electrical elements such as capacitors and resistors can be used to imitate the raw impedance data. The feature of the fitting can be controlled by arranging the elements of the equivalent circuits in a particular order and modifying the magnitudes of the various elements of the equivalent circuit. After capturing the best imitated fitting of raw impedance results for bare and coated CRS metal substrates, the variation in the magnitudes of the different elements of the equivalent circuits can be utilized to explain and compare the electrochemical behavior and the corrosion resistance properties of bare and coated CRS substrates.

All electrochemical impedance experiments were conducted in a 1-L temperature-controlled 3.5% NaCl solution using a three-electrode configuration, and all experiments were carried out in triplicate in order to confirm the repeatability of the raw impedance results. In the three-electrode configuration, silver/silver chloride electrode was used as the reference electrode, graphite rod was used as the auxiliary electrode, and bare or coated CRS metal substrates were used as the working electrodes. After immersing the working electrode in the 3.5% NaCl solution electrolyte, the potential of the working electrode was allowed to stabilize for 1 h before conducting the impedance studies. Once raw impedance results for bare and coated CRS substrates were collected, the equivalent circuit depicted in

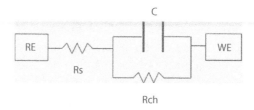

Figure 14.8 Equivalent circuits used to fit raw electrochemical impedance data.

Figure 14.8 was used to fit the raw impedance data. It is worth mentioning that the specific type and order of the various elements of the circuit were selected in order to deliver the finest fitting for raw impedance data for all testing samples. In this circuit, R_s represents the resistance of the temperature-controlled electrolyte solution, R_{ch} is the charge transfer resistance of bare or coated CRS metal substrates, while CPE represents a constant phase element.

Raw impedance data and the fitting results are depicted in Figure 14.9, which is known in the corrosion industry as Nyquist plots. In these plots, the real and the imaginary parts of the impedance results are presented, and the variation in the impedance behavior of the bare and coated metal substrates can be used to evaluate the corrosion protection properties of the prepared protective coatings. Generally, an increase in the size of the impedance semicircle behavior represents an enhancement in corrosion mitigation property. The results presented in the Nyquist plots demonstrate that CRS metal substrates can be protected from corrosion in a chloride-rich environment by applying epoxy protective coating on the metal substrate. In addition, it can be clearly observed that the incorporation of GO as a filler in polymeric matrix may have excelled the corrosion protection property of the resin. However, it was interesting to observe that the corrosion resistance property of the E/GO composite coating can be further enhanced by chemical modification of the surface of GO sheets. This can be seen as a significant enhancement in corrosion mitigation

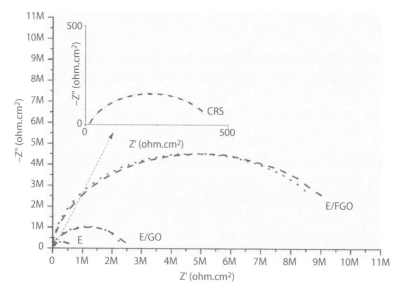

Figure 14.9 Nyquist plots for bare CRS and epoxy-, E/GO-, and E/FGO-coated CRS substrates.

properties of E/FGO over epoxy and E/GO composite coatings, where the enhancement is manifested in a substantial rise in the magnitude of the real part of impedance data at the lowest recorded range of frequency.

Raw impedance data reported in the Nyquist plots were used to carry qualitative analysis on the electrochemical behavior as well as the corrosion resistance properties of bare and coated CRS substrates. In addition to the qualitative analysis conducted on the raw impedance results, the fitting data can be used to carry quantitative analysis in order to evaluate the electrochemical behavior and the corrosion protection properties of the prepared protective composite coatings. This quantitative analysis was conducted by examining the variations in the magnitudes of the various elements of the equivalent circuit depicted in Figure 14.8, which was utilized to fit the raw impedance results. It is worth mentioning that the elements used to configure the equivalent circuit and the unique arrangement of these elements have been selected in order to deliver the finest fitting data for the raw impedance results. The magnitudes of the various elements of the equivalent circuit are reported in Table 14.2 and the triplicate measurements of impedance for all samples facilitate the calculation of the standard deviation of some elements of the equivalent circuit. Charge transfer resistance is an important parameter that can be utilized to explain the electrochemical behavior of bare and coated metal substrates. Furthermore, the variation in the charge transfer resistances of bare and coated metal samples can be used to evaluate the corrosion resistance properties of the protective coatings. The reported magnitudes of the charge transfer resistances in this study demonstrate the possibility of slowing down the corrosion process of CRS metal substrates in a corrosive environment by applying epoxy protective coating on the surface of the CRS substrate. In addition, the incorporation of GO in the epoxy resin may enhance the corrosion resistance property of the hosting polymer resin. Moreover, the corrosion protection performance of E/GO composite coatings can be further excelled by surface functionalization of GO sheets with the amino group. This can be clearly observed as an increase in the charge transfer resistances from 432.8 ohm cm^2 for the CRS substrate to 6.2×10^5 ohm cm^2 for epoxy, 2.6×10^6 ohm cm^2 for pristine E/GO, and 9.8×10^6 ohm cm^2 for E/FGO-coated CRS substrates. Finally, it was interesting to observe that in addition to the advanced corrosion protection property of E/FGO, the functionalization of GO sheets magnifies the reliability of the protective composite coating. This can be observed as a significant attenuation in the standard deviation of the charge transfer resistance for E/FGO compared to epoxy and E/GO composite coatings.

Table 14.2 Electrochemical corrosion parameters obtained from equivalent circuit for EIS raw measurements for CRS and epoxy-, E/GO-, and E/FGO-coated CRS in 3.5% NaCl solution.

Sample	R_S Ω cm^2	C F	R_{ch} $\Omega.cm^2$	$R_{ch, STD}$ Ω cm^2
CRS	18.1	4.2×10^{-4}	432.8	2
Epoxy	18.2	6.1×10^{-11}	6.2×10^5	150
E/GO	18.3	1.5×10^{-10}	2.6×10^6	390
E/FGO	18.0	6.1×10^{-11}	9.8×10^6	160

Bode plot is another approach that can be utilized to represent the electrochemical impedance behavior of bare and coated metal substrates and compare the corrosion protection properties of different protective coatings. In bode plots, the logarithm of impedance modulus (|Z|) is presented over the logarithm of the entire range of frequencies, and the variation of the impedance modulus at the lowest recorded frequency can be used to compare the corrosion resistance properties of the different protective composite coatings. Bode plots for all samples are depicted in Figure 14.10a, while Figure 14.10b represents the phase plots for raw impedance data. The results depicted in the Bode plots illustrate the advantage of the utilization of GO sheets in the epoxy coating in order to extend the corrosion resistance property of epoxy. Furthermore, the results demonstrate that the utilization of FGO as a filler in the polymeric matrix will remarkably enhance the corrosion protection property, and this can be observed as an increase in the logarithm of impedance modulus at the lowest frequency range from 2.7 ohm cm² for the CRS substrate to 5.7 ohm cm² for epoxy, 6.3 ohm cm² for pristine E/GO, and 6.9 ohm cm² for E/FGO-coated CRS substrates.

The impedance study illustrates the possibility of extending the life span of CRS metal substrates by covering the metal substrates with epoxy protective composite coating. In addition, the impedance results illustrate the opportunity of enhancing the corrosion resistance property of epoxy by incorporating GO sheets as a filler in the composites. The positive impact of the incorporation of GO sheets in the epoxy resin on the corrosion mitigation property can be attributed to the shielding property of GO sheets [13], which may act as a barrier that prolongs the pathways corrosive agents follow to reach the interface between the underneath CRS metal substrates and the protective composite coatings. Furthermore, the various approaches used to evaluate the impedance behavior and the corrosion protection performance of the different protective coatings illuminate the benefit of surface modification of GO sheets before the incorporation of the sheets in the epoxy resin. The remarkable enhancement in the corrosion resistance property of E/FGO composite coatings over other

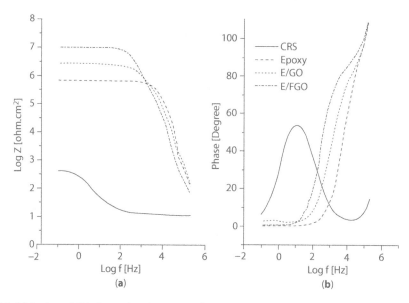

Figure 14.10 (a) Bode and (b) phase plots for CRS and epoxy-, E/GO-, and E/FGO-coated CRS substrates.

protective coatings can be attributed to the superior degree of dispersion of the FGO filler in the resin matrix, as depicted in the TEM images, and this may further increase the tortuosity of pathways for corrosive agents such as oxygen, moisture, and chloride ions to reach the surface of the coated CRS metal substrates.

14.3.5 Potentiodynamic Polarization

Potentiodynamic measurement is another electrochemical technique that can be utilized to explain the electrochemical behavior as well as the corrosion resistance property of bare and coated metal substrates. In this approach, a three-electrode configuration was used with bare or coated CRS metal substrates as the working electrode, silver/silver chloride (Ag/AgCl) electrode as the reference electrode, a graphite rod as the auxiliary electrode, and a 1-L temperature-controlled 3.5% NaCl solution as testing electrolyte. In potentiodynamic measurements, the potential of the working electrode is shifted over a certain range of potential difference versus the constant potential of the reference electrode in order to observe the anodic and cathodic behaviors of the bare and coated metal substrates. Here, too, the potential of the working electrode was allowed to stabilize for 1 h after immersing the working electrode in the electrolyte and the potential of the working electrode was noted as the open circuit potential (OCP). Potentiodynamic measurements were conducted in triplicate in order to examine the reproducibility of the data. Moreover, a new electrolyte solution was prepared for each experiment since potentiodynamic measurement is a destructive electrochemical test that might introduce corrosion residue to the electrolyte. In this study, the potentiodynamic measurements were conducted by scanning the potential of the working electrodes in the potential range of −0.5 V to 0.5 V starting from the OCP of the corrosion cell at a constant rate of 20 mV/min.

It should be noted that the study focused on the particular area where the potential of the working electrode alternates between anodic and cathodic behaviors. The shift in the potential of the working electrode between anodic and cathodic behavior is usually presented with the associated current transfer from or to the working electrode in a plot known in corrosion studies as the Tafel plots as depicted in Figure 14.11. The representation of Tafel plots facilitates the extraction of significant corrosion parameters such the corrosion potential (E_{corr}) and the corrosion current (I_{corr}). The extracted corrosion parameters from the potentiodynamic measurements are reported in Table 14.3. In addition, the triplicate measurements were utilized to conduct statistical analysis on the corrosion parameters and compute the standards deviations of E_{corr} ($E_{corr,STD}$) and I_{corr} ($I_{corr,STD}$) as described in the table.

In addition to the extracted corrosion parameters from the Tafel plots, the Stern-Geary equation can be utilized to compute the polarization resistance (R_p) for bare and coated CRS substrates as defined in Equation 14.3. Furthermore, the extracted corrosion parameters can also be utilized to compute the corrosion protection efficiency (P_{EF}) of the various prepared protective composite coatings using Equation 14.4.

$$R_P = \frac{(b_a \times b_c)}{2.303 \times (b_a + b_c) \times I_{corr}} \tag{14.3}$$

$$P_{EF}[\%] = (1 - I_{corr}/I^\circ_{corr}) \times 100 \tag{14.4}$$

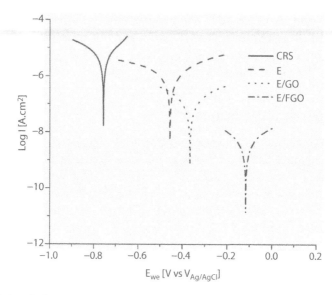

Figure 14.11 Tafel plots for bare CRS and epoxy-, E/GO-, and E/FGO-coated CRS substrates.

Table 14.3 Electrochemical corrosion parameters obtained from potentiodynamic measurements for bare CRS and epoxy-, E/GO-, and E/FGO-coated CRS in a 3.5% NaCl solution.

Sample	E_{corr} mV vs. Ag/AgCl	$E_{corr,STD}$ V vs. Ag/AgCl	I_{corr} μA/cm²	$I_{corr,STD}$ μA/cm²	b_a	b_c	R_p Ω cm²	P_{EF} %
CRS	−751	0.1	6.9	0.001	161.5	268.5	6.3	–
Epoxy	−529	2.3	0.52	0.002	292.8	397.2	139.5	92.4
E/GO	−362	12.5	0.09	0.009	218.6	207.5	509.2	98.7
E/FGO	−122	3.9	0.007	0.001	272.3	251.1	8033.6	99.9

In Equation 14.3, b_a and b_c represent the anodic and the cathodic slops of the Tafel plots, and the intersection between the extrapolated linear portion of these slops facilitates the extraction of E_{corr} and I_{corr}, while the protection efficiencies of the protective coatings were calculated using the corrosion currents of the bare ($I^°_{corr}$) and coated (I_{corr}) CRS metal substrates. Moreover, a quantitative analysis can be conducted on the extracted and computed corrosion parameters from potentiodynamic measurements, where a drop in the magnitude of I_{corr} and an increase in the E_{corr} and R_p signify advance in the corrosion protection properties of protective composite coatings.

The collected and extracted results from the potentiodynamic measurements indicate the decent corrosion protection property of epoxy coating on the CRS metal substrate in corrosive medium. However, the results clearly indicate the possibility of enhancement in this noble corrosion resistance property of epoxy resin by incorporating a small loading of GO sheets as a filler, and this can be observed as a drop in the corrosion current and a positive shift in both corrosion potential and protection efficiency. Moreover, the study demonstrates that the surface functionalization of GO sheets can significantly improve the corrosion protection property, and this can be seen as a further attenuation in the corrosion current and a further positive shift in corrosion potential and protection efficiency for the E/FGO-coated CRS metal substrate. Finally, an interesting finding can be observed from a statistical analysis of the various collected corrosion parameters. Despite the enhanced corrosion protection property of E/GO, statistical analysis illustrates the poor repeatability of the collected corrosion parameters for E/GO samples, and this can be observed as high magnitudes of $E_{corr,STD}$ and $I_{corr,STD}$. However, statistical analysis indicates that the improved corrosion protection property of E/FGO was united with notable reproducibility of the corrosion parameters as indicated by the small magnitudes of $E_{corr,STD}$ and $I_{corr,STD}$. The improvement in the corrosion resistance property of E/FGO as well as the noble repeatability of the collected corrosion parameters can be attributed to a high degree of dispersion of FGO sheets in the E/FGO composite coatings. Here, too, the barrier property of FGO sheets might shield corrosive agents from penetrating the protective composite coating and reaching the coating/metal substrate interface, where corrosion may take place.

14.3.6 Thermal Stability and UV Degradation

Thermal stability of a protective composite coating is a vital property that needs to be assessed in order to examine the durability of the protective coating against thermal influences. The impact of the incorporation of GO and FGO sheets in the thermal stability of the epoxy resin is examined using TGA and DSC techniques. In particular, TGA was utilized to evaluate the influences of the incorporation of the fillers in the thermal degradation property of epoxy, while DSC was utilized to investigate the impacts of the GO and FGO sheets in the glass transition temperature (T_g) of the composite. The thermal degradation behaviors of the prepared composite coatings were examined in the temperature range 25–800°C using a heating rate of 10°C/min, and the results are depicted in Figure 14.12. TGA results illustrate that the incorporation of GO sheets in the epoxy resin slightly increased the onset

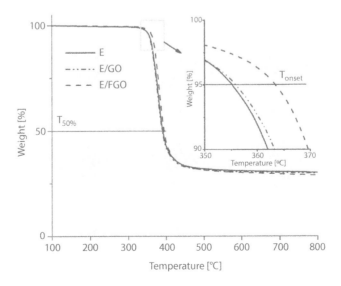

Figure 14.12 TGA thermograms of epoxy, E/GO, and E/FGO composite coatings.

temperature (T_{onset}), which can be described as the temperature at which the polymer composites degrade by 5 wt.% of the original weight. In addition, the results demonstrate that the functionalization of GO sheets contributes a further enhancement in the thermal stability of the E/FGO composite coatings. This can be observed as a further increase in T_{onset}, which increased from 355°C for epoxy to 355.9°C for E/GO and 363.5°C for E/FGO.

DSC analysis was conducted in the temperature range 25–200°C using a heating rate of 10°C/min and the results are depicted in Figure 14.13. The utilization of GO sheets as a filler in the E/GO composite coating marginally increased the T_g of the epoxy resin from 79.5°C to 81.6°C, whereas the incorporation of FGO as a filler delivers a significant rise in T_g to 86.4°C. The thermal analysis results obtained for epoxy, E/GO, and E/FGO are summarized in Table 14.4.

Resistance of UV degradation is another property of the prepared composite coatings that was assessed according to the ASTM-D4587 standard using an accelerated weathering tester. The UV degradation test was conducted by continuously exposing the prepared composite coatings to an alternating UV cycle at 60 ± 2°C for 8 h and a condensation cycle at 50 ± 2°C for 4 h over 30 days. The surface morphology of the epoxy, E/GO, and E/FGO composites was examined by SEM at the end of the exposure period after coating the samples with gold using the sputtering technique for 120 s as depicted in Figure 14.14.

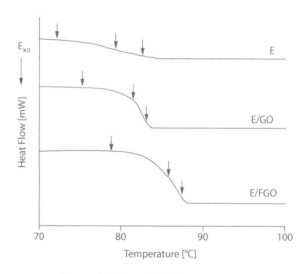

Figure 14.13 DSC thermograms of epoxy, E/GO, and E/FGO composite coatings.

Table 14.4 Thermal analysis results for epoxy, E/GO, and E/FGO composites.

Sample	Initial weight mg	T_{onset} °C	$T_{50\%}$ °C	Residue %	T_g °C
Epoxy	34.9	355	389.5	28.8	79.5
E/GO	34.8	355.9	391.3	28.7	81.6
E/FGO	34.9	363.5	394.3	28.9	86.4

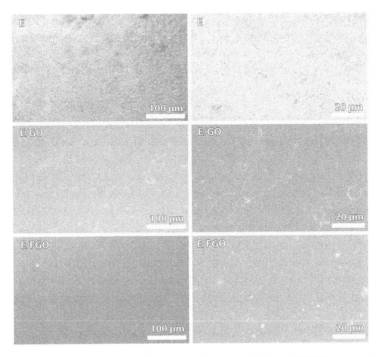

Figure 14.14 SEM images of post-UV degradation tests of epoxy, E/GO, and E/FGO composite coatings.

The post-UV degradation test results demonstrate severe damages on the surface of epoxy coating, which were manifested in widely spread pits and cracks, whereas minor damages were observed on the surface of E/GO, where some cracks were observed after the UV exposure period. Despite the fact that the incorporation of GO in the E/GO composite protective coating enhanced the UV degradation resistance property of the epoxy resin, the cracks observed on the surface of E/GO might grow deeper to reach the surface of the underneath CRS substrate. Such damage in the surface of the protective coating might cause a complete failure of the protective coating and expose the surface-coated metal substrate at specific areas. The migration of corrosive agents through these cracks on the surface of E/GO coating might cause a dangerous form of corrosion that can be difficult to detect or evaluate, which is pitting corrosion. On the other hand, the post-UV degradation images of the E/FGO composite coating show no sign of damages, which indicates a significant enhancement in the UV degradation resistance property of epoxy. The UV degradation test confirms that the incorporation of FGO as a filler in epoxy resin may extend the life span of the underneath metal substrate in outdoor application, where the coated metal substrate can be exposed to UV light. The observed advanced UV degradation resistance of E/FGO over E/GO composite coatings can be attributed to the superior degree of dispersion of FGO in the epoxy matrix compared to the agglomerated GO sheets in the E/GO as presented in TEM analysis.

The positive influences of the incorporation of GO and FGO in the glass transition temperature of epoxy resin can be attributed to the role of the fillers in restricting the mobility of the polymer chains in the amorphous phase of the resin [20]. Moreover, the enhancement in thermal and UV degradation of epoxy after the incorporation of the filler can be attributed to the interaction between the functional group on the surface of GO and FGO such as hydroxyl and epoxide groups and particularly the amino group on FGO with the epoxy resin.

14.3.7 Impact Resistance

The resistance of protective composite coating to deformation due to sudden impact is another initial property that needs to be assessed in addition to corrosion resistance, thermal stability, and UV degradation. In particular, the examination of the resistance of a coating to sudden impact is required to evaluate the durability of the coating in applications, where the coating might be exposed to impacts. The impact resistance tests were conducted following the procedures described in the ASTM D2795 standard. The influences of GO sheets and graphene materials on the mechanical properties of epoxy composites were reported in previous studies [21–24]. However, the main objective of this study is to evaluate the influences of the incorporation of GO and FGO on the resistance to deformation due to the sudden impact of E/GO and E/FGO composite coatings. The elevations at which the prepared protective composite coatings fail are reported in Figure 14.15.

The results depicted in the figure clarify that the incorporation of GO in the epoxy resin might enhance the impact of the polymer resin to sudden impact. Moreover, this enhancement in the impact resistance of epoxy can be furthered by surface modification of GO. Here, too, the positive impacts of the incorporation of GO and FGO in the impact resistances of E/GO and E/FGO composite coatings can be attributed to the interaction between the functional group on the surfaces of GO and FGO with the epoxy resin, which might increase the toughness of the epoxy composite.

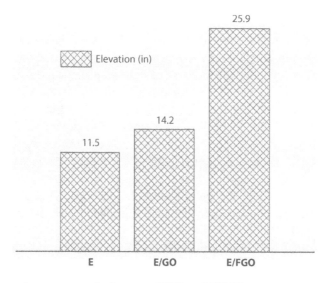

Figure 14.15 Impact resistance test results for epoxy, E/GO, and E/FGO composite coatings.

14.4 Conclusion

Functional graphene oxide sheets were prepared by attaching an amino functional group to GO sheets using a simple one-step synthesis procedure. The FGO sheets were incorporated in epoxy to the prepared E/FGO composite coating using an *in situ* pre-polymerization technique before exploring the composite coating as a protective coating on the CRS metal substrate. The study concluded that the corrosion protection properties of epoxy as well as E/GO composites can be excelled by chemical modification of GO sheets. Furthermore, the finding confirms the possibility of enhancing significant properties such as the thermal stability, UV degradation, and impact resistances of epoxy and E/GO composites by the utilization of FGO sheets as a filler in the hosting polymer resin. The advances in the various protection properties of epoxy resin after the incorporation of FGO sheets can be attributed to the superior degree of dispersion of the FGO filler in the composite coating as well as the possible interactions between the amino and other functional groups on the surface of FGO with the hosting polymer resin.

References

1. Shen, G.X., Chen, Y.C., Lin, C.J., Corrosion protection of 316 L stainless steel by a TiO_2 nanoparticle coating prepared by sol-gel method. *Thin Solid Films*, 489, 130–136, 2005.
2. Cecchetto, L., Delabouglise, D., Petit, J.P., On the mechanism of the anodic protection of aluminium alloy AA5182 by emeraldine base coatings. Evidences of a galvanic coupling. *Electrochim. Acta*, 52, 3485–3492, 2007.
3. Moretti, G., Guidi, F., Grion, G., Tryptamine as a green iron corrosion inhibitor in 0.5 M deaerated sulphuric acid. *Corros. Sci.*, 46, 387–403, 2004.
4. Jiang, M.-Y., Wu, L.-K., Hu, J.-M., Zhang, J.-Q., Silane-incorporated epoxy coatings on aluminum alloy (AA2024). Part 1: Improved corrosion performance. *Corros. Sci.*, 92, 118–126, 2015.
5. Jiang, M.-Y., Wu, L.-K., Hu, J.-M., Zhang, J.-Q., Silane-incorporated epoxy coatings on aluminum alloy (AA2024). Part 2: Mechanistic investigations. *Corros. Sci.*, 92, 127–135, 2015.
6. Mert, B.D., Corrosion protection of aluminum by electrochemically synthesized composite organic coating. *Corros. Sci.*, 103, 88–94, 2016.
7. Hiromoto, S., Self-healing property of hydroxyapatite and octacalcium phosphate coatings on pure magnesium and magnesium alloy. *Corros. Sci.*, 100, 284–294, 2015.
8. Liu, L. and Xu, J., A study of the erosion–corrosion behavior of nano-Cr_2O_3 particles reinforced Ni-based composite alloying layer in aqueous slurry environment. *Vacuum*, 85, 687–700, 2011.
9. Dhoke, S.K., Khanna, A.S., Sinha, T.J.M., Effect of nano-ZnO particles on the corrosion behavior of alkyd-based waterborne coatings. *Prog. Org. Coatings*, 64, 371–382, 2009.
10. Li, J. *et al.*, *In-situ* AFM and EIS study of a solventborne alkyd coating with nanoclay for corrosion protection of carbon steel. *Prog. Org. Coatings*, 87, 179–188, 2015.
11. Chang, K.C. *et al.*, Advanced anticorrosive coatings prepared from electroactive polyimide/graphene nanocomposites with synergistic effects of redox catalytic capability and gas barrier properties. *Express Polym. Lett.*, 8, 243–255, 2014.
12. Alhumade, H., Abdala, A., Yu, A., Elkamel, A., Simon, L., Corrosion inhibition of copper in sodium chloride solution using polyetherimide/graphene composites. *Can. J. Chem. Eng.*, 94, 896–904, 2016.

13. Alhumade, H., Yu, A., Elkamel, A., Simon, L., Abdala, A., Enhanced protective properties and UV stability of epoxy/graphene nanocomposite coating on stainless steel. *Express Polym. Lett.*, 10, 1034, 2016.
14. Liu, S., Gu, L., Zhao, H., Chen, J., Yu, H., Corrosion resistance of graphene-reinforced waterborne epoxy coatings. *J. Mater. Sci. Technol.*, 32, 425–431, 2016.
15. Chang, C.H. et al., Novel anticorrosion coatings prepared from polyaniline/graphene composites. *Carbon*, 50, 5044–5051, 2012.
16. Xu, Z. and Buehler, M.J., Geometry controls conformation of graphene sheets: Membranes, ribbons, and scrolls. *ACS Nano*, 4, 3869–3876, 2010.
17. Yu, Z. et al., Preparation of graphene oxide modified by titanium dioxide to enhance the anti-corrosion performance of epoxy coatings. *Surf. Coatings Technol.*, 276, 471–478, 2015.
18. Yang, Z. et al., Liquid-phase exfoliated fluorographene as a two dimensional coating filler for enhanced corrosion protection performance. *Corros. Sci.*, 103, 312–318, 2016.
19. Zheng, L., Wang, R., Young, R.J., Deng, L., Yang, F., Hao, L., Jiao, W., Liu, W., Control of the functionality of graphene oxide for its application in epoxy nanocomposites. *Polymer*, 54, 6437–6446, 2013.
20. Liao, K.-H., Aoyama, S., Abdala, A.A., Macosko, C., Does graphene change T_g of nanocomposites? *Macromolecules*, 47, 8311–8319, 2014.
21. Wan, Y.-J. et al., Covalent polymer functionalization of graphene for improved dielectric properties and thermal stability of epoxy composites. *Compos. Sci. Technol.*, 122, 27–35, 2016.
22. Rafiee, M.A. et al., Enhanced mechanical properties of nanocomposites at low graphene content. *ACS Nano*, 3, 3884–3890, 2009.
23. Chandrasekaran, S. et al., Fracture toughness and failure mechanism of graphene based epoxy composites. *Compos. Sci. Technol.*, 97, 90–99, 2014.
24. Bortz, D.R., Heras, E.G., Martin-Gullon, I., Impressive fatigue life and fracture toughness improvements in graphene oxide/epoxy composites. *Macromolecules*, 45, 238–245, 2011.

15
Supramolecular Graphene-Based Systems for Drug Delivery

Sandra M.A. Cruz[1]*, Paula A.A.P. Marques[1] and Artur J.M. Valente[2]

[1]*TEMA, Department of Mechanical Engineering, University of Aveiro, Aveiro, Portugal*
[2]*CQC, Department of Chemistry, University of Coimbra, Coimbra, Portugal*

Abstract

Cancer is now considered as a public health problem worldwide with increasing incidence and mortality. Along with innovations on the early detection, numerous efforts have also been made to improve the therapy. Cancer is a generic term for a large group of diseases that can affect any part of the body. This disease is heterogeneous and complex, hindering the clinical outcomes of new therapies. Among the typical treatments, chemotherapy is an effective drug treatment designed to kill cancer cells in individuals with various forms of carcinoma. However, its clinical benefits are limited such as several side effects, development of drug resistance, nonspecific, nonmolecular treatment, since this treatment uses chemical agents to destroy all dividing cells. Hoping to overcome, at least, some of these drawbacks, several drug delivery systems (DDS) have been developed in the past decades aiming at cancer-targeted delivery and controlled and sustained release of therapeutic agents inside the lesion.

The interface between nanotechnology and nanomedicine has contributed to the development of numerous drug career nanoplatforms in the last decades. Among them, graphene oxide (GO) and its derivatives attracted much attention due to their surprising properties: excellent biocompatibility, physiological stability, high specific surface area enriched with oxygen functionalities, cost–benefit, and scalable production make it an excellent candidate for pharmaceutical applications. The combination of GO and cyclodextrins (CD) has emerged as a new nanoplatform to DDS, with special relevance for cancer treatment. These two components acting as one system enhance drug-loading capacity and respond to different pH out/in cancer cells. Moreover, the hemocompatibility problems that can arise from the nonspecific interactions between GO sheets and blood components, promoting several types of precipitates, are minimized by the surface functionalization of GO with hydrophilic materials, which is the case of CD molecules. CD are cyclic oligosaccharides with a hydrophilic outer surface and a hydrophobic cavity and thus is a good candidate to perform the surface functionalization of GO. The hydrophobic character of CD's cavity can induce host–guest supramolecular interactions with hydrophobic molecules, e.g., drugs, improving the properties of the guest molecule, such as solubility enhancement and stability improvement. Additionally, CD acts as an efficient drug carrier, providing a controlled and sustained release, avoiding undesirable toxic effects.

In this chapter, the last advances in the GO-CD nanocomposites as anticancer DDS will be presented, with particular focus on their synthesis, biocompatibility, and drug release profiles.

Keywords: Graphene oxide, cyclodextrin, drug delivery, cancer, chemotherapy

Corresponding author: sandracruz@ua.pt

15.1 Introduction

Cancer is one of the major public healthcare problems worldwide with increasing incidence and mortality [1]. Despite the significant progress attained in medical technology, the mortality by cancer is above what was expected; this results in an increasing demanding for further research on cancer treatment. Nowadays, surgery, chemotherapy, and radiotherapy are the most common treatments. In several cases, surgery is not able to completely remove the primary tumor, but alternative therapies imply severe toxic side effects to healthy cells due to their nonspecificity to cancer cells [2], besides drug resistance, harsh tumor microenvironment that hinders drug penetration, and dose-limiting toxicity. These are the common issues related to the inefficiency of monotherapy [3]. The use of the combination of two or more classical drugs has started. However, this combination is only a mixture leading to treatment uncertainty [1].

In the past few years, the nanotechnology applied to medicine, named nanomedicine, developed various types of nanoparticles (NPs): polymeric micelles, liposomes, dendrimers, carbonaceous-based materials, etc. [2, 4, 5]. Therapy of cancer through diagnosis, imaging, and theranostics is one of the purposes of such developments. Numerous efforts have been made to create stimuli-responsive DDS with not only excellent *in vivo* pharmacokinetic profiles and tumor reversion ability but also enhanced cell uptake and/or highly selective drug controlled in the affected region, triggered by a certain stimulus [6, 7]. Moreover, DDS must be capable of anchoring drugs with different hydrophilicity, since the majority of anticancer drugs have low solubility in aqueous media [2, 8, 9]. In that way, it is possible to aspire for a decrease of drug side effects and an increase in treatment efficiency.

Graphene-based nanomaterials, due to their high surface area, biocompatibility, and versatile chemistry are good promising carriers for drug delivery [10]. The oxidative derivative of graphene, graphene oxide (GO), has received the attention of the scientific community due to its significant oxygen content, which allows the growth of chemical structures at the surface. Moreover, GO retains much of the properties of the highly valued nanomaterial pristine graphene, being easier to prepare and process and cheaper [11].

GO's hydrophilicity combined with the amphiphilic properties of macrocycles, such as cyclodextrins (CDs), has been exploited to carry both hydrophobic and hydrophilic drugs, since the majority of anticancer drugs are hydrophobic. However, as mentioned before, the combination of two drugs is expected to be more effective.

In this book chapter, we intend to contribute to the explanation on how nanotechnology can be useful in cancer treatment by using the DDS based on GO and CD composite-based materials. First, the performance as DDS of individual entities (GO and CD) will be described. Afterward, the potentialities and the latest developments of GO-CD as DDS will be highlighted.

15.2 Graphene Oxide and Cyclodextrin: Entities Applied in Drug Delivery

15.2.1 Graphene Oxide

Graphene is a single layer of sp^2-hybridized carbons with a large specific surface area and remarkable electrical, mechanical, and optical properties [10–12]. The hydrophilic derivative of graphene, GO, has a lower content of C (40%–60%) in favor of the presence of oxygen groups (hydroxyl, epoxy, and carboxyl) distributed in the graphene carbon network originating sp^3 domains and making it dispersible in aqueous media, and consequently in the physiological environment [4]. The large GO's surface area and oxygen functionalities allow easy functionalization, high drug-loading efficiency, and good dispersion [13].

The nature of the organic groups and sp^3/sp^2 ratio is strongly dependent on the type of protocols adopted for GO preparation and the source of graphite used. Chemical exfoliation is the most used technique to produce GO. Such method was started to be exploited in 1859, when Brodie made the first attempt to produce "graphite oxide" using chemical oxidation [14]. The addition of potassium chlorate to a mixture of graphite and nitric acid produced a material with mainly hydrogen, oxygen, and carbon. By further oxidation, 40 years later, Staudenmaier (1989) achieved a 2:1 C:O ratio in graphite oxidation [15]. However, in the late 1950s, Hummers [16] used powerful oxidizing agents (potassium permanganate) and strong acids (sulfuric and nitric acids) to separate the graphene layers from a source of graphite and achieve similar levels of oxidation as those obtained by Staudenmaier. The so-called "Hummers method", with many variations, is the current one to produce large-scale GO. Along with the chemical process, mechanical stirring or sonication is used to break the bonds between the carbon layers. The latter allows for a faster and more effective cleavage, as the cavitation of bubbles generated by ultrasonic fields produces shockwaves that break apart the graphite flakes. The exfoliation (Figure 15.1) is accomplished due to the strength of interactions between water and the oxygen-containing (epoxide and hydroxyl) functionalities introduced into the basal plane during oxidation. The hydrophilicity leads water to readily intercalate between the sheets and disperse them as individuals [11].

Quantity and/or type of oxygen groups at the GO surface can be controlled by its reduction, resulting in the well-known reduced GO (rGO). GO can be treated by chemical reduction (such as hydrazine, hydroquinone, sodium borohydride, and ascorbic

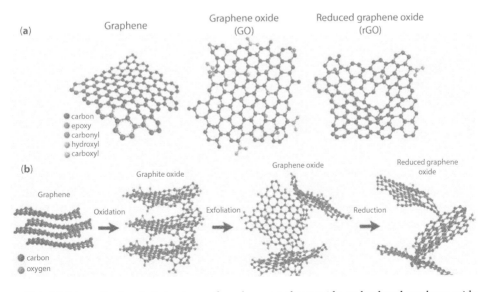

Figure 15.1 (a) Schematic chemical structures of graphene, graphene oxide, and reduced graphene oxide. (b) Route of graphite to reduce graphene oxide. Reprinted from Ref. [26] InTechOpen. Open access.

acid) or a thermal or UV process [2, 17, 18]. Indeed, oxygen functionalities and the lateral size of these graphene derivatives can modulate the kinetics and capacity of molecular adsorption [19, 20]. The heterogeneous electron transfer kinetics are directly related with the graphene derivatives' oxygen content and density of defect sites, which plays a key role in its functionalization efficiency [21].

Due to its properties, graphene derivatives have been widely used as nanocarriers for the delivery of drugs [2, 4, 6, 13, 19, 22, 23] and genes [4, 19, 24, 25].

Doxorubicin (DOX) is an intrinsically fluorescent anticancer drug and is, probably, the most widely used and studied drug using graphene-based delivery systems. DOX and other hydrophobic molecules can be loaded simply by noncovalent π–π stacking and hydrophobic interactions owing to its aromatic structure [27]. It has been reported that such drug-loading mechanism onto GO surface is more efficient in the case of DOX than other nanocarriers [28]. Due to the presence of carboxylic, epoxide, and hydroxide groups on its surfaces, various drugs can also be loaded at GO surface via covalent conjugation, hydrogen bonding, and electrostatic interaction [29–31]. For example, hydrogen bond can be formed between each -OH group of GO and DOX or the same group of GO and the -NH$_2$ group of the drug, at neutral pH [32].

The anthracycline antitumor drugs, where DOX is included, have aromatic rings and amino groups, which provides its high loading onto GO surface [27, 32–36]. The delivery of DOX carried by GO showed higher efficacy than free DOX against breast cancer MCF-7 cells [37]. Physically adsorbed DOX on GO surfaces has a pH-responsive release (particularly under acidic conditions).

Although graphene-based systems can be used as DDS, its biocompatibility remains a concern for the scientific community. The functionalization of graphene and its derivatives with stabilizers prevents graphene aggregation under physiological conditions, improving biocompatibility. Some stabilizers used are synthetic polymers, surfactants, natural polysaccharides, and proteins. Some studies have already shown that stabilized graphene can be a DDS owing to not only relatively low toxicity but also stability in the circulation and ability to load anticancer drugs [23].

Dai and co-workers [38] demonstrated that aromatic and hydrophobic drugs such as camptothecin and its analogs are able to bind to the graphene surface through noncovalent van der Waals interactions. The solubility of the camptothecin analog SN38 is two to three orders of magnitude greater than the prodrug SN38 free, after modified GO with poly(ethylene glycol) (PEG). Several studies have reported that PEG grafting can reduce the cytotoxicity of GO, resulting in increased biocompatibility and physiological stability [27, 39]. However, the dependence on the commonly used PEG nature (number of branches) also influences the viability and cell uptake speed [40]. Other examples of graphene surface polymeric modification was the study performed by Fan et al. [41] where a GO–sodium alginate (SA) conjugate was used as a carrier for DOX. In this case, the drug was loaded via π–π stacking and hydrogen bond interactions. A combination of several polymers was also studied to attempt to increase the efficacy of graphene derivative drug loading: PEG with sodium alginate [42] or low-molecular-weight polyethylenimine (PEI) [43, 44]; the latter is considered one of the most efficient nonviral gene delivery vector.

Several types of nano-formulations, including liposomes, have been developed in an effort to minimize the side effects of the anthracycline antitumor drugs [45, 46]. The surfactants are also used in GO modification to attain high drug-loading efficiency. Hydroxyethyl cellulose-neutral and hydroxyethyl cellulose-anionic surfactants, via noncovalent attachment, were shown to improve the stability and dispersion of a GO-DOX composite under physiological conditions [47].

Targeted delivery is another approach to minimize the side effects of anticancer drugs. DOX delivery has also been achieved, for example, by decorating rGO with antibodies or receptors. The specific cancer cell ligand lactobionic acid was used to functionalize GO, and this complex can specifically target cancer cells overexpressing asialoglycoprotein (ASGPR) receptors and inhibit cancer cells [44, 48]. An excellent DOX release efficiency was also attained with folic acid (FA) combined with rGO. FA-rGO showed specific targeting to MDA-MB 231 cancer cells (expressing the FA receptor) [49]. Miao *et al.* synthesized rGO coated by cholesteryl hyaluronic acid (CHA) where DOX loading capacity was fourfold greater than that of rGO [50]. Moreover, the colloidal stability of CHA-rGO and safety *in vivo* were higher compared to rGO. The results showed that drug delivered by CHA-rGO was significantly increased compared with free DOX and rGO-DOX. Tumor weights were reduced up to 14.1 (±0.1)% in mice treated with CHA-rGO/DOX when compared with untreated ones (Figure 15.2).

Miao W. *et al.* studied rGO modified with poly-L-lysine (PLL) and conjugated with anti-HER2 antibody to promote the targeting delivery of DOX to cancer cells nucleus [51]. Cellular uptake results showed that the internalization of these nanocarriers into MCF7/HER2 cells was much higher than the carriers without anti-HER2 antibody. The excellent cell uptake is due to the combination of specific antibody and conjugation of the cell-penetrating peptide, PLL, improving antitumor efficiency.

Drug nanocarriers can be modulated to be external stimuli responsive and only respond in the presence of such conditioning, improving the efficiency of delivery. For example, a pH-responsive supramolecular polymeric shell around a mesoporous silica-coated magnetic GO (Fe_3O_4@GO@$mSiO_2$) was used, in a controlled manner, to deliver DOX into cancerous tissue. The GO composite gained magnetic field sensitivity by the presence of Fe_3O_4 and is able to deliver DOX into target sites when external magnetic fields are applied [52]. Other external stimuli can be used to enhance the therapeutic functions of systems when applied from an external source: light (photothermal therapy, PTT) and temperature [53, 54]. In the case of PTT, GO acts as a therapeutic agent since it responds to near-infrared (NIR) irradiation. GO strongly absorbs light in the NIR range (700–900 nm), commonly called "therapeutic window," and is

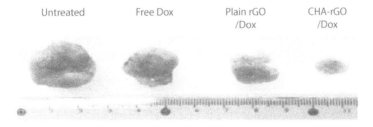

Figure 15.2 Anticancer effects of CHA-rGO nanophysisorplexes. KB tumor-bearing mice were intravenously treated with Dox alone or in complex with plain rGO or CHA-rGO (Dox, 2 mg/kg) every 3 days beginning on day 7. On day 24, tumor tissues were excised for visualization. Adapted with permission from Miao *et al.* [50]. Copyright (2018) Elsevier.

noninvasive, is harmless, and has skin-penetrating irradiation [55]. There are other studies where graphene/GO is conjugated with inorganic particles to load anticancer drugs and release it upon NIR stimulus [56–62].

Other types of functionalization can attribute their endogenous responsive character to the drug carrier [6]. Such response could be from the lesion environment, which is the target of drug delivery. Changes in pH are the most exploited since it is well-known that the tumor environment is more acidic than the whole cellular tissue. Under low pH values, hydrophobic drugs such as DOX can be protonated and the π–π stacking and hydrophobic interactions between drug molecules and the graphene surface become weaker, releasing the drug [63–65].

Levels of glutathione (GSH) are the main regulator of cellular redox environment [66]. Excess of GSH normally increases antioxidant capacity and oxidative stress, while deficiency leads to an increased susceptibility of oxidative stress. Therefore, GSH levels can be used as stimulus to trigger drug release from nanocarriers. An example is a study where DOX was stacking via π–π interaction to the nanoconjugates formed by nano-GO (NGO) modified via disulfide linkage with methoxy polyethylene glycol (mPEG), yielding NGO-SS-mPEG [67]. The disulfide bond of NGO-SS-mPEG, in the presence of intracellular GSH, was broken to rapidly release DOX, enhancing the efficacy of chemotherapy (Figure 15.3).

Some biomolecules, such as peptides [68], proteins [69], or nucleic acids [70–74] (Figure 15.4), can be delivered by graphene and its derivatives. The ring structure of nucleobases allows π–π stacking interactions of the highly hydrophilic nucleic acids with GO. Such combination promotes GO as a gene delivery system.

He *et al.* [71] synthesized GO-based multicolor fluorescent DNA nanoprobe, which allows rapid, sensitive, and selective detection of DNA targets in solution analyzing the interactions between DNA molecules and GO. This sensor is able to differentiate the secondary structure of DNA (i.e., the single- or double-stranded DNA), and when it is complemented with the use of functional nucleic acid structures, for example, aptamers, it can detect other analytes. Such sensor constitutes an advance to targeted gene delivery.

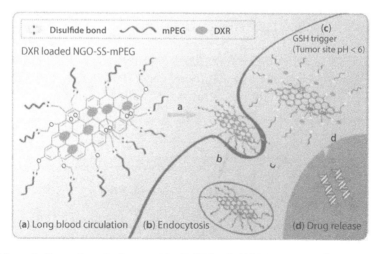

Figure 15.3 Schematic illustration of redox-sensitive DOX-loaded NGO-SS-mPEG for cancer cell chemotherapy. (a) PEG-shielded NGO with disulfide linkage for prolonged blood circulation. (b) Endocytosis of NGO-SS-mPEG in tumor cells via the EPR effect. (c) GSH trigger to induce PEG detachment. (d) Rapid drug release to kill cancer cells. Adapted with permission from Yang *et al.* [6]. Copyright (2018) Elsevier.

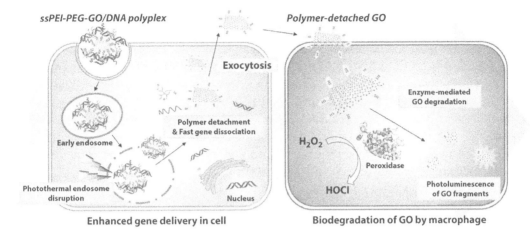

Figure15.4 Schematic illustration of the overall progress of ssPEG-PEI-GO/DNA polyplex from cellular uptake, photothermally enhanced gene delivery, and fast gene release in cancer cell to enzyme-mediated biodegradation and its monitoring in macrophage. Reprinted with permission from Ref. [56]. Copyright (2018) American Chemical Society.

As was referred before, PEG is widely used as a surface modifier to improve biocompatibility and physiological stability of nanomaterials for use in biological and medical applications. PEG-GO and PEG-rGO were loaded with ssRNA in a comparative study of loading and release [75]. Computational simulations (Figure 15.5) reveal that π–π stacking interactions between RGO and ssRNA are much stronger

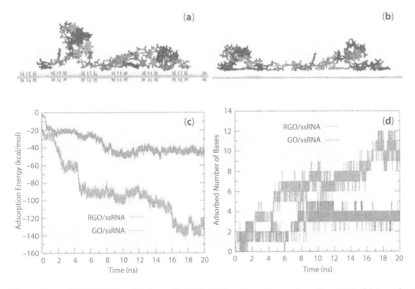

Figure 15.5 Snapshots of ssRNA adsorbed on the GO (a) in side view and on the RGO (b) in side view after the 20-ns MD simulations. The red color in Figure 15.4a and b represents oxygen-containing groups of the GO substrate; the blue, green, brown, and yellow colored parts stand for the bases of A, U, C, G, in ssRNA, respectively. (c) Attractive vdW interaction energy and (d) the number of the adsorbed bases in a nucleic acid sequence in the ssRNA–RGO and ssRNA–GO hybrid systems as functions of simulation time. Adapted from Zhang et al. [75]. Reproduced from Ref. [75] with permission from The Royal Society of Chemistry.

than those between GO and ssRNA, which is in agreement with the experimental results.

PEG was also used and combined with branched polyethylenimine (bPEI) via disulfide linkage to modify GO and design a delivery carrier (ssPEG-PEI-GO). The plasmid DNA (pDNA) efficiently interacts with the nanocarrier to form a stable complex through electrostatic interactions. After cellular uptake, ssPEG-PEI-GO/pDNA has the advantage of being able to easily escape from endosomes by photothermal conversion of GO upon NIR and subsequent photothermally induced endosome disruption. Into the cell, the intracellular environment enables polymer dissociation and, consequently, rapid gene delivery showing efficient gene transfection with low toxicity [74].

Multiple drug resistance (MDR) of cancer cells, a well-known problem in chemotherapy, can be overcome with short interfering RNA (siRNA), which induces specific silencing of targeted protein [76]. Polyethylenimine (PEI)-functionalized graphene oxide (PEI-GO) was reported as a carrier to sequential delivery of Bcl-2-targeted siRNA and DOX. It was demonstrated that such nanocarrier was effective in this simultaneous delivery, enhancing significantly therapeutic effects [77]. Another study using GO functionalization with low-molecular-weight BPEI was conducted by Kim *et al.* [73]. Although bPEI-GO has high gene delivery efficiency and cell viability, the photoluminescent properties (see Figure 15.6) of GO were increased, in a synergetic way, through

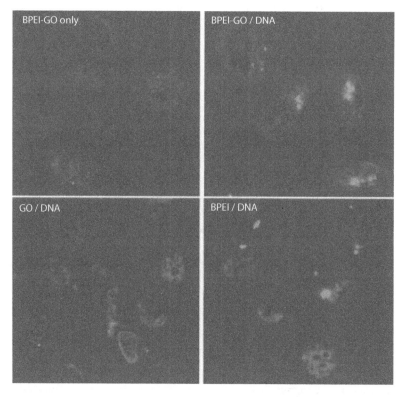

Figure 15.6 Confocal fluorescence microscopic images of a PC-3 cell line treated with BPEI–GO, DNA complexes with BPEI–GO, GO, and BPEI. In this study, pDNA was labeled with TOTO-3, and nuclei were stained with DAPI; a 488-nm laser was then applied to observe photoluminescence. Reprinted with permission from Ref. [73]. Copyright (2018) American Chemical Society.

its conjugation with bPEI, which give the opportunity of bPEI-GO to act, simultaneously, as fluorescent probe. These properties can be useful in gene delivery and bioimaging as well.

The dual functionality of former gene delivery reported systems can be encountered also in drug delivery systems based on GO. Fe_3O_4-PEG-GO was prepared for both magnetic imaging and drug delivery. High loaded levels of DOX was attained; the hybrid showed good physiological stability and cell viability. Simultaneously, the nanocomposite improved magnetic resonance imaging (MRI) contrast compared to bare Fe_3O_4 NPs [78]. In a similar approach, GO grafted with dendrimers and functionalized with gadolinium diethylene triaminepentaacetate (Gd-DTPA), and an antibody targeted to prostate stem cell antigen (PSCA) was employed for targeting and MR imaging of cancer cells overexpressing PSCA [79].

Live imaging and drug release are probably the dual function most explored in DDS because it is easier to observe the internalization of released drug. GO-based nanocarrier conjugated with folate can target delivery of anticancer drugs and self-monitor both *in vitro* and *in vivo* as labeled fluorescein protein. The cancer cells' apoptosis can be visualized by confocal fluorescence imaging [22]. A similar system was fluorescent manganese-doped zinc sulfide (ZnS/Mn) nanocrystals covalently attached to GO-PEG, which is capable of drug delivery and cell labeling [80].

15.2.2 Cyclodextrin

Cyclodextrins (CDs) are a group of natural cyclic oligomers formed by glucopyranose unities. The synthesis of cyclodextrins was first reported by Villiers [81], in 1891, as a consequence of an enzymatic degradation of starch, but these molecules become attractive to scientific community only after two reference works: the first by Schardinger [82], in 1930, and the second by Szejtli [83], in 1975. In between, Freudenberg and Meyer-Delius have reported the first accurate chemical structure of CDs [84]. Cyclodextrins are formed by glucopyranose units connected through α(1–4) ether linkages. The most common natural cyclodextrins are the α-, β-, and γ-cyclodextrins having six, seven, and eight glucopyranose units, respectively (Figure 15.7), formed through α(1–4) ether linkages. Due to chair conformation of glucopyranose units, CD has a truncated cone or torus shape [85] with internal diameter cavities ranging from 5.7 to 9.5 Å, wide end, from α- to γ-CDs, respectively, and 7.9 Å height. The structure of CDs is characterized

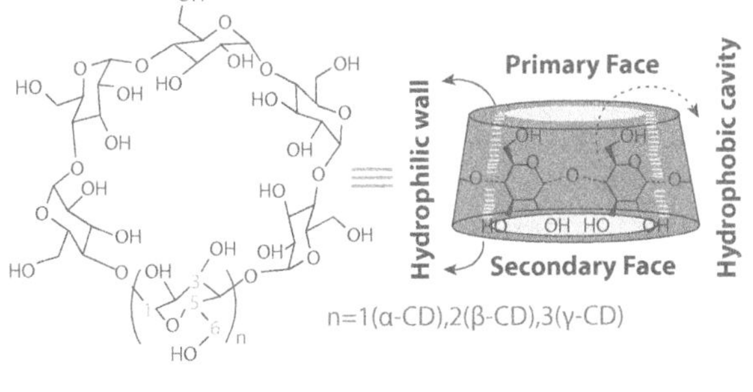

Figure 15.7 Spatial arrangement of cyclodextrin. Reproduced from Ref. [116] with permission from The Royal Society of Chemistry.

by the existence of hydroxyl groups oriented to outside the cavity: the primary hydroxyl groups and the secondary ones are localized at the narrow and wider edges of the truncated cone, respectively, resulting in a hydrophilic outer surface [86–88]. On the other hand, the C–H bonds are directed inward, leading to a hydrophobic cavity. The inner cavity has a high electron density due to nonbonding electron pairs of the glycosidic oxygen bridges, giving it some Lewis base character. The described spatial arrangement of cyclodextrin functional groups results in a molecule that combines hydrophobic, at the cavity, and hydrophilic, at the outer surface, characteristics. Such amphiphilic property of CD allows it to form host–guest supramolecular complexes. The hydrophobic molecules that can be entrapped by CD cavity include drugs [89–97], polymers [98, 99], inorganic salts [100–104], surfactants [87, 105–109], and dyes [110–115].

The interaction between CDs and guest molecules is, in general, driven by noncovalent interactions (such as, for example, van der Waals, hydrophobic, electrostatic, and charge transfer interactions), metal coordination, and hydrogen bonding [87]. The mechanism of guest threading into the hydrophobic cavity, in aqueous solutions, is accompanied by, simultaneously, dehydration of both cavity and guest molecules, making the process, in general, entropy-driven. This process depends on both CD cavity and guest sizes, geometry of guest molecules [117, 118] and interactions between water–water and water–CD inside the cavity [119, 120].

Host–guest complexes of CD may improve some properties of the guest molecule: solubility enhancement [121–124], stability improvement [124–126], control of volatility and sublimation, and physical separation of incompatible compounds [127–129]. Such properties, accompanied by its nontoxicity in humans, makes CD molecules unique to be applied in several different industries: chemical synthesis and catalysis [130–134], analytical chemistry [135–137], corrosion coatings [138–140], wastewater and soil treatment [141–145], pharmaceutical and biomedicals [122, 146–150], cosmetics [151], food technology [152, 153], and textile [154–156].

As was mentioned before, one major drawback of some drugs is its poor solubility in aqueous media, due to their hydrophobic character, generally characterized by their high octanol–water partition coefficients. To overcome this problem, CD has been used to entrap the drugs, and the results obtained show that these cyclic oligomers act as efficient drug carriers with a controlled and sustained release, avoiding undesirable toxic effects [88, 157]. However, this can be limited by the CD's own solubilities in water. The solubility in water of the α-, β-, and γ-CD is 13%, 2%, and 26% [weight by weight (w/w)], respectively [158]. From these values, it becomes clear that there is no trend between the solubility in water and the number of glucopyranose unities. Furthermore, the most commonly used cyclodextrin is β-CD, with its easy synthesis, and it is also the least soluble. The huge number of applications of β-CD can be justified by several factors: easy synthesis and low price, and the size of its internal cavity matches those of a large number of guest molecules (such as those containing aromatic groups or alkyl chains). Consequently, it is of great importance, for many applications, to deal with cyclodextrins with the same cavity volume as β-CD but with higher solubility in water, which can be obtained by functionalization. For example, hydroxypropyl-β-CD (HPβCD) has an aqueous solubility of approximately 60% (w/w) [88, 159]. Others examples of CD chemically modified are sulfobutyl ether-β-cyclodextrin and randomly methylated-β-cyclodextrin, whose solubilities are greater than 500–600 mg/mL. Also, the polymerized CDs epichlorohydrin-β-cyclodextrin and carboxy methyl epichlorohydrin β-cyclodextrin show improved aqueous solubility (more than 500 and 250 mg/mL, respectively) [160].

Along with solubility enhancement, the optimum drug loading and more effective release are the major challenges that nanotechnology researchers must face. Self-assembly of CD able part of such challenges since these structures can carry drug efficiently. Many CD-based polymeric nanosystems have been synthesized to this purpose [161] and can be classified as micelles, uni/multilamellar vesicles, nanospheres, nanocapsules, nanogels, nanoreservoirs, or CDplexes (Figure 15.8). Some examples of the application of such type of nanoassembly systems for the encapsulation and delivery of anticancer drug load/release are presented in Table 15.1.

Normally, CD micelles are only dependent on the its concentration [162]. These nano-sized colloidal particles (ranging from about 5 to 2000 nm) have a lipophilic central part and a hydrophilic outer part with a single hydrophobic core where the drugs are entrapped. Therefore, the harmful side effects and drug degradation are minimized, showing a better therapeutic profile to micelles [163].

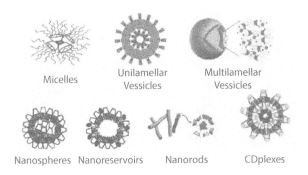

Figure 15.8 Schematic representation of some cyclodextrin nanoassemblies and supramolecular architectures. Reproduced from Ref. [161] with permission from Elsevier.

Table 15.1 Cyclodextrin-based carriers of anticancer drugs. Adapted from Gidwani *et al.* Copyright © 2015 Bina Gidwani and Amber Vyas [160]. Open access.

Drug	Cyclodextrin	Nanocarrier prepared	Outcome	Reference
Doxorubicin	γ-CD	Liposomes	Increased retention in tumor cells	[173]
Curcumin	HP-γ-CD	Liposomes	Improvement in therapeutic efficacy	[174]
Camptothecin	β-CD	Nanosponges	Improvement in therapeutic efficacy and reduction in toxic effects	[179]
Paclitaxel	β-CD	Nanosponges	Prolonged shelf life	[181]
Doxorubicin	β-cyclodextrin-based star copolymers	Micelles	Enhanced drug release	[182]
β-Lapachone	α-CD	Polymeric millirods	Sustained drug release	[183]

Amphiphilic CD complexes, with hydrophobic and hydrophilic tails, can be prepared using additional polymers or surfactants [164–170]. Both hydrophilic and hydrophobic drugs can be incorporated by lamellar vesicles due to their amphiphilic character: the hydrophilic drug stays in the aqueous phase and hydrophobic one remains in the lipid bilayer, retaining drugs en route to their destination [160, 171, 172]. The performance of CD-based liposomes was studied by Arima et al. [173], by using γ-CD/DOX supramolecular complexes encapsulated in PEGylated liposomes. The results showed retardation in tumor growth, increase in drug retention, and higher survival rate, after intravenous injection in BALB/c mice bearing colon-26 tumor cells. The complex 2-HP-γ-CD/curcumin was a potential assembly to treat breast cancer after the results obtained *in vitro* and *in vivo* [174].

CD-based nanosponges are hyper-branched and porous structures that use active carbonyl compounds as cross-linkers [175–178]. The anticancer drug camptothecin was efficiently loaded into a CD nanosponge and showed long release profiles besides good stability in phosphate buffer (pH 7.4) and plasma [179]. In other study using Paclitaxel (PTX) loaded into a CD nanosponge, the oral bioavailability of such drug increased 2.5-fold compared with marketed Taxol® [180].

The modification with polymers allows the synthesis of a broad range of supramolecular systems, including gels. Nanogels have the same properties as the gels but at nanoscale: they can host and protect drug molecules, and their release can be regulated by the incorporation of stimuli-responsive conformations or biodegradable bonds into the polymer network. Moya-Ortega et al. [148] have reviewed the CD-based nanogels applied to biomedical and pharmaceutical issues.

Some research groups have faced a new challenge: to explore graphene-based nanocomposites as responsive DDS. In particular, they focused much of their work on the development of pH-responsive systems since the variation of pH-tumor tissues is well known and can be easily followed. Damage reduction of normal tissues and bioavailability improvement of the drugs are the advantages of using such type of DDS [116].

DOX was successfully loaded into pH-responsive micelles using the host–guest interaction between benzimidazole-terminated poly(ethylene glycol) (PEG-BM) and cyclodextrin-modified poly(L-lactide) (CD-PLLA). The release rate showed that the model drug was rapidly released from the supramolecular micelles as the acidic environment was reduced from 7.4 to 5.5 [184] (Figure 15.9). Moreover, these supramolecular micelles showed higher tumor inhibition efficacy and reduced systemic toxicity compared to free DOX after intravenous injection into nude mice.

Other CD-based pH-responsive supramolecular matrices used as vectors for antineoplastic drugs to tumor tissues were prepared by Cai et al. 3-(3,4-dihydroxyphenyl)propionic acid (DHPA)-functionalized CD was conjugated onto the surfaces of hollow mesoporous silica nanoparticles (HMSNs) and subsequently PEG-grafted adamantane (ADA) was linked to HMSNs-β-CD driven by the host–guest interaction of Ada and CD. DOX was efficiently loaded into HMSNs-β-CD/Ada-PEG and released within the tumor environment responding to pH stimuli, thus inducing cell apoptosis and inhibiting tumor growth, while toxic side effects are minimal [185].

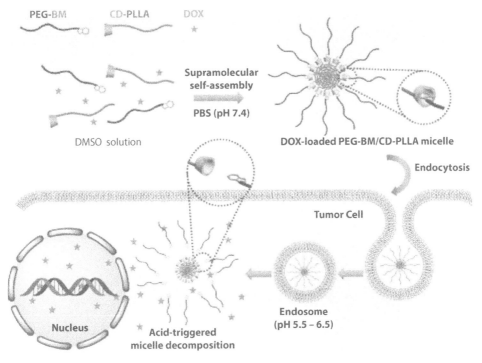

Figure 15.9 Schematic illustration of the formation and triggered drug release process for supramolecular micelles in response to the intracellular microenvironment. Reprinted (adapted) with permission from Ref. [184]. Copyright (2018) American Chemical Society.

Contrary to pH stimulus, the concentrations of redox substances in tumors are extremely low and change very suddenly. Although some attempts are being made for using CD [186, 187], controlled specific redox molecular mechanisms are very difficult to achieve using the abovementioned reasons.

CDs can be applied to gene delivery due to the binding affinity between them and nucleic acids [189, 190]. For example, a gene carrier composed of adamantane–functionalized folic acid (FA-ADA), an adamantane–functionalized poly(ethylene glycol) derivative (PEG-ADA), and cyclodextrin-grafted low-molecular-weight branched polyethylenimine (PEI-CD) was successfully developed. The gene carrier was based on host–guest interactions and could be efficient and show low toxicity for pDNA delivery to target cells [191].

Drug and gene co-delivery is a promising synergetic cancer therapeutic once the former can eliminate the damaged cells and the latter is able to change nucleic acid code responsible to cells proliferation. Assembled vesicles of hyperbranched-linear supramolecular amphiphile were synthesized to simultaneously load DNA and DOX hydrochloride (DOX·HCl). Nanovesicles were formed by hyperbranched-linear supramolecular amphiphile self-assembled amine compounds attached to cCD-centered hyperbranched polyglycerol (CD-HPG-TAEA) and linear adamantane-terminated octadecane (C18-AD) (Figure 15.10). The vesicle cleaved under pH 5 and cumulative drug release was around 80%; simultaneously, gene delivery also occurs in the nuclei [188].

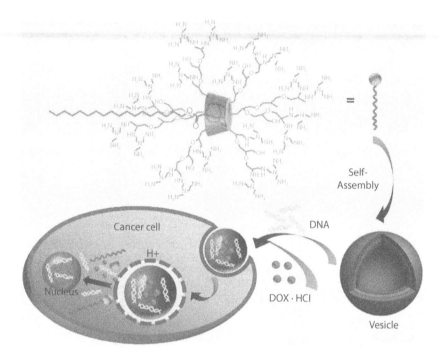

Figure 15.10 Illustration of nanovesicle self-engineering driven by host–guest interaction between AD-C18 and CDHPG-TAEA for gene delivery and drug encapsulation. Reprinted with permission from Ref. [188]. Copyright 2018 American Chemical Society.

15.3 GO-CD Nanocomposites as Drug Delivery Systems

The inadequate single-drug therapy can contribute to the chemo-resistance and tumor relapse [192]. Therefore, there is an urgent need for the development of methodologies that can overcome this barrier to drug formulation along with delivery problems, such as non-specific uptake and poor aqueous solubility [193].

As it was mentioned before, some DDS can respond to external stimulus to deliver the carried drug. Beyond this feature, GO can also interact with specific functional groups of drug molecules improving drug load capacity. CD has the great capacity to form inclusion complexes by host–guest interaction with some drugs with poor aqueous solubility, besides responding to external stimuli, particularly to pH changes. The conjugation of CD with GO would combine their advantages, and it is expected to maximize, in a synergetic way, their potentials in drug delivery [192].

15.3.1 Strategies of Preparation of GO-CD

The combination of CD and GO can be achieved by various routes using the attachment of these two compounds with drug(s) without any excipient through functionalization or forming a gel using both components. In the latter, the drug can be loaded first on the CD cavity and GO is added to promote cross-linking, or the drug is stacked onto the GO surface and CD is added to create the linkages that favor gel formation.

Pourjavadi et al. [194] prepared a nanostructure by noncovalently grafting β-cyclodextrin-graft-hyperbranched polyglycerol onto the edges of graphene ([RGO@(β-CD-g-HPG)n]) through host–guest interactions. Such functionalization of rGO confers good solubility in neutral aqueous solutions and was stable for long periods of time. DOX was previously loaded onto rGO surface to obtain [DOX-RGO@(β-CD-g-HPG)n].

The anticancer drug camptothecin (CPT) was used as a drug model to study the release and biocompatibility performance of the water-soluble nanocarrier based on β-CD functionalized GO. GO nanosheets were functionalized by grafting β-CD units via an amine–epoxy reaction, and the product was then modified with hyaluronated adamantane (HA-ADA) chains by noncovalent interaction of β-CD cavity with the adamantyl group [195]. CPT was attached via π–π stacking interaction between the planar GO surface and the aromatic ring of the drug molecule [193]. The inclusion of HA-ADA chains has the advantage of recognizing the HA receptor-expressing tumor cells in cancer metastasis, enhancing the oncological treatment using this aqueous soluble supramolecular carrier.

A similar DDS was obtained by GO supramolecular surface modification with folic acid (FA) through a synthetic bifunctional molecule that contains a planar porphyrin moiety as a binding group and an adamantane moiety that is encapsulated in the cavity of the CD. Also, here, DOX was loaded into a matrix driven by π–π interactions between the aromatic ring of the drug and porphyrin and GO. The presence of adamantane-grafted porphyrin and FA-modified β-cyclodextrin contributes to the integrity of final supramolecular entity based on graphene through folate-receptor-positive malignant cells and DOX release [196].

The efficiency of chemotherapeutics can be increased by delivering two chemical drugs. In the work developed by Wu *et al.* [192], DOX and topotecan (TPT) were the drugs used to test the new nanocarrier platform. DOX was bound to adamantane carboxylic acid (ADA-COOH) through covalent linkage to form ADA-DOX; GO was modified by ethylenediamino-β-CD (EDA-CD) to form GO-β-cyclodextrin (GO-CD). GO-CD was assembled with ADA-DOX via host–guest interaction while TPT was loaded onto GO through π–π stacking interaction.

Magnetic functionalization is another possible approach to prepare drug carriers with the purpose of enhancing DOX release into cancer cells. Two types of mesoporous silica-coated magnetic GO, Fe_3O_4@GO@$mSiO_2$, were synthesized [197] and modified by 3-aminopropyltriethoxysilane, methyl acrylate, and pentaethylene hexamine, to create dendrimer-like structures. α-CD was applied to the assembling structure, functioning to store the drug and preventing its release at the pH of healthy tissues.

A drug delivery material (rGO-C_6H_4-COOH) was synthesized using rGO linked covalently with *p*-aminobenzoic acid. PEI was grafted to enhance the higher water solubility of rGO-C_6H_4-COOH, and simultaneously, biotin was conjugated with PEI to enhance the targeting. β-CD was also introduced into the DDS, not only to reduce the cytotoxicity of PEI and rGO-C_6H_4-COOH but also to form a host–guest interaction water-insoluble drug (in this case, DOX). rGO-C_6H_4-COOH-NH-PEI-CD-biotin can act simultaneously as a delivery and as a theranostic material for cancer therapy, without cytotoxic effects on normal cells [198].

Other theranostics and drug carriers were prepared by Ko *et al.* [199]. Amine-functionalized GQDs (GQD-NH_2), nanocarrier labeled with herceptin (HER) and β-CD, were developed for breast cancer treatment. Here, GQDs provide diagnostic effects by emitting the color blue. Amine groups of GQD-NH_2 conjugate to the hydroxyl groups of β-CD through a facile 1,10-carbonyldiimidazole (CDI) coupling reaction. The active target HER was added to GQD-βCD via an amide bond between a carboxylic group of HER and one amine group of GQDs. The presence of HER provides an active targeting to

HER2-overexpressed breast cancer to enhance accumulation in the cancer cells. DOX was loaded into the cavity of β-CD by host–guest chemistry (Figure 15.11).

Nanospheres of GO and gelatin were successfully prepared, where PTX-HPβCD was incorporated. Here, PTX was included in the HPβCD cavity by a host–guest mechanism, whereas PTX-HPβCD was anchored to GO by ester bonds, hydrogen bonding, π-π stacking, and strong van der Waals interaction [200].

Another approach to obtaining nanospheres to carry PTX was shown by Tan *et al.* [201]. Carboxylated GO (GO-COOH) was modified using HPβCD to obtain a GO-COO-HPβCD nanohybrid (GN). PTX was loaded into GN by mixing the drug with ethanolic and GN aqueous solutions in alkaline medium. Glutaraldehyde (GA) acts as a cross-linker to form hydrogel nanospheres. At this stage, the aldehyde groups at both ends of GA reacted with the hydroxyl groups in GN, which, along with PTX, curled up together.

The oral route is the most comfortable for the patients, so ideally the drugs should be integrated on a system that can release the drug and be stable under different media encountered through the digestive tract. However, injectable hydrogels are also being exploited because drugs can be easily integrated into the hydrophilic polymer water solution and the polymer sol can become a gel *in situ* at the target sites [202]. Three-dimensional (3D) porous structures could be obtained by hydrogels that can be responsive to external stimuli to release the drug; furthermore, these 3D polymer networks can be easily synthesized and can be biocompatible and biodegradable. There are several approaches to producing GO/CD-based hydrogels where the main purpose was the incorporation of anticancer drugs.

A hydrogel based on GO was synthesized using poly(vinyl alcohol) (PVA), which can form physical cross-linked hydrogel composites. The 3D structure was obtained by mixing a suspension of GO and a PVA solution that was violently shaken until the gel is formed. PVA aqueous solution was prepared with the anticancer drug camptothecin (CPT) encapsulated on acetalated-β-cyclodextrin (Ac-β-CD) nanoparticles synthesized through a single

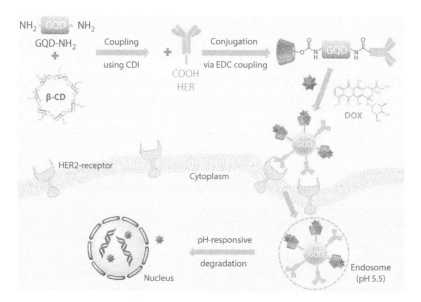

Figure 15.11 Preparation of DOX-loaded HER-labeled GQD-based nanocarriers and its drug release via cellular uptake for active targeting of breast cancer cells [199]. Published by The Royal Society of Chemistry.

oil-in-water (o/w) emulsion technique. By scanning electron microscopy (SEM), the typical porous cross-linked structure of hydrogel was observed and the wrinkle structure of GO was easy to find. The porous structure became more homogeneous with a smaller pore size when nanoparticle concentration is higher, which can be explained by the presence of more cross-linking sites [203].

DOX hydrochloride and CPT were incorporated in a hydrogel formed by GO and α-CD, mediated by Pluronic F-127. The drugs were dissolved and added to the Pluronic copolymer functionalized GO or rGO solutions, which were then mixed with the α-CD solution. The hydrogel was formed spontaneously after a certain period of time by supramolecular assembly of α-CD and PEO block of Pluronic F-127 due to the penetration of PEO chains into the cavities of cyclodextrin driven by hydrogen bonding and hydrophobic interactions [204].

A 3D regular porous structure was obtained by self-assembly of GO and pseudopolyrotaxane (PPR) formed by low-MW linear poly(ethylene glycol) (PEG) incorporating a series of α-cyclodextrin. Linear high-molecular-weight (MW) PEG or its copolymers and α-CD are the precursor of hydrogels for biomedical and pharmaceutical applications, which have no need for chemical cross-links [205, 206]. Low-MW PEG and α-CD only provide the cross-links and cannot provide a network because of the PEG segments [207–209]. Here, the presence of adjacent PPR contributes to the network formation. GO can also be used as a supramolecular building block for, e.g., hydrogels, providing improved properties such as high strength and biocompatibility [210–212]. GO surface was modified with the pyrene–poly(ethylene glycol) (Py–PEG) conjugate polymer by the strong π–π interactions between them. GO-Py-PEG layered aggregates are induced by parallel arrangement of GO sheets (Figure 15.12). The anticancer drug DOX and α-CD were added simultaneously in the GO–Py–PEG solution. When α-CD is introduced into the system, PEG chains thread into the cavities of a series of α-CD via the well-known host–guest inclusion interaction originating from the rigid necklace-like PPR supramolecular structure. The rigid structure of hydrogel was rapidly obtained by the strong hydrogen bond interactions occurring between the rigid PPR [213].

The advantages of having a DDS that carries a combination of anticancer drugs and at the same time is responsive to enhance targeted delivery are already mentioned. All these features were attained by the hydrogel produced by Ha *et al.* [214]. The proposed GO-based

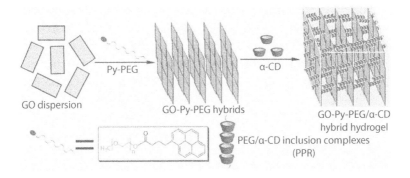

Figure 15.12 Schematic representation of the supramolecular hybrid hydrogel preparation strategy based on the self-assembly of Py–PEG-modified GO and a-CD. Reproduced from Ref. [213] with permission from The Royal Society of Chemistry.

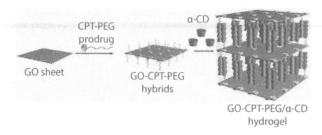

Figure 15.13 Schematic representation of the hybrid hydrogel preparation based on the host–guest interactions between prodrug-modified GO and a-CD. Reproduced from Ref. [214] with permission from The Royal Society of Chemistry.

hydrogel has an organized microstructure, is NIR light-responsive, and carries/releases hydrophilic and hydrophobic drugs. Due to the noncovalent hydrophobic interactions and π–π stacking, CPT can attach to the surface of GO by mixing a camptothecin-low-MW poly(ethylene glycol) (CPT-PEG) prodrug with GO water solution. Here, the process of hydrogel formation is very similar to a previously described process. When α-CD was added to GO-CPT-PEG solution under sonication, the solution became gradually blurred and then a homogeneous hydrogel (GO-CPT-PEG/α-CD) was formed after incubation at room temperature (Figure 15.13). The water-soluble anticancer drug 5-FU, which can be combined with CPT to enhance therapeutic advantages, can be loaded when GO hydrogel is highly hydrated. The 5-FU loaded GO hydrogel exhibited a typical dual-phase behavior for co-delivery of CPT and 5-FU [214].

15.3.2 Biocompatibility

The cytotoxicity of biomedical materials proposed to act as a DDS is a key point in assessing the feasibility of such material for use in medicine. Although this is a parameter that limits its future use, some of the studies using GO and CD exemplified above do not present biocompatibility studies. This is an important issue that needs to be addressed simultaneously with the DDS development.

Viability cell studies are mainly obtained by using model cell lines traditionally from mouse or human origin. For instance, Yang et al. [196] made cellular toxicity tests using OCT-1 cell lines (mouse osteoblasts, folate receptor-negative) as model cells. They show that DOX/GO-CD supramolecular assembly was almost nontoxic to normal cells during 24 h. The relative cellular viability of this DOX-loaded DDS reached 97% whereas free DOX only reached 57%. The morphology of cancer and normal cells in the presence of DO/GO-CD was also analyzed comparing the effect of free DOX and DOX/GO-CD on OCT-1 and HeLa cancer cells. The damages induced by free DOX and DOX-loaded DDS in cancer cells are similar, but the toxic effects of free DOX toward normal cells are higher than that of DOX-loaded DDS. These results are promising because the therapeutic effect against cancer cells seems to be similar using free DOX or is loaded on the GO-CD supramolecular assembly, but the latter presents lower toxicity to normal cells. The strong affinity between folic acid and folate receptor on the surface of GO-CD can be the reason for such results because this affinity favors the drug uptake of cancer cells. This process was not possible with OCT-1 since this cell line is folate receptor-negative.

HeLa cell lines, a type of human cervical carcinoma cell, was also used in toxicity assessment of GN/PTX [201] and TPT/GO-CD/AD-DOX [192] DDS. In the first case, cytotoxicity was evaluated using several materials: PTX, GN/PTX, and GN, during 24, 48, and 72 h. As shown in Figure 15.14, GN has no toxic effect on HeLa cell growth, while PTX/GN was very toxic against the same cells compared with PTX and GN. These results indicate that GN can be used as an efficient DDS and cause no toxicity against cells.

A very similar study was presented by Wu *et al.* [192], where it was possible to see that GO-CD/AD and only GO have little impact in cell viability. In some cases, cell viability is studied using two types of cells: cancer and healthy cells (Figure 15.15). This comparison was made in the study performed by Zhang *et al.* [193] using MDA-MB-231 cancer cells, a type of human breast cancer cells with abundant HA receptors being overexpressed on its surface. These authors not only prove that GO-CD–HA-ADA maintains cell viability but also conclude that CPT@GO-CD–HA-ADA displayed better anticancer activity. Cell nontoxicity was confirmed using GO-CD–HA-ADA, CPT, and CPT@GO-CD–HA-ADA against normal fibroblasts NIH3T3. With CPT@GO-CD–HA-ADA, the relative cellular viability of normal fibroblasts was 82.5%, which was quite high compared to that with free CPT (63.0%). These results demonstrated that the carrier developed in this study could be a promising safe DDS or applied in another biomedical device.

BT-474 and MCF-7 are other types of cancer cell lines of human breast cancer that are used in research. Active targeting efficiency complements biocompatibility for enhanced anticancer efficacy and reduced side effects. MCF-7 and BT-474 are HER2-negative and -positive breast cancer cell lines, respectively. In the recent study where GQD-comp DDS was prepared with GQD and CD, the functionalization with herceptin (HER), an antibody specific for anti-proliferation of HER2 that is overexpressed in breast cancer cells, was a strategy to treat this type of tumor [199]. Cell viability of MCF-7 and BT-474 cells was assessed in the presence of GQD-comp at different concentrations (0, 20, 100, 200, 300, 500 mg mL^{-1}) during 48 h. The results showed that GQD-comp as a DDS is very promising since it showed low cytotoxicity for MCF-7 (>95% up to 500 mg mL^{-1}), whereas the viability of BT-474 gradually decreased with increasing concentration of GQD-comp.

The *in vitro* cytotoxicity of GO-CPT-PEG/α-CD, a hydrogel based on α-CD and CPT-GO-modified GO, was developed using MTT assays in A549 lung cancer cell lines [214]. The overall cytotoxicity of the CPT-PEG and CPT-PEG/a-CD hydrogels was first evaluated

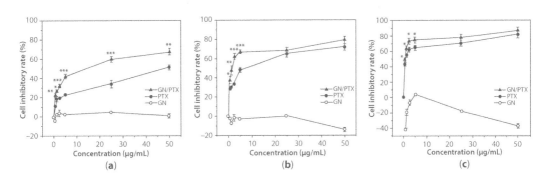

Figure 15.14 Cell inhibitory rate of GN/PTX and PTX against HeLa cells after culturing for 24 h (a), 48 h (b), and 72 h (c). The results represent the mean ± SD, $n \geq 3$ (***$p < 0.001$, **$p < 0.01$, *$p < 0.05$ versus cells with PTX). Reprinted from Ref. [201].

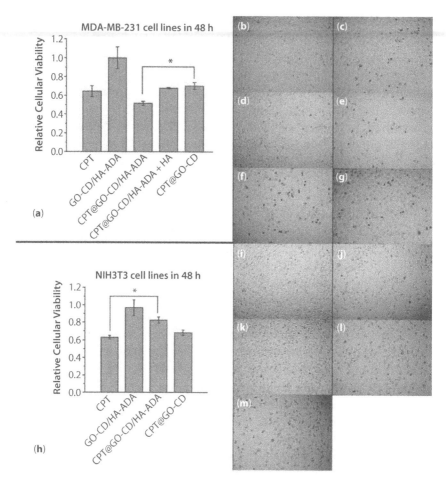

Figure 15.15 Relative cellular viability and cell photos of (a–g) MDA-MB-231 and (h–m) NIH3T3 cell lines after the treatment with blank (b and i), CPT (c and j), GO-CD–HA-ADA (d and k), CPT@GO-CD–HA-ADA (e and l), CPT@GO-CD–HA-ADA with an excess of HA (f), and CPT@GO-CD (g and m), respectively, in 48 h incubation ([CPT] = 1.0 mM). The statistically significant differences were indicated with asterisks ($p < 0.05$). Reproduced from Ref. [193] with permission from The Royal Society of Chemistry.

and compared with CPT as a free drug. Figure 15.16a shows that CP, CPT-PEG, and CPT-PEG/α-CD hydrogels have a minimal effect on the cytotoxicity, proving that the CPT-PEG prodrug in the hydrogel still possesses release ability.

15.3.3 Drug Release Profiles

Controlled and rapid release of drugs is essential to achieve successful applications of DDS.

In the DDS abovementioned based on GO and CD, it is possible to find three types of release studies: (a) experiments in phosphate-buffered saline (PBS) solution at 37°C, (b) different values of pH ranging from 7.5 to 5.0 to include simulated conditions of normal human tissue and tumor microenvironment, (c) and *in vivo* experiments.

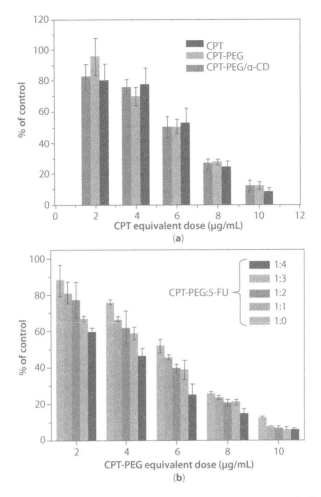

Figure 15.16 *In vitro* cytotoxicity of free CPT, CPT-PEG prodrug, CPT-PEG/α-CD hydrogel (a) and GO-CPT-PEG/α-CD hydrogel laden with different amount of 5-FU (b) to A549 lung cancer cells determined by MTT assay. Reproduced from Ref. [214] with permission from The Royal Society of Chemistry.

In the experiments of drug release in PBS solutions at 37°C, PTX-GO [200] and GO or rGO with α-CD [204] hydrogels showed different release rates of PTX and CPT, respectively. In the first, two PTX contents were loaded onto the nanospheres, being the one with higher content, which has more quantity released. In the second study, a comparison between GO and rGO systems was made. However, it was concluded that the controlled release occurs despite using GO or rGO systems.

In the group (b) of release studies, the pH-responsive character was studied. A profile of the main results obtained by several DDS based on GO and CD can be seen in Table 15.2. pH sensitivity implied the potential of the delivery systems in the acidic tumor extracellular fluids [215].

In the systems GO-CD–HAADA [190] and Ac-β-CD/GO [200], CPT was control released over 10 h and 4 days, respectively. With GO-CD–HAADA, the rate of drug release was much faster at pH 5.7, and this pH-responsive releasing behavior could definitely inhibit the growth and reproduction of tumor cells in cancer cell environments. From Ac-β-CD/GO, CPT was gradually released in the pH 7.5 solution in 20 h, and only 15%–20% was gradually released in the pH 5.5 solution in the same time period (Figure 15.17).

PTX was the drug tested in Tan et al. [198] with GO-COO-HP-β-CD systems. The drug release studies were carried out at pH 7.4, 6.5, and 5.0 at 37°C. PTX was released continuously over 150 h, and 70.2%, 65.1%, and 58.9% of total PTX were released at pH 5.0, 6.5, and 7.4. It can be concluded that drug release was dependent on the pH medium, with a lower pH (such as tumor tissues) being the most favorable.

DOX was the model drug tested in all the other systems referenced in Table 15.2. With GO–Py–PEG [213] and GQD+βCD [199], release of DOX was above 70%, at pH 5.0 or 5.5, attaining 100% with the system Fe3O4@GO@mSiO2 [197]. Besides being pH-responsive, GQD+βCD behavior also depends on temperature stimulus. When the temperature was raised from 25 to 37°C, at pH 5.5, approximately 60% of cumulated DOX was increased within 28 h (Figure 15.18). This dual stimuli-responsive drug release profile occurs because the formation of β-CD inclusion complexes with hydrophobic drugs is strongly affected by pH and temperature [216].

In the work reported by Wu et al. [192], they present the release profiles of dual GO-CD/AD. This particular dual-drug system was tested in simulated conditions of normal human tissue (pH 7.4) and tumor microenvironment (pH 5.9). Over 95 h, the cumulative release percentages of DOX and TPT were 70.3% and 77.6% at pH 5.9, respectively. These results show that π-π stacking interaction and host-guest bond were both sensitive to pH values. The rupture of these noncovalent bonds is faster due to pH decrease, which leads to faster TPT and DOX release.

Ha et al. [214] tested the release capacity of developed GO-CPT-PEG/α-CD hydrogel under in vitro and in vivo conditions. As the abovementioned DDS, this hydrogel also has dual-loading drugs, CPT and 5-FU. In the first stage, 5-FU was released from the hydrogel during 24 h at an almost constant rate, while the release of CPTPEG was sustained for more

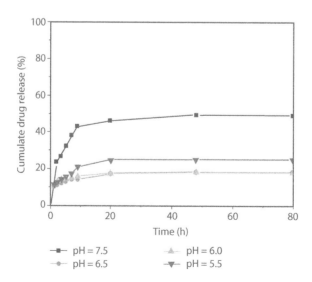

Figure 15.17 Cumulate CPT release behaviors of injectable hydrogel in different medium at 37°C. Copyright © 2016 Yuanfeng Ye and Xiaohong Hu [203]. Open access.

Table 15.2 Drug release profile from several DDS based on GO and CD.

DDS	Drug	pH	Main results	Ref.
GO-CD–HAADA	CPT	PBS (7.2), 5.7	Drug release profile is much faster at pH 5.7 than at pH 7.2 (during 10 h), which is promising to inhibit the growth and reproduction of tumor cells in cancer cell environments.	[193]
Ac-β-CD/GO	CPT	5.5–7.5	Initial drug release (10%) is independent of the pH medium. At pH 7.5, drug release rate is much higher than in other pH solutions. After 4 days, no more CPT was released.	[203]
GO-COO-HP-β-CD	PTX	7.4, 6.5, 5	PTX was released continuously over 150 h, and 70.2%, 65.1%, and 58.9% of total PTX were released at pH 5.0, 6.5, and 7.4.	[201]
GO-CD/AD	DOX and TPT	7.4, 5.9	Over 95 h, the cumulative release percentages of DOX and TPT were 18.9% and 20.9% at pH 7.4, and <70.3% and 77.6% at pH 5.9, respectively.	[192]
rGO-C6H4-CO-NH-PEI-NH-CO-biotin	DOX	7.4 (PBS), 5.5 (ABS)	DOX release rate was faster in PBS solution than that in ABS solution during the early stages, but slower at more than 14 h.	[198]
GO–Py–PEG	DOX	7.4, 5.0	Drug release was almost constant at pH 7.4 or 5.0, with the total release amount of 90% during 45 h, at pH 7.4 and 75% during 70 h, at pH 5.0.	[213]
GQD+βCD	DOX	7.4, 5.5	About 70% of DOX was released from DL-GQD at pH 5.5, while only 20% of DOX was released at pH 7.4 within 28 h.	[199]
Fe3O4@GO@mSiO2	DOX	7.4, 5.5	The drug release was about 100% at pH 5.5 (endosomal pH) during 48 h; but it was zero at pH 7.4.	[197]

than 6 days, in PBS (pH 7.4), at 37°C (Figure 15.19b). The release of 5-FU occurs in a diffusion-controlled manner due to the regular pores of hydrogel. The break of hydrogel bonds could be the reason for controlled and much slower release of CPT. In the second stage, GO-CPT-PEG/α-CD hydrogel suffered NIR light irradiance and CPT-PEG/α-CD was also irradiated for comparison. As shown in Figure 15.19c, 5-FU and CPT-PEG were released from the CPT-PEG/α-CD hydrogel slowly upon NIR light irradiation, while they were more rapidly released from the GO-CPT-PEG/a-CD hydrogel, under the same conditions.

To investigate the NIR-triggered drug release behavior *in vivo*, Chinese KunMing mice bearing H22 ascites sarcoma were intratumorally injected with the CyN-PEG

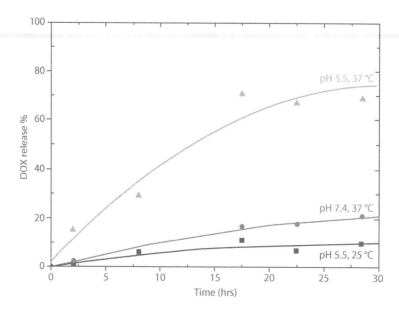

Figure 15.18 *In vitro* release profile of DOX from DL-GQD in PBS under different conditions: pH 5 at 37°C, pH 7.4 at 37°C, and pH 5 at 25°C [199] Published by The Royal Society of Chemistry.

(near-infrared fluorescent IR-780 dye) loaded GO-CPT-PEG/α-CD hydrogel, and *in vivo* whole-body fluorescence images were taken on an *in vivo* imaging system (IVIS). The fluorescence area in the tumor position increased obviously after 20 min of irradiation (right panel), while it almost remained unchanged in the case of no NIR light irradiation (left panel) (Figure 15.19d). The result indicates that CyN-PEG was minimally released from the hydrogel when no NIR light was irradiated but once NIR light irradiation was administered, its release quantity largely increased. These data suggest that NIR light irradiation could efficiently trigger the gel–sol transition of the GO-CPT-PEG/α-CD hydrogel, thus leading to drug release from the hydrogel *in vivo*.

Another *in vivo* drug release study is reported in the work of Yang *et al.* [196]. The experiments were carried out on a BALB/c nude mice model that contained HeLa cancer cells. The mice with tumors were divided into three groups, one of which was untreated (the control) and two were injected with DOX or assembly 1/2/DOX/GO by the tail vein. The untreated group showed a volume growth of the tumor. Growth tumor was suppressed in the DOX group, with a tumor growth inhibition of 46% by day 20. However, this mice group showed body weight and low survival because of the high toxicity of DOX toward normal cells and tissues. From day 21, mice started to die, with less than 30% of the mice surviving after day 27. In the group treated with 1/2/DOX/GO, the tumor growth inhibition was 71% on day 20. Moreover, after 4 weeks, mice were still alive. 1/2/DOX/GO was accumulated in malignant tumor tissues inhibiting its growth due to the specific binding between folic acid present in DDS and the folate receptor displayed on the surface of tumor cells. Besides that, the targeting effect maintained the low level of assembly in normal tissues, thereby reducing side effects.

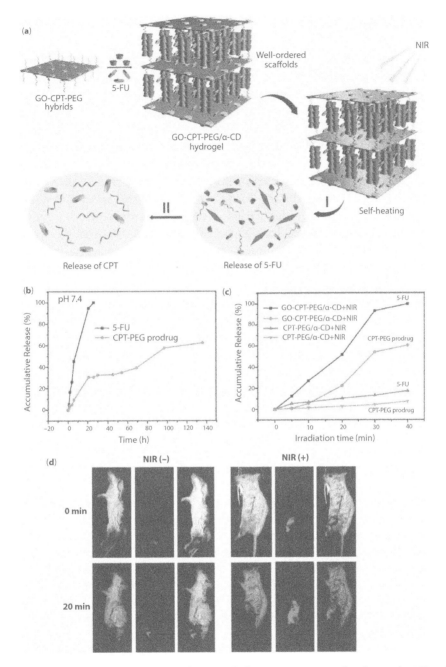

Figure 15.19 (a) Schematic representation of the NIR light-responsive cascade release of 5-FU and CPT. 5-FU and CPT-PEG prodrug release kinetics of the GO-CPT-PEG/a-CD hydrogel (b) and the GO-CPT-PEG/a-CD hydrogel or the native CPT-PEG/a-CD hydrogel triggered by NIR light irradiation (c) in PBS at pH 7.4 and 371°C. (d) *In vivo* NIR light-triggered dye release from the GO-CPT-PEG/a-CD hydrogels. Whole-body NIR fluorescence images of Chinese KunMing mice bearing H22 ascites sarcoma following a single intratumoral injection of the CyN-PEG loaded GO-CPT-PEG/a-CD hydrogel. Images were taken at 0 min and 20 min using an *in vivo* whole animal fluorescence imaging system (excitation at 625 nm, emission at 700 nm). For each panel, images from left to right show the bright field, CyN-PEG channel, and merge of the two images. Reproduced from Ref. [214] with permission from The Royal Society of Chemistry.

15.4 Concluding Remarks

As a shining star in material science, GO and its derivatives possess potential applications in various areas. Among them, the application of GO to DDS has attracted ever-increasing interest in recent years. Most drugs, in particular those with anti-cancer activity, are hydrophobic as a consequence of its aromatic structure; consequently, they can be loaded through π-π stacking in DDS. GO has a high surface area, and π-conjugated domains with functional groups that allow oxygen to its surface modification, characteristics that endow this nanomaterial an excellent capacity for drugs immobilization. The favorable biocompatibility with low toxicity promoted the interest of several research groups to explore GO's application as a molecular carrier for *in vitro* and *in vivo* drug delivery. To increase the drug load ability and simultaneously decrease the amount of the nanocarrier needed for the delivery, the association of GO with CD, another interesting group of molecules with a significant potential as drug carriers, is being explored. This biphasic GO-CD system presents an ultrahigh drug-loading ability and allows the simultaneous filling of different drugs, solving the problem of drug solubility and thus increasing treatment efficiency in a synergetic way. Some of the reported GO-CD-Drugs systems are also inside a stimulus-responsive hydrogel that increases the targeted release of drugs which is critical to overcome the drugs instability and poor dissolution in the gastrointestinal tract due to the differents values of pH in its different regions. The possibility of simultaneously delivering a highly localized amount of different drugs to tumor cells without affecting the surrounding healthy tissues still needs to be explored concomitantly with careful biocompatibility screening studies before clinical trials are carried out. The possibility of the GO-CD platform as a potential DDS was reviewed and represents a promising strategy with respect to classical cancer therapies like external and internal radiotherapy, chemotherapy, or surgery.

References

1. Zhang, B., Song, Y., Wang, T., Yang, S. *et al.*, Efficient co-delivery of immiscible hydrophilic/hydrophobic chemotherapeutics by lipid emulsions for improved treatment of cancer. *Int. J. Nanomed.*, 12, 2871–2886, 2017.
2. Zhou, Q., Zhang, L., Wu, H., Nanomaterials for cancer therapies. *Nanotechnol. Rev.*, 6, 2017.
3. Zhang, R.X., Wong, H.L., Xue, H.Y., Eoh, J.Y., Wu, X.Y., Nanomedicine of synergistic drug combinations for cancer therapy—Strategies and perspectives. *J. Control. Release*, 240, 489–503, 2016.
4. Lam, P.-L., Wong, W.-Y., Bian, Z., Chui, C.-H., Gambari, R., Recent advances in green nanoparticulate systems for drug delivery: Efficient delivery and safety concern. *Nanomedicine*, nnm-2016-0305, 2017.
5. Davis, M.E., Chen, Z.G., Shin, D.M., Nanoparticle therapeutics: An emerging treatment modality for cancer. 7, 771–782, 2008.
6. Yang, K., Feng, L., Liu, Z., Stimuli responsive drug delivery systems based on nano-graphene for cancer therapy. *Adv. Drug Delivery Rev.*, 105, 228–241, 2016.
7. Das, D., Ghosh, P., Ghosh, A., Haldar, C. *et al.*, Stimulus-responsive, biodegradable, biocompatible, covalently cross-linked hydrogel based on dextrin and poly(*N*-isopropylacrylamide) for *in vitro/in vivo* controlled drug release. *ACS Appl. Mater. Interfaces*, 7, 14338–14351, 2015.
8. Costa, D., Valente, A.J.M., Miguel, M.G., Queiroz, J., Plasmid DNA hydrogels for biomedical applications. *Adv. Colloid Interface Sci.*, 205, 257–264, 2014.

9. Costa, D., Valente, A.J.M., Miguel, M.G., Queiroz, J., Gel network photodisruption: A new strategy for the codelivery of plasmid DNA and drugs. *Langmuir*, 27, 13780–13789, 2011.
10. Reina, G., González-Domínguez, J.M., Criado, A., Vázquez, E. *et al.*, Promises, facts and challenges for graphene in biomedical applications. *Chem. Soc. Rev.*, 4400–4416, 2017.
11. Marques, P., Gonalves, G., Cruz, S., Almeida, N. *et al.*, *Adv. Nanocomposite Technol.*, A. Hashim (Ed.), InTech, Rijeka, 2011.
12. Cruz, S., Girão, A., Gonçalves, G., Marques, P., Graphene: The missing piece for cancer diagnosis? *Sensors*, 16, 137, 2016.
13. Ma, N., Zhang, B., Liu, J., Zhang, P. *et al.*, Green fabricated reduced graphene oxide: Evaluation of its application as nano-carrier for pH-sensitive drug delivery. *Int. J. Pharm.*, 496, 984–992, 2015.
14. Brodie, B.C., On the atomic weight of graphite. *Proc. R. Soc.* London, 10, 11–12, 1859.
15. Staudenmaier, L., Verfahren zur Darstellung der Graphitsäure. *Ber. Dtsch. Chem. Ges.*, 31, 1481–1487, 1898.
16. Hummers, W.S. and Offeman, R.E., Preparation of graphitic oxide. *J. Am. Chem. Soc.*, 80, 1339–1339, 1958.
17. Park, S., An, J., Jung, I., Piner, R.D. *et al.*, Colloidal suspensions of highly reduced graphene oxide in a wide variety of organic solvents. *Nano Lett.*, 9, 1593–1597, 2009.
18. Zhu, Y., Murali, S., Cai, W., Li, X. *et al.*, Graphene and graphene oxide: Synthesis, properties, and applications. *Adv. Mater.*, 22, 3906–3924, 2010.
19. Teradal, N.L. and Jelinek, R., Carbon nanomaterials in biological studies and biomedicine. *Adv. Healthcare Mater.*, 6, 1–36, 2017.
20. Lee, J., Yim, Y., Kim, S., Choi, M.H. *et al.*, In-depth investigation of the interaction between DNA and nano-sized graphene oxide. *Carbon N. Y.*, 97, 92–98, 2016.
21. Morales-Narváez, E., Baptista-Pires, L., Zamora-Gálvez, A., Merkoçi, A., Graphene-based biosensors: Going simple. *Adv. Mater.*, 29, 2017.
22. Tian, J., Luo, Y., Huang, L., Feng, Y. *et al.*, Pegylated folate and peptide-decorated graphene oxide nanovehicle for *in vivo* targeted delivery of anticancer drugs and therapeutic self-monitoring. *Biosens. Bioelectron.*, 80, 519–524, 2016.
23. Shim, G., Kim, M.-G.G., Park, J.Y., Oh, Y.-K.K., Graphene-based nanosheets for delivery of chemotherapeutics and biological drugs. *Adv. Drug Delivery Rev.*, 105, 205–227, 2016.
24. Chen, B., Liu, M., Zhang, L., Huang, J. *et al.*, Polyethylenimine-functionalized graphene oxide as an efficient gene delivery vector. *J. Mater. Chem.*, 21, 7736, 2011.
25. Teimouri, M., Nia, A.H., Abnous, K., Eshghi, H., Ramezani, M., Graphene oxide–cationic polymer conjugates: Synthesis and application as gene delivery vectors. *Plasmid*, 84–85, 51–60, 2016.
26. Jimenez-Cervantes, E., López-Barroso, J., Martínez-Hernández, A.L., Velasco-Santos, C., *Recent Adv. Graphene Res.*, P. Nayak (Ed.), InTech, 2016.
27. Sun, X.M., Liu, Z., Welsher, K., Robinson, J.T. *et al.*, Nano-graphene oxide for cellular imaging and drug delivery. *Nano Res.*, 1, 203–212, 2008.
28. Shen, A.J., Li, D.L., Cai, X.J., Dong, C.Y. *et al.*, Multifunctional nanocomposite based on graphene oxide for *in vitro* hepatocarcinoma diagnosis and treatment. *J. Biomed. Mater. Res. Part A*, 100 A, 2499–2506, 2012.
29. You, P., Yang, Y., Wang, M., Huang, X., Huang, X., Graphene oxide-based nanocarriers for cancer imaging and drug delivery. *Curr. Pharm. Des.*, 21, 3215–3222, 2015.
30. Bikhof Torbati, M., Ebrahimian, M., Yousefi, M., Shaabanzadeh, M., GO-PEG as a drug nanocarrier and its antiproliferative effect on human cervical cancer cell line. *Artif. Cells Nanomed. Biotechnol.*, 45, 0–6, 2017.

31. Zhao, X., Yang, L., Li, X., Jia, X. et al., Functionalized graphene oxide nanoparticles for cancer cell-specific delivery of antitumor drug. *Bioconjugate Chem.*, 26, 128–136, 2015.
32. Yang, X., Zhang, X., Liu, Z., Ma, Y., Huang, Y., Chen, Y., High-efficiency loading and controlled release of doxorubicin hydrochloride on graphene oxide. *J. Phys. Chem. C*, 112, 17554, 2008.
33. Mahdavi, M., Rahmani, F., Nouranian, S., Molecular simulation of pH-dependent diffusion, loading, and release of doxorubicin in graphene and graphene oxide drug delivery systems. *J. Mater. Chem. B*, 4, 7441–7451, 2016.
34. Zhang, Q., Li, W., Kong, T., Su, R. et al., Tailoring the interlayer interaction between doxorubicin-loaded graphene oxide nanosheets by controlling the drug content. *Carbon*, 51, 164–172, 2013.
35. Wu, S., Zhao, X., Cui, Z., Zhao, C. et al., Cytotoxicity of graphene oxide and graphene oxide loaded with doxorubicin on human multiple myeloma cells. *Int. J. Nanomed.*, 9, 1413–1421, 2014.
36. Wu, S., Zhao, X., Li, Y., Du, Q. et al., Adsorption properties of doxorubicin hydrochloride onto graphene oxide: Equilibrium, kinetic and thermodynamic studies. *Materials (Basel).*, 6, 2026–2042, 2013.
37. Wu, J., Wang, Y., Yang, X., Liu, Y. et al., Graphene oxide used as a carrier for adriamycin can reverse drug resistance in breast cancer cells. *Nanotechnology*, 23, 355101, 2012.
38. Liu, Z., Robinson, J.T., Sun, X., Dai, H., PEGylated nano-graphene oxide for delivery of water insoluble cancer drugs (b). *J. Am. Chem. Soc.*, 130, 10876–10877, 2008.
39. Yang, K., Zhang, S., Zhang, G., Sun, X. et al., Graphene in mice: Ultrahigh *in vivo* tumor uptake and efficient photothermal therapy. *Nano Lett.*, 10, 3318–3323, 2010.
40. Vila, M., Portolés, M.T., Marques, P.A.A.P., Feito, M.J. et al., Cell uptake survey of pegylated nanographene oxide. *Nanotechnology*, 23, 465103, 2012.
41. Fan, L., Ge, H., Zou, S., Xiao, Y. et al., Sodium alginate conjugated graphene oxide as a new carrier for drug delivery system. *Int. J. Biol. Macromol.*, 93, 582–590, 2016.
42. Zhao, X., Liu, L., Li, X., Zeng, J. et al., Biocompatible graphene oxide nanoparticle-based drug delivery platform for tumor microenvironment-responsive triggered release of doxorubicin. *Langmuir*, 30, 10419–10429, 2014.
43. He, Y., Zhang, L., Chen, Z., Liang, Y. et al., Enhanced chemotherapy efficacy by co-delivery of shABCG2 and doxorubicin with a pH-responsive charge-reversible layered graphene oxide nanocomplex. *J. Mater. Chem. B*, 3, 6462–6472, 2015.
44. Lv, Y., Tao, L., Annie Bligh, S.W., Yang, H. et al., Targeted delivery and controlled release of doxorubicin into cancer cells using a multifunctional graphene oxide. *Mater. Sci. Eng. C*, 59, 652–660, 2016.
45. O'Brien, M.E.R., Wigler, N., Inbar, M., Rosso, R. et al., Reduced cardiotoxicity and comparable efficacy in a phase III trial of pegylated liposomal doxorubicin HCl (CAELYX™/Doxil®) versus conventional doxorubicin for first-line treatment of metastatic breast cancer. *Ann. Oncol.*, 15, 440–449, 2004.
46. Gabizon, A., Shmeeda, H., Barenholz, Y., Pharmacokinetics of pegylated liposomal doxorubicin: Review of animal and human studies. *Clin. Pharmacokinet.*, 42, 419–436, 2003.
47. Zhang, Q., Chi, H., Tang, M., Chen, J. et al., Mixed surfactant modified graphene oxide nanocarriers for DOX delivery to cisplatin-resistant human ovarian carcinoma cells. *RSC Adv.*, 6, 87258–87269, 2016.
48. Pan, Q., Lv, Y., Williams, G.R., Tao, L. et al., Lactobionic acid and carboxymethyl chitosan functionalized graphene oxide nanocomposites as targeted anticancer drug delivery systems. *Carbohydr. Polym.*, 151, 812–820, 2016.
49. Park, Y.H., Park, S.Y., In, I., Direct noncovalent conjugation of folic acid on reduced graphene oxide as anticancer drug carrier. *J. Ind. Eng. Chem.*, 30, 190–196, 2015.

50. Miao, W., Shim, G., Kang, C.M., Lee, S. et al., Cholesteryl hyaluronic acid-coated, reduced graphene oxide nanosheets for anti-cancer drug delivery. *Biomaterials*, 34, 9638–9647, 2013.
51. Zheng, X.T., Ma, X.Q., Li, C.M., Highly efficient nuclear delivery of anti-cancer drugs using a bio-functionalized reduced graphene oxide. *J. Colloid Interface Sci.*, 467, 35–42, 2016.
52. Pourjavadi, A., Mazaheri Tehrani, Z., Jokar, S., Chitosan based supramolecular polypseudorotaxane as a pH-responsive polymer and their hybridization with mesoporous silica-coated magnetic graphene oxide for triggered anticancer drug delivery. *Polymer (United Kingdom)*, 76, 52–61, 2015.
53. Zhu, S., Li, J., Chen, Y., Chen, Z. et al., Grafting of graphene oxide with stimuli-responsive polymers by using ATRP for drug release. *J. Nanopart. Res.*, 14, 2012.
54. Pan, Y., Bao, H., Sahoo, N.G., Wu, T., Li, L., Water-soluble poly(N-isopropylacrylamide)-graphene sheets synthesized via click chemistry for drug delivery. *Adv. Funct. Mater.*, 21, 2754–2763, 2011.
55. Gonçalves, G., Vila, M., Portolés, M.-T.T., Vallet-Regi, M. et al., Nano-graphene oxide: A potential multifunctional platform for cancer therapy. *Adv. Healthcare Mater.*, 2, 1072–1090, 2013.
56. Song, J., Yang, X., Jacobson, O., Lin, L. et al., Sequential drug release and enhanced photothermal and photoacoustic effect of hybrid reduced graphene oxide-loaded ultrasmall gold nanorod vesicles for cancer therapy. *ACS Nano*, 9, 9199–9209, 2015.
57. Kurapati, R. and Raichur, A.M., Near-infrared light-responsive graphene oxide composite multilayer capsules: A novel route for remote controlled drug delivery. *Chem. Commun.*, 49, 734–736, 2013.
58. Chen, J., Liu, H., Zhao, C., Qin, G. et al., One-step reduction and PEGylation of graphene oxide for photothermally controlled drug delivery. *Biomaterials*, 35, 4986–4995, 2014.
59. Kim, H. and Kim, W.J., Photothermally controlled gene delivery by reduced graphene oxide-polyethylenimine nanocomposite. *Small*, 10, 117–126, 2014.
60. Shi, J., Wang, L., Zhang, J., Ma, R. et al., A tumor-targeting near-infrared laser-triggered drug delivery system based on GO@Ag nanoparticles for chemo-photothermal therapy and X-ray imaging. *Biomaterials*, 35, 5847–5861, 2014.
61. Wang, Y., Wang, K., Zhao, J., Liu, X. et al., Multifunctional mesoporous silica-coated graphene nanosheet used for chemo-photothermal synergistic targeted therapy of glioma. *J. Am. Chem. Soc.*, 135, 4799–4804, 2013.
62. Tang, Y., Hu, H., Zhang, M.G., Song, J. et al., An aptamer-targeting photoresponsive drug delivery system using "off–on" graphene oxide wrapped mesoporous silica nanoparticles. *Nanoscale*, 7, 6304–6310, 2015.
63. Kavitha, T., Haider Abdi, S.I., Park, S.-Y., pH-Sensitive nanocargo based on smart polymer functionalized graphene oxide for site-specific drug delivery. *Phys. Chem. Chem. Phys.*, 15, 5176, 2013.
64. Zhang, L., Xia, J., Zhao, Q., Liu, L., Zhang, Z., Functional graphene oxide as a nanocarrier for controlled loading and targeted delivery of mixed anticancer drugs. *Small*, 6, 537–544, 2010.
65. Hu, H., Yu, J., Li, Y., Zhao, J., Dong, H., Engineering of a novel pluronic F127/graphene nanohybrid for pH responsive drug delivery. *J. Biomed. Mater. Res. Part A*, 100 A, 141–148, 2012.
66. Ballatori, N., Krance, S.M., Notenboom, S., Shi, S. et al., Glutathione dysregulation and the etiology and progression of human diseases. *Biol. Chem.*, 390, 191–214, 2009.
67. Wen, H., Dong, C., Dong, H., Shen, A. et al., Engineered redox-responsive PEG detachment mechanism in PEGylated nano-graphene oxide for intracellular drug delivery. *Small*, 8, 760–769, 2012.
68. Shim, G., Lee, J., Kim, J., Lee, H.-J. et al., Functionalization of nano-graphenes by chimeric peptide engineering. *RSC Adv.*, 5, 49905–49913, 2015.

69. Shen, H., Liu, M., He, H., Zhang, L. et al., PEGylated graphene oxide-mediated protein delivery for cell function regulation. *ACS Appl. Mater. Interfaces*, 4, 6317–6323, 2012.
70. Lu, C.-H., Yang, H.-H., Zhu, C.-L., Chen, X., Chen, G.-N., A graphene platform for sensing biomolecules. *Angew. Chem. Int. Ed.*, 48, 4785–4787, 2009.
71. He, S., Song, B., Li, D., Zhu, C. et al., A graphene nanoprobe for rapid, sensitive, and multicolor fluorescent DNA analysis. *Adv. Funct. Mater.*, 20, 453–459, 2010.
72. Varghese, N., Mogera, U., Govindaraj, A., Das, A. et al., Binding of DNA nucleobases and nucleosides with graphene. *ChemPhysChem*, 10, 206–210, 2009.
73. Kim, H., Namgung, R., Singha, K., Oh, I.K., Kim, W.J., Graphene oxide-polyethylenimine nanoconstruct as a gene delivery vector and bioimaging tool. *Bioconjugate Chem.*, 22, 2558–2567, 2011.
74. Kim, H., Kim, J., Lee, M., Choi, H.C., Kim, W.J., Stimuli-regulated enzymatically degradable smart graphene-oxide-polymer nanocarrier facilitating photothermal gene delivery. *Adv. Healthcare Mater.*, 5, 1918–1930, 2016.
75. Zhang, L., Wang, Z., Lu, Z., Shen, H. et al., PEGylated reduced graphene oxide as a superior ssRNA delivery system. *J. Mater. Chem. B*, 1, 749–755, 2013.
76. Elbashir, S.M., Harborth, J., Lendeckel, W., Yalcin, A. et al., Duplexes of 21-nucleotide RNAs mediate RNA interference in cultured mammalian cells. *Nature*, 411, 494–498, 2001.
77. Zhang, L., Lu, Z., Zhao, Q., Huang, J. et al., Enhanced chemotherapy efficacy by sequential delivery of siRNA and anticancer drugs using PEI-grafted graphene oxide. *Small*, 7, 460–464, 2011.
78. Chen, W., Wen, X., Zhen, G., Zheng, X., Assembly of Fe_3O_4 nanoparticles on PEG-functionalized graphene oxide for efficient magnetic imaging and drug delivery. *RSC Adv.*, 5, 69307–69311, 2015.
79. Guo, L., Shi, H., Wu, H., Zhang, Y. et al., Prostate cancer targeted multifunctionalized graphene oxide for magnetic resonance imaging and drug delivery. *Carbon*, 107, 87–99, 2016.
80. Dinda, S., Kakran, M., Zeng, J., Sudhaharan, T. et al., Grafting of ZnS:Mn-doped nanocrystals and an anticancer drug onto graphene oxide for delivery and cell labeling. *ChemPlusChem*, 81, 100–107, 2016.
81. Villiers, A., Sur la fermentation de la fécule par l'action du ferment butyrique. *C. R. Acad. Sci.*, 112, 536–538, 1891.
82. Schardinger, F., Über thermophile bakterian aus verschiedenen speisen und milch, sowie über einige umsetzungsprodukte derselben in kohlenhydrathaltigen nährlösungen, darunter krystallisierte polysaccharide (dextrin) aus stärke. *Z. Untersuch. Nahr. Genussm.*, 6, 865–80, 1930.
83. Szejtli, J. and Bankyelod, E., Inclusion complexes of unsaturated fatty-acids with amylose and cyclodextrin. *Starke*, 27, 368–376, 1975.
84. Freudenberg, K. and Meyer-Delius, M., Über die Schardinger—Dextrine aus Stärke. *Ber. Dtsch. Chem. Ges. (A B Ser).*, 71, 1596–1600, 1938.
85. Loftsson, T. and Brewster, M.E., Pharmaceutical applications of cyclodextrins. 1. Drug solubilization and stabilization. *J. Pharm. Sci.*, 85, 1017–1025, 1996.
86. Saenger, W., Jacob, J., Gessler, K., Steiner, T. et al., Structures of the common cyclodextrins and their larger analogues—Beyond the doughnut. *Chem. Rev.*, 98, 1787–1802, 1998.
87. Valente, A.J.M.M. and Söderman, O., The formation of host–guest complexes between surfactants and cyclodextrins. *Adv. Colloid Interface Sci.*, 205, 156–176, 2014.
88. Teixeira, R.S., Veiga, F.J.B., Oliveira, R.S., Jones, S.A. et al., Effect of cyclodextrins and pH on the permeation of tetracaine: Supramolecular assemblies and release behavior. *Int. J. Pharm.*, 466, 349–358, 2014.
89. Figueiras, A., Sarraguça, J.M.G., Carvalho, R.A., Pais, A.A.C.C., Veiga, F.J.B., Interaction of omeprazole with a methylated derivative of β-cyclodextrin: Phase solubility, NMR spectroscopy and molecular simulation. *Pharm. Res.*, 24, 377–389, 2007.

90. Figueiras, A., Sarraguça, J.M.G., Pais, A.A.C.C., Carvalho, R., Veiga, J.F., The role of L-arginine in inclusion complexes of omeprazole with cyclodextrins. *AAPS PharmSciTech*, 11, 233–240, 2010.
91. Santos, C.I.A.V., Esteso, M.A., Sartorio, R., Ortona, O. *et al.*, A comparison between the diffusion properties of theophylline/β-cyclodextrin and theophylline/2-hydroxypropyl-β-cyclodextrin in aqueous systems. *J. Chem. Eng. Data*, 57, 1881–1886, 2012.
92. Bom, A., Bradley, M., Cameron, K., Clark, J.K. *et al.*, A novel concept of reversing neuromuscular block: Chemical encapsulation of rocuronium bromide by a cyclodextrin-based synthetic host. *Angew. Chem. Int. Ed.*, 41, 265–271, 2002.
93. Fernandes, C.M., Carvalho, R.A., Pereira da Costa, S., Veiga, F.J.B., Multimodal molecular encapsulation of nicardipine hydrochloride by β-cyclodextrin, hydroxypropyl-β-cyclodextrin and triacetyl-β-cyclodextrin in solution. Structural studies by 1H NMR and ROESY experiments. *Eur. J. Pharm. Sci.*, 18, 285–296, 2003.
94. Junquera, E. and Aicart, E., Potentiometric study of the encapsulation of ketoprophen by hydroxypropyl-β-cyclodextrin. Temperature, solvent, and salt effects. *J. Phys. Chem. B*, 101, 7163–7171, 1997.
95. Junquera, E. and Aicart, E., A fluorimetric, potentiometric and conductimetric study of the aqueous solutions of naproxen and its association with hydroxypropyl-β-cyclodextrin. *Int. J. Pharm.*, 176, 169–178, 1999.
96. Michel, D., Chitanda, J.M., Balogh, R., Yang, P. *et al.*, Design and evaluation of cyclodextrin-based delivery systems to incorporate poorly soluble curcumin analogs for the treatment of melanoma. *Eur. J. Pharm. Biopharm.*, 81, 548–556, 2012.
97. Gerola, A.P., Silva, D.C., Jesus, S., Carvalho, R.A. *et al.*, Synthesis and controlled curcumin supramolecular complex release from pH-sensitive modified gum-arabic-based hydrogels. *RSC Adv.*, 5, 94519–94533, 2015.
98. Hashidzume, A. and Harada, A., Recognition of polymer side chains by cyclodextrins. *Polym. Chem.*, 2, 2146, 2011.
99. Martínez-Tomé, M.J., Esquembre, R., Mallavia, R., Mateo, C.R., Formation and characterization of stable fluorescent complexes between neutral conjugated polymers and cyclodextrins. *J. Fluoresc.*, 23, 171–180, 2013.
100. Buvari, A. and Barcza, L., Complex formation of inorganic salts with β-cyclodextrin. *J. Inclusion Phenom. Mol. Recognit. Chem.*, 379–389, 1989.
101. Norkus, E. and Vaitkus, R., Interaction of lead(II) with b-cyclodextrin in alkaline solutions. 337, 1657–1661, 2002.
102. Ribeiro, A., Esteso, M., Lobo, V., Valente, A. *et al.*, Interactions of copper (II) chloride with β-cyclodextrin in aqueous solutions. *J. Carbohydr. Chem.*, 25, 173–185, 2006.
103. Ribeiro, A.C.F., Lobo, V.M.M., Valente, A.J.M., Simões, S.M.N. *et al.*, Association between ammonium monovanadate and β-cyclodextrin as seen by NMR and transport techniques. *Polyhedron*, 25, 3581–3587, 2006.
104. Kurokawa, G., Sekii, M., Ishida, T., Nogami, T., Crystal structure of a molecular complex from native β-cyclodextrin and copper(II) chloride. *Supramol. Chem.*, 16, 381–384, 2004.
105. Carlstedt, J., Bilalov, A., Krivtsova, E., Olsson, U., Lindman, B., Cyclodextrin-surfactant coassembly depends on the cyclodextrin ability to crystallize. *Langmuir*, 28, 2387–2394, 2012.
106. Haller, J. and Kaatze, U., Octylglucopyranoside and cyclodextrin in water. Self-aggregation and complex formation. *J. Phys. Chem. B*, 113, 1940–1947, 2009.
107. Jiang, L., Yan, Y., Huang, J., Versatility of cyclodextrins in self-assembly systems of amphiphiles. *Adv. Colloid Interface Sci.*, 169, 13–25, 2011.
108. Sehgal, P., Mizuki, T., Doe, H., Wimmer, R. *et al.*, Interactions and influence of α-cyclodextrin on the aggregation and interfacial properties of mixtures of nonionic and zwitterionic surfactants. *Colloid Polym. Sci.*, 287, 1243–1252, 2009.

109. Carvalho, R.A., Correia, H.A., Valente, A.J.M., Söderman, O., Nilsson, M., The effect of the head-group spacer length of 12-s-12 gemini surfactants in the host-guest association with β-cyclodextrin. *J. Colloid Interface Sci.*, 354, 725–732, 2011.
110. Jeong, S., Kang, W.Y., Song, C.K., Park, J.S., Supramolecular cyclodextrin-dye complex exhibiting selective and efficient quenching by lead ions. *Dyes Pigm.*, 93, 1544–1548, 2012.
111. Kyzas, G.Z., Lazaridis, N.K., Bikiaris, D.N., Optimization of chitosan and β-cyclodextrin molecularly imprinted polymer synthesis for dye adsorption. *Carbohydr. Polym.*, 91, 198–208, 2013.
112. Lao, W., Song, C., You, J., Ou, Q., Fluorescence and β-cyclodextrin inclusion properties of three carbazole-based dyes. *Dyes Pigm.*, 95, 619–626, 2012.
113. Zhao, J., Wang, J., Yu, C., Guo, L. et al., Prognostic factors affecting the clinical outcome of carcinoma ex pleomorphic adenoma in the major salivary gland. 1–8, 2013.
114. Mohanty, J., Bhasikuttan, A.C., Nail, W.M., Pal, H., Host–guest complexation of neutral red with macrocyclic host molecules: Contrasting pKa shifts and binding affinities for cucurbit[7]uril and β-cyclodextrin. *J. Phys. Chem. B*, 110, 5132–5138, 2006.
115. García-Río, L., Leis, J.R., Mejuto, J.C., Navarro-Vázquez, A. et al., Basic hydrolysis of crystal violet in β-cyclodextrin/surfactant mixed systems. *Langmuir*, 20, 606–613, 2004.
116. Liao, R., Lv, P., Wang, Q., Zheng, J. et al., Cyclodextrin-based biological stimuli-responsive carriers for smart and precision medicine. *Biomater. Sci.*, 5, 1736–1745, 2017.
117. Nilsson, M., Valente, A.J.M., Olofsson, G., Söderman, O., Bonini, M., Thermodynamic and kinetic characterization of host–guest association between bolaform surfactants and α- and β-cyclodextrins. *J. Phys. Chem. B*, 112, 11310–11316, 2008.
118. García-Río, L., Mejuto, J.C., Rodríguez-Dafonte, P., Hall, R.W., The role of water release from the cyclodextrin cavity in the complexation of benzoyl chlorides by dimethyl-β-cyclodextrin. *Tetrahedron*, 66, 2529–2537, 2010.
119. De Brauer, C., Germain, P., Merlin, M.P., Energetics of water/cyclodextrins interactions. *J. Inclusion Phenom.*, 44, 197–201, 2002.
120. Pajzderska, A., Czarnecki, P., Mielcarek, J., Wasicki, J., 1H NMR study of rehydration/dehydration and water mobility in β-cyclodextrin. *Carbohydr. Res.*, 346, 659–663, 2011.
121. Murtaza, G., Solubility enhancement of simvastatin: A review. *Acta Pol. Pharm. Drug Res.*, 69, 581–590, 2012.
122. Loftsson, T. and Brewster, M.E., Cyclodextrins as functional excipients: Methods to enhance complexation efficiency. *J. Pharm. Sci.*, 101, 3019–3032, 2012.
123. Singh, A., Worku, Z.A., Van den Mooter, G., Oral formulation strategies to improve solubility of poorly water-soluble drugs. *Expert Opin. Drug Delivery*, 8, 1361–1378, 2011.
124. Yuan, C., Du, L., Jin, Z., Xu, X., Storage stability and antioxidant activity of complex of astaxanthin with hydroxypropyl-β-cyclodextrin. *Carbohydr. Polym.*, 91, 385–389, 2013.
125. Kim, S., Cho, E., Yoo, J., Cho, E. et al., β-CD-mediated encapsulation enhanced stability and solubility of Astaxanthin. *J. Appl. Biol. Chem.*, 53, 559–565, 2010.
126. Yuan, C., Jin, Z., Xu, X., Zhuang, H., Shen, W., Preparation and stability of the inclusion complex of astaxanthin with hydroxypropyl-β-cyclodextrin. *Food Chem.*, 109, 264–268, 2008.
127. Chun, J.Y., You, S.K., Lee, M.Y., Choi, M.J., Min, S.G., Characterization of β-cyclodextrin self-aggregates for eugenol encapsulation. *Int. J. Food Eng.*, 8, 2012.
128. Wang, G., Wu, F., Zhang, X., Luo, M., Deng, N., Enhanced TiO_2 photocatalytic degradation of bisphenol E by β-cyclodextrin in suspended solutions. *J. Hazard. Mater.*, 133, 85–91, 2006.
129. Polyakov, N.E., Khan, V.K., Taraban, M.B., Leshina, T.V. et al., Complexation of lappaconitine with glycyrrhizic acid: Stability and reactivity studies. *J. Phys. Chem. B*, 109, 24526–24530, 2005.
130. Afkhami, A. and Khajavi, F., Effect of β-cyclodextrin, surfactants and solvent on the reactions of the recently synthesized Schiff base and its Cu(II) complex with cyanide ion. *J. Mol. Liq.*, 163, 20–26, 2011.

131. Hu, J., Huang, R., Cao, S., Hua, Y., Unique structure and property of cyclodextrin and its utility in polymer synthesis. 1–15, 2008.
132. Li, J., Tang, Y., Wang, Q., Li, X. *et al.*, Chiral surfactant-type catalyst for asymmetric reduction of aliphatic ketones in water. *J. Am. Chem. Soc.*, 134, 18522–18525, 2012.
133. Faugeras, P.A., Boëns, B., Elchinger, P.H., Brouillette, F. *et al.*, When cyclodextrins meet click chemistry. *Eur. J. Org. Chem.*, 4087–4105, 2012.
134. Gref, R. and Duchêne, D., Cyclodextrins as "smart" components of polymer nanoparticles. *J. Drug Delivery Sci. Technol.*, 22, 223–233, 2012.
135. Oka, Y., Nakamura, S., Uetani, Y., Morozumi, T., Nakamura, H., Determination of SDS using fluorescent γ-cyclodextrin based on TICT in aqueous solution. *Anal. Sci.*, 28, 973–978, 2012.
136. Oka, Y., Nakamura, S., Morozumi, T., Nakamura, H., Triton X-100 selective chemosensor based on β-cyclodextrin modified by anthracene derivative. *Talanta*, 82, 1622–1626, 2010.
137. Zhu, G., Yi, Y., Chen, J., Recent advances for cyclodextrin-based materials in electrochemical sensing. *TrAC, Trends Anal. Chem.*, 80, 232–241, 2016.
138. Zheng, S. and Li, J., Inorganic–organic sol gel hybrid coatings for corrosion protection of metals. *J. Sol-Gel Sci. Technol.*, 54, 174–187, 2010.
139. Antonijevic, M.M., Inhibitory action of non toxic compounds on the corrosion behaviour of 316 austenitic stainless steel in hydrochloric acid solution: Comparison of chitosan and cyclodextrin (vol 7, pg 6599, 2012). *Int. J. Electrochem. Sci.*, 7, 9042, 2012.
140. Chen, T. and Fu, J., An intelligent anticorrosion coating based on pH-responsive supramolecular nanocontainers. *Nanotechnology*, 23, 2012.
141. Karim, Z., Adnan, R., Husain, Q., A β-cyclodextrin-chitosan complex as the immobilization matrix for horseradish peroxidase and its application for the removal of azo dyes from textile effluent. *Int. Biodeterior. Biodegrad.*, 72, 10–17, 2012.
142. Crini, G., Recent developments in polysaccharide-based materials used as adsorbents in wastewater treatment. *Prog. Polym. Sci.*, 30, 38–70, 2005.
143. Crini, G., Peindy, H.N., Gimbert, F., Robert, C., Removal of C.I., Basic Green 4 (Malachite Green) from aqueous solutions by adsorption using cyclodextrin-based adsorbent: Kinetic and equilibrium studies. *Sep. Purif. Technol.*, 53, 97–110, 2007.
144. Guo, H., Zhang, J., Liu, Z., Yang, S., Sun, C., Effect of Tween80 and β-cyclodextrin on the distribution of herbicide mefenacet in soil-water system. *J. Hazard. Mater.*, 177, 1039–1045, 2010.
145. Viglianti, C., Hanna, K., De Brauer, C., Germain, P., Removal of polycyclic aromatic hydrocarbons from aged-contaminated soil using cyclodextrins: Experimental study. *Environ. Pollut.*, 140, 427–435, 2006.
146. Stella, V.J. and Rajewski, R.A., Cyclodextrins: Their future in drug formulation and delivery. *Pharm. Res.*, 14, 556–567, 1997.
147. Tomatsu, I., Peng, K., Kros, A., Photoresponsive hydrogels for biomedical applications. *Adv. Drug Delivery Rev.*, 63, 1257–1266, 2011.
148. Moya-Ortega, M.D., Alvarez-Lorenzo, C., Concheiro, A., Loftsson, T., Cyclodextrin-based nanogels for pharmaceutical and biomedical applications. *Int. J. Pharm.*, 428, 152–163, 2012.
149. Messner, M., Kurkov, S.V., Jansook, P., Loftsson, T., Self-assembled cyclodextrin aggregates and nanoparticles. *Int. J. Pharm.*, 387, 199–208, 2010.
150. Kurkov, S.V. and Loftsson, T., Cyclodextrins. *Int. J. Pharm.*, 453, 167–180, 2013.
151. Auzély-Velty, R., Self-assembling polysaccharide systems based on cyclodextrin complexation: Synthesis, properties and potential applications in the biomaterials field. *C. R. Chim.*, 14, 167–177, 2011.
152. Fang, Z. and Bhandari, B., Encapsulation of polyphenols—A review. *Trends Food Sci. Technol.*, 21, 510–523, 2010.

153. Astray, G., Gonzalez-Barreiro, C., Mejuto, J.C., Rial-Otero, R., Simal-Gándara, J., A review on the use of cyclodextrins in foods. *Food Hydrocolloids*, 23, 1631–1640, 2009.
154. Vivod, V. and Jaus, D., β-Cyclodextrin as retarding reagent in polyacrylonitrile dyeing. *Dyes & Pigments*, 74, 642–646, 2007.
155. Lisa, G., Cyclodextrins' Applications in the Textile Industry. *Cellulose Chemistry and Technology*, 42, 103, 2008.
156. Voncina, B. and Vivo, V., Eco-Friendly Text. Dye. *Finish*, InTech, 2013.
157. Castronuovo, G. and Niccoli, M., Thermodynamics of inclusion complexes of natural and modified cyclodextrins with propranolol in aqueous solution at 298 K. *Bioorg. Med. Chem.*, 14, 3883–3887, 2006.
158. Davis, M.E. and Brewster, M.E., Cyclodextrin-based pharmaceutics: Past, present and future. *Nat. Rev. Drug Discovery*, 3, 1023–1035, 2004.
159. Qi, Z.H. and Sikorski, C.T., Intell. Mater. Control. Release, vol. 728, S.M. Dinh, J.D. DeNuzzio, A.R. Comfort (Eds.), pp. 113–130, *American Chemical Society*, 1999.
160. Gidwani, B. and Vyas, A., A comprehensive review on cyclodextrin-based carriers for delivery of chemotherapeutic cytotoxic anticancer drugs. *Biomed. Res. Int.*, 2015, 2015.
161. Adeoye, O. and Cabral-Marques, H., Cyclodextrin nanosystems in oral drug delivery: A mini review. *Int. J. Pharm.*, 531, 521–531, 2017.
162. Loftsson, T., Self-assembled cyclodextrin nanoparticles and drug delivery. *J. Inclusion Phenom. Macrocycl. Chem.*, 80, 1–7, 2014.
163. Oerlemans, C., Bult, W., Bos, M., Storm, G. et al., Polymeric micelles in anticancer therapy: Targeting, imaging and triggered release. *Pharm. Res.*, 27, 2569–2589, 2010.
164. Sallas, F. and Darcy, R., Amphiphilic cyclodextrins—Advances in synthesis and supramolecular chemistry. *Eur. J. Org. Chem.*, 957–969, 2008.
165. Zhang, J. and Ma, P.X., Cyclodextrin-based supramolecular systems for drug delivery: Recent progress and future perspective. *Adv. Drug Delivery Rev.*, 65, 1215–1233, 2013.
166. Roux, M., Perly, B., Djedaïni-Pilard, F., Self-assemblies of amphiphilic cyclodextrins. *Eur. Biophys. J.*, 36, 861–867, 2007.
167. Bonnet, V., Gervaise, C., Djedaïni-Pilard, F., Furlan, A., Sarazin, C., Cyclodextrin nanoassemblies: A promising tool for drug delivery. *Drug Discovery Today*, 20, 1120–1126, 2015.
168. Sun, T., Ma, M., Yan, H., Shen, J. et al., Vesicular particles directly assembled from the cyclodextrin/UR-144 supramolecular amphiphiles. *Colloids Surf., A*, 424, 105–112, 2013.
169. Sun, T., Yan, H., Liu, G., Hao, J. et al., Strategy of directly employing paclitaxel to construct vesicles. *J. Phys. Chem. B*, 116, 14628–14636, 2012.
170. Ma, M., Guan, Y., Zhang, C., Hao, J. et al., Stimulus-responsive supramolecular vesicles with effective anticancer activity prepared by cyclodextrin and ftorafur. *Colloids Surf., A*, 454, 38–45, 2014.
171. Zerkoune, L., Angelova, A., Lesieur, S., Nano-assemblies of modified cyclodextrins and their complexes with guest molecules: Incorporation in nanostructured membranes and amphiphile nanoarchitectonics design. *Nanomaterials*, 4, 741–765, 2014.
172. Aktaş, Y., Yenice, I., Bilensoy, E., Hincal, A.A., Amphiphilic cyclodextrins as enabling excipients for drug delivery and for decades of scientific collaboration: Tribute to a distinguished scientist, French representative and friend—A historical perspective. *J. Drug Delivery Sci. Technol.*, 30, 261–265, 2015.
173. Arima, H., Hagiwara, Y., Hirayama, F., Uekama, K., Enhancement of antitumor effect of doxorubicin by its complexation with β-cyclodextrin in pegylated liposomes. *J. Drug Targeting*, 14, 225–232, 2006.
174. Dhule, S.S., Penfornis, P., Frazier, T., Walker, R. et al., Curcumin-loaded γ-cyclodextrin liposomal nanoparticles as delivery vehicles for osteosarcoma. *Nanomed. Nanotechnol. Biol. Med.*, 8, 440–451, 2012.

175. Trotta, F. and Cavalli, R., Characterization and applications of new hyper-cross-linked cyclodextrins. *Compos. Interfaces*, 16, 39–48, 2009.
176. Ansari, K.A., Vavia, P.R., Trotta, F., Cavalli, R., Cyclodextrin-based nanosponges for delivery of resveratrol: *In vitro* characterisation, stability, cytotoxicity and permeation study. *AAPS PharmSciTech*, 12, 279–286, 2011.
177. Trotta, F., Zanetti, M., Cavalli, R., Cyclodextrin-based nanosponges as drug carriers. *Beilstein J. Org. Chem.*, 8, 2091–2099, 2012.
178. Castiglione, F., Crupi, V., Majolino, D., Mele, A. et al., Vibrational dynamics and hydrogen bond properties of β-CD nanosponges: An FTIR-ATR, Raman and solid-state NMR spectroscopic study. *J. Inclusion Phenom. Macrocycl. Chem.*, 75, 247–254, 2013.
179. Swaminathan, S., Pastero, L., Serpe, L., Trotta, F. et al., Cyclodextrin-based nanosponges encapsulating camptothecin: Physicochemical characterization, stability and cytotoxicity. *Eur. J. Pharm. Biopharm.*, 74, 193–201, 2010.
180. Torne, S.J., Ansari, K.A., Vavia, P.R., Trotta, F., Cavalli, R., Enhanced oral paclitaxel bioavailability after administration of paclitaxel-loaded nanosponges. *Drug Delivery*, 17, 419–425, 2010.
181. Mognetti, B., Barberis, A., Marino, S., Berta, G. et al., In vitro enhancement of anticancer activity of paclitaxel by a cremophor free cyclodextrin-based nanosponge formulation. *J. Inclusion Phenom. Macrocycl. Chem.*, 74, 201–210, 2012.
182. Liu, T., Li, X., Qian, Y., Hu, X., Liu, S., Multifunctional pH-disintegrable micellar nanoparticles of asymmetrically functionalized β-cyclodextrin-based star copolymer covalently conjugated with doxorubicin and DOTA-Gd moieties. *Biomaterials*, 33, 2521–2531, 2012.
183. Wang, F., Blanco, E., Ai, H.U.A., Boothman, D.A., Gao, J., Modulating β-lapachone release from polymer millirods through cyclodextrin complexation. *J. Pharm. Sci.*, 95, 2309–2319, 2006.
184. Zhang, Z., Lv, Q., Gao, X., Chen, L. et al., pH-responsive poly(ethylene glycol)/poly(L-lactide) supramolecular micelles based on host-guest interaction. *ACS Appl. Mater. Interfaces*, 7, 8404–8411, 2015.
185. Liu, J., Luo, Z., Zhang, J., Luo, T. et al., Hollow mesoporous silica nanoparticles facilitated drug delivery via cascade pH stimuli in tumor microenvironment for tumor therapy. *Biomaterials*, 83, 51–65, 2016.
186. Sun, T., Shu, L., Shen, J., Ruan, C. et al., Photo and redox-responsive vesicles assembled from Bola-type superamphiphiles. *RSC Adv.*, 6, 52189–52200, 2016.
187. Liu, J., Xu, L., Jin, Y., Qi, C. et al., Cell-targeting cationic gene delivery system based on a modular design rationale. *ACS Appl. Mater. Interfaces*, 8, 14200–14210, 2016.
188. Yang, B., Dong, X., Lei, Q., Zhuo, R. et al., Host–guest interaction-based self-engineering of nano-sized vesicles for co-delivery of genes and anticancer drugs. *ACS Appl. Mater. Interfaces*, 7, 22084–22094, 2015.
189. Lai, W.-F., Cyclodextrins in non-viral gene delivery. *Biomaterials*, 35, 401–411, 2014.
190. Mellet, C.O., Fernández, J.M.G., Benito, J.M., Cyclodextrin-based gene delivery systems. *Chem. Soc. Rev.*, 40, 1586–1608, 2011.
191. Liao, R., Yi, S., Liu, M., Jin, W., Yang, B., Folic-acid-targeted self-assembling supramolecular carrier for gene delivery. *ChemBioChem*, 16, 1622–1628, 2015.
192. Wu, H., Peng, J., Wang, S., Xie, B. et al., Fabrication of graphene oxide-β-cyclodextrin nanoparticle releasing doxorubicin and topotecan for combination chemotherapy. *Mater. Technol.*, 30, 242–249, 2015.
193. Zhang, Y.-M., Cao, Y., Yang, Y., Chen, J.-T., Liu, Y., A small-sized graphene oxide supramolecular assembly for targeted delivery of camptothecin. *Chem. Commun. (Camb).*, 50, 13066–13069, 2014.
194. Pourjavadi, A., Eskandari, M., Hosseini, S.H., Nazari, M., Synthesis of water dispersible reduced graphene oxide via supramolecular complexation with modified β-cyclodextrin. *Int. J. Polym. Mater. Polym. Biomater.*, 66, 235–242, 2017.

195. Liu, J., Chen, G., Jiang, M., Supramolecular hybrid hydrogels from noncovalently functionalized graphene with block copolymers. *Macromolecules*, 44, 7682–7691, 2011.
196. Yang, Y., Zhang, Y.M., Chen, Y., Zhao, D. *et al.*, Construction of a graphene oxide based noncovalent multiple nanosupramolecular assembly as a scaffold for drug delivery. *Chem. Eur. J.*, 18, 4208–4215, 2012.
197. Pourjavadi, A., Tehrani, Z.M., Shakerpoor, A., Dendrimer-like supramolecular nanovalves based on polypseudorotaxane and mesoporous silica-coated magnetic graphene oxide: A potential pH-sensitive anticancer drug carrier. *Supramol. Chem.*, 28, 624–633, 2016.
198. Wei, G., Dong, R., Wang, D., Feng, L. *et al.*, Functional materials from the covalent modification of reduced graphene oxide and β-cyclodextrin as a drug delivery carrier. *New J. Chem.*, 38, 140–145, 2014.
199. Ko, N.R., Nafiujjaman, M., Lee, J.S., Lim, H.-N. *et al.*, Graphene quantum dot-based theranostic agents for active targeting of breast cancer. *RSC Adv.*, 7, 11420–11427, 2017.
200. He, Y., Chen, D., Xiao, G., Hydroxypropyl-β-cyclodextrin functionalized graphene oxide nanospheres as unmodified paclitaxel carriers. *Asian J. Chem.*, 26, 6005–6009, 2014.
201. Tan, J., Meng, N., Fan, Y., Su, Y. *et al.*, Hydroxypropyl-β-cyclodextrin–graphene oxide conjugates: Carriers for anti-cancer drugs. *Mater. Sci. Eng. C*, 61, 681–687, 2016.
202. Hu, X., Ma, L., Wang, C., Gao, C., Gelatin hydrogel prepared by photo-initiated polymerization and loaded with TGF-b1 for cartilage tissue engineering. *Macromol. Biosci.*, 9, 1194–1201, 2009.
203. Ye, Y. and Hu, X., A pH-Sensitive injectable nanoparticle composite hydrogel for anticancer drug delivery. *J. Nanomater.*, 2016, 1–8, 2016.
204. Hu, X., Li, D., Tan, H., Pan, C., Chen, X., Injectable graphene oxide/graphene composite supramolecular hydrogel for delivery of anti-cancer drugs. *J. Macromol. Sci. Part A Pure Appl. Chem.*, 51, 378–384, 2014.
205. Li, J., Harada, A., Kamachi, M., Sol–Gel Transition during inclusion complex formation between α-cyclodextrin and high molecular weight poly(ethylene glycol)s in aqueous solution. *Polym. J.*, 26, 1019–1026, 1994.
206. Li, J., Ni, X., Leong, K.W., Injectable drug-delivery systems based on supramolecular hydrogels formed by poly(ethylene oxide)s and α-cyclodextrin. *J. Biomed. Mater. Res.*, 65A, 196–202, 2003.
207. Li, J., Li, X., Ni, X., Wang, X. *et al.*, Self-assembled supramolecular hydrogels formed by biodegradable PEO-PHB-PEO triblock copolymers and α-cyclodextrin for controlled drug delivery. *Biomaterials*, 27, 4132–4140, 2006.
208. Ren, L., He, L., Sun, T., Dong, X. *et al.*, Dual-responsive supramolecular hydrogels from water-soluble PEG-grafted copolymers and cyclodextrin. *Macromol. Biosci.*, 9, 902–910, 2009.
209. Ha, W., Yu, J., Song, X., Zhang, Z. *et al.*, Prodrugs forming multifunctional supramolecular hydrogels for dual cancer drug delivery. *J. Mater. Chem. B*, 1, 5532–5538, 2013.
210. Mao, S., Lu, G., Chen, J., Three-dimensional graphene-based composites for energy applications. *Nanoscale*, 7, 6924–6943, 2015.
211. Li, C. and Shi, G., Three-dimensional graphene architectures. *Nanoscale*, 4, 5549, 2012.
212. Jiang, L. and Fan, Z., Design of advanced porous graphene materials: From graphene nanomesh to 3D architectures. *Nanoscale*, 6, 1922–1945, 2014.
213. Ha, W., Yu, J., Chen, J., Shi, Y., 3D graphene oxide supramolecular hybrid hydrogel with well-ordered interior microstructure prepared by a host–guest inclusion-induced self-assembly strategy. *RSC Adv.*, 6, 94723–94730, 2016.

214. Ha, W., Zhao, X.-B., Jiang, K., Kang, Y. et al., A three-dimensional graphene oxide supramolecular hydrogel for infrared light-responsive cascade release of two anticancer drugs. *Chem. Commun.*, 52, 3–6, 2016.
215. Wang, L., Ren, K., Wang, H., Wang, Y., Ji, J., pH-sensitive controlled release of doxorubicin from polyelectrolyte multilayers. *Colloids Surf. B*, 125, 127–133, 2015.
216. Stella, V.J., Rao, V.M., Zannou, E.A., Zia, V., Mechanisms of drug release from cyclodextrin complexes. *Adv. Drug Delivery Rev.*, 36, 3–16, 1999.

16

Polymeric Nanocomposites Including Graphene Nanoplatelets

Ismaeil Ghasemi* and Sepideh Gomari

Plastic Department, Iran Polymer and Petrochemical Institute, Tehran, Iran

Abstract

Polymer nanocomposites based on carbonaceous nanomaterials, especially graphene nanoplatelets, have attracted a great deal of attention in recent years. Graphene as a 1-nm-thick layer of carbon atoms has very unique properties. The main challenge for using graphene nanosheets in polymers is obtaining a good dispersion in the matrix. After describing methods of graphene production, various strategies for graphene functionalization are reviewed. Then, different routes of polymer/graphene nanocomposite preparation are demonstrated. These nanocomposites show improved thermal, electrical, mechanical, and gas barrier properties. The crystallization behavior, electrical conductivity, mechanical property, gas barrier property, thermal conductivity, and rheological behavior of polymer/graphene nanocomposites are reviewed. Hybrid nanocomposites including graphene and other nanofillers are discussed. Finally, the application of polymer/graphene nanocomposites in the future using these nanocomposites is elucidated.

Keywords: Graphene, polymeric nanocomposite, functionalization, crystallization, electrical conductivity, rheological behavior, physical properties

16.1 Introduction

Graphene as a new class of carbon allotropes was first introduced by Novoselov *et al.* in 2004 [1]. Graphene is a 1-nm-thick lattice of carbon atoms with sp^2 hybridization, which is the building block of graphite. The typical structure of graphite and graphene is shown in Figure 16.1. Among carbon allotropes, buckyball and carbon nanotubes (CNTs) are known as 0-D and 1-D, respectively, while graphene nanoplatelets (GnPs) are characterized as 2-D nanosheets. Unique properties such as high electrical conductivity (up to 6000 S/cm), superior heat conductivity (5000 W/(m K)), excellent Young's modulus (1 TPa), and tensile strength (130 GPa) are reported for a single-layer graphene, which nominates this material as the strongest available material [2].

There are different methods to make graphene nanosheets, including bottom-up and top-down approaches. The bottom-up approach initiates from carbon atoms to produce a single-layer graphene, which consisted of chemical vapor deposition (CVD), arc

---

*Corresponding author: i.ghasemi@ippi.ac.ir

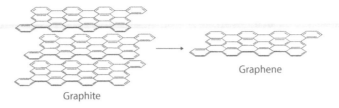

Figure 16.1 Typical structure of graphite and a single-layer graphene.

discharge, epitaxial growth on SiC, chemical conversion, reduction of CO, unzipping carbon nanotubes, and self-assembly of surfactants. Each aforesaid method has its own merits and demerits. For example, CVD and epitaxial growth on SiC produce small amounts of graphene, while free defect and large nanosheets can be obtained. On the other hand, the commencement of the top-down approach is chemical treatment of graphite. Mechanical cleavage, direct sonication, electrochemical, and superacid dissolution methods have been used to separate graphite layers. Some oxygen-containing groups (carboxyl, hydroxyl, and epoxide groups) may be created on the graphene surface in these methods.

The most favorable method for the large-scale production of graphene starts from exfoliation of graphite using strong oxidation procedures such as Hummer's method to produce graphene oxide (GO). Subsequently, GO would reduce to GnP using thermal reduction or chemical reduction. In the thermal reduction method, GO is rapidly heated to 1000°C for 30 s under inert atmosphere, which leads to production of thermally reduced graphene (TRG), while in chemical reduction, chemically reduced graphene (CRG) is produced using hydrazine, dimethylhydrazine, and sodium borohydride followed by hydrazine, hydroquinone, or UV-irradiated TiO_2 [2].

16.2 Functionalization of Graphene Nanosheets

The main challenge for using graphene nanosheets in polymers is obtaining a good dispersion in the matrix. Enhancement of the properties can be reached if proper exfoliation is achieved. Due to strong interactions between graphene layers including van der Waals and π–π interactions, achieving nanocomposites containing single-layer graphene is very difficult. The main strategy to improve the dispersion of these nanoparticles is functionalization with chemical groups. The existence of chemical groups on the graphene surface, which are compatible with the polymeric matrix, causes an increase in the filler–matrix interactions in comparison to filler–filler interactions. In other words, matrix-compatible chemical groups on the surface of nanosheets inhibit the restacking of the separated nanosheets.

Chemical functionalization of graphene can be classified into two main methods including covalent and noncovalent modifications. Covalent functionalization usually leads to structural change in graphene nanosheets from sp^2 to sp^3 hybridization with reduction of electronic conjugation. This reaction can be accomplished both on the surface and at the end of the sheets. The covalent modification consisted of the following methods: nucleophilic substitution, electrophilic substitution, condensation, and addition [3].

16.2.1 Covalent Modification

Nucleophilic substitution is a suitable method for mass production of functionalized GnP, which the epoxide groups react with $-NH_2$ groups of modifiers. This is because of the high rate of epoxide group reactions at ambient temperatures. It should be considered that ambient temperature is only applied for primary amines with short chains. In this method, GO was first prepared by Hummer's method and then reacted with NH_2-containing molecules. Kuila et al. prepared GO modified with dodecyl amine through nucleophilic substitution at room temperature [4], while for long-chain amine like octadecyl amine (ODA), higher temperatures are required for progressing of the reaction [5].

In electrophilic substitution, the hydrogen atoms are replaced by an electrophilic group. The most common example of this reaction is embedment of aryl diazonium salt on the surface of GO. Lomeda et al. prepared dodecyl benzene sulfonate (SDBS)-wrapped GO and then functionalized with aryl diazonium salt to obtain organosoluble nanosheets [6]. The modified nanosheets were easily dissolved in DMF, DMAc, and NMP at concentrations up to 1 mg/ml.

The nomination of the condensation method originated from the fact that a chemical reaction between two molecules has occurred and a larger molecule is obtained, producing a small molecule such as H_2O as by-product. The condensation method is the most prevalent way to functionalize graphene and is able to graft isocyanate, diisocyanate, amine compounds, and alkyl lithium reactants onto graphene. By grafting these reactants, the hydrophobicity of the graphene surface is reduced because of the formation of amide bonds (where reacted with carboxylic acid groups) and carbamate ester linkages (where reacted with hydroxyl groups). Salavagione et al. [7] modified GO by reacting between carboxylic acids of GO and hydroxyl groups of PVA in the presence of dicyclohexyl carbodiimide (DCC) and 4-dimethyl amino pyridine (DMAP) as catalysts. The obtained functionalized graphene nanosheets were then reduced by hydrazine hydrate, which were soluble in DMSO and water. The schematic of this reaction is shown in Figure 16.2.

The addition method involves the combination of one molecule with another to form a single larger molecule with no other by-product. The main limitation for this reaction is the existence of multiple bounds such as molecules with carbon–carbon double bonds, or with triple bonds. In this method, most works have been accomplished via cycloaddition reaction, which is a class of addition reaction. For example, chemically converted graphene

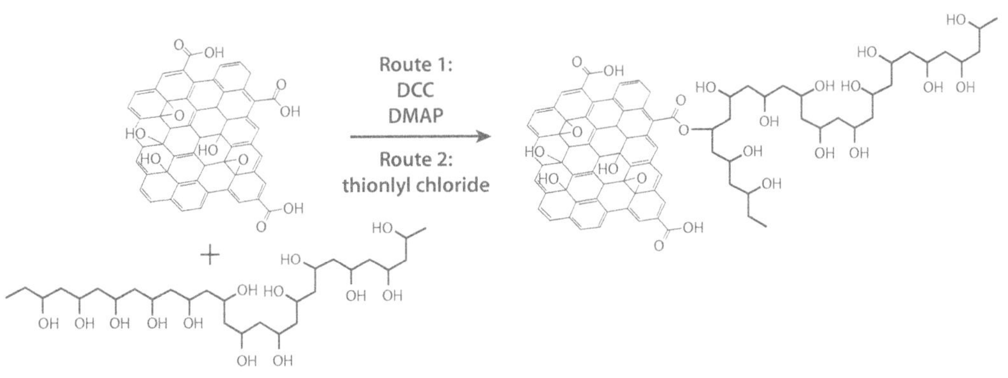

Figure 16.2 Schematic illustration of the esterification of graphite oxide with PVA [7].

was reacted by aryne cycloaddition under mild conditions by Zhong et al. [8]. The obtained functionalized graphene was homogeneously dispersed in ethanol. Choi et al. prepared a functionalized epitaxial graphene by the cycloaddition of azidotrimethylsilane [9]. They explained the reaction mechanism by removing N_2 and then [2 + 1] electrophilic cycloaddition or biradical pathway between nitrene and graphene. Since the addition method is easily applicable, many kinds of functional molecules can be grafted onto graphene.

16.2.2 Noncovalent Modification

There is a large body of reports for modification of carbon-based nanomaterials by non-covalent functionalization, which implies the efficiency and feasibility of this method. Noncovalent modifications need the physical adsorption of some molecules onto graphene surface via hydrophobic, van der Waals, or electrostatic forces. The reduction of aqueous dispersion GO is usually accomplished by hydrazine hydrate, which results in aggregation and agglomeration of graphene sheets. To avoid inevitable agglomeration, reduction of GO is done in the presence of a suitable surfactant.

As a well-known surfactant, SDBS is frequently used in the modification of carbon-based nanomaterials. Reduction of graphene in the presence of SDBS leads to production of surfactant-wrapped graphene sheets, which shows good dispersibility in water without considerable sacrificing of electrical conductivity [10]. Poly(sodium 4-styrenesulfonate) (PSS) was used by Stankovich et al. for noncovalent modification of graphene nanoplatelets [11]. They reported successful reduction, exfoliation, and high water dispersibility of GO by this method. Bai et al. used sulfonated polyaniline (SPANI) as a surface modifier for noncovalent modification of graphene due to its good electrical conductivity, electrochemical activity, and water solubility [12]. They reported good water dispersibility (>1 mg/ml), satisfactory electrical conductivity (30 S/m), and unique electrochemical properties for SPANI-functionalized graphene sheets. A zwitterionic surfactant was used by Keramati et al. for surface modification of graphene nanoplatelets [13]. XRD results demonstrated full exfoliation of nanosheets obtained by using a zwitterionic surfactant.

16.3 Preparation Methods of Polymeric Nanocomposites

Since the final properties of nanocomposites is affected by the dispersion status of GnP in the matrix, the selection and condition of the preparation method have a great importance. The exfoliation of graphene nanoplatelets into few layers and even down to single layers would provide the advantage of large surface area per unit volume of these nanoplatelets. Hence, it is of crucial interest to select a proper method for achieving the fully exfoliated state of nanoplatelets. During preparation of the nanocomposites, restacking of nanosheets is inevitable due to existence of van der Waals interactions among them. As mentioned earlier, functionalization of nanosheets could be a good strategy to prevent the restacking.

Similar to preparation methods of nanocomposites including other nanoparticles (organoclay, CNT, nano SiO_2, etc), there are three main methods to synthesize nanocomposites containing GnP: *in situ* polymerization, solution blending, and melt blending.

In the *in situ* polymerization technique, GnP is dispersed in the monomer or monomer solution and the polymerization reaction is initiated by heat or radiation. During the polymerization reaction, the formed macromolecules can diffuse between the nanosheets and facilitate the establishment of exfoliation structure. Poly(vinyl acetate) intercalated graphite oxide nanocomposite was prepared via an *in situ* intercalative thermally polymerization reaction by Liu *et al.* [14]. In another work, poly(methyl methacrylate) (PMMA)/GO nanocomposites were prepared by a novel method utilizing a macroazoinitiator [15]. Lee *et al.* produced nanocomposites of waterborne polyurethane with TRG nanosheets by this method [16].

The solution blending method has been used for the production of GO- or GnP-containing nanocomposites. Graphene is exfoliated in a solvent in which the polymer is soluble via mechanical stirring or ultrasonication. Then, the polymer adsorbs on the graphene nanosheets, followed by solvent removal. It should be considered that prolonged time and high ultrasonic power may cause the breakage of nanosheets. The better dispersion of nanosheets can be obtained for TRG in comparison to CRG in this method. The main reason for this behavior is the wrinkled structure of TRG that prevents the restacking, while for CRG, the flattened structure and strong interfacial interactions lead to more probable restacking [17]. Nanocomposites based on water-soluble polymers, such as poly(ethylene oxide) (PEO) [18] or poly(vinyl alcohol) (PVA) [19], have been prepared via this method because of the easy exfoliation of GO in water. For nonpolar polymers that dissolve in aprotic solvents, GO should be functionalized with organic moieties such as isocyanate or amine. Ren *et al.* functionalized GO with dodecyl amine and prepared high-density polyethylene (HDPE) nanocomposites by solution mixing method [20].

Melt mixing is more popular than other aforesaid methods because of its scalability in available industrial apparatuses like extrusion. However, there are some limitations for using this method. First, the most functionalized groups are unstable at the processing temperatures. In typical mixing temperatures of 200°C, oxygen-containing groups undergo thermal degradation. For this reason, TRG has been usually used in the melt mixing method instead of GO and other types of functionalized GnP [21]. However, Reghat *et al.* used the melt mixing method to produce nanocomposites based on PLA/GO at 175°C and reported a good dispersion of nanosheets [22]. Difficult feeding due to the low bulk density of graphene is the second challenge of melt mixing. At high loading of GnP, the homogenous feeding could hardly be achieved. The third limitation is related to the breakage of platelets under intense shear forces in melt mixing operations. From the published reports in the literature, one can find the better dispersion of graphene in the solution compared to the melt mixing method. A direct comparison of these two methods has been done by Kim *et al.* for polyurethane/graphene nanocomposites [17].

16.4 Crystallization Behavior of Polymer/Graphene Nanocomposites

Most of the used polymers as the matrix in polymer/graphene nanocomposites show a semicrystalline nature. A study of crystallization behavior is important because of its direct role in different properties of nanocomposites like mechanical performance. The crystallization process in polymers usually consisted of two steps, nucleation and crystal growth, which are affected by the presence of graphene nanosheets. When polymer samples are crystallized from the melt, at the initial stage, the spherulites grow outward until they impinge on their neighbors and stop growth at the intersection. Then, the secondary crystallization starts when polymer crystallizes in the remaining interlamellar regions.

16.4.1 Isothermal Crystallization Kinetics

Studies on the kinetics of polymer crystallization in the presence of GnP have been considered by researchers to understand the effect of GnP on the time development of crystallinity. There are several different kinds of experimental methods to evaluate crystallization kinetics, including differential scanning calorimetry (DSC), small-angle X-ray scattering (SAXS), vibrational spectroscopy, nuclear magnetic resonance (NMR), and polarized optical microscopy (POM). Nevertheless, DSC in isothermal and non-isothermal conditions and POM are the most common and most frequently used by the researchers [23, 24]. Generally, in isothermal experiments by DSC, the polymer melt is abruptly cooled (supercooled by liquid nitrogen) into the vicinity of crystallization temperature and held at this temperature for monitoring of crystallization process. This procedure leads to obtaining the overall rate of crystallization.

The same trend of the effect of GnP on the overall rate of crystallization was not observed and one can find contradictory results in the literature. This contradiction originated from the polymer structure and dispersion status of the GnP in the matrix. It seems that these parameters can affect the two steps of crystallization (nucleation and growth). The overall crystallization rate is the product of crystal growth rate and nucleation rate. The polymer–filler interaction could suppress polymer chain mobility to reduce the overall crystallization rate. On the other hand, well-dispersed nanoparticles can act as heterogeneous nucleating agents to enhance the nucleation rate. However, in some cases, nanoparticles have reduced the nucleating efficiency and anti-nucleating behavior has been observed [23]. Xu $et\ al.$ prepared isotactic polypropylene (iPP)/graphene oxide nanocomposites by the solution coagulation method and studied isothermal crystallization by DSC and POM [25]. They reported that the induction period and half-crystallization time of nanocomposites were greatly reduced during the isothermal crystallization process (Figure 16.3). In addition, considerable crystallite nucleation was detected at a very low loading of GO due to the adequate surface area. They fitted their data on the Lauritzen-Hoffman secondary nucleation theory [26] and confirmed the enhancement of nucleation. The Lauritzen-Hoffman equation is given in Equation 16.1:

$$G = G_0 \exp\left[-\frac{U}{R(T_0 - T_\infty)}\right] \exp\left[-\frac{K_g}{T_c \Delta T f}\right] \qquad (16.1)$$

where G is the spherulitic growth rate, U is the activation energy for transferring of segments into the crystallites, K_g is the nucleation constant, ΔT is the undercooling ($T_m^0 - T_c$), and T_m^0 is the equilibrium melting point; f is the factor $2T_c/(T_m^0 + T_c)$, which denotes the change in heat of function as the temperature is decreased below T_m^0, R is the gas constant and $T_\infty = T_g - 30K$ is the temperature at which all segment mobility is frozen and viscosity approaches an infinite value.

The isothermal crystallization of poly(L-lactic acid) (PLLA) was studied in the presence of GO and functionalized GnP [24, 27]. Wang $et\ al.$ prepared nanocomposites based on PLLA with the addition of 0.5, 1 and 2 wt.% GO via solution method. They reported that the overall crystallization rate was increased without changing the crystal structure and crystallization mechanism. The overall crystallization rate passed through a maximum at 1 wt.% of GO, which could be related to the existence of some aggregation at high GO loading. Despite GO acting as a nucleating agent at lower contents, the nucleation density reduced

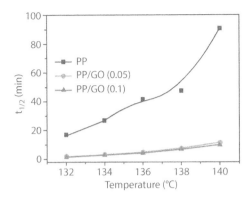

Figure 16.3 $t_{1/2}$ plots of neat PP, PP/GO (0.05wt.%), and PP/GO (0.1 wt.%) nanocomposites against different T_c [25].

at higher loadings because of the formation of aggregates, which was confirmed by POM images (Figure 16.4). Moreover, a lower overall crystallization rate was observed at a higher range of crystallization temperatures (123 to 138°C).

The effect of GnP functionalization on the isothermal crystallization behavior was studied by Manafi et al. [24]. In this study, PLA was grafted on the surface of GnP after oxidation and acylation. The kinetics of isothermal crystallization was investigated at different temperatures (115, 120, 125, and 130°C), and the obtained data were fitted on the Avrami model (Equation 16.2) [28].

$$X_t = 1 - \exp(-kt^n) \qquad (16.2)$$

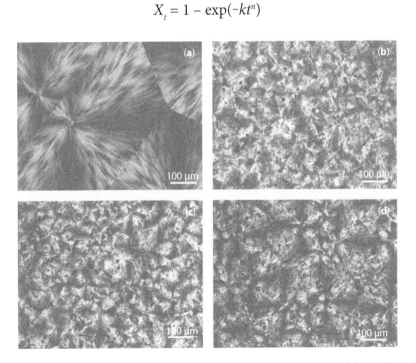

Figure 16.4 POM images of neat PLLA and its nanocomposites crystallized at 138°C; (a) neat PLLA, (b) PLLA/GO(0.5), (c) PLLA/GO(1), and (d) PLLA/GO(2).

The Avrami parameters accompanied by the half-time of crystallization ($t_{1/2}$) are depicted in Table 16.1. It can be found that samples containing functionalized graphene (FGnP) show lower $t_{1/2}$ at the same crystallization temperature. In the Avrami equation, n is dependent on the mechanism of nucleation and growth geometry of the crystals and k is the crystallization rate constant that involves both nucleation and growth rate parameters. They calculated $n = 2.4$, which means the circular diffusion on a lamellar structure. As seen in Table 16.1, k values for nanocomposites containing FGnP are larger than those of samples containing GnP nanoparticles at the same holding temperature. This means a higher crystallization rate for PLA in the nanocomposites in the presence of FGnP.

Since a decrease in rate of crystallization was reported in some studies due to the presence of functionalized GnP [23], some efforts were focused on the incorporation of chain promoter agents in the nanocomposites. For example, Liu and coworkers used PEG as a chain promoter for PLA/GnP nanocomposites [29]. A masterbatch of PEG and graphene was prepared by the freeze-drying method and incorporated into PLA via the solution method. Their results showed that the crystallinity and crystallization rate of PLA are greatly improved in the presence of this masterbatch. In another work, Xu et al. used PEG chemically grafted onto GnP as a simultaneous heterogeneous nucleation

Table 16.1 Summary of Avrami kinetic parameters for isothermal crystallization of PLA/GnP and PLA/FGnP nanocomposites at different holding temperatures [24].

Samples	Holding temperature (°C)	n	k (min^{-1})	$t_{1/2}$ (min)
PLA/GnP(0.5)	115	2.2	0.160	1.9
	120	2.6	0.120	2.2
	125	2.7	0.100	2.5
	130	2.3	0.109	2.7
PLA/FGnP(0.5)	115	2.3	0.200	1.5
	120	2.3	0.196	1.5
	125	2.0	0.130	2.7
	130	2.5	0.070	4.0
PLA/GnP(1)	115	2.1	0.180	1.8
	120	2.2	0.160	2.0
	125	2.2	0.110	2.9
	130	2.0	0.060	5.8
PLA/FGnP(1)	115	2.4	0.200	1.4
	120	2.3	0.202	1.5
	125	2.0	0.230	1.5
	130	2.5	0.099	2.8

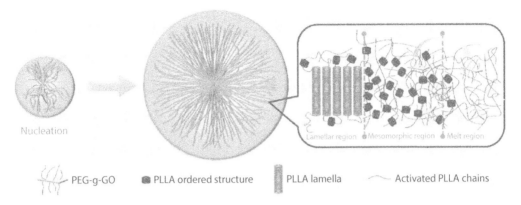

Figure 16.5 Crystallization development in PLA/PEG-g-GO nanocomposites [30].

agent and chain mobility promoter [30]. The performance mechanism of PEG-g-GO is schematically shown in Figure 16.5. Since PEG molecules are not able to crystallize with PLA, it is expected that the PEG-g-GO does not contribute in the growth front into the newly formed mesomorphic layer once the current region solidifies. They also reported smaller activation energy (U) according to Lauritzen-Hoffaman theory for samples containing 1 wt.% of PEG-g-GO (1887 J/mol) in comparison to those of neat PLA (2809 J/mol) and PLA/GO (2938 J/mol). The smaller U means that the molecular chains need less activation energy for moving across the interphase of crystalline/amorphous regions.

16.4.2 Non-Isothermal Crystallization Kinetics

The non-isothermal crystallization condition is more similar to the real situation in polymer processing. This method for investigating crystallization is usually accomplished using DSC in three steps, including heating, cooling, and reheating. The sample is heated higher than its melting point and kept for a few minutes to remove thermal and stress history, after which it was cooled to the correct temperature and again reheated above the melting temperature. This procedure can be done at different heating and cooling rates. The useful data are extracted from cooling and reheating steps: onset of crystallization temperature (T_{onset}), initial slope of the crystallization (S_i), full width at the half height maximum of the crystallization peak (FWHM), crystallization enthalpy (H_c), exotherm crystallization temperature (T_c), melting point (T_m), fusion enthalpy (H_m), and degree of crystallinity (X_c). It is found that the mentioned thermal properties can be affected by the presence of GnP nanoplatelets [31, 32].

Tarani *et al.* studied non-isothermal crystallization of HDPE in the presence of GnP [31]. The effect of graphene size (5, 15, and 25 μm in diameter) on crystallization behavior was investigated using a modified Avrami model. Their results showed that the overall crystallization rate, the activation energy, and the fold surface-free energy of the HDPE polymer were changed after the addition of nanofiller in different sizes. GnP incorporation leads to a greater T_c of HDPE, while the smaller diameter of GnP resulted in a more efficient heterogeneous nucleation and higher crystallization rate. The obtained data from the modified

Avrami model stated that the primary crystallization contains crystal growth until overlapping of lamellar stacks. The secondary stage, including the filling of the spherulites' gaps, shows a much slower rate than the primary stage.

Nanocomposites of poly(3-hexylthiophene) (P3HT) containing reduced graphene oxide (rGO) were prepared through *in situ* reduction of graphene oxide in the presence of P3HT [33]. The non-isothermal crystallization behaviors of nanocomposites were studied using Avrami, Ozawa, and Mo models. T_c and X_c of P3HT remarkably increased by incorporation of rGO. It was found that rGO has two contradictory roles: first, it acts as a nucleating agent to promote the crystallization of P3HT; second, it restricts the chain mobility of P3HT to retard the crystallization.

The non-isothermal crystallization kinetics of *in situ* polymerized nylon 6/GnP was studied by Zhang *et al.* [34]. The results according to the modified Avrami equation showed that, at lower cooling rates (at 5, 10, and 20°C/min), the crystallization rate of the nylon 6/graphene nanocomposites decreased, while at higher cooling rates (40°C/min), a higher rate of crystallization was observed for nanocomposites (as reported in Table 16.2).

Table 16.2 Non-isothermal kinetic parameters of nylon 6/graphene nanocomposites based on the modified Avrami equation at different cooling rates [34].

Sample	Cooling rate (°C/min)	$\log Z_t$	n	$\log Z_c$	Z_c
Nylon 6	40	1.96	2.39	0.049	1.119
	20	1.59	2.47	0.0795	1.201
	10	1.15	2.58	0.113	1.297
	5	0.40	2.54	0.08	1.202
Nylon/GnP(0.1)	40	2.08	2.45	0.052	1.127
	20	1.25	2.33	0.0625	1.155
	10	0.58	2.49	0.058	1.143
	5	0.18	2.32	0.036	1.086
Nylon/GnP(0.5)	40	2.29	2.30	0.0573	1.141
	20	1.37	2.24	0.0685	1.171
	10	0.88	2.32	0.088	1.225
	5	0.17	2.21	0.034	1.081
Nylon/GnP(1)	40	2.09	2.26	0.0523	1.128
	20	1.10	2.19	0.055	1.135
	10	0.83	2.25	0.083	1.211
	5	0.30	2.21	0.06	1.148

They claimed that this observation is related to the balance of two contradicting effects of graphene on the crystallization rate. They believed that at higher cooling rates, the positive effect played a leading role. However, at lower cooling rates, the negative effect of graphene is dominant.

Gomari *et al.* studied the non-isothermal crystallization of PEO/PEG-g-GnP (FGnP) nanocomposites with/without LiClO$_4$ salt for use as electrolyte in Li-ion batteries [23]. Modified Avrami and combined Avrami–Ozawa equations were used and revealed that the Avrami exponent values did not change in the presence of PEG-g-GnP, which means that the nucleation mechanism and crystal growth are not affected. However, the half-time of crystallization increased by the addition of PEG-g-GnP in both PEO and PEO:LiClO$_4$ systems. The effect of PEG-g-GnP on the parameters of non-isothermal crystallization such as T_c, ΔH_c, $t_{1/2}$, and kinetic rate constant (Z_c) based on the modified Avrami equation for PEO and PEO:LiClO$_4$ systems is shown in Figure 16.6. They concluded that there is a more considerable effect of PEG-g-GnP on the crystallization behavior of PEO:LiClO$_4$ in comparison to PEO.

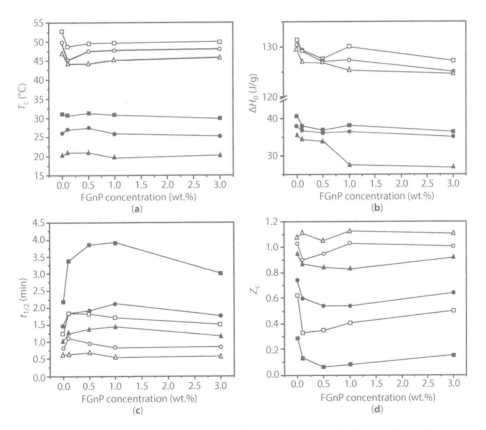

Figure 16.6 (a) Crystallization temperature, (b) crystallization enthalpy, (c) half-time of crystallization, and (d) kinetic rate constant based on modified Avrami equation as a function of FGnP concentration for PEO and nanocomposite samples (open symbols) and SPE and nanocomposite electrolytes (solid symbols) at 2°C/min (squares), 5°C/min (circles) and 10°C/min (triangles) [23].

16.5 Electrical Conductivity

Due to its highly inherent electrical conductivity, GnP can improve electrical conduction through the formation of electron pathways. The dispersion status of GnP plays a main role in enhancement of electrical conduction and determines the minimum required GnP content to achieve percolation threshold. In comparison to other-carbon based fillers, GnP is able to change insulating polymers to conductive materials at very low loadings. It should be noted that although chemical functionalization of GnP improves dispersion quality, due to the destruction of carbon structure of GnP, a lower electrical conductivity is obtained. In other words, the electrical conductivity is balanced by two factors: functionalization (negative effect) and dispersion (positive effect), in which the dispersion factor is usually dominant to increase the overall electrical conductivity. Since the dispersion of GnP is affected by the preparation method, it is found that the solution method leads to better dispersion and thus higher conductivity. The electrical conduction is also affected by the polymeric matrix and it is found that the thermoset polymers show a higher conductivity in comparison to thermoplastic materials at the same loadings [35].

At the percolation threshold, an interconnected network of nanosheets is formed through the polymeric matrix and electrical conductivity is abruptly increased to the ordinary values of conductive materials ($\approx 10^{-3}$ S/cm). There are two mechanisms for electron transport in the matrix: tunneling and contacting. The tunneling mechanism usually occurs when the electrons can transmit between two adjacent graphene sheets that are sufficiently close to each other and are separated by a thin layer of polymer. In the contact mechanism, the direct pathway for electrons is created by the physical contact between GnP nanosheets. At low concentrations of GnP, tunneling is the dominant mechanism because the number of graphene sheets is inadequate for physical contacting. On the other hand, at higher concentrations and above the percolation threshold point, electrical conduction is governed by contact mechanism. Both aforesaid mechanisms are affected by the dispersion state and specific surface area of GnP. The aspect ratio of graphene nanosheets should be taken into account for the determination of the percolation threshold. It is clear that with an increase of aspect ratio, the lower percolation threshold concentration is obtained.

The percolation threshold value is usually determined by the fitting of experimental data on the power law equation as follows:

$$\sigma_c = \sigma_f [(\phi - \phi_c)/(1 - \phi_c)]^t \quad (16.3)$$

where σ_f is the conductivity of the filler, φ is the filler volume fraction, φ_c is the percolation threshold, and t is the universal critical exponent. Conductivity of nanocomposite (σ_c) is plotted against filler volume fraction (φ). In addition, $\log \sigma_c$ versus $\log(\varphi - \varphi_c)$ is also drawn, where t and φ_c can be calculated. Typical graphs are shown in Figure 16.7.

In recent years, the production of conductive polymeric nanocomposites including GnP has attracted great research interest. Conductive nanocomposites based on a variety of polymers such as PET [37], PA6 [38], PVDF [39], PS [40], PI [41], PU [17], and HDPE [42] including reduced GO have been prepared.

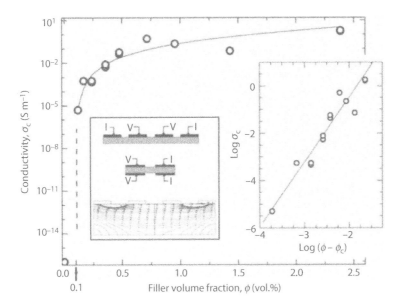

Figure 16.7 Electrical conductivity of polystyrene/phenyl isocyanate-functionalized graphene nanocomposites as a function of filler volume fraction; inset plot shows the power law dependence of conductivity [36].

The most reported data for electrical percolation threshold are under 1 vol.% [17, 37, 38, 40, 41]. However, in some studies, higher values are also obtained [39, 42]. One of the lowest reported percolation thresholds (0.1 vol.%) belongs to Stankovich's work, which synthesized PS/isocyanate-treated GO by solution blending [36]. In this work, treated GO was reduced by dimethylhydrazine. An extremely high aspect ratio and very good homogeneous dispersion of graphene sheets were conducted to such a low percolation threshold.

Another low percolation threshold was observed by Zheng *et al.* [38]. They reported 0.41 vol.% percolation threshold for nanocomposites based on PA6 and thermally reduced GO (rGO). Reduction of GO simultaneously occurred with *in situ* polymerization of ε-caprolactam. A very high conductivity of 0.028 S/cm was measured at 1.64 vol.% of rGO. This result was also attributed to the high aspect ratio, large specific surface area, and homogeneous dispersion of the rGO nanosheets in the matrix. It can be concluded that for achieving low electrical percolation threshold in polar polymers, using rGO is necessary. A comparison between the performance of TRG and exfoliated graphite (EG) in the electrical conductivity of nanocomposites based on PVDF was accomplished by Ansari and Giannelis [39]. Their results showed that a lower percolation threshold (2 wt.%) was obtained for PVDF/TRG nanocomposites compared to PVDF/EG samples (above 5 wt.%), as shown in Figure 16.8.

Using graphene nanosheets in elastomers for enhancing the electrical property has also been reported. For example, Araby *et al.* prepared SBR/GnP by solution mixing [43]. As shown in Figure 16.9, the breakdown of the insulating nature of cured SBR was achieved in the range of 5–7 vol.%. These relatively high values of percolation threshold originated from the high resistive nature of elastomer (in comparison to thermoplastics) and the unsuitable dispersion of GnP.

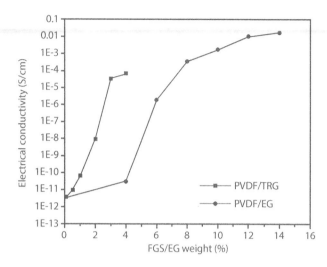

Figure 16.8 Electrical conductivity of PVDF/TRG and PVDF/EG nanocomposites [39].

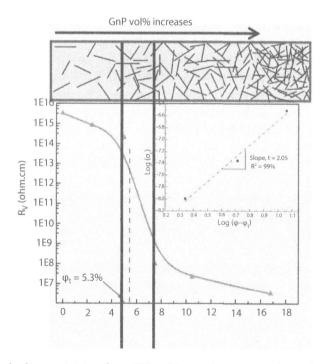

Figure 16.9 Electrical volume resistivity of neat SBR and its cured nanocomposites with GnP [43].

Electrically conductive nanocomposites of PET/graphene were prepared via melt compounding by Zhang *et al.* [37]. In this study, graphene nanosheets were prepared by oxidation of pristine graphite and then thermal reduction and exfoliation. They reported a low percolation threshold of 0.47 vol.% and a high electrical conductivity of 2.11 S/m by only 3.0 vol.% of graphene. This low percolation threshold may be attributed to the uniform dispersion of graphene nanosheets because of the good interaction between the oxygen-containing groups on the graphene surface and the active polar groups of PET. Low- and high-magnification TEM micrographs of PET/TRG (3 vol.%) are shown in Figure 16.10. In the low-magnification image (Figure 16.10a), an interconnected network throughout the matrix can be observed. On the other hand, in the high-magnification image (Figure 16.10b), well-dispersed nanosheets are observable. Wrinkled and overlapped graphene sheets are detectable in the image, which link individual graphene sheets to result in high electrical conductivity.

Qi *et al.* compared the electrical conductivity of PS/MWCNT and PS/TRG nanocomposites, which were prepared by solution method and formed by compression molding [40]. They found that the electrical conductivity of PS/TRG samples was two to four orders of magnitude higher than that of PS/MWCNT. They also added PLA (40 wt.%) to the compounds for the formation of a double percolated network. The percolation threshold of the PS/PLA/TRG samples was ~0.075 vol.%, which is ~4.5 times lower than that of PS/TRG. The main reason for this behavior is selective localization of nanoparticles in the PS phase, which causes the formation of a networked structure at relatively lower graphene concentrations. The percolation thresholds of these samples and localization of graphene are shown in Figure 16.11a and b, respectively. The strategy of the double percolated network was also followed in the case of PS/PMMA/functionalized graphene by Mao *et al.* [44].

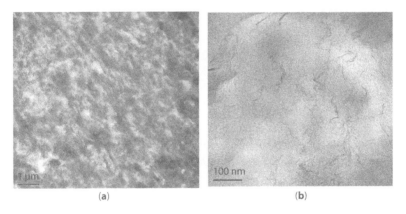

Figure 16.10 Two different magnifications of TEM images for PET/TRG (3 vol.%) nanocomposites [37].

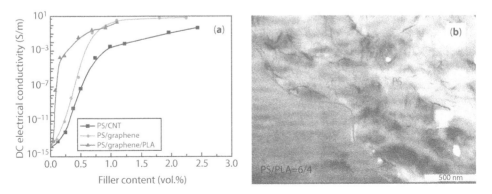

Figure 16.11 (a) Electrical conductivity against filler volume fraction for neat PS and its nanocomposites. (b) TEM image of PS/PLA (6/4) nanocomposite with ~0.46 vol.% (~1.0 wt.%) graphene [40].

16.6 Mechanical Properties

It has been reported that the modulus of graphene is around 1 TPa and its strength is about 130 GPa in non-defect conditions [2]. Indeed, graphene is the strongest material in nature. Although the presence of defects in the graphene surface during functionalization is inevitable, the modulus of the graphene containing defects is yet strong enough for the reinforcing of polymeric matrix (elastic modulus of CRG sheets is still as high as 0.25 TPa [45]).

Tensile properties including elastic modulus, tensile strength, and elongation at break are usually measured to evaluate the mechanical properties. It is noteworthy that the elastic modulus is usually increased by addition of graphene nanoplatelets. Interestingly, the increment of modulus in elastomeric materials is more pronounced. The main reason for the enhancement of modulus by incorporation of the graphene nanosheets is its higher modulus and independence of the modulus to the adhesion of the interface. In other words, since the modulus is determined in low elongations (<5%), the interface cannot play a main role. On the other hand, tensile strength and elongation at break are intensively affected by the nature of the filler–matrix interactions. An approach for improving filler–matrix interaction is the chemical functionalization of the nanoparticles with proper molecules, which show good affinity toward the matrix.

Generally, the mechanical reinforcement in polymeric nanocomposites could be obtained if these conditions have been achieved: (1) a homogenous dispersion of nanofiller in the matrix and (2) a strong interaction between nanofiller and matrix. The homogenous dispersion of graphene and its exfoliation into thin sheets or even single nanosheets provide an important advantage of large specific area of these nanosheets in the polymer. Furthermore, the strong interfacial adhesion ensures a good load transfer from polymer chains into the graphene nanosheets.

Wang *et al.* prepared nanocomposites based on PA6 and two types of graphene: neat graphene and modified graphene with 3-aminopropyltriethoxysilane [46]. The variation of tensile strength and modulus of the samples is shown in Figure 16.12. As can be seen, the elastic modulus was increased gradually in both nanocomposites. However, in samples containing functionalized graphene, a higher increment was detected. In the samples with neat graphene, the tensile strength was decreased due to incompatibilities between GnP and PA6 molecules. In this case, there is inadequate load transfer because of the large amount of microvoids or stress concentrated points between GnP and PA6. Nevertheless, samples with functionalized graphene showed higher tensile strength compared to neat PA6, except for high loadings, which could be related to higher probability of aggregation formation. The different interactions between FGnP–matrix and neat graphene–matrix are schematically shown in Figure 16.12c and d. Better interaction between FGnP and matrix has inhibited the formation of microvoids.

Tensile properties of polyimide/functionalized graphene (FGS) were investigated by Luong *et al.* [41]. *In situ* polymerized polyimide was obtained in the presence of chemically modified graphene with ethyl isocyanate. Stress–strain curves and derived data are demonstrated in Figure 16.13. Because of the good adhesion between FGS and the polymeric

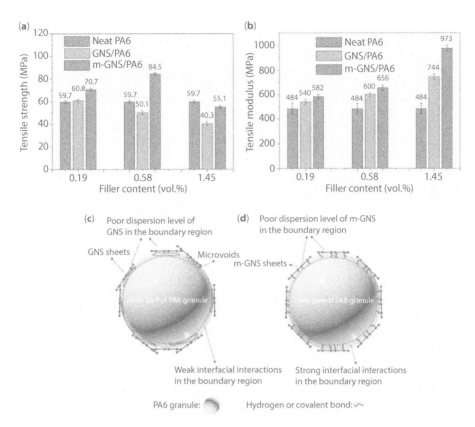

Figure 16.12 (a) Tensile strength and (b) tensile modulus of PA6 with neat (GNS) and modified (m-GNS) graphene; (c) mechanistic model of reinforcement for PA6/GNS and (d) model of PA6/m-GNS nanocomposites [46].

matrix and the addition of only 0.38 wt.% of FGS, a dramatic increase in Young's modulus was obtained (from 1.8 GPa to 2.3 GPa, which is approximately 30% improvement compared to that of pure polymer). Moreover, the tensile strength was increased from 122 MPa to 131 MPa.

Gomari *et al.* reported the tensile properties of polyethylene oxide (PEO) films with GnP and functionalized GnP with PEG (FGnP) [47]. The samples were produced by solution method and stress–strain curves are illustrated in Figure 16.14. Both neat PEO and nanocomposite samples showed yield behavior. The tensile strength of nanocomposites containing 1 wt.% of nanoplatelets increased by 148% versus 29%, toughness increased by 466% versus 165%, and elongation at break increased by 186% versus 87% for FGnP against GnP addition.

Steurer *et al.* prepared nanocomposites based on PA6, PP, PC, and SAN with TRG and studied the mechanical properties [48]. It was observed that by incorporation of TRG, the elastic modulus and elongation at break were increased and decreased, respectively, in comparison to neat resins.

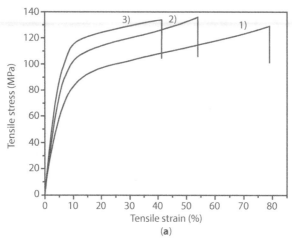

Figure 16.13 (a) Typical stress–strain curves of neat PI (1) and PI/FGS nanocomposite with 0.38 wt.% (2) and 0.75 wt.% of graphene (3); (b) summary of mechanical properties of nanocomposite films [41].

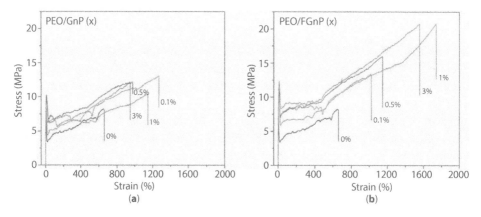

Figure 16.14 Stress–strain curves of neat PEO and its nanocomposites: (a) PEO/GnP(x), (b) PEO/FGnP(x) [47].

16.7 Gas Barrier Properties

It is well known that the addition of planar nanofillers such as organoclay and graphene causes more resistance against gas diffusion through polymers. Indeed, the planar filler creates longer tortuous pathways for penetration of gas molecules. Barrier properties of nanocomposite are controlled by the following factors: the nature of polymer matrix, filler properties (aspect ratio, loading, and intrinsic barrier property of filler), and dispersion status of the filler. In addition, the orientation of the filler, the interface of filler/matrix, and the crystallinity of the matrix could also influence the barrier resistance.

The most increments in barrier properties by addition of graphene have been reported for PET, EVOH, and PVA, which are intrinsically impenetrable. The effect of graphene nanosheets on the oxygen barrier property of PET was studied by Al-Jabareen *et al.* [49]. They reported that by incorporation of 1.5 wt.% of GnP, oxygen permeability reduced by 99% (0.1 cc/m^2/day/atm). This behavior is attributed to the well-dispersed condition of GnP and the higher degree of crystallinity due to the presence of GnP. The variation of oxygen transmission rate (OTR) and the degree of crystallization versus GnP content are shown in Figure 16.15a. Figure 16.15b shows a typical SEM image of the fractural surface of the PET/GnP (1.5 wt.%), which implies that GnP flakes aligned parallel to the plane of the film.

In another study on PET/graphene nanocomposites, Shim *et al.* found that the functionalization of GO by alkyl and alkyl ether groups can enhance the gas barrier property compared to un-functionalized graphene [50]. Oxygen flux rate and permeability coefficient of PET with GO and fGO are shown in Figure 16.16.

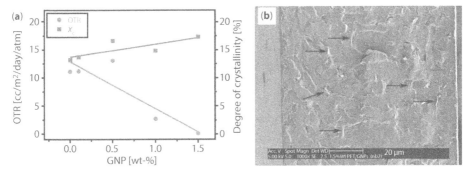

Figure 16.15 (a) OTR and degree of crystallinity as a function of GnP content in nanocomposite films, (b) SEM images of fractured surfaces of PET/GnP (1.5 wt.%) films [49].

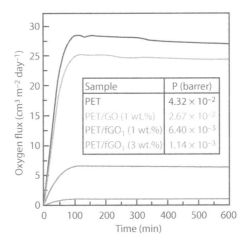

Figure 16.16 Oxygen flux rate and oxygen permeability coefficient of PET, PET/GO, and PET/fGO [50].

EVOH has good potential applications as packaging materials to extend the shelf life of foods and medicines. Yang et al. reported that the permeability coefficient of O_2 for EVOH/TRG nanocomposite films decreased to 8.517×10^{-15} cm^3 cm cm^{-2} s^{-1} Pa^{-1}, which is nearly 1671 times lower than that of virgin EVOH films [51]. They fitted experimental data on the well-known models (Nielsen [52] and Cussler theory models [53]) to analyze the oxygen transmission rate through the films as shown in Figure 16.17. As can be seen, the Cussler theory well adopted to their experimental data. In this model, the nanoplatelets are expected to be a completely oriented array throughout the entire polymer matrix.

The improvement of gas barrier properties of elastomers (such as IIR [54], XNBR [55], SBR [56], and PDMS [57]) by incorporation of graphene nanosheets was also investigated. Ha and coworkers prepared aminopropyl terminated telechelic PDMS/GO and examined gas permeability against common gases like H_2, N_2, CO_2, CH_4, and O_2 [57]. The obtained results are shown in Figure 16.18, which reveals that the addition of only 3.55 vol.% (8 wt.%) GO into the PDMS matrix caused a reduction of 99.9% in gas permeability for all aforesaid gases. Furthermore, the gas selectivity of nanocomposite samples was improved in comparison to neat PDMS for CO_2/N_2 and CO_2/CH_4. The adoption of experimental data on the Nielsen and Cussler models accompanied with SEM analysis was also performed. Based on these results, they proposed that gas permeation obeys different models depending on the concentration and alignment of the nanofiller. As can be seen in Figure 16.18, at low concentration, the modified Nielsen (random) is the proper model; at medium GO contents, the modified Nielsen (aligned) is fitted, and at high loadings, the modified Cussler is suitable.

The comparison between the gas barrier properties of IIR/TRG and IIR/nanoclay was accomplished by Sadasivuni et al. [54]. They reported that OTR for neat IIR, IIR/nanoclay (5phr), and IIR/TRG(5phr) is equal to 38.4, 35.6, and 28.4 ml/m^2/24 h, respectively. This observation was related to higher aspect ratio (130 versus 108) and higher specific surface area (2630 versus 750 m^2/g) of graphene.

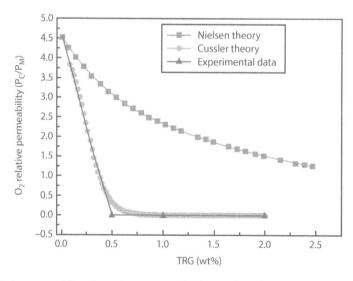

Figure 16.17 Relative permeability plots of experimental values and predicted values from Nielsen and Cussler theories for EVOH/TRG(0.5 wt.%) [51].

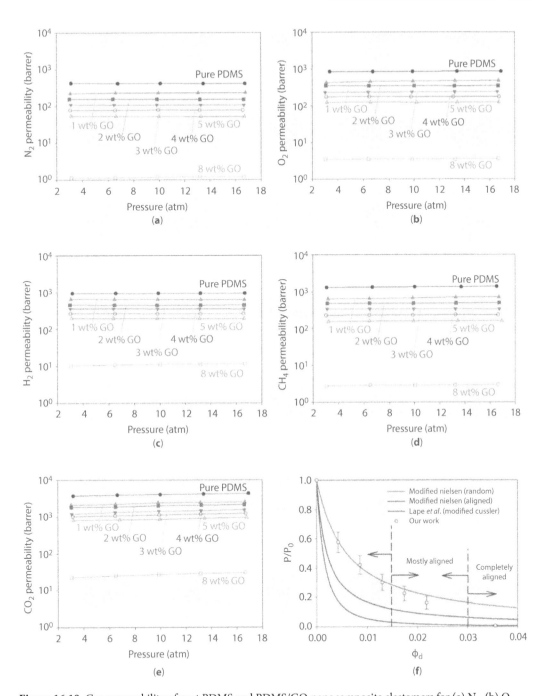

Figure 16.18 Gas permeability of neat PDMS and PDMS/GO nanocomposite elastomers for (a) N_2, (b) O_2, (c) H_2, (d) CH_4, and (e) CO_2; Relative permeability coefficient as a function of GO volume fraction divided into three different regions depending on the state of GO alignment and concentration of GO. Data points are the average relative permeability coefficient (R) values of all gases (N_2, O_2, H_2, CH_4, and CO_2) while the error bars are standard deviations of these data [30].

16.8 Thermal Conductivity

One of the other significant properties of graphene is its high thermal conductivity. It has been reported that single-layer graphene has a thermal conductivity of 4840–5300 W/mK. This value is much higher than that of polymers (0.1–1 W/mK) and even than other carbon-based fillers such as CNT and CNF (≈3000 W/mK) [2]. However, it should be noted that the enhancement of thermal conductivity is not as much as that of electrical conductivity upon the addition of graphene nanosheets. This is because of the lower difference between the thermal conductivity of graphene and polymers (0.1 versus 5000 W/mK) in comparison to their difference in electrical conductivity (6000 versus 10^{-15} S/cm). The obtained thermal conductive polymers can be used in tharmal pastes, thermal activated shape memory polymers, and electronic devices.

Heat transport in solid materials can be carried out by acoustic phonons. Acoustic phonons and their two-dimensional (2D) transport dominate the thermal conductivity of graphene and graphite at room temperature. Tuning the heat flow in graphene or graphene-based nanocomposites could be done through phonon scattering by substrates, edges, or interfaces. It is noteworthy that the outstanding thermal properties of graphene could lead to many interesting applications such as thermal management of polymeric nanocomposites.

The thermal conductivity can be determined using following equation:

$$K = \alpha \times \rho \times C \tag{16.4}$$

where K is thermal conductivity (W/mK), α is thermal diffusivity (mm²/s), ρ is density (g/cm³), and C is specific heat (J/g·K). Among these parameters, the measurement of α is more difficult, which is commonly evaluated by the laser flash technique. In this method, one side of the sample is irradiated by the heat pulse of laser. Heat transmission in thickness direction is measured by an infrared camera. Thermal diffusivity is calculated using the following equation [58]:

$$\alpha = \frac{1.38 L^2}{\pi^2 t_{1/2}} \tag{16.5}$$

where $t_{1/2}$ is the time to reach the half of the maximum temperature on the other side of the sample, π is the input power of the laser, and L is the thickness of the sample.

There are a few reports on the thermal conductivity of the polymer/graphene nanocomposites. Araby *et al.* prepared SBR/graphene nanocomposites by solution method [43]. They reported that thermal conductivity of neat SBR by incorporation of 24 vol.% of graphene increases from 0.177 to 0.480 W/mK (shown in Figure 16.19). As mentioned before, the enhancement of thermal conductivity is far lower than that of electrical conductivity due to high interfacial resistance, which is known as Kapitza resistance, and lower difference between thermal conductivity of graphene and polymer in comparison to that of electrical conductivity.

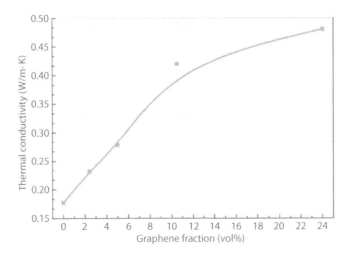

Figure 16.19 Thermal conductivity of neat SBR and its cured nanocomposites with GnP [43].

16.9 Rheology

It is well known that the rheological behavior and processability of polymers are strongly related to each other. Indeed, rheological properties indicate melt processing behavior in different processing methods of polymers such as extrusion and injection molding. In the case of nanocomposites, since the rheological properties are affected by size, shape, structure, and surface of the nanofillers, one can find the dispersion status of the nanosized disperse phase by measuring rheological parameters.

Both steady shear and dynamic oscillatory shear measurement techniques could be employed to understand the rheological behavior of materials. The general trend of viscosity variation of nanocomposites at low shear rate is increasing by filler content. In this region, solid-like behavior is often observed, which is ascribed to the physical jamming or formation of a solid network through the polymeric matrix. To assess the solid-like behavior of the nanocomposites, the slope of storage modulus (G′) at the terminal region is measured. The common trend for neat polymers in this region is that G′ and G″ are proportional to ω^2 and ω, respectively. However, for nanocomposites, G′ and G″ slopes deviate from the aforesaid values and nonterminal behavior is observed. Sabzi et al. determined the slope of G′ curves and plotted them as a function of two types of graphene concentration, as shown in Figure 16.20 [59]. As can be seen, a faster decrease is observed for NO_2-type graphene in comparison to xGn, which is related to a higher specific surface area and a much stronger network structure for this type of graphene.

At high shear rates, pseudoplasticity is usually detected. The main reason for this behavior is orientation of nanofillers in the direction of melt flow. A typical viscosity–frequency curve of polypropylene with different contents of graphene is shown in Figure 16.21.

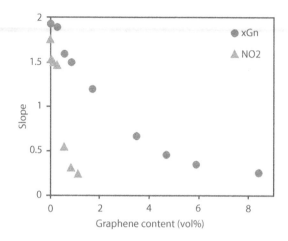

Figure 16.20 The slope of low frequency region of G′ curves versus graphene concentration [59].

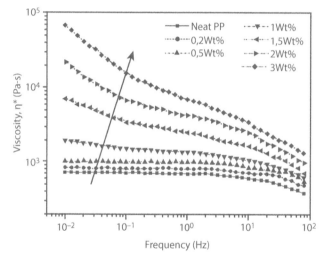

Figure 16.21 Complex viscosity (η*) versus frequency for neat PP and its nanocomposites with different graphene contents at 200°C [60].

Incorporation of graphene nanosheets is usually accompanied by an increase in elasticity. The intersection of two moduli (G′ and G″) can be an indication of the elastic nature of nanocomposites. At this point, which is called crossover frequency, G′ and G″ are equal. It is assumed that at the crossover point, the rheological transition from a viscoelastic solid (where G′ > G″) to a viscoelastic liquid (where G′ < G″) occurs. It can be concluded that with the increase of graphene content, the crossover frequency decreases, which implies the enhancement of elasticity by the addition of graphene. Basu *et al.* prepared nanocomposites based on general-purpose polystyrene and graphene via the *in situ* polymerization method and determined the crossover frequency and its corresponding relaxation time, which are shown in Table 16.3 [61]. As can be seen, the crossover frequency of samples decreases with increasing graphene content, which shows more pseudo-solid-like behavior of polymer melt by the addition of nanographene. From this table, one can find an increase of relaxation time with an increase of graphene content. It is well established that the relaxation time shows the

Table 16.3 Crossover frequencies and characteristic relaxation times for polystyrene/graphene nanocomposites [61].

Graphene content	0	0.25	0.5	0.75	1.00	1.50	2.00	2.50
Crossover freq. (s^{-1})	36.902	24.708	0.6608	0.8827	0.8523	1.4423	3.1264	0.1843
Relaxation time λ (s)	0.027	0.040	1.513	1.133	1.173	0.693	0.319	5.426

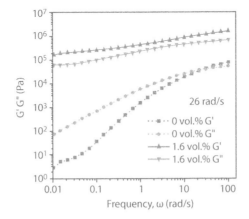

Figure 16.22 Crossover frequencies of G' and G" for neat PMMA and PMMA/graphene (1.6 vol.%) [62].

required time by the polymeric chains to relax during processing operation. The importance of relaxation time is that it could reflect the presence of residual stress in the sample. The Deborah number is defined as the ratio of polymer processing time to the longest relaxation time. If the Deborah number is greater than 1, the polymeric chains have adequate time for relaxation and thus there is no residual stress in the sample. Since graphene nanosheets restrict the chain mobility of the samples, complete relaxation becomes longer.

In some cases, the effect of graphene nanosheets on elasticity is very strong such that no crossover point is observed. In other words, G' is higher than G" in a whole range of frequency. For example, in neat PMMA, the crossover frequency appears at 26 rad/s, while PMMA/graphene nanocomposites represent a dominant elastic response (G' > G"), as shown in Figure 16.22 [62].

Another important rheological parameter is percolation threshold. The rheologocal behavior is generally changed before/after this point. Different methods are used to determine the percolation threshold of the filled polymers. One approach is based on the total excluded volume (V_{ex}) of randomly oriented discs. In this approach, the percolation threshold can be determined by:

$$\varphi_c = 1 - \exp\left(-\frac{\langle V_{ex} \rangle t}{\pi r}\right) \quad (16.6)$$

where t and r are thickness and radius of the discs, respectively.

In another approach, the elastic modulus (G'_0) of the nanocomposite can be correlated to volume fraction of the nanoparticle at low frequencies. This correlation can be presented in power law models as follows:

$$G'_0 \propto (\varphi - \varphi_{cG'})^\upsilon \tag{16.7}$$

where the exponent υ is the elasticity exponent, which is attributed to stress-bearing mechanisms. Zhang et al. produced a PMMA nanocomposite including three types of graphene with different C/O ratios [62] and calculated the rheological percolation threshold. As can be seen in Figure 16.23, the lowest percolation threshold is related to nanocomposites containing Graphene-13.2 (0.3 vol.%). In this figure, the number in front of graphene represents the C/O ratio. It seems that better dispersion has been obtained for graphene with the higher C/O ratio due to higher interfacial interaction and better polarity matching.

It is of great interest to compare the rheological and electrical percolation thresholds in polymeric nanocomposites. Generally, there are three classes, including $\varphi_{c\,rheo} < \varphi_{c\,elec}$, $\varphi_{c\,rheo} \approx \varphi_{c\,elec}$, $\varphi_{c\,rheo} > \varphi_{c\,elec}$. When $\varphi_{c\,rheo} > \varphi_{c\,elec}$, the electrical percolation threshold is reached by direct contact between graphene nanosheets and the formation of a conductive network. However, rheological percolation is only reached when a rigid network is formed through the matrix. Thus, it can be concluded that the graphene content is not high enough to enhance the rigidity of the matrix. In the case of $\varphi_{c\,rheo} < \varphi_{c\,elec}$, the graphene nanosheets are not directly in contact with each other such that no interconnected network is formed through the matrix. However, the adsorbed polymer chains onto the graphene surface would enhance the effective filler volume fraction and act as bridges between two adjacent nanosheets to facilitate the percolation. In this case, the affinity of the polymer chains into the graphene surface plays a dominant role. It should be noted that rheological percolation is reached when the distance between two neighboring nanosheets becomes lower than the critical value. This critical value is between the polymer entanglement distance and twice the radius of gyration (which depends on the temperature, nature, and molecular weight

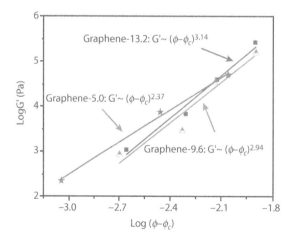

Figure 16.23 Storage modulus at $\omega = 0.1$ rad/s for PMMA containing different graphene types versus the difference between graphene fraction and rheological percolation threshold [62].

of the polymer). When the graphene–graphene distance is smaller than the reptation tube diameter of neat polymer, which is explained by reptation theory, the movement of polymer chains becomes rigid. Since the critical value for rheological percolation is generally smaller than the critical value of electrical percolation, one can find that $\varphi_{c\,rheo}$ is smaller than $\varphi_{c\,elec}$ in most cases.

16.10 Hybrid Nanocomposites Including Graphene and Other Nanofillers

Hybrid nanocomposites are made by a combination of two or more different types of fillers in the same matrix. Indeed, one of the used fillers should be in nanoscale and another filler can be micro/nano-sized. These hybrid nanocomposites represent a range of properties that cannot be achieved using a single kind of filler. Hybridization of the graphene nanosheets with other fillers can be a good choice for utilizing new properties. The mechanical and barrier properties and electrical and thermal conductivity can be enhanced by hybridization of graphene with different reinforements.

Chaharmahali *et al.* studied the effect of GnP on the physical, mechanical, and morphological properties of PP/bagase composites [63]. The bagase contents were 15 and 30 wt.%, while the used GnP was in the range of 0.1-1 wt.%. Their results showed that the incorporation of GnP has a considerable effect on the physical and mechanical properties such that the composite filled with 30 wt.% bagase and 0.1 wt.% GnP showed 22.5% increment in tensile strength, 29% higher tensile modulus, 6.8% increment in flexural strength, and 30% higher flexural modulus in comparison to the sample without graphene.

Patole *et al.* prepared self-assembled graphene/carbon nanotube/polystyrene hybrid nanocomposites and studied their thermal, mechanical, and electrical properties [64]. The samples were perepared via water-based *in situ* microemulsion polymerization, which is schematically shown in Figure 16.24. As shown in this figure, the self-assembled nanocomposites were used as a filler in polystyrene. They concluded that due to the gap bridging of CNT between the graphene sheets coated with polymer nanoparticles, the electrical property was enhanced. CNT bridging could be detected by TEM microscopy, and typical micrographs are shown in Figure 16.24b and c.

Hsiao *et al.* used TRG–silica hybrid fillers to enhance the thermal conductivity of epoxy resin while keeping the electrical resistivity [65]. GO–silica nanosheet "sandwiches" were

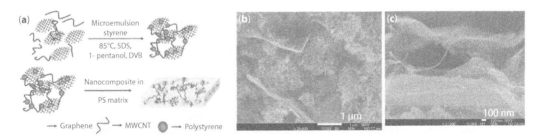

Figure 16.24 (a) Synthesis procedure of PS/CNT/graphene, (b and c) two different magnification of SEM images of PS/CNT/graphene nanocomposites [64].

synthesized using a sol-gel process and then thermally reduced at 700°C under Ar atmosphere. The steps of the sample preparation are shown in Figure 16.25.

The variation of electrical and thermal conductivities of the samples is illustrated in Figure 16.26. As can be seen, the thermal conductivity of nanocomposites is considerably improved by the addition of 1 wt.% of TRG–silica such that 61% higher thermal conductivity than the neat matrix is obtained. On the other hand, in Figure 16.26b, there is no change in electrical conductivity at this filler loading. The main reason for this behavior may be attributed to the hindering effect of silica layer on the creation of electrical paths. In other words, the silica layer covers the surface of the TRG and leads to the formation of 3D phonon transport channels that are able to enhance thermal conductivity.

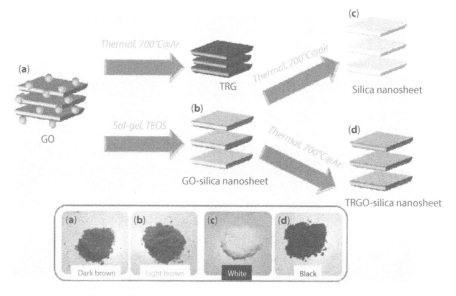

Figure 16.25 Schematic diagram of preparation procedure for (a) GO, (b) GO–silica, (c) silica, and (d) TRG–silica nanosheets [65]. (The color of each product is also illustrated at the bottom box.)

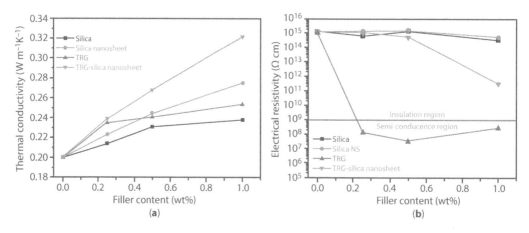

Figure 16.26 (a) Thermal conductivities and (b) electrical resistivities of nanocomposites containing three various fillers [65].

16.11 Applications of Polymer/Graphene Nanocomposites

Due to the extraordinary properties of graphene, novel applications in the field of optical, electrical, thermal, or barrier properties are created. The use of graphene-based polymeric nanocomposites have been reported in the gas barrier films, sensors, supercapacitors, photovoltaic devices, Li-ion batteries, water treatment, drug delivery systems, and tissue engineering.

It is necessary to protect some devices or materials against oxygen and moisture. For example, Li-ion batteries, fuel cells, and some other electronic devices are sensitive to oxygen and moisture, and their applications would deteriorate in the presence of these gases. On the other hand, in food packaging, the inhibition of gas and moisture diffusion is very critical. In comparison to the metal thin films (such as aluminum foil), polymeric films show better flexibility and transparency. However, polymeric films are inherently permeable to gases and vapors, which makes them unsuitable for packaging applications. Graphene as a plate-like nanofiller with a non-porous nature can effectively enhance the barrier property of polymeric films. Indeed, graphene nanosheets make a tortuous and longer pass for diffusing of gas molecules to significantly decline the permeability of polymer nanocomposite.

Graphene-containing nanocomposites based on sulfonated tetrafluoroethylene copolymer (Nafion), polypyrrole (PPy), polyaniline (PANI), and poly(isobutylene-co-isoprene) have been used in sensing applications. Nafion/reduced graphene nanocomposites showed high sensitivity toward metal ions such as lead and cadmium [66]. PANI/graphene or graphene oxide was used for sensing H_2 gas and methanol, respectively [67, 68]. *In situ* polymerized PPy/GO hydrogels were used to sense the ammonia gas, and this hydrogel showed 40% increase in sensivity compared to neat PPy [69].

Conducting polymers, such as polyaniline (PANI), polypyrrole (PPy), and polythiophene (PTh) and their derivatives, are a good candidate for high-performance energy devices. In recent years, polymeric nanocomposites based on conjugated polymers and graphene have been extensively used as electrode material of supercapacitors. Supercapacitors are energy storage devices that are capable of delivering high-power densities. Since graphene has a high specific surface area, it can improve the interactions between pseudocapacitive polymers and electrolyte to help electric double-layer generation. Graphene functionalization provides a higher interfacial interaction toward the polymer matrix and supplies a good synergetic effect for energy storage.

Solar cells have attracted great interest as a cost-effective power source in recent years. A polymer solar cell is a type of photovoltaic device that uses conductive polymers for absorption of light and charge transport to convert sunlight into electricity. In spite of the weak photoelectric properties of graphene, the formation of heterojunction nanostructures could improve the properties of pristine graphene. Due to the high electrical conductivity of graphene, it could greatly accept photoinduced charge carriers and enhance the rate of electron transfer of semiconductor conduction band. This would increase the efficiency of solar cells. Grapheme/poly(3,4-ethylenedioxythiophene):polystyrenesulfonate (graphene/PEDOT:PSS) nanocomposite film was electrodeposited on fluorine-doped tin oxide conductive substrate and used as a counter electrode in dye-sensitized solar cells [70]. The nanocomposite film showed low charge-transfer resistance on the electrolyte/electrode interface and high catalytic activity for the reduction of triiodide to iodide.

Lithium ion batteries (LIBs) are very useful devices to store and supply electricity for a long time with excellent advantages in capacity and cyclic stability. Thus, they have found many applications as the primary power sources in different electric vehicles. The performance of LIBs are very dependent on the electrode materials, i.e., anode and cathode. Conventional LIBs use graphite and lithium-containing transition metal oxides as anode and cathode, respectively. However, new researches have been focused on novel electrode materials with superior capacity. Graphene shows high electrochemical activity, high charge-carrier mobility, and strong mechanical properties and is developed as a new electrode material in LIBs, specially flexible LIBs [71, 72]. In addition, different electroactive polymers such as PPy and PANI have been investigated as nanocomposite polymer-based electrodes [73]. PVDF/graphene nanocomposites were also used as a conductive adhesive layer between active material and current collector of LIB to improve the electrochemical performance of anode [74].

Water filtration is one of the major applications of graphene-based nanocomposites. GO has a great adsorption capacity against heavy metal ions, synthetic dyes, and some other organic compounds due to its large amount of oxygen functional groups. Since GO easily disperses in water, the separation of GO sheets from the aqueous media after the adsorption process is really difficult. A proper method to facilitate the separation process is to blend them with polymers and make polymeric nanocomposites. In this context, several attempts have been made on the nanocomposites based on GO and biopolymers such as alginate and chitosan [75]. The presence of GO would simultaneously improve the adsorption capacity and mechanical performance of these polymers.

Application of graphene in drug delivery systems has been reported by researchers. In these studies, the proper polymers are grafted on the surface of GO and used as a nanocargo for drug delivery. Grafting of poly(N-vinyl caprolactam), poly(ethylene glycol), and polyethylenimine on GO have been used in drug delivery systems [76].

Nanocomposites based on biodegradable and biocompatible polymers such as poly(propylene carbonate) (PPC), polycaprolactone (PCL), and polylactic acid (PLA) have been considered for application in the field of tissue engineering [77]. In general, incorporation of high electrical conductive nanosized fillers (such as graphene) into polymers could lead to electrically trigged growth of cells. In other words, formation of conductive substrate facilitates cell growth via electrical stimulation. The biocompatibility of the biopolymer/graphene nanocomposite is retained, and the mechanical property and electrical conductivity are dramatically enhanced. Biodegradable foams and an environmental-friendly foaming process are a good choice for the production of porous materials that are applicable in tissue engineering scaffolds. The presence of graphene in these foams can overcome thermomechanical weakness and expand the range of applications for these scaffolds.

References

1. Novoselov, K.S., Geim, A.K., Morozov, S.V., Jiang, D., Zhang, Y., Dubonos, S.V., Grigorieva, I.V., Firsov, A.A., Electric field effect in atomically thin carbon films. *Science*, 306, 666–669, 2004.

2. Kim, H., Abdala, A.A., Macosko, C.W., Graphene/polymer nanocomposites. *Macromolecules*, 43, 6515–6530, 2010.
3. Kuila, T., Bose, S., Mishra, A.K., Khanra, P., Kim, N.H., Lee, J.H., Chemical functionalization of graphene and its applications. *Prog. Mater. Sci.*, 57, 1061–1105, 2012.
4. Kuila, T., Bose, S., Hong, C.E., Uddin, M.E., Khanra, P., Kim, N.H., Lee, J.H., Preparation of functionalized graphene/linear low density polyethylene composites by a solution mixing method. *Carbon*, 49, 1033–1037, 2011.
5. Li, W., Tang, X., Zhang, H., Jiang, Z., Yu, Z., Du, X.-S., Mai, Y.-W., Simultaneous surface functionalization and reduction of graphene oxide with octadecylamine for electrically conductive polystyrene composites. *Carbon*, 49, 4724–4730, 2011.
6. Lomeda, J.R., Doyle, C.D., Kosynkin, D.V., Hwang, W.-F., Tour, J.M., Diazonium functionalization of surfactant-wrapped chemically converted graphene sheets. *J. Am. Chem. Soc.*, 130, 16201–16206, 2008.
7. Salavagione, H.J., Gomez, M.A., Martínez, G., Polymeric modification of graphene through esterification of graphite oxide and poly(vinyl alcohol). *Macromolecules*, 42, 6331–6334, 2009.
8. Zhong, X., Jin, J., Li, S., Niu, Z., Hu, W., Li, R., Ma, J., Aryne cycloaddition: Highly efficient chemical modification of graphene. *Chem. Commun.*, 46, 7340–7342, 2010.
9. Choi, J., Kim, K., Kim, B., Lee, H., Kim, S., Covalent functionalization of epitaxial graphene by azidotrimethylsilane. *J. Phys. Chem. C*, 113, 9433–9435, 2009.
10. Chang, H., Wang, G., Yang, A., Tao, X., Liu, X., Shen, Y., Zheng, Z., A transparent, flexible, low-temperature, and solution-processible graphene composite electrode. *Adv. Funct. Mater.*, 20, 2893–2902, 2010.
11. Stankovich, S., Piner, R.D., Chen, X., Wu, N., Nguyen, S.T., Ruoff, R.S., Stable aqueous dispersions of graphitic nanoplatelets via the reduction of exfoliated graphite oxide in the presence of poly (sodium 4-styrenesulfonate). *J. Mater. Chem.*, 13, 16, 155–158, 2006.
12. Bai, H., Xu, Y., Zhao, L., Li, C., Shi, G., Non-covalent functionalization of graphene sheets by sulfonated polyaniline. *Chem. Commun.*, 1667–1669, 2009.
13. Keramati, M., Ghasemi, I., Karrabi, M., Azizi, H., Sabzi, M., Dispersion of graphene nanoplatelets in polylactic acid with the aid of a zwitterionic surfactant: Evaluation of the shape memory behavior. *Polym. Plast. Technol. Eng.*, 55, 1039–1047, 2016.
14. Liu, P., Gong, K., Xiao, P., Xiao, M., Preparation and characterization of poly (vinyl acetate)-intercalated graphite oxide nanocomposite. *J. Mater. Chem.*, 10, 933–935, 2000.
15. Jang, J.Y., Kim, M.S., Jeong, H.M., Shin, C.M., Graphite oxide/poly (methyl methacrylate) nanocomposites prepared by a novel method utilizing macroazoinitiator. *Compos. Sci. Technol.*, 69, 186–191, 2009.
16. Lee, Y.R., Raghu, A.V., Jeong, H.M., Kim, B.K., Properties of waterborne polyurethane/functionalized graphene sheet nanocomposites prepared by an *in situ* method. *Macromol. Chem. Phys.*, 210, 1247–1254, 2009.
17. Kim, H., Miura, Y., Macosko, C.W., Graphene/polyurethane nanocomposites for improved gas barrier and electrical conductivity. *Chem. Mater.*, 22, 3441–3450, 2010.
18. Matsuo, Y., Tahara, K., Sugie, Y., Structure and thermal properties of poly(ethylene oxide)-intercalated graphite oxide. *Carbon*, 35, 113–120, 1997.
19. Hirata, M., Gotou, T., Horiuchi, S., Fujiwara, M., Ohba, M., Thin-film particles of graphite oxide 1: High-yield synthesis and flexibility of the particles. *Carbon*, 42, 2929–2937, 2004.
20. Ren, P., Wang, H., Huang, H., Yan, D., Li, Z., Characterization and performance of dodecyl amine functionalized graphene oxide and dodecyl amine functionalized graphene/high-density polyethylene nanocomposites: A comparative study. *J. Appl. Polym. Sci.*, 131, 39803-39812, 2014.
21. Kim, H. and Macosko, C.W., Processing–property relationships of polycarbonate/graphene composites. *Polymer*, 50, 3797–3809, 2009.

22. Reghat, M., Ghasemi, I., Farno, E., Azizi, H., Namin, P.E., Karrabi, M., Investigation on shear induced isothermal crystallization of poly (lactic acid) nanocomposite based on graphene. *Soft Mater.*, 15, 103–112, 2017.
23. Gomari, S., Ghasemi, I., Esfandeh, M., Effect of polyethylene glycol-grafted graphene on the non-isothermal crystallization kinetics of poly(ethylene oxide) and poly(ethylene oxide): Lithium perchlorate electrolyte systems. *Mater. Res. Bull.*, 83, 24–34, 2016.
24. Manafi, P., Ghasemi, I., Karrabi, M., Azizi, H., Ehsaninamin, P., Effect of graphene nanoplatelets on crystallization kinetics of poly(lactic acid). *Soft Mater.*, 12, 433–444, 2014.
25. Xu, J.-Z., Liang, Y.-Y., Huang, H.-D., Zhong, G.-J., Lei, J., Chen, C., Li, Z.-M., Isothermal and nonisothermal crystallization of isotactic polypropylene/graphene oxide nanosheet nanocomposites. *J. Polym. Res.*, 19, 9975, 2012.
26. Hoffman, J.D. and Miller, R.L., Kinetic of crystallization from the melt and chain folding in polyethylene fractions revisited: Theory and experiment. *Polymer*, 38, 3151–3212, 1997.
27. Wang, H. and Qiu, Z., Crystallization kinetics and morphology of biodegradable poly (l-lactic acid)/graphene oxide nanocomposites: Influences of graphene oxide loading and crystallization temperature. *Thermochim. Acta*, 527, 40–46, 2012.
28. Avrami, M., Kinetics of phase change. I General theory. *J. Chem. Phys.*, 7, 1103–1112, 1939.
29. Liu, C., Ye, S., Feng, J., Promoting the dispersion of graphene and crystallization of poly (lactic acid) with a freezing-dried graphene/PEG masterbatch. *Compos. Sci. Technol.*, 144, 215–222, 2017.
30. Xu, J.-Z., Zhang, Z.-J., Xu, H., Chen, J.-B., Ran, R., Li, Z.-M., Highly enhanced crystallization kinetics of poly (l-lactic acid) by poly (ethylene glycol) grafted graphene oxide simultaneously as heterogeneous nucleation agent and chain mobility promoter. *Macromolecules*, 48, 4891–4900, 2015.
31. Tarani, E., Wurm, A., Schick, C., Bikiaris, D.N., Chrissafis, K., Vourlias, G., Effect of graphene nanoplatelets diameter on non-isothermal crystallization kinetics and melting behavior of high density polyethylene nanocomposites. *Thermochim. Acta*, 643, 94–103, 2016.
32. Li, C., Vongsvivut, J., She, X., Li, Y., She, F., Kong, L., New insight into non-isothermal crystallization of PVA–graphene composites. *Phys. Chem. Chem. Phys.*, 16, 22145–22158, 2014.
33. Yang, Z. and Lu, H., Nonisothermal crystallization behaviors of poly(3-hexylthiophene)/reduced graphene oxide nanocomposites. *J. Appl. Polym. Sci.*, 128, 802–810, 2013.
34. Zhang, F., Peng, X., Yan, W., Peng, Z., Shen, Y., Nonisothermal crystallization kinetics of *in situ* nylon 6/graphene composites by differential scanning calorimetry. *J. Polym. Sci. Part B Polym. Phys.*, 49, 1381–1388, 2011.
35. Zhang, J., Mine, M., Zhu, D., Matsuo, M., Electrical and dielectric behaviors and their origins in the three-dimensional polyvinyl alcohol/MWCNT composites with low percolation threshold. *Carbon*, 47, 1311–1320, 2009.
36. Stankovich, S., Dikin, D.A., Dommett, G.H.B., Kohlhaas, K.M., Zimney, E.J., Stach, E.A., Piner, R.D., Nguyen, S.T., Ruoff, R.S., Graphene-based composite materials. *Nature*, 442, 282–286, 2006.
37. Zhang, H.-B., Zheng, W.-G., Yan, Q., Yang, Y., Wang, J.-W., Lu, Z.-H., Ji, G.-Y., Yu, Z.-Z., Electrically conductive polyethylene terephthalate/graphene nanocomposites prepared by melt compounding. *Polymer*, 51, 1191–1196, 2010.
38. Zheng, D., Tang, G., Zhang, H.-B., Yu, Z.-Z., Yavari, F., Koratkar, N., Lim, S.-H., Lee, M.-W., *In situ* thermal reduction of graphene oxide for high electrical conductivity and low percolation threshold in polyamide 6 nanocomposites. *Compos. Sci. Technol.*, 72, 284–289, 2012.
39. Ansari, S. and Giannelis, E.P., Functionalized graphene sheet—Poly(vinylidene fluoride) conductive nanocomposites. *J. Polym. Sci. Part B Polym. Phys.*, 47, 888–897, 2009.

40. Qi, X.-Y., Yan, D., Jiang, Z., Cao, Y.-K., Yu, Z.-Z., Yavari, F., Koratkar, N., Enhanced electrical conductivity in polystyrene nanocomposites at ultra-low graphene content. *ACS Appl. Mater. Interfaces*, 3, 3130–3133, 2011.
41. Luong, N.D., Hippi, U., Korhonen, J.T., Soininen, A.J., Ruokolainen, J., Johansson, L.-S., Nam, J.-D., Seppälä, J., Enhanced mechanical and electrical properties of polyimide film by graphene sheets via *in situ* polymerization. *Polymer*, 52, 5237–5242, 2011.
42. Fim, F., de C., Basso, N.R.S., Graebin, A.P., Azambuja, D.S., Galland, G.B., Thermal, electrical, and mechanical properties of polyethylene–graphene nanocomposites obtained by *in situ* polymerization. *J. Appl. Polym. Sci.*, 128, 2630–2637, 2013.
43. Araby, S., Meng, Q., Zhang, L., Kang, H., Majewski, P., Tang, Y., Ma, J., Electrically and thermally conductive elastomer/graphene nanocomposites by solution mixing. *Polymer*, 55, 201–210, 2014.
44. Mao, C., Zhu, Y., Jiang, W., Design of electrical conductive composites: Tuning the morphology to improve the electrical properties of graphene filled immiscible polymer blends. *ACS Appl. Mater. Interfaces*, 4, 5281–5286, 2012.
45. Gómez-Navarro, C., Burghard, M., Kern, K., Elastic properties of chemically derived single graphene sheets. *Nano Lett.*, 8, 2045–2049, 2008.
46. Wang, P., Chong, H., Zhang, J., Lu, H., Constructing 3D graphene networks in polymer composites for significantly improved electrical and mechanical properties. *ACS Appl. Mater. Interfaces*, 9, 22006–22017, 2017.
47. Gomari, S., Esfandeh, M., Ghasemi, I., All-solid-state flexible nanocomposite polymer electrolytes based on poly(ethylene oxide): Lithium perchlorate using functionalized graphene. *Solid State Ionics*, 303, 37–46, 2017.
48. Steurer, P., Wissert, R., Thomann, R., Mülhaupt, R., Functionalized graphenes and thermoplastic nanocomposites based upon expanded graphite oxide. *Macromol. Rapid Commun.*, 30, 316–327, 2009.
49. Al-Jabareen, A., Al-Bustami, H., Harel, H., Marom, G., Improving the oxygen barrier properties of polyethylene terephthalate by graphite nanoplatelets. *J. Appl. Polym. Sci.*, 128, 1534–1539, 2013.
50. Shim, S.H., Kim, K.T., Lee, J.U., Jo, W.H., Facile method to functionalize graphene oxide and its application to poly(ethylene terephthalate)/graphene composite. *ACS Appl. Mater. Interfaces*, 4, 4184–4191, 2012.
51. Yang, J., Bai, L., Feng, G., Yang, X., Lv, M., Zhang, C., Hu, H., Wang, X., Thermal reduced graphene based poly(ethylene vinyl alcohol) nanocomposites: Enhanced mechanical properties, gas barrier, water resistance, and thermal stability. *Ind. Eng. Chem. Res.*, 52, 16745–16754, 2013.
52. Nielsen, L.E., Models for the permeability of filled polymer systems. *J. Macromol. Sci.*, 1, 929–942, 1967.
53. Cussler, E.L., Hughes, S.E., Ward, W.J., III, Aris, R., Barrier membranes. *J. Memb. Sci.*, 38, 161–174, 1988.
54. Sadasivuni, K.K., Saiter, A., Gautier, N., Thomas, S., Grohens, Y., Effect of molecular interactions on the performance of poly(isobutylene-co-isoprene)/graphene and clay nanocomposites. *Colloid Polym. Sci.*, 291, 1729–1740, 2013.
55. Kang, H., Zuo, K., Wang, Z., Zhang, L., Liu, L., Guo, B., Using a green method to develop graphene oxide/elastomers nanocomposites with combination of high barrier and mechanical performance. *Compos. Sci. Technol.*, 92, 1–8, 2014.
56. Xing, W., Tang, M., Wu, J., Huang, G., Li, H., Lei, Z., Fu, X., Li, H., Multifunctional properties of graphene/rubber nanocomposites fabricated by a modified latex compounding method. *Compos. Sci. Technol.*, 99, 67–74, 2014.

57. Ha, H., Park, J., Ando, S., Kim, C., Bin, Nagai, K., Freeman, B.D., Ellison, C.J., Gas permeation and selectivity of poly (dimethylsiloxane)/graphene oxide composite elastomer membranes. *J. Memb. Sci.*, 518, 131–140, 2016.
58. Xie, H., Cai, A., Wang, X., Thermal diffusivity and conductivity of multiwalled carbon nanotube arrays. *Phys. Lett. A*, 369, 120–123, 2007.
59. Sabzi, M., Jiang, L., Liu, F., Ghasemi, I., Atai, M., Graphene nanoplatelets as poly(lactic acid) modifier: Linear rheological behavior and electrical conductivity. *J. Mater. Chem. A*, 1, 8253–8261, 2013.
60. El Achaby, M., Arrakhiz, F., Vaudreuil, S., el Kacem Qaiss, A., Bousmina, M., Fassi-Fehri, O., Mechanical, thermal, and rheological properties of graphene-based polypropylene nanocomposites prepared by melt mixing. *Polym. Compos.*, 33, 733–744, 2012.
61. Basu, S., Singhi, M., Satapathy, B.K., Fahim, M., Dielectric, electrical, and rheological characterization of graphene-filled polystyrene nanocomposites. *Polym. Compos.*, 34, 2082–2093, 2013.
62. Sim, L.H., Gan, S.N., Chan, C.H., Yahya, R., ATR-FTIR studies on ion interaction of lithium perchlorate in polyacrylate/poly(ethylene oxide) blends. *Spectrochim. Acta Part A Mol. Biomol. Spectroscopy*, 76, 287–292, 2010.
63. Chaharmahali, M., Hamzeh, Y., Ebrahimi, G., Ashori, A., Ghasemi, I., Effects of nano-graphene on the physico-mechanical properties of bagasse/polypropylene composites. *Polym. Bull.*, 71, 337–349, 2014.
64. Patole, A.S., Patole, S.P., Jung, S.-Y., Yoo, J.-B., An, J.-H., Kim, T.-H., Self assembled graphene/carbon nanotube/polystyrene hybrid nanocomposite by *in situ* microemulsion polymerization. *Eur. Polym. J.*, 48, 252–259, 2012.
65. Hsiao, M.-C., Ma, C.-C.M., Chiang, J.-C., Ho, K.-K., Chou, T.-Y., Xie, X., Tsai, C.-H., Chang, L.-H., Hsieh, C.-K., Thermally conductive and electrically insulating epoxy nanocomposites with thermally reduced graphene oxide–silica hybrid nanosheets. *Nanoscale*, 5, 5863–5871, 2013.
66. Li, J., Guo, S., Zhai, Y., Wang, E., High-sensitivity determination of lead and cadmium based on the Nafion-graphene composite film. *Anal. Chim. Acta*, 649, 196–201, 2009.
67. Al-Mashat, L., Shin, K., Kalantar-zadeh, K., Plessis, J.D., Han, S.H., Kojima, R.W., Kaner, R.B., Li, D., Gou, X., Ippolito, S.J., Graphene/polyaniline nanocomposite for hydrogen sensing. *J. Phys. Chem. C*, 114, 16168–16173, 2010.
68. Konwer, S., Guha, A.K., Dolui, S.K., Graphene oxide-filled conducting polyaniline composites as methanol-sensing materials. *J. Mater. Sci.*, 48, 1729–1739, 2013.
69. Bai, H., Sheng, K., Zhang, P., Li, C., Shi, G., Graphene oxide/conducting polymer composite hydrogels. *J. Mater. Chem.*, 21, 18653–18658, 2011.
70. Yue, G., Wu, J., Xiao, Y., Lin, J., Huang, M., Lan, Z., Fan, L., Functionalized graphene/poly(3,4-ethylenedioxythiophene): Polystyrenesulfonate as counter electrode catalyst for dye-sensitized solar cells. *Energy*, 54, 315–321, 2013.
71. Liu, Y., Wang, W., Gu, L., Wang, Y., Ying, Y., Mao, Y., Sun, L., Peng, X., Flexible CuO nanosheets/reduced-graphene oxide composite paper: Binder-free anode for high-performance lithium-ion batteries. *ACS Appl. Mater. Interfaces*, 5, 9850–9855, 2013.
72. Liang, J., Zhao, Y., Guo, L., Li, L., Flexible free-standing graphene/SnO_2 nanocomposites paper for Li-ion battery. *ACS Appl. Mater. Interfaces*, 4, 5742–5748, 2012.
73. Zhao, Y., Huang, Y., Wang, Q., Graphene supported poly-pyrrole (PPY)/Li_2SnO_3 ternary composites as anode materials for lithium ion batteries. *Ceram. Int.*, 39, 6861–6866, 2013.
74. Lee, S. and Oh, E.-S., Performance enhancement of a lithium ion battery by incorporation of a graphene/polyvinylidene fluoride conductive adhesive layer between the current collector and the active material layer. *J. Power Sources*, 244, 721–725, 2013.

75. Platero, E., Fernandez, M.E., Bonelli, P.R., Cukierman, A.L., Graphene oxide/alginate beads as adsorbents: Influence of the load and the drying method on their physicochemical-mechanical properties and adsorptive performance. *J. Colloid Interface Sci.*, 491, 1–12, 2017.
76. Liu, Z., Robinson, J.T., Sun, X., Dai, H., PEGylated nanographene oxide for delivery of water-insoluble cancer drugs. *J. Am. Chem. Soc.*, 130, 10876–10877, 2008.
77. Sayyar, S., Murray, E., Thompson, B.C., Gambhir, S., Officer, D.L., Wallace, G.G., Covalently linked biocompatible graphene/polycaprolactone composites for tissue engineering. *Carbon*, 52, 296–304, 2013.

17
Graphene Oxide–Polyacrylamide Composites: Optical and Mechanical Characterizations

Gülşen Akın Evingür[1]* and Önder Pekcan[2†]

[1]*Faculty of Engineering, Piri Reis University, Tuzla-Istanbul, Turkey*
[2]*Faculty of Engineering and Natural Sciences, Kadir Has University, Cibali-Istanbul, Turkey*

Abstract

Graphene oxide (GO) is a two-dimensional carbon material with similar one-atom thickness, and is a light material having extremely high strength and thermal stability [1]. Thus, GO is an efficient filler for the enhancement of the electrical, mechanical, and thermal properties of composite materials [2]. We focused on GO as a nanofiller in polyacrylamide hydrogels and GO–PAAm composites to investigate the optical and mechanical properties of the composites in this chapter. Gelation, fractal analysis, and optical energy band gap measurements of the composites were performed by UV–Vis and fluorescence spectroscopy techniques. The sol-gel phase transition and its universality were monitored and tested as a function of GO contents. The geometrical distribution of GO during gelation was presented by the fractal analysis. The fractal dimension of the composite gels was estimated based on the power law exponent values using scaling models. UV–Vis spectroscopy was used to investigate the behavior of optical band gap of GO–PAAm composites. On the other hand, mechanical measurements were employed to determine toughness and compressive modulus of the polymer composites before and after swelling. The behavior of compressive modulus was explained by the theory of rubber elasticity.

Keywords: Graphene oxide, composite, gelation, fractal analysis, optical band gap, elasticity

17.1 Introduction

A composite is a mixture of two or more different materials producing a product that has different properties from its constituents. There are many polymer composite systems, including biopolymer-clay and hydrogel composites, which are interesting materials due to their adsorption [3] and tissue engineering capacities [4]. Polyacrylamide (PAAm) hydrogels are produced by free-radical cross-linking polymerization of acrylamide with N,N'methylenebis(acrylamide) (BIS), where the polymerization occurs in aqueous solutions of the monomers. When doped with nanomaterials, they are a class of composite hydrogels. In recent years, studies on the optical and electrical properties of composite gels have attracted much attention in view of their applications in optical devices [5].

---

*Corresponding author: gulsen.evingur@pirireis.edu.tr
†Corresponding author: pekcan@khas.edu.tr

On the other hand, graphene oxide (GO) is a monolayer of graphite oxide, and is a two-dimensional carbon material with similar one-atom thickness but with a large number of hydrophilic oxygenated functional groups [1]. GO is a light material having extremely high strength and thermal stability. Therefore, GO is an efficient filler for the enhancement of the electrical, mechanical, and thermal properties of composite materials [2]. GO possesses tunable electronic properties. The band gap of GO can be tailored by controlling coverage, arrangement, and relative ratio of the epoxy and hydroxyl groups. The optical absorption of GO is dominated by the $\pi-\pi^*$ transitions, which typically give rise to an absorption peak between 225 and 275 nm. The insulating nature of regular GO also limits its applications in electronic devices and energy storage [6]. It is also used in such applications as nanoelectronic devices, gas sensors, supercapacitors, tissue engineering, and drug delivery systems [2].

Graphene/polyacrylamide composites that are pH responsive via noncovalent interaction have been reported [7]. The resultant complexes show a reversible pH-responsive property although polyacrylamide itself does not possess such characteristics. Monitoring the gelation satisfies altering the structure and kinetics of the gelation. Fluorescence technique requires monitoring the gelation by using a fluorescence probe. Emission and excitation spectra, emission intensity, and lifetimes of fluorescence probes [8–10] are measured to investigate polarity [11] and viscosity [12] of the systems. Recently, using pyranine as an intrinsic fluoroprobe, the universality of composite gels could be described by classical and percolation models during the sol-gel phase transition in PAAm–sodium alginate (SA) [13], PAAm–kappa (κ)-carrageenan [14], PAAm–N-isopropylacrylamide (NIPA) [15], PAAm-multiwalled carbon nanotube composites (MWNTs) [16], and PAAm–poly(N-vinyl pyrrolidone) (PVP) [17]. In this transition, the cross-linked monomers (called clusters) react between them to produce larger molecules up to a sol-gel transition point [18]. Flory–Stockmayer's classical theory and percolation theory have been used for modeling the sol-gel phase transition [18–20].

The fractal analysis is a bridge between the microstructure and macroscopic properties of gels [21]. Fractal structure is usually provided to define the complexity of cross-linked molecules. The complexity in gel systems is described by the fractal dimension (D_f). Fractal analysis is widely used in studies on the microstructure of biomacromolecules, such as proteins, carbohydrate polysaccharides, and DNA [22]. The rheological properties and fractal dimensions of flaxseed gum gels were studied on various ionic strength values ranging between 0 and 1000 mM [22]. The calculated fractal dimensions of the gels were 2.06–2.49 or 1.42–2.18 based on the model selected and the ionic strength applied. The viscoelastic property and scaling behavior of acid (glucono-δ-lactone) induced soy protein isolate (SPI) gels were performed for various ionic strengths and protein contents [23]. Rheological analysis and confocal laser scanning microscopy analysis were applied to estimate the fractal dimensions, D_f, of the gels, and the values were found to vary between 2.319 and 2.729. The scaling behavior and fractal analysis of basil seed gum cross-linked with sodium trimetaphoshate have been investigated by rheological small amplitude oscillatory shear measurements [24]. The D_f values lay well within the range of fractal dimension values (1.5–2.8) reported for protein gels.

The clusters that are formed have a well-defined fractal dimensionality or compactness below, at, and above the percolation transition point depending on the mechanism that produces the larger molecules from the smaller ones [21]. In this case, there are at least two fractal dimensionalities that are obtained for the aggregates below the transition point and do not coincide with the fractal dimensionality of the percolation clusters below the percolation threshold.

The determination of band gap in materials is important in the semiconductor and nanomaterial industries [25]. Band gap refers to the difference in energy between the top of the valence band filled with electrons and the bottom of the conduction band devoid of electrons. The band gap energy of insulators is large (>4 eV), but lower for semiconductors (<3 eV) [26]. Optical properties and the electronic structure of amorphous germanium are obtained from photon energies between 0.08 and 1.6 eV [27]. The comparison of a theory with the experimental results leads to an estimate of the localization of the conduction band wave functions. Experimental evidence about the states in the gap of chalcogenide glasses is discussed [28]. The concentration of the states was estimated from the optical and photo emission measurement. Synthesis and optical characterization of poly(glycidylmethacrylate-co-methlacrylate) copolymers were investigated to calculate the optical constants, n and k [29]. The energy band gap of the copolymer was decided between 3.421 and 3.519 eV as a function of cross-linker, EGDMA concentration, and monomer weight. Transport and optical properties of silica (SiO_2)-polypyrrole nanocomposites were characterized by Fourier transform infrared spectrometer, UV spectrometer, and scanning electron microscope [30]. Optical absorption spectra reveal that the $\pi-\pi^*$ transition of polypyrrole shifts from 3.9 eV to 4.58 eV with an increase of silica content. The synthesis, optical, thermal, and electrochemical properties of low-band-gap copolymers were studied by NMR, UV–Vis spectrometer, thermal gravimetric analysis, and cyclic voltammeter, respectively [31]. The copolymers have a small optical band gap of 1.3–1.4 eV. Optical properties and electrical conductivity of polypyrrole–chitosan composite thin films were investigated to determine the optical transition characteristics and energy band gap of the composite films [32]. The optical band gap was obtained within 1.30–2.32 eV. The optical band gap of CdSe nanostructural films was estimated by using Tauc's model and absorption spectrum fitting method [33]. The energy band gap was calculated as 3.93 eV, 3.58 eV, and 2.52 eV, and also the width of the tail of localized states was found as 1.07 eV, 1.05 eV, and 2.32 eV, depending on deposition time in 6, 8, and 24 h, respectively. The TiO_2-graphene oxide nanocomposite was prepared by thermal hydrolysis of suspension with graphene oxide nanosheets and titania proxy complex [34]. The photocatalytic activity of the nanocomposites was performed under UV and visible light. When the band gap of pure TiO_2 is 3.20 eV, the band gap of the nanocomposite is reduced from 3.15 eV. The reduction of graphene oxide with glucose, fructose, and ascorbic acid was prepared at room temperature [35]. Their optical properties were determined by a UV–Vis spectrometer, and the optical band gap of GO can be reduced and tuned effectively from 2.70 eV to 1.15 eV. Polypyrrole nanoparticles were decorated by a reduced graphene oxide nanocomposite layer [36]. Optical band gap and thermal diffusivity were characterized by using a UV–Vis spectrometer and the photoacoustic technique, respectively. The optical band gap was in the range between 3.580 eV and 3.853 eV. Reduced graphene oxide (RGO)/polyacrylamide (PAAm) nanocomposite was prepared to study the adsorption kinetics of Pb (II) and methylene blue [37]. PAAm chains were grafted onto RGO sheets to enhance the dispersion property of RGO in aqueous solution and improve the adsorption capacity of RGO. Graphene oxide (GO) is added into poly(acrylamide) (PAAm) hydrogels to modify their mechanical and thermal properties [38]. The uniform polymer network of the GO–N′–methylenebisacrylamide (BIS) gels could distribute stress evenly on each chain, and the GO–BIS gels exhibit a tough tensile behavior.

On the other hand, as the mechanical properties of the hydrogels are often poor, it limits the performance of the gels in some applications. The addition of nanofillers such as clay-carbon nanotubes [39] has also been used to enhance the hydrogel mechanical strength and toughness [40, 41]. Elastic properties of poly(N-isopropylacrylamide) and polyacrylamide hydrogels were determined at various temperatures and cross-linker concentrations [42]. Young's moduli were found to be dependent of temperature though the cross-linker concentration. Elastic properties of highly cross-linked polyacrylamide gels were studied by N,N'-methylenebisacrylamide as cross-linking agent [43]. The modulus of PAAm gels increased with comonomer concentration at a fixed cross-linker agent.

The swelling characteristics and mechanical properties of gelatin–polyacrylamide interpenetrating networks have been reported [44]. The semi-IPN presented a greater elastic modulus when compared to the cross-linked PAAm hydrogel. The values of apparent cross-linking density were determined from the mechanical compression measurements at temperatures from 25 to 40°C. The equilibrium swelling and the plateau elastic modulus of a family of hydrogels made by the polymerization of acrylamide with itaconic acid or some of its esters were investigated as a function of composition and degree of cross-linking to produce materials with satisfactory swelling and elastic properties [45]. The mechanical properties of the GO/polyvinylalcohol(PVA) were observed to be improved [2], compared to pure PVA hydrogels, by a 132% increase in tensile strength with the addition of 0.8 wt.% of GO. The mechanical and thermal properties of GO-BIS (N,N-methylenebisacrylamide) gels vary by changing GO and BIS content [38]. The tensile strength of GO–BIS gels improved by increasing the GO content. The nanocomposites prepared by PAAm–GO nanocomposites are synthesized via *in situ* free-radical polymerization and investigated for the type and content of cross-linkers [46]. The microstructure of nanocomposites was characterized with TEM, DSC, ATR-FTIR, and XRD. The swelling properties of GO–polyacrylic acid–co-acrylamide nanocomposites were studied by FTIR, X-ray diffraction, DMA, field emission scanning electron microscopy, and optical microscopy [1]. The swelling capacity of the nanocomposites was compared with pure acrylamide with respect to pH behavior. On the other hand, GO was used as a drug binding effector in the anticancer drug in konjac-glucomannan/sodium alginate [47]. Controlled-release behavior of the hydrogels was studied by using FTIR and SEM. Tough graphene oxide composite hydrogels were prepared by using graphene peroxide as polyfunctional initiating and cross-linking centers [48]. The mechanical properties of the healed hydrogels were performed with respect to the healing time, temperature, GO content, and chemical cross-linker on the hydrogels. Therefore, the healing properties and the healing mechanism of the composite gels were revealed. GO/sodium alginate/PAAm ternary nanocomposite hydrogel was presented to improve mechanically [49].

We have studied the elastic percolation of the swollen polyacrylamide (PAAm)–multiwall carbon nanotube (MWNTs) composite [50], which indicates that the compressive elastic modulus increases dramatically up to 1 wt.% MWNT by increasing the nanotube content and then the elastic modulus decreases, presenting a critical MWNT value, showing that there is a sudden change in material elasticity. The effect of κC content on the composites was studied [51]. PAAm–κC composites were mentioned experimentally to decide the critical exponent of elasticity [52]. The elastic properties of the PAAm–κC composites are highly dependent on κC content, which directly affects the interactions between PAAm and κC monomers in the composites. Such monomer interactions will play a critical role in the load transfer and interfacial bonding that determine the elastic properties of the composites. Lastly, composites formed from PAAm–MWNTs were prepared via free-radical cross-linking copolymerization with various amounts of MWNTs varying in the range between 0.1 and 50 wt.% [53]. It is observed that elastic modulus increased when temperature is increased from 30°C to 60°C. Toughness, however, presented the reverse behavior versus temperature compared to the elastic modulus.

In this chapter, *in situ* fluorescence experiments are mentioned during the copolymerization and fractal analysis of AAm with various GO contents. The gel fraction was modeled by the classical and percolation theories, and the fractal analysis was applied to decide D_f's of the composite gels, respectively. Fractal dimensions of the composite gels were estimated based on the power law exponent values using scaling models. On the other hand, UV–Vis spectroscopy was used to investigate the optical band gap of the GO–PAAm composite. The main purpose of this part is to study and calculate the optical band gaps of the composite and compare their changes with the toughness of the composites as a function of GO content. Lastly, the mechanical behavior of the composites was investigated. This behavior was determined by using compressive technique and modeled by the theory of rubber elasticity.

17.2 Theoretical Considerations

17.2.1 Universality

Hydrogels have received considerable attention for the sol-gel phase transition process. Bethe lattice is a special lattice to define the sol-gel phase transition on the closed loop [18–20]. On the other hand, an alternative theory is the lattice percolation theory. Monomers occupied the sites of a periodic lattice [54, 55]. A bond between these lattice sites is formed randomly with probability p. The infinite cluster starts to form in the thermodynamic limit. p_c is called the percolation cluster in polymer language and also defined as the percolation threshold for a certain bond content. The critical exponents in these two theories are different from each other because of their universality. The exponents γ and β are for the weight average degree of polymerization, DP_w, and the gel fraction G (average cluster size, S_{av}, and the strength of the infinite network P_∞ in percolation language) near the percolation threshold. They are defined as

$$DP_w \propto (p_c - p)^{-\gamma} \quad p \to p_c^- \qquad (17.1)$$

$$G \propto (p - p_c)^{\beta} \quad p \to p_c^+ \qquad (17.2)$$

where the Flory–Stockmayer theory gives $\beta = \gamma = 1$, independent of the dimensionality while the percolation studies based on computer simulations give γ and β around 1.7 and 0.43 in three dimensions [54, 55].

In our case, near the percolation threshold, the critical point $|p - p_c|$ is linearly proportional to the $|t - t_c|$ and the fluorescence intensity from the bonded pyranines monitors the DP_w and G for below and above the gel point, respectively [56, 57], where t_c is defined as the gel point. When $t < t_c$, the maximum fluorescence intensity, I_{max}, measures the weight average degree of polymers. On the other hand, if $t > t_c$, the corrected intensity $I_{max} - I_{ct}$ measures only the gel fraction G. As a result, Equations 17.1 and 17.2 can be summarized.

$$I_{max} \propto DP_W = C_+(t_c - t)^{-\gamma} \quad t \to t_c^- \quad (17.3)$$

$$I_{ct} \propto DP_W = C_-(t_c - t)^{-\gamma'} \quad t \to t_c^+ \quad (17.4)$$

$$I' = I_{max} - I_{ct} \propto G = B(t - t_c)^{\beta} \quad t \to t_c^+ \quad (17.5)$$

where C_+, C_- and B are the critical amplitudes. I_{ct} is the intensity from finite clusters distributed through the infinite network. γ, γ', and β are the critical exponents [54, 55, 58, 59]. The ratio of the critical amplitudes was discussed by [54, 55, 60].

17.2.2 Fractal Analysis

After the decided gel point, the junction points of the final heterogeneous network structure occurred. When GO content increases, the density of the junction points without changing the distance between the monomers is increased [21].

Because, the density of monomers is proportional to the GO content in the cross-linked GO–PAAm composite gels, the following relation between fluorescence intensity at 427 nm and GO content is suggested [21]

$$I_{427nm} = (GO)^{Df} \quad (17.6)$$

Here, D_f is the fractal dimension of the composite gel.

17.2.3 Optical Energy Band Gap

17.2.3.1 Tauc's Model

The band gap in insulators/semiconductors is classified into direct and indirect band gaps. If the k vector (momentum) of the minimal energy state in conduction band and maximal energy state in the valence band are the same, it is called a "direct gap". If they are different, it is called an "indirect gap". The optical band gap (E_g) in an amorphous semiconductor is determined by Tauc [27, 28] (Equation 17.7)

$$[\alpha h\nu]^{2/n} = K(h\nu - E_g) \quad (17.7)$$

where K is a constant, hv is the energy of photon, and n represents the nature of the transition, which may have different values such as 1/2, 2, 3/2, or 3 for allowed direct, allowed indirect, forbidden direct, and forbidden indirect transitions, respectively.

The absorption coefficient α(v) at various wavelengths is calculated from the absorbance (A) and thickness (t) of the sample (Equation 17.8)

$$\alpha(v) = (2.303/t)A(v) \qquad (17.8)$$

The optical band gap (E_g) can be obtained by extrapolation of the linear portion of the plot $(\alpha h v)^{2/n}$ versus hv. For a direct transition n = 1, the equation becomes (Equation 17.9)

$$[\alpha h v]^2 = K(hv - E_g) \qquad (17.9)$$

17.2.3.2 Tail of Absorption Edge

It has been suggested that Urbach's rule may be used to relate the absorption coefficient to the incident photon energy [33] as

$$\alpha(v) = \alpha_0 \exp(hv/E_{tail}) \qquad (17.10)$$

where α is a constant and E_{tail} is the width of the tail of localized states (Urbach energy) corresponding to the optical transition between localized states adjacent to the valence band and extended state in the conduction band.

In Absorption Spectrum Fitting (ASF), from the absorption versus wavelength graph, the cutoff wavelength can be determined from the minimum value of absorbance, and then

$$E_{tail} = 1239.83/\lambda_{cut\text{-}off} \qquad (17.11)$$

was employed to calculate the E_{tail} values.

17.2.4 Elasticity

The stress/strain ratio is not constant at large deformations that the stress–strain relation becomes nonlinear for biological gels. The deviation from linearity is a product-dependent characteristic. In this nonlinear region, products can still be characterized by a modulus [61]. Comparison of theory with experiments and relationship between stress, strain, and molecular constitution of polymer networks were developed by Erman and Flory [62]. The compression measurements were mentioned through plots of the applied stress, τ (τ = f/A with f being the acting force and A the cross-section of the undeformed swollen specimen) versus $\lambda - \lambda^{-2}$, where $\lambda = L/L_0$ is the relative deformation caused by compression of the length of the swollen sample along the direction of the stress (L), with respect to the initial length (L_0) of the swollen but undistorted sample. Typical curvature at the smallest range of deformation is generally assigned to the imperfect geometry of the surface of the gel specimen [63]. The elastic modulus was determined by linear regression, as the slope of such plots as given in Equation 17.12.

$$S \approx \frac{(f/A)}{\alpha - \alpha^{-2}} \qquad (17.12)$$

17.3 Experiment

17.3.1 Preparation of PAAm–GO Composites

Composite gels were prepared by using 2M AAm (Acrylamide, Merck) with various amounts 5–50 (μl) of GO (graphene oxide, Graphenea) content at room temperature [64]. AAm, the linear component; BIS (N,N'-methylenebisacrylamide, Merck), the cross-linker; APS (ammonium persulfate), the initiator; and TEMED (tetramethylethylenediamine, Merck), the accelerator were dissolved in distilled water. The initiator and pyranine (8-hydroxypyrene-1,3,6-trisulfonic acid, trisodium salt, Sigma Aldrich) concentrations were kept constant at 7×10^{-3} M and 4×10^{-4} M, respectively. The solution was stirred (200 rpm) for 15 min to achieve a homogenous solution as 14 ml. All samples were deoxygenated by bubbling nitrogen for 10 min just before the polymerization process. After preparing the solutions, each pre-composite gel solution of 4 ml was poured into a quartz cell and put into the sample holder of the spectrometer for gelation, optical band gap, and fractal analysis as 4 ml. Also, the solution of 6 ml was poured into a plastic syringe for the mechanical measurement.

17.3.2 Fluorescence Measurement

The fluorescence intensity measurements were carried out using the Perkin Elmer LS 55 fluorescence spectroscopy. All measurements were made at the 90° position and slit widths were kept at 10 nm. When the sample was excited at 340 nm, which is the excitation wavelength for pyranine, the fluorescence intensities at 427 nm and 512 nm give the information about the binding pyranine to the monomers and free pyranine molecules, respectively. These wavelengths were monitored as a function of polymerization time.

17.3.3 UV Measurement

After the gelation process, the optical properties of the composites were studied using a UV–Vis spectrometer. A Schimadzu 1800 UV–Vis spectrometer was used to obtain the optical absorption data, and the distilled water was used as the reference solution for optical absorption measurements.

17.3.4 Mechanical Measurement

Compression measurements were carried out using an Instron 3345 testing machine attached with a 500-N force transducer as shown in Figure 17.1. The diameter of each disc

Figure 17.1 Photograph of compressive load cell.

was measured using a digital caliper, and all of the measurements were conducted a minimum of three times.

Composites prepared with various amounts of GO contents were cut into discs with a 10-mm diameter and 4-mm thickness as two samples used for before and after swelling, respectively. A probe with a flat end was used to compress the sample. Compression measurements were performed at a speed of 0.1 mm/min, a probe size of 10 cm, and up to 40% deformation ratio at 30°C before swelling.

On the other hand, the samples were maintained in distilled water at 30°C to achieve swelling equilibrium for the compression measurements of swollen samples. A final wash of all samples with distilled water was carried out for 1 week at room temperature to remove unreacted repeated units and to allow the gel to achieve swelling equilibrium. Compression measurements were carried out in the same method as above.

17.4 Results and Discussion

The sol-gel transition was performed by steady state fluorescence (SSF) for GO–PAAm composite gels with various GO contents. The universality of the transition can be tested as a function of gel fraction and time. The fluorescence spectrum of polymerization of GO–PAAm composite gels at different times was performed.

During the course of GO–PAAm polymerization, while the intensity of the 427-nm peak increased, the intensity of the 512-nm peak decreased. Figure 17.2a presents the fluorescence intensity of free pyranines in the composite at the 512-nm peak, I_{512nm}, as a function of the reaction time for 20- and 30-μl GO contents, respectively. First, the fluorescence intensity of the free pyranines decreased, then increased up to gel point, and then decreased to zero at the end of the reaction for 20- and 30-μl GO contents. Figure 17.2b shows the fluorescence intensities from the bonded pyranine against the gelation time for 20- and 30-μl GO contents. Since the maxima of the spectra, I_{427nm}, corresponded to bonded pyranine, the polymerization has progressed. The fluorescence spectra I_{427nm} as a function of time in Figure 17.2b were then used to evaluate the critical behavior of the sol-gel phase transition.

Figure 17.3a presents intensity curve during gelation of 30-μl GO contents in the composite gels. The curve was depicted by circle data representing the mirror symmetry I_{ms} of the intensity according to the axis perpendicular to time axis at $t = t_c$. The intensity from the clusters above the gel point was calculated as $I_{ct} = \dfrac{C_-}{C_+} I_{ms}$. Thus, the intensity below the symmetry axis monitors the average cluster size for $t < t_c$ and is given in Equations 17.3 and 17.4.

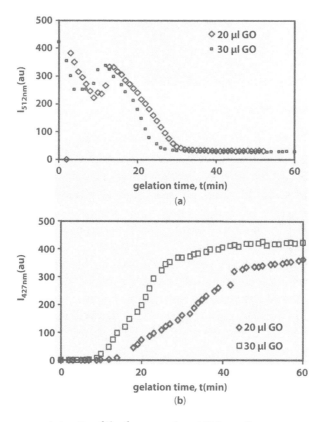

Figure 17.2 (a) Fluorescence intensity of the free pyranine at 512 nm, I_{512nm}, versus gelation time, and (b) fluorescence intensity variation of the pyranine, bonded to the PAAm versus gelation time for 20 μl and 30 μl GO contents of the composites, respectively.

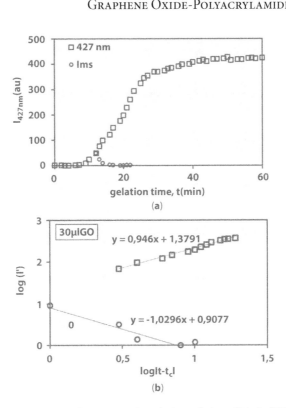

Figure 17.3 (a) Fluorescence intensity with respect to time during gelation of 30 μl of GO in the composites. The curve depicted by black circle data represents the mirror symmetry I_{ms} of the intensity with respect to the axis perpendicular to time axis at $t = t_c$. (b) Double logarithmic plot of the intensity I' versus time curves above t_c, 30 μl of GO content ($C_-/C_+ = 0.28$), respectively. The β exponent was determined from the slope of the straight line.

$I' = I_{427nm} - I_{ct}$, given in Equation 17.5, monitors the growing gel fraction for $t > t_c$. The t_c values are summarized in Table 17.1 together with the other parameters. Figure 17.3b represents the log–log plots of the typical intensity–time data below and above the gel point, t_c for the 30-μl GO content composite ($C_-/C_+ = 0.28$), where the slope of the straight line, close to the gel points, gives γ and β exponents, respectively.

The produced γ and β values together with t_c are listed in Table 17.1 for various GO contents. Here, we have to note that β and γ exponents as seen in Table 17.1 strongly support that AAm–GO composite gels during gelation obey the percolation picture (<25 μl GO contents), but classical results were produced above (>25 μl GO contents) content composite gels.

I_{427nm} against GO contents are given in Figure 17.4 for the GO in three different times (20, 30, and 50 min). In Figure 17.4, it is seen that I_{427nm} intensity increased till a certain GO content by indicating heterogeneities occurring during gelation. The fluorescence intensity above the gel point must be proportional to the number of pyranine molecules effectively surrounded by the percolating network [56]. After gel point, the junction points of the final heterogeneous network structure occur. When GO content increases, the density of the junction points without changing the distance between the monomers is increased [21].

Because the density of monomers is proportional to the GO content in the cross-linked GO–PAAm composite gels, the following relation between fluorescence intensity at 427 nm and GO content is suggested by Equation 17.6. In Figure 17.5, the log–log plot of I_{427nm}

Table 17.1 Experimentally measured parameters for GO–PAAm composites.

Model	Acrylamide [M]	GO (μl)	C^-/C^+	t_c (min)	β	γ
Percolation	2	0	1.0	5	1.04	–
			0.37		0.59	
			0.28		0.58	
			0.23		0.52	
			0.1		0.45	
		5	1.0	11	0.62	1.83
			0.37		0.61	1.83
			0.28		0.61	1.83
			0.23		0.60	1.83
			0.1		0.60	1.83
		8	1.0	14	0.60	1.83
			0.37		0.47	1.83
			0.28		0.45	1.83
			0.23		0.44	1.83
			0.1		0.43	1.83
		10	1.0	15	0.66	1.74
			0.37		0.68	1.74
			0.28		0.66	1.74
			0.23		0.65	1.74
			0.1		0.68	1.74
		15	1.0	16	0.62	1.76
			0.37		0.61	1.76
			0.28		0.60	1.76
			0.23		0.60	1.76
			0.1		0.66	1.76

(*Continued*)

Table 17.1 Experimentally measured parameters for GO–PAAm composites. (*Continued*)

Model	Acrylamide [M]	GO (µl)	C^-/C^+	t_c (min)	β	γ
		20	1.0	18	0.54	1.76
			0.37		0.53	1.76
			0.28		0.53	1.76
			0.23		0.52	1.76
			0.1		0.59	1.76
		25	1.0	19	0.66	1.88
			0.37		0.57	1.88
			0.28		0.55	1.88
			0.23		0.53	1.88
			0.1		0.50	1.88
Classical	2	30	1.0	10	1.05	1.02
			0.37		0.95	1.02
			0.28		0.94	1.02
			0.23		0.93	1.02
			0.1		0.94	1.02
		40	1.0	13	1.04	1.05
			0.37		0.92	1.05
			0.28		0.91	1.05
			0.23		0.90	1.05
			0.1		0.98	1.05
		50	1.0	16	1.00	1.01
			0.37		0.95	1.01
			0.28		0.95	1.01
			0.23		0.94	1.01
			0.1		0.95	1.01

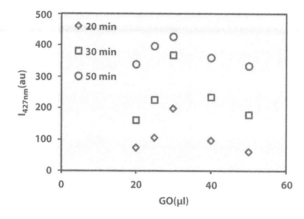

Figure 17.4 Plots of fluorescence intensities, I_{427nm} versus GO at various gelation times 20, 30, and 50 min, respectively.

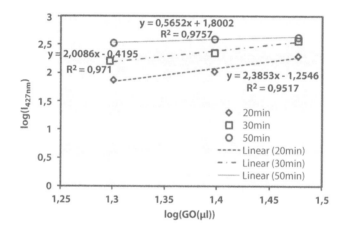

Figure 17.5 Log-log plots of I_{427nm} versus GO and their fits for 20, 30, and 50 min, respectively.

intensities is plotted versus GO content for various gelation times (20, 30, and 50 min), respectively, from which D_f values are produced from the slope of linear curves.

The produced fractal dimension is plotted versus gelation time in Figure 17.6 where it is seen that fractal dimension decreases as the gelation time is increased. Gelation at 20 min gives fractal dimension as 2.38, which is very close to 3D percolation cluster with 2.52 [58, 61]. Then, fractal dimension goes to diffusion-limited clusters with $D_f = 1.4$ and then lines up to Von Koch curve with random interval with $D_f = 1.14$ [21].

The UV–Vis spectra of the PAAm–10 μl GO composite between 300 and 900 nm are shown in Figure 17.7. Absorption maxima of the composite were located at 314, 364, 377, and 400 nm. The absorption bands at about 350–380 nm can be assigned to the lowest π–π* transitions.

For allowed direct transition, one can plot $(\alpha h\nu)^2$ versus $h\nu$, as shown in Figure 17.8, and extrapolate the linear portion of it to $\alpha = 0$ value to obtain the corresponding band gap.

The zoom of Figure 17.8 between 3 and 3.09 eV for the PAAm–10 μl GO composite is given in Figure 17.9. It was found that the band gap decreases with the increase in the values

GRAPHENE OXIDE-POLYACRYLAMIDE COMPOSITES 531

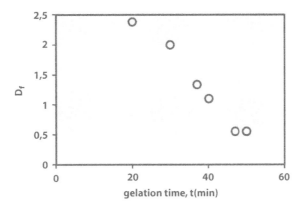

Figure 17.6 Plot of fractal dimension, D_f, against gelation time.

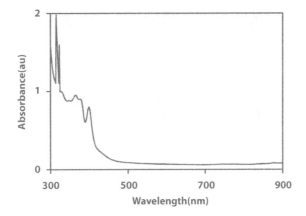

Figure 17.7 The absorbance of PAAm–10 μl GO composite between 300 and 900 nm.

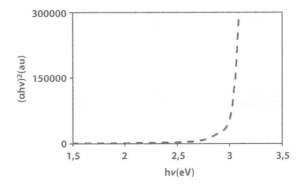

Figure 17.8 Plot of $(\alpha h\nu)^2$ versus $h\nu$ for the PAAm–10 μl GO composite to determine optical band gap.

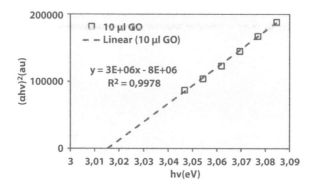

Figure 17.9 Calculation of energy band gap for PAAm–10 μl GO composite. (The zoom of Figure 17.2 between 3 and 3.09 eV.)

of GO content in the range 0–8 μl and then increases with further increases in GO content. The fundamental absorption edge of UV–Vis spectra can be correlated with the optical band gap E_{gap} by Tauc's relation as given in Equation 17.9.

The estimated optical band gaps for the composite of the GO content are listed in Table 17.2. The optical band gap is weakly dependent on GO content, except for the critical value of the 8-μl GO content sample.

The width of the tail of localized states was calculated from the Absorption Spectrum Fitting (ASF) method. The resulting spectrum obtained on the PAAm–10 μl GO composite is shown in Figure 17.10. The spectral data recorded showed the strong cutoff at 425 nm where the absorbance value is minimum. The cutoff wavelength and width of the tail of localized states for all GO content in the composite are summarized in Table 17.2. E_{tail} is the energy that yields an indication of depth-of-tail levels extending into the forbidden energy gap below the absorption edge. The larger the value of E_{tail}, the greater is the compositional,

Table 17.2 Optical energy band gap values of GO–PAAm composites for various GO contents from Tauc's model and ASF procedure, respectively.

GO (μl)	E_{gap} (eV)	λ_{cutoff} (nm)	E_{tail} (eV)
0	3.016	423	2.93
4	3.018	422	2.93
5	3.015	424	2.92
8	3.007	428	2.89
10	3.015	425	2.91
15	3.018	422	2.93
25	3.022	421	2.94
40	3.023	417	2.96

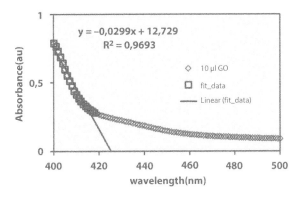

Figure 17.10 The absorbance of PAAm–10 μl GO composite between 400 and 600 nm to decide cutoff wavelength and calculate E_{tail}.

topological, or structural disorder [5]. The energies E_{tail} obtained by using Equation 17.11 for pure PAAm and the composites doped with GO are listed in Table 17.2, and it can be seen that the value of E_{tail} declines with rising content of GO in the PAAm for low GO content. The E_{tail} value for pure PAAm is 2.93 eV and increases for the high GO content region of the composite. This indicates that the composite has greater compositional and structural disorder as compared to the pure PAAm [5].

The gap value for PAAm–GO composites lies within the range 3.016–3.023 eV, as given in Figure 17.11a. The width of the tail of localized states shown in Figure 17.11b shows a decrease from 2.93 to 2.89 and then an increase from 2.89 to 2.96, producing a critical value at 8 μl of GO content. The values of E_{gap} and E_{tail} change weakly due to addition of GO in the PAAm.

This change has been mentioned for all the studied composites of various compositions. The change in the optical band gap of PAAm doped with GO can be explained in terms of a rise in the degree of disorder in the structure of the composite, and this is responsible for increasing the localized energy level concentration and, hence, reflects a decrease in the optical band gap. However, further rising the GO content, 8–40 μl, caused low concentration of localized energy levels. The presence of a low concentration of localized states employed an increase in the optical band gap, as it is evidenced by the shifting of cutoff wavelengths in the range 417–423 nm (given Table 17.2) for low and high concentrations, respectively.

The compressive load F (N) and compressive extension (mm) curves before and after the swelling processes for 5, 8, and 20 μl of GO content composites at 30°C are shown in Figure 17.12a and b, respectively. It is seen in Figure 17.12a that the force of the (5 μl) GO composite before swelling is around 4 N when the compressive extension is around 6 mm. On the other hand, the interactions between the monomers of the composites in the swollen state are much weaker than those in the collapsed state. In other words, the composite was easily affected by deformations after the swelling process as given in Figure 17.12b. The reason for this can be thermodynamically explained in that a decrease in length brings about an increase in entropy because of changes of the end-to-end distances of the network chains in GO–PAAm composites [50].

Typical strain–stress curves for GO–PAAm composites with different (5, 8, and 20 μl) GO content are presented before and after swelling in Figure 17.13a and b, respectively.

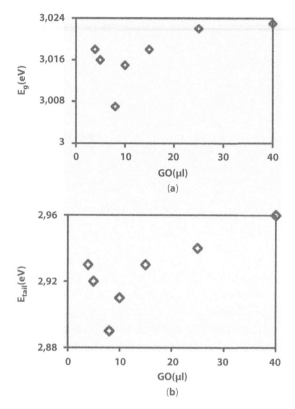

Figure 17.11 (a) The variation of the optical band gap with GO content and (b) the E_{tail} versus GO content, respectively.

The curves of 5 μl of GO contents in GO–PAAm composites before swelling is found to possess approximately five times larger modulus than the composite after swelling, which is the minimum value for the composite. It is also seen in Figure 17.13b that the composite presents a smaller strain of around 2% after swelling. On the other hand, the strain exceeds 100% when the GO content is 5 μl in GO–PAAm composites before swelling. In this case, it appears that an increase in the compressive force of the medium hinders the movement of free radicals, monomers, and enhances the possibility of chain transfer between GO and PAAm molecules.

As can be clearly seen, the shear modulus of GO–PAAm composites before and after swelling decreases with the increase of GO contents above 8 μl GO, as shown in Figure 17.14a. The shear modulus after swelling is found to be higher than the modulus before swelling above 8 μl GO content. On the other hand, the toughness, U_T, of the composite after swelling is found to be lower than the toughness after swelling between 8 and 50 μl GO content, as presented in Figure 17.14b. These findings are not surprising. It is obvious that toughness of the composite gel with water must be less than that of the gel without water. It is also well known that the swelling properties of composite gels mainly affect the shear modulus and toughness of the composite gels, because, basically, composite assembly on the basis of intermolecular hydrogen bond or other noncovalent interactions was constructed in the presence of the GO and PAAm networks, which further affected swelling performances of the composites [40]. It was deduced that the GO content could

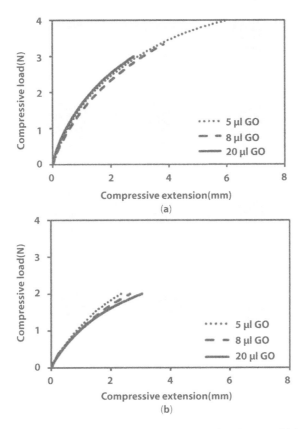

Figure 17.12 Compressive load F (N) versus compressive extension (mm) curves (a) before, and (b) after the swelling processes for 5, 8, and 20 µl of GO contents at 30°C, respectively.

act as a multifunctional cross-linker to form more junctions in the GO–PAAm composite and increase the cross-link density, leading to the reduction of swelling capacity [47]. On the other hand, the elasticity of the PAAm–MWNT composite was previously studied by us as a function of wt.% MWNT content [40, 53]. In PAAm–MWNT systems, the modulus increased dramatically up to 1 wt.% MWNT by increasing nanotube content and then decreased, presenting a critical MWNT value, indicating that there is a sudden change in the material elasticity.

17.5 Conclusion

In this chapter, polyacrylamide–graphene oxide (GO) composites were prepared by free-radical cross-linking copolymerization. The fractal analysis of the composites was analyzed in various GO contents during gelation and was investigated by using fluorescence technique. The fluorescence intensity at 427 nm was increased during gelation when the intensity at 512 nm was decreased. The fluorescence intensity of bonded pyranine to the composites shows that the gel fraction of the composites can be measured near the sol-gel phase transition. The gel fraction was modeled by the classical and percolation theories and the critical exponent, β and γ, agreed with the percolation theory

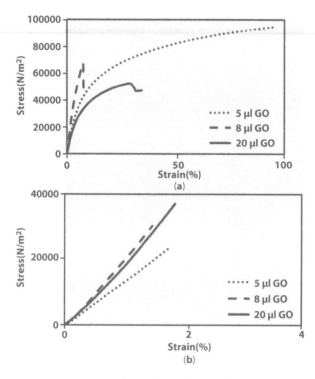

Figure 17.13 Stress versus strain curves (a) before and (b) after swelling processes for 5, 8, and 20 µl of GO contents at 30°C, respectively.

below 25 µl of GO contents and observed by classical theory 25 µl of GO contents. The analysis was applied to estimate the D_f values of the composites. Fractal dimensions of the composites were estimated based on the power law exponent values using scaling models. In addition, here we aimed to present the geometrical distribution of GO during gelation [66].

PAAm–graphene oxide (GO) composites were characterized by UV–Vis spectroscopy to monitor the optical band gap of GO–PAAm composites [65]. The behavior of the optical band gap was explained by the Tauc model and the Absorption Spectrum Fitting (ASF) method, which presents the width of band tail for the composites. It is important to note that the optical band gap of the composite is much higher for low GO (below 8 µl of GO) and high GO (above 8 µl of GO) content regions of the composite, producing a critical value at 8 µl of GO content. The decrease in the band gap (E_g) for low GO content, 0–8 µl, may be explained as a result of an increase in the degree of disorder in the structure of the composite because, as we know from the measurement of elasticity, the toughness of the composite is much higher for low GO (below 8 µl of GO) and high GO (above 8 µl of GO) content regions of the composite, suggesting a critical value at 8 µl of GO content. However, further increasing the GO content, 8–40 µl, caused low concentration of localized energy levels. The presence of a low concentration of localized states employed an increase in the optical band gap because the toughness of the composite increased dramatically, up to 8 µl of GO, by increasing GO content. Here, it is interesting to mention

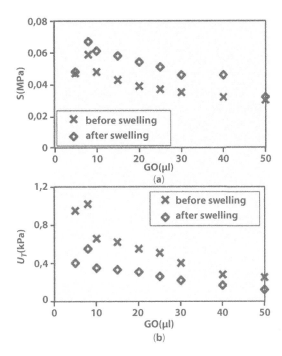

Figure 17.14 Effect of various GO content on the (a) shear modulus (S) and (b) toughness (U_T) of GO–PAAm composites before and after the swelling processes at 30°C, respectively.

that the critical value (8 µl) of GO content appears at the same position in all three cases, namely, optical band gap, E_{gap}, the width of the tail, E_{tail}, and toughness (U_T), all presenting the same critical GO value. In other words, low U_T values provide high E_{gap} and E_{tail} values, predicting the strong relationship between the physical nature and the electronic behavior of the GO–PAAm composite.

A compressive testing technique was used to determine the mechanical behavior of the GO–PAAm composite for various GO content before and after the swelling process. This technique was employed to measure force versus compression, i.e., stress versus strain, for the composite. Investigations indicated that shear modulus can be improved between 5 and 50 µl of GO content. It was observed that the mechanical properties of GO–PAAm composites are highly dependent on the GO content. The behavior of shear modulus was explained by the theory of rubber elasticity. It is important to note that the shear modulus of the composite is much lower for low GO (below 8 µl of GO) and high GO (above 8 µl of GO) content regions of the composite, suggesting a critical value at 8 µl of GO content. It is observed that the lowest shear modulus was achieved at 50 µl of GO content both before and after the swelling process. On the other hand, the highest shear modulus was obtained at 8 µl of GO at 30°C. After the swelling process, the shear modulus and toughness of the GO–PAAm composite decreases with increasing GO content similarly to the case before the swelling process. It was identified that the whole stress relaxation of GO–PAAm composites is composed of three contributions: relaxation observed commonly for elastomer, breakdown of physical cross-links, and swelling-induced relaxations. Therefore, the composite might have potential applications in many areas, such as biomedical, construction, and tissue engineering.

References

1. Huang, Y., Zeng, M., Ren, J., Wang, J., Fan, L., Xu, Q., Preparation and swelling properties of graphene oxide/poly (acrylic acid-co-acrylamide) super-absorbent hydrogel nanocomposites. *Colloids Surf., A*, 401, 97, 2012.
2. Zhang, L., Wang, Z., Xu, C., Li, Y., Gao, J., Wang, W., Lui, Y., High strength graphene oxide/polyvinyl alcohol composite hydrogels. *J. Mat. Chem.*, 2, 10399, 2011.
3. Kabiri, K. and Zohuriaan-Mehr, M.J., Superabsorbent hydrogel composites. *Polym. Adv. Technol.*, 14, 6, 438, 2003.
4. Park, H., Temenoff, J.S., Tabata, Y., Caplan, A.I., Mikos, A.G., Injectable biodegradable hydrogel composites for rabbit marrow mesenchymal stem cell and growth factor delivery for cartilage tissue engineering. *Biomaterials*, 28, 21, 3217, 2007.
5. Rawat, A., Mahavar, H.K., Chauhan, S., Tanwar, A., Singh, P.J., Optical band gap of polyvinylpyrrolidone (polyacrylamide blend film thin films). *Indian J. Pure App. Phys.*, 50, 100, 2012.
6. Li, F., Jiang, X., Zhao, J., Zhang, S., Graphene oxide: A promising nanomaterial for energy and environmental applications. *Nano Energy*, 16, 488, 2015.
7. Ren, L., Liu, T., Guo, J., Guo, S., Wang, X., Wang, W., A smart pH responsive graphene/polyacrylamide complex via noncovalent interaction. *Nanotechnology*, 21, 335701, 2010.
8. Barrow, G.M., *Introduction to Molecular Spectroscopy*, McGraw-Hill, New York, NY, 1962.
9. Birks, J.B., *Photophysics of Aromatic Molecules*, Wiley, Interscience, New York, NY, 1971.
10. Herculus, D.M., *Fluorescence and Phoshoperence Analysis*, Wiley Intersicence, New York, NY, 1965.
11. Jager, W.F., Volkers, A.A., Neckers, D.C., Solvatochromic fluorescent probes for monitoring the photopolymerization of dimethacrylates. *Macromolecules*, 28, 8153, 1995.
12. Vatanparast, R., Li, S., Hakala, K., Lemmetyinen, H., Monitoring of curing of polyurethane polymers with fluorescence method. *Macromolecules*, 33, 438, 2000.
13. Evingür, G.A., Tezcan, F., Erim, F.B., Pekcan, Ö., Monitoring of gelation of PAAm–sodium alginate (SA) composite by fluorescence technique. *Phase Transitions*, 85, 6, 530, 2012.
14. Aktaş, D.K., Evingur, G.A., Pekcan, Ö., Universal behaviour of gel formation from acrylamide-carregeenan mixture around the gel point: Fluorescence study. *J. Biomol. Struct. Dyn.*, 24, 1, 83, 2006.
15. Aktaş, D.K., Evingur, G.A., Pekcan, Ö., Steady state fluorescence technique for studying phase transitions in PAAm–PNIPA mixture. *Phase Transitions*, 82, 1, 53, 2009.
16. Aktaş, D.K., Evingur, G.A., Pekcan, Ö., Critical exponents of gelation and conductivity in multiwalled carbon nanotubes doped polyacrylamide gels. *Compos. Interfaces*, 17, 301, 2010.
17. Evingür, G.A., Kaygusuz, H., Erim, F.B., Pekcan, Ö., Gelation of PAAm- PVP composites: A fluorescence study. *Int. J. Mod. Phys. B*, 28, 20, 1450122 (11 pages), 2014.
18. Flory, P.J., Molecular size distribution in three dimensional polymers. I. Gelation, *J. Am. Chem. Soc.*, 63, 3083, 1941; Molecular size distribution in three dimensional polymers. II. Trifunctional branching units, *J. Am. Chem. Soc.*, 63, 3091, 1941; Molecular size distribution in three dimensional polymers. III. Tetrafunctional branching units. *J. Am. Chem. Soc.*, 63, 3096, 1941.
19. Stockmayer, W., Theory of molecular size distribution and gel formation in branched chain polymers. *J. Chem. Phys.*, 11, 45, 1943.
20. Stockmayer, W., Theory of molecular size distribution and gel formation in branched polymers II. General cross-linking. *J. Chem. Phys.*, 12, 125, 1944.
21. Pietronero, L. and Tosatti, E., *Fractals in Physics*, Elsevier Science Publishers B.V, 1986.
22. Wang, Y., Li, D., Wang, L.-J., Wu, M., Özkan, N., Rheological study and fractal analysis of flaxseed gum gels. *Carbohydr. Polym.*, 86, 594, 2011.

23. Bi, C., Li, D., Wang, L., Adhikari, B., Viscoelastic properties and fractal analysis of acid-induced SPI gels at different ionic strength. *Carbohydr. Polym.*, 92, 98, 2013.
24. Rafe, A. and Razavi, S.M.A., Scaling law, fractal analysis and rheological characteristics of physical gels cross-linked with sodium trimetaphosphate. *Food Hydrocolloids*, 62, 58, 2017.
25. Schimadzu, Measurements of band gap in compound semiconductors band gap determination from diffuse reflectance spectra. *Spectrophotometric Analysis*, A428.
26. Dharma, J. and Pisal, A., *Simple Method of Measuring the Band Gap Energy Value of TiO_2 in the powder form using UV/Vis/NIR Spectrometer*, Perkin Elmer Center, Shelton CT USA, Application Note.
27. Tauc, J., Optical properties and electronic structure of amorphous germanium. *Phys. Stat. Solids*, 15, 627, 1966.
28. Tauc, J. and Menth, A., States in the gap. *J. Non Cryst. Solids*, 8–10, 569, 1972.
29. Khafagi, M.G., Salem, A.M., Essawy, H.A., Synthesis and optical characterization of poly (glycidylmethacrylate-co-butylacrylate) copolymers. *Mater. Lett.*, 58, 3674, 2004.
30. Dutta, K. and De, S.K., Transport and optical properties of SiO_2–polypyrrole nanocomposites. *Solid State Commun.*, 140, 167, 2006.
31. Kminek, I., Vyprachticky, D., Kriz, J., Dybal, J., Cimrova, V., Low band gap copolymers containing thienothiadiazole units: Synthesis, optical, and electrochemical properties. *J. Polym. Sci. Part A: Polym. Chem.*, 48, 2743, 2010.
32. Abdi, M.M., Ekramul Mahmud, H.N.M., Chuah Abdullah, L., Kassim, A., Ab. Rahman, M.Z., Ying Chyi, J.L., Optical band gap and conductivity measurements of polypyrrole–chitosan composite thin films. *Chin. J. Polym. Sci.*, 30, 9, 2012.
33. Ghobadi, N., Band gap determination using absorption spectrum fitting procedure. *Int. Nano Lett.*, 3, 2, 2013.
34. Stengl, V., Bakardjieva, S., Grgar, T.M., Bludska, J., Kormunda, M., TiO_2–graphene oxide nanocomposite as advanced photocatalytic materials. *Chem. Cent. J.*, 7, 41, 2013.
35. Velasco-Soto, M.A., Perz-Garcia, S.A., Quintana, J.A., Cao, Y., Nyborg, L., Jimenez, L.L., Selective bang gap manipulation of graphene oxide by its reduction with mild reagents. *Carbon*, 93, 967, 2015.
36. Sadrolhosseini, A.R., Optical band gap and thermal diffusivity of polypyrole nanoparticles decorated reduced graphene oxide nanocomposite layer. *J. Nanomat.*, 1949042 (8 pages), 2016.
37. Yang, Y., Xie, Y., Pang, L., Li, M., Song, X., Wen, J., Zhao, H., Preparation of reduced graphene oxide/polyacrylamide nanocomposite and its adsorption of Pb(II) and methylene blue. *Langmuir*, 29, 10727, 2013.
38. Shen, J., Yan, B., Li, T., Long, Y., Li, N., Ye, M., Study on graphene oxide based polyacrylamide composite hydrogels. *Composites Part A*, 43, 1476, 2012.
39. Haraguchi, K. and Li, H.J., Mechanical properties and structure of polymer clay nanocomposite with high clay content. *Macromolecules*, 39, 1898, 2006.
40. Evingür, G.A. and Pekcan, Ö., Effect of multiwalled carbon nanotube (MWNT) on the behavior of swelling of polyacrylamide–MWNT composites. *J. Reinf. Plast. Compos.*, 33, 13, 1199, 2014.
41. Das, S., Irin, F., Ma, L., Bhattacharia, S.K., Hedden, R.C., Green, M.J., Rheology and morphology of pristine graphene/polyacrylamide gels. *ACS Appl. Mater. Interfaces*, 5, 8633, 2013.
42. Matzelle, T.R., Geuskens, G., Kruse, N., Elastic properties of poly(*N*-isopropylacrylamide) and poly(acrylamide) hydrogels studied by scanning force microscopy. *Macromolecules*, 36, 2926, 2003.
43. Baselga, J. and Hernandez-Fuentes., I., Pierola, M.A., Llorente. F., Elastic properties of highly cross-linked polyacrylamide gels. *Macromolecules*, 20, 3060, 1987.

44. Kaur, H. and Chatterji, P.R., Interpenetrating hydrogel networks. 2. Swelling and mechanical properties of the gelatin-polyacrylamide interpenetrating networks. *Macromolecules*, 23, 4868, 1990.
45. Valles, E., Durando, D., Katime, I., Mendizabal, E., Puig, J.E., Equilibrium swelling and mechanical properties of hydrogels of acrylamide and itaconic acid or its esters. *Polym. Bull.*, 44, 109, 2000.
46. Liu, R., Liang, S., Tang, X.Z., Yan, D., Li, X., Yu, Z.Z., Tough and highly stretchable graphene oxide/polyacrylamide nanocomposite hydrogels. *J. Mat. Chem.*, 22, 14160, 2012.
47. Wang, J., Liu, C., Shuai, Y., Cui, X., Nie, L., Controlled release of anticancer drug using graphene oxide as a drug-binding effector in konjac glucomannan/sodium alginate hydrogels. *Colloids Surf., B*, 113, 223, 2014.
48. Liu, J., Song, G., He, C., Wang, H., Self-healing in tough graphene oxide composite hydrogels. *Macromol. Rapid Commun.*, 34, 1002, 2013.
49. Fan, J., Shi, Z., Lian, M., Li, H., Yin, J., Mechanically strong graphene oxide/sodium alginate/polyacrylamide nanocomposite hydrogel with improved dye adsorption capacity. *J. Mat. Chem.*, 1, 7433, 2013.
50. Evingur, G.A. and Pekcan, Ö., Elastic percolation of swollen polyacrylamide (PAAm)–multiwall carbon nanotubes composite. *Phase Transitions*, 85, 553, 2012.
51. Evingur, G.A. and Pekcan, Ö., Kinetics models for the dynamical behaviors of PAAm–κ-carrageenan composite gels. *J. Bio. Phys.*, 41, 37, 2015.
52. Evingur, G.A. and Pekcan, Ö., Superelastic percolation network of polyacrylamide (PAAm)–kappa carrageenan (kC) composite. *Cellulose*, 20, 1145, 2013.
53. Evingur, G.A. and Pekcan, Ö., Temperature effect on elasticity of swollen composite formed from polyacrylamide (PAAm)–multiwall carbon nanotubes (MWNTs). *Engineering*, 4, 619, 2012.
54. Stauffer, D., Coniglio, A., Adam, M., Gelation and critical phenomena. *Adv. Polym. Sci.*, 44, 103, 1982.
55. Stauffer, D. and Aharony, A., *Introduction to Percolation Theory*, 2nd ed., Taylor and Francis, London, 1994.
56. Yılmaz, Y., Erzan, A., Pekcan, Ö., Critical exponents and fractal dimension at the sol-gel phase transition via *in situ* fluorescence experiments. *Phys. Rev. E*, 58, 7487, 1998.
57. Yılmaz, Y., Erzan, A., Pekcan, Ö., Slow regions percolate near glass transition. *Euro. Phys. J. E*, 9, 135, 2002.
58. Sahimi, M., *Application of Percolation Theory*, Taylor and Francis, London, 1994.
59. de Gennes, P.G., *Scaling Concepts in Polymer Physics*, Cornell University Press, Ithaca, 1988.
60. Aharony, A., Universal critical amplitude ratios for percolation. *Phys. Rev. B*, 22, 400, 1980.
61. Holdt, S.L. and Kraan, S., Bioactive compounds in seaweed: Functional food applications and legislation. *J. App. Phys. Coll.*, 23, 543, 2011.
62. Erman, B. and Flory, P.J., Relationships between stress, strain, and molecular constitution of polymer networks. Comparison of theory with experiments. *Macromolecules*, 15, 806, 1982.
63. Valencia, J., Baselga, J., Pierola, I.F., Compression elastic modulus of neutral, ionic, and amphoteric hydrogels based on N-vinylimidazole. *J. Polym. Sci. B: Polym. Phys.*, 47, 1078, 2009.
64. Evingur, G.A. and Pekcan, Ö., Mechanical properties of graphene oxide–polyacrylamide composites before and after swelling in water. *Polym. Bull.*, 75, 4, 1431–1439, 2018.
65. Evingur, G.A. and Pekcan, Ö., Optical band gap of PAAM–GO composites. *Compos. Struct.*, 183, 212, 2018.
66. Evingur, G.A. and Pekcan, Ö. *et al.*, Gelation and fractal analysis of graphene oxide-polyacrylamide composite gels. *Phase Transitions*, 85, 530–541, 2012.

18

Synthesis, Characterization, and Applications of Polymer/Graphene Oxide Composite Materials

Carmina Menchaca-Campos[1]*, César García-Pérez[1], Miriam Flores-Domínguez[1], Miguel A. García-Sánchez[2], M.A. Hernández-Gallegos[3], Alba Covelo[3] and Jorge Uruchurtu-Chavarín[1]

[1]*Centro de Investigación en Ingeniería y Ciencias Aplicadas, IICBA-UAEM, Cuernavaca, Mor. México*
[2]*Departamento de Química, Universidad Autónoma Metropolitana-Iztapalapa, Ciudad de México*
[3]*Centro de Ingeniería de Superficies y Acabados, Fac. de Ing. UNAM, Ciudad de México*

Abstract

In this work, examples of the fabrication and characterization of composite materials based on graphene oxide and polymers are demonstrated. The composite materials were deposited on different substrates and examples of their electrochemical performance and characteristics were evaluated in different electrolytes using different techniques, showing their potential applications. The chemical and structural properties of these materials were characterized using different techniques, such as scanning electron microscopy (SEM), Fourier transformed infrared spectroscopy, X-ray diffraction (R-X), ultraviolet–visible spectroscopy (UV–Vis), Raman spectroscopy, etc., in order to determine the interactions between the materials of which the composites are made of. The covalent functionalization of graphene oxide allows the polymer to be grafted. Their characterization suggests high level of interaction and integration between the components of the composite materials. This allows the enhancement of the mechanical properties and thermal and electronic conductivity, improving the materials performance for different applications.

Keywords: Graphene, graphene oxide, sol-gel, electrospinning, coatings, electrochemical, energy applications

18.1 Introduction

Developments in materials chemistry are being applied nowadays in energy systems, nanoscience and technology research. Functionalized materials, i.e., polymeric nanocomposites and hybrid metal-organic porous structure developments, are available now for various applications, for example, in hydrogen storage, carbon dioxide capture, toxic gases–fuels, and hydrocarbon molecules separation [1–3].

Graphene is a bi-dimensional allotropic carbon separated from graphite by Geim, Novoselov and coworkers in 2004 [4–6] and has attracted too much attention in research

Corresponding author: cmenchaca@uaem.mx

because its unique electronic structure and fascinating optical, thermal, chemical, and mechanical properties [7–15]. Due to its excellent mechanical properties, graphene can be used for generate composite materials; for example, its combination with a polymeric matrix can improve those properties. Furthermore, the reinforcement of fibers in this form improves their properties for diverse applications, such as thermal conductivity, among others [16].

Graphene is the construction block of other carbon materials already well known. It is a bi-dimensional material, a sheet composed of carbon atoms tied together in sp^2 hybridization, ordered in regular hexagons extending all along with one atom width. If the sheet is rolled up, simple wall carbon nanotubes are obtained, various sheets together form multiple-wall nanotubes, and graphene as a football is known as fullerans [17].

There are different ways to prepare graphene starting from abundant and cheap mineral graphite, with different advantages and disadvantages. The simplest method to obtain graphene is by mechanical exfoliation of graphite against an abrasive surface (generally Si/SiO) and yielding the particles through the known adhesive tape collection [18, 19]. Through this easy procedure, it is possible to obtain large-sized monolayer graphene sheets (up to 0.2 mm) with a high structural quality and excellent electronic properties. It is a giant aromatic macromolecule with excellent thermal, electrical, optical, and mechanical properties, conducting heat and electricity, in two dimensions.

Unfortunately, the low yielding rendered and the laborious processes of layers separation made impractical the above mentioned method. Nowadays, more alternative and refined methods are being developed to prepare graphene to lower production costs and increase the quantity produced of this material. The most promising methods to produce graphene include chemical vapor phase deposit (CVD), epitaxial growth in electrically insulated silica carbide surfaces (SiC), and chemical processing of graphite oxide obtaining graphene oxide (GO) sheets (highly oxidative graphene) [20–22].

Graphite oxide is a compound of carbon, oxygen, and hydrogen in variable ratios, commonly obtained by treating graphite with strong oxidizers ($KMnO_4$) in acid medium (H_2SO_4). After the oxidation process, a material possessing a layered structure consisting of piled-up sheets of GO is obtained [23].

Rigorously, "oxide" is a wrong name, since graphite is not a metal [1]. The bulk material disperses in basic solutions yielding monomolecular sheets, known as graphene oxide (GO) by analogy to graphene, the single-layer form of graphite. Graphene oxide sheets have attracted substantial interest not only as a possible intermediate for the manufacture of graphene but also as a constituent of composite and hybrid systems. It renders particular properties to such systems, apart from having hydrophilic properties unlike graphene, with the latter being hydrophobic. Typically, it preserves the layer structure of the parent graphite, but the layers are buckled and the interlayer spacing is about two times larger (~0.7 nm) than that of graphite. Also, GO layers are about 1.1 ± 0.2 nm thick [2, 3, 7, 8]. The edges of each layer are terminated with carboxyl and carbonyl groups. The detailed structure is still not understood because of its disorder and disorganized layers packing [24].

One of the methods used to separate the layers of graphite consist of an aggressive oxidation process, which functionalizes the surroundings and certain places on the graphene surface, mainly those in which defects do exist. As a consequence, oxygenated organic functions can be attached to those places, promoting attraction with polar species and solvents and repulsion with the hydrophobic regions of the graphene layer [25]. The existence of organic functions attached to the surface and periphery of the graphene layers made its covalent union with other chemical or biochemical species possible, making it susceptible to being used in different technological areas. This modification causes layers to lose their planarity, promoting their separation [26].

Graphene oxide was prepared 150 years ago by B. C. Brodie and improved a century after by W. S. Hummers and R. E. Offerman [23], involving a strong oxidizer substance as mentioned before, in sulfuric acid medium. After oxidation, a laminar structure of GO is obtained. These layers are highly oxygen functionalized and thus are highly hydrophilic [27]. These oxygenated groups and water adsorbed molecules increase the laminar distance considerably; therefore, interaction energy among layers diminishes and GO is easily exfoliated in aqueous media. Colloidal suspensions of GO monolayer sheets are obtained, and these are stabilized, thanks to the electrostatic repulsion produced by negative charges acquired during dispersion due to ionization of some functional groups present [28].

The GO structure is a non-stoichiometric material formed by aromatic regions separated by aliphatic regions containing large proportion of hydroxyl and epoxy groups over basal planes with just a small amount of carbonyl and carboxyl groups present on the sheets' edges (Figure 18.1). Due to the presence of these groups, GO is electrically insulating, limiting its applicability in many instances. To reduce this, treatments are necessary to obtain conducting sheets [25].

Different methods do exist to perform controlled reduction such as chemical and electrochemical, and the most extended involves hydrazine (H_2N-NH_2) as reducing agent. After reduction, it is possible to perform an added thermal treatment in the range between 150 and 1100°C to increase the reduction efficiency, improving the sheets' structural quality. Due to the hazards and toxicity of hydrazine, nowadays more safe reduction agents can use to substitute it [29–31].

Reducing GO sheets' conductivity is greatly increased, in the order of three to four times the original, probably as a consequence of the original graphitic structure restoration. But the restoration is not fully achieved and oxygenated functionalities introduced in the oxidation phase are not fully eliminated from the GO sheets. Even though the carbon/oxygen atomic relation increases after reduction from values around 2 for GO, up to 10 or above for the hydrazine reduction used, relatively low values reached indicate that the oxygen presence continues to be significant. This affects the low structural quality of the sheets obtained. For all these, the different types of graphene obtained are far from the high-quality sheets obtained from mechanical exfoliation. The easily produced and processed mass turns

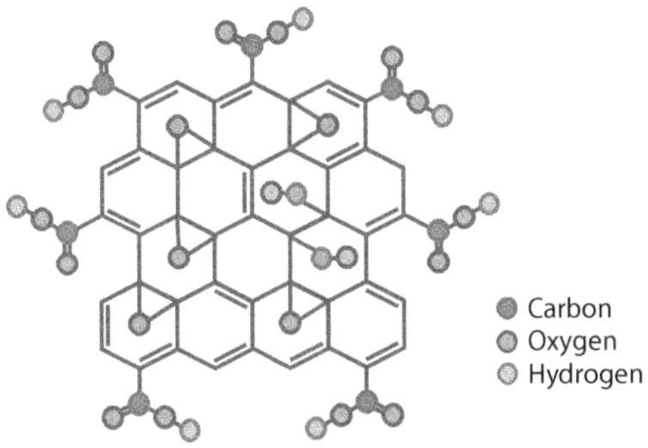

Figure 18.1 GO structure.

the chemically modified graphene (CMG) to an ideal match candidate for a large number of new materials applications including electronic conductors composites with high mechanical resistance, flexible conducting coatings for screens, gas molecular sensors, etc. [26].

A polymer is a large molecule or macromolecule composed of many repeated subunits. Because of their broad range of properties, both synthetic and natural polymers play an essential role in daily life. Polymers range from synthetic plastics such as polystyrene to natural biopolymers such as DNA and proteins that are fundamental to biological structure and function. Polymers, both natural and synthetic, are synthesized via polymerization of many small molecules, known as monomers. Consequently, their large molecular mass relative to small-molecule compounds produces unique physical properties, including toughness, viscoelasticity, and a tendency to form glasses and semicrystalline structures rather than crystals [1–3].

Polymer fibers possess a long length in relation to their diameter ratio. The incorporation of loading nanomaterials in electrospun fibers improves their properties, rendering a series of new applications, in particular mechanically improved hybrids and composite materials. Properties and characteristics of special importance for this type of fibers are the small diameter, great surface area and low size pores, crucial characteristics for catalysis, filtration and adsorption applications of the systems. Functionalized materials including graphene and GO-polymeric nanocomposites and the development of porous hybrid and organometal structures are becoming available for diverse applications, for example, hydrogen storage, carbon dioxide trapping, fuel and toxic gases, and hydrocarbon separation [14, 15].

18.2 Graphene Oxide Synthesis

There are many ways to obtain graphene; it can be produced by micro-mechanical exfoliation of highly ordered pyrrolitic graphite, epitaxial growth, chemical vapor deposition, and the reduction of graphene oxide (GO). However, GO is the derivative that we seek, because of the functionalized groups around the molecule that allow the interaction or bonding with polymers or any other kind of molecule that we need to attach to the monolayer or graphene.

In general, in the case of our work, GO is obtained by two methods: first, the modified Hummers method, followed by electrochemical reduction, and the second, by mechanical exfoliation of graphite followed by heat and oxidative treatments, in order to obtain the oxygenated functional groups needed.

18.2.1 Synthesis of GO by the Modified Hummers Method

Graphene oxide is synthetized by direct oxidation of graphite dust, using the modified Hummers method [32, 33]. Graphite is previously exfoliated in order to completely oxidize to graphite oxide. This procedure takes place with H_2SO_4, $K_2S_2O_8$, and P_2O_5 at varying temperatures. When this oxidative mixture makes contact with graphite, bubbles begin to form, indicating the intercalation reaction, ending their formation around 30 min after. Then, the mixture is diluted in water and left to rest all night long. The next day, the mixture must be filtered and rinsed to eliminate any quantity of acid. The solid obtained is kept in a desiccator in order to maintain dry.

After that, graphite oxide obtained is placed in sulfuric acid in an ice bath, because of the exothermic reaction. Then, with $KMnO_4$ as an oxidant agent, in order to functionalize the graphite sheets with oxygenated functional groups, an increase in the interplanar space along the chain axis is necessary, that is, from 3.4 Å (graphite) to 6.25–7.5 Å (GO) [34]. This oxidation avoids stacking of graphene, facilitating its dispersion in aqueous media as well as in polar organic solvents due to the rise in its polarity. After total dissolution, the reaction must continue in hot water bath for 2 h. Distilled water is added to the mixture in a cold bath in order to avoid raising the temperature due to the exothermic reaction. Water diminishes the reactivity of the mixture. After several dilutions, H_2O_2 at 30% is added. The mixture became bright yellow, and 1 day after, the superficial material is decanted. The remaining mixture is rinsed with 10 wt.% HCl and then with distilled water. The remaining solid is filtered, dialyzed, centrifuged, and lyophilized. Then, graphene oxide is ready.

18.2.2 Synthesis of Graphene Oxide by Exfoliation Method

The GO was synthesized starting from the mechanical exfoliation of graphite, and different treatments are performed to obtain the best GO conditions. These include high-temperature treatment calcination in a muffle with a constant oxygen flux, in order to remove organic material impurities, contaminants, and undesirable functional groups, and oxidation of graphitic species that maintain the GO layers bonded. The treatment is at 700°C for a period of 2 h. This procedure allows for a greater separation of graphite sheets. Afterward, this material is left to cool at room temperature and then subjected to a chemical treatment consisting of ultrasonic agitation for 3 h in formic acid, KOH-NaOH basic, and hydrogen peroxide solution at 60°C. The generated particles during this process are separated, and the floating lighter particles were washed, dried at ambient temperature, and characterized. The best conditions are for the H_2O_2 chemical treatment to obtain the GO [1].

18.2.3 Electro-Reduction of GO

Once the graphene oxide is obtained, it is placed in the convenient electrode and mixed with the material required, to tailor a composite or hybrid material, and electrochemically reduced, in order to eliminate functional groups and make it more conductive and suitable, for example, in energy applications.

In these studies, electrochemical techniques are used, namely, polarization curves (CP potentiodynamic and CA potentiostatic) and cyclic voltammetry (CV). Electrochemical conditions depend on the specific system under study. First of all, a CP test of the base material is done, normally from 0 to 3 mV to find the corrosion potential and, from this point, establish the reduction conditions in order to reduce the oxygenated functional groups present in the GO. Reduction takes place in a hydrazine environment. Constant potential offers a major control on the parameters applied (potential and reduction time). This technique yields high-quality GO [35].

Cyclic voltammetry has been used to reduce the functional groups that decorate the GO sheets and as a tool to determine the reactions that occur in the electrodes like reversible and irreversible processes. Also, this technique can be used to calculate the specific capacitance of electrodes [36, 37].

18.3 Graphene Oxide Characterization

SEM Graphene Oxide Characterization. Figure 18.2 presents the SEM micrographs characterizing the mechanically exfoliated graphite and graphene particles, subjected to temperature and different solutions in ultrasound (Figure 18.2a to d). The acid or basic treatments deliver aggregate solids of smaller, disordered, and slightly folded layers of graphene oxide. In thermally treated samples, a porous structure is found, due to impurities present in the original graphite, and eliminated during the thermal treatment at 700°C (Figure 18.2a).

After the thermal treatment, to facilitate from the porous structure the separation of the graphite sheets, a subsequent chemical treatment in acid, basic, and peroxide solutions (formic acid, KOH, NaOH, or H_2O_2), and with ultrasonic vibrations is tested (Figure 18.2b, c, and d, respectively). The procedure in peroxide solution renders a solid with lower dimensions in the graphene oxide sheets, in which planar or slightly folded separated GO layers are obtained over the surface. In Figures 18.3 and 18.4, a SEM general view of the surface and a higher magnification show the effect of this treatment that rendered folded sheets thinner and varied in size.

A folded particle sample is characterized using SEM and EDX and is shown in Figure 18.4. The chemical composition of a GO thin sheet characterization shows the presence of carbon and oxygen in its structure, confirming the carbon constituted structure and the presence of oxygen groups attached to it. The low oxygen percentage is due to the weak peroxide oxidation process, compared to other chemical oxidation procedures, such as the Hummers

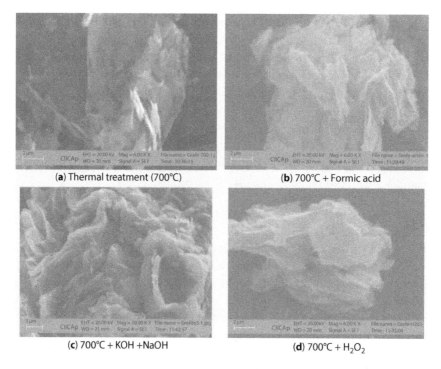

Figure 18.2 SEM micrographs of mechanical exfoliated graphite. (a) Thermal treatment (700°C), (b) Thermal treatment (700°C) + formic acid, (c) Thermal treatment (700°C) + KOH + NaOH, and (d) Thermal treatment (700°C) + H_2O_2.

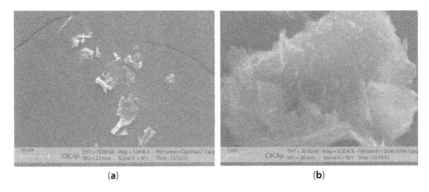

Figure 18.3 SEM micrographs of GO. (a) General view, (b) 6 kX detail view.

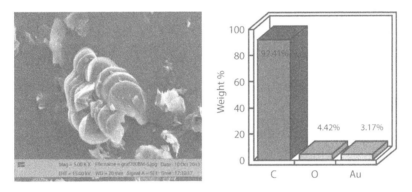

Figure 18.4 GO: SEM and EDX characterization.

process. The O/C ratio obtained in our procedure is 0.048 between the ratios for pristine graphite (0.014) and GO (0.582) [38, 39].

Analyses with SEM (Figure 18.5a) and FT-IR (Figure 18.5b) are performed for the GO particles showing better conditions, obtained in peroxide solution. The material structural changes were observed and compared to the mechanically exfoliated graphite *blank*. The spectrum obtained for mechanically exfoliated graphite after chemical treatment in the peroxide solution showed characteristic bands related to GO, around 1628 cm^{-1}, corresponding to the stretching vibration of the C=C bond, assigned to the π-bonds that form the extended conjugated and aromatic layer of graphene.

Other bands correspond to the characteristic vibration (around 1050 cm^{-1}) of C–O bonds, as well as the bands (3000–3500 cm^{-1} and 1419 cm^{-1}) associated to the stretching and bending of the OH bonds. Also, an 880 cm^{-1} band was observed, associated with vibrations of the epoxy group, although small bands near the 1700 cm^{-1} region are related to the presence of carbonyl (C=O) or carboxyl (–COO) groups formed at the edges of the graphite plate layers. The most characteristic features in the FT-IR spectrum of GO are the adsorption bands corresponding to the C=O carbonyl stretching at 1733 cm^{-1}, the O–H deformation vibration at 1412 cm^{-1}, the C–OH stretching at 1226 cm^{-1}, and the C–O stretching at 1053 cm^{-1}. The O–H stretches appear at 3400 cm^{-1} as a broadband intense signal; the resonances at 1621 cm^{-1} are assigned to the vibrations of the adsorbed water molecules and

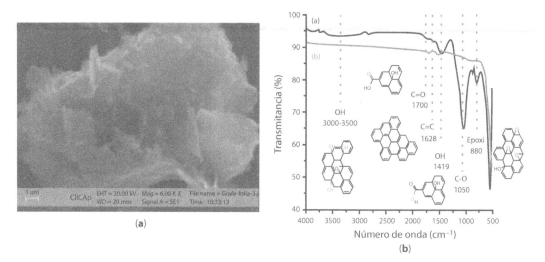

Figure 18.5 Analysis with (a) SEM and (b) FT-IR, for the GO particles showing better conditions, obtained in peroxide solution.

overlapped with the C=C skeletal vibrations of nonoxidized graphitic domains, according to the literature [40].

These results confirm the best condition for GO sheet formation. The presence of oxygen in the GO layer is accompanied by loss of planarity of the affected carbons, which, combined with the more hydrophilic nature of these groups, induces the penetration of water solvent molecules and makes the separation of some layers possible.

18.4 Applications of Polymer/Graphene Oxide Composite Materials

At the present time, the applications of thin films and coatings based on hybrid or nanocomposites for corrosion protection are becoming quite successful. These developments take advantage of properties of the individual components, which in combination yield materials with enhanced new properties.

A potential use of graphene oxide is as a corrosion coating additive to limit diffusion of aggressive species through it. The material planar form could diminish diffusion, decreasing the effective pathway of ionic species. This could have repercussions in corrosion protection coatings and materials with barrier properties.

18.4.1 Corrosion Protection Coating Application

Graphite is a bi-dimensional carbon layer bonded by van der Waals forces [41]. When this layer is separated, a mono-atomic carbon sheet is formed, called graphene. Due to its unique structural and electrical properties, as well as its large specific surface area [42], graphene has been applied into composite materials [40, 43], polymer functionalization devices [44, 45], and electrochemical/corrosion applications [46–48], among others [49, 50]. There are different traditional methods to obtain graphene sheets according to the purposes of research and uses such as mechanical procedure, chemical synthesis, and electrochemical extraction [50–52]. The latter has been widely developed by Alanyalioglu *et al.* [50], which consisted of intercalation of an anionic surfactant into the graphite layers via cathodic and anodic polarization. The final graphene flakes are dissolved in alkaline solution. This methodology is commonly used when nonagglomerated graphene layers are needed. On the other hand, creation of graphene by using chemical vapor deposit (CVD) has become the largest and most extensive methodology to deposit large area surfaces [53].

In the electrochemical field, graphene has been extensively used as a barrier layer or coating on metals against corrosion phenomena without altering or adding extra weight to the substrate under different temperature and oxygen concentration conditions [54]. Graphene is considered to act as an ionic barrier for metals that in conjunction with topcoat coatings provide a multiprotective system under extreme aggressive conditions [55, 56]. However, graphene in conjunction with functionalized graphene oxide (with silanes) has been proven to be an effective anticorrosive alternative for different metals such as steel substrate [57], since the final surface material is a hydrophobic or superhydrophobic material suitable for corrosion protection [58].

For cases where graphene or graphene oxide is incorporated into coatings as nanosheets, nanoparticles, or nanofillers, these elements can act both as corrosion inhibitors and as barrier layers. Thus, an efficient corrosion protection will depend on the well-balanced properties of hydrophobicity, high surface protection area, and good adhesion of the loaded graphene/polymer matrix with the metal substrate.

18.4.1.1 Corrosion Protection Properties of Sol-Gel Coatings Reinforced with Graphene Nanoparticles on Aluminum

Sol-gel coatings based on siloxanes have extensively demonstrated their effective protection properties to prevent or delay corrosion on metallic substrates mainly on alkaline and neutral media [59, 60] in different metals such as aluminum, steel, zinc, and magnesium substrates. Since sol-gel coatings do not contain chromate-based chemical compounds, their effectiveness relies on the composition of the inorganic and organic precursors and the morphological properties [61].

The sol-gel coatings can act either as single layers or as part of a multilayer top-coat system. Their average low thickness (1–10 μm) allows the incorporation of fibers, clays, or pigments to improve their physical, mechanical, and electrochemical properties. For this reason, the incorporation of graphene into a sol-gel matrix as an effective barrier to prevent penetration of corroded species may result in a promising way to enhance durability of the coatings without increasing the thickness of or adding extra weight to the system. The most common way to incorporate nanoparticles into a sol-gel matrix is during the condensation and hydrolysis reactions. Afterward, by using dip-coating, spin-coating, spraying, or electro-deposition procedures, the sol coating is deposited on metallic substrates [62]. The evolution of the sol-gel coatings for corrosion control is reported elsewhere [62].

18.4.1.1.1 Hybrid Sol-Gel Coating/Graphene Preparation

An example of the positive effect of the incorporation of graphene into sol-gel coatings as a corrosion protective layer, Figure 18.2 shows the electrochemical impedance results of a coated aluminum sample in NaCl 0.1 M after 2800 h of testing. However, before explaining these results in detail, it is convenient to describe the graphene/sol-gel formulation. A high-purity graphite rod was subjected to heat treatment at 700°C for 2 h in order to eliminate occluded gas, to burn impurities, and to induce the expansion of the graphite. Afterward, the electrochemical exfoliation procedure described by Alanyalioglu et al. was applied [50, 52]. This consisted of exfoliating the treated graphite rod by applying an oxidation potential of +2.0 V for 12 h in a sodium dodecyl sulfate solution (SDS) followed by a cathodic reduction at −1.0 V for 2 h. Finally, the final solution is centrifuged with acetone in a ratio 1.1 at 4000 rpm for 20 min. The graphene sheets were collected and washed in methanol solution.

The electrochemical setup for the exfoliation procedure consists of the graphite rod as a working electrode and two platinum wires as reference and counter electrodes. This graphene/methanol mixture dissolved in polyvinylalcohol (PVA) was deposited on AA2024-T333 aluminum samples via electrospinning, which consists of a commercial plastic syringe attached to a commercial injection pump. The metallic needle was connected to the positive terminal of a high-voltage power supply while the ground terminal was electrically connected to the aluminum sample. The electrospinning conditions were as follows: 10 kV, a distance between the collector and the needle of 3 cm, and a flow rate of 2.2 μl/min for 20 min. After graphene deposition, the sample was covered with the sol-gel coating by means of spin-coating procedure.

For the hybrid sol-gel coating, two different precursors, organic and inorganic, are prepared. The organic sol consists of mixing 2-propanol with 3-glycidoxypropyltrimethoxisilane and nitric acid solution at pH 5, whereas the inorganic counterpart was synthesized by mixing ethyl aceto-acetate with tetra-n-propoxy zirconium also with acidic nitric solution. Finally, both sols were mixed and mechanically and ultrasonically stirred for 1 h.

18.4.1.1.2 Characterization—Morphology

Figure 18.6 shows the SEM/EDX assessment of graphene particles at different magnifications from 1000× up to 50,000×. From these results, it is evident that electrochemical exfoliation produces graphene nanoparticles rather than graphene oxide since the main elemental signal peaks mainly belong to carbon. The micrographs also reveal that graphene contains superimposed layers due to different textures and color contrast along the whole sheets as seen in Figure 18.6c; thus, the average graphene particle size varies from 0.5 μm downward in accordance to higher magnifications. In addition, Raman analysis was carried out on these particles in order to confirm whether the Raman signal belongs to graphene or not. As seen in Figure 18.6d, two signals at 1593 (G peak) cm^{-1} and 2700 cm^{-1} (2D band) correspond to graphene orbital sp^2 [63, 64]. The low ratio of the 2D versus G peak intensities as well as the asymmetrical shape of the 2D peak indicate that the sample does not correspond to a single monolayer as evidenced by SEM micrographs.

Figure 18.7 shows a cross-section micrograph of the entire system. The different layers on the AA2024-T3 substrate can be seen. The PVA/graphene nanofibers generated an average homogenous layer of 3.2 μm in thickness, whereas the sol-gel coating had a nominal thickness of 15.8 μm. Both layers are well formed and thus clearly distinguished along the sample.

In Figure 18.8, it is seen that the final impedance (resistance) at low frequencies is around 2×10^7 ohm-cm^2. The phase angle results (Figure 18.8b) shows at least three identifiable

APPLICATIONS OF POLYMER/GO MATERIALS 551

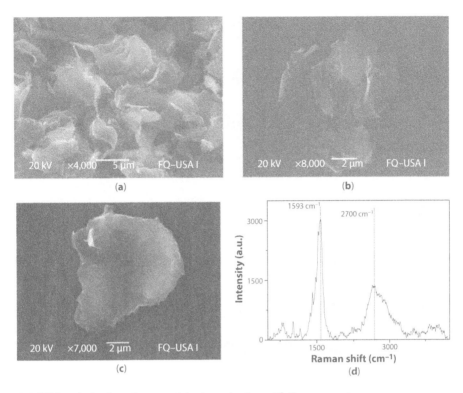

Figure 18.6 SEM analysis of graphene particles in conjuction with Raman spectroscopy.

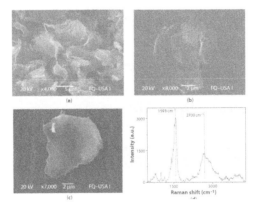

Figure 18.7 Cross-section micrograph of aluminum/PVA-graphene/sol-gel.

different time constants as a function of frequency. The higher frequency one, located in the order of kHz, is associated with the dielectric properties of the sol-gel coating, while the second and third constants are ascribed to the graphene sheets layer and to the oxides/hydroxides present on the metal/graphene interface, respectively. In order to quantify the sol-gel resistance (R_c), it is necessary to estimate the magnitude of the diameter of the first semicircle where the dielectric properties are given. As seen in Figure 18.8a (enlarged area), the R_c varies alternatively from 200 to 800 kΩ cm^2 during 2800 h of testing, which represents changes on the sol-gel surface.

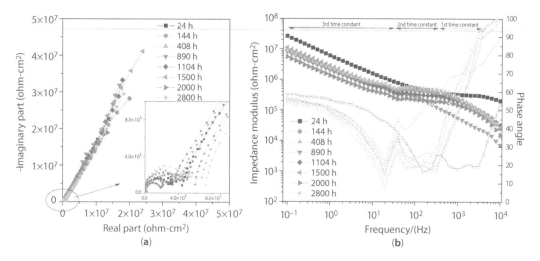

Figure 18.8 EIS spectra of aluminum/PVA-graphene/sol-gel in 0.1 M NaCl during 2880 h. (a) Nyquist and (b) Bode and phase angle.

The Nyquist diagrams indicate that the general behavior of the entire system corresponds to a capacitive response with high protection properties as seen at low frequencies. This stable behavior indicates that the graphene acts as an effective barrier layer at the metal interface providing a long-lasting protection effect. Therefore, the combination and interaction of graphene with the hybrid sol-gel system provide a very positive effect of the corrosion protection of aluminum in saline media.

These anti-corrosion properties are related to the good adhesion of the system. The sol-gel has alkoxide hydrolyzable silanol groups that interact with the metal hydroxyls to produce metal–oxygen–aluminum bonds [65], which, in conjunction with the hydrogen bonds of the carboxyl groups of carboxyl–graphene nanosheets, generates stronger molecular interactions. This adherence improvement is reflected on the undamaged coating at longer exposure times as evidenced on EIS results and supported by atomic force microscopy assessment after immersion test (Figure 18.9).

The final appearance was very smooth with a low roughness profile. It is seen that after immersion in NaCl for 2280 h, the sol-gel coating did not generate agglomerates or voids, which means that the PVA loaded with graphene sheets induced additional anchoring bonds to the aluminum substrate. The average roughness was around 347.2 nm.

Figure 18.9 AFM assessment of coated AA2024-T3 after immersion in NaCl for 2880 h.

18.4.1.2 Graphene Oxide–Nylon Coating System for Steel

Research has been done trying to find useful recycling applications with added value, for scrap metals and polymer discards. Corrosion protection coatings are an important industrial painting and coating sector, because metallic corrosion occurs in almost any environment and accounts for economic losses in the order of 4%–6% of the gross national product in every country. Corrosion can be defined as an undesirable degradation or deterioration of metals and alloys. This definition can also be applied to nonmetals such as ceramics, glass, concrete, etc. [2].

For example, bipolar plates in combustion cells are commonly made with low corrosion and good surface contact resistance graphite compounds [3]. Nevertheless, manufacturing, permeability, and durability against vibration are unfavorable compared to metals. On the other hand, metallic plates suffer from corrosion and form a passive film causing a reduction in contact resistance and contamination of the catalyst and ionomer [14, 15, 24]. A possible solution to these inconveniences is coating the plates to avoid corrosion and passive film formation, to improve charge transport and energy transfer through the combustion cell [66–68].

Nylon is a common low-cost polymer widely used, with excellent mechanical properties such as strength, stiffness, hardness, and toughness coming from its mechanical resistance due to the attraction of their chains because of hydrogen bonds and cross-linking. A possible application for commercial nylon 6-6 is electrospun fibers and functionalized graphene oxide (Ny/FGO) producing a corrosion protection composite coating over a metallic substrate, through electrochemical procedures [69].

18.4.1.2.1 Nylon 6-6 Electrospun Film Procedure

The electrospinning process was discovered by Formhals [70], consisting of three fundamental setup parts: high-voltage power source, a syringe with a metallic needle for the polymeric solution to be stored, and an earthly connected collector (screen). After polymeric solution preparation, the syringe is filled up and electrically charged with the high voltage (between 1 and 30 kV), and the needle tip is placed at a distance between 5 and 30 cm from the collector. The process proceeds because the applied electric force over the polymeric solution defeats the surface tension, producing a solution flow directed toward the screen, and the electrically charged fibers after solvent evaporation are deposited over the collector. Jia *et al.* [71] and Wang *et al.* [72] reported the possibility of obtaining electrospun nanofibers from polymeric mixtures, incorporating nanoparticles, mentioning the potential of reinforcement for composite materials. This procedure was adopted in some of the examples presented along this chapter.

Coating film samples were prepared electrospinning nylon (Ny) fibers under environmental room temperature (25°C) with 1.20 g nylon 6-6 dissolved in 7 ml of formic acid, and this mixture is left under gentle agitation for approximately 12 h. Electrospinning is carried out using a power source and a dosage syringe, at 12 kV voltage and a tip–collector distance of 12 cm and a flow rate of 0.2 ml/h (Figure 18.10). An electrospun nylon 6-6 (Ny)/GO coating was formed after a few hours and collected over a porous silica (because it is a conductive material) plate used as screen [73].

18.4.1.2.2 Electrodes Preparation

Two different electrospun Ny/GO coatings are prepared from a polymeric solution, consisting of 90% formic acid, 0.36% or 2% by weight of GO, and the rest of Ny. To functionalize the

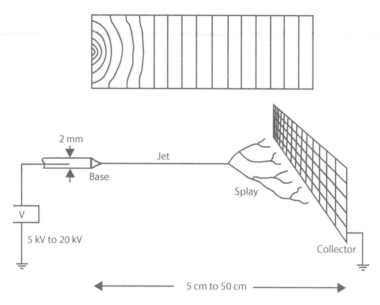

Figure 18.10 Electrospinning scheme.

GO to form the composite coating, the system is electrochemically treated, preparing electrochemical cells using porous silica covered with electrospun Ny/GO films as electrodes. Polarization curves were performed to determine the best electrochemical oxidation/reduction conditions under different acidic (H_2SO_4), alkaline (KOH-NaOH), and peroxide (H_2O_2) solutions. Further oxidation and reduction by electrochemical procedures to obtain Ny/FGO composite are carried out.

18.4.1.2.3 Polarization Curves

To obtain the best electrochemical conditions for the Ny/FGO coating formation, polarization curves were obtained under acid, neutral, and basic solutions to observe the best oxidation–reduction conditions. According to the polarization curves, both electrochemical kinetic reactions were affected by the different solutions. The higher oxidation condition is obtained in KOH and peroxide solutions, while for the reduction, reaction peroxide or H_2SO_4 solutions are greater.

After obtaining the best electrochemical conditions, the procedure adopted was to oxidize in alkaline and reduce in acid solutions, adding up a few drops of hydrazine to suppress the oxygen reduction reaction, improving the efficiency of the reduction process to obtain the Ny/GO composite bonding taking place during this process [1].

18.4.1.2.4 Electrochemical Coating Formation

Potentiostatic oxidation and reduction curves (±1000 mV) using KOH (pH 12) or H_2SO_4 (pH 2) are applied to further oxidize carbon species and then to reduce them over the porous silica substrate containing either the electrospun Ny/GO composite film or the electrospun nylon film covered with a layer of GO, to form the Ny/GO film composite coating in both cases.

18.4.1.2.5 Coating Characterization

Functionalized graphene oxide and nylon fiber samples are characterized using SEM EDX, UV–visible, and FTIR techniques, and the results were presented previously by the authors [69].

These conditions are used to form the composite electrospun Ny/GO coating. The nylon film thickness formed during 3 h was established and determined through SEM analysis. An average around 9.5 μm film thickness of nylon fibers is obtained through the electrospinning deposition. In the composite sheet area, the presence of large and well-defined nylon fibers forms an intricate and compact mesh that covers the substrate. Because of its size, the presence of the GO layers cannot easily be seen [1].

18.4.1.2.6 Electrochemical Impedance Spectroscopy

Electrochemical impedance measurements are performed to evaluate the Ny/GO at 0.36 and 2% by wt. electrospun composite coating under different Na_2SO_4 concentration solutions (Table 18.1) and the total impedance values for the Ny/GO film sample for different solution concentrations are presented. In general, an inverse relation was obtained for the overall impedance as a function of solution concentration, reflecting coating performance [74]. This is due to the difficulty of aggressive species diffusing through the coating as well as modifying the electron discharge of the cathodic reaction, therefore modifying the metal degradation [75, 76]. This also reflects the effects of the solution over the porous silica substrate.

Capacitance values obtained for the Ny/GO film suggest a charge storage capacity for the coating condition. The capacitance obtained for the Ny/GO0.36% electrospun sample presents higher values and therefore greater charge storage. On the other hand, total impedance values are greater for the 2% GO coating condition. A proposal is made of the functionalized graphene oxide, and polymeric fiber association occurs by means of the dipole–dipole and hydrogen bridge interactions.

18.4.2 Storage Energy Applications

Since prehistoric ages, humanity has been seeking different ways to store energy for survival purposes. However, increase in technological usage has demonstrated that, nowadays, storage energy devices must be cheaper, portable, sustainable, and of high-quality

Table 18.1 Electrochemical impedance parameters of the Ny/GO as a function of solution concentration.

Na_2SO_4 concentration	Z_T(ohm-cm^2) Ny/GO 0.36% electrospun	C_{dl} (F/cm^2) Ny/GO 0.36% electrospun	Z_T(ohm-cm^2) Ny/GO 0.36% electrospun	C_{dl} (F/cm^2) Ny/GO 0.36% electrospun
0.01 M	1.5E4	2.66E−5	3.5E7	1.06E−7
0.1 M	8E3	4.48 E−5	3.0E4	8.58 E−7
1.0 M	7E3	4.49 E−5	2.0E4	7.47 E−7

capacity compared to before. Supercapacitors (electrochemical capacitors) use materials with more capacitance area than the normal capacitors and have been successfully used in some commercial electronic devices, but its use can be expanded to other technological applications.

A longer life, high efficiency to charge–discharge process, cyclability (>1E6 cycles), and fast energy deliverance under extreme temperature conditions are searched for supercapacitors (electrochemical capacitors), without losing their charge capacity and with the presence of less toxic components including carbon-based porous materials, transition metal oxides, and conducting polymers. All of these possess advantages and disadvantages; therefore, for the preparation of electrodes, tendency is directed toward the use of composite electrodes, combining beneficial aspects and compensating for the limitations of each individual material [69].

Some nanoporous materials based on graphene oxide–polymer–electron donor dopants show potential as supercapacitor. The combination of graphene oxide and polymers showed resistive properties, and when electronic donors were included, the conductivity properties were greatly increased. These systems include the use of graphene oxide combined with nylon-porphyrin (electrospun and in paste form) and were presented by Garcia and coworkers [69]. They performed and showed electrochemical and physicochemical characterization for both systems.

Porphyrin has shown the ability to protonate/deprotonate in a reversible way, and their nanocomposite materials were characterized for the first time for supercapacitor applications. The results obtained are encouraging to pursue the best conditions in the development of these materials for supercapacitor applications.

18.4.2.1 Graphene Oxide–Nylon–Porphyrin System

Following the procedure adopted and presented above, a three-component composite material was manufactured according to similar reasoning. Among aromatic molecules, porphyrins (H_2P) are modified or substituted aromatic tetrapyrrole macrocyclic compounds exhibiting a wide range of interesting coordinations: catalytic, medical, photoelectrical, and medical properties suitable to be used in high-tech devices. Porphyrins are of main interest in electronics due to their rich electronic/photonic properties (including charge transport, energy transfer, light absorption, or emission) [77]. The relatively easy synthesis and purification of substituted tetra-phenyl porphyrins ($H_2(S)TPP$) make them attractive for preparation of diverse technological systems.

Porphyrin, formed by four pyrrole rings bonded through *methane* (=CH) bridges, forms a planar and highly conjugated macrocycle with four central nitrogen atoms that confers high complexation character on it. Synthetic porphyrin complexes involves almost all metallic elements, and the central space of the molecule can only accommodate ions having an atomic radius smaller than 0.201 nm [77]. Larger ions would be located outside of the molecular plane. Both peripheral and remaining pyrrole hydrogens, as well as those localized on the =CH bridges, can be substituted by different chemical groups to render a family of different compounds.

The majority of polymeric materials are limited in their technological applications for their high electrical resistivity, plastic deformation, low conductivity, and thermal stability. Nevertheless, porphyrins and graphene oxide exhibit excellent properties that are seen to level off these shortcomings, incorporating them into polymeric matrices to form a composite or hybrid material.

Nylon is a common technological important polymer widely utilized, related to its complex morphological changes under different conditions and its low cost. It has excellent mechanical conditions such as strength, stiffness, hardness, and toughness coming from its mechanical resistance due to the attraction of their chains because of hydrogen bonds and cross-linking. In different areas of research, the synthesis or manufacture of composite and hybrid materials has been developed with the purpose of taking advantage of the particular characteristics of the materials that are part of them, as well as their new properties, resulting from the interactions between the different materials. An example is the nylon/$H_2T(p-NH_2)PP$/GO compound, which, according to its properties [69], makes it attractive to be used in energy applications such as fuel cells, capacitors, solar cells, etc.

18.4.2.2 Nylon/$H_2T(p-NH_2)PP$ System Preparation

To form the nylon/porphyrin composite, two solutions are prepared. Solution A: Composed of a mixture of hexamethylene-diamine and sodium hydroxide in distilled water at 6.1% and 2.3%, respectively. Solution B: Prepared with 2 ml of adipoyl chloride dissolved in 22 ml of chloroform and the quantity desired of reacting porphyrin, which were 5 and 100 mg of $H_2T(p-NH_2)PP$. Once the solutions were prepared, hexamethylene-diamine solution "A" is poured carefully into adipoyl chloride and the porphyrin solution "B". Immediately, the separation of solutions and the formation of nylon/$H_2T(p-NH_2)PP$ compound are observed, just at the interface of the two liquid volumes. The polymer formed at the interface is removed very slowly using tweezers, allowing contact and reaction of the two solutions to facilitate composite nylon/$H_2T(p-NH_2)PP$ formation. Finally, the compound is washed with distilled water to remove traces of reagents and allowed to dry in an oven at 80°C for 12 h.

18.4.2.3 Nylon/$H_2T(p-NH_2)PP$/GO Compound Preparation

To form the nylon/$H_2T(p-NH_2)PP$ composite, the compound was dissolved in formic acid, and graphene oxide was added and the mixture is ultrasonically treated at a temperature of 60–65°C for 12 h.

The process starts with the preparation of a polymer solution. In the case of the nylon/GO, the weight percent of GO were 25% or 50%. The concentrations of porphyrin used for the composite nylon/$H_2T(p-NH_2)PP$ were 5 or 100 mg, which correspond to 0.1% or 1% by weight of the total compound mix. A concentration of GO of 25% or 50% is used for the formation of nylon/GO films, and for the nylon/$H_2T(p-NH_2)PP$ ratio, the percent of $H_2T(p-NH_2)PP$ were 0.1 or 1%. Additionally, for the nylon/$H_2T(p-NH_2)PP$/GO systems, the weight percent of GO is set at 25% and those for $H_2T(p-NH_2)PP$ are set at 0.1% or 1%, using electrospinning times of 5 min, 1 h, and 2 h for both cases, being the screen or collector stainless steel, where the polymer fibers are deposited. All compounds were dissolved in formic acid and left stirring for 12 h. The syringe (3 ml) is prepared by cutting and grinding the bevel part of the needle, which is electrically charged when connected to the power source. Detailed experimental procedure can be found in a previous work [69]. The experimental parameters are presented in Table 18.2.

Table 18.2 Experimental parameters set up for electrospinning.

Composite	Solvent	Charge (kV)	Flow rate (µl/min)	Viscosity (cp)	Tip–collector distance (cm)
Nylon	Formic acid	12	0.3	122.24	15
Nylon/GO	Formic acid	12	0.4	102.16	12
Nylon/$H_2T(p-NH_2)$PP	Formic acid/chloroform	12	0.4	101.86	12
Nylon/$H_2T(p-NH_2)$PP/GO	Formic acid/chloroform	13	0.2	106.96	12

18.4.2.4 Characterization

18.4.2.4.1   Scanning Electron Microscopy

SEM characterization of Tetrakis-(para-aminophenil) porphyrin (Ny/$H_2T(p-NH_2)$PP/GO) shows a homogeneous distribution of GO sheets at 25% and 50% by wt. concentrations over the polymeric matrix Ny/$H_2T(p-NH_2)$PP. A complex network of the whole composite can also be seen (Figure 18.11). The presence of a functionalized GO sheet, and the possible union among Ny/$H_2T(p-NH_2)$PP/GO composite fibers around it suggest the presence of hydrophilic groups over the surface, making the interaction with the amide group (-CO-NH) from the composite fibers possible, confirmed from FTIR characterization.

18.4.2.4.2   Fourier Transformation Infrared

The FTIR spectrum of free base species (Figure 18.12) presents one band at around 3300 cm^{-1} and another one at 960 cm^{-1}, ascribed to the NH bond stretching and bending frequencies of NH_2 substituents and of the central nitrogens of the macrocyclic porphyrin free bases. The bands located in the range from 2850 to 3150 cm^{-1} are attributed to C–H bond vibrations of the benzene and pyrrole rings. Bands located at around 1490 to 1650 cm^{-1} are assigned to C=C vibrations and those located at around 1350 and 1272 cm^{-1} are due to –C=N and C–N stretching vibrations. Signals at around 1800 to 1900 cm^{-1} as well as the

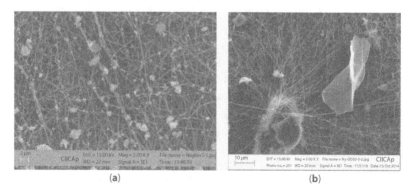

Figure 18.11 Homogeneous distribution of GO sheets over the polymeric matrix Ny/$H_2T(p-NH_2)$PP: (a) general view and (b) detailed view.

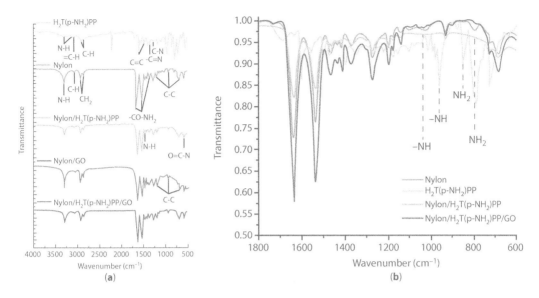

Figure 18.12 FTIR of nylon/$H_2T(p-NH_2)PP$ (a) spectra and (b) amplified section.

bands at about 800 cm^{-1} and 750 cm^{-1} are attributed to C–H bond bending vibrations of para-substituted phenyls [78].

Pure nylon 66 FTIR spectra bands at 3314 and 3221 cm^{-1} and those arising at 1450 cm^{-1} and 750 cm^{-1} can be observed and assigned to the stretching, deformation, and wagging vibrations of N–H bonds. The bands at 2946 and 2867 cm^{-1} are associated to the CH_2 stretching vibrations. The C=O stretching vibrations are observed at around 1717 cm^{-1}. The stretching, asymmetric deformation, and wagging of NH amide groups are observed at 1654, 1547, and 1376 cm^{-1}, respectively. The bands located around 1140 cm^{-1} can be attributed to CO–CH symmetric bending vibration combined with CH_2 twisting. Bands at 936 and 600 cm^{-1} are associated with the stretching and bending vibrations of C–C bonds, and the band at 583 cm^{-1} can be due to O=C–N bending. Additionally, the bands appearing at 936 and 1140 cm^{-1} are associated to the crystalline and amorphous structures of nylon 66, respectively [79].

Characteristic bands in the spectra of nylon/$H_2T(p-NH_2)PP$ and nylon/$H_2T(p-NH_2)PP$/GO 25% or 50% by weight composites, as well as the nylon/GO samples, can be observed in Figure 18.12a. Comparison of the different compounds with the base material (Ny) in the amplified spectra (Figure 18.12b) amplifies small bands that disappear at 1100, 1030, and 800 cm^{-1}, corresponding to primary (NH_2) and secondary (-NH-) amines; likewise, the disappearance of other bands in the spectra of the different compounds is observed. The porphyrin spectrum bands associated with the -NH group appears at 966 cm^{-1} wavelength and those attributed to the stretching vibrations of NH_2 substituents are located at around 850 and 793 cm^{-1}.

The disappearance of the aforementioned bands could be caused by the interaction between the base material (nylon 66), porphyrin ($H_2T(p-NH_2)PP$), or inclusively by the reaction of periphery amine groups of these last species, and the functional groups present on the periphery of the GO sheets, mainly carboxyl (COOH) or carbonyl (C=O) groups [69].

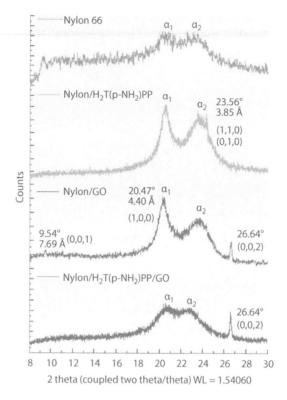

Figure 18.13 X-R diffraction patterns of Ny, Ny/H$_2$T(p-NH$_2$)PP, Ny/GO, and Ny/H$_2$T(p-NH$_2$)PP/GO.

18.4.2.4.3 X-Ray Diffraction

The X-ray diffraction patterns (Figure 18.13) of Ny, Ny/H$_2$T(p-NH$_2$)PP, Ny/GO, and Ny/H$_2$T(p-NH$_2$)PP/GO samples show the same diffraction pattern of a predominant amorphous material with some crystallinity. Bands at around 20.47° and 23.56° correspond to the reflection of (100) and (010,110) of the α phase of Ny crystals oriented in a triclinic cell. The α$_1$ phase corresponds to the distance between adjacent chains of Ny, interacting through hydrogen bonding, while the α$_2$ phase is attributed to the distance between lamellae of polymer. The couple of bands attributed to the α phase are more intense for the Ny/H$_2$T(p-NH$_2$)PP sample than for the pristine Ny. This difference could be attributed to a slight crystallinity increment induced by the incorporation of the porphyrin in the polyamide network [78]. The X-R diffraction patterns of the compounds Ny/GO and Ny/H$_2$T(p-NH$_2$)PP/GO show bands at 26.64° and 9.54° diffraction 2θ angle; for the case of composite Ny/GO, bands are characteristic of graphene oxide.

18.4.2.4.4 Ultraviolet–Visible Nylon/H$_2$T(p-NH$_2$)PP/GO Characterization

In the UV–visible absorption spectra of the Ny/H$_2$T(p-NH$_2$)PP/GO system, an absorption peak at 280 nm assigned to the π-π* transition of the aromatic C=C bond is observed (Figure 18.14), present in the structure of GO as well as in that of porphyrin. Also, an absorption peak at 427 nm representing Soret or B band, characteristic of porphyrins, is observed overlapped in the spectrum of the GO [80].

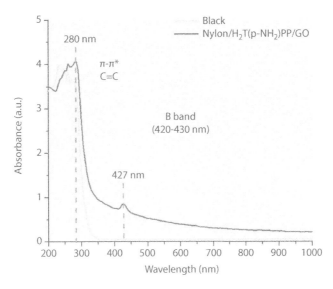

Figure 18.14 UV–Visible absorption spectra of the Ny/H$_2$T(p-NH$_2$)PP/GO system.

18.4.2.5 Electrochemical Evaluation

18.4.2.5.1 EIS of Nylon/H$_2$T(p-NH$_2$)PP/GO Composites

Using electrochemical impedance spectroscopy (EIS), composite coatings over stainless steel substrate are evaluated in H$_2$SO$_4$ acid solution. After no significant change is detected with increasing concentration of GO from 25% to 50%, it is decided to use 25% of GO and to only vary the concentration of 5 mg to 100 mg of H$_2$T(p-NH$_2$)PP as well as electrospinning times from 5 min to 1 and 2 h. In the presence of both GO and H$_2$T(p-NH$_2$)PP, a significant decrease in the total impedance of approximately six orders of magnitude is obtained, compared with the blank (stainless steel) sample and therefore greatly increasing metal dissolution.

The EIS evaluation for the Ny/H$_2$T(p-NH$_2$)PP/GO composite at different electrospinning times is presented in Figure 18.15, showing impedance as a function of the different concentration of H$_2$T(p-NH$_2$)PP, causing higher concentrations and longer electrospinning times, inducing lower total impedance modulus, compared to the metal substrate (Figure 18.15a). The phase angle shows the possible presence of three distinctive time constants for all cases: the first one is due to the coating with low resistance values, the second one is due to a charge transfer process and metallic dissolution, and the third is associated to mass transport. In the nylon/H$_2$T(p-NH$_2$)PP(100 mg)/GO composite (Figure 18.15b) when the electrospinning time rose from 5 min to 2 h, the total impedance of the Bode plot and phase angle response changed, due to the change and decrease of the values in the time constants. Therefore, for binary systems, the metal substrate is protected, while for the nylon/H$_2$T(p-NH$_2$)PP/GO ternary system, the coating is not protective [81].

The capacitance values obtained for the different systems suggest a charge in storage capacity for the coating condition. The capacitance obtained for the Ny/H$_2$T(p-NH$_2$)PP(100 mg)/GO system presents higher values and therefore greater charge storage, which could be appropriate for diverse applications. The capacitance values obtained for the different systems suggest a loaded storage capacity for the coating. Figure 18.16 presents the pore

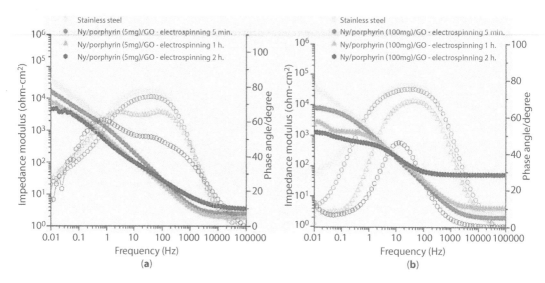

Figure 18.15 EIS Bode and phase angle electrochemical evaluation of the (a) nylon/$H_2T(p-NH_2)PP$(5 mg)/GO (25%) and (b) nylon/$H_2T(p-NH_2)PP$(100 mg)/GO (25%) in a 1 M H_2SO_4 solution at different electrospinning times.

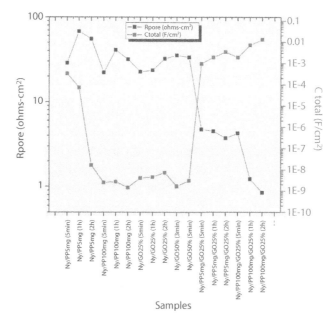

Figure 18.16 Pore resistance and total capacitance as a function of system coating composition.

resistance and total capacitance of the samples, as a function of diverse coating preparation. It can be seen that both parameters present opposite trends. This means that for a decrease in pore resistance, an increase in total capacitance was obtained.

In electrospun Ny in the presence of $H_2T(p-NH_2)PP$ or GO, the barrier effect predominates, presenting higher pore resistance values obtained [82], independent of the electrospinning times. The $H_2T(p-NH_2)PP$ with the combined presence of GO decreased further the pore resistance values, eventually triggering the total capacitance values due to the possible synergistic effect of $H_2T(p-NH_2)PP$ and GO. This effect promotes mass transfer through the porous coating enhanced by the three-component interaction. This means that between minor pore resistance values, the surface is more porous, allowing more surface of contact, promoting the diffusion and charge storage in the metal surface double layer and complex film network. This is confirmed by the voltammetric curves presented below.

18.4.2.5.2 Cyclic Voltammetry of Carbon Cloth Electrodes

With voltammetric tests, the specific capacitance of the nylon/$H_2T(p-NH_2)PP$/GO 80% by weight composite over a carbon cloth was determined, through the current originated by an electron transfer reaction, which occurs on the surface of the electrode as a function of potential. Figure 18.17 shows the specific capacitance with respect to a potential range applied for the Ny/$H_2T(p-NH_2)PP$/GO composite compound, which has a capacitive voltammetric response. It can be seen that the specific capacitance is a function of the sweep rate; at lower scan rates, higher specific capacitance values were obtained for this system. This is because at a lower sweep rate, there is enough time for the phenomena to show itself and be revealed such as charge and mass transfer, which take place in the double electrochemical layer, present between the electrolyte and the surface of the electrode. Nevertheless, capacitance values obtained are below those reported in the literature [10, 83–85].

According to the EIS, the combined effect of $H_2T(p-NH_2)PP$/GO present in the Ny/$H_2T(p-NH_2)PP$/GO coating reduced the total impedance similar to stainless steel values, suggesting that the compounds' interaction promote metal dissolution, decreasing the effect of the coating as a physical barrier. Pore resistance values are low due to good ionic transport properties reflecting the type of coating formed using the electrospinning coating application. Low electrospinning times of nylon/$H_2T(p-NH_2)PP$/GO coating formation present this ionic effect, while longer times decreased further the impedance values probably due to a greater contact surface.

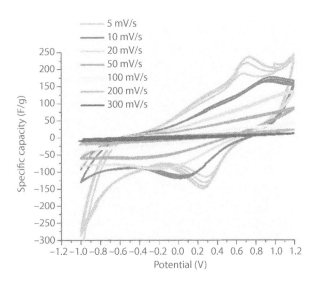

Figure 18.17 Cyclic voltammetry of nylon/$H_2T(p-NH_2)PP$/GO (80% GO) compound at different scan rates.

In recent investigations, proton permeability of graphene was proposed [86]. To reduce this barrier and increase permeability, graphene can be coated with a discontinuous layer of some catalytic metal (platinum), which acts as a catalyst for the passage of hydrogen ions. This same phenomenon is observed in the tests of the nylon/$H_2T(p-NH_2)PP$/GO system due to the synergistic action between the GO and $H_2T(p-NH_2)PP$ structure, where central nitrogen atoms of porphyrin can be reversibly protonated and thus act as receptor and promoter of the passage of protons through the coating. Another phenomenon that reinforces the theoretical explanation [86] is the reduction of the total impedance about two orders of magnitude with respect to the stainless steel (blank), as can be seen in the Bode impedance diagram (Figure 18.15b), indicating the greatly increased ionic conductivity of the system.

The voltammetric results presented suggest a slow reaction process taking place that is reflected in the specific capacitance properties of the coating formed. As it is presented in Figure 18.17, the voltameter rate shows an effect in the shape of the curve, where for higher rates, the voltammogram is flattened, while at lower rates, it tends to have a rectangular shape presenting two peaks associated to the anodic and cathodic reactions and an increase in the specific capacitance [85].

18.4.3 Water Solar Heater Application

In solar thermal collectors, efficiency is directly related to heat loss and heat transfer by conduction and radiation. Heat transfer is more rapid if the temperature difference between the collector surface and the environment is larger. The same goes for the collector surface and the fluid.

The most important part of the solar collector, which determines its efficiency, is the absorber, which is the part of the system where the incident solar energy is translated into heat and transmitted to a fluid medium, such as water. The absorption plates of conventional solar collectors are being traditionally made of metal (stainless steel, aluminum, copper); this makes the system expensive and heavy. Currently, studies have focused on replacing metal materials with less expensive materials without sacrificing the thermal efficiency of the system.

Polymeric materials offer potential advantages over the metallic materials currently in use, such as reduced cost of materials and manufacture, resistance to corrosion, as well as better assembly with other components. The major disadvantage of using polymer materials as a collector absorber is their low thermal conductivity compared to the metal ones. In order to increase thermal conductivity of polymeric materials, different investigations have been made on the development of composite materials with aggregates and of highly thermal conductivity materials such as graphite, carbon black, carbon fibers, metal, and metal oxide microparticles [87–89]. One of the carbon based aggregates is graphene, which recently gathered significant attention due to its fascinating properties, such as high thermal conductivity, good mechanical strength, light weight, low cost, and fair dispersion in polymer matrices [90–95].

Different polymeric materials were used in the manufacture of solar water heater components (cover, water storage tank, absorber plates or tubes, etc.) [96–99]. Polypropylene (PP) is one of the most important and commonly used thermoplastic materials, and its properties have been enhanced by incorporation of various aggregates to obtain improved mechanical, thermal, and electrical conductivity properties, among others [100–105]. In addition, PP composites have also been proven to be suitable to manufacture solar water heater collectors [106, 107].

Graphene oxide is added as a filler in the polypropylene (PP) matrix for the purpose of improving thermal conductivity and to obtain a compound to be used as heat absorber collector's material for solar water heaters.

18.4.3.1 Graphene Oxide–Polypropylene (GO/PP) Composite

Synthesis of composite material is based on a matrix of recycled polypropylene (PP), waste material from the automotive industry, and GO synthesized from graphite via the modified Hummers method (see above).

The material composite is prepared from a blend of GO (10 wt.%) and PP to obtain a homogenous polymer–filler compound. Recycled polymer is ground and then passed through a 1-mm-mesh sieve, obtaining a uniform particle size of PP. Afterward, GO is added to PP powder and blended for 10 min, obtaining a homogeneous mixture. The resulting powder is molded using the compression molding principle, heating the sample in a furnace to a temperature above the melting point of the polymeric matrix (220°C) for 2 h, at a constant pressure according to the thickness required, to allow the material melt inside the mold.

Finally, the composite material is cooled in the mold until it reaches a solid state, conforming to the shape of the mold. The specimens obtained had a circular form with dimensions of 100 mm diameter and 5 mm thickness (see Figure 18.18).

Thermal conductivity and mechanical material resistance properties are important for materials to be used in water solar heater applications. Thermal conductivity is the capacity to transfer thermal energy (heat) by imposing a temperature gradient. Material resistance is the capacity to sustain the mechanical stress applied.

A guarded hot plate apparatus is used to determine the thermal conductivity of composite material. The analysis is conducted in accordance with the ASTM E1225-99 standard [108]. The operation principle of this method is based on heat transfer using a conduction technique under stable state conditions between a cool and a hot plate with a guard [109]. A temperature gradient is established between the two plates to produce a steady, unidirectional, and uniform heat flow through the specimens (Figure 18.19). Under these conditions, the thermal conductivity, λ, is determined according to the following relation:

$$\lambda = \varphi \frac{d}{S\Delta T} \quad (18.1)$$

where φ is the power supplied to the metering section (W), S is the area of the specimens (m²), ΔT is the mean difference in temperature between the plates (K), and d is the mean thickness of the specimens (m).

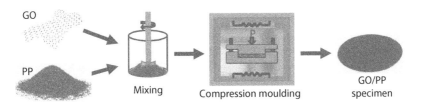

Figure 18.18 Process to obtain the GO/PP compound specimens.

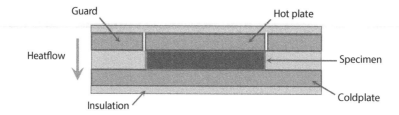

Figure 18.19 Cross-section of a guarded hot plate apparatus configuration with a specimen to measure thermal conductivity.

Thermal conductivity is obtained from 3-mm-thick and 100-mm-diameter circular shaped GO/PP composite material samples. Tests are performed under ambient conditions. The established temperatures are 25°C and 35°C for the cold and hot plates, respectively. Steady-state condition is reached after 7 h, and half an hour after, the temperature gradient is measured. Thermal conductivity is obtained from Equation 18.1.

Results (Figure 18.20) show that thermal conductivity for GO/PP composites is 2.816 W/m·K, representing an enhancement of about 2.85 times, as compared with that of recycled PP (0.988 W/m·K). This demonstrates that the addition of GO increases the thermal conductivity of the composite material proposed.

The GO added as aggregate to the polymeric material, as well as gaining in thermal conductivity, changes the mechanical properties too, such as the Young's modulus.

For tensile test, samples are machined from molded plates of GO/PP to obtain specimens with standardized dimensions. Mechanical properties are measured through universal testing machine and tests are conducted according to the ASTM D638 Standard [110]. The tests are performed until the final fracture occurs.

The tensile results obtained are given in Figure 18.21, for stress–strain curves for each tested material (compound and recycled). The first section of the curve (without stress) is related to the initial adjustment required by the test. In the next section, a linear behavior is observed, which represents the elastic behavior of the materials tested and allows Young's modulus E to be obtained using Equation 18.2. The third section, corresponding

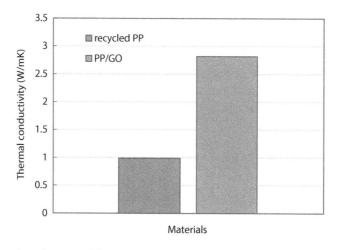

Figure 18.20 Thermal conductivity of the GO/PP compared with the recycled PP.

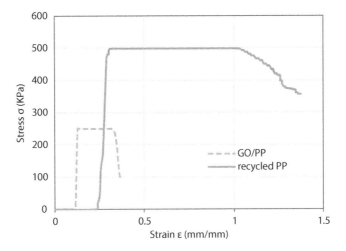

Figure 18.21 Representative stress–strain curves of GO/PP and PP materials.

to a constant strain increased, is indicative of plastic deformation, until fracture of the specimen is reached. As seen in Figure 18.21, the GO/PP compound curve presents a greater slope, which means greater E, signifying a more fragile behavior than the recycled PP.

$$E = \frac{\Delta \sigma}{\Delta \varepsilon} \quad (18.2)$$

where σ is stress and ε is strain.

The results (Figure 18.22) show that for GO/PP, the elastic modulus increased approximately by 2.85 times as compared with that of recycled PP. The compound mechanical properties are enhanced by the presence of GO, due to its very high mechanical strength and Young's modulus.

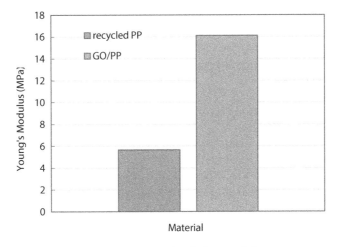

Figure 18.22 Young's modulus of the GO/PP compared with the recycled PP.

Based on the results of the thermal conductivity and tension tests, the GO (wt. 10%)/PP composite material is an excellent alternative to be used as absorbing material of the solar heat collector, since it has significantly improved thermal conductivity and mechanical rigidity. Furthermore, the use of this material reduces the total weight of the heater, thus diminishing costs and mitigation of environmental pollution from polymer wastes.

References

1. Menchaca-Campos, C., García-Pérez, C., Castañeda, I., García-Sánchez, M.A., Guardián, R., Uruchurtu, J., Nylon/graphene oxide electrospun composite coating. *J. Polym. Sci.*, 2013, 1, 2013.
2. Licona-Sánchez, T.deJ., Álvarez-Romo, G.A., Mendoza-Huizar, L.H., Galán-Vidal, C.A., Palomar-Pardavé, M., Romero-Romo, M., Herrera-Hernández, H., Juárez-García, J.M., Uruchurtu, J., Nucleation and growth kinetics of electrodeposited sulfate-doped polypyrrole: Determination of the diffusion coefficient of SO_4^{2-} in the polymeric membrane. *J. Phys. Chem. B*, 114, 9737, 2010.
3. Wang, H. and Turner, J.A., Ferritic stainless steels as bipolar plate material for polymer electrolyte membrane fuel cells. *J. Power Sources*, 128, 193, 2004.
4. Geim, A.K. and Novoselov, K., The rise of graphene. *Nat. Mater.*, 6, 183, 2007.
5. Novoselov, K.S., Geim, A.K., Morozov, S.V., Jiang, D., Katsnelson, M.I., Grigorieva, I.V., Two-dimensional gas of massless Dirac fermions in graphene. *Nature*, 438, 197, 2005.
6. Gómez, N.C., Meyer, J.C., Sundaram, R.S., Chuvilin, A., Kurasch, S., Burghard, M., Kern, K., Kaiser, U., Atomic structure of reduced graphene oxide. *Nano Lett.*, 10, 1144, 2010.
7. Reddy, K.R., Jeong, H.M., Lee, Y., Raghu, A.V., Synthesis of MWCNTs-core/thiophene polymer-sheath composite nanocables by a cationic surfactant-assisted chemical oxidative polymerization and their structural properties. *J. Polym. Sci., Part A: Polym. Chem.*, 48, 1477, 2010.
8. Khan, M.U., Reddy, K.R., Snguanwongchai, T., Haque, E., Gomes, V.G., Polymer brush synthesis on surface modified carbon nanotubes via *in situ* emulsion polymerization. *Colloid Polym. Sci.*, 294, 1599, 2016.
9. Gnädinger, F., Middendorf, P., Fox, B., Interfacial shear strength studies of experimental carbon fibres, novel thermosetting polyurethane and epoxy matrices and bespoke sizing agents. *Compos. Sci. Technol.*, 133, 104, 2016.
10. Hassan, M., Reddy, K.R., Haque, E., Minett, A.I., Gomes, V.G., High-yield aqueous phase exfoliation of graphene for facile nanocomposite synthesis via emulsion polymerization. *J. Colloid Interface Sci.*, 410, 43, 2013.
11. Han, S.J., Lee, H., Jeong, H.M., Kim, B.K., Raghu, A.V., Reddy, K.R., Graphene modified lipophilically by stearic acid and its composite with low density polyethylene. *J. Macromol. Sci. Part B: Phys.*, 53, 1193, 2014.
12. Choi, S.H., Kim, D.H., Raghu, A.V., Reddy, K.R., Lee, H., Yoon, K.S., Properties of graphene/waterborne polyurethane nanocomposites cast from colloidal dispersion mixtures. *J. Macromol. Sci. Part B: Phys.*, 51, 197, 2012.
13. Lee, Y.R., Kim, S.C., Lee, H., Jeong, H.M., Raghu, A.V., Reddy, K.R., Kim, B.K., Graphite oxides as effective fire retardants of epoxy resin. *Macromolecular Research*, 19, 66, 2011.
14. Reddy, K.R., Sin, B.C., Ryu, K.S., Noh, J., Lee, Y., *In situ* self-organization of carbon black–polyaniline composites from nanospheres to nanorods: Synthesis, morphology, structure and electrical conductivity. *Synth. Met.*, 159, 1934, 2009.
15. Mehta, V. and Cooper, J.S., Review and analysis of PEM fuel cell design and manufacturing. *J. Power Sources*, 144, 32, 2003.

16. Sing, V., Joung, D., Zhai, I., Das, S., Khondaker, S.L., Seal, S., Graphene based materials past, present and future. *Prog. Mat. Sci.*, 56, 1178.
17. Prezhdo, O.V., Graphene—The ultimate surface material. *Surf. Sci.*, 605, 1607, 2011.
18. Yang, R.T., *Chemistry and Physics of Carbon*, vol. 19, P.A. Thrower (Ed.), pp. 163–210, Marcel Dekker, New York, 1984.
19. Paredes, J.I., Martínez Alonso, A., Tascón, J.M., Multiscale imaging and tip-scratch studies reveal insight into the plasma oxidation of graphite. *Langmuir*, 23, 8932, 2007.
20. Lahaye, J. and Ehrburger, P., *Fundamental Issues in Control of Carbon Gasification Reactivity*, Kluwer Academic Publishers, Dordrecht, 1991.
21. Kobayashi, Y., Fukui, K.I., Enoki, T., Kusakabe, K., Observation of zigzag and armchair edges of graphite using scanning tunneling microscopy and spectroscopy. *Phys. Rev. B*, 71, 193406, 2006.
22. Ishigami, M., Chen, J.H., Cullen, W.G., Fuhrer, M.S., Williams, E.D., Atomic structure of graphene on SiO_2. *Nano Lett.*, 7, 1643, 2007.
23. Leconte, N., Moser, J., Ordejón, P., Tao, H., Lherbier, A.I., Bachtold, A., Alsina, F., Sotomayor, T.C., Charlier, J.C., Roche, S., Damaging graphene with ozone treatment: A chemically tunable metal–insulator transition. *ACS Nano*, 4, 4033, 2010.
24. Borup, R. and Vanderborgh, N., Design and testing criteria for bipolar plate materials for pem fuel cell applications. *MRS. Proc.*, 393, 151, 1995.
25. Clark, D.T., Cromarty, B.J., Dilks, A., A theoretical investigation of molecular core binding and relaxation energies in a series of oxygen-containing organic molecules of interest in the study of surface oxidation of polymers. *J. Polym. Sci., Part A: Polym. Chem.*, 16, 3173, 1978.
26. Paredes, J.I., Martínez-Alonso, A., Tascón, J.M.D., Scanning probe microscopies for the characterization of porous solids: Strengths and limitation. *Stud. Surf. Sci. Catal.*, 144, 1, 2002.
27. Lerf, A., He, H., Forster, M., Klinowski, J., Structure of graphite oxide revisited. *J. Phys. Chem. B*, 102, 4477, 1998.
28. Zhang, G., Sun, S., Yang, D., Dodelet, J.P., Sacher, W., The surface analytical characterization of carbon fibers functionalized by H_2SO_4/HNO_3 treatment. *Carbon*, 46, 196, 2008.
29. McCarley, R.L., Hendricks, S.A., Bard, A., Controlled nanofabrication of highly oriented pyrolytic graphite with the scanning tunneling microscope. *J. Phys. Chem.*, 96, 10089, 1992.
30. Wei, Z., Wang, D., Kim, S., Kim, S.Y., Hu, Y., Yakes, M.K., Laracuente, A.R., Dai, Z., Marder, S.R., Berger, C., King, W.P., de Heer, W.A., Sheehan, P.E., Riedo, E., Nanoscale tunable reduction of graphene oxide for graphene electronics. *Science*, 328, 1373, 2010.
31. Schrier, J., Helium separation using porous graphene membranes. *J. Phys. Chem. Lett.*, 1, 2284, 2010.
32. Albañil-Sánchez, L., *Preparación de Nanocompuestos Nylon/Grafeno: Electrohilado y Microestructura*, M. Sci. Thesis, Universidad del Autónoma del Estado de Morelos, México, 2010.
33. Hummers, W.S., Jr. and Offeman, R.E., Preparation of graphitic oxide. *J. Am. Chem. Soc.*, 80, 1339, 1958.
34. Wang, G., Yang, J., Park, J., Gou, X., Wang, B., Liu, H., Yao, J., Facile synthesis and characterization of graphene nanosheets. *J. Phys. Chem.*, 112, 8192, 2008.
35. Peng, X.P., Liu, X.X., Diamond, D., Lau, K.T., Synthesis of electrochemically-reduced graphene oxide film with controllable size and thickness and its use in supercapacitor. *Carbon*, 49, 3488, 2011.
36. Liu, C., Teng, Y., Liu, R., Luo, S., Tang, Y., Chen, L., Cai, Q., Fabrication of graphene films on arrays for photocatalytic application. *Carbon*, 49, 5312, 2011.
37. Kauppila, J., Kunnas, P., Damlin, P., Viinikanoja, A., Kvarnström, C., Electrochemical reduction of graphene oxide films in aqueous and organic solutions. *Electrochim. Acta*, 89, 84, 2013.

38. Lopez, V., Sundaram, R.S., Gomez-Navarro, C., Olea, D., Burghard, M., Gomez-Herrero, J., Zamora, F., Kern, K., Chemical vapor deposition repair of graphene oxide: A route to highly-conductive graphene monolayers. *Adv. Mater.*, 21, 4683, 2009.
39. Mattevi, C., Eda, G., Agnoli, S., Miller, S., Mkhoyan, K.A., Celik, O., Mastrogiovanni, D., Granozzi, G., Garfunkel, E., Chhowalla, M., Evolution of electrical, chemical, and structural properties of transparent and conducting. *Adv. Funct. Mater.*, 19, 2577, 2009.
40. Stankovich, S., Dikin, D.A., Dommett, G.H.B., Kohlhaas, K.M., Zimney, E.J., Stach, E.A., Piner, R.D., Nguyen, S.T., Ruoff, R.S., Graphene-based composite materials. *Nature*, 442, 282, 2006.
41. Zou, H., Wu, S., Shen, J., Polymer/silica nanocomposites: Preparation, characterization, properties, and applications. *Chem. Rev.*, 108, 3893, 2008.
42. Novoselov, K.S., Geim, A.K., Morozov, S.V., Jiang, D., Zhang, Y., Dubonos, S.V., Grigorieva, I.V., Firsov, A.A., Electric field effect in atomically thin carbon films. *Science*, 306, 666, 2004.
43. Paton, K.R., Varrla, E., Backes, C., Smith, R.J., Khan, U., O'Neill, A., Boland, C., Lotya, M., Istrate, O.M., King, P., Higgins, T., Barwich, S., May, P., Puczkarski, P., Ahmed, I., Moebius, M., Pettersson, H., Long, E., Coelho, J., O'Brien, S.E., McGuire, E.K., Sanchez, B.M., Duesberg, G.S., McEvoy, N., Pennycook, T.J., Downing, C., Crossley, A., Nicolosi, V., Coleman, J.N., Scalable production of large quantities of defect-free few-layer graphene by shear exfoliation in liquids. *Nat. Mater.*, 13, 624, 2014.
44. Gu, Z., Zhang, L., Li, C., Preparation of highly conductive polypyrrole/graphite oxide composites via *in situ* polymerization. *J. Macromol. Sci., Part B*, 48, 1093, 2009.
45. Ramanathan, T., Abdala, A.A., Stankovich, S., Dikin, D.A., Herrera-Alonso, M., Piner, R.D., Adamson, D.H., Schniepp, H.C., Chen, X., Ruoff, R.S., Nguyen, S.T., Aksay, I.A., Prud'Homme, R.K., Brinson, L.C., Functionalized graphene sheets for polymer nanocomposites. *Nat. Nanotechnol.*, 3, 327, 2008.
46. Chang, K.C., Hsu, M.H., Lu, H.I., Lai, M.C., Liu, P.J., Hsu, C.H., Ji, W.F., Chuang, T.L., Wei, Y., Yeh, J.M., Liu, W.R., Room-temperature cured hydrophobic epoxy/graphene composites as corrosion inhibitor for cold-rolled steel. *Carbon*, 66, 144, 2014.
47. Hernandez, M., Genesca, J., Ramos, C., Bucio, E., Bañuelos, J.G., Covelo, A., Corrosion Resistance of AA2024-T3 coated with graphene/sol-gel films, in: *Solid State Phenomena, Corrosion and Surface Engineering*, vol. 227, J. Michalska and M. Sowa (Eds.), pp. 115–118, Scientific.Net, Trans Tech Publications, 2015.
48. Lin, Y.M., Jenkins, K.A., Valdes-Garcia, A., Small, J.P., Farmer, D.B., Avouris, P., Operation of graphene transistors at gigahertz frequencies. *Nano Lett.*, 9, 422, 2008.
49. Liu, X., Xiong, J., Lv, Y., Zuo, Y., Study on corrosion electrochemical behavior of several different coating systems by EIS. *Prog. Org. Coat.*, 64, 497, 2009.
50. Alanyalioglu, M., Segura, J.J., Oró-Solè, J., Casañ-Pastor, N., The synthesis of graphene sheets with controlled thickness and order using surfactant-assisted electrochemical processes. *Carbon*, 50, 142, 2012.
51. Wang, G., Zhang, L., Zhang, J., A review of electrode materials for electrochemical supercapacitors. *Chem. Soc. Rev.*, 41, 797, 2012.
52. Kaniyoor, A., Baby, T.T., Ramaprabhu, S., Graphene synthesis *via* hydrogen induced low temperature exfoliation of graphite oxide. *J. Mater. Chem.*, 20, 8467, 2010.
53. Robin, J., Ashokreddy, A., Vijayan, C., Pradeep, T., Single- and few-layer graphene growth on stainless steel substrates by direct thermal chemical vapor deposition. *Nanotechnology*, 22, 165701, 2011.
54. Bunch, J.S., Verbridge, S.S., Alden, J.S., Van der Zande, A.M., Parpia, J.M., Craighead, H.G., McEuen, P.L., Impermeable atomic membranes from graphene sheets. *Nano Lett.*, 8, 2458, 2008.

55. Ramezanzadeh, B., Ahmadi, A., Mahdavian, M., Enhancement of the corrosion protection performance and cathodic delamination resistance of epoxy coating through treatment of steel substrate by a novel nanometric sol-gel based silane composite film filled with functionalized graphene oxide nanosheets. *Corros. Sci.*, 109, 182, 2016.
56. Ikhe, A.B., Kale, A.B., Jeong, J., Reece, M.J., Choi, S.H., Pyo, M., Perfluorinated polysiloxane hybridized with graphene oxide for corrosion inhibition of AZ31 magnesium alloy. *Corros. Sci.*, 109, 238, 2016.
57. Lee, C.Y., Bae, J.H., Kim, T.Y., Chang, S.H., Kim, S.Y., Using silane-functionalized graphene oxides for enhancing the interfacial bonding strength of carbon/epoxy composites. *Composites Part A*, 75, 11, 2015.
58. Yin, B., Fang, L., Hu, J., Tang, A.Q., Wei, W.-H., He, J., Preparation and properties of super-hydrophobic coating on magnesium alloy. *Appl. Surf. Sci.*, 257, 1666, 2010.
59. Figueira, R.B., Silva, C.J.R., Pereira, E.V., Organic–inorganic hybrid sol–gel coatings for metal corrosion protection: A review of recent progress. *J. Coat. Technol. Res.*, 12, 1, 2015.
60. Hernandez, M., Covelo, A., Menchaca, C., Uruchurtu, J., Genesca, J., Characterization of the protective properties of hydrotalcite on hybrid organic-inorganic sol-gel coatings. *Corrosion*, 70, 828, 2014.
61. Hernandez, M., Inti-Ramos, O., Guadalupe-Bañuelos, J., Bucio, E., Covelo, A., Correlation of high-hydrophobic sol-gel coatings with electrochemical and morphological measurements deposited on AA2024. *Surf. Inter. Anal.*, 48, 670, 2016.
62. Menchaca-Campos, C., Uruchurtu, J., Hernández-Gallegos, M., Covelo, A., García-Sanchez, M.A., Smart protection of polymer-inhibitor doped systems, in: *Intelligent Coatings for Corrosion Control*, A. Tiwari, L. Hihara, J. Rawlins (Eds.), pp. 447–454, Elsevier, 2015.
63. Nemes-Incze, P., Magda, G., Kamarás, K., Biró, L.P., Crystallographic orientation dependent etching of graphene layers. *Phys. Status Solidi C*, 7, 1241, 2010.
64. Sonde, S., Giannazzo, F., Raineri, V., Rimini, E., Nanoscale capacitive behaviour of ion irradiated graphene on silicon oxide substrate. *Phys. Status Solidi B*, 247, 907, 2010.
65. Álvarez, D., Collazo, A., Hernández, M., Nóvoa, X.R., Pérez, C., Characterization of hybrid sol–gel coatings doped with hydrotalcite-like compounds to improve corrosion resistance of AA2024-T3 alloys. *Prog. Org. Coat.*, 67, 152, 2010.
66. Obreja, V.V.N., On the performance of supercapacitors with electrodes based on carbon nanotubes and carbon activated material—A review. *Physica E*, 40, 2596, 2008.
67. Wu, F.C., Tseng, R.L., Hu, C.C., Wang, C.C., Physical and electrochemical characterization of activated carbons prepared from firewoods for supercapacitors. *J. Power Sources*, 138, 351, 2004.
68. Kierzek, K., Frackowiak, E., Lota, G., Gryglewicz, G., Machnikowski, J., Electrochemical capacitors based on highly porous carbons prepared by KOH activation. *Electrochim. Acta*, 49, 515, 2004.
69. García-Pérez, C., Menchaca-Campos, C., García-Sánchez, M.A., Pereyra-Laguna, E., Rodríguez-Pérez, O., Uruchurtu-Chavarín, J., Nylon/porphyrin/graphene oxide fiber ternary composite, synthesis and characterization. *Open J. Compos. Mater.*, 7, 146, 2017.
70. Formhals, A., Process and apparatus for preparing artificial threads. US Patent 1975504, assigned to Richard Schreiber Gastell and Anton Formhals, 1934.
71. Jia, Y., Gong, J., Gu, X., Kim, H., Dong, J., Shen, X., Fabrication and characterization of poly (vinyl alcohol)/chitosan blend nano fibers produced by electrospinning method. *Carbohydr. Polym.*, 1, 7, 2006.
72. Wang, H., Lu, X., Zhao, Y., Wang, C., Preparation and characterization of ZnS: Cu/PVA composite nanofibers via electrospinning. *Mater. Lett.*, 60, 2480, 2006.

73. Soto-Quintero, A., Uruchurtu Chavarín, J., Cruz Silva, R., Bahena, D., Menchaca, C., Electrospinning smart polymeric inhibitor nanocontainer system for copper corrosion. *ECS Trans.*, 36, 119, 2011.
74. Kendig, M. and Scully, J., Basic aspects of electrochemical impedance application for the life prediction of organic coatings on metals. *Corrosion*, 46, 22, 1990.
75. Khanna, A.S., Totlani, M.K., Singh, S.K., *Corrosion and Its Control*, Elsevier, Amsterdam, The Netherlands, 1998.
76. Reneker, D.H. and Chun, I., Nanometre diameter fibres of polymer, produced by electrospinning. *Nanotechnology*, 7, 216, 1996.
77. García-Sánchez, M.A., Rojas-González, F., Menchaca-Campos, E.C., Tello-Solís, S.R., Quiroz-Segoviano, R.I.Y., Diaz-Alejo, L.A., Salas-Bañales, E., Campero, A., Crossed and linked histories of tetrapyrrolic macrocycles and their use for engineering pores within sol-gel matrices. *Molecules*, 18, 588, 2013.
78. Díaz-Alejo, L.A., Menchaca-Campos, E.C., Uruchurtu-Chavarín, J., Sosa-Fonseca, R., García-Sánchez, M.A., Effects of the addition of ortho- and para-NH2 substituted tetraphenylporphyrins on the structure of nylon 66. *Int. J. Polym. Sci.*, 2013, 323854, 2013.
79. Starkweather, H.W. and Moynihan, R.E., Density, infrared absorption, and crystallinity in 66 and 610 nylons. *J. Polym. Sci. Part A*, 22, 363, 1956.
80. Smith, K.M., *Porphyrins and Metalloporphyrins*, Elsevier Scientific Publishing, Amsterdam, The Netherlands, 1976.
81. Njoku, D.I., Cui, M., Xiao, H., Shang, B., Li, Y., Understanding the anticorrosive protective mechanisms of modified epoxy coatings with improved barrier, active and self healing functionalities: EIS and spectroscopic techniques. *Sci. Rep.*, 7, 15597, 1, 2017.
82. Uruchurtu Chavarin, J., Electrochemical investigations of the activation mechanism of aluminum. *Corrosion*, 47, 472, 1991.
83. Qu, G., Cheng, J., Li, X., Yuan, D., Chen, P., Chen, X., Wang, B., Peng, H., Supercapacitors: A fiber supercapacitor with high energy density based on hollow graphene/conducting polymer fiber electrode. *Adv. Mater.*, 28, 3646, 2016.
84. Cakici, M., Reddy, K.R., Alonso-Marroquin, F., Advanced electrochemical energy storage supercapacitors based on the flexible carbon fiber fabric-coated with uniform coral-like MnO_2 structured electrodes. *Chem. Eng. J.*, 309, 151, 2017.
85. Wang, H., Maiyalagan, T., Wang, X., Review on recent progress in nitrogen-doped graphene: Synthesis, characterization, and its potential applications. *ACS Catal.*, 2, 781, 2012.
86. Hu, S., Lozada-Hidalgo, M., Wang, F.C., Mishchenko, A., Schedin, F., Nair, R.R., Hill, E.W., Boukhvalov, D.W., Katsnelson, M.I., Dryfe, R.A.W., Grigorieva, I.V., Wu, H.A., Geim, A.K., Proton transport through one atom thick crystals. *Nature*, 516, 227, 2014.
87. Stankovich, S., Synthesis of graphene-based nanosheets via chemical reduction of exfoliated graphite oxide. *Carbon*, 45, 1558, 2007.
88. Veca, L.M., Polymer functionalization and solubilization of carbon nanosheets. *Chem. Commun. Camb.*, 18, 2565, 2009.
89. Balandin, A.A., Superior thermal conductivity of single-layer graphene. *Nano Lett.*, 8, 902, 2008.
90. Luo, W., Cheng, C., Zhou, S., Zou, H., Liang, M., Thermal, electrical and rheological behavior of high-density polyethylene/graphite composites. *Iran. Polym. J.*, 24, 573, 2015.
91. Breuer, O. and Sundararaj, U., Big returns from small fibers: A review of polymer/carbon nanotube composites. *Polym. Compos.*, 25, 630, 2004.
92. Coleman, J.N., Khan, U., Blau, W.J., Gunko, Y.K., Small but strong: A review of the mechanical properties of carbon nanotube–polymer composites. *Carbon*, 44, 1624, 2006.

93. Kuilla, T., Bhadra, S., Yao, D., Kim, N.H., Bose, S., Lee, J.H., Recent advances in graphene based polymer composites. *Prog. Polym. Sci.*, 35, 1350, 2010.
94. Goli, P., Legedza, S., Dhar, A., Salgado, R., Renteria, J., Balandin, A.A., Graphene enhanced hybrid phase change materials for thermal management of Li-ion batteries. *J. Power Sources*, 248, 37, 2014.
95. Renteria, J.D., Strongly anisotropic thermal conductivity of free-standing reduced graphene oxide films annealed at high temperature. *Adv. Funct. Mater.*, 25, 4664, 2015.
96. Ariyawiriyanan, W., Meekaew, T., Yamphang, M., Tuenpusa, P., Boonwan, J., Euaphantasate, N., Chungpaibulpatana, S., Thermal efficiency of solar collector made from thermoplastics. *Energy Procedia*, 34, 500, 2013.
97. de la Peña, J.L. and Aguilar, R., Polymer solar collectors. A better alternative to heat water in Mexican homes. *Energy Procedia*, 57, 2205, 2014.
98. Dorfling, C., Hornung, C.H., Hallmark, B., Beaumont, R.J.J., Fovargue, H., Mackley, M.R., The experimental response and modelling of a solar heat collector fabricated from plastic microcapillary films. *Sol. Energy Mater. Sol. Cells*, 94, 1207, 2010.
99. Ango, D.A.M., Medale, M., Abid, C., Optimization of the design of a polymer flat plate solar collector. *Sol. Energy*, 87, 64, 2013.
100. Logakis, E., Pollatos, E., Pandis, C., Peoglos, V., Zuburtikudis, I., Delides, C.G., Structure–property relationships in isotactic polypropylene/multi-walled carbon nanotubes nanocomposites. *Compos. Sci. Technol.*, 70, 328, 2010.
101. Song, M.Y., Cho, S.Y., Kim, N.R., Jung, S.H., Lee, J.K., Yun, Y.S., Alkylated and restored graphene oxide nanoribbon-reinforced isotactic-polypropylene nanocomposites. *Carbon*, 108, 274, 2016.
102. Feng, C.P., Ni, H.Y., Chen, J., Wang, W., Facile method to fabricate highly thermally conductive graphite/PP composite with network structures. *ACS Appl. Mater. Inter.*, 8, 19732, 2016.
103. Zha, J.X., Li, T., Bao, R.Y., Bai, L., Liu, Z.Y., Yang, W., Yang, M.B., Constructing a special 'sosatie' structure to finely dispersing MWCNT for enhanced electrical conductivity, ultra-high dielectric performance and toughness of iPP/OBC/MWCNT nanocomposites. *Compos. Sci. Technol.*, 139, 17, 2017.
104. Zhang, D.L., Zha, J.W., Li, C.Q., Li, W.K., Wang, S.J., Wen, Y.Q., Dang, Z.M., High thermal conductivity and excellent electrical insulation performance in double-percolated three-phase polymer nanocomposites. *Compos. Sci. Technol.*, 144, 36, 2017.
105. Yang, J.L., Huang, Y.J., Lv, Y.D., Li, S.R., Wang, Q., Li, G.X., The synergistic mechanism of thermally reduced graphene oxide and antioxidant in improving the thermo-oxidative stability of polypropylene. *Carbon*, 89, 340, 2015.
106. Povacz, M., Wallner, G.M., Grabmann, M.K., Beißmann, S., Grabmayer, K., Buchberger, W., Lang, R.W., Novel solar thermal collector systems in polymer design–Part 3: Aging behavior of PP absorber materials. *Energy Procedia*, 91, 392, 2016.
107. Kim, S., Kissick, J., Spence, S., Boyle, C., Design, analysis and performance of a polymer–carbon nanotubes based economic solar collector. *Sol. Energy*, 134, 251, 2016.
108. Standard Test Method for thermal conductivity of solid by the guarded comparative longitudinal heat flow technique, ASTM E1225-99, 1999.
109. Standard Test Method for Steady-State Thermal Properties by means of the Guarded-Hot-Plate apparatus, ASTM C-177-97, 1993.
110. Standard Test Method for Tensile Properties of Plastics, ASTM D638. 638-03, 2008.

Index

3D-graphene, 171, 173, 188–190

Ab initio design, 171
Absorption, 270
Additive manufacturing, 41
Adhesion, 426, 431
Alloy/de-alloy, 325
Alpha-sublattice, 187
Aluminum, 549–552, 564
Amorphorus polycrystalline compound, 354
Amorphous silicon (A-Si), 354
Amorphous structure, 346
Annealing, 272
Anode, 325
Anthocyanin counts, 369
Anthracycline antitumor drugs, 446
Applications, 234, 235, 237, 239, 244, 248, 252, 254
Arrhenius equation, 192
Aspect ratio, 492–493, 499–500

Ball milling, 38
Band gap, 265
Batteries, 324
Beta-sublattice, 187
Binding energy, 175–179, 182, 186, 192, 197, 200
Breaking strength, 300
Bridge contact bonds, 184–185
Brinell hardness, 308
Bulk density, 485

Cadmium telluride (CdTe), 354
Capacity, 324
Carbon, 265
Carbon nanotube (CNT), 28
Carbon/graphene composite, 325
Carbyne, 171–187, 190, 195–201
Carbynophene, 171, 173, 185–186, 188, 190–192, 199

Ceramic, 147–169
Ceramic matrix composites (CMCs), 27
Characterization and dispersion, 428–431
Characterization and sensor test, 400–401
Chemical composition studies of GO and rGO, 404–405
Chemical exfoliation, 343
Chemical processes, 346
Chemical stability, 343
Chemical vapor deposition, 207, 208, 209
Chitosan, 240, 242, 243, 251
CIE chromaticity, 266
Clad forming, 45
CNT, 267
Coatings, 544, 548–550, 552–555, 561–564
Cold isostatic pressing (CIP), 30
Colloid solution, 263
Composite, 327
Compressive strength, 311, 312
Concentrations, 269
Conductive nanocomposites, 492, 494
Consolidation, 40
Contact bonds, 171, 184–186, 192, 199
Copper indium gallium selenite (CIGS), 354
Corrosion, 545, 548–549, 552–554
   inhibitors, 549
   protection, 548–549, 552–553
Corrosion resistance, 48
Coulombic efficiency, 324
Covalent bonding, 273
Covalent modification
   addition method, 483
   condensation method, 483
   electrophilic substitution, 483
   nucleophilic substitution, 483
Crystallization rate, 486–491
Cumulene, 175–176, 180–181
Cyclic loading treatments, 136–137, 141–142
Cyclic stability, 328
Cyclodextrin, 451

D peak, 267
Defect density, 205, 212
Device fabrication, 400
DFT calculations, 179, 181, 194, 197
Discharge capacity, 336
Dispersion, 147, 150–155, 157, 161, 163, 164
Dispersion status, 484, 486, 492, 499, 503
Dopants, 264
DOS, 184, 186–191
Dynamic mechanical analysis, 244, 246, 250

Elastic modulus, 297, 302, 310, 316, 496–497, 506
Elasticity, 504–506
Elasticity modulus, 174, 177–178, 185, 191
Electric arc spray, 44
Electric conductivity, 289, 290–293, 299, 309
Electrical conductivity, 234, 235, 238–240, 246, 247, 343, 481, 492–495, 502, 507–508
Electrical properties, 148, 155, 158, 165
Electrical resistivity, 295, 296, 303
Electro-brush plating, 44
Electrochemical, 541, 545, 548–550, 553–554, 563
  exfoliation, 550
  reduction, 544–545, 554
  techniques, 545, 549, 555–556, 561–562
Electrochemical deposition, 42
Electrochemical properties, 328
Electrodeposition, 30
Electron density, 181
Electrophoretic deposition (EPD), 71
Electrospinning, 550, 553–554, 557–558, 561–563
Electrospun fibers, 544, 553
  films, 553
  Ny/GO, 553–556
Elementary cell, 173, 189–190
Elongation, 300
Emission spectra, 266
Endurance, 271
Energy, 543, 553, 555–556, 565
  applications, 545, 555, 557
  devices, 555
  solar, 564
  storage, 555, 563
  systems, 541
Enthalpy, 184, 187, 189–190, 199
Enzymes, 213, 224
Epoxy, 242, 243, 245, 249, 252, 255
ES bandgap, 180–182, 191–192, 195–197, 199
Even–odd effect, 175, 178, 197, 200

Exfoliation, 207, 210
Experiemental procedure
  material characterization, 424–425
  material synthesis, 424
Extrusion, 30

Fe_2O_3, 263
Fermi level, 186–187, 189
Few-layer graphene (FLG), 64
Flame spray, 44
Flow stress, 314
Fluctuation model, 171, 192, 195
Friction stir processing, 45
Fringe regions, 126–127
Functional groups, 270, 342
Functionalization, 235, 239, 244, 246, 247, 249, 254

G peak, 267
Gas sensing mechanism, 409–410
Gas sensor studies, 410–414
Graphene, 28, 147–169, 171–174, 179, 181, 183–190, 192, 196, 198–199, 261, 326, 444, 541–552, 564
  functionalized, 541–546, 548–550, 553, 555–557, 558–559
  oxide, 541–546, 548–550, 553, 555–557, 560, 565
Graphene nanoplatelet (GNP), 64
Graphene oxide (GO), 28, 205, 206, 261, 444–445, 517–522, 525–537
  absorbance, 531–533
  elasticity, 523, 533–537
  fluorescence measurement, 524, 526–527
  fractal analysis, 518, 522, 530–532
  mechanical measurement, 517, 524–525
  optical energy band gap, 519, 522, 532–534
  sol-gel, 525–529
  tail of absorption edge, 523
  Tauc's model, 522–523
  universality, 521–522, 525–530
  UV measurement, 524
Graphene-based composites
  graphene–nanoparticle composites, 15–16, 18
  graphene–polymer composites, 11, 13–14
Graphite, 203, 204, 205, 207, 343

Half-time of crystallization, 488, 491
Hexagonal phase, 171, 189

High velocity oxyfuel spray (HVOF), 44
HOMO, 265
Honeycomb network, 267
Hot isostatic pressing (HIP), 30
Hot-press sintering, 39
Hummer's method, 344, 355
Hybrid, 326
Hybrid materials, 224
Hybrids, 266
Hydroxyapatite (HA), 71
Hydroxyl groups, 205, 218

Initial strain, 115, 134–141
Impact resistance, 427, 442
Impedance, 426, 433–437
In situ polymerization, 53
Interatomic bonds, 171, 175, 177–178, 197
Intercalation/de-intercalation, 325
Interconnected network, 492, 495, 506
Interfacial bonding state, 142
Interfacial interaction, 485, 497, 506, 509
Interfacial properties, 121–123, 134–136, 140, 141–143
Interfacial shear strength, 122, 123, 128, 132–134, 139–140, 142
Interfacial shear stress, 115, 121, 127–129, 131, 134, 139
ITO layer, 273
ITO phones, 267
I-V measurements, 272

K point, 267

Large-sized graphene, 123
Laser sintering, 47
Layer by layer assembly, 53
Lifetime, 171–172, 192, 195–200
Limit strain, 123, 127–128, 131–132, 138–139, 142
Lithium ion batteries, 324
Lone pair, 270
Lower unoccupied molecular orbital (LUMO), 359
LUMO, 265

Macroporous structures, 205
Magnetic resonance imaging, 213
Materials
  composite, 541–542, 544, 548, 553, 556, 564–566, 568
  hybrid, 541–542, 544–545, 548–550, 552, 557
  nano, 544
Material synthesis, 424
Mechanical exfoliation, 328
Mechanical properties, 148, 153–156, 158, 159, 166, 167
Melt blending, 51
Melting and solidification, 30
Memeristor, 271
Metal matrix composites (MMCs), 27
Metal oxide, 325
Metal sulfide, 325
Metal to ligand charge transfer (MCLT), 367
M-graphene, 171–172, 187–190, 199
Microhardness, 305, 307
Microwave irradiation, 344
Molecular-level mixing, 45
Morphological and elemental analysis of the ZnO NSs-rGO hybrids, 405–407
Morphological and elemental studies of GO and rGO, 403–404
Morphological studies of the ZnO NSs, 402–403
Multishape, 249, 252

N719 dye, 358
Nanocomposite, 541, 544, 548, 556
Nanofibers, 240, 243, 246
Nanomaterial, 261
Nanoparticle, 255, 549–550, 553
Nanoscrolls, 264
Nanostructure, 171–172, 195–196, 199, 326
Negative bias, 273
Noncovalent modification, 484
Nonvolatile memory, 275
Nucleating agent, 486, 490
Nucleation, 485–486, 488, 491
Ny/GO system, 553–555, 557–560, 562
Ny/H2T(p-NH2)PP system, 557–560, 562
Ny/H2T(p-NH2)PP/GO system, 556–564

Ohmic conduction, 278
On/off current ratio, 277
Optical properties, 261
Orientation, 499, 503
Oxygen functional group, 270

Pandanus amaryllifolius, 368
Plasma spray, 44
Peak position, 116–120, 123–125, 135–138

Percolation threshold, 492–495, 505–507
Peripheral carboxylate group, 205
Permeabilty coefficient, 499
Phase change, 273
Phonon mode, 269
Phosphor, 269
Photocatalytic activity, 224
Photoluminescence, 265
Poly(lactic acid) (PLA), 54, 246, 248
Poly(methyl methacrylate) (PMMA), 54
Poly(N-isopropylacrylamide), 252
Poly(propylene carbonate), 240, 242, 243, 251
Poly(vinyl acetate), 240, 246, 249
Poly(vinyl alcohol), 240, 243, 251
Poly(ε-caprolactone), 239, 240, 243, 246, 250, 252
Polyacrylamide, 243, 248, 517, 518, 520, 521, 524–537
Polycarbonate, 50
Polycrystalline silicon, 354
Polyethylene, 50
Polyethylene glycol (PEG), 65
Polyimide, 50
Polymer, 50, 541–542, 544, 548–549, 553, 555–558, 564–566, 568
Polymer matrix composites (PMCs), 27
Polypropylene, 50
Polystyrene (PS), 52
Polytetrafluoroethylene (PTFE), 75
Polyurethane, 50, 238–240, 242–249, 252, 255
Polyyne, 175, 179
Positive bias, 276
Potentiodynamic polarization, 426, 437–439
Powder metallurgy, 30
Processing routes for graphene composites, 99, 100
   in situ polymerization/crystallization, 102, 103
   layer-by-layer assembly, 103, 104
   melt bending/mixing, 100, 101
   other processing techniques, 105
   solution bending/mixing, 101, 102
      chemical reduction, 105
      colloidal processing, 105, 106
      powder processing, 106
      sol-gel methods, 105
Pulse electroplating, 43

Quantum yield, 265
Quantum-espresso, 172, 200

Raman 2D shift to strain coefficient, 119, 120, 124, 125
Raman scattering, 116
Raman spectra, 266
Raman spectroscopy, 115–117, 120–123, 129, 134, 136, 141–143
Rate capability, 328
RBLM mode, 267
Reduced graphene oxide (rGO), 58, 336, 445–446
Relative critical length, 127–129, 131–134, 138–139
RESET, 275
Resistive switching, 271
Retention, 271
Rolling, 30

SAED, 263
Scale effect, 178
Scrolling, 262
SEM, 263
Sensing, 206, 213, 214, 215, 219, 224
SET, 272
Shape memory effect, 236, 238, 240, 242, 247–254
Shape memory polymer, 234–255
Shockley–Queisser limit, 354
Sintering, 30, 150, 151, 153–155, 161, 162, 165, 167, 168
Size effect, 126, 129, 131, 133–134, 142
Small-angle neutron scattering (SANS), 70
Sol–gel, 549–552
Solid-like behavior, 503–504
Solution compounding, 52
Sonication, 275
Sp2, 261
Sp2 bond, 187
Sp3, 271
Sp3 bond, 187, 189
Spark plasma sintering (SPS), 30
Spherulite, 485, 490
Spin coating, 272
Steel, 549, 553, 557, 561–564
Storage capacity, 555, 561
Strain distribution, 122, 126, 129–131, 134, 136, 139, 141
Strain gradient, 126, 127, 129, 131
Strain stable region, 126
Strength and stability, 171, 174

Structural studies of GO, rGO, ZnO NSs, and ZnO NSs-rGO hybrids, 407–408
Supercapactors, 556
Surface area, 343, 484, 486, 492–493, 500, 503, 509
Surface roughness, 122, 135
Synergistic effect, 213, 216, 218, 222
Synthesis of bare reduced graphene oxide (rGO) nanosheets, 399–400
Synthesis of graphene
  bottom-up approach, 5
  top-down approach, 3, 4
Synthesis of ZnO nanostructures
  microflowers (MFs), 398
  microspheres (MSs), 399
  microurchins (MUs), 399
  nanocapsules (NCs), 398
  nanoflakes (NFls), 398
  nanograins (NGs), 398
  nanoparticles (NPs), 397
  nanorods (NRs), 398
Synthesis of ZnO NSs-rGO hybrids, 400

TCLC, 279
TEM, 265
Temperature coefficient of resistance, 296
Temperature memory effect, 251, 252
Tensile strength, 297, 310, 311, 312, 316
Thermal conductivity, 297, 299, 315, 502, 507–508
Thermal fluctuation, 171, 192–193, 196
Thermal spray, 44
Thermal stability, 427, 437–440
Thermomechanical stability, 192

Transition metal oxides, 325
Tribological, 155, 162, 163, 165, 169

Ultimate tensile strength, 48, 311
Ultrahigh molecular weight polyethylene (UHMWPE), 75
UPS analysis, 265
UV degradation, 427, 441

Valence band, 265
Van der Waals, 147, 152, 154
Van der Walls attraction, 264
Vicker's hardness, 289, 291, 292, 297, 302, 316

Wear resistance, 49
Weight loss, 426–427, 432–433
Wrapping, 264
Wrinkeled structure, 485, 495

Yield strength, 47, 288, 300, 310, 312
Young's modulus, 288, 300

Zinc oxide nanocomposites (GZNC), 77
  graphene filled polymer composites, 92, 94
    graphene filled polymers, 94
    layered graphene polymers, 94
    polymer-functionalized graphene nanosheets, 94
  graphene bioactive composites, 99
  graphene colloids and coatings, 97, 98
  graphene nanostructure composites, 95
  hybrid graphene/microfiber composites, 95–97

CPSIA information can be obtained
at www.ICGtesting.com
Printed in the USA
BVHW060206140619
550972BV00017B/85/P